Advances in Intelligent Systems and Computing

Volume 1153

The series "Advances in Intelligent Systems and Computing" contains publications on theory, applications, and design methods of Intelligent Systems and Intelligent Computing. Virtually all disciplines such as engineering, natural sciences, computer and information science, ICT, economics, business, e-commerce, environment, healthcare, life science are covered. The list of topics spans all the areas of modern intelligent systems and computing such as: computational intelligence, soft computing including neural networks, fuzzy systems, evolutionary computing and the fusion of these paradigms, social intelligence, ambient intelligence, computational neuroscience, artificial life, virtual worlds and society, cognitive science and systems, Perception and Vision, DNA and immune based systems, self-organizing and adaptive systems, e-Learning and teaching, human-centered and human-centric computing, recommender systems, intelligent control, robotics and mechatronics including human-machine teaming, knowledge-based paradigms, learning paradigms, machine ethics, intelligent data analysis, knowledge management, intelligent agents, intelligent decision making and support, intelligent network security, trust management, interactive entertainment, Web intelligence and multimedia.

The publications within "Advances in Intelligent Systems and Computing" are primarily proceedings of important conferences, symposia and congresses. They cover significant recent developments in the field, both of a foundational and applicable character. An important characteristic feature of the series is the short publication time and world-wide distribution. This permits a rapid and broad dissemination of research results.

**** Indexing: The books of this series are submitted to ISI Proceedings, EI-Compendex, DBLP, SCOPUS, Google Scholar and Springerlink ****

More information about this series at http://www.springer.com/series/11156

Aboul-Ella Hassanien · Ahmad Taher Azar ·
Tarek Gaber · Diego Oliva · Fahmy M. Tolba
Editors

Proceedings of the International Conference on Artificial Intelligence and Computer Vision (AICV2020)

Springer

Editors
Aboul-Ella Hassanien
Information Technology Department
Cairo University, Faculty of Computers and Information
Giza, Egypt

Tarek Gaber
Faculty of Computers and Information
Suez Canal University
Ismailia, Egypt

Fahmy M. Tolba
Faculty of Computer and Information Sciences
Ain Shams University
Cairo, Egypt

Ahmad Taher Azar
Faculty of Computers and Information
Benha University
Banha, Egypt

Diego Oliva
Departamento de Ciencias Computacionales
Universidad de Guadalajara, CUCEI
Guadajalara, Jalisco, Mexico

ISSN 2194-5357 ISSN 2194-5365 (electronic)
Advances in Intelligent Systems and Computing
ISBN 978-3-030-44288-0 ISBN 978-3-030-44289-7 (eBook)
https://doi.org/10.1007/978-3-030-44289-7

This Springer imprint is published by the registered company Springer Nature Switzerland AG
The registered company address is: Gewerbestrasse 11, 6330 Cham, Switzerland

Aboul Ella Hassanien dedicated this work to the Scientific Research Group in Egypt members and my wife Nazaha Hassan El Saman

Preface

The first International Conference on Artificial Intelligence and Computer Vision (AICV2020), which took place in Cairo, Egypt, from April 8–9, 2020, is an international conference covering research and development in the field of artificial intelligence and computer vision. The first edition of AICV2020 is organized by the Scientific Research Group in Egypt (SRGE). It is organized to provide an international forum that brings together those who are actively involved in the areas of interest, to report on up-to-the-minute innovations and developments, to summarize the state of the art, to exchange ideas and advances in all aspects of artificial intelligence and computer vision. The conference proceedings has seven major tracks:

- Artificial intelligence
- Swarm-based optimization
- Mining and data analysis
- Deep learning and applications
- Machine learning and applications image processing and computer Vision
- Intelligent systems and applications
- Intelligent networks

All submissions were reviewed by three reviewers on average, with no distinction between papers submitted for all conference tracks. We are convinced that the quality and diversity of the topics covered will satisfy both the attendees and the readers of these conference proceedings. We express our sincere thanks to the plenary speakers and international program committee members for helping us to formulate a rich technical program. We would like to extend our sincere appreciation for the outstanding work contributed over many months by the Organizing Committee: Local Organization Chair and Publicity Chair. We also wish to express our appreciation to the SRGE members for their assistance. We would like to emphasize that the success of AICV2020 would not have been possible without the support of many committed volunteers who generously contributed their time, expertise, and resources toward making the conference an unqualified success.

Finally, thanks to the Springer team for their support in all stages of the production of the proceedings. We hope that you will enjoy the conference program.

Aboul-Ella Hassanien
Ahmad Taher Azar
Tarek Gaber
Diego Oliva
Fahmy M. Tolba

Organization

Honorary Chair

Fahmy M. Tolba	Faculty of Computer and Information Sciences, Ain Shams University, Cairo, Egypt

International Advisory Board

Igor Farkas	Slovak Republic
Hassan Hussien Soliman	Mansoura University, Egypt
Vijayan Sugumaran	USA
Siddhartha Bhattacharyya	India
Maged ElKamary	Egypt
Abdelkrim Haqiq	Morocco
Swagatam Das	India
Ahmed Mohamed Hamad	Egypt
Fatos Xhafa	Spain
Vaclav Snasel	Czech Republic
Ivan Zelinka	Czech Republic
Janusz Kacprzyk	Poland
Tai-hoon Kim	Korea

General Chair

Aboul Ella Hassanien, Egypt

Conference Co-chairs

Ashraf Darwish, Egypt
Ahmad Taher Azar, Egypt

Program Chairs

Diego Oliva	Departamento de Ciencias Computacionales, Universidad de Guadalajara, CUCEI, Guadajalara, Jalisco, Mexico
Tarek Gaber	Faculty of Computers and Information, Suez Canal University, Ismailia, Egypt

Publicity Chairs

Nour Moustafa	Australia
Mohamed M. ElMogy	Mansoura University, Egypt
Manik Sharma	India
Nour Mahmoud	Egypt
Jyotir Moy Chatterjee	India
Mohamed Hamed	Egypt

Technical Program Committee

Nour Moustafa	Australia
Samir Elmougy	Mansoura University, Egypt
Magdy Zakarya	Mansoura University, Egypt
Rania Elgohary	Ain Shams University, Egypt
Jatinderkumar R. Saini	SICSR, India
Azza Ahmed	Imam Abdulrahman Bin Faisal University, KSA
Anjali Awasthi	Concordia University, Canada
Abdelkrim Haqiq	Hassan 1st University, Morocco
A. V. Senthil Kumar	Hindusthan College of Arts and Science, India
Benjamin Apraku Gyampoh	Ghana
R. S. Ajin	India
Brian Galli	Long Island University, USA
Camelia Pintea	TU Cluj-Napoca, Romania
Chakib Bennjima	University of Sousse, Tunisia
Christos Volos	Aristotle University of Thessaloniki, Greece
Faisal Talib	Aligarh Muslim University, India
Hajar Mousannif	Cadi Ayyad University, Morocco
Irene Mavrommati	Hellenic Open University, Greece
Jaouad Boumhidi	Sidi Mohammed Ben Abdellah University (USMBA), Morocco
Jesus Manuel Munoz-Pacheco	Autonomous University of Puebla, Mexico
Jihene Malek	Higher Institute of Applied Sciences and Technology Sousse, Tunisia

Jin Xu	Behavior Matrix LL, USA
Kusuma Mohanchandra	Dayananda Sagar College of Engineering, India
Laura Romero	University of Seville, Spain
Mariem Ben Abdallah	Monastir University, Tunisia
Marius Balas	Aurel Vlaicu University of Arad, Romania
Mario Pavone	University of Catania, Italy
Mohamed Khalgui	University of Carthage, Tunisia
Muaz A. Niazi	COMSATS Institute of Information Technology, Pakistan
Nickolas S. Sapidis	University of Western Macedonia, Greece
Nilanjan Dey	Techno India College of Technology, India
Nizar Banu, P. K.	B.S. Abdur Rahman University, India
Nizar Rokbani	University of Sousse, Tunisia
Peter Gczy	National Institute of Advanced Industrial Science and Technology (AIST), Japan
Philip Moore	University College Falmouth, UK
Valentina Balas	Aurel Vlaicu University of Arad, Romania
Viet-Thanh Pham	Hanoi University of Science and Technology, Vietnam
Salvador Hinojosa	Universidad Complutense de Madrid, España
Alberto Ochoa	Universidad Autonoma de Ciudad Juarez, Mexico
Jorge Ruiz-Vanoye	Universidad Autonoma del Estado de Hidalgo, Mexico
Marco Perez-Cisneros	Universidad de Guadalajara, Mexico
Erik Cuevas	Universidad de Guadalajara, Mexico
Daniel Zaldivar	Universidad de Guadalajara, Mexico
Valentin Osuna-Enciso	Universidad de Guadalajara, Mexico
Eman Nashnush	University of Salford, Greater Manchester, UK
Tooska Dargahi	University of Salford, Greater Manchester, UK
Sana Belguith	University of Salford, Greater Manchester, UK
Santosh More	University of Salford, Greater Manchester, UK
Julian Bass	University of Salford, Greater Manchester, UK
Omar Alani	University of Salford, Greater Manchester, UK
Kaja Mohideen	PSG College of Technology, Anna University, India
Fernando Fausto	Universidad de Guadalajara, México
Adrián González	Universidad de Guadalajara, México

Local Arrangement Committee

Mohamed Abdelfatah (Chair), Egypt
Moura Rafat, Egypt
Heba Aboul Ella

Contents

Intelligent Networks

Artificial Intelligence

Artificial Intelligence-Based Plant's Diseases Classification

Lobna M. Abou El-Maged[1,4](✉), Ashraf Darwish[2,4], and Aboul Ella Hassanien[3,4]

[1] Computer Science Department, MET High Institute, Mansoura, Egypt
lobna_acd@hotmail.com
[2] Faculty of Science, Helwan University, Cairo, Egypt
[3] Faculty of Computers and Information, Cairo University, Giza, Egypt
[4] Scientific Research Group in Egypt (SRGE), Cairo, Egypt
http://www.egyptscience.net

Abstract. In this paper, we proposed an artificial intelligence model for plants diseases classification based on convolutional neural network (CNN). The proposed model consists of three phases; (a) preprocessing phase, which augmented the data and balanced the dataset; (b) classification and evaluation phase based on pre-train CNN VGG16 and evaluate the results; (c) optimize the hyperparameters of CNN using Gaussian method. The proposed model is tested on the plant's images dataset. The dataset consists of nine plants with thirty-three cases for diseased and healthy plant's leaves. The experimental results before the optimization of pre-trained CNN VGG16 achieve 95.87% classification accuracy. The experimental results improved to 98.67% classification accuracy after applied the Gaussian process for optimizing hyperparameters.

Keywords: Plants diseases · Deep learning · Convolution neural network · Hyper parameter optimization · Gaussian process

1 Introduction

Machine learning (ML) and artificial intelligence (AI) have become dominant problem-solving techniques in many areas of research. Deep learning (DL) is a machine learning approach that involves training neural networks with many feed-forwarded layers on large data sets [1–3]. It has been used in many fields like; image recognition, voice recognition, and also other huge applications that deal with the analysis of large volumes of data including self-driving cars and natural language processing systems [1–3].

The DL is differing than the traditional machine learning in how features are extracted. Traditional machine learning approaches use feature extraction algorithms as a preceding stage for the classifier, but in DL the features are learned and represented hierarchically in multiple levels. So, DL is better for machine learning than the traditional machine learning approaches [4]. There are many types of DL and the most commonly used is the Convolutional Neural Networks (CNN) [3]. The CNN has many architectures that concerning the identification of plant diseases; AlexNet [5], Overfeat [6], AlexNetOWTBn [7], GoogLeNet [8], and VGG [9]. CNN has a number parameters called hyperparameters,

A.-E. Hassanien et al. (Eds.): AICV 2020, AISC 1153, pp. 3–15, 2020.
https://doi.org/10.1007/978-3-030-44289-7_1

these parameters determine the network structure, these parameters are variables such as; the number of hidden layers and the learning rate, the values of the hyperparameters are set before training the network [10]. In this paper, the Gaussian optimization was adapted to find the optimal hyperparameters in a fine-tuning CNN.

On the other hand, the agricultural sector is one of the most important sectors of the economy for all countries; therefore it is one of the elements of sustainable development. Plant diseases significantly affect crop productivity and agricultural wealth. The simple farmer cannot easily identify plant diseases, but he goes to an expert in plant diseases, which in turn may be specialized only in diseases of fruits or vegetables or even in one type of them.

This paper proposed an intelligent system based on deep learning technology to classify different types of plant diseases for vegetables and fruits. This system may be a substitute for the human expert, which achieves the speed and accuracy in the identification and treatment of plant diseases quickly before the spread of the disease. Also, get optimal performance for the identification of the plants' diseases.

The rest of this paper is present as follows. Section 2 describes the basics and background. Section 3 presents the Materials and methods. Section 4 discusses the experimental results. Finally; Sect. 5 presents the conclusion.

2 Preliminaries

This section presents the basics of CNN, VGG16, fine-tuning, and Gaussian optimization.

2.1 Convolutions Neural Network

CNN is an evolution of traditional artificial neural networks and it is network architecture for deep learning [1, 4, 11]. They are used to object recognition and image classifications. A CNN is like a multilayer neural network; it consists of convolutional layers and then followed by fully connected layers. Each layer generates a successive abstraction of the input data with a high level, called a feature map, which saves essential unique information. Amazing performance could be achieve in modern CNNs by employing a very deep hierarchy of layers [11]. The numbers of layers used in DL range from five to more than a thousand [1, 5–9].

In the image analysis tasks, the CNN can be trained to do image classification, object detection, image segmentation and finally image processing [12, 13]. The CNN architecture is consist of two main layers; layers for feature extraction and layers for classification [4]. The feature extraction layers are convolution layer and max-pooling layer:

Convolution Layer: The learnable kernels are used for convolving feature maps from previous layers. The output of the kernels goes through a linear or non-linear activation function specified as a sigmoid, inflated tangent, Softmax, rectified additive, and sameness functions to make the feature maps. Each of the output feature maps can be conjunctive with much than one input feature map.

Suppose

$$LR_i^l = f(\sum\nolimits_{j \in Mi} LR^{l-1} * k_{ji}^l + B_i^l) \quad (1)$$

Where LR_i^l the product of the present layer is, LR_i^{l-1} is the product of the previous layer, k_{ji}^l is the kernel for the present layer, and B_i^l are biases for the presentation layer. Fi represents a selection of input maps. An additive bias B is given for each output map.

Max-Pooling Layer: This bed perfect distribution the dimension of the feature maps depending on the filler of the land distribution mask [4].

The classification layer uses the extracted features from a previous convolutional layer to classify each class. The layers are fully connected in the classification layer. The final layer is represented as vectors with scalar values which are passed to the fully connected layers. The fully connected feed-forward neuronal layers are misused as a soft-max classification layer [4]. Still, in most cases, two to quaternion layers feature been observed in assorted architectures including LeNet, AlexNet, and VGG Net [4].

Soft-max function calculates the probabilities distribution of different events. The soft-max function calculates the probabilities for each class over all possible classes. The calculated probabilities determining the target class for the present inputs [13].

CNN has a number of hyperparameters. The hyperparameters determine the structure of the network such as the number of hidden layers and the learning rate, the hyperparameters are set before the network training [10]. We need to optimize the hyperparameter to reduce the effort of the human necessary for applying machine learning and to enhance the performance of machine learning algorithms. The hyperparameter can be a real value such as "learning rate", integer value such "number of layers", binary, or categorical such as "choice of optimizing" [14].

For enhancing the deep networks training and to avoid the over-fitting, a lot of techniques were proposed. Typical entirety allow pot normalization (BN), and dropout [12]. To enhance the expressive commonwealth of CNNs, non-linear activation functions were well-studied, like as ReLU, ELU, and SELU. Also, a lot of realm peculiar techniques were also mature to add fine-tune networks on the special application [12].

2.2 VGG16 Architecture

VGG16 is the CNN model proposed by Simonyan and Zisserman [9]. The model operated on a dataset contains over 14 million images categories of 1000 classes. It achieves 92.7% as a test accuracy in ImageNet dataset. VGG16 achieves improvement over AlexNet by replacing large kernel-sized filters with 3 × 3 kernel-sized filters one after another. NVIDIA Titan Black GPU's was used to train VGG16 for weeks.

In VGG16, the image with a fixed size 224 × 224 RGB is an input to the convolution layer. There is a stack of convolution layers where the image passes through them using filters with a very small receptive field: 3 × 3. In one of the configurations, it also utilizes 1 × 1 convolution filters, which can be seen as a linear transformation of the input channels then followed by non-linearity transformation. The stride is set to 1 pixel; the spatial padding of convolution. The input of the layer is such that the spatial resolution is preserved after convolution, i.e. the padding is 1-pixel for 3 × 3 convolution layers. After that, the spatial pooling is achieved by 5 max-pooling layers, which follow some of the convolution layers, not all ones. Max-pooling is performed over a 2 × 2 pixel window, with stride 2. After the stack of convolutional layers, three Fully-Connected (FC) layers come. The first two FC layers have 4096 channels each, the third one performs 1000-way ILSVRC classification, so it contains 1000 channels one for each class. The final layer is the soft-max layer. The fully connected layers' configuration is the same all networks. All hidden layers are prepared by the rectification (ReLU) non-linearity. It is also noted that none of the networks (except for one) contain Local Response Normalisation (LRN), such normalization does not improve the performance on the ILSVRC dataset, but leads to more complexity for time and memory [9, 15].

The CNN architecture can be reused for new datasets which differ than its own dataset, the new datasets may vary in size or nature of images. So; we can use fine-tuning deal with this situation.

Fine-tuning tries to improve the effectiveness or efficiency of a process or function by making small modifications in the used CNN architecture to improve or optimize the outcome [16].

2.3 Gaussian Optimization Method

The optimization method is an iterative algorithm has two main parts, the probabilistic surrogate model and the acquisition function to determine which next point to be evaluated. The alternate modeling is fitted to all observations of the reference use made so far in each iteration. Then the acquisition function uses the prophetic dispersion of the probabilistic copy to determine the utility of various candidate points, trading off exploration and exploitation. Compared to evaluating the dear operate, the acquisition role is to compute and can thence be well optimized [14, 17]. The expected improvement (EI):

$$E[I(\lambda)] = E[\max(f^{*}\min - Y, 0)] \quad (2)$$

EI can be computed in the closed-form if the model prediction Y at configuration λ according to a normal distribution:

$$E[I(\lambda)] = \left(f^{*}_{min} - \mu(\lambda)\right)\Phi\left(\frac{f^{*}_{min} - \mu(\lambda)}{\sigma}\right) + \sigma\,\emptyset\left(\frac{f^{*}_{min} - \mu(\lambda)}{\sigma}\right) \quad (3)$$

3 Materials and Methods

3.1 Plant's Image Dataset

The dataset of the plant's diseases images is a "kaggle-dataset" [18], it consists 159.984 images of diseases and healthy leaves' plants present in Fig. 1; 128,028 of plants leaves images for training and 31,956 images for testing. The images are present diseased and healthy plant's leaves. The dataset is characterized by distinguishing images which may be taken at different angles and different backgrounds. Table 1 shows that the data set contains thirty-three categories of nine plants leaves; apple, cherry, corn, grape, peach, Pepper, potato, Strawberry and tomato.

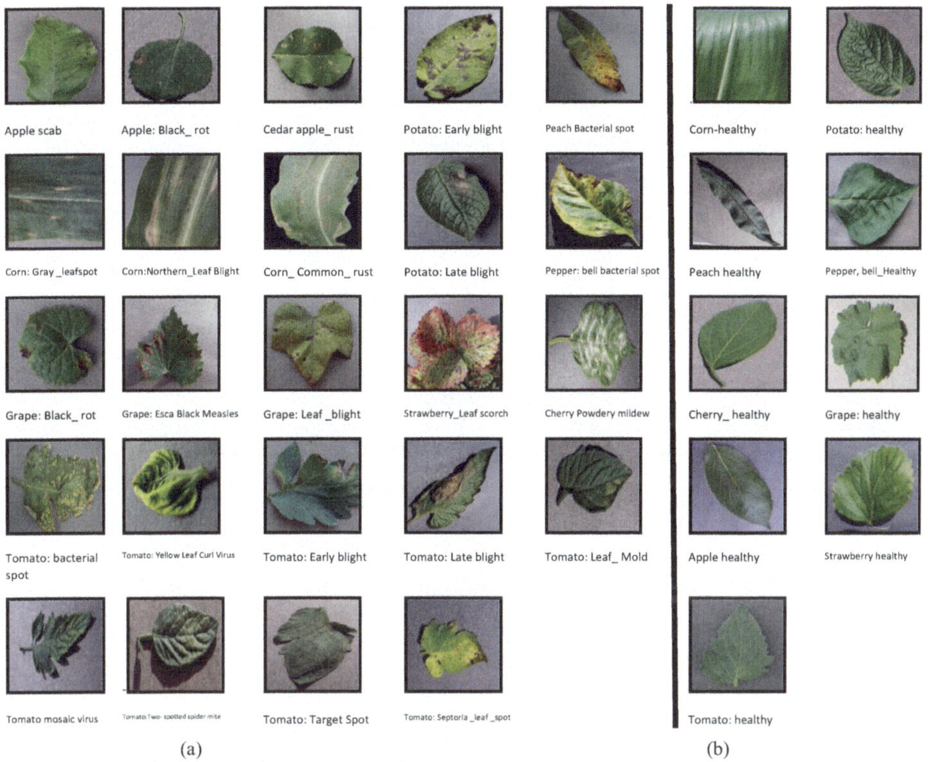

Fig. 1. a. sample of diseased plants' leaves b. sample of healthy plants' leaves1

3.2 The Proposed Plant's Diseases Classification Model

The AI proposed model is consisting of three main phases (a) preprocessing phase, (b) classification and evaluation phase and (c) hyperparameter optimization using Gaussian process phase. Each phase is composed of a number of steps as presented in Fig. 2. In the AI proposed model; the image of the plants are gained from the data set, after executing the preprocessing phase, the Boolean variable "optimize" is set true, then process classification and evaluation phase executed, the output of this phase is the evaluation of the results. If the results are satisfying and the value of the "optimize" variable is "false", the algorithm will end otherwise, the optimization phase will execute a number of n times. The detail of each process is given below.

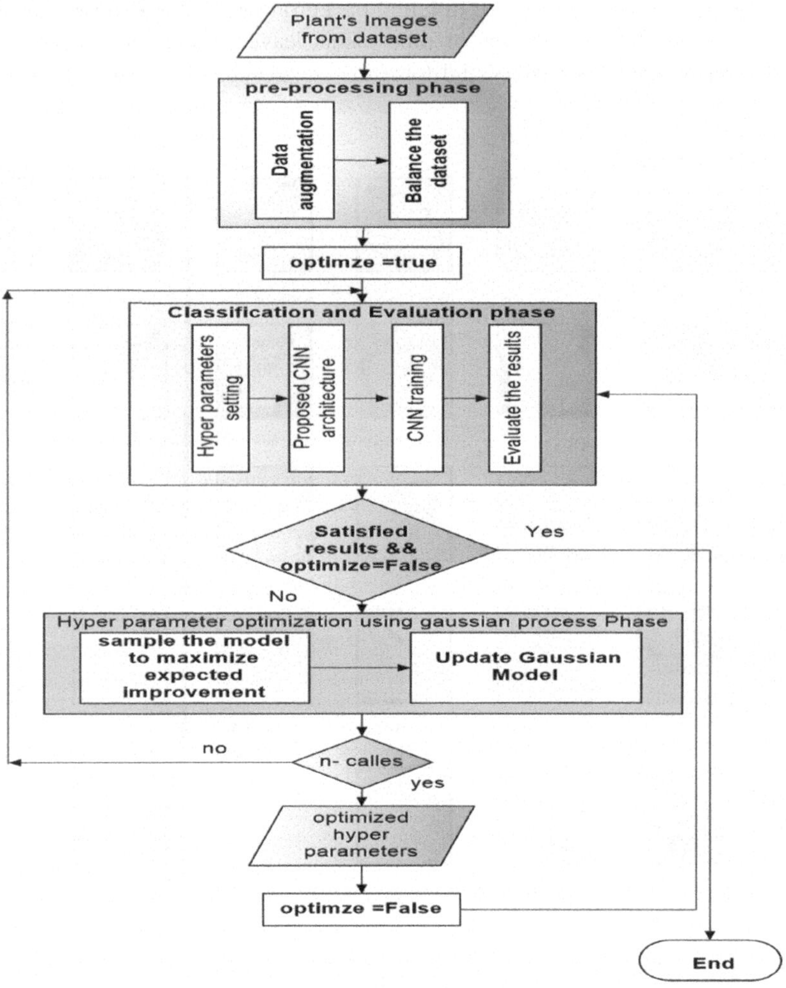

Fig. 2. The proposed Artificial Intelligent plant's diseases classification model

3.2.1 Preprocessing Phase

Preprocessing phase is used for processing and formatting images for use in the next stages, it consisting of two processes; data augmentation and balanced the dataset as in Fig. 2. The detail of each process is given below.

***Data Augmentation*:** The data was augmented to avoid overfitting and present more varieties in the data set. The augmentation for data was done in five different methods. The augmentation methods are; rotate right 30°, rotate left 30 & + 90°, flip horizontally about Y-axis and shear.

Balance Dataset: The dataset showed in the Table 1 is imbalance dataset, the imbalanced lead to inaccurate results, so we calculate the class weight for the categories in the dataset to use it later for our building model. The class weight will be, for example, the first weight is '1.92135642' which presents 2017 images for the apple scab is less than third weight '4.40165289' presents 880 images for the cedar-apple rust.

Table 1. Plants' images dataset description

#	Plant	Disease	Train samples	Test samples
1	Apple	Scab	2017	505
2		Black rot	1988	496
3		Cedar apple rust	880	202
4		Healthy	5264	1316
5	Cherry	Healthy	2736	680
6		Powdery mildew	3368	840
7	Corn	Cercospora leaf spot Gray leaf spot	1644	408
8		Common rust	3816	952
9		Healthy	3720	928
10		Northern Leaf Blight	3125	788
11	Grape	Black rot	3776	944
12		Esca_(Black Measles)	4428	1104
13		Healthy	1356	336
14		Leaf blight_(Isariopsis_Leaf_Spot)	3444	860
15	Peach	Bacterial _spot	7352	1837
16		Healthy	1152	228
17	Pepper	Bacterial _spot	3193	796
18		Healthy	4725	1180
19	Potato	Early _blight	3200	800
20		Healthy	488	120
21		Late _blight	3200	800
22	Strawberry	Healthy	1460	364
23		Leaf scorch	3552	884
24	Tomato	Bacterial spot	6808	1700
25		Early blight	3200	800
26		Healthy	5089	1272
27		Late blight	6109	1524
28		Leaf Mold	3048	760
29		Septoria leaf spot	5668	1416
30		Two- spotted spider mite	5364	1340
31		Target Spot	4496	1120
32		Mosaic virus	1196	296
33		Yellow leaf curl virus	17144	2484
Total			**128,028**	**38836**

3.2.2 Classification and Evaluation Phase

The classification and evaluation phase is consists of 4 processes; hyperparameter setting, proposed CNN architecture, CNN Training, and evaluation the results. The detail of each step is given below.

Hyperparameter Setting: This process is responsible for setting values of the hyperparameters, this work uses three types of hyperparameters; learning rate, drop_out value, and activation function type.

Proposed CNN Architecture: The proposed architecture depends on the usage of a pre-train CNN VGG16 architecture which is a very convincing way since it has been used previously and has given good results. The proposed architecture is consists of two main steps; the first one is to create a CNN VGG16 then the second one is to make some adaptation in the architecture to fit the dataset. Figure 3 presents the proposed CNN architecture.

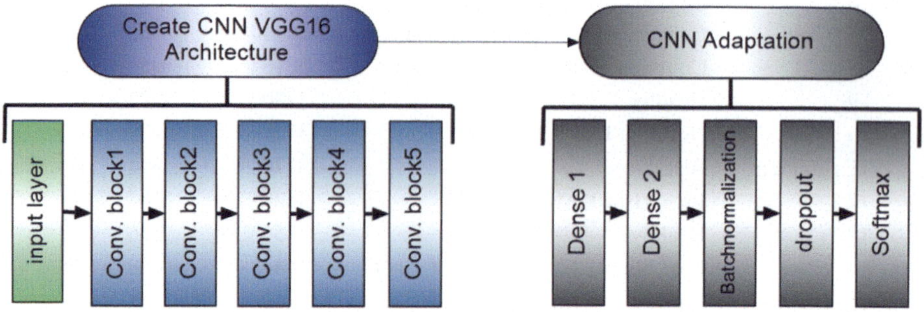

Fig. 3. CNN proposed architecture

As seen in the early section, the VGG16 is a CNN architecture used to classify 1000 classes, it consists of 16 layers. In the proposed architecture, the convolutions layers in the VGG16 is used, then modifying the rest architecture by adding two dense layers, BatchNormalization layer, Dropout layer to prevent overfitting, and finally the classification layer.

The CNN Training Process: In the training process the fine-tuning is used Fig. 4. The Fine-Tuning is optimized both the weights of the new classification layers that

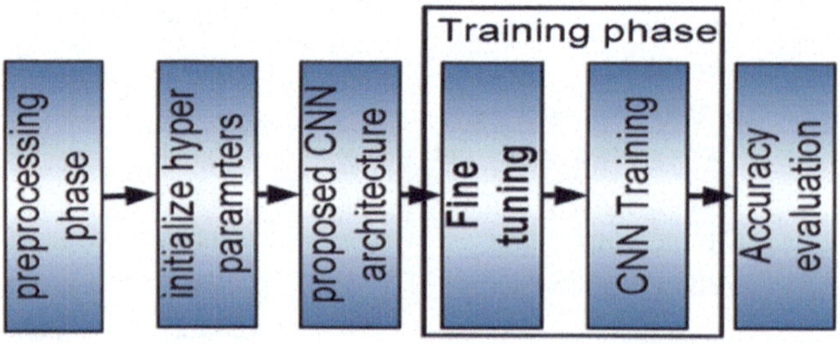

Fig. 4. Proposed model with fine – tuning

have been added and also all of the layers from the VGG16 model [16]. The training process is based on the hyper-parameter values which sat in the previous process.

Evaluation the Results: This process is responsible for evolution the accuracy of the proposed AI model, the evaluation results used to decide a road chart of the model as seen in Fig. 2.

3.3 Hyper Parameter Optimization for CNN Using Gaussian Process Phase

The optimization method is used for two reasons; the first one is to get the optimal hyper_ parameters' values and the other is to provide the human effort in trying to get the optimal hype_ parameters' values by doing the system with this task.

The proposed model setting parameters used the learning rate = 1e−6 with Adam optimizer, drop out rate = 0.2 and activation function is 'relu' as initial values. In the Gaussian method; the model is sampled with expected improvement, then update the Gaussian model according to the classification accuracy, after that check if the number of iteration calls is maximum or not to decide whether re-execute the model or not.

After finishing the n calls iteration, it will give optimized hyperparameters.

4 Experiments Results and Discussion

The experiments are done using tensor flow and Keras with GPU google colab. The experiments are done in three experiments. In the first experiment is implementing the proposed model without optimization. In the second experiment, perform the optimization process using the Gaussian method. In the third experiment, implement the proposed model after optimization.

4.1 Experiment (I): Without Optimization

In this experiment, the structure of VGG16 is modified by adding two dense layers, adding a BatchNormalization layer, adding a Dropout layer, and finally a classification layer. In this experiment, we improved both the weights of the new classification layers and all layers of the VGG16 model.

The hyper_parameters used are learning rate = 6−1e, drop_out = 0.2 and the activation function is "relu". Figure 5. shows the Test-set classification accuracy: 95.87%.

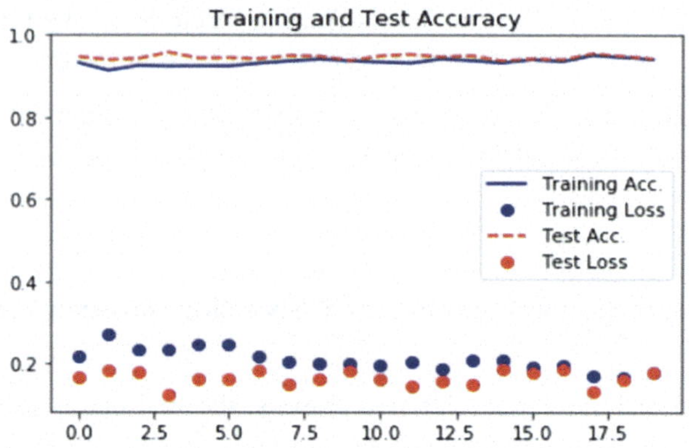

Fig. 5. Training history of CNN network before hyperparameter optimization

4.2 Experiment (II): Hyper_Parameter Optimization Using Gaussian Process

In this experiment, Gaussian process optimization is used with EI and with minimum 11 iteration calls. The hyper-parameters are ‘learning_rate’ within range (from low = 1e−6 to high = 1e−2), ‘drop_out’ within range (from low = 0.1 to high = 0.9) and finally ‘activation’ with categories (‘relu’ and ‘sigmoid’). The fitness function is the function that creates and trains a neural network with the given hyper-parameters, and then evaluates its performance on the data set. The function then returns the so-called fitness, which is the negative classification accuracy on the validation set. It is negative because the performance minimization instead of in Fig. 6. For example test accuracy is about 91% when learning rate = 6.8.99, Drop_out = 0.63 and activation function is “sigmoid”. It is illustrated that the best values are; learning rate = 1e−5 with droup out = 0.3 and activation function is ‘relu’.

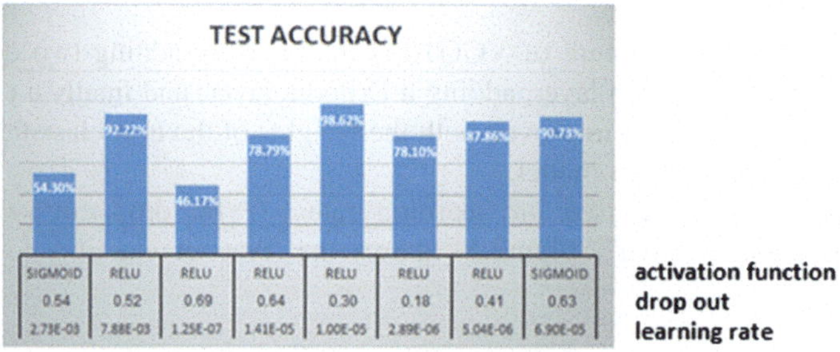

Fig. 6. Test accuracies during optimization process

4.3 Experiment (III): Hyperparameters Optimization

In this experiment, the CNN architecture is run with new parameters such as learning rate = 1e−5, drop out = 0.3 and activation function = 'relu'. It gives Test-set classification accuracy: 98.67%. Figure 7 shows the historical training.

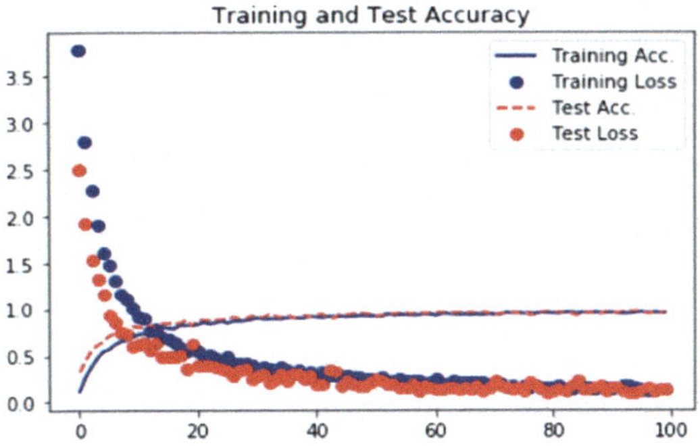

Fig. 7. Training history of CNN network of proposed AI model.

The results as shown in Table 2 shows that the total test accuracy is improved with 2.8%. At the plants level, test accuracy for most plants are improved but test accuracy for a small number of plants doesn't change such as cherry and corn. It is noted that the plant, whose accuracy has not changed, has fairly high accuracy.

Table 2. Accuracy test results

- #	Plant	Disease	Test accuracy before hyper parameters optimization	Test accuracy after hyper parameters optimization
1	Apple	Scab	91.47%	94.44%
2		Black rot	96.58%	100.00%
3		Cedar apple rust	97.33%	99.55%
4		Healthy	99.69%	100.00%
5	Cherry	Healthy	98.21%	98.21%
6		Powdery mildew	98.68%	98.68%
7	Corn	Cercospora leaf spot Gray leaf spot	95.15%	98.04%
8		Common rust	99.79%	100.00%
9		Healthy	92.89%	96.70%
10		Northern Leaf Blight	99.89%	99.89%
11	Grape	Black rot	97.88%	97.88%
12		Esca_(Black Measles)	96.24%	99.73%
13		Healthy	97.93%	100.00%
14		Leaf blight_ (Isariopsis_Leaf_Spot)	99.41%	100.00%
15	Peach	Bacterial _spot	99.89%	99.89%
16		Healthy	92.91%	95.49%

(*continued*)

Table 2. (*continued*)

- #	Plant	Disease	Test accuracy before hyper parameters optimization	Test accuracy after hyper parameters optimization
17	Pepper	Bacterial _spot	97.86%	98.99%
18		Healthy	98.74%	99.75%
19	Potato	Early _blight	98.27%	99.38%
20		Healthy	93.20%	98.24%
21		Late _blight	90.08%	98.33%
22	Strawberry	Healthy	99.55%	99.66%
23		Leaf scorch	96.15%	100.00%
24	Tomato	Bacterial spot	95.49%	99.00%
25		Early blight	86.90%	99.59%
26		Healthy	96.72%	98.62%
27		Late blight	91.31%	98.42%
28		Leaf mold	97.30%	98.80%
29		Septoria leaf spot	97.52%	99.55%
30		Two- spotted spider mite	90.63%	98.21%
31		Target spot	97.97%	99.11%
32		Mosaic virus	84.80%	92.23%
33		Yellow leaf curl virus	97.22%	99.84%
Total			95.87%	98.67%

5 Conclusion and Future Work

Agriculture plays an important role in the economic development for many countries so classification the plant's diseases are very important. Deep learning mimics human thinking, it has been used extensively in the last decade. VGG16 is an architecture of CNN. Using fine-tuning CNN VGG16 model by adding several dense layers, BatchNormalization layer, Dropout layer and finally the classification layer, this lead to accuracy 95.87%. Automated hyperparameters optimization is important to help the human by reducing his effort necessary for applying machine learning. It used to enhance the performance of the algorithm for machine learning. The Gaussian method can be used for hyperparameters optimization. It uses the predictive distribution of the probabilistic model. The Gaussian method identifies the importance of different candidate points, trading off exploration and exploitation, this function is cheap. Using Gaussian process optimization with EI illustrated that the best values are; learning rate = 1e–5 with drop_out = 0.3 and activation function is 'real'. Using fine-tuning with new parameters: learning rate = 1e−5, drop_out = 0.3 and activation function = 'relu', It gives accuracy 98.76%. I.e. Gaussian process optimization helps in improving accuracy from 95.87% to 98.67%. In future work, we will use the swarm algorithm for hyperparameters optimization as a trial to enhance accuracy.

References

1. Ferentinos, K.P.: Deep learning models for plant disease detection and diagnosis. Comput. Electron. Agric. **145**, 311–318 (2018)
2. Zhang, X., et al.: Identification of maize leaf diseases using improved deep convolutional neural networks. In: IEEE Xplore Digital Library, Digital Object Identifier (2018). https://doi.org/10.1109/access.2018.2844405
3. Barbedo, J.G.A.: Factors influencing the use of deep learning for plant disease recognition. Biosyst. Eng. **72**, 84–91 (2018)
4. The History Began from AlexNet: A Comprehensive Survey on Deep Learning Approaches. https://arxiv.org/abs/1803.01164
5. Krizhevsky, A., Sutskever, I., Hinton, G.E.: ImageNet classification with deep convolutional neural networks. In: Proceedings of 25th International Conference on Neural Information Processing Systems, Lake Tahoe, Nevada, 03–06 December 2012, vol. 1, pp. 1097–1105 (2012)
6. Sermanet, P. et al.: Overfeat: integrated recognition, localization and detection using convolutional networks. arXiv:1312.6229 (2013)
7. Krizhevsky, A.: One weird trick for parallelizing convolutional neural networks. arXiv:1404.5997 (2014)
8. Szegedy, C., Liu, W., Jia, Y., Sermanet, P., Reed, S., Anguelov, D., et al.: Going deeper with convolutions. In: Proceedings of the IEEE Conference on Computer Vision and Pattern Recognition (2015)
9. Simonyan, K., Zisserman, A.: Very deep convolutional networks for large-scale image recognition. arXiv:1409.1556 (2014)
10. https://towardsdatascience.com/what-are-hyperparameters-and-how-to-tune-the-hyperparameters-in-a-deep-neural-network-d0604917584a. Accessed May 2019
11. LeCun, Y., et al.: Gradient based learning applied to document recognition. In: Proceedings of the IEEE, November 1998
12. Wang, T., Li, B.: Data dropout: optimizing training data for convolutional neural networks. arXiv:1809.00193v2 (2018)
13. Islam, M.S. et al.: InceptB: a CNN based classification approach for recognizing traditional bengali games. In: 8th International Conference on Advances in Computing and Communication (ICACC-2018). Elsevier (2018)
14. Feurer, A., Hutter, F.: Automated machine learning: methods, systems, challenges. In: Chapter1: Hyperparameter Optimization. Springer (2019)
15. https://neurohive.io/en/popular-networks/vgg16/. Accessed March 2019
16. Hussain, M., Bird, J.J., Faria, D.R.: Advances in Computational Intelligence Systems. In: UKCI 2018, AISC 840, pp. 191–202 (2019)
17. Rasmussen, C.E., Williams, C.K.I.: Gaussian processes for machine learning. The MIT Press (2006). www.GaussianProcess.org/gpml. ISBN 026218253X. c, Massachusetts Institute of Technology
18. https://storage.googleapis.com/kaggle-datasets/53092/100927/plantdisease.zip

Machine Learning for Apple Fruit Diseases Classification System

Atrab A. Abd El-aziz[1,5(✉)], Ashraf Darwish[2,5], Diego Oliva[3,5], and Aboul Ella Hassanien[4,5]

[1] Faculty of Computers and Information, Kafrelsheikh University, Kafr El-Sheikh, Egypt
atrab.ahmed@gmail.com
[2] Faculty of Science, Helwan University, Cairo, Egypt
[3] Departamento de Ciencias Computacionales, Universidad de Guadalajara, CUCEI, Guadalajara, Jal, Mexico
[4] Faculty of Computers and Artificial Intelligence, Cairo University, Giza, Egypt
[5] Scientific Research Group in Egypt (SRGE), Cairo, Egypt
http://www.egyptscience.net

Abstract. There are a growing demand and an urgent need for fruits due to the increase in the world population. Fruit diseases cause a devastating problem in production losses worldwide. The healthy recognition of fruits and apples is an important issue for the economic and agricultural fields. In this paper, a recognition system for apple fruit diseases detection is proposed and experimentally validated. The K-Means based segmentation technique is applied. In regards to performance enhancement, different features extraction techniques are applied and classified using Support Vector Machine, K-NN, Multi-Class Support Vector Machine, and Multi-Label KNN (ML-KNN). The proposed model can significantly support accurate detection and automatic classification of apple fruit diseases. The average accuracy of diseases classification is achieved up to 97.5% and up to 99% for apple health classification.

Keywords: Multi-Class Support Vector Machine · Multi-Label KNN · Fruit · K-means · Agriculture

1 Introduction

Early examination of fruit infections can help decrease fatalities and may anticipate the additional extent of infections. Automatic recognition of fruit infections is a popular challenge in machine learning due to its important purposes in agriculture. In this regards, intelligent automatic examinations of fruit diseases are considered a critical need [1]. There are several types of fruit infections that influence the quality and production of fruits. However, numerous efforts have been made in this area. The techniques of recognition of health and infections of the fruit continue being a challenge due to the great irregularity of the types of infection and the normal inconsistency in the texture and color of the different categories of fruits. Apples are one of the most consumed fruits in the world [2]. The diseases of the apple can generate great losses in

A.-E. Hassanien et al. (Eds.): AICV 2020, AISC 1153, pp. 16–25, 2020.
https://doi.org/10.1007/978-3-030-44289-7_2

quality and quantity of harvest. In this sense, it is crucial to recognize rotten fruits early [3]. Therefore, in this paper, different feature extractors based on textures and colors properties are used. Firstly, the KNN and SVM are used for normal and rotten apple recognition and then MSVM and ML-KNN are applied for apple diseases recognition. The fusion of features vectors from different descriptors is applied to further enhance recognition accuracy.

The rest of this paper is organized as follows. Section 2 briefly discusses the literature review. Section 3, the proposed system is explained in detail. Section 4 explains the experimental results. Finally, Sect. 5 describes the conclusion.

2 Literature Review

Traditional recognition approaches were based on the colors characteristics of apple fruit. The main problems with these methods are low performance. Unay et al. [4] have developed an apple binary classification framework based on MLP segmentation with statistical features methods and a variety of classifiers such as fuzzy C-means, fuzzy k-NN, and SVM. The evaluated accuracy was 93.5%. Dubey et al. [5] have implemented a framework for the classification of fruit diseases using Complete Local Binary Patterns (CLBP), Local Binary Patterns (LBP), Color coherence vector (CCV), Global color histogram (GCH) as a feature extractor and a multi-class support vector machine (MSVM) as a classifier. The accuracy of the model reached 93% using (CLBP).

Ashok et al. [6] have implemented a framework for apple (defective, non-defective) classification. They tested different features extractor such as (Mean boundary gradient, mean intensity, Fourier descriptors filler blank, invariant moments) and the classification was based on Probabilistic Neural Network (PNN) with 86.52% accuracy. An improved contribution [7] based on texture features using histograms for the recognition of fruit diseases with MSVM. They reported an accuracy of 97%.

Another contribution by the same author [8] they implemented more feature extractors include a combination of features (color, shape, and texture) using (CCV), (GCH), CLBP, LBP and Zernike Moments (ZM) with MSVM as a classifier. They reported 95, 9% by combining CCV, CLBP and ZM features. |Other related work done by authors in [9], that uses binary classification and using different feature extraction methods including histogram of oriented gradients (HOG), Law's Texture Energy (LTE), Gray level co-occurrence matrix (GLCM). They implemented many classifiers such as logistic regression, SVM, Linear Discriminant and k-NN. The reported performance is 98.9% using SVM.

3 Apple Diseases Classification System

The familiar diseases in apple fruit are caused by fungal or infection. These common diseases include apple (rot, scab, and blotch). In this paper, the recognition system is divided into two parts: the first one is binary classification system for the apple (Good or Rotten) recognition, the second one is a multi-classification system to detect the type of disease of infected apple. The feature extraction is performed by using many features of image based on (color and texture), and then are classified using many classification algorithms. Four primary phases are included in the proposed system namely image preprocessing, segmentation, feature extraction, and classification, the principal architecture illustrated in Fig. 1.

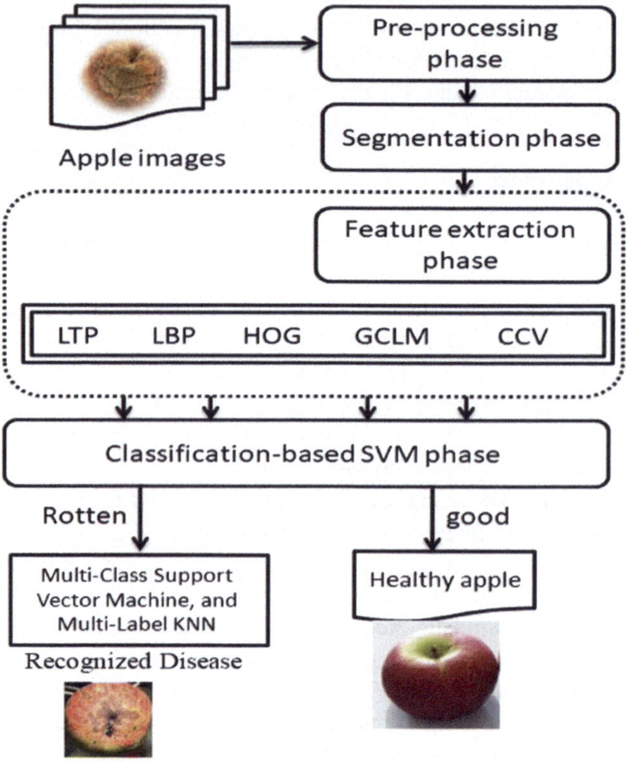

Fig. 1. Apple classification model

3.1 Preprocessing Phase

The first step is to normalize the images so that they have the same size. Preprocessing involves converting images from RGB color space to L*a*b* Color Space. The processing time is intended to reduce since using L*a*b* color space as the information is stored in only two channels in this space.

3.2 Apple Segmentation Phase

Image segmentation is one of the methods frequently used to classify an image accurately. Divide an image into regions so that the pixels have great similarity in each region and a high contrast between regions. Within the proposed framework, the K-means based approach is used for apple segmentation by distinguishing the infected portion.

The K-means algorithm segments the component into a K number by reducing the sum of squares of distances between the input objects and the corresponding area [10]. K-means calculate the centroid as follows:

$$c_i = \frac{1}{|S_i|} \sum\nolimits_{x_j \in s_i} X_j \tag{1}$$

Where S_i is the number of instances in cluster i and i = 1, 2, …, K. The fitness function is defined as follows:

$$fit = \sum\nolimits_{i=1}^{k} \sum\nolimits_{x_j \in s_i} d(c_i, X_i) \tag{2}$$

Where d (a, b) indicates squared Euclidean.

3.3 Feature Extraction Phase

The combination of features based on texture and color is applied to improve system performance. The features are derived from the healthy and defective images of the apple. In the proposed system, the texture-based feature descriptors such as LTP, LBP, HOG, GLCM, and color-based descriptors such as CCV are applied.

3.3.1 Local Ternary Patterns (LTP)

LTP has been proved to have great discriminative power for texture classification since it is resistant to illumination effects and invariant to monotonic gray-level transformations. The value of LTP of a pixel (i, j) is calculated by [11]:

$$LTP_{(P,R)}^{Pattern}(i,j) = \sum\nolimits_{p=0}^{P-1} s_{pattern}(g_p)2^p \tag{3}$$

$$S(P) = \begin{cases} -1 & if\, g_c - k \\ 0 & if\, p \geq g_c - k\, and\, P \geq g + k \\ 1 & if\, p > g_c - k \end{cases} \tag{4}$$

Where P is the neighbor's numbers, p is the neighboring pixel and gp is its grey level, R is the circle radius and gc is its grey level value, k is the neighbor threshold constant.

3.3.2 Local Binary Pattern (LBP)

LBP is an efficient texture-based descriptor which applied to encode pixel-based information based on their different mapping tables. LBP value of pixel (i, j) is computed by [11]:

$$S(P) = \begin{cases} -1 & P > 0 \\ 1 & P \leq 0 \end{cases} \tag{5}$$

LTP and LBP are similar in architecture since using a number of neighbors and a circle. The principal variation is that LBP thresholds the pixels into two values (0 and 1) while LTP thresholds the pixels into three values (−1, 0 and 1).

3.3.3 Histogram of Oriented Gradients (HOG)

HOG descriptor is used to extract the orientated gradients of the image by calculating orientation histograms in local region of edge intensity. The HOG descriptor is used to detect apple defects by dividing the image into equal cells and combining adjacent cells into a larger block, then calculates horizontal and vertical gradients; finally, superimposed slides to build the HOG features of the window. The HOG descriptor is calculated by [12]:

$$n = (Width - BlocKW/Step - 1) * (Height - BlockH/Step + 1) \tag{6}$$

3.3.4 Gray Level Co-occurrence Matrix (GLCM)

GLCM is a texture-based feature statistical method. The Gray level, intensity, and the frequency of the sequence of different gray levels co-occurrence of the image are tabulated. These tabulated data provides the measure of fluctuation of pixels intensity and is used for the estimation of texture feature. It gives the measure of fluctuation in intensity of pixels [13].

3.3.5 Color Coherence Vector (CCV)

CCV is a method of representation of characteristics based on color that represents the coherence of color as the amount at which the pixels of the image of the same color are the parts of a huge and uniform color area. In the CCV calculations, the components connected in the image are classified, then CCV discovers two-color histograms, finally, the two histograms are concatenated to form a single histogram [14].

3.4 Apple Classification Phase

K-Nearest Neighbor (K-NN): It is a supervised learning algorithm that is implemented for both classification and regression. The algorithm classifies data sets into several classes by calculating Euclidean and Hamming distance. KNN classifies data through the k-neighbors and most neighbors describe the data classified as output. Let feature vector is represented by x = (x1, x2, …, xm), then the distance between sample xi and xj is defined by the following formula [15]:

$$d(x_i, x_j) = \sqrt{\sum_{l=1}^{n} (x_i^l - x_j^l)} \tag{7}$$

Support Vector Machine: The SVM classifier is a supervised learning algorithm. SVM is a better choice for binary classification which is a method used of linearly separable problems. The core of SVM is to find a separating hyper-plane by [16],

$$g(x) = W^T * X + W_0 \tag{8}$$

Where W^T is the vector of weights and X is the input features. SVM maximizes the boundaries between different classes to ensure classification accuracy. It focuses on the minimization of structural risk in the learning process. It has a good ability to generalize in characteristics of non-linear, high-dimension and small data sets. In this article, an SVM linear classifier with the kernel function 'RBF' is applied for the classification of good and rotten apple [17].

4 Experimental Results

The results are divided into two parts. (1) Results using the binary classification of the apple as healthy or rotten, and (2) results of classification of multiple classes from the apple data set that has three types of diseases (rot, blotch and scab). Data sets in [8] are used as a benchmark which involves four different kinds of apple: Normal (80), Blotch (80), rot (80), and scab (80) with a total of 320 apple fruit samples as shown in Figs. 2 and 3. In this paper, all experiments is done using a MATLAB (2016 b) on a PC operating with Intel 2.7 GHz, 64-bit Windows 10 and 8 GB RAM.

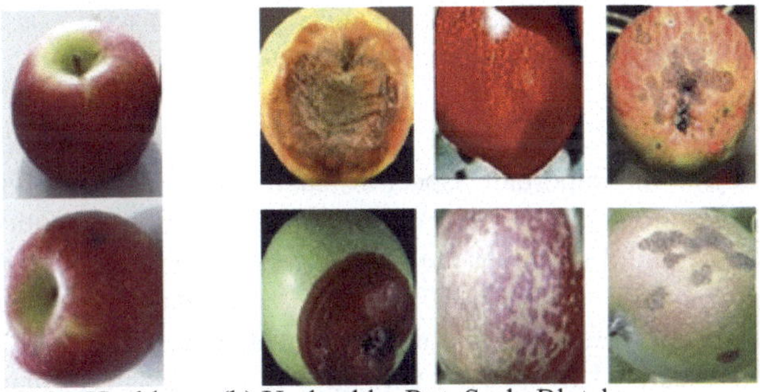

(a) Healthy (b) Un-healthy Rot, Scab, Blotch

Fig. 2. Samples of the image data sets

The investigations of some feature descriptors based on textures and colors derived from apple images with respect to the K-NN and SVM classifiers are applied in the search to find the best category of apple. We divide data into (70) samples per class for training with total (280) sample and remaining samples for testing. The accuracy of the proposed method is determined by:

$$Accuracy(\%) = \frac{Total\ number\ of\ correctly\ Classified\ Samples}{Total\ Number\ of\ Samples} * 100 \tag{9}$$

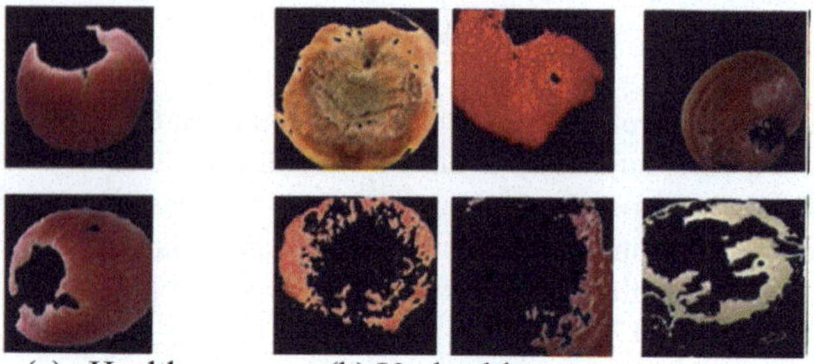
(a) Healthy (b) Un-healthy

Fig. 3. K-means segmentation results

The performance of different feature extractors is evaluated by applying two different classifiers as shown in Table 1. Experiments show that GLCM does not work well and the reported accuracy is lower for KNN and SVM; However, HOG works well with KNN, LTP works best with SVM with 100% accuracy recognition.

Table 1. Normal and abnormal classification accuracy

Classifier	LTP	LBP	HOG	GLCM	CCV
K-NN	98.4%	99%	**99.06%**	95.6%	96.65%
SVM	**100%**	75%	98.04%	75.06%	99.8%

Class accuracy is an important aspect when it comes to classifying apple fruit disease. This information indicates the classes that need more attention in resolving confusions. Tables 2 and 3 represent the accuracy for each class of apple disease of 3 classes using (HOG, LTP, CCV, LBP and GLCM) descriptors with ML-KNN and MSVM. Also, a fusion of features is done by concatenating the LTP, CCV and LBP features together. Clearly, LBP and GLCM have the lowest performance with SVM where the merged functions work well.

Table 2. Classification accuracy of apple diseases using ML-KNN

ML-KNN						
Classifier	LTP	LBP	HOG	GLCM	CCV	LTP+CCV+LBP
Blotch	95%	96.2%	95%	96.2%	96.2%	**96.2%**
Rot	93.7%	93.7%	91.2%	**96.2%**	95%	**98.7%**
Scab	96.2%	97.5%	93.7%	98.7%	98.7%	**98.75%**
Average accuracy	95%	95.8%	93.3%	97%	96.6%	**97.9%**

Table 3. Classification accuracy of apple diseases using MSVM

MSVM						
Classifier	LTP	LBP	HOG	GLCM	CCV	LTP+CCV+LBP
Blotch	95%	77.5%	92.5%	67.5%	91.2%	**95%**
Rot	90%	78.5%	90%	68%	87%	**95%**
Scab	95%	52.5%	93.7	61.5%	75%	**93.7%**
Average accuracy	93.3%	69%	92%	51%	84.5%	**95.4%**

Apple's disease classification is performed well using ML-KNN on MSVM with an average accuracy of 97.7%, where SVM performs well in binary classification. Sequentially, ML-KNN resulting average accuracy of feature fusion with ML-KNN is 97.9% and 95.4% with MSVM with 70 training samples for each class.

The experiments shows that GLCM and LBP don't perform well with SVM in binary classification. One possible explanation is that, GLCM and LBP features have only texture information, they don't considers color and neighboring information; however CCV and LTP performs better than GLCM and LBP in both texture space and color space as LTP is resistant to illumination effects and CCV uses frequency of each color in coherent and incoherent regions separately.

From Tables 2 and 3, it is clear that in multiclass classifications, fusing LTP, CCV and LBP features yield better results than separate feature extractors because both LTP, CCV and LBP uses the neighboring, texture and color information of each pixel in the image. Both LTP and LBP are robust to illumination differences, so fusing them with CCV are computationally more efficient and enhanced the classification accuracy. The comparisons of proposed approach with previous contributions using the same dataset [5, 8] are illustrated in Figs. 4 and 5. As shown the proposed model performs better than previous contributions in terms of accuracy.

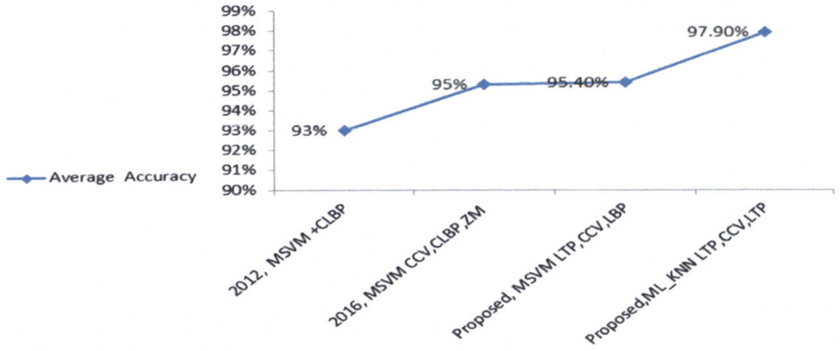

Fig. 4. Average recognition comparison between previous works and proposed model classification models

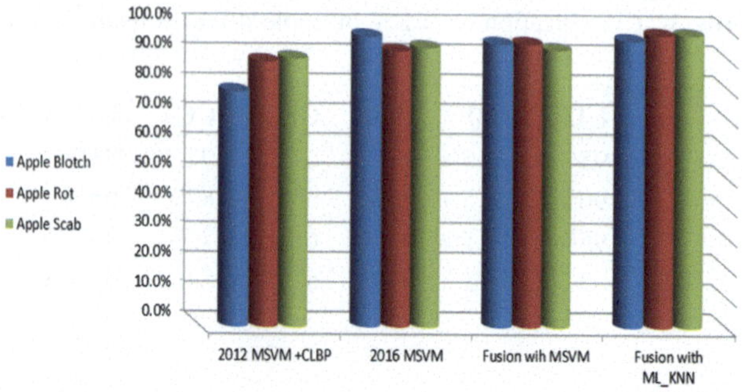

Fig. 5. Diseases recognition accuracy between previous works and proposed model classification models

5 Conclusions

In this paper, an automated intelligent system for the recognition of apple fruits health and diseases is introduced. The apple segmentation is performed using K-means-based image segmentation approach for a region of interest extraction. The main contribution of the paper is the recognition of apple health (normal or rotten) and the enhancement of apple diseases (blotch, scab, and rot) classification for early treatment using color and texture properties. The experiments concentrated on the analysis of different feature descriptors performance with KNN, ML-KNN, SVM, and MSVM classifiers. The HOG, LTP, LBP, GLCM, CCV and fusion of (LTP, CCV, and LBP) feature extraction descriptors are applied. According to the result, the percentage of accuracy is enough to recognition apple diseases. The proposed model is used for apple health recognition and infected apple detection. Experiments point to the improvements and vulnerability of the proposed methods for the problem of classification of apple diseases. According to the classification results, the healthy apples are simply distinct corresponded to the infected apple fruits plus the fusion of the color and texture-based features outperform separate types. The fusion improved average accuracy rate to reach 97.9%. In the future, further work includes leaf feature extraction for plant classification.

References

1. Dubey, S.R., Jalal, A.S.: Application of image processing in fruit and vegetable analysis: a review. J. Intell. Syst. **24**, 405–424 (2015)
2. Roberts, M.J., Schimmelpfennig, D., Ashley, E., Livingston, M.: The value of plant disease early-warning systems: a case study of USDA's soybean rust coordinated framework. United States Department of Agriculture, Economic Research Service, Economic Research Report No. 18. Available via (2006)
3. Rocha, A., Hauagge, D.C., Wainer, J., Goldenstein, S.: Automatic fruit and vegetable classification from images. Comput. Electron. Agric. **70**, 96–104 (2010)
4. Unay, D., et al.: Automatic grading of bi-colored apples by multispectral machine vision. Comput. Electron. Agric. **75**(1), 204–212 (2011)

5. Dubey, S.R., Jalal, A.S.: Detection and classification of apple fruit diseases using complete local binary patterns. In: Proceedings of the 3rd International Conference on Computer and Communication Technology, Allahabad, India, pp. 346–351 (2012)
6. Ashok, V., Vinod, D.S.: Automatic quality evaluation of fruits using probabilistic neural network approach. In: 2014 International Conference on Contemporary Computing and Informatics (IC3I). IEEE (2014)
7. Dubey, S.R., Jalal, A.S.: Fruit disease recognition using improved sum and difference histogram from images. Int. J. Appl. Pattern Recognit. **1**(2), 199–220 (2014)
8. Dubey, S.R., Jalal, A.S.: Apple disease classification using color, texture and shape features from images. Signal Image Video Process. **10**(5), 819–826 (2016)
9. Singh, S., Singh, N.P.: Machine learning-based classification of good and rotten apple. In: Recent Trends in Communication, Computing, and Electronics, pp. 377–386. Springer, Singapore (2019)
10. Hrosik, R.C., et al.: Brain image segmentation based on firefly algorithm combined with K-means clustering. Stud. Inform. Control **28**, 167–176 (2019)
11. Tan, X., Triggs, B.: Enhanced local texture feature sets for face recognition under difficult lighting conditions, pp. 168–182. INRIA & Laboratoire Jean Kuntzmann, 655 avenue de l'Europe, Montbonnot 38330, France (2007)
12. Wei, Y., et al.: Multi-vehicle detection algorithm through combining Harr and HOG features. Math. Comput. Simul. **155**, 130–145 (2019)
13. Thu, W.M., et al.: GLCM and LTP based classification of food types (2018)
14. Pass, G., Zabih, R., Miller, J.: Comparing images using color coherence vectors. In: Proceedings of the fourth ACM International Conference on Multimedia, New York, USA, pp. 65–73 (1997)
15. Sun, Z., et al.: Mutual information based multi-label feature selection via constrained convex optimization. Neurocomputing **329**, 447–456 (2019)
16. Gola, J., et al.: Objective microstructure classification by support vector machine (SVM) using a combination of morphological parameters and textural features for low carbon steels. Comput. Mater. Sci. **160**, 186–196 (2019)
17. Mao, T., et al.: Defect recognition method based on HOG and SVM for drone inspection images of power transmission line. In: 2019 International Conference on High Performance Big Data and Intelligent Systems (HPBD&IS). IEEE (2019)

Experimental Modeling of Hexapod Robot Using Artificial Intelligence

Abdelrahman Sayed Sayed[1,2(✉)], Amr Ahmed Mohamed[2], Ahmed Magd Aly[1,2], Youssef Mohamed Hassan[2], Abdallah Mahir Abdulaziz[2], Hossam Hassan Ammar[1,2], and Rafaat Shalaby[2,3]

[1] Smart Engineering Systems Research Center (SESC), Nile University, Sheikh Zayed City, Giza 12588, Egypt

[2] School of Engineering and Applied Sciences, Nile University Campus, Sheikh Zayed District, Juhayna Square, 6th of October City, Giza 12588, Egypt

{ab.sayed,a.ahmed,a.magd,y.aly,a.maher,hhassan,rshalaby}@nu.edu.eg

[3] Faculty of Electronic Engineering, Menofia University, Menofia, Egypt

Abstract. Hexapod Robots gave us the opportunity to study walking robots without facing problems such as stability and expensive custom made hardware. It has a great deal of flexibility in moving over different terrains even if some legs become malfunctioned or facing some difficulties in movement. In this study the kinematic analysis of CH3-R 18DOF Hexapod Robot is discussed where each leg contains three revolute joints in order to mimic the structure of a spider. To develop the overall kinematic model of CH3-R robot, direct and inverse kinematic analyses for each leg have been considered where the Denavit-Hartenberg (D-H) conventions will be used to perform the forward kinematic analysis of the six-legged robot while the inverse kinematics are obtained by simplifying the architecture of the robot into 7 modules, a hexagon trunk and 6 limbs between the trunk and the ground. A neural network (NN) model was employed in order to compare and decide the angles prediction, training time and overall performance with the ones obtained from the analytical solution of the inverse kinematics. The data points used for training the two models and testing their performance was acquired from giving the robot an initial and final position and recording the change in the coordinates and time by using an IMU sensor.

Keywords: Hexapod · Kinematic analysis · Neural networks

1 Introduction

Legged walking robots became the main goal of many studies and researches as the main role of legged robots is to simulate the characterstics of some biological inspired organisms [1]. The Hexapod multi-legged robot have powerful potentials compared to tracked or wheeled robots. Legged robots have the ability to adapt to complex external environments and deal with dangerous or unreachable tasks, whereas tracked and wheeled vehicles are only effective with obstacles

A.-E. Hassanien et al. (Eds.): AICV 2020, AISC 1153, pp. 26–36, 2020.
https://doi.org/10.1007/978-3-030-44289-7_3

that are smaller than half the wheel diameter. Moreover, legged robots have the ability to find the local horizontal footholds in regional steep terrain, as they can climb extreme angles [2]. The available types of legged robots ranges between single-legged hoppers, bipedals, tripods, quadrupeds, hexapods, octopods and other types having more limb legs [3–5]. Hexapod robots demonstrated distinctive advantages, such as walking on unstable or irregular terrain, and having more mobility, flexibility and stability than tracked and wheeled robots [6–8]. The applications include but are not limited to exploration, mining, search and rescue and industrial environments on earth and below. There are also many studies about hexapod robot models, especially in gait planning and control, as in [9] that describes a central pattern generator to generate the gait or in [10] that implements a free based gait on posture control. The problem in the control of such mobile systems is defining the proper application of parameters according to the planned goals to be achieved. So, defining the sampling rate, which controller to be used, how much accuracy is required by the locomotion pattern, the specifications of the actuators and sensors are all considered to be necessary steps, but they do not always outcome a trivial solution. In most of the cases the mathematical models of such systems are complex and non-linear, and finding a proper controller for such systems require the application of a sort of linearization techniques in order to reach a linear simple plant for the robot [11]. The rest of the paper is organized as follows, in Sect. 2 the kinematic model of the hexapod is discussed, Sect. 3 discusses the Neural Network structure and its parameters, Sect. 4 shows the discussion and results. Finally, Sect. 5 is reserved for the conclusion.

2 Kinematic Model of Hexapod

The typical hexapod robots types can be classified into rectangular and hexagonal configuration as shown in Fig. 1. The rectangular configuration is inspired from walking insects with six legs distributed symmetrically along two sides where each side has three legs, this configuration is fast in forward direction but not handy in moving or turning sideways. The hexagonal configuration has

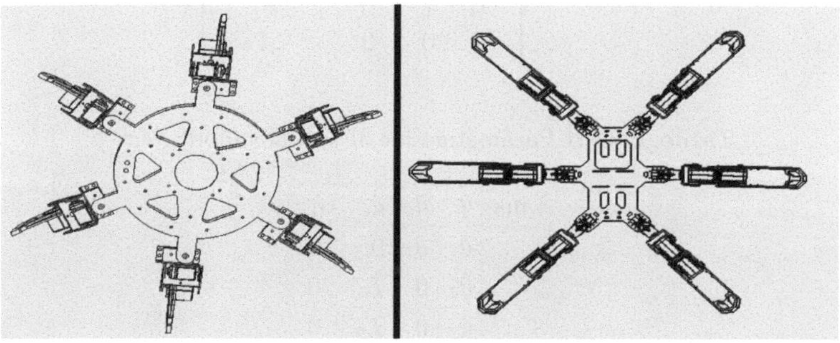

Fig. 1. Two common types of hexapod robots

six legs disturbed in an axisymmetric way around the body. The advantage of the hexagonal configuration is being more flexible in achieving the same walking speeds in all directions. The ch3-r hexapod robot compared to other hexapod robots is that its legs can rotate around the center of the body [12–17].

2.1 Forward Kinematics Problem

Each single leg of the hexapod can be regarded as a serial manipulator where its base is fixed on the trunk and its end effector is on the ground. The Denavit-Hartenberg(D-H) method is applied for one leg, so we can write the transformation matrix for each joint based on the frames shown in Fig. 2.

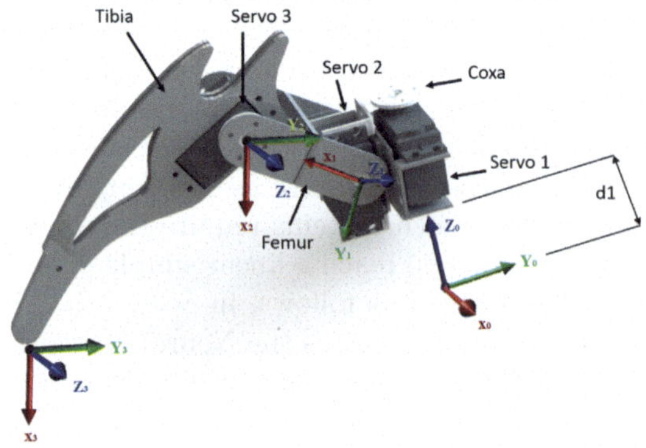

Fig. 2. Frame assignments for finding D-H parameters

Table 1 shows the D-H parameters that include link angle θ_i, link distance d_i, link length a_i and link twist α_i. The transformation matrices between the legs coordinates can be obtained as:

$$T_{femur}^{coxa} = \begin{bmatrix} cos\theta_1 & 0 & sin\theta_1 & L_1 cos\theta_1 \\ sin\theta_1 & 0 & -cos\theta_1 & L_1 sin\theta_1 \\ 0 & 1 & 0 & d_1 \\ 0 & 0 & 0 & 1 \end{bmatrix} \quad (1)$$

Table 1. D-H Parameters for 3DOF hexapod's leg

Link	θ_i	d_i	a_i	α_i
1	θ_1	d_1	0	0
2	θ_2	0	L_1	90
3	θ_3	0	L_2	0
4	0	0	L_3	0

$$T_{tibia}^{femur} = \begin{bmatrix} cos\theta_2 & -sin\theta_2 & 0 & L_2cos\theta_2 \\ sin\theta_2 & cos\theta_2 & 0 & L_2sin\theta_2 \\ 0 & 0 & 1 & 0 \\ 0 & 0 & 0 & 1 \end{bmatrix} \quad (2)$$

$$T_{base}^{tibia} = \begin{bmatrix} cos\theta_3 & -sin\theta_3 & 0 & L_3cos\theta_3 \\ sin\theta_3 & cos\theta_3 & 0 & L_3sin\theta_3 \\ 0 & 0 & 1 & 0 \\ 0 & 0 & 0 & 1 \end{bmatrix} \quad (3)$$

From Eqs. (1), (2) and (3) the homogeneous transformation matrices of the legs links can be obtained as:

$$T_{base}^{coxa} = T_{femur}^{coxa} T_{tibia}^{femur} T_{base}^{tibia} \quad (4)$$

where P_x, P_y and P_z refers to the position matrix between the foot tip coordinate and the body coordinate as :

$$P_x = L_1c\theta_1 + L_2c\theta_1c\theta_2 - L_3c\theta_1s\theta_2s\theta_3 + L_3c\theta_1c\theta_2c\theta_3 \quad (5)$$

$$P_y = L_1s\theta_1 + L_2s\theta_1c\theta_2 + L_3s\theta_1c\theta_2c\theta_3 - L_3s\theta_1s\theta_2s\theta_3 \quad (6)$$

$$P_z = d_1 + L_2s\theta_2 + L_3c\theta_2s\theta_3 + L_3s\theta_2c\theta_3 \quad (7)$$

2.2 Inverse Kinematics Problem

The forward kinematic problem is solved in the previous section but the reverse operation is always interesting and if it is possible for the leg to reach a position in space without any singularities, then it is also possible to obtain the joint angles for the given position. There are two main methods for solving the inverse kinematics which are the Numerical solutions but because of their iterative nature they are much slower than the Closed-form solutions which are based on analytical expressions or the solution of a polynomial having a degree 4 or

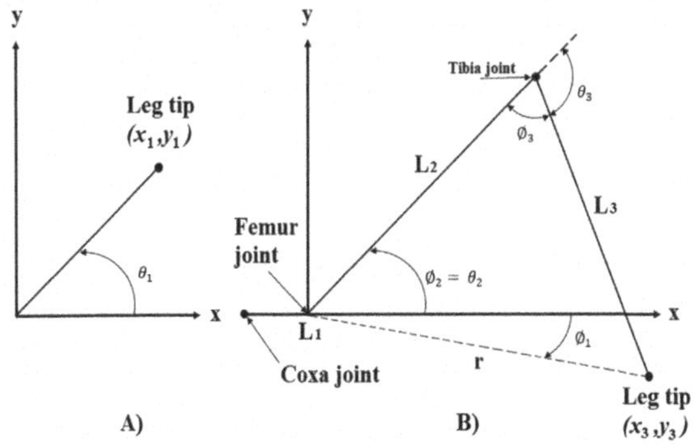

Fig. 3. 2D triangle with vertices of link frames origins. Adopted from [21]

less [18–20]. The following equations show how to solve the inverse kinematics based on the triangle with vertices in the origins of the coxa, femur and tibia frames as shown in Fig. 3.

$$coxaangle\theta_1 = tan^{-1}(\frac{Y_4}{X_4}) \tag{8}$$

$$TrueX = \sqrt{X^2 + Y^2} - L_{coxa} \tag{9}$$

$$Im = \sqrt{(TrueX)^2 + Z^2} \tag{10}$$

Get femur angle above the horizon

$$q_1 = tan^{-1}\frac{Z}{TrueX} \tag{11}$$

Apply the law of cosine in the triangle abc

$$q_2 = cos{-1}\frac{\sqrt{Lf^2 + Im^2 - Lt^2}}{2 * Lf * Im} \tag{12}$$

and obtain tibia angle from femur

$$\varphi = cos{-1}\frac{\sqrt{Lf^2 + Im^2 - Lt^2}}{2 * Lf * Lt} \tag{13}$$

$$tibiaangle\theta_3 = \varphi - 180 \tag{14}$$

After obtaining the inverse kinematic solution at the initial position as shown in Fig. 4 and Table 2, we simulated the gait of the initial position on each leg as shown in Fig. 5.

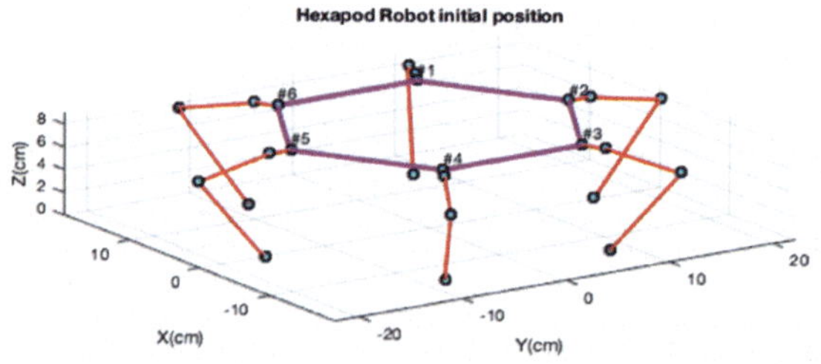

Fig. 4. Hexapod robot initial position simulated in MATLAB

Table 2. Initial position leg joint angles

Leg angles	coxa	femur	tibia
Leg_1	60	10.98	118.96
Leg_2	0	10.98	118.96
Leg_3	−59.99	10.98	118.96
Leg_4	−119.99	10.98	118.96
Leg_5	−179.99	10.98	118.96
Leg_6	120	10.98	118.96

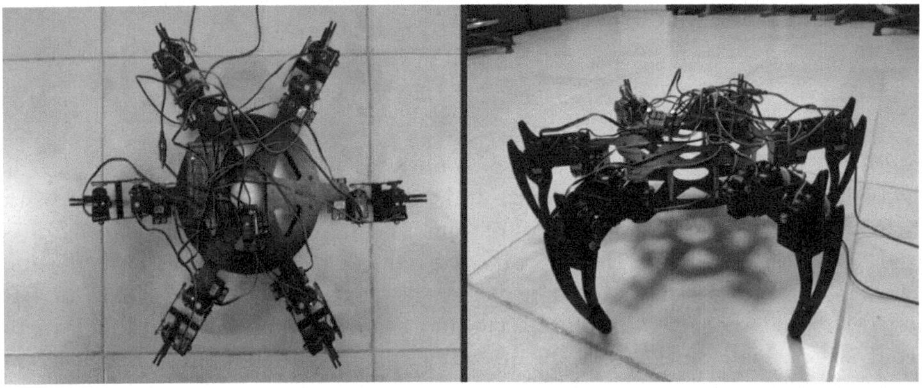

Fig. 5. Hexapod robot initial position

3 Neural Network Structure

In order to get the inverse kinematic solution of this robot, four neural network models were designed. Each model has 6 inputs, as the angular speeds of each upper joint, and one output, which is either the angular or linear velocity of the robot's body. It is a feed-forward neural network model with back propagation. It has 6 input neurons and one output neuron, in addition to a hidden layer that consists of 10 neurons. All the neurons were established to use the tangent sigmoid as a transfer function for their output. The first two models were trained with the Levenberg-Marquardt Algorithm (LMA) using MATLAB's NN toolbox. This algorithm is usually used to solve non-linear least squares problems. Before LMA was implemented, either gradient descent or Gauss-Newton were used as training functions. Nevertheless, they both had some disadvantages; thus LMA combined them together in one algorithm. As it turned out that if back-propagation was used with gradient descent, the algorithm will act slowly and will not reach to the optimal solution. On the other hand, Gauss-Newton algorithm have the highest probability to give an optimal solution. The Hessian matrix for this algorithm is expressed as follows:

$$H = J^T J \tag{15}$$

$$g = J^T e \tag{16}$$

where J represents the jacobian matrix while e represents the net error vector. The LMA then behaves as Newton is expressed as shown in the following equation:

$$X_{K+1} = X_K - [J^T J + \mu I]^{-1} J^T e \tag{17}$$

where X_{K+1} represents the new weight calculated as gradient and X_K, which is the current weight using Newton function [22]. The other two models were trained using Bayesian Regularization Algorithm (BR) using MATLAB's NN toolbox. In the training processes the following term is usually used as a measure of errors in the network results:

$$F = E_D(D|\omega, M) = \frac{1}{N}\sum_{i=1}^{n}(\hat{t_i} - t_i) \tag{18}$$

where E_D represents the mean sum of squares for the network errors and D is the training set that consists of inputs and targets while M is the neural network with its hidden layers, activation functions and its overall architecture. E_D is used to avoid over-fitting. However, it may reduce the variance while increasing the bias, thus BR is used to reduce both of them. The regularization in BR networks adds another additional term and then uses an objective function to avoid large weights which results in a much smoother mapping. The weights therefore can be updated using Bayes' rule:

$$P(\omega|D, \alpha, \beta, M) = \frac{P(D|\omega, \beta, M) \cdot P(\omega|\alpha, M)}{P(D|\omega, \beta, M)} \tag{19}$$

where α and β are hyperparameters that need to be estimated function parameters. According to Bayes' rule, the network will be a probabilistic framework because of the inclusion of probability distribution of network weights [23]. In order to train the models, a v-rep simulation was implemented to form the training set. In this simulation, the angles of the legs were recorded and exported to an excel sheet along with the angular and linear velocities of the robot's body. All the information about the characteristics of the used neural network model are listed in Table 3.

Table 3. Characteristics of the used neural networks model

Property	Value
Number of layers	3 (Input, hidden and output)
Neurons and activation functions at layer 1, 2	(6,10) Sigmoidal
Neurons and activation functions at layer 3	(1) Linear
Training algorithm	Levenberg-Marquardt and Bayesian regularization
Training error goal	1e−20
Max epochs allowed	100000

4 Experimental Results and Discussion

The output of the neural network's model for the linear and angular velocities succeeded to mimic the actual output in real life with a root mean square error (RMSE) of 0.0437 for Levenberg-Marquardt and 9.0124 for Bayesian regularization both are for the linear velocity results. On the other hand, for the angular velocity the RMSE was 0.0405 for Levenberg-Marquardt model and 7.4839 for Bayesian regularization model. The four estimated velocity outputs and their actual counterparts are shown in Figs. 6 and 7. A validation set was used to show if the system can predict the correct outputs for inputs that were not used in the training. Tables 4 and 5 show in details the statistical analysis of the output results from the neural networks' models.

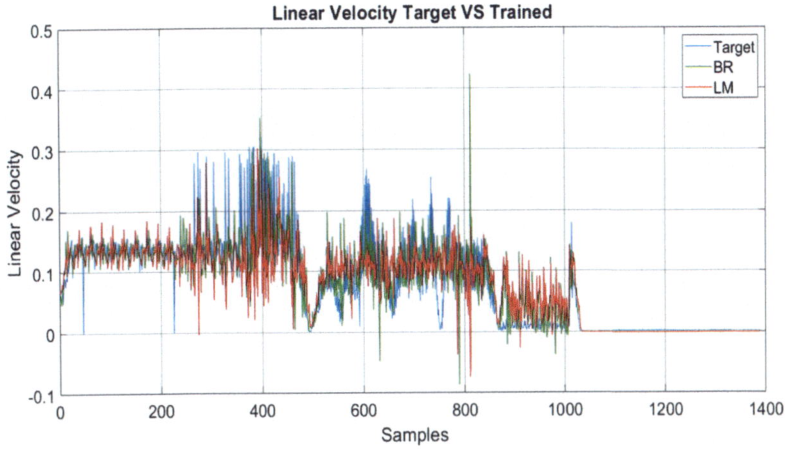

Fig. 6. NN testing's output for linear velocities with LM and BR vs the actual output

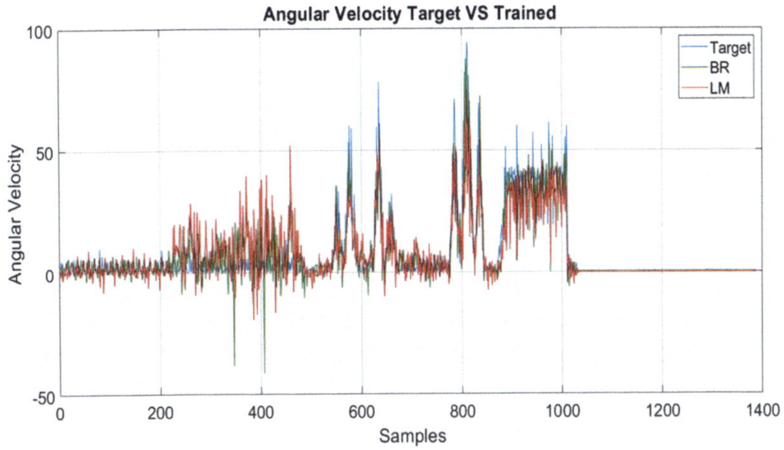

Fig. 7. NN testing's output for angular velocities with LM and BR vs the actual output

Table 4. Statistical analysis for NN model with Levenberg-Marquadrt training function

	Linear velocity	Angular velocity
Mean	0.0792	7.7332
SD	0.0603	12.3269
Mean absolute error (MAE)	0.0266	5.1638
Root means squared error (RMSE)	0.0437	9.0124
Normalized Root means squared error (NRMSE)	0.5533	1.1506
Pearson Rp	0.8007	0.7889
Pearson R^2p	0.6411	0.6224
Spearman Rs (S-Rs)	0.8138	0.5990

Table 5. Statistical Analysis for NN model with Bayesian regularization training function

	Linear velocity	Angular velocity
Mean	0.0793	7.6490
SD	0.0651	13.1370
Mean absolute error (MAE)	0.0238	4.3564
Root means squared error (RMSE)	0.0405	7.4839
Normalized Root means squared error (NRMSE)	0.5128	0.9555
Pearson Rp	0.8335	0.8598
Pearson R^2p	0.6947	0.7392
Spearman Rs (S-Rs)	0.8295	0.5992

Figure 8 shows the performance plots with the mean error decreasing gradually.

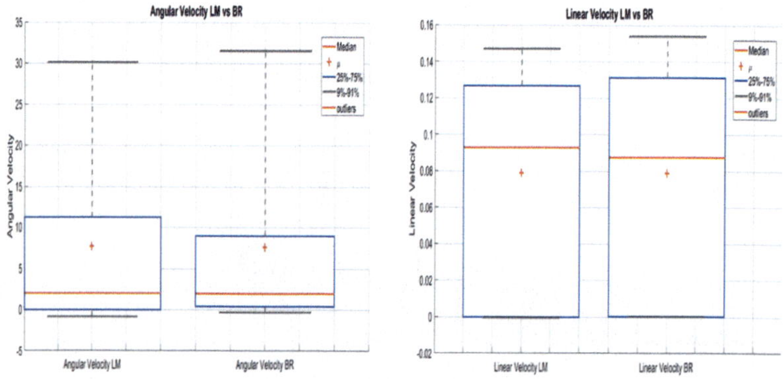

Fig. 8. Angular linear velocities training performance with LM and BR

5 Conclusion

In the present study, the inverse kinematics solution for CH3-R hexapod robot is obtained by using the Denavit-Hartenberg (D-H) parameters assuming that each separate leg is a serial open chain manipulator. The simulated results were compared with the actual robot in terms of the robot initial position, linear and angular velocities. A neural network algorithm was implemented in order to simplify the mathematical chaos of hexapod's model. The required output from the neural network was the solution of the inverse kinematics of the robot or in other words the required motor angles that would be feed to the robot to reach the required position. That implemented performed very well in estimating the actual angular and linear velocities of the robot. However, as it can be seen from the discussed results, the Levenberg-Marquardt training function was more accurate than the Bayesian regularization function in this application. That can be concluded from the fact that the Levenberg's model took much less epochs to reach to the best performance while Bayesian's model was 10 times higher than that. On the other hand, the Spearman correlation factor for Levenberg's model was also much better than that of the Bayesian's model as shown in Tables 4 and 5. For the future work, adaptive neuro-fuzzy inference system (ANFIS) can be implemented to predict the outputs and it is power lies in its ability to be optimized with metaheuristic techniques and genetic algorithms.

References

1. Oliveira, I., Barbosa, R., Silva, M.: Modelling, trajectory planning and control of a quadruped robot using matlab®/simulink™. In: Ollero, A., Sanfeliu, A., Montano, L., Lau, N., Cardeira, C. (eds.) ROBOT 2017: Third Iberian Robotics Conference, pp. 756–767. Springer, Cham (2018)
2. Burkus, E., Odry, P.: Autonomous hexapod walker robot "szabad(ka)". In: 2007 5th International Symposium on Intelligent Systems and Informatics, pp. 103–106, August 2007
3. Chen, L., Zhong, G., Liu, Z., Deng, H.: A novel hexapod robot: design and mobility analysis. In: 2016 IEEE International Conference on Mechatronics and Automation, pp. 2094–2098, August 2016
4. Sun, J., Ren, J., Jin, Y., Wang, B., Chen, D.: Hexapod robot kinematics modeling and tripod gait design based on the foot end trajectory. In: 2017 IEEE International Conference on Robotics and Biomimetics (ROBIO), pp. 2611–2616, December 2017
5. Raibert, M.H., Tello, E.R.: Legged robots that balance. IEEE Expert **1**, 89–89 (1986)
6. Stoica, A., Carbone, G., Ceccarelli, M., Pisla, D.: Cassino hexapod: experiences and new leg design. In: 2010 IEEE International Conference on Automation, Quality and Testing, Robotics (AQTR), vol. 3, pp. 1–6, May 2010
7. Wang, Z., Ding, X., Rovetta, A., Giusti, A.: Mobility analysis of the typical gait of a radial symmetrical six-legged robot. Mechatronics **21**, 1133–1146 (2011)
8. Cao, Q., van Rijn, A.T., Poulakakis, I.: On the control of gait transitions in quadrupedal running. In: 2015 IEEE/RSJ International Conference on Intelligent Robots and Systems (IROS), pp. 5136–5141, September 2015

9. Ya-xin, Y., Bo, J.: Gait generation and control of a hexapod walking robot. Open Electr. Electron. Eng. J. **8**, 335–341 (2014)
10. Suzuki, Y., Yuki, M., Inagaki, S.: Adaptive walking of hexapod robots based on FCP gait control considering postural stabilization. In: The Proceedings of JSME annual Conference on Robotics and Mechatronics (Robomec), vol. 2018, pp. 2P2–I02, December 2018
11. Azar, A.T., Sayed, A.S., Shahin, A.S., Elkholy, H.A., Ammar, H.H.: PID controller for 2-DOFs twin rotor MIMO system tuned with particle swarm optimization. In: Hassanien, A.E., Shaalan, K., Tolba, M.F. (eds.) Proceedings of the International Conference on Advanced Intelligent Systems and Informatics, pp. 229–242. Springer, Cham (2020)
12. Muñoz, P., Castaño, B., R-Moreno, M.: Simulation of the hexapod robot ptinto walking on irregular surfaces. Int. J. Simul. Modell. **14**, 5–16 (2015)
13. Pa, P.S., Wu, C.M.: Design of a hexapod robot with a servo control and a man-machine interface. Robot. Comput.-Integr. Manuf. **28**(3), 351–358 (2012)
14. Chen, J., Liu, Y., Zhao, J., Zhanghe, H., Jin, H.: Biomimetic design and optimal swing of a hexapod robot leg. J. Bionic Eng. **11**, 26–35 (2014)
15. Zhao, J., Zhang, H., Liu, Y., Yan, J., Zang, X., Zhou, Z.: Development of the hexapod robot HITCR-II for walking on unstructured terrain. In: 2012 IEEE International Conference on Mechatronics and Automation, pp. 64–69, August 2012
16. Faal, S.G., Chen, F., Tao, W., Agheli, M., Tasdighikalat, S., Onal, C.D.: Hierarchical Kinematic Design of Foldable Hexapedal Locomotion Platforms. J. Mech. Robot. **8**(1), 011005 (2015)
17. Hodoshima, R., Watanabe, S., Nishiyama, Y., Sakaki, A., Ohura, Y., Kotosaka, S.: Development of ASURA I: Harvestman-like hexapod walking robot—approach for long-legged robot and leg mechanism design. In: 2013 IEEE/RSJ International Conference on Intelligent Robots and Systems, pp. 4669–4674, November 2013
18. Spong, M.W., Hutchinson, S.A., Vidyasagar, M.: Robot modeling and control. IEEE Control Syst. **26**(6), 113–115 (2006)
19. George Lee, C.S.: Robot arm kinematics, dynamics, and control. Computer **15**(12), 62–80 (1982)
20. Korein, J.U., Badler, N.I.: Techniques for generating the goal-directed motion of articulated structures. IEEE Comput. Graph. Appl. **9**, 71–81 (1982)
21. Ollervides, J., Orrante-Sakanassi, J., Santibáñez, V., Dzul, A.: Navigation control system of walking hexapod robot. In: 2012 IEEE Ninth Electronics, Robotics and Automotive Mechanics Conference, pp. 60–65, November 2012
22. Du, Y.-C., Stephanus, A.: Levenberg-Marquardt neural network algorithm for degree of arteriovenous fistula stenosis classification using a dual optical photoplethysmography sensor. Sensors **18**, 2322 (2018)
23. Kayri, M.: Predictive abilities of Bayesian regularization and Levenberg-Marquardt algorithms in artificial neural networks: a comparative empirical study on social data. Math. Comput. Appl. **21**, 1–11 (2016)

A Systematic Review of the Factors Affecting the Artificial Intelligence Implementation in the Health Care Sector

Shaikha F. S. Alhashmi[1], Muhammad Alshurideh[2,3], Barween Al Kurdi[4], and Said A. Salloum[1,5(✉)]

[1] Faculty of Engineering and IT, The British University in Dubai, Dubai, UAE
[2] Faculty of Business, University of Jordan, Amman, Jordan
[3] Management Department, University of Sharjah, Sharjah, UAE
[4] Marketing Department, Amman Arab University, Amman, Jordan
[5] Research Institute of Sciences and Engineering, University of Sharjah, Sharjah, UAE
ssalloum@sharjah.ac.ae

Abstract. A systematic review for twenty-three research studies published between (2015–2018) was reviewed and analyzed deeply in order to answer the research question for the critical success factors for implementing artificial intelligence (AI) projects within the health sector. Afterwards, different constructs were elicited from the previous targeted studies to end up with a list of the most widely used external factors that are relied upon frequently during the process of adopting the technology. The study provide a critical discussion about what does (AI) means and how it can be linked to use the Technology Acceptance Model (TAM) to discuss the willingness to use (AI) systems by individuals.

Keywords: Systematic review · Artificial Intelligence (AI) · Technology Acceptance Model (TAM) · Health sector

1 Introduction

(AI) technology is growing rapidly to become an effective contribution of monitoring patients and management control in the health domain. Advanced technology is serving various operations within the health industry such as, data management system, prediction of future diseases and examination patients' health conditions. Employing (AI) practices can affect people health conditions in the developed countries, supporting the society by predicting future illnesses and fighting against medication shortage. Therefore, a need arises to analyze some of the factors that are considered critical and successful in implementing these projects. The aim is to define what the critical success factors for implementing (AI) projects for health care sector. Moreover, the study aims to bring and analysis some of what previous published studies utilized of different theoretical frameworks in different fields to assist in building a proper background of AI and to check which model that encompasses the most external factors such as, managerial, organizational, operational, strategic and IT infrastructure factors

A.-E. Hassanien et al. (Eds.): AICV 2020, AISC 1153, pp. 37–49, 2020.
https://doi.org/10.1007/978-3-030-44289-7_4

that can be used to study the acceptance and adoption of the (AI) projects in healthcare [1, 2]. Additionally, Massive quantity of research studies relating to the critical success factors for Information Technology (IT) projects in various fields has been reviewed [3–6]. Accordingly, it is found that none of the literature review papers has explored all the following critical success factors in health domain such as, managerial, operational, strategic, IT infrastructure and organizational factors in one unique paper. In addition, no single empirical research paper focused its potential on establishing (AI) projects in the healthcare sector within the emerging markets contexts. However, other research papers focuses on CSFs in IT projects for instance, Enterprise Resource Planning (ERP), Business Intelligence (BI) and Data Management Information Systems (DMIS). Besides, educational field is not an exceptional, various projects such as, e-learning is employing (AI) knowledge to foster the educational sector.

2 Background

2.1 Study Importance and Contribution

Possibilities in monitoring patient's conditions will be manageable with effectiveness since the adoption and acceptance of (AI) projects will expand towards assisting individual's health history. Acceptance and adoption of (AI) projects critical success factors are dependable on the way it's operating and the provided advantages in the respective domain within healthcare. Referring to [7], (AI) served in different roles within various sectors globally and these roles have been investigated by other authors for different studies according to their disciplines. According to [8], stated that nurses in hospitals are making diagnoses and noting down doctor's orders and playing an important role in adopting Electronic Medical Record in their environment. Nurses' attitudes toward accepting or rejecting (AI) projects in their environment are important to adopt the defined (AI) project. Along as, there is no even a single previous study that addressed the five factors (managerial, operational, strategic, IT infrastructure and organizational) for implementing (AI) projects in the healthcare sector. Current paper evaluates these factors practices as far as adoption or rather acceptance of the (AI) projects in healthcare are concerned [9]. The major rationale behind this paper is to other organizations and institutions to design and initiating different (AI) strategies in the healthcare sector. Accordingly, this paper contribution will add knowledge to the existing literature and demonstrate (AI) projects critical success factors. Furthermore, this paper describes extensively the CSF of (AI) projects in health sector and the structure way to understand these factors and prove its significance in enhancing the health care domain for governments' interests.

2.2 Problem Definition

Multiple research studies illustrate that (AI) concepts have been employed in various projects within different domains such as, education, business and (IT). In addition, some governments started pushing organization in having (AI) projects in their environment for serving people in a better and faster way. For example, UAE strategy for

2020 is to drive the growth by using (AI) applications for better government performances and implementing innovative productive environments. For instance, [10] pointed out that the acceptance and adoption of (AI) projects in health sector is practically depended on doctors and physicians' perceptions. Beside, previous research papers did not study extremely the five factors managerial, operational, strategic, IT infrastructure and organizational together in single research paper, the influences of physicians believe for (AI) projects execution. [10] resulted three significant factors for acceptance of ICT projects in health domain, compatibility, Perceived Usefulness and Perceived Ease of Use. Therefore, current paper analyzes the external factors employed extensively for acceptance of (AI) concepts in health projects [11]. Furthermore, an evaluation is conducted for the most widely used external factors considered to have an influence on (AI) projects, managerial, operational, strategic, IT infrastructure and organizational factors.

2.3 The Aim of Research

According to [12], the influence of AI in the healthcare system is growing rapidly to bring changes in the sector. Precisely, medicine field is experiencing major developments in both big data and sophisticated machine learning techniques [12]. As far as [13] is concerned, one of such developments is the way machine learning is being used currently to give different indispensable tools that are necessary for analyzing medical data intelligently. It is, therefore, true that (AI) is playing significant roles in the healthcare sector by giving solutions to biomedical problems [14]. Existing paper purpose is to explore out the main critical success factors (managerial, operational, strategic, IT infrastructure and organizational) for implementing (AI) projects within the healthcare environment. Therefore, the main contribution of this paper is to bridge the existing gap of not studying the above five factors within (AI) projects for health care sector and it will be covered by investigate, explore and propose critical success factors within health sector and the relation between these CSF with (AI) projects.

3 Literature Review

(AI) is a branch of computer science that simulates human processes into computer systems [15]. Means that computer systems are made to behave and act like human beings. "Nowadays, more and more computer technology and (AI) take part in disease diagnosis" [16] but "our possible relations to AI persons could be more complicated than they first might appear" [17]. It means that advancing technology is impacting human life directly in such a way that almost all aspect of the people is bound to the impacts of the technologies. However, the way people relate with the computer systems can be said to be much more complicated than it is literally assumed in the public domain. Nonetheless, service delivery in the current health care sector is prone to succumb to the effects of digital disturbance as information technologies and internet influence the way physicians carry out their daily professional tasks while attending to their clients. For example, [18] assert that the process of managing information is time-consuming for healthcare providers and implementation of relevant projects such as electronic health record systems allow clinicians to document the patient care records without much difficulty and again, free. According to [19], implementation of (AI) projects in health care results in improving the sector. Specifically, (AI) has proved to be significant in different healthcare practices such as a neural network, data mining, deep learning and machine learning [20]. Referring to (AI) it's considered a technology on itself and this technology has been known for making the work of different personnel not only easier but also manageable within short time. While implementing (AI) projects in the health domain, some enhancement opportunities could be resulted from the process of having accurate patients 'data and translate it to a clear useful information.

It is due to such benefits that (AI) projects are gaining much popularity in the health care sector. [21] state that in spite of the fact that benefits of implementing AI projects in healthcare have received much acknowledgment in the medical domain; there are some instances in which these implementations fail because of the rejections fostered by stakeholders in the field. Besides, the practice of rejection has prompted creation and testing different technology adoption models to give a clear prediction of the sources of rejections that some of the (AI) projects get during their implementation stages [22]. Some of these technology adoption models are Theory of Reasoned Action (TRA), Diffusion of Innovation Theory (DOI), Technology Acceptance Model (TAM), Uniform Theory of Acceptance and Use of Technology (UTAUT), Theory of Planned Behavior (TPB), and Social Cognitive Theory (SCT) [23]. Among the listed technology adoption theories, TAM that has been employed extensively in the information system to determine the acceptance of the technology among desired users [24]. The main reason for conducting this particular study is that there is a big gap existing in the current literature that related to the implementation of (AI) in the health care and the acceptance of those projects by the users. The existing literature only focuses on the acceptance of the technology projects in general but it does not narrow down to (AI) projects in the health domain exactly. Therefore, the main research question that this study is trying to answer is what are the critical success factors for implementing (AI) projects for health care sector using the TAM? Unlike other typical information technology projects, (AI) project deals with simulation of the human processes by machines [25, 26], hence their uniqueness. On the other hand, implementation of the (AI) projects in healthcare provide a great uniqueness.

3.1 Artificial Intelligence Projects in the Health Sector

The concept of (AI) in the health domain is bound to the probabilities patient's health improvements and cost reduction in the health care sector. Different organizations in the health industry are integrating the provisions of machine learning as they make both better and faster diagnoses comparing to human beings. Having a closer look at the way of (AI) being adoption will lead different governmental and no-governmental bodies to adapt (AI) and bring more investments to accept and adapt such new concept. For example, Vistdubai.com (2018) states that Dubai Health Authority (DHA) has already put in place a strategy which uses both (AI) and robotics to make the processes of healthcare automatic with an aim of meeting the rising local and regional demands. The strategy has an objective of ensuring the success of the (DHA) mission that states,

"Harnessing creativity and innovation by generating and transforming ideas into reality to deliver value-based and sustainable healthcare services through engaging all partners and stakeholders" [27]. Indeed, providing services through (AI) as well as maximizing integration of the (AI) concepts into the medical services is one of the five themes laid out in the UAE (AI) strategy. Currently, there are numerous projects of (AI) that have been implemented in healthcare to achieve the desired outcomes of improving patients' results as well as reducing the costs of healthcare. For instance, IBM Watson is a healthcare technology that has the ability to understand natural language (IBM 2018). Computer systems can also give responses to questions if prompted to do so and this system is therefore used to dig deep into the relevant data of the patient as well as other sources of data that are available. According to [28], one of the domains where predictions are appreciated with the importance they deserve is medicine. (AI) systems have excellence in conducting predictive analytics. According to [29], machine learning algorithms predict the hospitalization risks as well as death chances based on either socio-economic or predictive data. It is even stated "predictive analytics for healthcare using machine learning is a challenging task to help doctors decide the exact treatments for saving lives" [30].

In the literature research conducted by [31], it was found out that machine learning is effective when predictive analytics involves a huge data set. But again, [32] link both prediction and treatment of the diseases such as heart attack to the extraction of medical data. In such circumstances, effective predictions and management of conditions call for daily monitoring [33]. Basing on Yamada and Kobayashi [34], health monitoring improves the quality of life as well giving required support for aging populations [35]. Moreover, (AI) projects in healthcare help in censoring data, which is helpful in the prognosis modeling of the diseases [36]. In addition, (AI) projects make it possible for the enhancement of the Computer-aided diagnosis (CAD) systems, which improves the computed tomography's feasibility as explained by [37]. For instance, [38] point out that the use of CAD has been active in the screening mammography for a long period. Mammography screening increases the rate of detecting breast cancer during the early stages of the condition's development [39]. From this discourse, it is shows that implementing (AI) projects results in the development of beneficial systems in the healthcare sector, ranging from making physicians work with a sense of simplicity and with a lot of ease of saving valuable resources such as time and energy [28, 40]. According to [41], the methods of the advanced (AI) within the healthcare sector have capabilities of bringing about the highest quality of care in three different ways. Ensuring that non-obvious is discovered, enhancing relationships relevant in the clinical settings and ensuring that timely are interventions are enabled.

The main reason for conducting this particular research study is that there is a big gap existing in the current literature between the implementation of (AI) projects in health care and the acceptance of those projects by the users. The existing literature only focuses on the critical success factors that allow to accept (AI) projects in general but does not narrow down to (AI) projects in the medical domain. Therefore, the main research question that this study is based on is what are the critical success factors that determine the success implementing of (AI) projects in the healthcare sector? Unlike other typical information technology projects, (AI) projects deal with simulation of the human processes by machines, hence their uniqueness.

3.2 The Technology Acceptance Model (TAM)

Recognitions of various technologies in different sectors, TAM has established itself as a strong component in facilitating the acceptance of these technologies. Based on the studies reviewed, system quality, computer self-efficacy, self-efficacy, satisfaction, subjective norm, trust, enjoyment and information quality are the most common external factors that are widely used [3, 42, 43]. Indeed, TAM is now a vital framework that is employed in different sectors to integrate the ever-advancing technology. For instance, [9] used TAM when coming up with various factors that influence the adoption as well as the implementation of Business Intelligence and Analytics Projects in Organizations. The Technology Acceptance Model is built on two distinct but related personal beliefs, perceived usefulness as well as perceived ease of use [44]. However, the two identified pillars of the TAM are known to be influenced by not only external but also system-specific factors called upon while the process of making predictions of the subject's attitude towards accepting and integrating the respective technology desired. It is also a common notion in the public domain that as much as the attitude of the technology users is influenced by the system-specific and external factors, the attitude itself influences the behavioral intention of the technology users to adopt and implement the proposed technology in all the sectors [45–49]. It is this fact that forms the ground for justifying the way system use is predicted in the actual settings. According to [50], the main objective of the establishment of the Technology Acceptance Model is to assist in determining readiness of users in accepting the information systems. The figure below shows the original TAM by [51].

A closer look at the TAM shows that the model [52–54] explicitly explores various determinant factors influencing both computer use and other related or rather pertinent technologies in different contexts. Specifically, the TAM provides a good platform of the description of technology acceptance influencers in distinct user groups and technologies [11, 55]. In general, its' clear that TAM has a single purpose, is to investigate the way external factors lead to influencing attitudes, intentions as well as beliefs through recognition of the few variables that brought out in previous studies. However, [56] assert that the process is done with consideration of not only cognitive but also emotional factors that dictate technology acceptance among user groups.

3.3 Linking Technology Acceptance Model (TAM) to Artificial Intelligence (AI) Projects

There are different themes in the existing literature capable of assessing how different user groups accept the implementation of (AI) projects in the health care sector. Since the Technology Acceptance Model has been effective in giving out the rationale for people accepting the technologies, it is considered to be a significant predictive framework in the medicine domain [1, 57, 58]. Despite the fact of TAM's origin is linked to the United States of America, the provisions and concepts of this robust predictive model has had a rapid widespread and it is now being employed in different parts of the globe within different contexts [59–61]. According to [58], external factors are considered extensively exhausted while extending the concepts of TAM to various corners of the world in the health care sector. After carrying out a systematic analysis of 23 studies published in the duration of four years (2015–2018), 56 external factors were identified and assessed. Only five out of the 56 external factors were considered to be widely used in fostering technology acceptance among the user groups. Include managerial, operational, organizational, strategic and IT infrastructure [7, 58, 62]. After obtaining relevant published studies, different constructs presented in the research studies were put together to come up with a list of the external factors relied upon frequently during the process of adopting the technology. In addition, factors that appeared in three or more studies were considered as the most widely used external factors in the whole idea of extending TAM during the technology acceptance process. The table below shows the distribution of various external factors obtained from different studies.

4 Research Methodology

Research methodology part presents the analysis of acceptance and adoption of (AI) health projects for this paper. This study relies on using the secondary data heavily to find what factors used by existing literature of the previously published studies then see what models or theories can be used. Data collection method in the current paper involves a carry out an assessment for a number of (AI) papers in the healthcare context. Precisely, a systematic analysis design include 23 published studies mainly on the adoption and acceptance of (AI) and IT have been reviewed. After conducting a systemic review of the credible sources, the main elements in each study were elicited and summarized in a planned manner. Based on that, after proper consideration, it can be witnessed that the TAM model can be sued properly and all its elements were used properly. Then, the research capitalized the TAM theoretical framework that contains of five fundamental constructs in the literature. The five core constructs of the (TAM) includes perceived ease of use, perceived usefulness, behavioral intention to use, attitude towards usage and the actual system use of AI projects in the healthcare. After reviewing a systematic approach for the 23 published research studies in the literature on the most widely used external factors for the extended TAM were identified in education, Information Technology, healthcare, and business fields. The research seeks to examine and analyze the critical success factors associated with the implementation of (AI) projects in health care sector. Understanding the attitude of project owners and associated employees in implementing (AI) projects for health sectors can assist in comprehending the critical success factors regarding project's strengths and weakness. As hypothesized, the attitude is significant in determining the level of health sector willingness to use (AI) systems in the treatment plans of their respective patients (Table 1).

Table 1. Twenty three studies were reviewed to identify external factors of TAM

Authors	Technology type	Country	User
[7]	Business	Not given	Customers
[63]	Healthcare	Canada	Physicians
[58]	Healthcare	Egypt	Physicians
[64]	Healthcare	Indonesia	Physicians
[60]	Healthcare	Iran	Physicians, nurses, and managers
[11]	Healthcare	Iran	Physicians
[2]	E-Commerce	Lebanon	Customers
[65]	Healthcare	Malaysia	Physicians
[42]	Healthcare	Malaysia	Doctors
[54]	Education	Mauritius	Teachers
[10]	Healthcare	Morocco	Physicians
[66]	Healthcare	Morocco	Physicians
[67]	E-learning	Saudi Arabia	Students
[55]	Healthcare	Spain	Healthcare users
[68]	Health tourism	Thailand	Physicians
[61]	E-learning	Thailand	Students
[62]	E-Government	Thailand	Citizens
[59]	Healthcare	Turkey	Physicians
[69]	E-learning	UAE	Students
[9]	Business	UAE	Customers
[70]	Education	UAE	Students
[57]	Healthcare	UAE	Physicians
[1]	Education	USA	Teachers

5 Conclusion and Future Work

The science of simulating machines to perform different human processes is called (AI), including learning, reasoning and solving problems. With advancing technology, (AI) is affecting almost every sector of human life, includes the healthcare sector that is an essential segment in the lives of human beings. Implementation of (AI) projects in healthcare is beneficial in numerous ways. For instance, it provides a good platform on which the patient's data is managed. The resultant information can be employed in predictive analytics, which in turn influences the monitoring process of the patients' conditions. It is also noted that previous research studies failed to link the acceptance of artificial projects in healthcare by the user groups. Created a literature gap that this study has an objective of bridging. Healthcare sector projects are much different from the other sector projects because they deal with delicate human life once implemented.

The aims of this study can be achieved by answering the following questions what are the critical success factors for implementing AI projects in health care sectors. To do so, large number of studies were reviewed to and brought to current situation of AI projects implementation process in the healthcare sector. Major emphasis was done on the implementation of the AI projects alongside other technological resources in the medical domain. Besides, various factors that are believed to influence the perceptions and concerning the use of AI in the health centers were explained. Then, the link between TAM and (AI) projects in the healthcare sector were mentioned. Methodology assist in gathering the required data from the literature to answer the research question for this paper along with the utilized research structure. Besides, literature review part has different aims and goals such as, reviewing systematically the most recent (AI) projects adoption systems findings that have fostered the rapid extension of the Technology Acceptance Model and identifying the most widely used external factors. Additionally, this paper has a goal of identifying various strengths of the precise relationship that might be in existence among these external factors that are widely used in promoting the variables of the Technology Acceptance Model in the implementation of (AI) projects within the health sector. After reviewing 23 research studies it illustrated that system quality, computer self-efficacy, self-efficacy, satisfaction, subjective norm, trust, enjoyment and information quality are the most widely used external factors for promoting the Technology Acceptance Model. Finally, none of the 23 research papers did address the five factors together in one research paper (managerial, technological, operational, strategic and IT infrastructure.

To understand the acceptance and adoption of (AI) projects in health domain a modified (TAM) model was employed in this paper. The implications for researchers to have practical application to their AI health project by confirming the critical external factors based on their organizational culture and their process, ensuring the best practice application to guarantee project success. The consequences thereby, is to improve the attitude towards (AI) projects for their business environment.

Acknowledgment. This work is a part of a dissertation submitted in fulfilment of MSc Informatics (Knowledge & Data Management) Faculty of Engineering & Information Technology At The British University in Dubai.

References

1. Fathema, N., Shannon, D., Ross, M.: Expanding the technology acceptance model (TAM) to examine faculty use of learning management systems (LMSs) in higher education institutions. J. Online Learn. Teach. **11**(2), 210–232 (2015)
2. Fayad, R., Paper, D.: The technology acceptance model e-commerce extension: a conceptual framework. Proc. Econ. Financ. **26**, 1000–1006 (2015)
3. Salloum, S.A., Al-Emran, M., Shaalan, K., Tarhini, A.: Factors affecting the E-learning acceptance: a case study from UAE. Educ. Inf. Technol. **24**, 1–22 (2018)
4. Salloum, S.A., Shaalan, K.: Adoption of e-book for university students. In: International Conference on Advanced Intelligent Systems and Informatics, pp. 481–494 (2018)
5. Salloum, S.A., Al-Emran, M., Khalaf, R., Habes, M., Shaalan, K.: An innovative study of e-payment systems adoption in higher education: theoretical constructs and empirical analysis. Int. J. Interact. Mob. Technol. **13**(6), 68–83 (2019)

6. Salloum, S.A., Al-Emran, M., Habes, M., Alghizzawi, M., Ghani, M.A., Shaalan, K.: Understanding the impact of social media practices on E-learning systems acceptance. In: International Conference on Advanced Intelligent Systems and Informatics, pp. 360–369 (2019)
7. Who, X.: Extending TAM: success factors of mobile marketing. Am. Acad. Sch. Res. J. **1**(1), 1–5 (2011)
8. Aldosari, B., Al-Mansour, S., Aldosari, H., Alanazi, A.: Assessment of factors influencing nurses acceptance of electronic medical record in a Saudi Arabia hospital. Inform. Med. Unlocked **10**(December 2017), 82–88 (2018)
9. Al-Adwan, A., Al-Adwan, A., Smedley, J.: Exploring students acceptance of e-learning using technology acceptance model in Jordanian universities. Int. J. Educ. Dev. Inf. Commun. Technol. **9**(2), 4 (2013)
10. Bennani, A.-E., Oumlil, R.: The Acceptance of ICT by geriatricians reinforces the value of care for seniors in Morocco. IBIMA Publ. J. African Res. Bus. Technol. J. Afr. Res. Bus. Technol. **2014**(2014), 1–10 (2014)
11. Nadri, H., Rahimi, B., Afshar, H.L., Samadbeik, M., Garavand, A.: Factors affecting acceptance of hospital information systems based on extended technology acceptance model: a case study in three paraclinical departments. Appl. Clin. Inform. **9**(02), 238–247 (2018)
12. Price, I.I., Nicholson, W.: Artificial intelligence in health care: applications and legal implications (2017)
13. Kononenko, I.: Machine learning for medical diagnosis: history, state of the art and perspective. Artif. Intell. Med. **23**(1), 89–109 (2001)
14. Shukla, S., Gupta, D.L., Prasad, B.R.: Comparative study of recent trends on cancer disease prediction using data mining techniques. Int. J. Database Theory Appl. **9**(9), 107–118 (2016)
15. Swarup: No title. Artif. Intell. Int. J. Comput. Corp. Res. **2**(4) (2012)
16. Jie, W., Hai-yan, L., Biao, C., Yuan, Z.: Application of educational data mining on analysis of students' online learning behavior. In: 2017 2nd International Conference onImage, Vision and Computing (ICIVC), pp. 1011–1015 (2017)
17. Lawrence, J., Palacios-González, D., Harris, C.: Artificial Intelligence. Cambridge Q. Healthc. Ethics **25**(02), 250–261 (2016)
18. Moen, H., et al.: Comparison of automatic summarisation methods for clinical free text notes. Artif. Intell. Med. **67**, 25–37 (2016)
19. Ziuziański, P., Furmankiewicz, M., Sołtysik-Piorunkiewicz, A.: E-health artificial intelligence system implementation: case study of knowledge management dashboard of epidemiological data in Poland. Int. J. Biol. Biomed. Eng. **8**, 164–171 (2014)
20. Jordan, M.I., Mitchell, T.M.: Machine learning: trends, perspectives, and prospects. Science **349**(6245), 255–260 (2015)
21. Wallis, L., et al.: A roadmap for the implementation of mHealth innovations for image-based diagnostic support in clinical and public-health settings: a focus on front-line health workers and health-system organizations. Glob. Health Action **10**(sup3), 1340254 (2017)
22. Najafabadi, M.M., Villanustre, F., Khoshgoftaar, T.M., Seliya, N., Wald, R., Muharemagic, E.: Deep learning applications and challenges in big data analytics. J. Big Data **2**(1), 1 (2015)
23. Maruping, L.M., Bala, H., Venkatesh, V., Brown, S.A.: Going beyond intention: integrating behavioral expectation into the unified theory of acceptance and use of technology. J. Assoc. Inf. Sci. Technol. **68**(3), 623–637 (2017)
24. Alshurideh, D.M.: Do electronic loyalty programs still drive customer choice and repeat purchase behaviour? Int. J. Electron. Cust. Relatsh. Manag. **12**(1), 40–57 (2019)
25. Mokyr, J.: The British Industrial Revolution: An Economic Perspective. Routledge, Abingdon (2018)
26. Mijwel, M.M.: History of Artificial Intelligence. Comput. Sci. Coll. Sci. 1–6 (2015)

27. Authority, D.H.: No title (2018)
28. Albu, A., Stanciu, L.: Benefits of using artificial intelligence in medical predictions. In: 2015 E-Health and Bioengineering Conference (EHB), pp. 1–4 (2015)
29. Panicacci, S., Donati, M., Fanucci, L., Bellin, I., Profili, F., Francesconi, P.: Population health management exploiting machine learning algorithms to identify high-risk patients. In: 2018 IEEE 31st International Symposium on Computer-Based Medical Systems (CBMS), pp. 298–303 (2018)
30. Charleonnan, A., Fufaung, T., Niyomwong, T., Chokchueypattanakit, W., Suwannawach, S., Ninchawee, N.: Predictive analytics for chronic kidney disease using machine learning techniques. In: 2016 Management and Innovation Technology International Conference (MITicon), p. MIT-80 (2016)
31. Nithya, B., Ilango, V.:"Predictive analytics in health care using machine learning tools and techniques. In: 2017 International Conference on Intelligent Computing and Control Systems (ICICCS), pp. 492–499 (2017)
32. Rajamhoana, S.P., Devi, C.A., Umamaheswari, K., Kiruba, R., Karunya, K., Deepika, R.: Analysis of neural networks based heart disease prediction system. In: 2018 11th International Conference on Human System Interaction (HSI), pp. 233–239 (2018)
33. Zhang, Q., Zhou, D., Zeng, X.: Hear the heart: daily cardiac health monitoring using Ear-ECG and machine learning. In: 2017 IEEE 8th Annual Ubiquitous Computing, Electronics and Mobile Communication Conference (UEMCON), pp. 448–451 (2017)
34. Yamada, Y., Kobayashi, M.: Detecting mental fatigue from eye-tracking data gathered while watching video: evaluation in younger and older adults. Artif. Intell. Med. (2018)
35. Library, I.X.D.: Cover art: customized web-based system for elderly people using elements of artificial intelligence [online]. Univ. Košice. IEEE Xplore Digit. Libr. (2019). https://ieeexplore.ieee.org/document/7733252. Accessed 20 January 2019
36. Chen, D., Goyal, G., Go, R., Parikh, S., Ngufor, C.: Predicting time to first treatment in chronic lymphocytic leukemia using machine learning survival and classification methods. In: 2018 IEEE International Conference on Healthcare Informatics (ICHI), pp. 407–408 (2018)
37. Nibali, A., He, Z., Wollersheim, D.: Pulmonary nodule classification with deep residual networks. Int. J. Comput. Assist. Radiol. Surg. **12**(10), 1799–1808 (2017)
38. Fotin, S.V., Yin, Y., Haldankar, H., Hoffmeister, J.W., Periaswamy, S.: Detection of soft tissue densities from digital breast tomosynthesis: comparison of conventional and deep learning approaches. In: Medical Imaging 2016: Computer-Aided Diagnosis, vol. 9785, p. 97850X (2016)
39. Ertosun, M.G., Rubin, D.L.: Probabilistic visual search for masses within mammography images using deep learning. in: 2015 IEEE International Conference on Bioinformatics and Biomedicine (BIBM), pp. 1310–1315 (2015)
40. Shaikhina, T., Khovanova, N.A.: Handling limited datasets with neural networks in medical applications: a small-data approach. Artif. Intell. Med. **75**, 51–63 (2017)
41. Vemulapalli, V., et al.: Non-obvious correlations to disease management unraveled by Bayesian artificial intelligence analyses of CMS data. Artif. Intell. Med. **74**, 1–8 (2016)
42. Baharom, F., Khorma, O.T., Mohd, H., Bashayreh, M.G.: Developing an extended technology acceptance model: doctors' acceptance of electronic medical records in Jordan. In: ICOCI (2011)
43. Razali, N.M., Wah, Y.B.: Power comparisons of shapiro-wilk, kolmogorov-smirnov, lilliefors and anderson-darling tests. J. Stat. Model. Anal. **2**(1), 21–33 (2011)
44. Marangunić, N., Granić, A.: Technology acceptance model: a literature review from 1986 to 2013. Univers. Access Inf. Soc. **14**(1), 81–95 (2015)

45. Al Dmour, H., Alshurideh, M., Shishan, F.: The influence of mobile application quality and attributes on the continuance intention of mobile shopping. Life Sci. J. **11**(10), 172–181 (2014)
46. Alshurideh, M., Salloum, S. A., Al Kurdi, B., Al-Emran, M.: Factors affecting the social networks acceptance: an empirical study using PLS-SEM approach. In: 8th International Conference on Software and Computer Applications (2019)
47. Alshurideh, M., Al Kurdi, B., Salloum, S.A.: Examining the main mobile learning system drivers' effects: a mix empirical examination of both the expectation-confirmation model (ECM) and the technology acceptance model (TAM). In: International Conference on Advanced Intelligent Systems and Informatics, pp. 406–417 (2019)
48. Alshurideh, M.T., Salloum, S.A., Al Kurdi, B., Monem, A.A., Shaalan, K.: Understanding the quality determinants that influence the intention to use the mobile learning platforms: a practical study. Int. J. Interact. Mob. Technol. **13**(11), 157–183 (2019)
49. Alshurideh, M.T., Shaltoni, A.M., Hijawi, D.S.: Marketing communications role in shaping consumer awareness of cause-related marketing campaigns. Int. J. Mark. Stud. **6**(2), 163 (2014)
50. Briz-Ponce, L., García-Peñalvo, F.J.: An empirical assessment of a technology acceptance model for apps in medical education. J. Med. Syst. **39**(11), 176 (2015)
51. Davis, F.D.: Perceived usefulness, perceived ease of use, and user acceptance of information technology. MIS Q. **13**(3), 319–340 (1989)
52. Altamony, H., Alshurideh, M., Obeidat, B.: Information systems for competitive advantage: implementation of an organisational strategic management process. In: Proceedings of the 18th IBIMA Conference on Innovation and Sustainable Economic Competitive Advantage: From Regional Development to World Economic, 9–10 May 2012, Istanbul, Turkey (2012)
53. Muk, A., Chung, C.: Applying the technology acceptance model in a two-country study of SMS advertising. J. Bus. Res. **68**(1), 1–6 (2015)
54. Teeroovengadum, V., Heeraman, N., Jugurnath, B.: Examining the antecedents of ICT adoption in education using an extended technology acceptance model (TAM). Int. J. Educ. Dev. ICT **13**(3), 4–23 (2017)
55. Solano-Lorente, M., Martínez-Caro, E., Cegarra-Navarro, J.G.: Designing a framework to develop eloyalty for online healthcare services. Electron. J. Knowl. Manag. **11**(1), 107–115 (2013)
56. Abu-Shanab, E., Al-Tarawneh, H.: The influence of social networks on high school students' performance. Int. J. Web-Based Learn. Teach. Technol. **10**(2), 49–59 (2015)
57. Alloghani, M., Hussain, A., Al-Jumeily, D., Abuelma'atti, O.: Technology acceptance model for the use of M-health services among health related users in UAE. In: 2015 International Conference on Developments of E-Systems Engineering (DeSE), pp. 213–217 (2015)
58. Emad, H., El-Bakry, H.M., Asem, A.: A modified technology acceptance model for health informatics (2016)
59. Basak, E., Gumussoy, C.A., Calisir, F.: Examining the factors affecting PDA acceptance among physicians: an extended technology acceptance model. J. Healthc. Eng. **6**(3), 399–418 (2015)
60. Safdari, R., Saeedi, M.G., Valinejadi, A., Bouraghi, H., Shahnavazi, H.: Technology acceptance model in health care centers of Iran. Int. J. Comput. Sci. Netw. Secur. **17**(1), 42 (2017)
61. Punnoose, A.C.: Determinants of intention to use eLearning based on the technology acceptance model. J. Inf. Technol. Educ. Res. **11**(1), 301–337 (2012)
62. Wangpipatwong, S., Chutimaskul, W., Papasratorn, B.: Understanding citizen's continuance intention to use e-government website: a composite view of technology acceptance model and computer self-efficacy. Electron. J. e-Govern. **6**(1), 55–64 (2008)

63. Strudwick, G.: Predicting nurses' use of healthcare technology using the technology acceptance model: an integrative review. CIN Comput. Inform. Nurs. **33**(5), 189–198 (2015)
64. Helia, V.N., Indira Asri, V., Kusrini, E., Miranda, S.: Modified technology acceptance model for hospital information system evaluation–a case study (2018)
65. San, A.N.C., Yee, C.J.: The modified technology acceptance model for private clinical physicians: a case study in Malaysia, Penang. Int. J. Acad. Res. Bus. Soc. Sci. **3**(2), 380 (2013)
66. Bennani, A.-E., Oumlil, R.: IT acceptance by nurses in morocco: application of a modified unified theory of acceptance and use of technology. IBIMA Bus. Rev. **2014**, 1 (2014)
67. Alharbi, S., Drew, S.: Using the technology acceptance model in understanding academics' behavioural intention to use learning management systems. Int. J. Adv. Comput. Sci. Appl. **5**(1), 143–155 (2014)
68. Phatthana, W., Mat, N.K.N.: The application of technology acceptance model (TAM) on health tourism e-purchase intention predictors in Thailand. In: 2010 International Conference on Business and Economics Research, vol. 1, pp. 196–199 (2011)
69. Salloum, S.A., Alhamad, A.Q.M., Al-Emran, M., Monem, A.A., Shaalan, K.: Exploring students' acceptance of e-learning through the development of a comprehensive technology acceptance model. IEEE Access **7**, 128445–128462 (2019)
70. Salloum, S.A.S., Shaalan, K.: Investigating students' acceptance of E-learning system in higher educational environments in the UAE: applying the extended technology acceptance model (TAM). The British University in Dubai (2018)

Machine Learning and Deep Learning Techniques for Cybersecurity: A Review

Said A. Salloum[1,2(✉)], Muhammad Alshurideh[3,4], Ashraf Elnagar[1,5], and Khaled Shaalan[2]

[1] Research Institute of Sciences and Engineering, University of Sharjah, Sharjah, UAE
ssalloum@sharjah.ac.ae
[2] Faculty of Engineering and IT, The British University in Dubai, Dubai, UAE
[3] Faculty of Business, University of Jordan, Amman, Jordan
[4] Management Department, University of Sharjah, Sharjah, UAE
[5] Department of Computer Science, University of Sharjah, Sharjah, UAE

Abstract. In this review, significant literature surveys on machine learning (ML) and deep learning (DL) techniques for network analysis of intrusion detection are explained. In addition, it presents a short tutorial explanation on every ML/DL method. Data holds a significant position in ML/DL methods; hence this paper highlights the datasets used in machine learning techniques, which are the primary tools for analyzing network traffic and detecting abnormalities. In addition, we elaborate on the issues faced in using ML/DL for cybersecurity and offer recommendations for future studies.

Keywords: Cyber security · Machine learning · Deep learning

1 Introduction

This paper takes into view the cyber security applications and presents the outcomes of a literature survey of machine learning (ML), deep learning (DL), and data mining (DM) methods. In addition, it explains the (ML/DL)/DM methods and their applications to cyber intrusion detection issues. Besides providing a set of comparison criteria for (ML/DL)/DM methods, the paper analyzes the complexity of different (ML/DL)/DM algorithms. Subsequently, best methods and a set of recommendations are given subject to the attributes of the cyber problem. The set of processes and technologies that are formulated for the prevention of unauthorized access, attack, change, or destruction of networks, computers, data and programs is referred to as the cyber security [1]. The network security systems and computer (host) security systems basically make up the cyber security systems. At least, a single antivirus software, firewall and an intrusion detection system (IDS) must be included in each of these entities [1]. In addition to identification, IDSs do discover and determine the unauthorized use, alteration, duplication and damage of information systems [2]. The internal intrusions (intra-organization attacks) and external intrusions (inter-organization attacks) are known as the security breaches. In sensing security threats, the artificial intelligent and machine learning based methodologies have become an essential part of our lives because of the rise in the ratio of

A.-E. Hassanien et al. (Eds.): AICV 2020, AISC 1153, pp. 50–57, 2020.
https://doi.org/10.1007/978-3-030-44289-7_5

cyber-attacks [3]. The standards of security play an important role in the provision of the best security applications keeping in view their implementation and security requirements. They are vital for cyber security research from this perspective, such as intrusion detection system. While machine and deep learning algorithms and particular datasets can be corroborated through some research studies, literature doesn't contain much content on the (ML/DL)/DM techniques for cyber security and security-related datasets.

While concentrating on the (ML/DL)/DM methods and their descriptions, the (ML/DL)/DM techniques are focused by this survey paper aiming at cyber security. Besides publication of several reviews [1, 3–7], many papers containing these methods have also been published [8, 9]. Compared to earlier reviews, the publications fulfilling certain criteria will be the major emphasis of our paper. The techniques, such as: "machine learning" or "deep learning" and cybersecurity along with "data mining" were used to perform Google Scholar queries. The highly cited papers containing well-known techniques were among the main concerns. Nevertheless, there was a realization that novel and innovative techniques would possibly be ignored; therefore a few of these articles were also selected. By and large, selection of papers was done with a view to incorporate at least one and preferably a few representative papers about each of the (ML/DL)/DM categories. The remainder of this paper is organized as follows: Sect. 2 offered a summary of some relevant literature. Section 4 overviews different (ML/DL)/DM methods used in cyber security. Section 4 discusses cybersecurity datasets for (ML/DL)/DM and Sect. 5 concludes the paper with a brief summary of the paper's key points and other closing remarks.

2 Literature Review

With the assistance of computer software, an exclusive way to solve fundamental scientific and engineering questions is provided by the machine learning applications [10]. During last twenty years, radical advancements have been witnessed in the field of machine learning with an easy access to the beginners [11]. A laboratory "black-box" environment has basically triggered the machine learning, which has then transformed into a practical application and commercial companies are progressively implementing it on a large scale [10, 11]. The software applications computer vision [12, 13], natural language processing [13–15], speech recognition [13–16], robot control [13, 17, 18] and other emerging applications are the innovations of machine learning [10]. For bringing improvements in user experience, to promote special offerings and suggest purchases [10], machine learning is used by major companies such as Amazon, Facebook and Google. Instead of programming the traditional input-process-output, it is far easier for machine learning developers to train a system. They achieve it by developing simulations of anticipated output [11]. Various industries have observed the impact of machine learning in the shape of data-intensive issues, such as, cybersecurity [10]. Some other fields ranging from biology [19–23] to cosmology to social science [24–30] can also take advantages of Machine learning for bringing remarkable transformations [11]. The experimental data can be processed and analyzed by machine learning in new ways [11]. Theoretically, we can better understand the concepts of "big data" by working on machine learning algorithms. Moreover, these technologies can be employed to improve associated performance metrics in vertical applications. There can be a great variation in the machine learning algorithms with respect to unique functions (e.g., logistic regression, linear Regression, Naive Bayes, decision trees, random forest, support vector machines, deep learning and Gradient Boosting algorithms). Nevertheless, machine learning introduces creative ways to analyze huge amounts of data with an aim to generate evolutionary approach. Moreover, greater optimization can be provided by the successive generations of algorithms [11]. The large volume of data can ideally be processed through machine learning and cybersecurity [11]. The platforms and networks are vulnerable to attack. The effectiveness of these attacks depend on the number of tools to scan and evaluate targets [10]. Machine learning is used by the adversaries to further increase their attacks. Recently, some journals have published few surveys on the security perspective of machine learning and artificial intelligence [4–6, 31, 32]. In addition, [4] emphasized on contemporary literature on Intrusion Detection for computer network security and Machine Learning Techniques used in Internet-of-Things. Subsequently, for network analysis of intrusion detection, [6] explained key literature surveys on machine learning (ML) and deep learning (DL) methods with an explanatory description of each ML/DL method.

During training and testing or inferring of machine learning from a data driven view, a comprehensive literature review regarding defensive techniques and security threats was presented by [5]. In addition, [31] briefed on security issues regarding artificial intelligence, especially the reinforcement and supervised learning algorithms. In contrast, updates on security issues and defensive methods in the life cycle of a machine learning-based system from training to inference were reviewed by [32].

3 Classification of Machine Learning Algorithms to Cybersecurity

3.1 Classical Machine Learning Techniques

Development of models receiving input data besides utilizing statistical analysis to forecast an output value within an suitable range is the key objective of ML [3]. The ML is one of the rapidly growing areas with comprehensive applications in the domain of computer science [3]. The supervised, unsupervised and Reinforcement Learning are among the classifications of ML algorithms. The well-known procedures used in the machine learning algorithms are none other than the supervised algorithms. Furthermore, regression and classification are the further sub-divisions of supervised algorithms. A number of machine learning algorithms are used in literature [33, 34]. The Logistic Regression [35], Decision Tree [36], Naive Bayes [35, 37, 38], SVM [35, 37, 39, 40], K-Means [38], KNN [37, 40] and Random Forest [41] are among the commonly used machine-learning algorithms.

3.2 Deep Learning Techniques

A new branch of machine learning, i.e., the deep learning has currently gained widespread recognition and the same has been used for intrusion detection. Moreover, traditional methods are outshined by the deep learning as per the findings of the studies [8]. For flow-based anomaly detection, a deep learning approach subject to a deep neural network has been utilized by the authors in [42]. According to the experimental outcomes, anomaly detection in software defined networks can also be performed through deep learning. In [43], using self-taught learning (STL) on the benchmark NSL-KDD dataset, a deep learning based approach is proposed in a network intrusion detection system. The technique is found to be more efficient in terms of its performance as compared to those discussed in previous studies. Nonetheless, the feature reduction ability of the deep learning is emphasized by this category of references. It implements classification through the traditional supervision model and deep learning methods are primarily used for pre-training. Applying the deep learning method to directly perform classification is an unusual task, and literature is devoid of the studies illustrating the performance in multiclass classification. The RNNs are believed to be the reduced-size neural networks according to [31]. In this article, three-layer RNN architecture was suggested for misuse-based IDs with 41 features as inputs and four intrusion categories as outputs. The ability of deep learning for modeling high-dimensional features is not exhibited by the reduced RNNs and the nodes of layers are partially connected. The performance of the model in the binary classification is not explored by the authors.

[8] learned to model an intrusion detection system on the basis of deep learning and a deep learning approach was recommended for intrusion detection using recurrent neural networks (RNN-IDS). Furthermore, the performance of the model in binary classification and multiclass classification was also investigated. It was discovered that the performance of the proposed model is affected by different learning rate and the number of neurons. For modeling a classification model with high accuracy, the RNN-IDS are found very suitable according to the experimental outcomes. In both multiclass and binary classification, it depicts high performance than the traditional machine learning classification methods. The RNN-IDS model offers a new research technique for intrusion detection and the accuracy of the intrusion detection is also enhanced.

4 Cybersecurity Datasets

Currently, data is prepared by various research groups both for their own analysis and for provision to community repositories [3]. Using machine learning and artificial intelligent research, the present security-related datasets are explained by this section.

4.1 KDD Cup 1999 Dataset

The data obtained from MIT Lincoln Labs entails BSM list files and tcpdump in addition to KDD Cup 1999 [3]. The data captured in DARPA'98 IDS evaluation program contributed to this particular dataset [11]. For assessment of intrusion detection systems, this dataset is also considered as benchmark data. One of the renowned data sets to assess performance of anomaly detection methods is none other than the KDD'99 [3]. Presently, KDD dataset is being used by various researchers [38, 44–46]

4.2 ISOT (Information Security and Object Technology) Dataset

The openly available botnets and normal datasets comprising of 1,675,424 total traffic flow collectively formulate the ISOT dataset. As far as malicious traffic in ISOT is concerned, the same was acquired from French chapter of honeynet project encompassing Storm and Waledac botnets [3, 35, 47, 48].

4.3 HTTP CSIC 2010 Dataset

Thousands of web requests automatically generated and developed at Information Security Institute of CSIC (Spanish Research National Council) are collectively known as HTTP CSIC 2010 dataset [3]. The web attack protection systems can be tested through this dataset. Nearly 6,000 normal requests and above 25,000 anomalous requests formulate this data. Moreover, HTTP requests are labeled as anomalous or normal [3]. The web detection in [49–51] is successfully accomplished through this dataset.

4.4 CTU-13 (Czech Technical University) Dataset

The mixture of seizures of 13 different malwares in a nonfictional network environment is referred to as the CTU-13 (Czech Technical University) dataset. This dataset aims to acquire real mixed botnet traffic. Foregoing in view, normal traffic is generated by verified normal hosts, while botnet traffic is generated by Infected hosts [3]. This dataset is a carefully labeled dataset capturing the procedures carried out in a controlled environment, which is one of the benefits of using it [39, 40, 48, 52].

4.5 UNSW-NB15 Dataset

In the Cyber Range Lab of the Australian Centre for Cyber Security (ACCS), the IXIA Perfect Storm tool has reportedly generated the UNSW-NB15 Dataset [3]. Nearly one hour of anonymized traffic traces from a DDoS attack in 2007 are included in this dataset [47, 49, 53]. This dataset entails nine types of major attacks, including Fuzzers, Backdoors, Analysis, Exploits, DoS, Reconnaissance, Generic, Worms and Shellcode [3].

5 Conclusion and Future Work

One of the key issues for national and international security is the safety of computer systems from cyber-attacks. With numerous machine learning techniques, a survey on security concerns has been presented in this article. Several datasets have been employed to conduct various researches. In addition, a significant role in protection of computer systems is performed by machine learning and artificial intelligence.

The literature review of ML/DL and DM methods used for cyber is explained by this paper. The example papers explaining various ML/DL and DM techniques in the cyber domain were carefully found and widespread classes of various datasets have been defined along with their advantages and disadvantages. We aim to generate new dataset in the future, which will be open to all.

References

1. Buczak, A.L., Guven, E.: A survey of data mining and machine learning methods for cyber security intrusion detection. IEEE Commun. Surv. Tutor. **18**(2), 1153–1176 (2015)
2. Mukkamala, S., Sung, A., Abraham, A.: Cyber security challenges: designing efficient intrusion detection systems and antivirus tools. In: Vemuri, V.R. (ed.) Enhancing Computer Security with Smart Technology 2006, pp. 125–163 (2005)
3. Yavanoglu, O., Aydos, M.: A review on cyber security datasets for machine learning algorithms. In: 2017 IEEE International Conference on Big Data (Big Data), pp. 2186–2193 (2017)
4. da Costa, K.A.P., Papa, J.P., Lisboa, C.O., Munoz, R., de Albuquerque, V.H.C.: Internet of Things: a survey on machine learning-based intrusion detection approaches. Comput. Netw. **151**, 147–157 (2019)
5. Liu, Q., Li, P., Zhao, W., Cai, W., Yu, S., Leung, V.C.M.: A survey on security threats and defensive techniques of machine learning: a data driven view. IEEE Access **6**, 12103–12117 (2018)
6. Xin, Y., et al.: Machine learning and deep learning methods for cybersecurity. IEEE Access **6**, 35365–35381 (2018)
7. Dua, S., Du, X.: Data Mining and Machine Learning in Cybersecurity. Auerbach Publications (2016)
8. Yin, C., Zhu, Y., Fei, J., He, X.: A deep learning approach for intrusion detection using recurrent neural networks. IEEE Access **5**, 21954–21961 (2017)
9. Xiao, L., Wan, X., Lu, X., Zhang, Y., Wu, D.: IoT security techniques based on machine learning (2018). arXiv Prepr. arXiv:1801.06275
10. Jordan, M.I., Mitchell, T.M.: Machine learning: Trends, perspectives, and prospects. Science (80-.) **349**(6245), 255–260 (2015)
11. Fraley, J.B., Cannady, J.: The promise of machine learning in cybersecurity. SoutheastCon **2017**, 1–6 (2017)
12. Alazab, M., Tang, M.: Deep Learning Applications for Cyber Security. Springer, Heidelberg (2019)
13. Li, J.: Cyber security meets artificial intelligence: a survey. Front. Inf. Technol. Electron. Eng. **19**(12), 1462–1474 (2018)
14. Jones, C.L., Bridges, R.A., Huffer, K.M.T., Goodall, J.R.: Towards a relation extraction framework for cyber-security concepts. In: Proceedings of the 10th Annual Cyber and Information Security Research Conference, p. 11 (2015)
15. McNeil, N., Bridges, R.A., Iannacone, M.D., Czejdo, B., Perez, N., Goodall, J.R.: Pace: pattern accurate computationally efficient bootstrapping for timely discovery of cyber-security concepts. In: 2013 12th International Conference on Machine Learning and Applications, vol. 2, pp. 60–65 (2013)
16. Zhang, Q., Man, D., Yang, W.: Using HMM for intent recognition in cyber security situation awareness. In: 2009 Second International Symposium on Knowledge Acquisition and Modeling, vol. 2, pp. 166–169 (2009)

17. Bonaci, T., Herron, J., Yusuf, T., Yan, J., Kohno, T., Chizeck, H.J.: To make a robot secure: an experimental analysis of cyber security threats against teleoperated surgical robots (2015). arXiv Prepr. arXiv:1504.04339
18. Hacioglu, U., Sevgilioglu, G.: The evolving role of automated systems and its cyber-security issue for global business operations in Industry 4.0. Int. J. Bus. Ecosyst. Strategy **1**(1), 1–11 (2019)
19. Alhashmi, S.F.S., Salloum, S.A., Abdallah, S.: Critical success factors for implementing artificial intelligence (AI) projects in Dubai government United Arab Emirates (UAE) health sector: applying the extended technology acceptance model (TAM). In: International Conference on Advanced Intelligent Systems and Informatics, pp. 393–405 (2019)
20. Darwish, A., Ezzat, D., Hassanien, A.E.: An optimized model based on convolutional neural networks and orthogonal learning particle swarm optimization algorithm for plant diseases diagnosis. Swarm Evol. Comput. **52**, 100616 (2020)
21. Abdelghafar, S., Darwish, A., Hassanien, A.E.: Intelligent health monitoring systems for space missions based on data mining techniques. In: Machine Learning and Data Mining in Aerospace Technology, pp. 65–78. Springer (2020)
22. Elsayad, D., Ali, A., Shedeed, H.A., Tolba, M.F.: PAGeneRN: parallel architecture for gene regulatory network. In: Data Analytics in Medicine: Concepts, Methodologies, Tools, and Applications, pp. 1052–1075. IGI Global (2020)
23. Pacheco, A.G.C., Ali, A.-R., Trappenberg, T.: Skin cancer detection based on deep learning and entropy to detect outlier samples (2019). arXiv Prepr. arXiv:1909.04525
24. Salloum, S.A., Al-Emran, M., Monem, A., Shaalan, K.: A survey of text mining in social media: facebook and twitter perspectives. Adv. Sci. Technol. Eng. Syst. J. **2**(1), 127–133 (2017)
25. Alomari, K.M., AlHamad, A.Q., Salloum, S.: Prediction of the digital game rating systems based on the ESRB. Opción **35**(19), 1368–1393 (2019)
26. Salloum, S.A., Al-Emran, M., Shaalan, K.: Mining social media text: extracting knowledge from facebook. Int. J. Comput. Digit. Syst. **6**(2), 73–81 (2017)
27. Salloum, S.A., Al-Emran, M., Abdallah, S., Shaalan, K.: Analyzing the Arab Gulf newspapers using text mining techniques. In: International Conference on Advanced Intelligent Systems and Informatics, pp. 396–405 (2017)
28. Salloum, S.A., Al-Emran, M., Shaalan, K.: Mining text in news channels: a case study from facebook. Int. J. Inf. Technol. Lang. Stud. **1**(1), 1–9 (2017)
29. Salloum, S.A., AlHamad, A.Q., Al-Emran, M., Shaalan, K.: A survey of Arabic text mining, vol. 740 (2018)
30. Salloum, S.A., Mhamdi, C., Al-Emran, M., Shaalan, K.: Analysis and classification of Arabic newspapers' facebook pages using text mining techniques. Int. J. Inf. Technol. Lang. Stud. **1**(2), 8–17 (2017)
31. Amodei, D., Olah, C., Steinhardt, J., Christiano, P., Schulman, J., Mané, D.: Concrete problems in AI safety (2016). arXiv Prepr. arXiv:1606.06565
32. Papernot, N., McDaniel, P., Sinha, A., Wellman, M.: Towards the science of security and privacy in machine learning (2016). arXiv Prepr. arXiv:1611.03814
33. Feily, M., Shahrestani, A., Ramadass, S.: A survey of botnet and botnet detection. In: 2009 Third International Conference on Emerging Security Information, Systems and Technologies, pp. 268–273 (2009)
34. Ben Salem, M., Hershkop, S., Stolfo, S.J.: A survey of insider attack detection research. In: Insider Attack and Cyber Security, pp. 69–90. Springer (2008)
35. Bhamare, D., Salman, T., Samaka, M., Erbad, A., Jain, R.: Feasibility of supervised machine learning for cloud security. In: 2016 International Conference on Information Science and Security (ICISS), pp. 1–5 (2016)

36. Gallagher, B., Eliassi-Rad, T.: Classification of http attacks: a study on the ECML/PKDD 2007 discovery challenge. Lawrence Livermore National Lab. (LLNL), Livermore, CA (United States) (2009)
37. Haddadi, F., Le Cong, D., Porter, L., Zincir-Heywood, A.N.: On the effectiveness of different botnet detection approaches. In: International Conference on Information Security Practice and Experience, pp. 121–135 (2015)
38. Xie, M., Hu, J., Slay, J.: Evaluating host-based anomaly detection systems: application of the one-class SVM algorithm to ADFA-LD. In: 2014 11th International Conference on Fuzzy Systems and Knowledge Discovery (FSKD), pp. 978–982 (2014)
39. Kato, K., Klyuev, V.: An intelligent DDoS attack detection system using packet analysis and support vector machine. In: IJICR, pp. 478–485 (2014)
40. Yusof, A.R., Udzir, N.I., Selamat, A.: An evaluation on KNN-SVM algorithm for detection and prediction of DDoS attack. In: International Conference on Industrial, Engineering and Other Applications of Applied Intelligent Systems, pp. 95–102 (2016)
41. Hasan, M.A.M., Nasser, M., Ahmad, S., Molla, K.I.: Feature selection for intrusion detection using random forest. J. Inf. Secur. **7**(03), 129 (2016)
42. Javaid, A., Niyaz, Q., Sun, W., Alam, M.: A deep learning approach for network intrusion detection system. In: Proceedings of the 9th EAI International Conference on Bio-inspired Information and Communications Technologies (formerly BIONETICS), pp. 21–26 (2016)
43. Tang, T.A., Mhamdi, L., McLernon, D., Zaidi, S.A.R., Ghogho, M.: Deep learning approach for network intrusion detection in software defined networking. In: 2016 International Conference on Wireless Networks and Mobile Communications (WINCOM), pp. 258–263 (2016)
44. Chowdhury, S., et al.: Botnet detection using graph-based feature clustering. J. Big Data **4** (1), 14 (2017)
45. Neethu, B.: Adaptive intrusion detection using machine learning. Int. J. Comput. Sci. Netw. Secur. **13**(3), 118 (2013)
46. Kozik, R., Choraś, M., Renk, R., Hołubowicz, W.: A proposal of algorithm for web applications cyber attack detection. In: IFIP International Conference on Computer Information Systems and Industrial Management, pp. 680–687 (2015)
47. Lippmann, R., Haines, J.W., Fried, D.J., Korba, J., Das, K.: The 1999 DARPA off-line intrusion detection evaluation. Comput. Netw. **34**(4), 579–595 (2000)
48. Saad, S., et al.: Detecting P2P botnets through network behavior analysis and machine learning. In: 2011 Ninth Annual International Conference on Privacy, Security and Trust, pp. 174–180 (2011)
49. Torrano-Gimenez, C., Perez-Villegas, A., Alvarez, G.: A self-learning anomaly-based web application firewall. In: Computational Intelligence in Security for Information Systems, pp. 85–92. Springer (2009)
50. Torrano-Gimenez, C., Pérez-Villegas, A., Álvarez, G., Fernández-Medina, E., Malek, M., Hernando, J.: An anomaly-based web application firewall. In: SECRYPT, pp. 23–28 (2009)
51. Nguyen, H.T., Torrano-Gimenez, C., Alvarez, G., Petrović, S., Franke, K.: Application of the generic feature selection measure in detection of web attacks. In: Computational Intelligence in Security for Information Systems, pp. 25–32. Springer (2011)
52. Hoque, N., Bhattacharyya, D.K., Kalita, J.K.: A novel measure for low-rate and high-rate DDoS attack detection using multivariate data analysis. In: 2016 8th International Conference on Communication Systems and Networks (COMSNETS), pp. 1–2 (2016)
53. Torrano-Giménez, C., Perez-Villegas, A., Alvarez Maranón, G.: An anomaly-based approach for intrusion detection in web traffic (2010)

Mining and Data Analysis

A Survey of Semantic Analysis Approaches

Said A. Salloum[1,2(✉)], Rehan Khan[1], and Khaled Shaalan[1]

[1] Faculty of Engineering and IT, The British University in Dubai, Dubai, UAE
[2] Research Institute of Sciences and Engineering, University of Sharjah, Sharjah, UAE
ssalloum@sharjah.ac.ae

Abstract. Semantics is a branch of linguistics, which aims to investigate the meaning of language. Semantics deals with the meaning of sentences and words as fundamentals in the world. Semantic analysis within the framework of natural language processing evaluates and represents human language and analyzes texts written in the English language and other natural languages with the interpretation similar to those of human beings. This study aimed to critically review semantic analysis and revealed that explicit semantic analysis, latent semantic analysis, and sentiment analysis contribute to the leaning of natural languages and texts, enable computers to process natural languages, and reveal opinion attitudes in texts. The future prospect is in the domain of sentiment lexes. The overall results of the study were that semantics is paramount in processing natural languages and aid in machine learning. This study has covered various aspects including the Natural Language Processing (NLP), Latent Semantic Analysis (LSA), Explicit Semantic Analysis (ESA), and Sentiment Analysis (SA) in different sections of this study. However, LSA has been covered in detail with specific inputs from various sources. This study also highlights the future prospects of semantic analysis domain and finally the study is concluded with the result section where areas of improvement are highlighted and the recommendations are made for the future research. This study also highlights the weakness and the limitations of the study in the discussion (Sect. 4) and results (Sect. 5).

Keywords: Natural language processing · Sentiment analysis · Explicit semantic analysis · Latent semantic analysis · Wikipedia linked based measures

1 Introduction

Documented data occurs in an ever-increasing number of researches, [1–4], and business situations, and is encountered in different contexts by different scientists and scholars [5–12]. [13] indicate that factoring rapid increase in data gathering, language processing, and information retrieval attract investigators pursuing high system performance [14–16] argue that researchers from a variety of disciplines have interests in evaluating abstracts, titles, and publication full-text bodies to categorize attributes including research topics, methods, and theories associated with the research nature. According to [17], many multidisciplinary fields including biochemistry, engineering, data mining, neuroscience, and bioinformatics [18] have emerged in the past numerous

A.-E. Hassanien et al. (Eds.): AICV 2020, AISC 1153, pp. 61–70, 2020.
https://doi.org/10.1007/978-3-030-44289-7_6

decades. Because multidisciplinary environments involve theories, techniques, and methods from several areas, it is not easy to understand all the research processes in these fields [17]. Hence, there should be a method of organizing documents and texts into groups based on their pertinent structures. The method, which researchers and scholars utilize to group texts based on the underlying structures is semantic analysis. [19] state that semantic analysis accords the exact meaning and interpretation of the dictionary meaning emanating from structures developed by syntactic assessment [20]. Semantics is a branch of linguistics, which aims to investigate the meaning of language and language exhibits a meaningful message due to semantic interaction with diverse linguistic categories, syntax, phonology, and lexicon [19]. In this regard, semantic analysis is concerned with the meaning of words and sentences as elements in the world. This assignment reviews critically – offers descriptive, analytical, and interpretative evaluations of semantics. To accomplish this, the assignment factors semantic classification for practical natural language processing [21], latent semantic analysis, explicit semantic analysis, and sentiment analysis. The analyses contribute to the facilitation of free communication and the understanding of the textual data, which can be applied in information systems. The main challenges of Machine learning and Artificial intelligence is learning from natural language and text. Any significant advancement in this area has solid effect on numerous applications varying from speech recognition, information and data retrieval, natural language processing and machine interpretation.

It is important to emphasize that this study has also deliberately considered some research papers published before 2006 because of some interesting findings by the authors. Nonetheless, the latest research papers have been specifically focused. Literature Review.

2 Literature Review

2.1 Technical Part

2.1.1 Natural Language Processing

Semantic analysis interprets language structures through natural language analysis. Natural language processing (NLP) is a theory-motivated scope of computational techniques for the automatic evaluation of representation of human language [22]. [23] indicate that the field of NLP is concerned with the building of computer software systems that process and analyze texts written in the English language and other languages with interpretation and generation capacities similar to those of human beings. [24] adds that NLP is a field of research and application that examines how computers can be utilized to comprehend and manipulate natural language speeches and texts to accomplish useful endeavors. The aim of NLP researchers is to learn how people understand and use language in order to develop effective tools and methods to make computers understand and affect natural languages and to perform the required tasks [24]. The extracted meanings can be utilized for various tasks including undertaking robotic actions, production of equivalent translation in a dissimilar natural language in the context of machine translation systems, and retrieving information from databases. Natural language processing system inventions are associated with the semantic analysis of linguistic structures. For instance, [25] present a semantic discourse presentation structure (SemDRS), which is a model that interprets linguistic structures outputs based on the natural language linguistic analysis framework. SemDRS performs the transformation by contrasting the input linguistic construction to an automatic program in such a way that SemDRS correctly capture the area-specific interpretation of the linguistic expression [25]. For the undefined words in the natural languages, [26] indicate that there is an invention that requires users to input synonyms existing in the dictionary so that they can process natural languages and understand their meanings. NLP is applicable in diverse fields of research including machine translation, user interfaces, multi-language information retrieval, expert systems, speech recognition, natural language text processing, and summarization, and artificial intelligence.

2.1.2 Latent Semantic Analysis

The intelligence and machine learning face challenges such as difficulties learning from the natural language and text, but latent semantic analysis can address the challenges [27]. The fundamental issue is to understand the denotation and utilization of words in an information driven manner, for e.g. from some given content without any prior knowledge of linguistics.

[17] argue that latent semantic analysis (LSA) is a theoretical framework of acquiring knowledge, induction, and representation. LSA is a systematic machine learning methodology for evaluating the associations and similarities of structures among terms and documents without depending on human experience, knowledge bases, past theoretical models, and semantic dictionaries [17]. [16] indicate that LSA was introduced in the late 1980s as an information retrieval approach developed to better the indexing of library and search engine performances. Later, LSA became a method and a theory for retrieving and representing word meaning including sorting words and category judgments. [28] opine that LSA is utilized because of its unparalleled capacity to reveal the conceptual content of the unstructured data based on different mathematical dimension reduction approaches that approximate the linear combinations of the interpretation of words and concepts. LSA utilizes the categorized words and concepts to permit subsequent in-depth analyses. Latent semantic analysis is important in the whole semantic analysis discourse. Latent Semantic Analysis (LSA) is a natural language processing method, which extracts ideas and concepts from the collection of terms to generate an arrangement of terminologies connected by the underlying assumption that words similar in interpretation occur in the analogous segments of the text [28]. LSA identified a model called "Probabilistic Latent Semantic Analysis (PLSA)" where the variable has a sound computable foundation and identified an applicable propagative model of the data [29]. Therefore, LSA addresses challenges of learning from the natural language and text since its goal is to locate data mapping that offers information beyond the level of lexical interpretation and reveals a semantic association between the parties of interest [29]. Because of its generality, it is a suitable and valuable analytical tool with a diverse range of applications in business and information systems to analyze textual data. Some of the areas LSA can be applied include quantitative literature reviews, evaluation of textual data in computer-mediated

communication, evaluation of client feedback interviews, and management of knowledge repositories.

The statistical understanding of LSA leads to the emergence of probabilistic latent semantic analysis (PLSA), which is a new model of LSA. The probabilistic variant of LSA presents a formidable statistical foundation that explains a solid generative framework of the data [29]. The goal of PLSA is to identify and distinguish between contrasting contexts of word utilization without relying on a dictionary. Semantic relatedness interventions seek to encourage computers to reason for written text. PLSA has two applications such as permitting the disambiguation of polysems – words with more than one meaning and projecting the topical similarities by aggregating together words that form a common context [13, 29]. The research paper claimed substantial performance could be obtained by using this technique. Latent semantic analysis can complement unsupervised learning for a different range of applications in information retrieval and text Learning. LSA and its new model, PLSA are critical in the evaluation of text and natural language to comprehend the meaning and utilization of words in different machine learning, artificial intelligence, and other applications.

2.1.3 Explicit Semantic Analysis

To process natural language, computers need access to domain-specific world knowledge and a huge volume of common sense, which can be provided by explicit semantic analysis. Unlike LSA that utilizes statistical co-occurrence information, explicit semantic analysis utilizes knowledge humans gather and organize [30]. Explicit semantic analysis assists text representation with huge amounts of world knowledge such as those of Wikipedia to enable computers to process natural language and reason about the semantic associations of natural language utterances [31]. Explicit semantic analysis is critical in the computation of words and text relatedness and is applicable in artificial intelligence and natural language processing.

The motivation behind semantic relatedness measures is to enable Wikipedia Link-based Measure (WLM) approach have been developed to extract the semantic relatedness measures from Wikipedia. The main difference between the WLM and the conventional method is that relatedness is defined by using hyperlink structures [31]. In this technique, links between articles were used instead of textual content to provide precise measures cost-effectively. This approach is similar to Explicit Semantic Analysis (ESA) as it uses the online encyclopedia for this purpose. ESA is still considered as the best measures in terms of accuracy and effectiveness; the authors were successful in matching the accuracy when the individual articles in Wikipedia is defined as a term. Future work may include applying WLM to different tasks to explore its effectiveness further.

2.1.4 Sentiment Analysis

It is not a new job for advertising researchers to derive the opinions and feelings of customers from a pool of qualitative data. Conventionally, the coding of manual content can perform such a task. Researchers thoroughly and comprehensively examine the data and independently develop their own categories of sentiments. The growing popularity of sentiment analysis can be due to its unique benefits: firstly, computer-aided feelings sentiment analysis is not only more effective than manual coding but also yield comparable results.

Opinion and attitudes in a language are fundamental aspects of semantic analysis, which sentiment analysis reveals [32]. Sentiment analysis encompasses a sequence of methodologies, tools and techniques, related to the detection and extraction of individual data including opinion and attitudes from languages [33]. The objective of sentiment analysis is to identify in what way sentiments are presented in scripts and if the expressions exhibit positive (favorable) and negative (unfavorable) views on a subject [34]. The analysis identifies sentiments expressions, polarity, and power of the expressions, and the association to the topic. The identified elements contribute to the subjective comprehension of texts, which is critical in revealing the meaning of language. Twitter is a social networking and microblogging site for users to post real-time tweets [35–38]. Tweets are brief posts, just 140 characters long. Given the nature of this micro-blogging platform, people use acronyms, make mistakes with spelling, use hashtags, and other expressions that convey their particular meaning. [39] used the twitter data in their research to analyze the sentiment analysis, POS features and kernel tree usage. Twitter data was preprocessed by using an acronym dictionary and emoticons. A glossary of emoticons was prepared by labeling 170 emoticons available on Wikipedia. Different types of labels were assigned to emoticons, such as positive, negative, neutral, etc. Unigram model was used as a baseline and reported an overall increase of 4% for two activities: binary versus negative and neutral versus negative versus positive.

In summation, the Sect. 2 of the assignment critically analyzed NLP in general and narrowed down to latent semantic, explicit semantic, and sentiment analyses. Whereas NLP analyzes texts written in the English language and other natural languages, LSA provides a framework for machine learning to learn from natural languages. The explicit semantic analysis presents domain-specific world knowledge for computers to process natural language, while sentiment analysis investigates opinions and attitudes in the text.

3 Future Prospects

Semantic analysis is primarily concerned with the meaning of language [32, 40]. Language shows a meaningful message since it involves semantic interactions with the dissimilar language groupings such as lexicon and syntax [41]. The technical part revealed a deeper technical content of the semantic analysis including aspects such as latent semantic, explicit semantic, and sentiment semantic within the framework of practical natural language processing. Because of this, future research within the area would be sentiment expressions, which sentiment analysis as a technique of semantic analysis identifies. Sentiment expressions motivate me because the meaning of words and sentences should not only end at the understanding of the text but also at the comprehension of how they contain emotionally loaded opinions. [34] provide a framework for sentiment analysis and define sentiment expressions. In this regard, the grammatical arrangement of words in a sentence is important since it encompasses the elements in the world that constitute the meaning. However, the element of sentiment expression must be clearly understood based on the interpretation of words such as verbs, adjectives, nouns, prepositions, and adverbs. For instance, [34] argue that sentiment expressions utilizing adjectives such as good denotes noun modification with positive polarity. For example, a good product is a sentiment expression, which shows a positive attribute of a product. Similarly, an adverb such as beautifully in the case of play beautiful modifies an adverb to project a positive sentiment expression [34]. Hence, sentiment expressions utilizing adverbs, nouns, and adjectives can be explained based on positive and negative polarity argumentation. Critically, sentiment expression presents the prospect for research undertaking within the semantic analysis because of the applicability of the concept. In the field of computer science, opinion mining is critical when testing the quality of products. Therefore, computer science students have the opportunity to capitalize on large volumes of data and documents to develop powerful functionalities for competitive advantage. A research area is a powerful tool for analyzing and detecting the favorable and unfavorable attributes of innovations and products for appropriate risk analysis and management. In this regard, the study aims to achieve the fundamental association between sentiment expressions and practical natural language processing, especially in the construction of systems that process texts written in conventional languages, natural languages, their interpretation, and contributions in assigning products and concepts the positive, negative, and neutral polarities. The methodology applicable to future research is the critical review of the literature about the topic area. The critical review method of literature offers an opportunity to assess the strengths and weaknesses of the published literature on the research area, analyze ideas and content, and relate them to other aspects of natural language processing. Fundamentally, the critical review methodology provides the basis for describing, analyzing, and interpreting the literature and extracts the presentation associated with the topic area. In conclusion, the future prospect is sentiment expressions. The motivation for the selection is to comprehend how words contain emotionally loaded opinions. The future prospect is critical since it can be applied in gathering data about product quality.

4 Discussion

Summarily, the report has been a critical review of over 12 published research papers on semantic analysis [13, 22, 30, 42–50]. Some papers have been scanned quickly to check the relevance of the topic and the interesting findings revealed by the authors. The analysis unearthed that natural language processing focuses on the development of computer software infrastructure that can evaluate texts composed in English and other natural languages. The computer software should be able to interpret and generate concepts similar to human beings. However, the ability of the NLP to extract the meaning from the text depends on representing the meaning in a style suitable for computer manipulation. Therefore, what I did in the report was to draw a connection between the NLP, which is a theory-motivated scope of computational methods for automatic evaluation and representation of human language with semantic analysis. This study was able to present the ambit of semantic analysis is the meaning of words and sentences, which NLP can rely on to explore how computers can be utilized to comprehend and manipulate natural speeches and texts for various applications such as machine translation, artificial intelligence, user interfaces, and expert systems.

Fundamentally, this study was able to reveal how SemDRS is a model, which performs semantic analysis based on the natural language linguistic analysis platform. The model can process natural languages containing unidentified words and present their meanings in a format that the computer can understand. Moreover, this study was able to capture the meaning and application of latent semantic analysis and how its statistical variance, PLSA generate model data to address challenges associated with the inability of artificial intelligence and machine learning to acquire the meaning and utilization of words in a data-controlled fashion. Finally, this study was able to explain how explicit semantic analysis assists in semantic relatedness of natural language and how sentiment analysis demonstrates favorable and unfavorable opinions for competitive analysis. Reflectively, it was interesting to discover that semantic analysis is an important aspect of natural language processing, which can be applied in computer science, artificial intelligence, and other domains of learning. It was a significant discovery because when the term semantic is introduced, the first conclusion is that it is limited to the meaning of words and sentences. After reviewing the research papers, it is concluded that the semantic analysis is a critical area in the field of machine learning since there must be a mechanism for enabling machines to interpret natural languages to domain-specific world knowledge. The interesting area, which this study did not have time to cover, was syntactic-semantic analysis, especially word-class function (WCF) as presented by [51]. Also, this research paper was published in the year 1997 and the paper was skimmed quickly. I think the word-oriented aspect of WCF would have been an enriching area of focus, particularly on how words contribute fundamentally to language comprehension. In conclusion, a summary of the text reflects the content in a simplified version. It reflects what the researcher did, and the personal views concerning the research. This area is critical since it also reveals areas not captured in the proposal and how they could have contributed to the overall enrichment of the composition.

5 Conclusion

Based on the presentation in the technical part, it is evident that semantics is necessary to process any natural language. [19] contend that in the absence of semantic analysis, the results of machine learning may be ambiguous. In this regard, semantics is a field of linguistics, which deals with the study and interpretation of the meaning and is critical in natural language processing. To solve ambiguities and other challenges associated with natural language processing, semantic classification is of great value. Hence, it is important to collect knowledge on how people comprehend and utilize language to develop appropriate methods and tools to make computers comprehend and manipulate natural languages to undertake tasks such as machine translation. Moreover, it is clear that sentiment analysis contributes to the building of accurate and reliable semantic analysis of words and sentences and latent semantic analysis is applicable in the management of knowledge repositories. Since probabilistic latent semantic analysis is based on sound statistical application, it can be employed to enable machine learning to handle the distinctions between the lexical and semantic level interpretations.

In summation, the critical review of sources related to semantic analysis revealed the weakness of machine learning in processing natural languages. Positively, the semantic analysis demonstrated that computer software with capacities such as those of human beings can be developed to process texts composed in natural languages and English and utilize them in machine translation. Hence, the proposal demonstrates the capacity of semantic analysis in the whole framework of natural language processing and the contribution in user interface investigations. Due to the time limitation, this study could not focus vertically or deep dive into specific domain. However, this study provides a robust foundation to proceed with the detailed research in coming months.

References

1. Alshurideh, M.T., Salloum, S.A., Al Kurdi, B., Monem, A.A., Shaalan, K.: Understanding the quality determinants that influence the intention to use the mobile learning platforms: a practical study. Int. J. Interact. Mob. Technol. **13**(11), 157–183 (2019)
2. Salloum, S.A.S., Shaalan, K.: Investigating students' acceptance of e-learning system in higher educational environments in the UAE: applying the extended technology acceptance model (TAM). The British University in Dubai (2018)
3. Salloum, S.A., Al-Emran, M., Shaalan, K., Tarhini, A.: Factors affecting the E-learning acceptance: a case study from UAE. Educ. Inf. Technol. **24**, 1–22 (2018)
4. Salloum, S.A., Alhamad, A.Q.M., Al-Emran, M., Monem, A.A., Shaalan, K.: Exploring students' acceptance of E-learning through the development of a comprehensive technology acceptance model. IEEE Access **7**, 128445–128462 (2019)
5. Mhamdi, C., Al-Emran, M., Salloum, S.A.: Text mining and analytics: a case study from news channels posts on Facebook, vol. 740 (2018)
6. Salloum, S.A., AlHamad, A.Q., Al-Emran, M., Shaalan, K.: A survey of Arabic text mining, vol. 740 (2018)
7. Salloum, S.A., Al-Emran, M., Abdallah, S., Shaalan, K.: Analyzing the Arab Gulf newspapers using text mining techniques. In: International Conference on Advanced Intelligent Systems and Informatics, pp. 396–405 (2017)
8. Salloum, S.A., Al-Emran, M., Shaalan, K.: Mining social media text: extracting knowledge from Facebook. Int. J. Comput. Digit. Syst. **6**(2), 73–81 (2017)
9. Salloum, S.A., Al-Emran, M., Monem, A.A., Shaalan, K.: Using text mining techniques for extracting information from research articles. In: Studies in Computational Intelligence, vol. 740. Springer (2018)
10. Salloum, S.A., Mhamdi, C., Al-Emran, M., Shaalan, K.: Analysis and classification of Arabic newspapers' Facebook pages using text mining techniques. Int. J. Inf. Technol. Lang. Stud. **1**(2), 8–17 (2017)
11. Salloum, S.A., Al-Emran, M., Shaalan, K.: Mining text in news channels: a case study from Facebook. Int. J. Inf. Technol. Lang. Stud. **1**(1), 1–9 (2017)
12. Salloum, S.A., Al-Emran, M., Monem, A.A., Shaalan, K.: A survey of text mining in social media: Facebook and Twitter perspectives. Adv. Sci. Technol. Eng. Syst. J **2**(1), 127–133 (2017)
13. Chien, J.-T., Wu, M.-S.: Adaptive Bayesian latent semantic analysis. IEEE Trans. Audio. Speech. Lang. Processing **16**(1), 198–207 (2007)

14. Chang, C.-H., Kayed, M., Girgis, M.R., Shaalan, K.F.: A survey of web information extraction systems. IEEE Trans. Knowl. Data Eng. **18**(10), 1411–1428 (2006)
15. Al-Emran, M., Mezhuyev, V., Kamaludin, A., Shaalan, K.: The impact of knowledge management processes on information systems: a systematic review. Int. J. Inf. Manage. **43** (July), 173–187 (2018)
16. Evangelopoulos, N., Zhang, X., Prybutok, V.R.: Latent semantic analysis: five methodological recommendations. Eur. J. Inf. Syst. **21**(1), 70–86 (2012)
17. Kou, G., Peng, Y.: An application of latent semantic analysis for text categorization. Int. J. Comput. Commun. Control **10**(3), 357–369 (2015)
18. Alhashmi, S.F.S., Salloum, S. A., Abdallah, S.: Critical success factors for implementing artificial intelligence (AI) projects in dubai government United Arab Emirates (UAE) health sector: applying the extended technology acceptance model (TAM). In: International Conference on Advanced Intelligent Systems and Informatics, pp. 393–405(2019)
19. Rajani, S., Hanumanthappa, M.: Techniques of semantic analysis for natural language processing – a detailed survey. Int. J. Adv. Res. Comput. Commun. Eng. **5**(2) (2016)
20. Farghaly, A., Shaalan, K.: Arabic natural language processing: challenges and solutions. ACM Trans. Asian Lang. Inf. Process. **8**(4), 14 (2009)
21. Liu, X., He, P., Chen, W., Gao, J.: Multi-task deep neural networks for natural language understanding. arXiv Prepr. arXiv1901.11504 (2019)
22. Cambria, E., White, B.: Jumping NLP curves: a review of natural language processing research. IEEE Comput. Intell. Mag. **9**(2), 48–57 (2014)
23. Mahesh, K., Nirenburg, S.: Semantic classification for practical natural language processing. In: Proceedings of Sixth ASIS SIG/CR Classification Research Workshop: An Interdisciplinary Meeting, pp. 116–139 (1995)
24. Chowdhury, G.G.: Natural language processing. Annu. Rev. Inf. Sci. Technol. **37**(1), 51–89 (2003)
25. Calcagno, M.V., Barklund, P.J., Zhao, L., Azzam, S., Knoll, S.S., Chang, S.: Semantic analysis system for interpreting linguistic structures output by a natural language linguistic analysis system. Google Patents, 13 February 2007
26. Katayama, Y., Nakanishi, K., Yoshiura, H., Hirasawa, K.: System for processing natural language including identifying grammatical rule and semantic concept of an undefined word. Google Patents, 28 April 1992
27. Khan, A., Baharudin, B., Lee, L.H., Khan, K.: A review of machine learning algorithms for text-documents classification. J. Adv. Inf. Technol. **1**(1), 4–20 (2010)
28. Hutchison, P.D., Daigle, R.J., George, B.: Application of latent semantic analysis in AIS academic research. Int. J. Account. Inf. Syst. **31**, 83–96 (2018)
29. Hofmann, T.: Unsupervised learning by probabilistic latent semantic analysis. Mach. Learn. **42**(1–2), 177–196 (2001)
30. Gabrilovich, E., Markovitch, S.: Computing semantic relatedness using Wikipedia-based explicit semantic analysis. In: IJcAI, vol. 7, pp. 1606–1611 (2007)
31. Witten, I.H., Milne, D.N.: An effective, low-cost measure of semantic relatedness obtained from Wikipedia links (2008)
32. Akerkar, R.: Natural language processing. In: Artificial Intelligence for Business, pp. 53–62. Springer (2019)
33. Mäntylä, M.V., Graziotin, D., Kuutila, M.: The evolution of sentiment analysis—a review of research topics, venues, and top cited papers. Comput. Sci. Rev. **27**, 16–32 (2018)
34. Nasukawa, T., Yi, J.: Sentiment analysis: capturing favorability using natural language processing. In: Proceedings of the 2nd International Conference on Knowledge Capture, pp. 70–77 (2003)

35. Abraham, A., Hassanien, A.-E., Snášel, V.: Computational Social Network Analysis: Trends. Tools and Research Advances. Springer, Heidelberg (2009)
36. Ghali, N., Panda, M., Hassanien, A.E., Abraham, A., Snasel, V.: Social networks analysis: tools, measures and visualization. In: Computational Social Networks, pp. 3–23. Springer (2012)
37. Salloum, S.A., Mhamdi, C., Al Kurdi, B., Shaalan, K.: Factors affecting the adoption and meaningful use of social media: a structural equation modeling approach. Int. J. Inf. Technol. Lang. Stud. **2**(3), 96–109 (2018)
38. Al-Maroof, R.S., Salloum, S.A., AlHamadand, A.Q.M., Shaalan, K.: A unified model for the use and acceptance of stickers in social media messaging. In: International Conference on Advanced Intelligent Systems and Informatics, pp. 370–381 (2019)
39. Agarwal, A., Xie, B., Vovsha, I., Rambow, O., Passonneau, R.: Sentiment analysis of Twitter data. In: Proceedings of the Workshop on Language in Social Media (LSM 2011), pp. 30–38 (2011)
40. Baets, W., Oldenboom, E., Hosken, C.: The potential of semantic analysis for business (education). Available SSRN 3364133 (2019)
41. Salloum, S.A., Al-Emran, M., Shaalan, K.: A survey of lexical functional grammar in the Arabic context. Int. J. Com. Net. Tech. **4**(3), 141–147 (2016)
42. Godbole, N., Srinivasaiah, M., Skiena, S.: Large-scale Sentiment analysis for news and blogs. Icwsm **7**(21), 219–222 (2007)
43. Delmonte, R. Pallotta, V.: Opinion mining and sentiment analysis need text understanding. In: Advances in Distributed Agent-Based Retrieval Tools, pp. 81–95. Springer (2011)
44. Al-Batah, M.S., Mrayyen, S., Alzaqebah, M.: Investigation of naive bayes combined with multilayer perceptron for arabic sentiment analysis and opinion mining. JCS **14**(8), 1104–1114 (2018)
45. Chopra, F.K., Bhatia, R.: A critical review of sentiment analysis. Int. J. Comput. Appl. **149** (10), 37–40 (2016)
46. Pak, A., Paroubek, P.: Twitter as a corpus for sentiment analysis and opinion mining. In: LREc, vol. 10, no. 2010, pp. 1320–1326 (2010)
47. Rahate, R.S., Emmanuel, M.: Feature selection for sentiment analysis by using svm. Int. J. Comput. Appl. **84**(5), 24–32 (2013)
48. Hofmann, T.: Probabilistic latent semantic analysis. In: Proceedings of the Fifteenth Conference on Uncertainty in Artificial Intelligence, pp. 289–296 (1999)
49. Medagoda, N., Shanmuganathan, S., Whalley, J.: A comparative analysis of opinion mining and sentiment classification in non-English languages. In: 2013 International Conference on Advances in ICT for Emerging Regions (ICTer), pp. 144–148 (2013)
50. Mukherjee, S., Bhattacharyya, P.: Sentiment analysis in Twitter with lightweight discourse analysis. In: Proceedings of COLING 2012, pp. 1847–1864 (2012)
51. Helbig, H., Hartrumpf, S.: Word class functions for syntactic-semantic analysis. In: Proceedings of the 2nd International Conference on Recent Advances in Natural Language Processing (RANLP 1997), pp. 312–317 (1997)

Mining Publication Papers via Text Mining: A Case Study

Ahmed S. Ibrahim(✉), Sally Saad, and Mostafa Aref

Faculty of Computer and Information Sciences, Ain Shams University, Cairo 11566, Egypt
{a.saeedibr,sallysaad,mostafa.aref}@cis.asu.edu.eg

Abstract. The amount of data that produced is increased day after day especially data as a text, so with this massive production it would be difficult to analyze or extract information to discover the patterns from the unstructured text. Text mining is used for availing the massive amount of knowledge that is in the text and deriving high quality information from the text automatically. This Process would save effort and time. Text mining considered as a subset of data mining where data mining is more generic. This paper proposes a methodology of mining a text for a case study related to publication papers. Some of text mining approaches will be introduced for mining the publication papers using machine learning (ML) and natural language processing (NLP) techniques. Describing each phase as following: First phase is keywords extraction using natural language processing techniques, second phase named entity recognition and last phase is document classification. The last two phases are using the ML techniques. Then a case study is built to simulate the system phases, showing what is the input and the output in each phase.

Keywords: Text mining · Data mining · Named Entity Recognition (NER) · Mining publication papers · Machine Learning (ML)

1 Introduction

At our current pace the amount of data is generated everyday is truly massive. There are tons bytes of data created each day as the byte is the unit of storage on electronic devices. Data here might be a number or text, so mining the data is an important phase to detect patterns in a random data and understand trends, patterns, correlations by using all that information. The limits between text mining and data mining are ambiguous. In text mining, natural language texts (unstructured text) is the most common type to be used for patterns extraction rather than from structured databases of facts. There are two types of text: structured text and unstructured text. Mining the unstructured text is considered a difficult process as the text won't be in a form of row and columns it would be in a form of document which contains set of paragraphs. So text mining might be a powerful process for analyzing text without prior knowledge of the accurate

A.-E. Hassanien et al. (Eds.): AICV 2020, AISC 1153, pp. 71–80, 2020.
https://doi.org/10.1007/978-3-030-44289-7_7

keywords or terms, discover invisible relationships and trends in the text, finding the themes and key concepts of the text. As all of publication papers documents contain many types of data but mostly text so mining the text would help a lot to find the undiscovered relationships among the papers that is related to specific paper and that result to save the effort and time for the researcher. Text Miming has mainly five approaches [3]. The first is information Retrieval approach, which is the activity of obtaining information resources (mostly documents) from a set of unstructured data collections that satisfies the information need. The second approach is Natural Language Processing especially used in pre-processing phase. The third approach is the information extraction. The fourth approach is unsupervised learning methods. Last approach is supervised learning methods, so the structure of the system consists of the machine learning algorithm and text mining techniques to build it. This paper is organized in the following manner: Sect. 2 for Related Work, Sect. 3 Proposed Methodology, Sect. 4 Case Study and finally followed by Conclusion and References.

2 Related Work

Liao [1] described the text mining process steps as following: collecting data, preprocessing, text diversion, extracting feature, pattern election and evaluation. Further more, there are several approaches used for text mining for example: clustering, natural language processing, information extraction, topic tracking, categorization, text summarization, and their application are surveyed.

Henriksson and Moen [4] discussed integrating a framework Medline biomedical database that helps to remove unuseful specifics and extract useful information. Integration of the framework are based on the following modules: recognizing named entities (NER), classifying text, generating and testing hypothesis, extracting abbreviations, relationship and synonym.

Sumathy and Chidambaram [5] discussed the type of text documents that might be structured documents, semi structured documents or unstructured documents and discussed the techniques of extracting valuable information. Giving a brief abstract for applications and tools that used to mine the text and viewing issues for mining the text. A generic framework has been presented in this paper would be visualized for mining concept as following: knowledge distillation phases and text refinement. There is a dependency between intermediate form of mining the entity representation and a particular field.

Khan et al. [10] presented a review of classification text using ML approaches and feature selection algorithm. Most commonly used methods that perform in a good manner for selecting features are the Gain and Chi square statistics. Most appropriate algorithms used for text classification are Support Vector Machine (SVM), Naive Bayes (NB) and K-Nearest Neighbor (KNN).

Roul and Varshneya [7] presented a method to identify the similarity within the document and discover information regarding specific subjects using the document TF-IDF (Term Frequency-Inverse Document Frequency) algorithm. Presents a method to gather the similar text using k-mean clustering technique for bottom up approach. However, there are two different approaches for mining web-based text, a top down and bottom up approach.

Chau et al. [6] used machine learning (ML) with feed-forward/back-propagation neural network for extracting meaningful entities for instance: person's name, location and organization. The system extracts main entities from police narrative reports. The system performed in a good manner for the following entities: (74.1% precision), narcotic drugs (85.4% precision), However, performed not good as expected for the following entities: Personal property, addresses.

Jai Sharma et al. described the recent efforts and contributions in text mining techniques. There are various techniques that used for mining text to extract information, analyze and discover the pattern from a text. Some of these techniques used the ML approach and some used the NLP approach. The first technique is text classification considered as a supervised technique. The second technique is clustering is an unsupervised technique. Third technique is the information extractions (IE), the documents are transformed into the structured databases where text mining techniques would be applied for extracting knowledge or interesting patterns. Fourth technique information visualization provides browsing abilities, in addition to a simple search by viewing large textual resources in a visual hierarchy or map. Fifth technique is NLP concerns with natural language generation (NLG) and natural language understanding (NLU).

3 Proposed Methodology

This paper presents Mining methodology for mining the publication papers. This methodology depends on using text mining techniques and machine learning approaches as mentioned above. The methodology goal is to search for scientific papers using similar keywords that used in a specific paper resulting from this process is a set of papers which might relate to each other and that would help in discovering the hidden relationships among papers, after that recognizing of the named entities for each papers which helps to find the relation among keywords, finally classifying all documents to specific class see Fig. 1.

3.1 Reading the Document

Reading documents is the first phase for mining the publication papers, those documents might be in any file extension i.e. (pdf, docx, txt, xml ... etc.). Most of document contains a variety of data however most data appear in the document. Would in a format of text and that what is needed for mining the papers.

3.2 Extracting Keywords

Is the process which identifies automatically keywords that best describe the subject of a document to the reader. Extraction of keywords in a manual way is

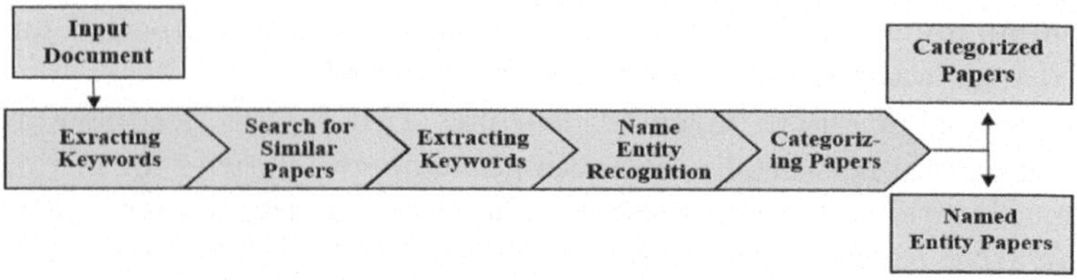

Fig. 1. Workflow of mining the scientific papers.

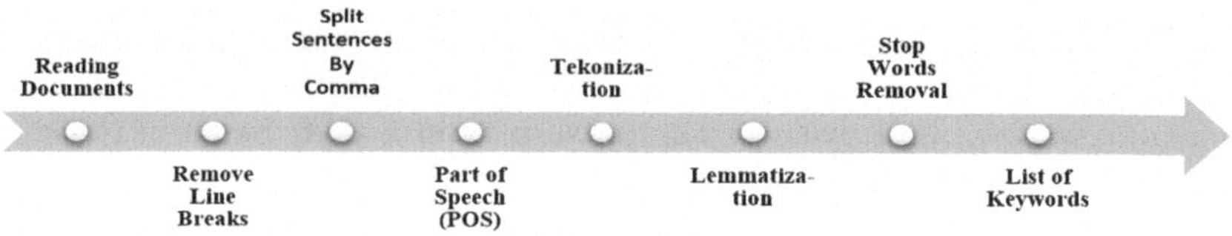

Fig. 2. Pre-processing phases for mining the publication papers.

very hard process and consumes time. Therefore, there is in need for automatically extracting keywords from documents. Keyword extraction, also referred as key phrase extraction, is an field of text mining that aims to identify the most useful and important keywords and phrases. To prepare papers for keywords extraction it is needed to pre-process for removing inconsistencies, redundancies words, so each document go through some phases to filter the words from document and gather only keywords with the highest effectiveness in the papers the following Fig. 2 shows the phases of pre-processing resulting a list of words that would be used for searching with a related topic.

- **Parts of Speeches (POS):** POS tagging sometimes referred to as (grammatical tagging) is the process of identifying a word into their POS, based on its context and definition. The word might have various POS based on the context of the word that is used, so it is not an explicit task. It is significant to realize that a word can sometimes be in more than one POS. For instance:
  1- Play can be a verb: he plays football everyday.
  2- also be a noun: that is a good play.
  After removing the line breaks and split the documents by the comma a list of sentences would be generated to be ready to get processed for next phase, Stanford POS techniques are used to find the POS of each word [8]. The result would be a list containing each word with its corresponding POS.

- **Tokenization:** The process of identifying the document as a set of words by splitting or breaking each sentence into a sequence of pieces (token) such as words, keywords, symbol, phrase. Sentences are splitted by white space, punctuation marks or line breaks.

- **Lemmatization:** Lemmatization in linguistics is the process of reducing the words to their base (root) so they are able to be analyzed as a single element, identified by the lemma of the word or dictionary form. In many languages, words could be appeared in various inflected forms. For instance, in English, the verb 'take' may appears as following: 'take', 'took', 'takes', 'taking'. The root form 'take', is called the lemma in the dictionaries. Word lexeme is the combination of the root form with the POS. Lemmatization relies on recognize the intentional POS and the meaning of the word in a sentence, also within the entire document or even bigger context surrounding that sentence (contiguous sentences). Lemmatization and stemming are close to each other. However, The distinction is that a stemmer operates on a word without taking into consideration the knowledge of the word context [11]. Determining POS tags using NLP Stanford POS tagger based on Probabilistic approaches [9] where those values are gained from a tagged corpus in a manual way. Based on Fig. 2 each word is lemmatize to its root depending on the context of the word in the sentence and that by using the POS generated by the previous phase, so the input would be the word and its corresponding POS. The output of this phase would be a list of words generated containing each words with its roots.

- **Removal of Stopping words:** The process of removing useless words that will not affect the meaning of the sentence those words called stop words such as a, an, but, and, of, the etc. also the search engine had been programmed to ignore or filtering those words.

3.3 Searching for Similar Papers

After the documents is pre-processed and represented in a list of keywords, then the frequency of each keyword is calculated using TF-IDF term, the list of keywords will be refine using a threshold value where the most repeated keywords would be taken only if its above or equal this threshold, otherwise these words would not considered as a keyword. The list of the most frequent keywords is used to search for scientific papers containing similar keywords using a search engine see Fig. 3, then a collection of documents that match those keywords are gathered for the determine the similarity phase. The next phase is to extract the keywords from each paper that already gathered to find the related papers see Fig. 4 using the cosine similarity measure [2]. A list of keywords is extracted for Each related document that will be used for another phase.

3.4 Named Entity Recognition (NER)

Entity identification (also is known as entity extraction or name entity recognition) is to recognize the main concepts in the document for example person, locations and organizations. Is one of important tasks to solve many problems in research field. Entities extraction can be considered as sub-task of information extraction that deals with unstructured or structured documents involves two tasks, first task is proper names identification in text. Second task is to classify these entities into a set of predefined groups of interest, for example: person's name, organization, location, address, date, time expression and concept. The

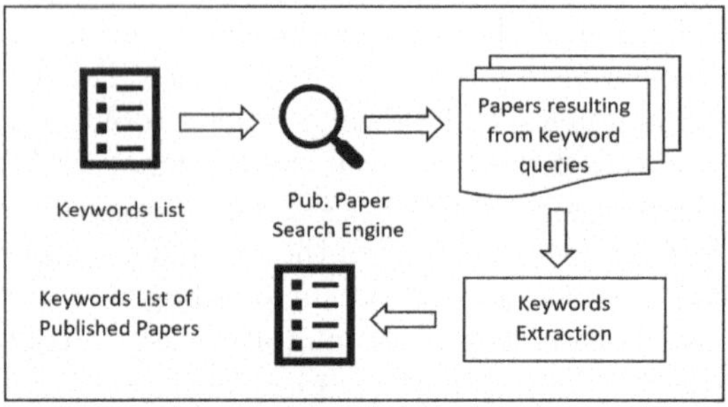

Fig. 3. Process of using extracted keywords to search for related papers.

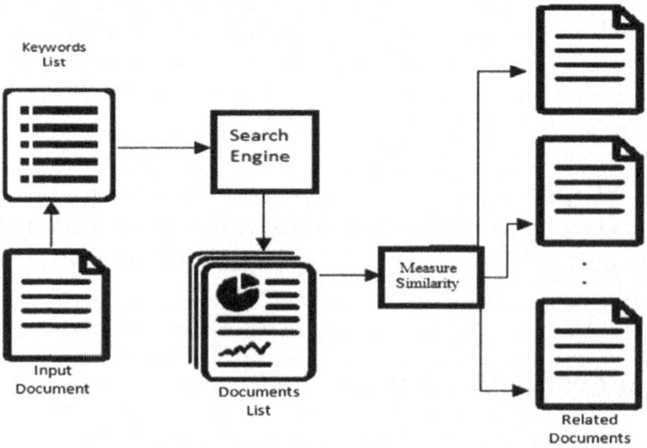

Fig. 4. The process of searching for related documents

NER is applied on all related papers that matches the input document. To apply this approach, tokenization and POS tagging to the sentence should be applied see Fig. 5. Then applying noun phrase chunking to using a regular expression consisting of rules that indicate how sentences should be chunked. Using this pattern = $NP : <DT>?<JJ>*<NN>$.

Fig. 5. Pre-processing phases before entity recognition.

3.5 Publication Papers Classification

The process of classifying the documents using ML is to assign a specific publication paper to specific class see Fig. 6. Document classification has two different methods: manual and automatic classification. In manual document classification, users translate the meaning of text, differentiate the relationships between concepts and the category of the documents. This gives the users an extra control over classification process. However, classifying manually would be costly and consume more time. Machine learning one of the main methods that used for automatic classification. There are three categories of learning: supervised, unsupervised and semi-supervised learning. All these types could be used in any classification problem; however, it differs in whether it uses labeled data or not. Document categorization is considered as one of the techniques of text mining [4]. The dataset is divided into two groups the first group for training the model and the second group for testing, 100 papers are collected to build the model 60% of the paper are for training and 40% of papers are for testing so, the input for testing model is a document from testing group the output would be the category of documents that belongs to.

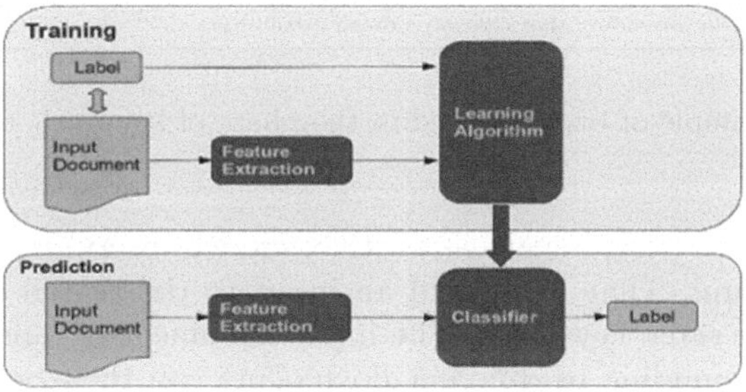

Fig. 6. Document classification model.

4 Case Study

This case study is going to examine mining publication papers system and the methods that applied using text mining techniques for analyzing unstructured contents. A Document contains a paragraph that used to build this study case and go through all phases in the methodology, to show what the input and the output in each phase See Fig. 1. In the initial phase the paragraph is read from the document and pre-processed base on the work-flow mentioned above as shown in Fig. 2. After reading the paragraph and pre-processed the keywords of the document is extracted from the paragraph and this after applying the part of speech, lemmatization and removing stopping words to reduce inconsistencies, redundancies words, so a list of keywords is generated of noun words only as the scientific terms mostly nouns see Table 1.

The second phase would determine the frequency of each keyword contained in extracted list. Then a list of most frequent keywords is generated by a threshold value between 4% to 7%. The list contains frequent keywords as following:

Table 1. A list of keywords extracted from a document

Promise	Learning	Model	Probability	Distribution	Kind	Data
Intelligence	Application	Success	Dropout	Maximum	Image	Learning
Algorithm	Likelihood	Audio	Estimation	Model	Unit	Estimation
Procedure	Waveform	Input	Gradient	Strategy	Difficulties	Speech
Class	Model	Difficulty	Symbol	Label	Impact	Benefit
Language	Success	Context	Unit	Corpora	Back-propagation	Computation

> The promise of deep learning is to discover rich, hierarchical models [2] that represent probability distributions over the kinds of data encountered in artificial intelligence applications, such as natural images, audio waveforms containing speech, and symbols in natural language corpora. So far, the most striking successes in deep learning have involved discriminative models, usually those that map a high-dimensional, rich sensory input to a class label [14, 22]. These striking successes have primarily been based on the backpropagation and dropout algorithms, using piecewise linear units [19, 9, 10] which have a particularly well-behaved gradient. Deep generative models have had less of an impact, due to the difficulty of approximating many intractable probabilistic computations that arise in maximum likelihood estimation and related strategies, and due to difficulty of leveraging the benefits of piecewise linear units in the generative context. We propose a new generative model estimation procedure that sidesteps these difficulties.

Fig. 7. Example of text the used in the phase of keywords extraction.

deep, learning, piece-wise, linear, generative, model, discriminative, probability, distribution, natural. That list would an input to the search engine to gather papers containing same keywords. The input document d1 and the list of documents ld2 corresponding to different documents can then be compared using standard similarity measures such as the cosine. The search is done by measuring the similarity between the input paper and other papers resulting a list of documents that related to same field of the input document. The most frequent keywords will an input to the search engine. Only the most 10 recent papers are taken for each keyword used in the searching phase. The output would be roughly 150 documents, however most of those documents aren't related to the input document. The next step is measuring the similarity between the input document with other documents, resulting 50 related documents see Fig. 4.

Third phase is Name Entity Recognition where the similar documents are going to be the input to this module, the objective of the NER is to define the main entities of the documents and this going to be the output of the module as shown in Fig. 8. Input of that module would be the document and the output would be a list of entities and each entity assigned to a list of words See Table 2.

Last phase is categorizing the papers based on their field using machine learning approach where that module is divided into two phases, the first phase is the training phase where the classifier gives a value to each pair belongs to $D \times C$. Where D is set of training documents $D = d_1,\ldots,d_n$ and C is a set of pre-defined categories $C = C_1,\ldots,\ C_m$. If the value is true, then document d_j is assigned a category c_i, so the input in the training phase going to be set of documents with their predefined category and the output is the model will be trained. The predefined categories that used are as following graphics,

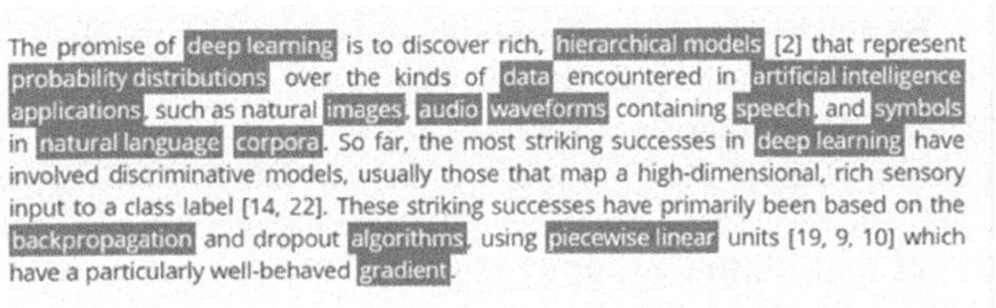
The promise of deep learning is to discover rich, hierarchical models [2] that represent probability distributions over the kinds of data encountered in artificial intelligence applications, such as natural images, audio waveforms containing speech, and symbols in natural language corpora. So far, the most striking successes in deep learning have involved discriminative models, usually those that map a high-dimensional, rich sensory input to a class label [14, 22]. These striking successes have primarily been based on the backpropagation and dropout algorithms, using piecewise linear units [19, 9, 10] which have a particularly well-behaved gradient.

Fig. 8. Name entity identification

Table 2. An example of entities found in the document

Entities	Number of keywords
Work:	0
Places:	0
Organization:	0
Concepts:	17
Events:	0
Persons:	0

security, artificial intelligence and hardware, for example the input would be the document in Fig. 7 to the prediction model to find the category of this document. The output would be artificial intelligence which is the category of the input document.

5 Conclusion

This paper introduced a methodology that concerns with mining the publication papers using text mining techniques and machine learning. The proposed system consists of the following features. Firstly, extracting the most frequent keywords using natural language processing methodologies by using a threshold value however, the threshold value would be generated automatically based on the number of words in each paper. Secondly, categorizing documents using machine learning. Lastly, identify the main theme for the documents using also machine learning. Many of the researches that using text mining involved in mining data for business wise to predict the changes in different sectors based on the given data. However, number of researches that concerns mining the scientific papers is very limited. Therefore, focusing on this point is very important where it has a great impact for researchers throughout saving efforts and time for researchers by collecting the most papers in a specific domain.

References

1. Liao, S.-H., Chu, P.-H., Hsiao, P.-Y.: Data mining techniques and applications-a decade review from 2000 to 2011. Expert Syst. Appl. **39**(12), 11 30–11 311 (2012)

2. Rajendra, R., Saransh, V.: A novel modified apriori approach for web document clustering. Int. J. Comput. Appl. **3**, 159–171 (2013)
3. Thilagavathi, K., Shanmuga, V.: A survey on text mining techniques. Int. J. Adv. Res. Comput. Sci. Robot. **2**(10), 41–50 (2014). ISSN: 2320 7345, in IJRCAAR
4. Henriksson, A., Moen, H., Skeppstedt, M., Daudaravicius, V., Duneld, M.: Synonym extraction and abbreviation expansion with ensembles of semantic spaces. J. Biomed. Semant. **5**(1), 1 (2014)
5. Sumathy, K., Chidambaram, M.: Text mining: concepts, applications, tools and issues-an overview. Int. J. Comput. Appl. **80**(4) (2013)
6. Chau, M., Xu, J.J., Chen, H.: Extracting meaningful entities from police narrative reports. In: Proceedings of the 2002 Annual National Conference on Digital Government Research, pp. 1–5. Digital Government Society of North America (2002)
7. Roul, R.K., Varshneya, S., Kalra, A., Sahay, S.K.: A novel modified apriori approach for web document clustering. In: Computational Intelligence in Data Mining, vol. 3, pp. 159–171. Springer, New Delhi (2015)
8. Cutting, D., Kupiec, J., Pederson, J., Sibun, P.: Practical part of speech tagger. In: Proceedings of the Third Conference on Applied Natural Language Processing (1992)
9. Lafferty, J., McCallum, A., Pereira, F.: Conditional random field: probabilistic model for segmenting and labeling sequence data. In: Proceedings of the Eighteenth International Conference on Machine Learning (2001)
10. Khan, A., Baharudin, B., Lee, L.H., Khan, K.: A review of machine learning algorithms for text-documents classification. J. Adv. Inf. Technol. **1**(1), 4–20 (2010)
11. Ratnaparkhi, A.: A maximum entropy model for part of speech tagging. In: Proceedings of the Conference on Empirical Methods in Natural Language Processing, pp. 133–142 (1996)

Medical Imbalanced Data Classification Based on Random Forests

Engy El-shafeiy[1,3(✉)] and Amr Abohany[2,3]

[1] Department of Computers Engineering, Faculty of Engineering, Mansoura University, Mansoura, Egypt
engy.elshafeiy@gmail.com

[2] Department of Information Systems, Faculty of Computer Science and Information, Kafrelsheikh University, Kafr El Sheikh, Egypt

[3] Scientific Research Group in Egypt, Cairo, Egypt
http://www.egyptscience.net

Abstract. This paper studies the problem of imbalance in medical datasets. Today, modern machine learning techniques are becoming increasingly popular for this type of problem, with examples in the areas of health and medical. One of the major difficulties with this technique is that the database handled is highly imbalance. Under-sampling and over-sampling techniques are used to work around this problem. In this paper, we apply random forests, which are combinations of decision trees fitted to subsamples of the data, built using under-sampling and over-sampling. At the end of the work, we compare fit metrics obtained in the various specifications of the models tested and evaluate their results inside and outside the sample. We observed that random forest techniques using imbalanced sub-samples smaller than the original sample presented the best performance among the random forests used and an improvement compared to that practiced in the medical dataset.

Keywords: Random forest · Imbalanced medical dataset · Under-sampling · Oversampling

1 Introduction

Rotation Forest (RF) is an ensemble classifier that incorporates decision tree algorithms and the theory of linear analysis. In the latest research, RF widely used in different areas with an impressive performance as opposed to the traditional Machine Learning (ML) algorithms, granted that a sufficient number of fundamental classifiers are available. The conventional RF, however, struggles from Linearly Inseparable Datasets (LIDs) classification [1]. Zhu et al. [2] proposed a minimal cost and strong efficiency approach for extracting approvals, responsive application programming interfaces, tracking events of the system and approval levels as main attributes and used the ensemble RF to create a model to determine whether or not the Mobile app is harmful. Hong et al. [3] investigated and compared the use of methods, such as AdaBoost, Bagging, and RF, for landslide sensitivity evaluation integrated with the J48 Decision Tree (JDT) Classifier. Wang et al. [4] proposed a novel Feature Weighted Rotation

A.-E. Hassanien et al. (Eds.): AICV 2020, AISC 1153, pp. 81–91, 2020.
https://doi.org/10.1007/978-3-030-44289-7_8

Forest (FWRF) algorithm handling protein-protein interaction data, which containing noise and highly dimensional. Pham et al. [5] proposed a hybrid algorithm that would combine the Radial Base Function (RBF) neural network classifier and the RF Ensemble, which are ML techniques to address classification problems. Decision Tree (DT) is integrating randomness into its building process and afterward creating a set out of it [6]. Gul et al. [7] proposed a collection of kNN classifier subsets, ESkNN, for a two-step classification process. Classifiers were selected depending on their results using out-of-sample accuracy. Sun et al. [8] presented a new DT series of models for imbalanced corporate credit assessment depend on a Synthesized Minority Over-Sampling Technique (SMOTE) and a Bagging Ensemble Learning Algorithm with Differential Sampling Rate (DSR) called DTE-SBD (Decision Tree Ensemble depend on SMOTE, Bagging and DSR). Lango et al. [9] studied the roughly balanced basic properties of baggage which may affect the efficiency of its classification. Chen et al. [10] introduced a novel approach based on the combination of Fisher's Linear Discriminating Function (FLDA) with Rotational Forest (RFLDA) and Bagging (BFLDA) assemblies for groundwater capacity evaluation in the Ningtiaota area of Shaanxi, China. Learning from imbalanced datasets is one of the main difficulties in ML. Although a regular classifier could have a better performance on a balanced dataset, its performance deteriorates significantly when implemented on an imbalanced dataset. This poor performance is rather problematic, particularly in the identification of the minority class, which is typically a class of interest. Over-sampling the minority class is one of the most promising solutions for imbalanced data learning [11]. Maldonado et al. [12] suggested a novel approach to the collection of features designed to address two key problems in ML, namely class inequality and high dimensionality. Piri et al. [13] introduced a new Synthetic Information Minority Oversampling (SIMO) algorithm that would exploit the SVM. Zhang et al. [14] studied the high-dimensional problems affecting the sampling process, whereas class - imbalanced problems interfere with the selection of features, and then implemented SVM and examined its benefits in addressing high-dimensional and class- imbalanced problems. The methods of Douzas et al. [15] are depending on local knowledge instead of on the distribution of the total minority class. Veganzones et al. [16] proposed a joint analysis of the degree of disparity, loss of results and methods of treatment. Tahan et al. [17] presented a novel Evolutionary Multi-Objective Discretization Algorithm for Binary Imbalanced Datasets (EMDIDs).

2 Background

2.1 Classifiers and Supervised Learning

This section introduces techniques that fall into the group of techniques formed by supervised training. The techniques presented here were used to compare with the technique proposed in this paper. Supervised training techniques need a set already to learn of classified instances (training set). The technique used for supervised training within this paper will be presented here.

Decision Tree (DT). DTs [19] are statistical models represented by a set of conditions that divide the learning sample into smaller and smaller parts, where we have the classes {0 = abnormal; 1 = normal}, the only necessary change in the tree algorithm corresponds to the classification criteria of the nodes and "pruning" of the tree. Therefore, for a node m representing the region R_m with N_m observations represents the proportion of class k observations at node m. Therefore, the observations in node m are classified for class $k(m), k\{0, 1\}$; for the majority class referring to node m.

$$p_{mk} = \frac{1}{N_m} \sum I(y_i = K)_{Xi \epsilon Rm} \tag{1}$$

The criterion used to perform the partitions is the utility of the attribute for the classification. By this criterion, a certain gain of information is applied to each attribute. The test attribute selected for the current node is the one that has the highest information gain, measured through some index. From this application, a new partition process starts. Traditionally the partition criterion is based on the Gini index, which measures the degree of heterogeneity of the data. This index on a given node is given by:

$$Index\,Gini = p(1 - p) \tag{2}$$

Where p is the relative frequency of class $k = 1$ at each node.

Bagging. The bootstrap method [19] aims to approximate the sample distribution of a statistic of interest using the observations. The bootstrap method is used to improve the accuracy of sample estimates through computational processes and is used in a range of areas and applications. In [19] the training data set is defined as $Z_i = (z_1, z_2, \ldots\ldots, z_n)$ where $Z_i = (x_i, y_i)$. The basic idea is to treat Z as the population and use resampling with replacement to build Z n size sub-samples. This step is performed a large number of times, producing B bootstrap samples $Z^1, \ldots\ldots, Z^B$. For each sample, the model is re-estimated and, thus, we obtain a sample of estimators, later used to build its distribution. Bagging uses bootstrap samples to build a collection of DTs that are combined into a single classifier. Given a training sample, we build B bootstrap samples, obtaining in each sample a DT $f(x)$. The bagging estimation is defined by the average of the predictions of all trees or the most voted. In the first case, we obtain an estimate of the probability while in the second, the majority class.

$$f_{bag}(x) = \frac{1}{B} \sum_{b=1}^{B} f^b(x) \tag{3}$$

The main idea of the bagging technique is to use the average of many unstable, but approximately non-biased, models to reduce the variance of the final predictor. The out-of-bag error is an unbiased estimator of the out-of-sample error and, for a z_i observation, uses only trees estimated using Z^b samples that did not contain this observation. More precisely, where C_i is the number of bootstrap samples that don't contain the observation z_i.

$$Err_{oob}\left(f_{bag}\right) = \frac{1}{N}\sum_{i=1}^{n}\frac{1}{C_i}\sum_{b:z_i \nexists Z^b} I\left(y_i \neq f^b(x_i)\right) \quad (4)$$

Random Forests. Random Forest was introduced by Breiman [20] and represents a bagging modification, in which a large collection of correlated trees is built, in which it considers the average of the results. The idea of random forest, demonstrated in its algorithm, is to improve the reduction of variance through bagging by reducing the correlation between trees, without greatly increasing the value of variance. This is achieved in the tree growth process through the random selection of the input variables. Besides, there is a facility to parallelize the process given the independence of the estimation of each tree [19]. As proposed in [20] the first step in the random forest is to select a bootstrap sample from the training sample. When the random forest is applied to an extremely imbalanced sample, there is a significant probability that few elements from the less frequent class will be selected, resulting in a tree with low performance to predict the less frequent class. A good way to solve this problem is to use a stratified bootstrap, that is, select samples with replacement within each class. This alternative, however, does not completely solve the imbalanced class problem. As presented in [21], for classification trees, artificially making classes in equal sizes through under-sampling of the majority class or over-sampling of the minority class is generally more effective concerning a certain measure of performance. However, there may be some negative points. The algorithm for the balanced random forest is shown below:

(A) For each iteration in a random forest, select a bootstrap sample from the minority class. Randomly select the same number of cases, with substitution, from the majority class.
(B) Induce a data classification tree to the maximum size, without pruning. The tree is induced with the CART algorithm [19], with the following modification: in each node, instead of searching through all the variables for the ideal division, just search through a set of variables selected at random.
(C) Repeat the two steps above for the desired number of times. Add the set forecasts and make the final forecast.

For this paper, a small change in item 1 was made. When selecting a bootstrap sample, weights were assigned to both classes and balancing the probability of each subset of the database is selected. Thus, at each bootstrap, the probability of selecting the observation for the minority class will be times greater than for the majority class. Therefore, the calculation was done as follows:

$$\eta = \int_{\frac{n-\Phi}{\Phi}}^{1} \begin{matrix} \textit{for majority class;} \\ \textit{for minority class;} \end{matrix} \quad (5)$$

Where N is the total observations of the training sample and is the total observations of the minority class. The number of variables randomly selected in item 2 was from $\sqrt{N-1}$.

2.2 Classification with Imbalanced Data

Sample balancing techniques, a very common technique when it comes to classes with low frequency is sampling and the methods that will be used in this paper are under-sampling and over-sampling. The purpose of both methods is to address the imbalance of the classes, making them less frequent class more observed, making it more frequent. Under-sampling consists of eliminating observations of the most frequent and over-sampling class consists of replicating the least frequent class [22]. Under-sampling, by eliminating observations from the most frequent class, can lose the effects of observations relevant to the estimation of the technique. In addition, oversampling, when replicating the less frequent class, can cause the model to overfit. Oversampling performs the replication of preexisting instances in the original dataset or synthetically generates new instances. The selection of the preexisting instances can occur randomly (e.g. Random Oversampling (ROS)) or directed by sub-concepts that compose the class. Regarding the generation of synthetic data, the interpolation technique is commonly used for this purpose, as is the case of the Synthetic minority oversampling technique [23].

Synthetic Minority Over-sampling Technique (SMOTE). SMOTE is a technique proposed by Chawla et al. [24] that generates artificial data through interpolation. It uses the KNN algorithm to create representatives between their k neighbors. The neighbor is multiplied by a random number between 0 and 1 and the new vector is applied over the element under consideration, selecting a point in the space, which falls on the line between the element and its neighbor. At this point, the new representative of the minority class is created. This step is represented by Eq. (6), in which x_i is the selected example, x_i, and its closest neighbor x_i' is the random number.

$$x_{novo} = x_i + \left(x_i' - x_i\right) * \gamma \tag{6}$$

3 Training the Resampling Technique

The results of the paper is divided as follows: initially, the specifications of the Technique are presented, detailing the division of the groups with different characteristics determined by the use or not of sampling techniques and the configurations used to estimate the Technique; in the second part of the paper we present the performance evaluation metrics considering each configuration applied in each group; and at the end of the paper, the configurations with the best performance found for each group will be presented according to the chosen metrics and the best method will be determined according to the established criteria.

The Random Forests Used in the Paper are Divided into Four Groups:

(A) Group 1: imbalanced sub-samples with the original sample size.
(B) Group 2: over-sampling. Sub-samples balanced with the original sample size.
(C) Group 3: imbalanced sub-samples smaller in size than the original sample.
(D) Group 4: under-sampling or over-sampling. Sub-samples balanced with a size smaller than the original sample, depending on the size of the sub-sample.

Within each group we use different configurations, varying the number of trees and the size of the sub-samples. Two forms of class attribution are considered: the majority of votes and proportion. Performance evaluation metrics the choice of the metric to measure the performance of the predictive model in the case of very imbalanced classes is critical for evaluating the results. In this work, we used the confusion matrix generated by the classification of normal to calculate statistics. The technique returns the probability of belonging to each class, different cut-off values, chosen a priori are used to construct the confusion matrix. From the confusion matrix presented in Table 1, we obtain the metrics True Positive (TP), False Positive (FP), True Negative (TN) and False Negative (FN). These quantities are used to calculate the performance metrics used at work. We use the following metrics, also adopted by [22, 25] and discussed in more detail in [26].

Table 1. Parameters matrix.

	Positive forecast	Negative forecast
Real positive	TP (True Positive)	FN (False Negative)
Real negative	FP (False Positive)	TN (True Negative)

$$True\,positive\,rate\,ou\,recall(Acc^{+}) = \frac{TP}{TP + FN'} \tag{7}$$

$$True\,negative\,rate(Acc^{-}) = \frac{TN}{TN + FP'} \tag{8}$$

$$G - mean(G) = (Acc^{+} \times Acc^{-})^{\frac{1}{2}}, \tag{9}$$

$$Weighted\,accuracy(W.Acc) = \beta Acc^{+} + (1 - \beta)Acc^{-}, \tag{10}$$

$$Precision(Prec) = \frac{TP}{TP + FN'} \tag{11}$$

$$F - measure(F) = \frac{2 \times Prec \times Acc^{+}}{Prec + Acc^{+}}, \tag{12}$$

The metrics above are often used to measure performance in binary classifier predictions. The Acc^{+} and Acc^{-} rates represent respectively the number of positive and negative cases that are correctly classified. The g-mean metric was proposed by [27] and represents the product of both rates (Acc^{+} and Acc^{-}). This metric indicates the balance between classification performances in the majority and minority classes. If there is a low rate of classification of positive cases, the g-mean value will be low, even with high Acc^{-}. The precision metric measures the proportion of false positives classified by the forecast, that is, the higher the precision, the lower the number of patients classified incorrectly in this class. Its interpretation indicates that the technique performed better for the positive class when it has a high value. Weighted accuracy is widely used when working with imbalanced classes [25]. For each application, the value is adjusted to values between 0 and 1. This metric is a weighting of Acc^+ and Acc^-, being important to measure the accuracy of the model and the balance between the respective rates. For any classifier, there is always a tradeoff between increasing the value of Acc^+ or Acc^-, and consequently the same applies for recall and accuracy. In the case of building analyzes on extremely imbalanced data, the minority class is often of great interest. In many applications, such as predict the disease or diagnostics, the choice of a high-precision Acc^+ classifier is paramount, while maintaining reasonable accuracy for Acc^-. In this paper, the balance between these metrics is important.

3.1 Evaluation and Comparison of the Proposed Technique Performance

In this paper, we selected eight medical datasets. The eight datasets are two classes' imbalanced medical datasets [28]. The imbalanced degree of the datasets chosen change from 0.061 (highly imbalanced) to 0.349 (only slightly imbalanced), where the imbalanced degree is defined as the ratio of the size of the abnormal class to that of the normal class. The details of the datasets are shown in Table 2.

Table 2. The medical dataset used.

Datasets	Imbalance degree	Size
Breast-cancer	0.297	286
Breast-Wisconsin	0.345	699
Diabetes	0.349	768
Hepatitis	0.206	155
Lymphography-normal-fibrosis	0.0405	148
New-thyroid1	0.162	215
New-thyroid2	0.162	215
Sick	0.061	3772

To evaluate the proposed technique performance, we compare it with SMOTE [24], UnderBagging [29], OverBagging [30], UnderOverBagging [30], Random Forests [20], Logistic Regression [18], Decision Tree [19] and UnderOver_random forests Technique as shown in Table 3.

In the breast cancer diagnostic dataset, the configurations that obtained the highest values in relation to the g-mean metric are shown in Table 4. For this metric, the group that obtained the best performance was group 3, with 2,00 trees, with a sub-sample size of 15,0, considering the tree probabilistic and 0.8 as the cutoff point. When comparing under-sampling with over-sampling, we see better performance in all metrics with the use of over-sampling. When comparing their configurations, under-sampling showed better results for a few trees and a smaller size of sub-samples. Group 1 showed results close to the group with the best performance, showing good results for forecasting normal (Tables 5 and 6).

Table 3. The average accuracy of our technique compared with other techniques chosen.

Datasets	Breast-cancer	Breast-Wisconsin	Hepatitis	Diabetes	Lymphography-normal-fibrosis	New-thyroid1	New-thyroid2	Sick
SMOTE	0.667597	0.754589	0.783255	0.885435	0.745899	0.889655	0.795243	0.963569
UnderBagging	0.778958	0.885678	0.798879	0.933564	0.896325	0.899547	0.899549	0.965294
OverBagging	0.798959	0.895529	0.799589	0.949874	0.899887	0.898979	0.898899	0.979877
UnderOverBagging	0.888957	0.909498	0.879455	0.949876	0.901478	0.902974	0.909875	0.975395
Random forests	0.839944	0.898779	0.895878	0.835985	0.859547	0.849575	0.839884	0.978959
Logistic regression	0.885323	0.858963	0.869458	0.875234	0.886324	0.898525	0.895447	0.986878
Decision tree	0.897985	0.875698	0.898755	0.889577	0.896576	0.905475	0.909785	0.986595
UderOver_random forests technique	0.939744	0.943159	0.929326	0.939343	0.927733	0.951353	0.957243	0.999999

Table 4. Our technique results considering the g-mean metric to the breast cancer diagnostic dataset.

Groups	Acc^+	Acc^-	G	$W.Acc$	$Prec$	F
1	0,868979	0,918315	0,840336	0,793868	0,372889	0,502236
2	0,852727	0,915968	0, 840393	0,777532	0,439418	0,387286
3	0,792727	0,925869	0,837394	0,789934	0,394214	0,519781
4 under-sampling	0,791399	0,895032	0,787884	0,732005	0,278158	0,398343
4 over-sampling	0,846948	0,921954	0,828500	0,775616	0,367943	0,493024

Table 5. Our technique results considering the weighted accuracy metric to the breast cancer diagnostic dataset.

Groups	Acc^+	Acc^-	G	$W.Acc$	$Prec$	F
1	0,758979	0,918315	0,840336	0,793868	0,372889	0,502236
2	0,635394	0,967679	0,787532	0,682275	0,439418	0,516165
3	0,923596	0,562922	0,824019	0,863484	0,117404	0,208301
4 under-sampling	0,801399	0,888032	0,889884	0,742005	0,278158	0,398343
4 over-sampling	0,848735	0,912888	0,825941	0,776760	0,347705	0,483947

Table 6. Our technique results considering the f-measure metric to the breast cancer diagnostic dataset.

Groups	Acc^+	Acc^-	G	$W.Acc$	$Prec$	F
1	0,661478	0,995036	0,757465	0,633736	0,877209	0,684698
2	0,757675	0,995114	0,744575	0,630415	0,796885	0,656158
3	0,795353	0,857600	0,820203	0,797327	0,257115	0,690701
4 under-sampling	0,599226	0,987838	0,642160	0,498995	0,795599	0,658609
4 over-sampling	0,699744	0,995243	0,733923	0,615760	0,922016	0,678387

One of the initial insights observed from the results is that there are no significant differences when the number of trees is increased. For probabilistic trees, the different cut points showed significantly different results, even when compared to classification trees. Regarding the size of sub-samples, differences in performance can be observed. In addition, in general, over-sampling performs better when compared to under-sampling.

4 Conclusion

This paper studied the problem of imbalance in medical datasets. Random forests, which are combinations of decision trees fitted to subsamples of the data, built using under-sampling and over-sampling. This paper Compared fit metrics obtained in the various specifications of the models tested and their results are evaluated inside and outside the sample. The results shown that random forest techniques using imbalanced sub-samples smaller than the original sample presented the best performance among the random forests used and an improvement compared to that practiced in the medical dataset.

References

1. Lu, H., et al.: Kernel principal component analysis combining rotation forest method for linearly inseparable data. Cogn. Syst. Res. **53**, 111–122 (2019)
2. Zhu, H.-J., et al.: DroidDet: effective and robust detection of android malware using static analysis along with rotation forest model. Neurocomputing **272**, 638–646 (2018)
3. Hong, H., et al.: Landslide susceptibility mapping using J48 decision tree with AdaBoost, bagging and rotation forest ensembles in the Guangchang area (China). Catena **163**, 399–413 (2018)
4. Wang, L., et al.: An improved efficient rotation forest algorithm to predict the interactions among proteins. Soft Comput. **22**(10), 3373–3381 (2018)
5. Pham, B.T., et al.: A hybrid machine learning ensemble approach based on a radial basis function neural network and rotation forest for landslide susceptibility modeling: a case study in the Himalayan area India. Int. J. Sediment Res. **33**(2), 157–170 (2018)
6. Lee, S.-J., et al.: A novel bagging C4.5 algorithm based on wrapper feature selection for supporting wise clinical decision making. J. Biomed. Inform. **78**, 144–155 (2018)
7. Gul, A., et al.: Ensemble of a subset of kNN classifiers. Adv. Data Anal. Classif. **12**(4), 827–840 (2018)
8. Sun, J., et al.: Unbalanced enterprise credit evaluation with DTE-SBD: decision tree ensemble based on SMOTE and bagging with differentiated sampling rates. Inf. Sci. **425**, 76–91 (2018)
9. Lango, M., Stefanowski, J.: Multi-class and feature selection extensions of roughly balanced bagging for unbalanced data. J. Intell. Inf. Syst. **50**(1), 97–127 (2018)
10. Chen, W., et al.: Novel hybrid integration approach of bagging-based fisher's linear discriminant function for groundwater potential analysis. Nat. Resour. Res. 1–20 (2019)
11. García, S., et al.: Dynamic ensemble selection for multi-class unbalanced datasets. Inf. Sci. **445**, 22–37 (2018)
12. Maldonado, S., López, J.: Dealing with high-dimensional class-unbalanced datasets: embedded feature selection for SVM classification. Appl. Soft Comput. **67**, 94–105 (2018)
13. Piri, S., Delen, D., Liu, T.: A synthetic informative minority over-sampling (SIMO) algorithm leveraging support vector machine to enhance learning from unbalanced datasets. Decis. Support Syst. **106**, 15–29 (2018)
14. Zhang, C., et al.: Research on classification method of high-dimensional class-unbalanced datasets based on SVM. Int. J. Mach. Learn. Cybern. **10**(7), 1765–1778 (2019)
15. Douzas, G., Bacao, F.: Effective data generation for unbalanced learning using conditional generative adversarial networks. Expert Syst. Appl. **91**, 464–471 (2018)
16. Veganzones, D., Séverin, E.: An investigation of bankruptcy prediction in unbalanced datasets. Decis. Support Syst. **112**, 111–124 (2018)
17. Tahan, M.H., Asadi, S.: EMDID: evolutionary multi-objective discretization for unbalanced datasets. Inf. Sci. **432**, 442–461 (2018)
18. Hosmer Jr., D.W., Lemeshow, S., Sturdivant, R.X.: Applied logistic regression, vol. 398. Wiley, Hoboken (2013)

19. Friedman, J., Hastie, T., Tibshirani, R.: The Elements of Statistical Learning, Springer Series in Statistics. vol. 1. no. 10. Springer, New York (2001)
20. Breiman, L.: Random forests. Mach. Learn. **45**(1), 5–32 (2001)
21. Drummond, C., Holte, R.C.: C4. 5, class imbalance, and cost sensitivity: why under-sampling beats over-sampling. In: Workshop on Learning from Imbalanced Datasets II, vol. 11. Citeseer, Washington DC (2003)
22. Burez, J., Van den Poel, D.: Handling class imbalance in customer churn prediction. Expert Syst. Appl. **36**(3), 4626–4636 (2009)
23. Weiss, G.M.: Mining with rarity: a unifying framework. ACM SIGKDD Explor. Newslett. **6**(1), 7–19 (2004)
24. Chawla, N.V., et al.: SMOTE: synthetic minority over-sampling technique. J. Artif. Intell. Res. **16**, 321–357 (2002)
25. Chen, C., Liaw, A., Breiman, L.: Using random forest to learn imbalanced data. Univ. Calif. Berkeley **110**(1-12), 24 (2004)
26. Bekkar, M., Djemaa, H.K., Alitouche, T.A.: Evaluation measures for models assessment over imbalanced data sets. J. Inf. Eng. Appl. **3**(10) (2013)
27. Kubat, M., Matwin, S.: Addressing the curse of imbalanced training sets: one-sided selection. In: ICML, vol. 97 (1997)
28. Hulse, J.V., Khoshgoftaar, T.M., Napolitano, A.: Experimental perspectives on learning from imbalanced data. In: ICML 2007 Proceedings of the 24th International Conference on Machine Learning, pp. 935–942, Corvalis, OR, USA (2007)
29. Sanz, J.A., et al.: An evolutionary underbagging approach to tackle the survival prediction of trauma patients: a case study at the hospital of Navarre. IEEE Access **7**, 76009–76021 (2019)
30. Wang, S., Yao, X.: Diversity analysis on imbalanced data sets by using ensemble models. In: IEEE Symposium on Computational Intelligence and Data Mining, pp. 324–331 (2009)

Mining in Educational Data: Review and Future Directions

Said A. Salloum[1,2(✉)], Muhammad Alshurideh[3,4], Ashraf Elnagar[1,5], and Khaled Shaalan[2]

[1] Research Institute of Sciences and Engineering, University of Sharjah, Sharjah, UAE
ssalloum@sharjah.ac.ae
[2] Faculty of Engineering and IT, The British University in Dubai, Dubai, UAE
[3] Faculty of Business, University of Jordan, Amman, Jordan
[4] Management Department, University of Sharjah, Sharjah, UAE
[5] Department of Computer Science, University of Sharjah, Sharjah, UAE

Abstract. One of the developing fields of the present times is educational data mining that pertains to developing methods that help in examining various kinds of data obtained from the educational field. A vital part is played by data mining in the education field, particularly when behavior is being assessed in an online learning setting. This is because data mining is capable of analyzing and identifying the hidden information regarding the data itself, which is very difficult and takes up a lot of time if performed manually. This review has the objective of examining the way data mining was handled by researchers in the past and the most recent trends on data mining in educational research, as well as to evaluate the likelihood of employing machine learning in the field of education. The various limitations inherent in the current research are examined and recommendations are made for future research.

Keywords: Educational data ming · Online learning · Machine learning

1 Introduction

The most advanced universities of present times frequently use data mining methods to examine the data collected and to extract information and knowledge to facilitate decision-making [1–6]. To offer significant understanding of the present research trends of the Education Data Mining (EDM), several review studies were carried out [7–11]. However, further studies are still needed to examine this issue from different angles. Previous research was found to disregard the examination of EDM in terms of electronic learning (E-learning) studies [12–14] from a variety of perspectives [15–18]. The purpose of Educational data mining is to develop techniques that analyze different kinds of data obtained from the field of education [19]. In addition, the field of educational data mining is evolving and is related to the development of techniques for examining the different kinds of data obtained from the educational domain [20, 21]. The purpose of the this review paper is to determine ways of applying data mining techniques to higher education system by providing the most widely used methods and the most relevant studies carried out in this field to date. The remaining parts of the paper include the

A.-E. Hassanien et al. (Eds.): AICV 2020, AISC 1153, pp. 92–102, 2020.
https://doi.org/10.1007/978-3-030-44289-7_9

following: A summary of some relevant literature is presented in related work, as well as the data mining techniques in educational systems are explained in Sect. 2.

2 Literature Review

Data mining is the most robust methodology used for assessing valuable information from the data warehouse [22]. Data mining is used for predicting hidden information through extraction method so that decision-making can be improved [23]. The use of data mining for educational activities has been extended on the basis of performance of students, staff and managerial decisions [19]. Another term that can be used for data mining is knowledge discovery through data [24, 25, 26]. Data mining is a multidimensional field comprising of different aspects, for example, learning, statistics, information technology, artificial intelligence, retrieval and visualization of data [27]. The education system has become more balanced due to the improved mining application [28]. The education field has seen a fast development of the idea of educational data mining (EDM) with respect to the various kinds of educational institutes [29]. On the other hand, academic analyst is specifically linked to institutional effectiveness and problems related to student performance [12, 13, 30–37]. Those areas that directly influence those studying at the institution are part of the EDM [38]. In this section, we provide a review of the last ten studies applied to data mining techniques in educational environment ranging from 2016 to 2019, as shown in Table 1.

Table 1. List of studies focused on exploring educational data mining.

Authors	Research problem	Technique	Approach
[39]	Prediction of students performance	Classification	Decision tree
[40]	Predicting academic performance of students	Classification and clustering	SVM, Naïve Bayes, Decision tree and Neural Network classifiers
[41]	Predicting students' academic procrastination	Classification and clustering	k-means clustering, ZeroR, OneR, ID3, J48, random forest, decision stump, JRip, PART, NBTree, and Prism
[42]	How data mining can help admission working process, and how data mining can predict the student's jobs	Association Rule Mining and Classification	ID 3 Decision Tree
[43]	Predicting students' performance	Clustering, classification, regression	Naive Bayes, Decision Tree, and Artificial Neural Network
[44]	Students' disposition analysis	Clustering	k-means
[45]	Predicting students' graduation	Classification	Multi-Layer Perceptron (MLP)

(*continued*)

Table 1. (*continued*)

Authors	Research problem	Technique	Approach
[46]	Student placement prediction	Clustering, classification, regression	J48, Naïve Bayes, Random Forest, and Random Tree, Multiple Linear Regression, binomial logistic regression, Recursive Partitioning and Regression Tree (rpart), conditional inference tree (ctree) and Neural Network (nnet) algorithms
[47]	Students' performance prediction- predicting final grades of students	Multiple regression analysis	Recurrent Neural Network (RNN)
[48]	Students' performance prediction	Classification	NaiveBayes, Bayesian Network, ID3, J48 and Neural Network
[49]	Predict dropout at the universities	Multiple regression analysis	Neural Network - Multilayer perceptron algorithms and radial basis function
[50]	Predicting students' academic performance	Classification	Decision tree classifiers and neural networks

2.1 Data Mining Techniques in Educational Systems

One of the elements of data mining is educational data mining, the key focus of which is on developing models for extracting hidden knowledge from the student's data, using which the academic performance of students may be enhanced. In the process of Education Data mining, raw material may be converted from various educational systems into valuable information that can be employed by teachers, students and their parents, educational researchers and the developers of educational software system. Educational data mining may also be considered as a new model that is part of the prevailing education system, which is able to generate positive interaction with different parts of the system. This will enable it to eventually attain the objective of enhancing teaching [51].

The definition of Educational Data Mining (EDM) is the application of techniques of traditional data mining to educational data analysis, with the objective of obtaining solutions to problems in the field of education [10]. There are certain EDM applications that include the formulation of e-learning systems [10, 52], clustering educational data [53] as well as making student performance predictions [54]. Several kinds of techniques are presently quite popular in educational data mining, which are part of the following categories: sequential pattern, clustering, prediction, classification, machine learning models and association rule analysis.

2.2 The Traditional Data Mining Techniques Applied in Educational Settings

Clustering is the most extensively known techniques for data mining [39, 55–62], followed by classification [42, 55, 63–65] sequential pattern [18, 58, 66], prediction [40, 67], and finally, association rule analysis [42, 55]. Going back to the year 1995 till the year 2005, majority of the studies on educational data mining often used the association rule analysis technique [11] because it involved a lesser degree of expertise compared to other techniques [68]. Nonetheless, at the start of the year 2005, the trend shifted as clustering and classification methods were often used for analysis by the researchers [9]. There are frequently a large number of outputs obtained for association rule, majority of which are not interesting and cannot be comprehended easily by those who are not experienced in data mining [69]. To select the correct algorithms, researchers should initially design the data and make it consistent with the required output [9]. When their study is small scale in nature, they can use clustering approach because data splitting that is required in classification approach is not needed in this technique [9]. In addition, the researchers are always able to compare with various algorithms using the same database that was used in [55]. This would help in determining if identical results would be attained when a distinct approach was used.

2.3 Machine Learning Applied to Learning Analytics and Educational Data Mining

It would be appropriate to consider Machine Learning (ML) as a component of artificial intelligence (AI). Machine learning is essentially the process through which a machine or model is provided access to data and is able to learn on its own. It was conceived by Arthur Samuel in 1959 that we should not have to teach computers; instead, we should allow them to learn themselves. To explain his theory, he came up with the term 'Machine Learning', which is now the standard term used to explain the computers' ability to learn on their own [70]. Machine Learning (ML) refers to the programming of computers to enhance a performance standard through example data or experience [71]. When a machine learning algorithm is implemented, it indicates that a model is implemented that outputs appropriate information, considering that input data has been given [72]. Another extensively used method in the field of Data Mining is Artificial Neural Network (ANN) [73, 74]. This method has been used in clustering [59, 75], regression [47, 76], classification [47, 76–80], time series forecasting [47, 79] and visualization. Though ANN has been used quite extensively in data mining, a significant aspect to point out is that there is a section of the data mining community that somewhat criticizes the fact that the models formulated using this paradigm cannot be interpreted [81]. These models do offer accurate predictions from the data; however, it is not easy to obtain human-interpretable rules that encapsulate their predictions [74]. At present, researchers and scientists are intrigued by the use of machine learning in the field of education. A few of the areas of interest include the following:

2.3.1 Predict Student Performance

Predicting student performance is an important use of machine learning. The machine learning model is used for "learning" about every student, which enables it to identify their shortcomings and determine ways through which they can improve, for example by attending more lectures or reviewing further literature [47, 54, 78, 80, 82–84]. Using these models many kinds of knowledge can be discovered such as clustering, association rules, and classifications [85]. The discovered knowledge can be used for prediction regarding enrolment of students [86] in a particular course, alienation of traditional classroom teaching model [87], detection of unfair means used in online examination, detection of abnormal values in the result sheets of the students, prediction about students performance and so on [85].

2.3.2 Use Unbiased Methods for Testing and Grading Students

It becomes possible to generate computerized adaptive assessments through machine learning [88, 89]. Teachers and students receive consistent feedback from the machine learning-based assessment regarding the way the student learns, the support that they require and the developments they are experience in terms of their learning objectives [72, 76, 90].

2.3.3 Enhance Retention

Retention rates are also enhanced through machine learning, for example learning analytics. When students "at risk" are determined, they can be approached by schools and universities to assist them in achieving success [91]. Prediction models that include all personal, social, psychological and other environmental variables are necessitated for the effective prediction of the retention rate of the students. The retention of students with high accuracy is beneficial for identify the students with low academic achievements initially. It is required that the identified students can be assisted more by the teacher so that their performance is improved in future [86].

2.3.4 Provide Support to Teachers and Institution Stuff

Using algorithms based on machine learning, it becomes easier to perform classification of the written assessment tests of students [92]. Intelligent analysis of assessment data assists in achieving a better understanding of student performance, the quality of the test and individual question [93].

3 Conclusion and Future Work

In this paper, the state of the art is evaluated in terms of EDM and surveys of the most pertinent studies till now in this field. Once all the published bibliography in the area of EDM has been gathered and reviewed, we chose the most significant studies of every author. After this, every study was categorized, not only on the basis of the kind of data and DM techniques employed, but also on the basis of the kind of educational activity they tackle. The introduction of EDM is considered as a forthcoming research domain that is relevant to various well-established research domains, such as classical data mining techniques, association, classification, regression, clustering and prediction.

In this study, new methods are examined and the latest studies in the field of Learning Analytics and Educational Data Mining that have used Deep Learning techniques are determined. It is believed that EDM is an imminent event that is going to alter the overall field of education. In the scope of future section, the different areas that require improvement are stated, for example: obtaining bigger datasets, including adaptable datasets, hybridizing the techniques used, improving the credibility of EDM and performing comparisons between the various methods. An idea of the way the findings of the studies performed on the suggested lines will enhance the EDM research is presented in future scope. In addition, other DM fields will benefit from the latest algorithms created. The study will also enhance the confidence of the EDM authorities to incorporate the recommendations put forward into the practical world systems. The system will be influenced by the solutions put forward in a way that the students, teachers and the administration are able to achieve the ideal outcomes from the system.

References

1. Saa, A.A., Al-Emran, M., Shaalan, K.: Factors affecting students' performance in higher education: a systematic review of predictive data mining techniques. Technol. Knowl. Learn. **24**, 567–598 (2019)
2. Salloum, S.A., AlHamad, A.Q., Al-Emran, M., Shaalan, K.: A survey of Arabic text mining, vol. 740 (2018)
3. Mhamdi, C., Al-Emran, M., Salloum, S.A.: Text mining and analytics: a case study from news channels posts on Facebook, vol. 740 (2018)
4. Hassanien, A.E., Darwish, A., El-Askary, H.: Machine Learning and Data Mining in Aerospace Technology. Springer, Cham (2020)
5. Hassanien, A.E.: Machine Learning Paradigms: Theory and Application. Springer, Cham (2019)
6. Ismail, F.H., Hassanien, A.E.: Extracting valuable associations among textural features of medical images. In: 2018 13th International Conference on Computer Engineering and Systems (ICCES), pp. 605–608 (2018)
7. Ahuja, R., Jha, A., Maurya, R., Srivastava, R.: Analysis of educational data mining. In: Harmony Search and Nature Inspired Optimization Algorithms, pp. 897–907. Springer (2019)
8. Sarra, A., Fontanella, L., Di Zio, S.: Identifying students at risk of academic failure within the educational data mining framework. Soc. Indic. Res. **146**(1–2), 41–60 (2019)
9. Mohamad, S.K., Tasir, Z.: Educational data mining: a review. Procedia-Soc. Behav. Sci. **97**, 320–324 (2013)
10. Baker, R.S.J.D., Yacef, K.: The state of educational data mining in 2009: a review and future visions. JEDM: J. Educ. Data Min. **1**(1), 3–17 (2009)
11. Romero, C., Ventura, S.: Educational data mining: a survey from 1995 to 2005. Expert Syst. Appl. **33**(1), 135–146 (2007)
12. Salloum, S.A., Alhamad, A.Q.M., Al-Emran, M., Monem, A.A., Shaalan, K.: Exploring students' acceptance of E-learning through the development of a comprehensive technology acceptance model. IEEE Access **7**, 128445–128462 (2019)
13. Alshurideh, M., Salloum, S.A., Al Kurdi, B., Al-Emran, M.: Factors affecting the social networks acceptance: an empirical study using PLS-SEM approach. In: 8th International Conference on Software and Computer Applications (2019)

14. Alshurideh, M.T., Salloum, S.A., Al Kurdi, B., Monem, A.A., Shaalan, K.: Understanding the quality determinants that influence the intention to use the mobile learning platforms: a practical study. Int. J. Interact. Mob. Technol. **13**(11), 157–183 (2019)
15. Mitrofanova, Y.S., Sherstobitova, A.A., Filippova, O.A.: Modeling smart learning processes based on educational data mining tools. In: Smart Education and e-Learning 2019, pp. 561–571. Springer (2019)
16. Menaka, M.S., Kesavaraj, G.: A study on e-learning system to analyse student performance using data mining (2019)
17. Cerezo, R., Bogarín, A., Esteban, M., Romero, C.: Process mining for self-regulated learning assessment in e-learning. J. Comput. High. Educ. **32**, 74–88 (2020)
18. Keskin, S., Şahin, M., Yurdugül, H.: Online learners' navigational patterns based on data mining in terms of learning achievement. In: Learning Technologies for Transforming Large-Scale Teaching, Learning, and Assessment, pp. 105–121. Springer (2019)
19. Fernandes, E., Holanda, M., Victorino, M., Borges, V., Carvalho, R., Van Erven, G.: Educational data mining: predictive analysis of academic performance of public school students in the capital of Brazil. J. Bus. Res. (2018)
20. Salloum, S.A., Al-Emran, M., Monem, A.A., Shaalan, K.: Using text mining techniques for extracting information from research articles. In: Studies in Computational Intelligence, vol. 740. Springer (2018)
21. Salloum, S.A., Al-Emran, M., Abdallah, S., Shaalan, K.: Analyzing the Arab gulf newspapers using text mining techniques. In: International Conference on Advanced Intelligent Systems and Informatics, pp. 396–405 (2017)
22. Salloum, S.A., Al-Emran, M., Shaalan, K.: Mining social media text: extracting knowledge from Facebook. Int. J. Comput. Digit. Syst. **6**(2), 73–81 (2017)
23. Salloum, S.A., Mhamdi, C., Al-Emran, M., Shaalan, K.: Analysis and classification of Arabic newspapers' Facebook pages using text mining techniques. Int. J. Inf. Technol. Lang. Stud. **1**(2), 8–17 (2017)
24. Cummins, M.R.: Nonhypothesis-driven research: data mining and knowledge discovery. In: Clinical Research Informatics, pp. 341–356. Springer (2019)
25. Salloum, S.A., Al-Emran, M., Monem, A.A., Shaalan, K.: A survey of text mining in social media: Facebook and Twitter perspectives. Adv. Sci. Technol. Eng. Syst. J **2**(1), 127–133 (2017)
26. Alomari, K.M., AlHamad, A.Q., Salloum, S.: Prediction of the digital game rating systems based on the ESRB (2019)
27. Arunachalam, A.S., Velmurugan, T.: Analyzing student performance using evolutionary artificial neural network algorithm. Int. J. Eng. Technol. **7**(2.26), 67–73 (2018)
28. Romero, C., Ventura, S., García, E.: Data mining in course management systems: moodle case study and tutorial. Comput. Educ. **51**(1), 368–384 (2008)
29. Sachin, R.B., Vijay, M.S.: A survey and future vision of data mining in educational field. In: 2012 Second International Conference on Advanced Computing & Communication Technologies, pp. 96–100 (2012)
30. Salloum, S.A., Shaalan, K.: Factors affecting students' acceptance of e-learning system in higher education using UTAUT and structural equation modeling approaches. In: International Conference on Advanced Intelligent Systems and Informatics, pp. 469–480 (2018)
31. Salloum, S.A., Al-Emran, M., Habes, M., Alghizzawi, M., Ghani, M.A., Shaalan, K.: Understanding the impact of social media practices on e-learning systems acceptance. In: International Conference on Advanced Intelligent Systems and Informatics, pp. 360–369 (2019)

32. Salloum, S.A., Mhamdi, C., Al Kurdi, B., Shaalan, K.: Factors affecting the adoption and meaningful use of social media: a structural equation modeling approach. Int. J. Inf. Technol. Lang. Stud. **2**(3), 96–109 (2018)
33. Salloum, S.A., Maqableh, W., Mhamdi, C., Al Kurdi, B., Shaalan, K.: Studying the social media adoption by university students in the United Arab Emirates. Int. J. Inf. Technol. Lang. Stud. **2**(3), 83–95 (2018)
34. Salloum, S.A.S., Shaalan, K.: Investigating students' acceptance of e-learning system in higher educational environments in the UAE: applying the extended technology acceptance model (TAM). The British University in Dubai (2018)
35. Habes, M., Alghizzawi, M., Khalaf, R., Salloum, S.A., Ghani, M.A.: The relationship between social media and academic performance: Facebook perspective. Int. J. Inf. Technol. Lang. Stud. **2**(1), 12–18 (2018)
36. Salloum, S.A., Al-Emran, M., Shaalan, K., Tarhini, A.: Factors affecting the E-learning acceptance: a case study from UAE. Educ. Inf. Technol. **24**, 509–530 (2019)
37. Al-Emran, M., Salloum, S.A.: Students' attitudes towards the use of mobile technologies in e-evaluation. Int. J. Interact. Mob. Technol. **11**(5), 195–202 (2017)
38. Kabakchieva, D.: Predicting student performance by using data mining methods for classification. Cybern. Inf. Technol. **13**(1), 61–72 (2013)
39. Durairaj, M., Vijitha, C.: Educational data mining for prediction of student performance using clustering algorithms. Int. J. Comput. Sci. Inf. Technol. **5**(4), 5987–5991 (2014)
40. Francis, B.K., Babu, S.S.: Predicting academic performance of students using a hybrid data mining approach. J. Med. Syst. **43**(6), 162 (2019)
41. Akram, A., et al.: Predicting students' academic procrastination in blended learning course using homework submission data. IEEE Access **7**, 102487–102498 (2019)
42. Rojanavasu, P.: Educational data analytics using association rule mining and classification. In: 2019 Joint International Conference on Digital Arts, Media and Technology with ECTI Northern Section Conference on Electrical, Electronics, Computer and Telecommunications Engineering (ECTI DAMT-NCON), pp. 142–145 (2019)
43. Sana, B., Siddiqui, I.F., Arain, Q.A.: Analyzing students' academic performance through educational data mining. 3c Tecnol. glosas innovación Apl. a la pyme **8**(29), 402–421 (2019)
44. Bharara, S., Sabitha, S., Bansal, A.: Application of learning analytics using clustering data Mining for Students' disposition analysis. Educ. Inf. Technol. **23**(2), 957–984 (2018)
45. Nurhayati, O.D., Bachri, O.S., Supriyanto, A., Hasbullah, M.: Graduation prediction system using artificial neural network. Int. J. Mech. Eng. Technol. **9**(7), 1051–1057 (2018)
46. Rao, K.S., Swapna, N., Kumar, P.P.: Educational data mining for student placement prediction using machine learning algorithms. Int. J. Eng. Technol. Sci. **7**(1.2), 43–46 (2018)
47. Okubo, F., Yamashita, T., Shimada, A., Ogata, H.: A neural network approach for students' performance prediction. In: LAK 2017, pp. 598–599 (2017)
48. Almarabeh, H.: Analysis of students' performance by using different data mining classifiers. Int. J. Mod. Educ. Comput. Sci. **9**(8), 9 (2017)
49. Alban, M., Mauricio, D.: Neural networks to predict dropout at the universities. Int. J. Mach. Learn. Comput. **9**(2), 149–153 (2019)
50. Feng, J.: Predicting students' academic performance with decision tree and neural network (2019)
51. Jie, W., Hai-yan, L., Biao, C., Yuan, Z.: Application of educational data mining on analysis of students' online learning behavior. In: 2017 2nd International Conference on Image, Vision and Computing (ICIVC), pp. 1011–1015 (2017)

52. Lara, J.A., Lizcano, D., Martínez, M.A., Pazos, J., Riera, T.: A system for knowledge discovery in e-learning environments within the European Higher Education Area-Application to student data from Open University of Madrid, UDIMA. Comput. Educ. **72**, 23–36 (2014)
53. Chakraborty, B., Chakma, K., Mukherjee, A.: A density-based clustering algorithm and experiments on student dataset with noises using Rough set theory. In: 2016 IEEE International Conference on Engineering and Technology (ICETECH), pp. 431–436 (2016)
54. Chauhan, N., Shah, K., Karn, D., Dalal, J.: Prediction of student's performance using machine learning (2019). SSRN 3370802
55. Pechenizkiy, M., Calders, T., Vasilyeva, E., De Bra, P.: Mining the student assessment data: lessons drawn from a small scale case study. In: Educational Data Mining 2008 (2008)
56. Shih, Y.-C., Huang, P.-R., Hsu, Y.-C., Chen, S.Y.: A complete understanding of disorientation problems in Web-based learning. Turkish Online J. Educ. Technol. **11**(3), 1–13 (2012)
57. Talavera, L., Gaudioso, E.: Mining student data to characterize similar behavior groups in unstructured collaboration spaces. In: Workshop on Artificial Intelligence in CSCL. 16th European Conference on Artificial Intelligence, pp. 17–23 (2004)
58. Perera, D., Kay, J., Koprinska, I., Yacef, K., Zaïane, O.R.: Clustering and sequential pattern mining of online collaborative learning data. IEEE Trans. Knowl. Data Eng. **21**(6), 759–772 (2008)
59. Dutt, A., Aghabozrgi, S., Ismail, M.A.B., Mahroeian, H.: Clustering algorithms applied in educational data mining. Int. J. Inf. Electron. Eng. **5**(2), 112 (2015)
60. Bogarín, A., Romero, C., Cerezo, R., Sánchez-Santillán, M.: Clustering for improving educational process mining. In: Proceedings of the Fourth International Conference on Learning Analytics and Knowledge, pp. 11–15 (2014)
61. Fernandes, E., Holanda, M., Victorino, M., Borges, V., Carvalho, R., Van Erven, G.: Educational data mining: predictive analysis of academic performance of public school students in the capital of Brazil. J. Bus. Res. **94**, 335–343 (2019)
62. Palomo-Duarte, M., Berns, A., Yañez Escolano, A., Dodero, J.-M.: Clustering analysis of game-based learning: worth it for all students? J. Gaming Virtual Worlds **11**(1), 45–66 (2019)
63. Ahmed, A.B.E.D., Elaraby, I.S.: Data mining: a prediction for student's performance using classification method. World J. Comput. Appl. Technol. **2**(2), 43–47 (2014)
64. Anjewierden, A., Kolloffel, B., Hulshof, C.: Towards educational data mining: using data mining methods for automated chat analysis to understand and support inquiry learning processes (2007)
65. Adebayo, A.O., Chaubey, M.S.: Data mining classification techniques on the analysis of student's performance. GSJ **7**(4), 45–52 (2019)
66. Kay, J., Maisonneuve, N., Yacef, K., Zaïane, O.: Mining patterns of events in students' teamwork data. In: Proceedings of the Workshop on Educational Data Mining at the 8th International Conference on Intelligent Tutoring Systems (ITS 2006), pp. 45–52 (2006)
67. Tiwari, A.K., Ramakrishna, G., Sharma, L.K., Kashyap, S.K.: Academic performance prediction algorithm based on fuzzy data mining. Int. J. Artif. Intelegence **8**(1), 26–32 (2019)
68. Merceron, A., Yacef, K.: Revisiting interestingness of strong symmetric association rules in educational data. In: Proceedings of the International Workshop on Applying Data Mining in e-Learning, Creete, Greece, pp. 3–12 (2007)
69. García, E., Romero, C., Ventura, S., Calders, T.: Drawbacks and solutions of applying association rule mining in learning management systems. In: Proceedings of the International Workshop on Applying Data Mining in e-Learning (ADML 2007), Crete, Greece, pp. 13–22 (2007)

70. Samuel, A.L.: Some studies in machine learning using the game of checkers. II—recent progress. IBM J. Res. Dev. **11**(6), 601–617 (1967)
71. Alpaydin, E.: Introduction to Machine Learning. MIT Press, Cambridge (2009)
72. Kučak, D., Juričić, V., Đambić, G.: Machine learning in education-a survey of current research trends. In: Annals of DAAAM and Proceedings, vol. 29 (2018)
73. Stahl, F., Jordanov, I.: An overview of the use of neural networks for data mining tasks. Wiley Interdiscip Rev. Data Min. Knowl. Discov. **2**(3), 193–208 (2012)
74. Coelho, O.B., Silveira, I.: Deep learning applied to learning analytics and educational data mining: a systematic literature review. In: Brazilian Symposium on Computers in Education (Simpósio Brasileiro de Informática na Educação-SBIE), vol. 28, no. 1, p. 143 (2017)
75. Vellido, A., Castro, F., Nebot, A.: Clustering educational data. In: Handbook of Educational Data Mining, pp. 75–92 (2010)
76. Li, J., Wong, Y., Kankanhalli, M.S.: Multi-stream deep learning framework for automated presentation assessment. In: 2016 IEEE International Symposium on Multimedia (ISM), pp. 222–225 (2016)
77. Gross, E., Wshah, S., Simmons, I., Skinner, G.: A handwriting recognition system for the classroom. In: Proceedings of the Fifth International Conference on Learning Analytics and Knowledge, pp. 218–222 (2015)
78. Guo, B., Zhang, R., Xu, G., Shi, C., Yang, L.: Predicting students performance in educational data mining. In: 2015 International Symposium on Educational Technology (ISET), pp. 125–128 (2015)
79. Tang, S., Peterson, J.C., Pardos, Z.A.: Deep neural networks and how they apply to sequential education data. In: Proceedings of the Third (2016) ACM Conference on Learning @ Scale, pp. 321–324 (2016)
80. Wang, L., Sy, A., Liu, L., Piech, C.: Deep knowledge tracing on programming exercises. In: Proceedings of the Fourth (2017) ACM Conference on Learning @ Scale, pp. 201–204 (2017)
81. Craven, M.W., Shavlik, J.W.: Using neural networks for data mining. Futur. Gener. Comput. Syst. **13**(2–3), 211–229 (1997)
82. Anozie, N., Junker, B.W.: Predicting end-of-year accountability assessment scores from monthly student records in an online tutoring system (2006)
83. Khan, I., Al Sadiri, A., Ahmad, A.R., Jabeur, N.: Tracking student performance in introductory programming by means of machine learning. In: 2019 4th MEC International Conference on Big Data and Smart City (ICBDSC), pp. 1–6 (2019)
84. Livieris, I.E., Drakopoulou, K., Tampakas, V.T., Mikropoulos, T.A., Pintelas, P.: Predicting secondary school students' performance utilizing a semi-supervised learning approach. J. Educ. Comput. Res. **57**(2), 448–470 (2019)
85. Yadav, S.K., Pal, S.: Data mining: a prediction for performance improvement of engineering students using classification, arXiv Prepr. arXiv:1203.3832 (2012)
86. Yadav, S.K., Bharadwaj, B., Pal, S.: Mining education data to predict student's retention: a comparative study, arXiv Prepr. arXiv:1203.2987 (2012)
87. Akinola, O.S., Akinkunmi, B.O., Alo, T.S.: A data mining model for predicting computer programming proficiency of computer science undergraduate students (2012)
88. Luckin, R., Holmes, W., Griffiths, M., Forcier, L.B.: Intelligence unleashed: an argument for AI in education (2016)
89. Meseguer-Brocal, G., Cohen-Hadria, A., Peeters, G.: DALI: a large dataset of synchronized audio, lyrics and notes, automatically created using teacher-student machine learning paradigm, arXiv Prepr. arXiv:1906.10606 (2019)
90. El-Alfy, E.-S.M., Abdel-Aal, R.E.: Construction and analysis of educational tests using abductive machine learning. Comput. Educ. **51**(1), 1–16 (2008)

91. Đambić, G., Krajcar, M., Bele, D.: Machine learning model for early detection of higher education students that need additional attention in introductory programming courses. Int. J. Digit. Technol. Econ. **1**(1), 1–11 (2016)
92. Celar, S., Stojkic, Z., Seremet, Z., Marusic, Z., Zelenika, D.: Classification of test documents based on handwritten student ID's characteristics. Procedia Eng. **100**, 782–790 (2015)
93. Pechenizkiy, M., Trcka, N., Vasilyeva, E., Van der Aalst, W., De Bra, P.: Process mining online assessment data. In: International Working Group on Educational Data Mining (2009)

Green Supply Chain Management and Firm's Performance: A Review

Sara Jasim[1,2], Muneer Al-Mubarak[1], and Allam Hamdan[2(✉)]

[1] Engineering Management and Systems Engineering Off-Campus Program, School of Engineering and Applied Science, The George Washington University, Washington, D.C., USA

[2] College of Business and Finance, Ahlia University, Manama, Bahrain
allamh3@hotmail.com

Abstract. As the world grows and develops, people become more aware of business operational processes that are extremely harmful to our environment. Recently, customers' demands, and governmental legislations have forced domestic and global organizations to implement environmentally sustainable practices in order to cure what has been damaged in the previous decades. This study focuses on the impact of green supply chain management practices on firm's performance. These practices are green purchasing, green packaging, green manufacturing, green design and green marketing.

Keywords: Green supply chain management · Practices · Performance

1 Introduction

People are extremely aware of all the operational and manufacturing processes that really damage the environment (Touboulic and Walker 2015). Global researchers are interested in studying the influence of embodying green practices in business operations (Al Khattab et al. 2015). Researchers have advanced the concept of supply chain management (SCM) to a more effective concept that is more concerned with environment (Touboulic and Walker 2015). During the last 40 years SCM became harmful to our environment as it needs more of strategic development to coordinate business. Responding to the demands of end customer, it is important to integrate business processes with green strategies (Green et al. 2012). This requires having green strategies that include; green purchasing, green manufacturing, green design, green packaging and green marketing (Islam et al. 2017). Each green strategy has its own uniqueness with different ways of implementation to improve firm's performance (Kenneth et al. 2012). Green purchasing is known as a cooperation between organizations and suppliers to build and develop sustainable products or services (Zhu et al. 2008). Green packaging is referred to the process of recycling and reusing materials to save resources and avoid wastes (Zhou 2014). Eco-design is defined as a manufacturing process that reduces consumption of resources and energy (Kenneth et al. 2012). Green manufacturing concerns with manufacturing processes and manages the flow of environmental wastes (Rusinko 2007). Green marketing promotes a product based on its environmental performance and is divided into four sets: market analysis, integrated

A.-E. Hassanien et al. (Eds.): AICV 2020, AISC 1153, pp. 103–111, 2020.
https://doi.org/10.1007/978-3-030-44289-7_10

communication practices, supplying, and pricing (Zampese et al. 2016). Firm's performance is concerned with several aspects such as economic, environment, and organization. Economic performance focuses on profitability and firm's ability to reduce the costs, environmental performance minimizes air emissions, wastes and toxic materials, and organizational performance monitors marketing performance (Kenneth et al. 2012). The green supply chain management (GSCM) term came to light in the last decades and few organizations are truly implementing it. The importance of environmental protection is increasing due to social awareness, governmental rules, legislations and global organizations competitiveness as well as the communities and charities that boost the ecological protection awareness. Since the SCM is consuming lots of energy, costs and resources, people and firms are more interested in a healthier environment and effective performance. Automobile firms are consuming lots of resources, materials and energy in order to create sufficient designs that satisfy their customers. The process of manufacturing and design is causing a serious burden over the environment and firms' profits. Firms that deliver shipments are also facing the same problem as they consume lots of energy on transportation and packaging processes. For such problems, GSCM can be the right solution that would recover all that damages with a proper implementation. This paper focuses on GSCM practices to increase awareness and maximize firm's performance. The study provides an insight on how firms are dealing with green practices and their influence on performance. The aim is to increase the awareness level among firms on the importance of GSCM practices in maximizing the performance by enhancing the existing environmental-friendly system that aims to protect environment and resources where at the same time increase the firm's performance level. The contribution of this paper is on measuring the impact of green practices on firms' performance. The next sections include literature review, presentation of a conceptual model with associated research hypotheses, elaboration on the study method used, demonstration of study analysis and discussion, and conclusions along with implications and future studies.

2 Literature Review

It is extremely important for organizations to balance between the economic and environmental performance due to the market competitiveness, regulatory and community pressures. As long as the pressure increases, firms have to implement a strategy that aims to reduce the ecological impact of their products and services. In order to apply the governmental rules toward the environment, the enterprises must success in establishing and re-examining their business purpose (Perotti et al. 2012). There is good relation between eco-design and firm's performance where eco-design is linked to market demand (Hasan 2013). Researchers suggested that merging green programs within business strategies will not only improve the core competencies but will also enhance its competitiveness (Khan and Qianli 2017). Implementing green design practices leads to an efficient utilization of energy and water. Firms are giving priority to ecological issues and incorporating green procedures in their processes to stay competitive and profitable in the market (Sarkar 2012). Supply networks are becoming complex and having lots of fragments (Touboulic and Walker 2015). At the same time, the social awareness of the surrounding ecological issues is growing very fast which highly preamble the introduction of the GSCM in late 1980's. The importance of GSCM is in its capability to handle various green activities that are related to SCM. Although, GSCM to do a lot with competitiveness, sustainability, and logistics, however, the main challenge in GSCM is the performance in its several forms like environmental, financial, economic and operational (Islam et al. 2017). Recently, it has become extremely important for organizations to balance between the economic and environmental performance due to the market competitiveness, regulatory and community pressures. As the pressure increases, the logistics will have to implement a strategy that aims to reduce the ecological impact of their products and services (Tachizawa et al. 2015). The green management practices include many approaches such as, eco-efficiency, environmental management systems and clear production (Bjorklund et al. 2012).

2.1 Green Supply Chain Management Practices

The pressure for greener and healthier environment grows up to reduce the negative impact (Fraj et al. 2011). The green supply chain practice (GSCP) ranges from green purchasing that flows from supplier to manufacturer to customer, which is referred as "closing the loop". According to Zhu and Sarkis (2007a), there are three-fold-effects: impact on environmental performance, impact on financial performance, and impact on operational performance. The internal environmental operations management practices of eco-efficiency are intended to convert the input of any services into an output through different steps, flows and processes (Hsu et al. 2016). Perotti et al. (2012) concluded that the level of adoption to GSCP is somehow limited in the economic, operational and ecological levels. Holt and Ghobadain (2009) said that manufacturers sensed a maximal pressure to enhance the environmental performance if the legislations, laws and internal drivers were applied. The green initiatives help firms reducing their handling costs and expensive inputs. Clear production is resulted by using the advanced technologies and innovative processes which directly improves the product quality. GSCM practices meet the environmental goals and enhance the public image of the firms (Sharma 2013). Firms need to have a unique selling technique for their products and services, but in all cases, protecting the environment is the most important objective for all firms (Lin et al. 2014). Several practices are getting under the GSCM umbrella such as green purchasing, green manufacturing, green packaging, green design, green marketing, reverse logistics, green information systems, green procurement, and green distribution (Srivastava 2007). Currently, (93%) of firms worldwide are implementing green strategies (Uddin and Khan 2015). In turn, the number of consumers who purchase sustainable goods is rising and the "green consumption" concept is gaining lots of attention (Punyatoya 2014). Among all the GSCM practices, buying materials is the most significant one because it measures the effectiveness of the firm's environmental management. Chien and Shih (2007) defined green purchasing as a purchasing process that entails environmental and social issues with the aim to reduce, reuse and recycle materials through the supply chain. Products and services are used effectively to minimize negative environmental impacts with regards manufacturing, transportation, use and, recycling (Sarkar 2012). Green purchasing is implemented by applying environmental criteria on products and services (Sarkar 2012). Using green purchasing creates a good brand name, lowers pollution, and saves costs (Ramakrishnan et al. 2015). On the other hand, Packaging is made from different materials which makes it easier to handle and protect the product. Green packaging supports the development of packaging sustainability and avoids any harm on human health and environment. Green packaging is all about recycling the source materials (Zhou 2014). Burning the trash leads to a new energy source, while using materials that decompose by itself will not cause pollutions to the environment. These practices save natural resources and reduces pollution (Besch and Pålsson 2016). Green packaging practice enables an efficient distribution and lowers environmental effects (Islam et al. 2017). Green packaging is praised due to its ability to reduce waste and improve quality of air. It considers the cost of materials, shipping, performance, convenience, compliance, and environmental effect (Liu et al. 2013). Manufacturing consumes lots of energy, materials and generates lots of wastes. Only a successful management system can minimize its negative effects. Green manufacturing is the new way of manufacturing goods. Like any other green practice, green manufacturing focuses on using resources efficiently and minimizing wastes (Namagembe et al. 2016). It is a system that integrates issues of products and process design with issues of manufacturing planning and control. Green manufacturing looks after identifying, quantifying, assessing, and managing environmental wastes with the aim of better environment and resource efficiency (Liu et al. 2013). Green design or eco-design improves product's outlook and is an environmentally friendly practice. It prevents the product from harming the environment during manufacturing, distribution and consumption. It requires designing products in a way that reduces consumption of energy, facilitates recycling, and reuses of materials efficiently and environmentally friendly (Zhu et al. 2008). The process of green design requires a budget that can be controlled. There are hundred millions of creative designs that can be made and only the budget can decide which design can fit. Being conscious of the budget limitation will simplify the work process. Lowering the cost, increasing the firm's profit and developing the design industry depend on utilizing the economic resources rationally (Yue et al. 2015). Green materials should be non-toxic, non-radioactive, easy recycling, high performance, paper materials, edible, fast and smooth decomposition (Besch and Pålsson 2016). The eco-system cycle of packaging materials used for the design is important. When the useful life ends for the used product and is back to manufacturing, the new product must be as good as the previous ones. At the end, if the materials used are safe, natural, and organic, returning them again to soil and adding fertilizers will be more beneficial than energy consumption (Rokka and Uusitalo 2008). Green marketing or ecological marketing focuses on solving ecological problems. All green practices including green marketing help firms achieving their objectives and satisfy human needs by using materials wisely without wastes (Peattie 2001). Green marketing has a wide range of activities including changing the product's process, modifying the product and changes in the packaging. It contributes in protecting the environment through minimizing the disposal of goods and services that would spoil the environment. Moving toward the green concept might be a bit costly, but later on it will prove its effectiveness. Furthermore, green products are utilizing green technologies to make harmless products to the environment. In order to maintain natural resources and keep up the greenness development, it is important to use clean technologies (Singhal and Singhal 2015).

3 Conceptual Model

Research hypotheses are developed based on a conceptual model that determines variables. Figure 1 illustrates the study framework that includes five independent variables: green purchasing, green packaging, green manufacturing, green design and green marketing, showing their influence on firm's performance, a dependent variable.

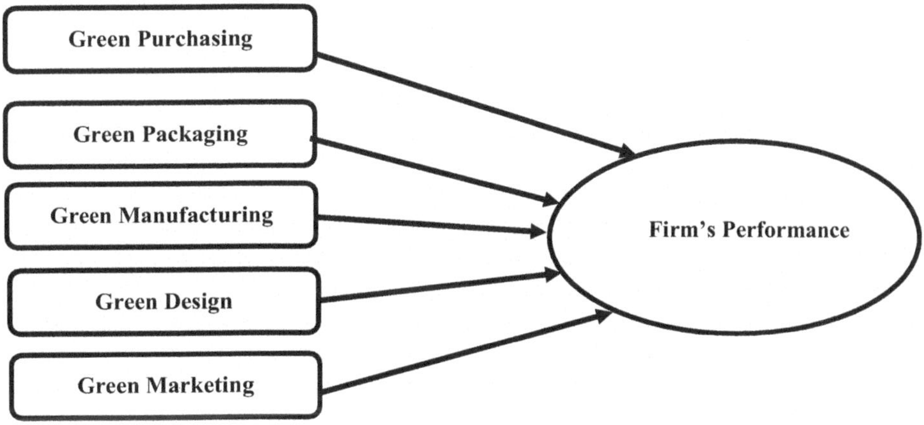

Fig. 1. Research conceptual model

3.1 Green Practices and Performance

Green purchasing encourages recycling without breaking down the system. Efficient implementation of green purchasing activities has reflected on costs saving, pollution control and firm's environmental performance (Kleindorfer et al. 2005). Green purchasing assures the market competitiveness, and works on enhancing firm's performance (Zailani et al. 2015). The crucial role of green purchasing is its ability to reduce waste by encouraging recycling activities without interrupting the system (Khan and Qianli 2017). Successful implementation of green purchasing system has controlled the cost of pollution along with some improvements on firm's performance and reputation in the market (Khan and Qianli 2017; Islam et al. 2017). Zailani et al. (2015) found a positive impact of green purchasing over firm's performance and saving in the environment from the danger of toxic materials. Sustainable purchasing could be seen in five areas which are supply chain management, ecological, external environment management, design operation management and environmental authentication (Yang et al. 2014). Thus, green purchasing is a reliable tool for controlling the environmental pollution, constructing a good green image and reputation (Khan and Qianli 2017). A significant improvement in packaging process was found when green packaging was used, as it reduces materials and increases customer's credibility (Tomar and Oza 2015). The implementation of green packaging improves productivity and positively influences firm's performance (Islam et al. 2017). Adopting best resources within the green manufacturing leads to a long-term competitive advantage and high quality products with the lowest cost. Green manufacturing technologies eliminate waste, reduce the manufacturing steps, produce efficient goods and minimize costs (Narasimhan and Schoenherr 2012). It improves firm's image and reputation (Khan and Qianli 2017). Firms integrate total quality management (TQM) and manufacturing practices with green manufacturing to come up with superior performance (Murovec et al. 2012). Green manufacturing influences competitive edges of suppliers (Subramanian and Gunasekaran 2014; Tseng and Chiu 2013). Baines et al. (2012) claimed that green manufacturing is having a substantial role in adopting GSCM practices as it decreases harmful manufacturing activities with few amount of waste in the overall process and improves efficiency and performance. Therefore, green manufacturing practice improves organizational performance (Khan and Qianli 2017; Islam et al. 2017; Tomar and Oza 2015). A sustainable product is measured based on its impact on the environment throughout its life. Introducing sustainable products' innovation with eco-design increases the manufacturing capabilities (Kleindorfer et al. 2005). A large number of firms implement green design practices in their system just to distinguish themselves form the markets' competitors (Lin et al. 2013). Büyüközkan and Çifçi (2012), said that implementation of eco-design practices can significantly improve products and processes. It protects the environment and enhances firm's sustainability (Zhu et al. 2007). Green design products have a positive impact on firm's performance where customers are willing to pay for sustainable products that save the environment (Lin et al. 2013). Green marketing is a new management term and can influence firm's performance (Zampese et al. 2016). Fraj et al. (2011) mentioned that firms' performance is improved when green marketing strategies are used. Baker and Sinkula (2005) observed a direct relationship between green marketing and new products success which in turns increases market shares.

4 Conclusion and Implications

Environment is the foundation of our health and without a free-pollutant environment, humanity will face many diseases and illnesses. This study was conducted in order to measure the level of green implementation in Bahrain on two different industries. It is obvious that Bahrain is in its early stages toward achieving business sustainability as not many firms are following the new concept. The situation in Bahrain is considered to be moderate when it comes to the green concept, a level which needs to be improved to be able to proceed to a world's development standard. The green practices that lead to a better performance vary in the study. Green purchasing shows a slight positive impact with no impact of green packaging on firm's performance. This could be attributed to the use of traditional way of purchasing, and the recycling costs. Green manufacturing shows high positive influence over the firm's performance especially when it comes to the automobile industry as it is extremely important to follow a green process in manufacturing the vehicles. The green design was found to have a negative impact on firm's performance as customers request to have over design products. Green marketing was found to have direct and positive impact over firm's performance where communication between firms and customers is enhanced. It is concluded that green manufacturing and green marketing are the most influential practices, however, these results vary from one country to another based on society's awareness level, firms' orientations and green habits. To conclude, "going green" is not an easy process for most of the firms as it requires deep examination of each operational aspect. One of the firm's responsibilities is to plan and implement green standards to avoid harmful practices of supply chain, a process which requires patience and lots of efforts.

4.1 Theoretical Implications

The research focuses on the green practices that enhance the overall firm's performance. It shows that a specific group of green practices can improve the situation inside and outside the firms. It also shows that due to cultural issues some practices are demanded more than others. While the study confirms the importance of green practices it shows the differences in weight which might be attributed to the background and level of maturity of the sample chosen in the country. In this regard, it was found that the green manufacturing was the most adopted practice that shapes the industries, followed by green marketing and green purchasing. When such green practices are measured and implemented well, firms can improve their performance and make different stakeholders happy.

4.2 Managerial Implications

Researchers are claiming that sustaining the SC will efficiently support many firms specially when it comes to manufacturing. Integrating the green practices in the SCM with the collaboration of suppliers and consumers will structure a very strong knowledge base for the GSCM to enhance performance. Working on the improvements of the SCM strategies will satisfy consumers' needs. Managers, therefore, are asked to hold the responsibility of implementing such strategies for better performance. If the SC is modified, then customers' demands are fulfilled, and overall performance is improved. Green practices can sustain the environment and form a better financial resource. This study encourages adopting and implementing the green concept in Bahrain. It contributes in understanding the current green situation of Bahraini firms. Firms are encouraged to focus on the most demanding practices to better efficiency and customer satisfaction.

References

Al Khattab, S.A., Abu-Rumman, A.H., Massad, M.M.: The impact of the green supply chain management on environmental-based marketing performance. J. Serv. Sci. Manag. **8**, 588–597 (2015)

Baines, T., Brown, S., Benedettini, O., Ball, P.: Examining green production and its role within the competitive strategy of manufacturers. J. Ind. Eng. Manag. **5**(1), 53–87 (2012)

Baker, W.E., Sinkula, J.M.: Environmental marketing strategy and firm performance: effects on new product performance and market share. J. Acad. Mark. Sci. **33**(4), 461–475 (2005)

Besch, K., Pålsson, H.: A supply chain perspective on green packaging development-theory versus practice. Packag. Technol. Sci. Int. J. **29**(1), 45–63 (2016)

Bjorklund, M., Martinsen, U., Abrahamsson, M.: Performance measurements in the greening of supply chains. Supply Chain. Manag. Int. J. **17**(1), 29–39 (2012)

Büyüközkan, G., Çifçi, G.: Evaluation of the green supply chain management practices: a fuzzy ANP approach. Prod. Plann. Control **23**(6), 405–418 (2012)

Chien, M.K., Shih, L.H.: An empirical study of the implementation of green supply chain management practices in the electrical and electronic industry and their relation to organizational performances. Int. J. Environ. Sci. **4**(3), 383–394 (2007)

Fraj, E., Martínez, E., Matute, J.: Green marketing strategy and the firm's performance: the moderating role of environmental culture. J. Strat. Mark. **19**(4), 339–355 (2011)

Uddin, S.M.F., Khan, M.: Exploring green purchasing behaviour of young urban consumers: empirical evidences from India. South Asian J. Glob. Bus. Res. **5**(1), 85–103 (2015). Emerald Insight

Green, K., Zelbst, P., Meacham, F.: Green supply chain management practices: impact on performance. Supply Chain Manag. Int. J. **17**(3), 290–305 (2012)

Hasan, M.: Sustainable supply chain management practices and operational performance. Int. J. Bus. Manag. **3**(1), 42–48 (2013)

Holt, D., Ghobadian, A.: An empirical study of green supply chain management practices amongst UK manufacturers. J. Manuf. Technol. Manag. **20**(7), 933–956 (2009)

Hsu, C.-C., Tan, K.-C., Zailani, M.S.H.: Strategic orientations, sustainable supply chain initiatives, and reverse logistics: empirical evidence from an emerging market. Int. J. Oper. Prod. Manag. **36**(1), 86–110 (2016)

Islam, S., Karia, N., Fauzi, F.B.A., Soliman, M.S.M.: A review on green supply chain aspects and practices. Manag. Marketing. Chall. Knowl. Soc. **12**(1), 12–36 (2017)

Khan, S.A.R., Qianli, D.: Impact of green supply chain management practices on firms' performance: an empirical study from the perspective of Pakistan. Environ. Sci. Pollut. Res. **24**, 16829–16844 (2017)

Kleindorfer, P.R., Singhal, K., Wassenhove, L.N.: Sustainable operations management. Prod. Oper. Manag. **14**(4), 482–492 (2005)

Lin, K.H., Zhang, Y.Q., Tsai, S.B., Wen, S.S., Wang, C.K.: Performance of green supply chain in manufacturing industry - the case of a PCB manufacturer. Appl. Mech. Mater. **687–691**, 4803–4806 (2014)

Lin, R.J., Tan, K.H., Geng, Y.: Market demand, green product innovation, and firm performance: evidence from Vietnam motorcycle industry. J. Clean. Prod. **40**(1), 101–107 (2013)

Liu, H., Ke, W., Kee Wei, K., Hua, Z.: Effects of supply chain integration and market orientation on firm performance: evidence from China. Int. J. Oper. Prod. Manag. **33**(3), 322–346 (2013)

Murovec, N., Erker, R.S., Prodan, I.: Determinants of environmental investments: testing the structural model. J. Clean. Prod. **37**, 265–277 (2012)

Namagembe, S., Sridharan, R., Ryan, S.: Green supply chain management practice adoption in Ugandan SME manufacturing firms. World J. Sci. Technol. Sustain. Dev. **13**(3), 154–173 (2016)

Narasimhan, R., Schoenherr, T.: The effects of integrated supply management practices and environmental management practices on relative competitive quality advantage. Int. J. Prod. Res. **50**(4), 1185–1201 (2012)

Peattie, K.: Towards sustainability: the third age of green marketing. Mark. Rev. **2**(2), 129–146 (2001)

Perotti, S., Zorzini, M., Cagno, E.: Green supply chain practices and company performance: the case of 3PLs in Italy. Int. J. Phys. Distrib. Logist. Manag. **42**(7), 640–672 (2012)

Punyatoya, P.: Linking environmental awareness and perceived brand eco-friendliness to brand trust and purchase intention. Glob. Bus. Rev. **15**(2), 279–289 (2014)

Ramakrishnan, P., Haron, H., Goh, Y.: Factors influencing green purchasing adoption for small and medium enterprises (SMEs) in Malaysia. Int. J. Bus. Soc. **16**(1), 39–56 (2015)

Rokka, J., Uusitalo, L.: Preference for green packaging in consumer product choices – do consumers care? Int. J. Consum. Stud. **32**(5), 516–525 (2008)

Rusinko, C.A.: Green manufacturing: an evaluation of environmentally sustainable manufacturing practices and their impact on competitive outcomes. IEEE Trans. Eng. Manag. **54**(3), 445–454 (2007)

Sarkar, A.N.: Green supply chain management: a potent tool for sustainable green marketing. Asia-Pac. J. Manag. **8**(4), 491–507 (2012)

Sharma, M.: A study on the concept of green supply chain management. J. Supply Chain Manag. Syst. **2**(1), 1–7 (2013)

Singhal, A., Singhal, P.: Exploratory research on green marketing in India. Int. J. Manag. Res. Rev. **5**(12), 1134–1141 (2015)

Srivastava, S.K.: Green supply-chain management: a state-of-the-art literature review. Int. J. Manag. Rev. **9**(1), 53–80 (2007)

Subramanian, N., Gunasekaran, A.: Cleaner supply-chain management practices for twenty-first century organizational competitiveness: practice-performance framework and research propositions. Int. J. Prod. Econ. **164**, 216–233 (2014)

Tachizawa, E., Gimenez, C., Sierra, V.: Green supply chain management approaches: drivers and performance implications. Int. J. Oper. Prod. Manag. **35**(11), 1546–1566 (2015)

Tomar, A., Oza, H.: Green supply chain management practices implementation and effect on organizational performance of ISO14001 certified manufacturing companies of India. Abhinav Int. Mon. Ref. J. Res. Manag. Technol. **4**(9), 21–26 (2015)

Touboulic, A., Walker, H.: Theories in sustainable supply chain management: a structured literature review. Int. J. Phys. Distrib. Logist. Manag. **45**(1/2), 16–42 (2015)

Tseng, M.L., Chiu, A.S.: Evaluating firm's green supply chain management in linguistic preferences. J. Clean. Prod. **40**, 22–31 (2013)

Yang, J., Dungrawala, H., Hua, H., Manukyan, A., Abraham, L., Lane, W., Mead, H., Wright, J., Schneider, B.L.: Cell size and growth rate are major determinants of replicative lifespan. Cell Cycle **10**(1), 144–155 (2014)

Yue, J., Jia, X., Dong, B.: The application of materials in green packaging. **722**, 34–37 (2015)

Zailani, S., Govindan, K., Iranmanesh, M., Shaharudin, M.R., Chong, Y.S.: Green innovation adoption in automotive supply chain: the Malaysian case. J. Cleaner Prod. **108**(A), 1115–1122 (2015)

Zampese, E., Moori, R., Caldeira, A.: Green marketing as a mediator between supply chain management and organizational performance. **17**(3), 183–211 (2016)

Zhou, F.: The theory of green packaging design and its application. Appl. Mech. Mater. **635–637**, 248–252 (2014)

Zhu, Q., Sarkis, J.: The moderating effects of institutional pressures on emergent green supply chain practices and performance. Int. J. Prod. Res. **45**(18–19), 4333–4355 (2007)

Zhu, Q., Sarkis, J.: Green supply chain management: pressures, practices and performance within the Chinese automobile industry. J. Cleaner Prod. **15**(2), 1041–1052 (2007a)

Zhu, Q., Sarkis, J., Lai, K.: Confirmation of a measurement model for green supply chain management practices implementation. Int. J. Prod. Econ. **111**(2), 261–273 (2008)

Role of Financial Technology FinTech: A Survey

Anjum Razzaque and Allam Hamdan(✉)

College of Business and Finance, Ahlia University,
Manama, Kingdom of Bahrain
allamh3@hotmail.com

Abstract. FinTech is not any new theory or a concept. But rather it is a phase which tends to bubble up every now and then as new forms of technologies revolutionize the world. FinTech has been applied in various sectors. Multiple scholars have investigated this phenomenon from the prism of the users in order to comprehend their perceptions of benefits and risks for using FinTech: predominantly revealing that perceived benefit and perceived risks affect intention to use of FinTech; and for raising awareness for FinTech users. This is a review of current literature and it proposes a conceptual model viable for future empirical assessment. Also, recommendations are made in this article for future research in relation to the model and the nine propositions of this study.

Keywords: FinTech · Perceived benefits · Perceived risks

1 Introduction

FinTech is unlimited to certain clear segments that prevails economies, e.g., financing, or business models, e.g., peer-to-peer lending; but instead includes the complete range of facilities and products conventionally provided by the financial services industry (Arner et al. 2015). Organizations utilize the best of technologies for improving their chances for a higher rate of competitive advantage through excellence in their services. "FinTech" blends finance and technology which automatically and efficiently delivered while showing off itself as being a cost effect technology. Also, such financial services make this current technology so attractive for its customers, particular the use of FinTech through the cell phone (Dahlberg et al. 2008) such that entrepreneurs and end-users can manage their financial services by the pre-developed algorithms hosted on the cloud. FinTech emerged recently by utilizing technologies in back-end financial servers for banks; covering a spectrum of innovations and processes on the front-end user interfaces to aid end-users to manage their finances (Venkatesh et al. 2003). With the advent of the mobile services, online payments are possible via humbler fund transfers by simply verifying bank account (Dahlberg et al. 2008). Such experiences have labeled FinTech as destructive for banks. Traditional banks have been observed to fall behind when it comes to delivering FinTech services. This slow-down of banks' performance in delivering FinTech has led to their downfall within the global competition caused from FinTech. FinTech is the ticket for startups who area able to provide the world with new FinTech services/solutions; thus, able to achieve a

A.-E. Hassanien et al. (Eds.): AICV 2020, AISC 1153, pp. 112–117, 2020.
https://doi.org/10.1007/978-3-030-44289-7_11

competitive advantage over the traditional banks (Chong 2013). Our study critiques current literature to comprehend what effect users' perception of benefits of using FinTech and the perception of risks of using fintech has on their intention to continue using FinTech.

2 Literature Review

The primary usage of technologies that are being used by banks and trading corporations were dependent on physical resources of transportation and storage. Moreover, markets were mainly restricted to a local scope. These limitations were increased with the rise of information and communications technology. As a result, the visual and the electric telegraph allowed the information to be transferred over larger areas; which led to neglecting the location. The telegraph's economic impacts were vital since it became well-known for its industrial development of modern societies. These relatable technologies lasted for the better part of the 20th century. Therefore, the rise of the technological communication in the late 20th century lead to the foundation of the financial technologies phase. It started depending on the electronic transactions between customers, markets, and financial institutions on an international level. The technology also spread towards the banking sector, which affected the banking value chain that progressed to four groups: (1) 3 Customers (retail, commercial, investment), (2) Channels (branches, brokers, social, mobile), (3) financial service providers (banking and non-banking institutions), and (4) interbank providers (exchanges and networks). FinTech 1.0 (1866–1967) was from the installation of the first transatlantic cable to the invention of the first ATM, finance and technology were combined to produce the first period of financial globalization. Technologies such as financial interlinkages across borders, allowing transmission of financial information, transactions and payments. FinTech 2.0 (1967–2008) was when FinTech remained internal to the sector and was conquered mainly by the traditional financial services industry that were using technology to provide financial products and services. This era witnessed the introduction of electronic payments and clearing systems, ATM machines and online banking. During the mid-1990s, the financial services industry became the single largest purchaser of IT – a position which remains to this day. FinTech 3.0 (2008-present) is since the global financial crisis ended, disruptive new startups and newly established technology companies have begun to deliver financial products and services directly to businesses and the general public and where new entrants tend to focus on a single-purpose solution, designed to offer an improved experience in just one product or service (Razzaque et al. 2020).

FinTech is transforming the delivery of financial services across the Middle East. FinTech solutions like the Sandbox support FinTech startups of the Kingdom of Bahrain. Regional banks are developing digital smart solutions in collaboration with Bahrain based firms such as Benefit Pay. FinTech offers earlier transactions than radically revolutionize the traditionally one. With FinTech one can request a balance sheet in real-time using s smart phone FinTech enabled (Razzaque et al. 2020). FinTech is a competitive strategy i.e., globally deployed in several banking institutions through its virtual payment gateways using smartphones; hence is changing the old financial school practice and is serving to advance the customers financial situations and outcomes, by promoting more advanced, i.e., FinTech associations traditional with online financial services (Razzaque et al. 2020).

There is a constant expansion in the investments pertaining to FinTech in order to advance technological breakthroughs in several arenas, e.g., smartphone networks, cloud computing, and data analytics (Castiglione et al. 2015; Lee and Kim 2015; Wen et al. 2013). FinTech has become a revolution in the current era due to its drivers, i.e., cost-saving conditions, innovative prospects, and the demands of the customer since the recent years (Wigglesworth 2016). However, there are challenges that have been faced due to the high adoption of FinTech. For instance, massive implementations of FinTech have led to complicated systems which need to be well integrated to work effectively and efficiently, and the unique demands raised from the expectations of using FinTech. The current research generated solution pertaining to security and privacy, etc. Within the financial sector, cyber concerns were initially a business process issue. But now this is a vital IT security issue as in the current research scholars are concerned with the investments of the banking sector pertaining to security issues which deem difficult to predicate as well as to prove or regulate (Wang et al. 2002; Banu and Nagaveni 2013; Lu et al. 2000). Furthermore, newer financial services a bundled with privacy concerns where implementing such solutions, e.g., insurance solutions, negatively affects the financial institutions (Elnagdy et al. 2016; Zhang and Soong 2004). These are serious challenging issues with another emerging FinTech issue being the web apps being attached online via financial frauds (Sharma and Panigrahi 2013). Traditions protocols for detecting frauds are statistically inclined and insufficient in detecting real-time security threats (Chandola et al. 2012).

As per Chong (2013), the open cell phone architecture with malware, and unencrypted confidential information easily get compromised, causing stealing of funds: a massive security concern. Regulators need to address transaction authorization, and the security of payment system. Though blockchain is not susceptible to attacks, some platforms suspend activities when compromised during hacking attacks which lead to financial theft: an epidemic for FinTech users. As per Al- Howard and Sheth (1969), operational risk is the most significant due to its strong correlation to other risks; resulting from inadequate or failed internal processes, people, and systems, or from external events. Three operational breakpoints are identified when digital payment processes fail. Financial provider must manage risks from the lens of security, data confidentiality, data system integrity, system availability and outsourcing. Such risks are closely associated with reputation risks and legal risks for banks which could bear legal consequences. Technological failure such as transaction delays from poor or scant Internet signal, or ill working devices, result in wrong data entry, system maintenance errors, execution/delivery/process management failures. With increasing atomization of processes, the number of technology-based breakpoints may rise if there is minimum awareness of their existence. Human errors, e.g., incorrect account number or funds amount entry. Most client errors are caused by poor process design. Following are the nine propositions formulated based on the argument critiqued in our study:

- P1, P2 and P3: Economic benefit, seamless transaction and convenience effect the perceived benefit of FinTech.
- P4, P5 and P6: Financial risk, legal risk, security risk, and operational risk effect the perceived risk of FinTech.
- P8 and P9: Perceived benefit bares and perceived risk effect on FinTech continuance intention.

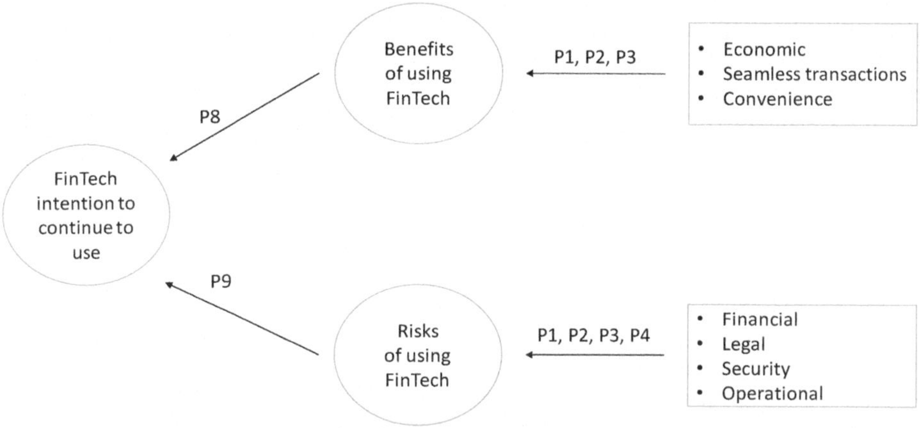

Fig. 1. Conceptual model

3 Conclusion

This study proposed a conceptual model to comprehend what perception the adaption of FinTech has on the financial industry, and this was based on a survey of current literature. Even though FinTech has gone through various phases of improvement, still there are challenges which have been posed in the current literature, and this review of current literature has personified the current challenges which question the continuing adoption of FinTech. There is a significant advancements in the field of financial services, which heavily contribute to the global discrimination of FinTech; particularly when it comes to the awareness of FinTech. This study's review of critiqued literature investigate the degree of acceptance or rejection of FinTech users, reflecting the extent of awareness of using the FinTech. Several variables significantly contribute to promote FinTech. This is from the perspective of the convenience and the ease of use of applying FinTech while keeping under consideration the non-financial risks, secure concerns when consuming FinTech on smartphone platforms. Once the FinTech firms raise such an awareness, they are able to tap into the desires of the end-users whom, in fact, claim for user-friendly means of using gaining the FinTech experience. Our review of literature evidences a turbulent FinTech landscape which in turn affects consumers' behaviors, which in turn force financial organizations to offer even a wide spectrum of financial solutions.

This study recommends that the financial firms support further studies to enhance awareness of FinTech. Financial firms can get involved by furnishing state of the art FinTech solutions using the Fig. 1 model of our study, viable for future empirical assessment within any sector that uses FinTech services; e.g., bankers. Also, companies planning to apply FinTech can investigate the use of their financial services so they will gather a deeper sympathetic of the perspectives of clients. Future research should also look at how FinTech technology can work when aligned with knowledge management processes, such as sharing of knowledge, and knowledge management tools, like the utilization of virtual environments for making the FinTech solutions awareness so for an improved adoptability of FinTech. Hence, studies should consider investigate areas investigated by scholars like (Al-Hashimi et al. 2019; Alshaikh et al. 2017; Razzaque 2019; Razzaque and Jalal-Karim 2010; Razzaque and Eldabi 2018).

References

Al-Hashimi, M., Razzaque, A., Hamdan, A., Reyad, S., Al-Sartawi, A.M., Madbouly, A.: The impact of internet addiction on Bahraini employees' performance. In: EMCIS 2019: 16th European Mediterranean and Middle Eastern Conference on Information Systems, 1–9 December 2019, Dubai, UAE (2019)

Alshaikh, I.Y., Razzaque, A., Alalawi, M.: Positive emotion and social capital affect knowledge sharing: case of the public sector of the Kingdom of Bahrain. In: European, Mediterranean and Middle Eastern Conference on Information Systems (EMCIS 2017 – 7–9 September 2017). Lecture Notes in Business Information Processing. Springer, Coimbra (2017)

Arner, D.W., Barberis, J., Buckley, R.P.: The Evolution of FinTech: A New Post-Crisis Paradigm, vol. 47. University of Hong Kong, Hong Kong (2015)

Banu, R., Nagaveni, N.: Evaluation of a perturbation-based technique for privacy. Inf. Sci. **232**, 437–448 (2013)

Castiglione, A., Pizzolante, R., De Santis, A., Carpentieri, B., Castiglione, A., Palmieri, F.: Cloud-based adaptive compression and secure management services for 3D healthcare data. Future Gener. Comput. Syst. **43–44**, 120–134 (2015)

Chandola, V., Banerjee, A., Kumar, V.: Anomaly detection for discrete sequences: a. IEEE Trans. Knowl. Data Eng. **24**(5), 823–839 (2012)

Chong, A.Y.: Understanding mobile commerce continuance intentions: an empirical analysis of Chinese consumers. J. Comput. Inform. Syst. **53**, 22–30 (2013)

Dahlberg, T., Mallat, N., Ondrus, J., Zmijewska, A.: Past, present and future of mobile payments research: a literature review. Electron. Commer. Res. Appl. **7**, 165–181 (2008)

Elnagdy, S., Qiu, M., Gai, K.: Cyber incident classifications using ontology-based. In: Proceedings of the 2nd IEEE International Conference of Scalable and Smart Cloud, pp. 301–306 (2016)

Howard, J., Sheth, J.: The Theory of Buyer Behavior. Wiley, New York (1969)

Lee, T., Kim, H.: An exploratory study on Fintech industry in korea: crowd funding case. In: Proceedings of 2nd International Conference on Innovative Engineering Technologies, Bangkok, Thailand, pp. 58–64 (2015)

Lu, Y., Roychowdhury, V., Vandenberghe, L.: Distributed parallel support vector. IEEE Trans. Neural Netw. **19**(7), 1167–1178 (2000)

Razzaque, A.: Collaborative innovation aids medical decisions in virtual community: a review of literature. In: Al-Sartawi, A., Hussani, K., Hannon, A., Hamdan, A. (eds.) Global Approaches to Sustainability Through Learning and Education. IGI Global publications (2019)

Razzaque, A., Cummings, R., Karolak, M., Hamdan, A.: The propensity to use FinTech: input from bankers in the Kingdom of Bahrain. J. Inform. Knowl. Manag. **19**(1), 20–45 (2020)

Razzaque, A., Jalal-Karim, A.: The influence of knowledge management on EHR to Improve the quality of health care services. In: European, Mediterranean and Middle Eastern Conference on Information Systems (EMCIS 2010), Abu-Dhabi, UAE (2010)

Razzaque, A., Eldabi, T.: Assessing the impact of physicians' virtual communities on their medical decision making quality. In: 51st Hawaii International Conference on System Sciences - HICSS-51. University of Hawaii at Manoa, Hawaii (2018)

Sharma, A., Panigrahi, P.: A review of financial accounting fraud detection based on data mining techniques. arXiv, 1309.3944 (2013)

Venkatesh, V., Norris, M.C., Davis, G.B., Davis, F.D.: User acceptance of information technology: towards a unified view. MIS Q. **27**(3), 425–478 (2003)

Wang, H., Mylopoulos, J., Liao, S.: Intelligent agents and financial risk monitoring systems. Commun. ACM **45**(3), 83–88 (2002)

Wen, S., Zhou, W., Zhang, J., Xiang, Y., Zhou, W., Jia, W.: Modeling propagation dynamics of social network worms. IEEE Trans. Parallel Distrib. Syst. **24**(8), 1633–1643 (2013)

Wigglesworth, R.: Fintech: search for a super-algo (2016). https://www.ft.com/content/5eb91614-bee5-11e5-846f-79b0e3d20eaf

Zhang, Y., Soong, B.: Performance evaluation of GSM/GPRS networks with channel re-allocation scheme. IEEE Commun. Lett. **8**(5), 280–282 (2004)

Applications of Integrated Management and Distribution Information in Power Supply System

Yanbo Liu[1(✉)], Yingjie Zang[2], Chao Sun[2], and Jingwen Sun[3]

[1] Dispatching and Control Center, State Grid Liaoning Electric Power Supply Co. Ltd., Shenyang 110004, China
lybstc0531@126.com

[2] Shenyang Institute of Engineering, Shenyang 110136, China

[3] State Grid Shenyang Electric Power Supply Company, Shenyang 100300, China

Abstract. Urban grid planning needs to focus on sound investment plans to improve the safety and reliability of power system power supply to meet the growing needs of users. Does not adversely affect the environment. The contradiction in urban power grid and urban planning is an important reason that plagues the rapid development of power grid enterprises. Therefore, under the trend of urbanized dogs, urban power grid construction planning and urban overall planning are required to match. Urban planning needs to consider urban power grid planning as an important component, and match urban development to achieve simultaneous development. The integrated management and distribution system is the platform that contains all the information processing. It can share information and improve the reliability and security of the grid operation. This paper first introduces the integrated system of urban problems and its concept and architecture, focusing on the overall design form and process information management model, system function structure and several aspects of the main functions of the integrated system.

Keywords: Comprehensive management · Distribution system · Integrated management

1 Introduction

The key to urban power grid design planning is economic development, which requires reasonable investment, can meet the requirements of the power grid, and improve power quality to meet the needs of urban residents for shops. The use of information technology can better meet the power needs of grid users. Taking the construction of distribution network information technology as an example, adopting the form of centralized distribution network inspection, eliminating defects, two tickets and other business areas, combining network production system management processes, forms, solidification processes, mobile terminal operation management, organization and realization of production information form distribution Network pilot [1]. With the availability of distribution network, it is closely related to daily operation and

A.-E. Hassanien et al. (Eds.): AICV 2020, AISC 1153, pp. 118–124, 2020.
https://doi.org/10.1007/978-3-030-44289-7_12

maintenance, improving information and standard operation (pdas, tablet) in the form of mobile terminals, and website operation support standardization tools to expand the field of information management, use of information technology, and distribution network. Standardization in the field of operations management. Through information processing, the supervision of on-site operation is strengthened, and work efficiency is greatly improved (Fig. 1).

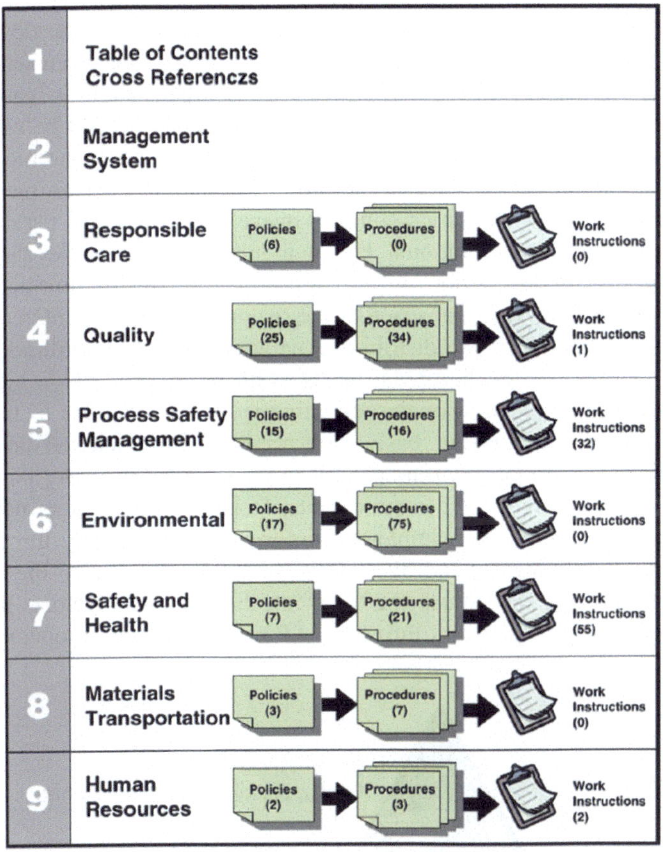

Fig. 1. Management system design model

Many of the management issues facing urban power grid planning have plagued the power system for a long time. In the most basic power supply and distribution enterprises, the main economic management system can be built to achieve independent operation, and when the staff arrives at the scene to fill in relevant information and input the computer to complete the operation, this can save a large part of the paper overhead. After the information is blocked, it cannot be fed back, resulting in failure to provide an effective service. The working method used today, its management efficiency is not high, it is not convenient to comprehensively monitor the management system, especially in the following aspects:

The system is inconvenient to carry and difficult to fill in and adjust. After completing the on-site work, you need to go back and upload into the distribution network production system. This method is not flexible and wastes time and there are many times when printing errors occur. After completing the on-site operational tasks, all task records are stored in paper form, which is not conducive to inspection and search, and is difficult to perform statistics and analysis.

In the process of urban planning, we should pay attention to the development of urban power grids. The development of power grids must keep pace with the development of cities. Whether the layout of power grids in urban spaces is reasonable is the main focus of power grid planning. In addition, technology should be emphasized in the planning process of power grids. And the rationality of the economic level makes the overall planning of the power grid more professional. The urban overall planning process needs to combine the city's comprehensive status quo and regional changes, and requires a scientific and rational layout to make the city's development more scientific. Both urban construction and urban power grid construction need to control the construction land and regional space of the region. The two are interconnected and coordinated. The stability of the distribution network requires coordinated control in many aspects, and the production team needs to fill in the production system log for each team. The printed version of the repetitive work system adds to the workload. During field operations, relevant basic data needs to be consulted and cannot be adapted to complex outdoor environments. It is inconvenient to carry paper materials and check that the site meets the requirements. In general, regulators need to obtain information in a timely manner, implement information technology through network production and management, hand over information management to the site, and use information technology to adjust the field of distribution network management. Improve the efficiency of [2] work through the form of information operation (Fig. 2).

Fig. 2. Document hierarchy

2 The Systems and Definitions of Integrated Management and Distribution

The key to using the modern information technology to establish an integrated power distribution network management system is to use the power information statistical system to integrate into the power system power distribution reverse side to assist in the planning, standardization and integration of enterprise information network construction. Through the leadership decision support portal integrated distribution network planning, power supply reliability for user management, engineering, electronic data transmission, power customers to provide comprehensive information query camp, auxiliary meter readings, near real-time data representation program analysis and other services and business development capabilities, I believe The mobile information integration system will be established in the camp [3].

In order to use the collected information to complete network supervision and control, improve monitoring and management, it is necessary to collect and process information uniformly to improve work efficiency. Power grid planning and urban planning are mutually reinforcing relationships. Urban planning guides the direction of grid planning. Power grid planning is also an important guarantee for urban planning. Coordinating the consistency between urban construction planning and grid planning development can greatly enhance social and economic benefits. In the next phase of power grid planning, we must do a good job in foundation work, load forecasting, and coordinate urban construction, systematically, standardize, and improve the specific planning of power grid planning, preliminary period, investment budget, etc., and build "safe, reliable, green, and efficient smart grid". The implementation plan is summarized below.

3 In the Form of Information Management and Process

(a) Start work orders, forms and distribution network production systems, and automatically receive order information management platforms.
(b) The system is divided into single or multiple tasks as required.
(c) The corresponding tasks are processed by the system and sent to the following website operators (PDAs, tablets) in the form of work tasks and worksheets.
(d) The on-site mobile terminal operator works on site and fills out the worksheet.
(e) The control personnel can conduct command and monitoring of the on-site tasks at any time, for example: work situation, work content, such as completion [4] (Fig. 3).

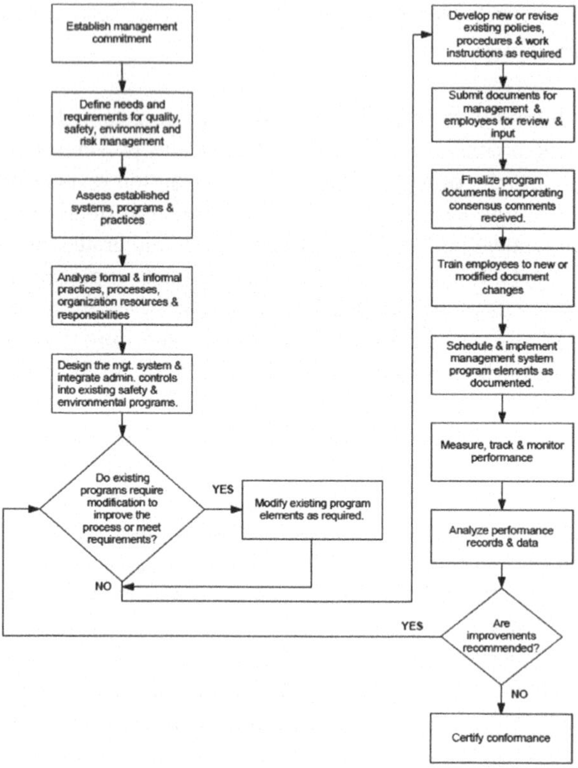

Fig. 3. Management system development and implementation process

4 Inspection Management

Inspection management, distribution line and equipment periodic inspection, special inspection, night inspection, fault inspection, supervision and inspection are divided into five categories.

(a) Regularly monitor the status of routes along the route and cover health care.
(b) Special inspections: all or part of the lines in bad weather (such as typhoons, storms, etc.), floods, fires, line overloads, external influences, abnormal operations or other special conditions (such as peak load periods) [5] an examination.
(c) Evening monitoring: If the grid is peaked or foggy or the weather is gloomy, check the line, heat and flashover discharge at the contact surface of the insulator switch.
(d) Fault trip: After the line fault, the fault condition and extent of the line cannot be determined.
(e) Monitoring access: Check the outdoor handheld PDA and use the card reader to check if the label has been checked in the automatic identification device and check the required technical content points. If there is unprocessed defect information, it should be timely [6, 7] (Fig. 4).

1	**Eliminate duplication/redundancy:** • Use same administrative controls to drive all program elements
2	**Ensure compliance with applicable regulatory requirements Including:** • OSHA PSM • EPA RMP • OSHA Safety and Health • EPA Environmental • DOT Materials Transportation • Other as applicable
3	**Ensure uniform compliance to Company/Industry standards** • ISO 9001 • ISO 14000 • QS-9000 • Responsible Care
4	**Clarify responsibilities and ownership for managing, performing and verifying the work**
5	**Maximize cost efficiencies / business results including the transition period for implementing change in the system**
6	**Achieve goals / objectives with minimum effort**
7	**Provide flexibility / adjustability within the new system to facilitate continuous improvement**
8	**Develop a management system where performance can be measured against**

Fig. 4. Example of management system objectives

Defect management, PDA system library download defects and background should be consistent. This number and the entire system should provide defect library functionality defects for incremental library updates. Off - PDA operation is used to eliminate defects, eliminate the defects of the construction unit personnel after receiving the task to eliminate defects, defect elimination using PDA data download [8, 9] to the scene for defect elimination. The tasks and defect information lost after the consumer PDA is cancelled, the ticket purchase and management work are eliminated according to the ticket purchase guide, and the relevant process information is completed and handed over to the PDA after the defect is eliminated. Libraries should add defect records: on-site PDA defect and defect classification staff operations are simple. Defects are recorded by the PDA, and the defective device should be clearly marked after it is added. The PDA camera is defected on site, and the attachment background image is uploaded to check the status of the device, eliminate defects before and after, and eliminate defects [10, 11].

5 Conclusion

Application integration business, in short, business process GIS data generation, sales, construction, etc., focusing on data transmission and processing to optimize power distribution and integrated management networks, and applying IT service industry to improve power network construction, It can improve the management level and reliability of power grid operation, give full play to the importance of distribution network, and build an integrated management and marketing system to enable grid users to be satisfied.

References

1. Lim, S., Cho, Y.: Development and validation study of the Korean genital image scale; focused on Korean married women. Int. J. Multimed. Ubiquit. Eng. **13**, 7–12 (2018)
2. Weon, I.S., Lee, S.G.: Velocity-based object detection in dynamic environment using YOLO-based deep learning algorithm. Int. J. Multimed. Ubiquit. Eng. **14**, 7–12 (2019)
3. Hong, C.: Development of COP-based position sensor. Int. J. Smart Device Appl. **4**, 29–34 (2016)
4. Park, J., Ryoo, H.Y.: Service classification of home robot. Int. J. Smart Home. **11**, 41–46 (2017)
5. Eoml, K., Choi, H., Won, K., Won, S., Jung, K.K.: Temperature history system using the NFC self-powered temperature sensor tag. Int. J. Smart Home **13**, 7–14 (2019)
6. Jeong, H.: An analysis on nursing students' perception of protective actions and management for medical information at hospital. Int. J. Reliable Inform. Assur. **4**, 19–24 (2016)
7. Fan, Y.F., Mohammed, A., George, M.L.: Gunshot detection system for national security. Int. J. Reliable Inform. Assur. **5**, 31–36 (2017)
8. Park, Y., Lim, Y.H., Chae, M.S.: A study on passwordless authentication technology and its effects. Int. J. Reliable Inform. Assur. **6**, 13–20 (2018)
9. Rao, L.G.: Key exposure in cloud data services. Int. J. Big Data Secur. Intell. **4**, 15–20 (2017)
10. Lee, J.S.: An analysis of SIFs for materials using artificial intelligence technique. Int. J. Bio-Sci. Bio-Technol. **9**, 39–44 (2017)
11. Shin, J.H., Lee, S.S., Kim, T.H.: Mechanical properties of non-cement matrix according to mixing method circulating fluidized bed combustion boiler fly ash and base on blast furnace slag and polysilicon. Asia-Pac. J. Model. Simul. Mech. Syst. Design Anal. **1**, 13–22 (2016)
12. Yang, S.J.: Feature-oriented vs. perception-oriented M-interactivity. Int. J. Comput. Graph. **7**, 1–6 (2017)
13. Kushwah, Y.S., Tomar, G.S., Bhadauria, S.S.: Performance comparison of microstrip high-pass filters for different dielectric substrates. Asia-Pac. J. Adv. Res. Electr. Electron. Eng. **1**, 9–20 (2017)

Swarm-Based Optimization and Applications

Vortex Swarm Optimization: New Metaheuristic Algorithm

Ahmed Sabry A. Elrahman(✉) and Hesham A. Hefny

Department of Computer Science, Faculty of Graduate Studies for Statistical Research, Cairo University, Giza, Egypt
ahmed.sabry200@yahoo.com

Abstract. This paper proposes a new swarm intelligence algorithm called vortex swarm optimization, inspired from vortex collective behavior where particles move around a common known center, they cover all the directions so they play as alarm system where any individual at any direction see the target (resources, predator…etc.) it informs other individuals to take an action. This type of collective behavior was seen in many creatures like (Bats and fishes). They use this type of collective behavior for feeding, access to resources, moving, information transmission, decision making, and protection from predators. This type is different from straight-ahead collective behavior (polarized collective behavior) where individuals move forward. The majority of swarm intelligence algorithms inspired their techniques from straight-ahead collective behavior, but vortex collective behavior had a few pieces of research so the importance of this research appears. The algorithm performs the two basics functions that any searching algorithm had (exploitation and exploration) by circling phase the algorithm will know the direction of the promising region, By swimming phase the algorithm will go to this region (exploration)and by attacking phase individuals will reduce their interaction range and reduce their velocity to get much closer to each other and best region found so far (exploitation), Algorithm skip local minimum by eliminating some individuals to a new regions to explore them and to change vortex direction to explore new directions. The algorithm was tested over ten benchmark functions and it competes with some other powerful algorithms.

Keywords: Vortex swarm optimization · Collective behavior · Optimization · Swarm intelligence · An optimization technique · Swarms · Microbial intelligence · Search · Artificial intelligence · Vortex

1 Introduction

Collective behavior is one of the most magnificent phenomena where a lot of particles move based on little information and simple rules. There are different major types of collective behavior one of them is "vortex" where particles move around empty core center, bat doughnut, duck swirl, and a fish torus are different names for collective vortex behavior. In this work, we will propose a new swarm intelligence algorithm called vortex swarm optimization to solve optimization problems based on collective vortex behavior.

A.-E. Hassanien et al. (Eds.): AICV 2020, AISC 1153, pp. 127–136, 2020.
https://doi.org/10.1007/978-3-030-44289-7_13

2 Vortex Collective Behavior

Schools of fish, birds' flocks, and insects' swarms their motion behavior are considered as collective behavior where a group of individuals has to make quick decisions about what actions they will perform in a dangerous and uncertain environment. Note that all actions are done synchronously [1, 2]. The particle behavior is completely determined by the collective effects of its neighbors so the behavior of a lonely particle will be completely different from a particle with one or more neighbors [3]. There are many reasons for organisms to perform collective behavior, like foraging, protection, finding mates, decision making, energy savings and migration as individual needs some navigational information's provided by others [4–6]. Vortex collective behavior is a type of collective behavior where particles move around a specific point. Many creatures adopt vortex collective behavior like fishes that use it as a protection system as when they face a predator they come together and a big vortex shape will be formed to scare the predator. They form this shape to protect themselves from the predator as fishes have a blind spot. By forming vortex shape they make an alarm system as all fishes in all directions so any fish see the predator will alarm others [4]. Collective vortex behavior is very important in collective decision making for bats and fishes as when there are not any consensuses about the direction that the group will move to they form vortex shape and be in the same place until enough numbers of individuals decide to break the vortex, go to specific direction and become temporary group leaders [4]. Some insects and animals obliged to adopt vortex collective behavior because they are unable to adapt to the environment [4, 11]. Bacteria also adopt vortex collective behavior, Paenibacillus vortex which is a kind of Bacteria and very smart form blue groups of bacteria called vortices that swarm around the common centers to search for food and escape from antibiotics [4, 7–9]. Vortex shape has some benefits for all individuals in swarm-like shoveler ducks which move in circles (vortex shape) to make nutrients rise from the water so they can reach to it [10]. Vortex shape is very important for some species as it provides access to resources to all individuals like oxygen, light, temperature [4].

3 The Proposed Vortex Swarm Optimization (VSO)

Vortex swarm optimization plays two roles foraging and attacking (exploration and exploitation). Foraging behavior is a behavior where particles' interaction range is static and the propulsion force is relativity big value. Particles cover almost directions around the common center and there is a mutation in one dimension for some random particles to make the vortex changes its direction to make all particles cover all directions around the common center (global best location) and explore new regions. When an algorithm finds a better solution around the best location, the individuals will swim in that direction and begin to repeat this step until either reaches the maximum swimming steps or there is no a better solution than the center, so the Algorithm will reach the best promising area. After awhile algorithm will play another role, attacking behavior (exploitation) where interaction range is decreasing every iteration and propulsion force is decreasing also so all individuals will get close to the best promising area to get better results. The VSO contains four phases as shown in Fig. 1. Also, Algorithm 1 shows the steps and processes of the VSO. Where M, MI, ATT, R, and RDR are the maximum value of propulsion force factor, the minimum value of propulsion force factor, attack iteration, the radius of interaction range and radius decreasing rate respectively.

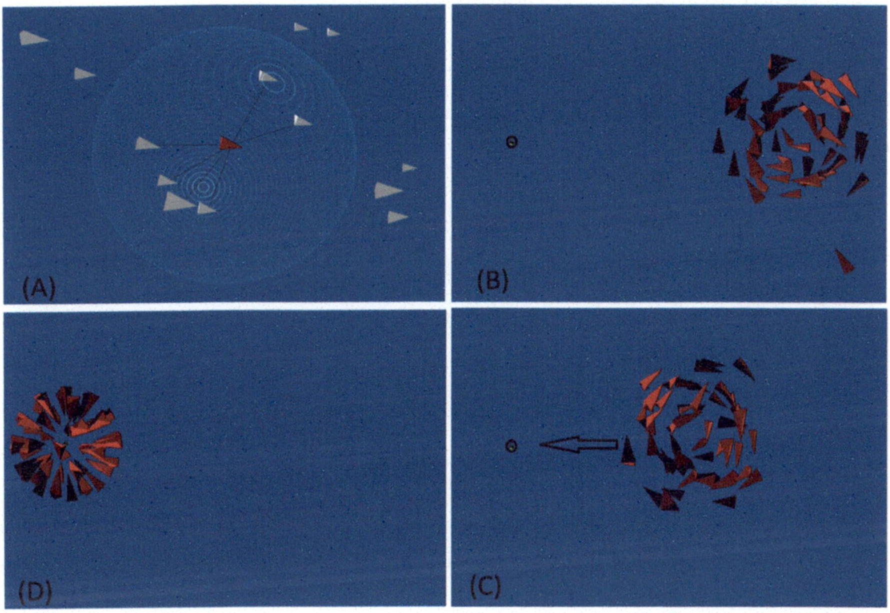

Fig. 1. The main components of the vortex swarm optimization algorithm (A) Determine neighborhoods phase (B) Circling phase (D) Attacking phase (C) Swimming phase

Algorithm 1:- vortex swarm optimization

Input:-
M, MI, ATT, R, RDR
Result:-
Global best position or near it
Method:-
InteractionRadius=R, W= M, Particles=initialize(Particles)
Evaluate (Particles)
C= position of a particle that has the best position
While Iteration ≤ MaxIteration **do**
    Circling (InteractionRadius; Particles; W; C)
    Swimming (Particles, Center, BestIterationFitness, BestIterationposition)
    IF Iteration ≥ ATT
        Attacking(Particles, InteractionRadius, W)
    End IF
End

Phase (1) Determine Neighborhoods: Any particle must determine its neighbors who located in its interaction range by calculating the distance between every particle and its location. Where the neighbor whose location within a circle (or sphere in 3d) its radius is R which is its interaction range radius, or in another word must satisfy the following condition:

$$\left|\overrightarrow{x_i} - \overrightarrow{x_j}\right| < R \tag{1}$$

Where $\overrightarrow{x_i}$ is the location of the particle and $\overrightarrow{x_j}$ is the position of the tested particle (i.e., candidate neighbor), R is the radius of the particle's interaction range.

Phase (2) Circling Phase: In this phase algorithm form vortex and based on the previous mechanisms algorithm can form it by making all particles follow three rules:

- Separation: This refers to collision avoidance with neighbors, with this rule any two particles cannot be in the same place at the same time.
- Alignment: which refer to velocity matching with neighbors in sensory range
- Attraction: particles are attracted to a specific point

In this phase all particles move in circles, the new position equal to the previous position plus the new velocity as follows:

$$\overrightarrow{x_i}(t+1) = \overrightarrow{x_i}(t) + \overrightarrow{v_i}(t+1) \tag{2}$$

Where $\overrightarrow{x_i}(t+1)$ is ith particle's new position, $\overrightarrow{x_i}(t)$ is the particle previous position and $\overrightarrow{v_i}(t+1)$ is the new particle's velocity which could be calculated by adding separated, alignment, attraction rules and propulsion force as the following formula:

$$\overrightarrow{v}(t+1) = s_i + A_i + C_i + w\overrightarrow{v}(t) \tag{3}$$

Where s_i is the result of applying separation rule, A_i is the result of applying alignment rule, C_i is the result of applying attraction rule, $w\overrightarrow{v}(t)$ represents individual propulsion force and w is propulsion force factor. Particles after form the vortex they need to change its direction to explore all directions as the alignment rule makes all particles circle in one direction, so we need some disturbance in our model to force particles to choose another direction (from right to left, up to down, right diagonal, left diagonal …etc.) to explore all directions as follows:

$$\overrightarrow{x_i}(t+1)_D = \begin{cases} \overrightarrow{x_i}(t+1)_D & \text{if } rand_{0,1} > ep \\ rand_{up,low} & \text{if } rand_{0,1} < ep \end{cases} \tag{4}$$

Where $\overrightarrow{x_i}(t+1)_D$ is the new position at random dimension D, $round_{up,low}$ is a random number between upper value and lower value, $rand_{0,1}$ is a random value between [0, 1] and ep is the elimination probability, so if particle's probability is lower than specific number (ep) a random dimension will be mutated and if particle's probability is higher than (ep) then the particle still as it is. Algorithm 2 illustrates the circling phase with separation, alignment and attraction rules.

(a) **Separation rule**

This rule responsible for avoiding collision between particles to avoid any two particles be in the same place at the same time to test the largest number of points Fig. 2(B) as follows:

$$s_i = -\sum_{j=1}^{n_r} \overrightarrow{x_j} - \overrightarrow{x_i} \tag{5}$$

Where x_i is the position of the ith particle and x_j is the position of the jth particle, n_r is the number of neighbors located in the ith particle sensory range.

(b) **Alignment**

This rule responsible for aligning particles in one direction, every particle changes its direction to the average directions of its neighbor's Fig. 2(A) as follows:

$$A_i = \frac{1}{n_r} * \sum_{j=1}^{n_r} \overrightarrow{v_j} \tag{6}$$

Where n_r is the number of neighbors located in the ith interaction range, V_j is the velocity of jth neighbor.

(c) **Attraction**

This rule responsible for attracting all particles to the best position found so far Fig. 2C as following:

$$c_i = c - \overrightarrow{x_i} \tag{7}$$

Where c is the attractant point (best position found so far).

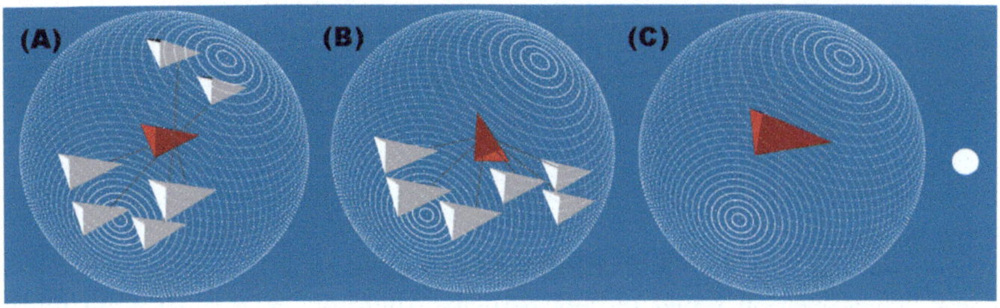

Fig. 2. (A) Alignment rule (B) Separation rule (C) Attraction rule.

```
Algorithm 2:- circling phase
Input:-
Particles; Radius;W;C
Result:-
Best Iteration Position
Method:-
For i ← 1 to ParticlesNmbers do
|       Neighbours ← GetNeighbours(i, Particles, Radius)  eq(1)
|       A ← Alignment(i, neighbours)                      eq(6)
|       S ← Seperation(i, neighbours)                     eq(5)
|       C ← Attraction (i, C)                             eq(7)
|       Velocity[i] ← W * Velocity[i] + A + S + C
|       Position[i] ← Position[i] + Velocity[i]
 End
 mutateSomeParticles(Particles)                           eq(4)
 Evaluate(Particles)
End
```

Phase (3) Swimming Phase: After performing circling phase and all particles are around specific point they begin to take the same direction of the center of vortex which determine by the difference between the position of the best particle (that its fitness is better than all particles and in the same time has fitness better than the center fitness (Global best)) and the center location (C). After performing the circling phase center will move towards the best position By the following equation:

$$\overrightarrow{x_c}(t+1) = \overrightarrow{x_c}(t) + \overrightarrow{C_i}(t+1) \tag{8}$$

Where $\overrightarrow{x_c}(t+1)$ is the center's new position, $\overrightarrow{x_c}(t)$ is the center's previous position and $\overrightarrow{C_i}(t+1)$ is the center's new velocity which equals to:

$$\overrightarrow{C_i}(t+1) = \text{Gbest} - C \tag{9}$$

Where $\overrightarrow{C_i}(t+1)$ is the velocity of the center, Gbest is the position of the best particle found so far and C is the center position. All particles will swim in the same direction as the following equation:

$$\overrightarrow{x_i}(t+1) = \overrightarrow{x_i}(t) + \overrightarrow{C_i}(t+1) \tag{10}$$

And repeat this step until a maximum number of swimming steps reached or there isn't any better position found by any particle, Algorithm 3, Fig. 1(C) are illustrate swimming phase.

```
Algorithm 3:- Swimming Phase
Input:
Particles, Center, BestIterationFitness, BestIterationposition
Result:
New positions of particles, New Center
Method:
While BestFitness < CenterFitness do
    CenterVelocity ← BestPosition − CenterPosition        eq(9)
    CenterPosition ← CenterPosition + CenterVelocity      eq(8)

    For i ← 1 to ParticlesNumbers do
    |    Position[i] ← Position[i] + CenterVelocity       eq(10)
    End

    BestFitness, BestPosition ← EvaluateParticles(ParticlesPositions)
    CenterFitness ← EvaluateCenter(CenterPosition)

    If NumOf Steps > MaximumSwimSteps then
    |    break
    |  Else
    |    NumOf Steps ← NumOf Steps + 1
    End If
End
```

Phase (4) Attacking: Any algorithm needs tow strategies (exploration and exploitation) here in this model we will reduce the propulsion force and interaction range to exploit the best region found so far.

After a while and when the algorithm reaches the best promising area it begins to exploit the best region to enhance the best solution by reducing particles' interaction range to get closer from each other by the following equation:

$$R(\mathrm{iter}) = \begin{cases} \mathrm{Val} & \mathrm{iter} < \mathrm{ATT} \\ R(\mathrm{iter}-1) \times (1-\mathrm{RDR}) & \mathrm{iter} \geq \mathrm{ATT} \end{cases} \tag{11}$$

Where $R(\mathrm{iter})$ is the current radius of particle's sensory range (interaction range), Val is the value of R before attacking iteration, $R\,(\mathrm{iter}-1)$ is the interaction radius of the previous iteration and RDR is the radius decreasing rate. And also the propulsion force of particles will decrease by decreasing propulsion force factor to get much closer to the best position found as follows:

$$w(\mathrm{iter}) = \begin{cases} M & \mathrm{iter} < \mathrm{ATT} \\ w(\mathrm{iter}-1) - \left(\frac{M-MI}{MT-ATT}\right) & \mathrm{iter} \geq \mathrm{ATT} \end{cases} \tag{12}$$

Where iter is the current iteration, M is the maximum propulsion force factor, MI minimum propulsion force factor, MT is the maximum iteration, ATT is the attack iteration which is followed by reducing the interaction radius and propulsion force. Figure 1(D) illustrates the attacking phase where interaction range and propulsion force are reduced so particles are close to each other and attacking the best region from all directions.

Algorithm 4:- Attacking phase
Input:- Particles; Radius; W **output:-** New value of (Radius; W) **Method:-** Update Radius() eq(11) Update W() eq(12) End:-

4 Results of Benchmark Functions

Here we will test our new algorithm on some benchmark functions Table 1, benchmark testing functions are divided to two groups unimodal [f1, f2, f6, f7, f8], multi-modal [f3, f4, f5, f9, f10] where unimodal have single optimum and multi-modal have more than one optimum but one is global optimum and the rest are local optimum, the results are compared with three well-known algorithms [12]: Dragonfly algorithm, particle swarm optimization, Genetic algorithm each algorithm run 30 times over all the functions with 10 Dimensions, 30 individuals perform algorithm and 500 iterations for each run, average and standard deviation of best solutions are calculated Table 2. The parameters for the experiment were as following:

Table 1. Benchmark functions.

Function number	Function	Dim	Range
F1	$\sum_{i=1}^{n} x_i^2$	10	[−100, 100]
F2	$\sum_{i=1}^{n} \lvert x_i \rvert + \prod_{i=1}^{n} \lvert x_i \rvert$	10	[−10, 10]
F3	$10n + \sum_{i=1}^{n} \left(x_i^2 - 10\cos(2\pi x_i)\right)$	10	[−5.12, 5.12]
F4	$1 + \sum_{i=1}^{n} \frac{x_i^2}{4000} \prod_{i=1}^{n} \cos\left(\frac{x_i}{\sqrt{i}}\right)$	10	[−600, 600]
F5	$-a.\exp\left(-b\sqrt{\frac{1}{n}\sum_{i=1}^{n} x_i^2}\right) - \exp\left(\frac{1}{n}\sum_{i=1}^{n}\cos(cx_i)\right) + a + \exp(1)$	10	[−32, 32]

(*continued*)

Table 1. (*continued*)

Function number	Function	Dim	Range
F6	$max_{i=1,\ldots n}\|x_i\|$	10	[−100, 100]
F7	$\sum_{i=1}^{n}\left[b\left(x_{i+1}-x_i^2\right)^2+(a-x_i)^2\right]$	10	[−5, 10]
F8	$\sum_{i=1}^{n} ix_i^4 + random[0.1)$	10	[−1.28, 1.28]
F9	$\sum_{i=1}^{D} u(x_i,10,100,4)+\frac{\pi}{D}\{10sin^2(\pi y_i) + \sum_{i=1}^{D-1}(y_i-1)^2\left[1+10sin^2(\pi y_{i+1})\right]+(y_D-1)^2\}$ $y_i = 1+\frac{1}{4}(x_i+1)$ $u(x_i,a,k,m)=\begin{cases}k(x_i-a)^m, & x_i>a,\\ 0, & -a\le x_i\le a,\\ k(-x_i,-a)^m, & x_i<-a,\end{cases}$	10	[−50, 50]
F10	$\sum_{i=1}^{D} u(x_i,5,10,4)+\frac{1}{10}\{sin^2(3\pi x_i) + \sum_{i=1}^{D-1}(x_i-1)^2\left[1+sin^2(3\pi x_{i+1})\right] + (x_D-1)^2\left[1+sin^2(2\pi x_D)\right]\}$ $\mathbf{u}(\mathbf{x_i},\mathbf{a},\mathbf{k},\mathbf{m})=\begin{cases}\mathbf{k}(\mathbf{x_i}-\mathbf{1})^{\mathbf{m}}, & \mathbf{x_i}>a,\\ \mathbf{0}, & -a\le \mathbf{x_i}\le a,\\ \mathbf{k}(-\mathbf{x_i},-\mathbf{1})^{\mathbf{m}}, & \mathbf{x_i}<-a,\end{cases}$	10	[−50, 50]

Table 2. Results.

Function	VSO		DA		PSO		GA	
	AVG	STD	AVG	STD	AVG	STD	AVG	STD
F1	**4.93E−43**	**2.69515E−42**	2.85E−18	7.16E−18	4.2E−18	1.31E−17	748.5972	324.9262
F2	**5.92119E−17**	**3.24317E−16**	1.49E−05	3.76E−05	0.003154	0.009811	5.971358	1.533102
F3	**0.132661208**	**0.344003018**	16.01883	9.479113	10.44724	7.879807	25.51886	6.66936
F4	**0.07639295**	**0.029420366**	0.193354	0.073495	0.083463	0.035067	7.719959	3.62607
F5	**7.49E−14**	**5.35562E−14**	0.23103	0.487053	0.280137	0.601817	9.498785	1.271393
F6	3.11E−03	2.04E−03	**0.000988**	**0.002776**	1748E−03	2515E−03	21.16304	2.605406
F7	420.0552	1841.004	**7.600558**	**6.786473**	63.45331	80.12726	133307.1	85,007.62
F8	**4.90E−05**	**3.47954E−05**	0.010293	0.004691	0.005973	0.003583	0.166872	0.072571
F9	**4.97E−32**	**1.41397E−32**	0.031101	0.098349	8.57E−11	2.71E−10	1858.502	5820.215
F10	**1.35E−32**	**2.2504E−34**	0.002197	0.004633	0.002197	0.004633	68,047.23	87,736.76

Sensory range = 1.e−4, max swim steps = 500, Attack Iteration = 300, elimination probability = 0.05, Maximum propulsion force = 0.2, Minimum propulsion force = 0.05, Decrease radius rate = 0.9.

5 Conclusion

In this paper, we proposed a new swarm intelligence algorithm called vortex swarm optimization, the proposed algorithm was tested over ten benchmark functions and compared with other two algorithms, the results showed that the proposed algorithm has competitive results with them, the result also showed that it is a promising field for future research.

References

1. Couzin, I.D., Krause, J.: Self-organization and collective behavior in vertebrates. Adv. Study Behav. **32**, 1–75 (2003)
2. Beach, T.W.: Thought transference (or what?) in birds. Saturday Rev. Polit. Lit. Sci. Art **152** (3960), 362 (1931)
3. Vicsek, T.: A question of scale. Nature **411**(6836), 421 (2001)
4. Delcourt, J., Bode, N.W.F., Denoel, M.: Collective vortex behaviors: diversity, proximate, and ultimate causes of circular animal group movements. Q. Rev. Biol. **91**(1), 1–24 (2016)
5. Colorado Zuluaga, G.J.: Why animals come together, with the special case of mixed-species bird flocks. Rev. EIA **19**, 49–66 (2013)
6. Couzin, I.D.: Collective cognition in animal groups. Trends Cogn. Sci. **13**(1), 36–43 (2009)
7. Kuchment, A.: The smartest bacteria on Earth. Sci. Am. **304**(6), 70–71 (2011)
8. Finkelshtein, A., et al.: Paenibacillus vortex—a bacterial guide to the wisdom of the crowd. In: BIOCOMMUNICATION: Sign-Mediated Interactions between Cells and Organisms, pp. 257–283 (2017)
9. Jacob, E.B., et al.: Bacterial linguistic communication and social intelligence. TRENDS in Microbiol. **12**(8), 366–372 (2004)
10. Bode, N.W.F., Delcourt, J.: Individual-to-resource landscape interaction strength can explain different collective feeding behaviors. PloS one **8**(10), e75879 (2013)
11. Couzin, I.D., Franks, N.R.: Self-organized lane formation and optimized traffic flow in army ants. Proc. Roy. Soc. London Ser. B: Biol. Sci. **270**(1511), 139–146 (2003)
12. Mirjalili, S.: Dragonfly algorithm: a new meta-heuristic optimization technique for solving single-objective, discrete, and multi-objective problems. Neural Comput. Appl. **27**(4), 1053–1073 (2016)

An Artificial Intelligence System for Apple Fruit Disease Classification Based on Support Vector Machine and Cockroach Swarm Optimization

Mohamed A. El-dosuky[1,4(✉)], Diego Oliva[2,4], and Aboul Ella Hassanien[3,4]

[1] Faculty of Computers and Information, Mansoura University, Mansoura, Egypt
mouh_sal_010@mans.edu.eg
[2] Universidad de Guadalajara CUCEI, Guadalajara, Jalisco, Mexico
[3] Faculty of Computers and Artificial Intelligence, Cairo University, Giza, Egypt
[4] Scientific research group in Egypt (SRGE), Cairo, Egypt

Abstract. This paper presents an artificial intelligent system for detection of apple diseases using Support Vector Machine (SVM) and Cockroach Swarm Optimization (CSO). This paper faces a challenge of content ambiguity. This challenge is mitigated by combining texture classification based on initial k-means clustering. The proposed system is able to extract useful features. There are 4 classes: scab disease, rot disease, blotch disease, and normal. The dataset consists of 320 apple images in total divided into 80 images for each class. SVM is actually a binary classifier. The paper shows how to extend it for multi-class classification. The performance of SVM strongly depends on proper choice of its parameters. CSO meta-heuristic performs fine-tunes for SVM parameters including nonlinear hyper planes, penalty parameter of error term, and degree which is used to find the hyper plane for splitting the data. Results show 94.65% accuracy in detecting the normal apple.

Keywords: Support Vector Machine · Cockroach Swarm Optimization · Texture classification · K-means clustering

1 Introduction

Artificial intelligence (AI) enables machines to simulate human intelligence. A particular application of AI is computer vision (CV).

Undoubtedly, fruit diseases reduce their quality and price. This paper focuses on apple diseases. A human can easily spot apple disease. There are 4 classes: (a) scab disease, (b) rot disease, (c) blotch disease, and (d) normal. An apple with scab is gray or brown. Rot has sunken brown spots. Blotch has irregular patterns. The *problem* that faces apple diseases recognition by computer is the challenge of content ambiguity. This challenge is mitigated by combining texture classification based on initial k-means clustering [1].

A.-E. Hassanien et al. (Eds.): AICV 2020, AISC 1153, pp. 137–147, 2020.
https://doi.org/10.1007/978-3-030-44289-7_14

This paper presents a proposed model for the detection of apple diseases using Support Vector Machine (SVM) [2] and Cockroach Swarm Optimization (CSO) [3] for parameter fine-tuning. This model is able to extract useful features. SVM is actually a binary classifier. The paper shows how to extend it for multi-class classification. The performance of SVM strongly depends on proper choice of its parameters. CSO meta-heuristic performs fine-tuning for SVM parameters.

Reference [3] shows that CSO quickly converges to solution with accurate results compared to Particle Swarm Optimization (PSO).

Most of related work regarding apple segmentation is carried out utilizing a mere threshold such as [4]. The threshold could be globally adaptive such as Otsu's algorithm [5]. The most employed model is Bayesian classifier such as [6] and [7]. Local Binary Pattern (LBP) is proposed [8].

It is a simple yet powerful tool for texture classification with rotation-invariance [9]. Complete Local Binary Pattern (CLBP) is presented in [10]. The climax of using LBP is in a notable previous work [11], in which image is segmented using K-Means before extracting some features, and finally classifying images.

Texture feature extraction can be based on difference and sum of neighboring pixels intensity values [12]. Recently, Adaptive Neuro-Fuzzy Interface System (ANFIS) is proved empirically to be more powerful in predicting apple skin [13].

Current state-of-the-art of fruits and vegetables classification has been developed for small datasets and limited classes, but there is a challenge in the hyper-dimensional features in real-life big datasets [14]. The increase in the dimensionality requires more training data [15]. For superior optimization for SVM, we may use meta-heuristics such as artificial bee colony algorithm (ABC) [16]. For a recent detailed survey of fruits and vegetables quality evaluation using computer vision, refer to [17].

The rest of this paper is presented as follows. Section 2 is dedicated for presenting K-mean clustering, SVM, and CSO. Section 3 is for presenting the proposed model. Validating the proposed model via experimentations and results are in Sect. 4. Section 5 is for conclusion and future directions.

2 Materials and Methods

This section reviews the key methods on which this paper depends, such as K-means clustering, Support Vector Machine (SVM), and Cockroach Swarm Optimization (CSO).

2.1 K-Means Clustering

K-means clustering [2] operates on m points in order to reducing the within-cluster sum of squares, a. k. a the sum of distances of each point in the group to k centers. Hence, the objective is to find arg_s min $\sum_{i=1}^{k} \sum x \in S_i \|x - \mu_i\|^2$. Where S are the speciments on which k-mean clustering operates and μ_i is mean of points in S_i.

The total number of clusters (k) must be set innitailly, and then repeat the following steps until the cluster members no longer change.

1. Randomly assign the cluster center:
   In the training group data, "randomly" find the K record as the initial seed (the center of the initial cluster)
2. Generate the initial cluster:
   Calculate the distance between each record to each random seed, then compare which random seed is closest to the record, then the record will be assigned to the nearest cluster center, and a cluster boundary will be formed, generating a set of members of the initial cluster
3. Generate a new better center:
   Recalculate the mass center of the cluster based on each case within the boundary, and replace the previous random seed with the new better center as the center of the group
4. Change the cluster boundary:
   After assigning a new better center, compare the distance between each record and the new cluster center again, and then reassign the cluster to which each case belongs based on the distance.

2.2 Support Vector Machines

Support Vector Machine (SVM) [2] is a method of supervised learning in machine learning. Based on some features, it finds an optimal hyper plane to separate the different categories of data.

The so-called optimal hyper plane is that the boundary between the two categories can reach the maximum, and the sample points closest to the boundary provide the most classified information of the SVM, which is called the Support Vector.

In the two-dimensional example, we hope to find a line that separates the points of the two classes, and we also want the line to be larger. The larger the margin, the better.

There are two main strategies for SVM on multivariate bifurcation:

a. One-against-rest: There are k SVMs, and the mth SVM can separate m^{th} from other. It will distinguish whether it belongs to a specific SVM.
b. One-against-one: For creating an SVM for any case, a total of $\frac{k(k-1)}{2}$ tests required.

The SVM can only distinguish between the errors used in the initial training. Assuming that the SVM was originally trained with a and b, it would naturally be quite good for data belonging to a or b, but if our test data is also mixed with other data, such as c, then we can hardly determine proper classes.

So, when classifying, we can first ask the classifier "1 to 5" (meaning it can answer "is the first or fifth class"), if it answers 5, we go left and ask "2" For the "5" classifier, if it is still said to be "5", we will continue to the left, so that we can continue to ask questions and get the classification results. We actually called 4 classifiers (if the number of categories is k, we will have $k - 1$ tests. The classification speed is fast, and there is no classification overlap.

2.3 Cockroach Swarm Optimization

The original cockroach swarm optimization (CSO) is introduced by Zhaohui and Haiyan [3]. Cockroaches secrete pheromones to inform other individuals of the nutritional status (cooperation), but they come out of it when food shortages (competition) occur. Cockroaches adapt to extreme environmental conditions and fluctuate the utilization of nutrients in these surroundings using various combinations of movement, habitat selection, etc. [18]. These types of swarm attributes are good for developing an optimization algorithm for research over a large area in limited computing time. CSO first step randomly creates a set of solution candidates. Furthermore, in each loop, CSO conducts 3 procedures, namely hunt swarming, dispersion, & ruthlessness.

In conducting hunt swarming procedure, cockroaches with the strongest power carry the best local solution candidate (P_i) and travel to the optimal global level (P_g) within the scope of his or her vision. It happens that a cockroach suddenly becomes stronger by finding a best solution candidate.

The dispersion procedure is performed frequently to ensure diversity. It performs a random step of all cockroaches.

The ruthlessness procedure mimics the phenomenon of eating weaker cockroaches when food resources are limited. It is implemented as replacing random individuals by the best individual.

CSO has the following steps [3]:

1. Initialize CSO's parameters (the_step, visual_scope, the_dimension, stopping_criteria).
2. Create n individuals (population; X_1, X_2, … X_n)
3. Search population for best local optimal solution within the visual_scope of the ith individual.
4. Conduct hunt-swarming and update P_g
5. if X_i is local optimum, then it moves to P_g via Eq. (1):

$$X_i := X_i + \text{the\_step.random\_number.}\left(P_g - X_i\right) \tag{1}$$

   Where random_number is within [0, 1];
   Otherwise, X_i moves to P_i (within his or her visual_scope) via Eq. (2):

$$X_i := X_i + \text{the\_step.random\_number.}(P_i - X_i) \tag{2}$$

6. Run dispersion procedure and update P_g.
7. Run ruthlessness procedure ($X_k = P_g$ or $X_k = 0$, where $k = 1, \ldots, n$).
8. Repeat steps 3–7 until termination criteria are met
9. Output last outcomes (global optima).

3 The Proposed Apple Fruit Disease Classification System

Figure 1 depicts the main classification model phases. Initially, preprocessing is carried out by converting the image to L*a*b. Then, K-means clustering is performed. Classification is done with SVM, with Cockroach Swarm Optimization fine tuning.

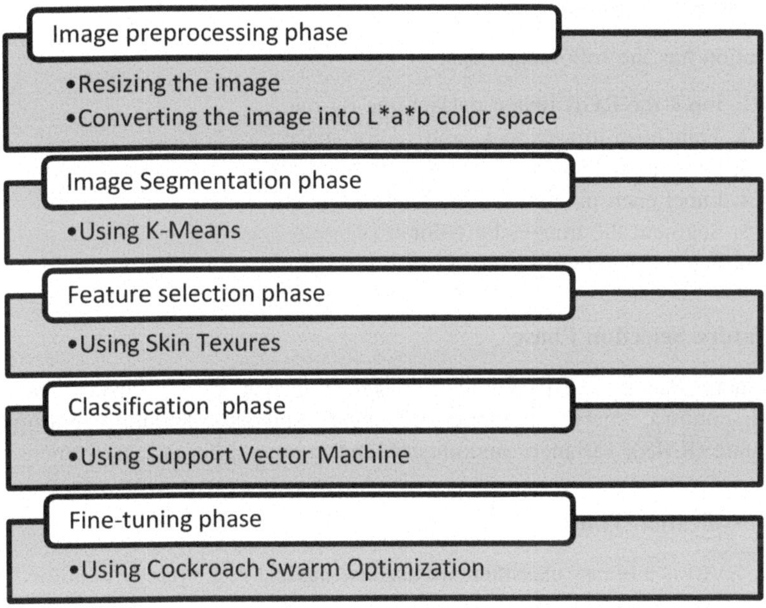

Fig. 1. The main apple fruit diseases classification model

Each of these phases will be discussed in the following subsections.

3.1 Preprocessing Phase

Imported images are resized to be of the same size. Images are scrutinized to get the range of normal apple as shown in Table 1.

Table 1. Apple skin range

	Min	Max
Redness component	126	165
Blueness component	119	208
Hue	0	0.9973

Hue range is narrowed to the range [0.01, 0.3889] for better results in detecting apple texture. Generally, your computer monitor uses RGB color space, but our eyes operate in the more accurate L*a*b* color space, but it is rarely used in everyday life because it is too complicated. In L*a*b color space, color mode consists of one Luminance and two color axes (a, b).

3.2 Apple Segmentation Phase

Segmentation has the following steps:

Step 1. Input the RGB image dataset one-by-one.
Step 2. Transform images to L*a*b* color space.
Step 3. Perform K-Means clustering.
Step 4. Label each pixel based on K-Means result
Step 5. Segment the images by color
Step 6. Output the segment having disease

3.3 Feature Selection Phase

Local Binary Pattern (LBP) could be used; however the following features are extracted: contrast, energy, homogeneity, mean, standard deviation, entropy, root-mean-square (RMS), variance, smoothness, and inverse difference movement (IDM).

3.4 Classification Phase

Basically SVM is a binary classifier; we use one-against-rest strategy to allow SVM on multiple classes, as shown in the following algorithm.

```
Input:
        Training dataset D,
        Total number of local models, k
        Hyper-parameter γ
        C tuning margin
Output:
        k local SVM models
Begin
        Create k clusters denoted D1, D2,...DK, corresponding centers are c1,
c2, ck
        For i ← 1 to k do
              lSVMi = SVM (D, γ, C)
        End for
        Return kSVM – model = { (c1, lSVM1), ..., (ck, lSVMk) }
End
```

3.5 Fine-Tuning Phase

CSO initially starts with generating a random quorum. While the stopping conditions are not met repeat the following steps.

1. Randomize coordinates of cockroaches
2. Measure fitness of cockroaches
3. *Swarming* of a quorum of cockroaches: If there is a good neighbor, then update the coordinates of the remaining cockroaches following that good neighbor.
4. *Dispersion* in presence of danger
5. *Ruthless component:* cockroaches eliminate each other in resource shortage.

CSO fine-tunes the following three SVM parameters:

- γ for nonlinear hyperplanes,
  γ values to try are [0.1, 1, 10, 100] (**4 values**)
- C which is a penalty parameter of error term,
  C values to try are [0.1, 1, 10, 100, 1000] (**5 values**)
- degree which is used to find the hyperplane to split the data
  degree values to try are [0, 1, 2, 3, 4, 5, 6] (**7 values**).

We generate a big number of SVM models (**140** = 4*5*7) with varying values of parameters. The choice between them is based on the fitness function which is based on SVM ability to classify.

4 Experimental Results and Discussion

Some parts of the implementation is done in Python, the rest is a menu-based MATLAB implementation is developed using R2015a Matlab on Microsoft Windows 8 Pro. The dataset consists of 320 apple images in total divided into 80 images for each class. The "normal" class consists of apples with no damage. Three classes are for diseases. An apple with "scab" is gray or brown, "rot" has sunken brown spots, and "blotch" has irregular patterns. A sample of the data set is shown in Fig. 2.

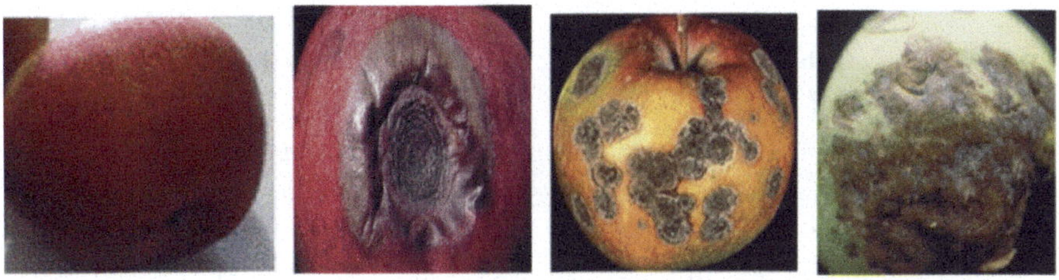

a) Normal b) rot c) scab d) blotch

Fig. 2. Sample dataset with representative of each class

Figure 3 shows how an image is clustered by k-mean clustering and k equals 4. Figure 4 shows result of image clustering of some arbitrary images.

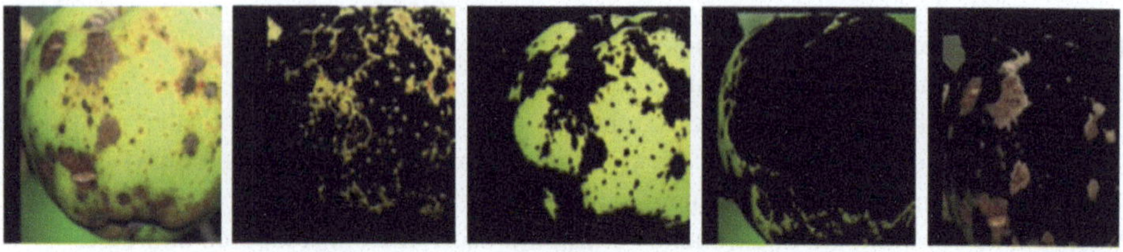

(a) Original (b) Iter-1 (c) Iter-2 (d) Iter-1 (e) Iter-4

Fig. 3. Clustering results

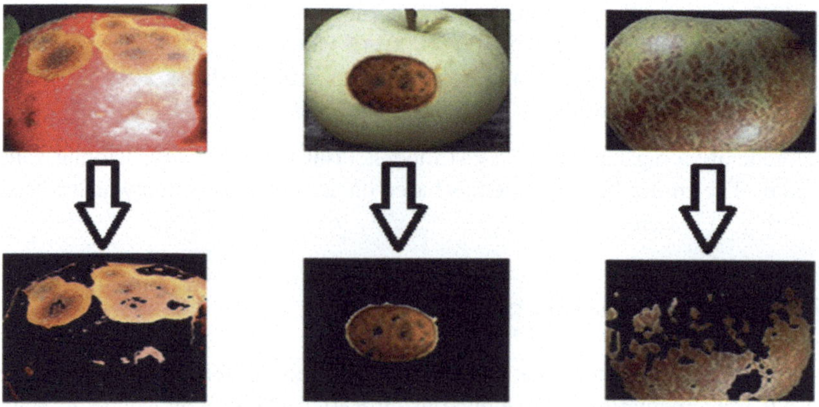

Fig. 4. Result of arbitrary images clustering

Table 2 shows accuracy for different dataset sizes at training. Note that the non-normal classes are intersecting as if these classes are overlapping.

Table 2. Training accuracy for different dataset sizes

Class	Size								
	10	15	20	25	30	35	40	45	50
Normal	94	94.5	95	95.5	98.5	98.5	98.5	99	100
Rot	75	78	80	84	85	88	91	90	89.5
Blotch	65	76.5	78	80	85	88	88	85	90
Scab	65	64	65.5	77.5	85	84	89	95	97

Testing results show 94.65% accuracy in detecting the normal apple. This result is better than previous work [11] which has accuracy 93%.

Non-normal classes have lower accuracy compared to normal class. For instance, rot class has an accuracy of 84.5% before optimization and 88.89% after optimization. Receiver operating characteristic (ROC) varies based on the testing data set. ROC for rot class against other classes is shown in Fig. 5.

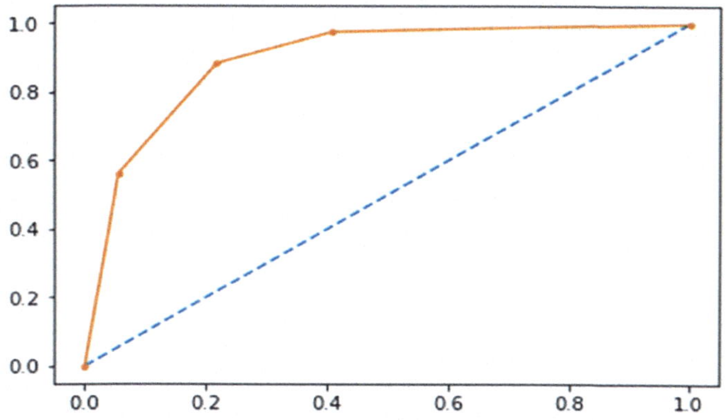

Fig. 5. ROC for "rot" class against other classes

Sensitivity and specifity of non-normal classes are shown in Table 3. Accuracy of non-normal classes both before and after optimization is shown in Table 4.

Table 3. Sensitivity and Specifity

Class	Measure	
	Sensitivity	Specifity
Rot	97	40
Blotch	89.84	57.1
Scab	92.1	42.86

Table 4. Testing accuracy before and after optimization

Class	Accuracy	
	Before	After
Rot	84.5	88.89
Blotch	81.7222	82.2
Scab	80.222	84.4

5 Conclusion and Future Directions

Classification is performed via support vector machine, with CSO fine-tuning. Testing results show 94.65% accuracy in detecting the normal apple. This gives a room for future enhancement. Another possible future direction is trying deep learning models such as Deep Belief Network and Auto-encoder. Another possible direction is to try other meta-heuristics other than CSO.

References

1. Jain, A.K.: Data clustering: 50 years beyond K-means. Pattern Recogn. Lett. **31**(8), 651–666 (2010)
2. Chang, C.-C., Lin, C.-J.: LIBSVM: a library for support vector machines. ACM Trans. Intell. Syst. Technol. (TIST) **2**(3), 27 (2011)
3. Zhaohui, C., Haiyan, T.: Cockroach swarm optimization. In: 2010 2nd International Conference on Computer Engineering and Technology (2010)
4. Mehl, P.M., Chao, K., Kim, M., Chen, Y.R.: Detection of defects on selected apple cultivars using hyperspectral and multispectral image analysis. Appl. Eng. Agric. **18**, 219–226 (2002)
5. Kim, M.S., Lefcourt, A.M., Chen, Y.R., Tao, Y.: Automated detection of fecal contamination of apples based on multispectral fluorescence image fusion. J. Food Eng. **71**, 85–91 (2005)
6. Leemans, V., Magein, H., Destain, M.F.: Defect segmentation on 'jonagold' apples using colour vision and a bayesian classification method. Comput. Electron. Agric. **23**, 43–53 (1999)
7. Kleynen, O., Leemans, V., Destain, M.F.: Development of a multi-spectral vision system for the detection of defects on apples. J. Food Eng. **69**, 41–49 (2005)
8. Ojala, T., Pietikäinen, M., Mäenpää, T.T.: Multi-resolution gray-scale and rotation invariant texture classification with local binary pattern. IEEE Trans. Pattern Anal. Mach. Intell. **24**(7), 971–987 (2002)
9. Ojala, T., Mäenpää, T., Pietikäinen, M., Viertola, J., Kyllönen, J., Huovinen, S.: Outex – new framework for empirical evaluation of texture analysis algorithm. In: Proceedings of International Conference on Pattern Recognition, pp. 701–706 (2002)
10. Guo, Z., Zhang, L., Zhang, D.: A completed modeling of local binary pattern operator for texture classification. IEEE Trans. Image Process. **19**(6), 1657–1663 (2010)
11. Dubey, S.R., Jalal, A.S.: Detection and classification of apple fruit diseases using complete local binary patterns. In: 2012 Third International Conference on Computer and Communication Technology, pp. 346–351. IEEE (2012)
12. Dubey, S.R., Jalal, A.S.: Robust approach for fruit and vegetable classification. Proc. Eng. **1**(38), 3449–3453 (2012)
13. Kus, Z.A., et al.: Estimation of the colour properties of apples varieties using neural network. Erwerbs-Obstbau **59**(4), 291–299 (2017)
14. Hameed, K., Chai, D., Rassau, A.: A comprehensive review of fruit and vegetable classification techniques. Image Vis. Comput. **1**(80), 24–44 (2018)
15. Rocha, A., Hauagge, D.C., Wainer, J., Goldenstein, S.: Automatic fruit and vegetable classification from images. Comput. Electron. Agric. **70**(1), 96–104 (2010)

16. Zhu, M., Hahn, A., Wen, Y.Q., Bolles, A.: Identification-based simplified model of large container ships using support vector machines and artificial bee colony algorithm. Appl. Ocean Res. **1**(68), 249–261 (2017)
17. Bhargava, A., Bansal, A.: Fruits and vegetables quality evaluation using computer vision: a review. J. King Saud Univ.-Comput. Inform. Sci. (2018)
18. Bell, W.J., Roth, L.M., Nalepa, C.A.: Cockroaches: Ecology, Behavior, and Natural History. Johns Hopkins University Press, Baltimore (2007)

Controlling Directed Particle Swarm Optimization for Delivering Nano-robots to Cancer Cells

Doaa Ezzat(✉), Safaa Amin, Howida A. Shedeed, and Mohamed F. Tolba

Scientific Computing Department,
Faculty of Computer and Information Sciences,
Ain Shams University, Cairo, Egypt
doaa.ezzat@gmail.com

Abstract. Cancer is one of the most dangerous diseases in this century. There are two traditional methods for treating cancer (radio-therapy and chemo-therapy). Using these methods causes very harmful side effects on healthy organs. Because of these side effects, doctors sometimes need to decrease the drug dose or delay the therapy. Some researchers proposed to use Nano-robots to deliver anti-cancer drugs to only cancer cells without touching the healthy tissues. One of the most recently proposed algorithms for delivering Nano-robots to cancer area is Directed Particle Swarm Optimization (DPSO) [1]. This algorithm can efficiently deliver all Nano-robots to the target area in a very short time. In this paper, we deeply discuss this algorithm and we also propose a new method for controlling this algorithm by tuning its direct step.

Keywords: Nano-robots · Cancer treatment · DPSO · Drug delivery · Direct step

1 Introduction

According to recent statistics, it was found that the number of cancer deaths largely increased in the last decades. Scientists expected that this number will be 13.1 million cancer deaths in 2030 [2]. Unfortunately, there is no completely safe therapy for treating this disease. The side effects caused by the traditional methods are very harmful on the uninfected tissues.

To limit these side effects, it was proposed to use a swarm of Nano-robots to attack tumor cells at the cellular level [3]. A Nano-robot is a programmable robot with Nano-scale. Its size can be from 0.1 to 100 nm (a Nano-meter = 10^{-9} m). A Nano-robot usually contains sensors, actuators, a power supply, a data transmission module and an on-board computer. The treatment is performed by injecting the Nano-robots in the blood vessel nearby the tumor region. Then, they try to travel through this vessel until they find tumor cells. After reaching the target area, they start to release the drug in a certain rate to destroy only the cancer cells that can be recognized by their pH value.

A.-E. Hassanien et al. (Eds.): AICV 2020, AISC 1153, pp. 148–158, 2020.
https://doi.org/10.1007/978-3-030-44289-7_15

During their journey inside the blood vessel, Nano-robots need to control and optimize their movement using a suitable algorithm like a swarm intelligence algorithm.

Particle Swarm Optimization (PSO) is an optimization algorithm inspired from the social behaviour of birds and fish. It was firstly proposed by Eberhart and Kennedy [4]. They proposed to use a swarm of particles moving in the search space. These particles try to reach the optimal solution. Each particle has a position and a velocity and represents one solution in the space. Each particle updates its position using the best position encountered by itself (*pbest*) and the best position encountered by all particles in the swarm (*gbest*).

PSO has many benefits like simplicity, insensitivity to scaling, efficiency and the ability to be parallelized. So, PSO is a very appropriate solution for the problem of Nano-robots delivery to the tumor area [5–7]. Researchers made a lot of modifications in PSO in order to improve its capability to solve this problem and many other problems [1]. However, a lot of research is needed to speed up the process of delivering Nano-robots to cancer area and to make sure that the whole swarm of Nano-robots is successfully delivered to this area. The speed of Nano-robots delivery is very significant especially because the Nano-robot has very limited power and this increases the importance of finding the shortest path. Also, if some of the Nano-robots get lost on their way to the target area, the drug dose will be not sufficient to destroy the complete tumor region. This may cause the remaining tumor cells, that can't be destroyed, to reproduce new infected cells; thus the treatment process fails.

To overcome these problems, a modification in PSO was proposed. This modification reduces the number of iterations needed for delivering Nano-robots to cancer area. This modified algorithm is called Directed PSO (DPSO) which can deliver the whole swarm of Nano-robots to the target area after a very few iterations [1]. This paper discusses DPSO in more details and proposes a new solution for controlling the direct step used in this algorithm. This direct step needs to be carefully tuned to improve the efficiency of DPSO.

2 Related Work

In 1961, Feynman was the first scientist that introduced the idea of nanotechnology [8]. In 1980's, Drexler began to discuss the possibility of manufacturing engineering systems that work at the molecular level [9]. Holland and Melhuish [10] investigated the ability of programmable agents with limited sensing, mobility and computation to perform a task. This task was to move toward a light source using simple rule-based algorithms.

From 1990's until now, a lot of work has been made in the field of using Nano-robots in detecting and treating diseases [11–17]. Cavalcanti et al. proposed many architectures for Nano-robots used in medical purposes [12–16, 18]. They introduced new hardware architecture for Nano-robots used in diagnosing brain aneurysm [12] and diabetes control [13]. They also presented a simulation of Nano-robots that have sensors for identifying medical target [14]. These Nano-robots have the ability to move with six-degree-of-freedom, and each Nano-robot contains a sensor for measuring chemical concentration without any communications between the Nano-robots.

Lewis and Bekey [19] applied swarm intelligence algorithms and chemical signalling techniques to Nano-robots for solving the problem of destroying tumor cells. Chandrasekaran and Hougen [20] proposed new control strategies that used swarm intelligence algorithms in medical Nano-robots. In these control strategies, each Nano-robot can measure the chemical signal, make a decision for the next move, and at last reach the target. This target is supposed to have the highest or the lowest chemical concentration. Also, in these control strategies, there was no communication between Nano-robots.

In 2013, Zhang et al. presented a cooperative control technique for Nano-robots that can measure the pH value by their sensors to deliver drugs to the cancer cells [21]. They introduced a distributed control algorithm where cooperation is done through robot-to-robot communications. In their control algorithm, the Nano-robots can sense and trace the change of pH values, until they reach the tumor centre that has the smallest pH value.

In 2014, Ahmed et al. proposed a novel strategy for Nano-robots to deliver drugs to the tumor cells [5]. They presented a modification in the Particle Swarm Optimization (PSO) algorithm to deliver the Nano-robots to the tumor area. Their algorithm was called Modified PSO (MPSO). They also introduced a modification in the obstacle avoidance algorithm for avoiding collision with blood cells. Then they combined this Modified Obstacle Avoidance (MOA) algorithm with the MPSO algorithm to control the movement of the Nano-robots inside the blood stream. In 2016, researchers improved the Bacterial Foraging Optimization Algorithm (BFOA) to find and destroy cancer cells through cooperation between Nano-robots [6]. In 2017, a new bio-detection framework called touchable computing was proposed for detecting cancer cells [7]. In the same year, Lambe used Q-learning algorithm for minimizing the path travelled by the Nano-robots to reach the target area [22].

In 2018, a modification was introduced in PSO to speed up the process of Nano-robots delivery to cancer area. This modification was called Directed Particle Swarm Optimization (DPSO) algorithm [1]. It can deliver a group of Nano-robots to the cancer area after very small number of iterations. A comparison was made between DPSO and four different optimization algorithms and the experiments proved that DPSO outperforms all these algorithms in reducing the number of iterations needed for delivering Nano-robots to the target [1]. In 2019, DPSO was combined with a quorum sensing algorithm to kill cancer cells and the experiments showed that DPSO can also reduce the number of iterations needed to destroy cancer cells [23].

3 Directed Particle Swarm Optimization Algorithm (DPSO)

When applying PSO to the problem of finding the cancer area using a swarm of Nano-robots, each Nano-robot in this swarm represents a solution in the solution space [1]. The fitness value of a solution represented by a Nano-robot at some position is the pH value measured by this Nano-robot at this position. Researchers found that the pH value equals 7.4 at the normal cells and this value continuously decreases until the Nano-robot enters the cancer area. In this area, they found that the pH value is less than or equals to 5.7 [21]. If the classical PSO is applied to our problem, then it will be as follows [4]:

1. Initialize Nano-robots with random positions and random velocities
2. Evaluate the pH value (the fitness value) sensed by each Nano-robot
3. For each Nano-robot, if its current pH value is smaller than the pH value sensed at its *pbest* position, then: *pbest* = its current position
4. Update the gbest to be the best position for the whole swarm
5. Update the velocity *V* of each Nano-robot by this equation:

$$V_i^{t+1} = V_i^t + C_1 R_1 \left(pbest_i^t - P_i^t\right) + C_2 R_2 \left(gbest^t - P_i^t\right) \tag{1}$$

6. Update the position *P* of each Nano-robot by this equation:

$$P_i^{t+1} = P_i^t + V_i^{t+1} \tag{2}$$

7. Re-evaluate the fitness values
8. If the target is reached, then terminate
9. Otherwise, go to step 3

Where P_i^t is the position of Nano-robot i at iteration t, V_i^t is the velocity of Nano-robot i at iteration t, C_1 and C_2 are constants indicate the importance of *pbest* and *gbest* respectively, R_1 and R_2 are random values.

In DPSO, Nano-robots use this classical PSO (discussed above) to update their velocities and positions until one of these Nano-robots reaches the target. Then all other Nano-robots in the swarm direct themselves toward the position of the one that reached the target. Then these Nano-robots move in this new direction with direct steps until all of them are delivered to the target [1]. Figure 1 shows the details of making a direct step.

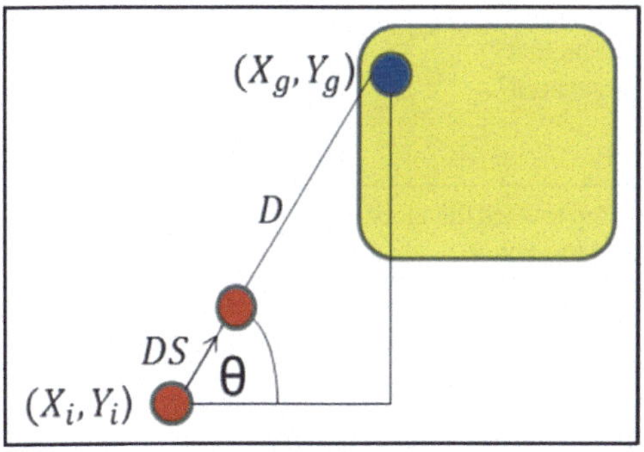

Fig. 1. Details of direct step

To make a direct step, let (X_i, Y_i) be the position of a Nano-robot *i* and (X_g, Y_g) the position of the global best Nano-robot that has already reached the target area. Then we can compute the angle θ between the Nano-robot *i* and the global best Nano-robot as follows:

$$\theta = \tan^{-1}\frac{(Y_g - Y_i)}{(X_g - X_i)} \tag{3}$$

Then for the iteration $t + 1$, the Nano-robot *i* updates its position to make a direct step of size equals *DS* using the following formulas:

$$X_i^{t+1} = X_i^t + \mathrm{DS}\ \cos\theta \tag{4}$$

$$Y_i^{t+1} = Y_i^t + \mathrm{DS}\ \sin\theta \tag{5}$$

The concept of the DPSO basically comes from the fact that the solution space has one target. This fact can be exploited when at least one Nano-robot is delivered to this target; this Nano-robot drags all other Nano-robots in the swarm to its position. In this way, the number of iterations needed to deliver all Nano-robots to cancer area can be considerably decreased. In other words, only one or two PSO iterations are needed for delivering at least one Nano-robot to the target area. Then some few direct steps are needed to deliver the rest of the swarm to this area [1]. Figure 2 shows the main steps of DPSO.

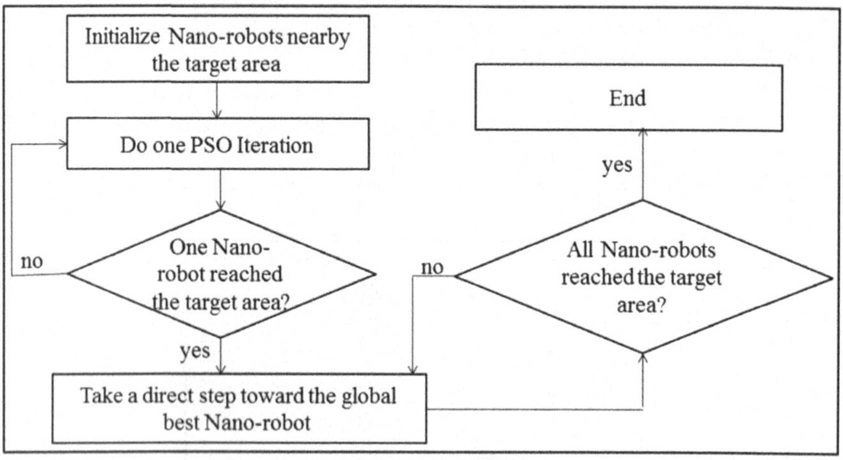

Fig. 2. Main steps of DPSO algorithm

Simulation experiments have shown that, in some cases, some Nano-robots get lost when using PSO and some other optimization algorithms. DPSO outperforms these algorithms in this aspect because it can guarantee the reachability of the target by all Nano-robots in the swarm. We will explain later in this paper the results of our experiments in more details.

4 Controlling the Direct Step

The speed efficiency of DPSO is affected by the size of the Direct Step *DS*. So, *DS* should be chosen carefully, because if this step size becomes too large, then the Nano-robot will oscillate around the target. And if this step size becomes too small, then the algorithm will be very slow.

If the distance between a Nano-robot *i* and the global best Nano-robot is *D* and *DS* is so larger than *D* (as shown in Fig. 3).Then the Nano-robot *i* will direct himself toward the global best Nano-robot and take a *DS* in this direction so it may go to the other side of the target area. Then it will turn himself to the global best Nano-robot and take another *DS* to return back to the original side of the target area. This may cause the oscillation of the Nano-robot around the target and therefore the Nano-robot *i* will take more steps until reaching the target area. So *DS* should be smaller than *D*. If *DS* is so smaller than *D* (as shown in Fig. 4), then the Nano-robot *i* will take a lot of steps until reaching the target area. So in this case, *DS* should be increased by a reasonable value.

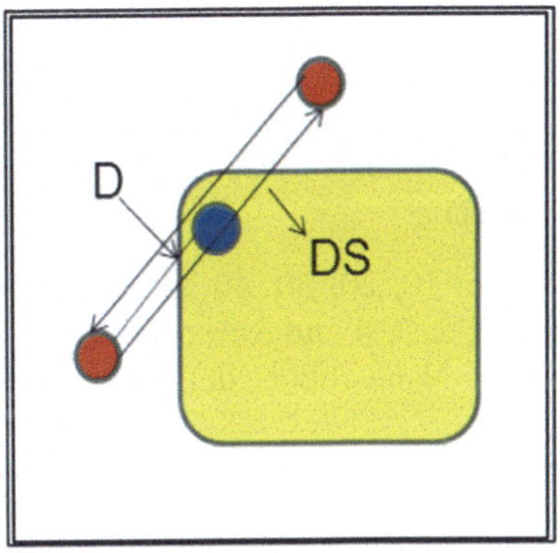

Fig. 3. DS is so larger than D

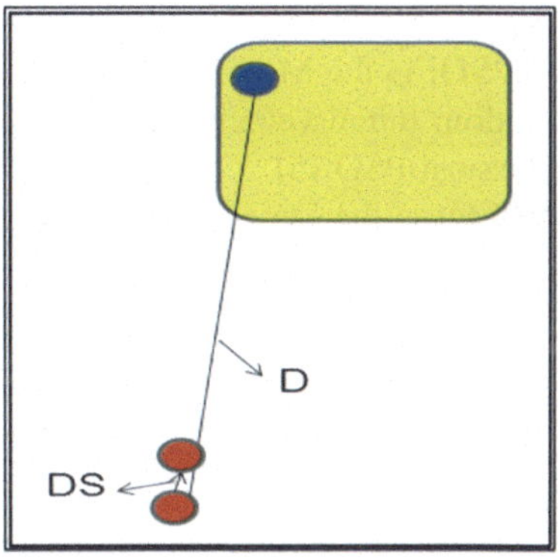

Fig. 4. DS is so smaller than D

To avoid these problems, we propose to determine the size of the direct step *DS* as follows: Firstly, compute *D* that is the distance between a Nano-robot *i* in the swarm and the global best Nano-robot using the following equation:

$$D = \sqrt{(X_g - X_i)^2 + (Y_g - Y_i)^2} \quad (6)$$

Then the size of the direct step *DS* can be controlled based on fuzzy logic as follows:

- If *D* is small then let *i* make a small *DS*.
- Else If *D* is medium then let *i* make a medium *DS*.
- Else If *D* is large then let *i* make a large.

In this way we can guarantee that *DS* is related to *D*. Thus, *DS* can't be too large or too small.

5 Simulation and Results

Our simulator was implemented using visual C++ to simulate the movement of the Nano-robots after locating them randomly nearby the tumor area. The simulation was applied using four different algorithms. To be unbiased, the same starting positions of Nano-robots were used for the four algorithms in each run. The four algorithms used in the simulation are as follows:

1. **PSO:** is the classical PSO algorithm (mentioned earlier)
2. **BFOA:** is the Bacterial Foraging Optimization Algorithm, which simulates the movement of bacteria. In this algorithm, each Nano-robot moves a unit step in a random direction, then it makes another unit step in the same direction if the fitness value is increased; otherwise it will take a unit step toward another random direction [6]
3. **MPSO:** is the Modified PSO algorithm, in which each Nano-robot moves with its random initial velocity for some period. After this period, MPSO behaves like the classical PSO [5]
4. **DPSO:** is the Directed PSO (discussed in this paper)

We conducted 10 experiments for each setup (using 10, 20, 30, 40 and 50 Nano-robots), and the averaged results are shown in Tables 1, 2, 3, 4 and 5 respectively. For each algorithm, these five tables show the number of Nano-robots delivered to the target area after 10, 20 and 30 iterations.

Table 1. The average number of nano-robots delivered using 10 nano-robots.

Number of iterations	PSO	BFOA	MPSO	DPSO
10	7	2	7	9
20	9	3	9	10
30	9	4	9	10

Table 2. The average number of nano-robots delivered using 20 nano-robots.

Number of iterations	PSO	BFOA	MPSO	DPSO
10	12	3	13	19
20	16	7	17	20
30	17	10	18	20

Table 3. The average number of nano-robots delivered using 30 nano-robots.

Number of iterations	PSO	BFOA	MPSO	DPSO
10	20	6	20	29
20	24	12	25	30
30	26	17	26	30

Table 4. The average number of nano-robots delivered using 40 nano-robots.

Number of iterations	PSO	BFOA	MPSO	DPSO
10	23	5	19	37
20	31	14	30	40
30	33	21	34	40

Table 5. The average number of nano-robots delivered using 50 nano-robots.

Number of iterations	PSO	BFOA	MPSO	DPSO
10	30	5	22	46
20	38	18	35	50
30	42	25	39	50

From these results, it can be concluded that each algorithm has a certain delivery rate for a certain range of Nano-robots. Table 6 states the percentage of the Nano-robots delivered to the target area after the first 10 iterations for each algorithm.

Table 6. The percentage of nano-robots delivered after the first 10 iterations

Number of nano-robots	PSO	BFOA	MPSO	DPSO
From 10 to 30	60%–70%	15%–20%	65%–70%	94%
From 40 to 50	58%–60%	10%–13%	44%–48%	92%

So, we can conclude that the efficiency of the four algorithms is affected (reduced) by increasing the number of Nano-robots but this effect (reduction) is very small for DPSO if it is compared with the other three algorithms. This proves the stability of DPSO.

In addition to this, the average number of iterations, at which all Nano-robots were delivered to the target using DPSO, is: 10 for 10 Nano-robots, 11 for 20 Nano-robots and 12 for 30 Nano-robots. In other words, we need only one extra iteration for delivering 10 extra Nano-robots to the target. But when using more than 30 Nano-robots (40 or 50), DPSO needs only 12 iterations to deliver the whole swarm to the target area.

The experiments showed that, some Nano-robots may be lost and can't be delivered to the target during the first 100 iterations. This phenomenon was observed in the experiments of PSO, BFOA and MPSO, but it was not observed in DPSO experiments. Table 7 compares the four algorithms on the basis of number of experiments conducted without loss in Nano-robots, using 10, 20, 30, 40 and 50 Nano-robots. In this table we assume that, if a Nano-robot can't reach the target before the completion of the first 100 iterations, then we will consider this Nano-robot is lost.

Table 7. Number of experiments without loss in nano-robots

Number of nano-robots	PSO	BFOA	MPSO	DPSO
10	7	0	8	10
20	3	0	8	10
30	2	0	4	10
40	0	0	0	10
50	2	0	0	10

In Table 7, we can notice that:

1. For PSO and MPSO, when using larger number of Nano-robots, the capability of conserving all Nano-robots is usually reduced
2. BFOA is the worst algorithm because all BFOA experiments have loss in Nano-robots
3. DPSO is the best algorithm because none of the Nano-robots was lost in all DPSO experiments

To prove the efficiency of DPSO with the controlled Direct Step (DS), we conducted 25 experiments using 10, 20, 30, 40 and 50 Nano-robots (5 experiments for each setup). These experiments were repeated using a small DS, a large DS and a controlled DS. The average number of iterations needed to deliver the whole swarm to the target area is computed in each setup. Table 8 describes the results of these experiments.

Table 8. The average number of iterations needed to deliver the whole swarm using DPSO with different sizes of DS.

Number of nano-robots	With small DS	With large DS	With controlled DS
10	37	14	11
20	36	17	12
30	39	21	11
40	45	19	12
50	46	21	12

This Table shows that using DPSO with a controlled DS always makes the algorithm faster than when using DPSO with a small or a large DS. When controlling DS, whatever the size of the Nano-robots swarm, DPSO needs only 11 or 12 iterations to deliver the whole swarm to the cancer area.

6 Conclusions

Cancer treatment is so challenging because chemical drugs should be delivered to cancer cells without affecting the healthy ones. Using Nano-robots is an excellent way to achieve this goal. It was proposed in many algorithms to deliver drugs to cancer cells using Nano-robots. This paper gives more details about a recent technique called Directed Particle Swarm Optimization (DPSO) that efficiently delivers a swarm of Nano-robots to the tumor area. It was found experimentally that DPSO is faster than PSO, BFOA and MPSO in Nano-robots delivery to the target area. Also, DPSO can guarantee the reachability of the target by all Nano-robots in the swarm. Finally, the size of the direct step used in DPSO should be chosen carefully. So, this paper also proposes a novel method to control the size of this direct step to improve the efficiency of DPSO.

References

1. Ezzat, D., Amin, S., Shedeed, H.A., Tolba, M.F.: Directed particle swarm optimization technique for delivering nano-robots to cancer cells. In: Proceedings of IEEE International Conference on Computer Engineering and Systems (ICCES 2018), Cairo, Egypt, pp. 80–84 (2018)
2. Devasena Umai, R., Brindha Devi, P., Thiruchelvi, R.: A review on DNA nanobots - a new technique for cancer treatment. Asian J. Pharm. Clin. Res. **11**(6), 61–64 (2018)
3. Venkatesan, M., Jolad, B.: Nanorobots in cancer treatment. In: International Conference on Emerging Trends in Robotics and Communication Technologies (INTERACT), pp. 258–264. IEEE (2010)
4. Eberhart, R., Kennedy, J.: A new optimizer using particle swarm theory. In: Proceedings of the Sixth International Symposium on Micro machine and Human Science, pp. 39–43 (1995)
5. Ahmed, S.: Nano-robotics control for biomedical applications (unpublished doctoral dissertation). Ain Shams University, Cairo, Egypt (2014)
6. Cao, J., Li, M., Wang, H., Huang, L., Zhao, Y.: An improved bacterial foraging algorithm with cooperative learning for eradicating cancer cells using nanorobots. In: Proceedings of the IEEE International Conference on Robotics and Biomimetics, Qingdao, China (2016)
7. Chen, Y., Shi, S., Yao, X., Nakano, T.: Touchable computing: computing-inspired bio-detection. IEEE Trans. Nanobiosci. **16**, 810–821 (2017)
8. Feynman, R.P.: There's plenty of room at the bottom: an invitation to enter a new field of physics. In: Gilbert, H.D. (ed.) Miniaturization, Reinhold (1961)
9. Drexler, K.E.: Engines of Creation: The Coming Era of Nanotechnology. Anchor Press, Norwell (1986)
10. Holland, O.E., Melhuish, C.R.: Getting the most from the least lessons for the nanoscale from minimal mobile agents. In: Proceedings of the Fifth International Workshop on Artificial Life, pp. 175–214 (1996)

11. Ummat, A., Sharma, G., Mavroidis, C., Dubey, A.: Biomedical Engineering Handbook. State of the Art and Future Challenges. CRC Press, London, Chapter Bio-Nanorobotics (2005)
12. Cavalcanti, A., Shirinzadeh, B., Fukuda, T., Ikeda, S.: Nanorobot for brain aneurysm. Int. J. Robot. Res. **28**, 558–570 (2009)
13. Cavalcanti, A., Shirinzadeh, B., Kretly, L.: Medical nanorobotics for diabetes control. Nanomedicine **4**, 127–138 (2008)
14. Cavalcanti, A., Shirinzadeh, B., Freitas, R., Hogg, T.: Nanorobot architecture for medical target identification. Nanotechnol **19**, 1–15 (2008)
15. Cavalcanti, A., Freitas, R.: Nanorobotics control design: a collective behavior approach for medicine. IEEE Trans. Nanobiosci. **4**, 133–140 (2005)
16. Cavalcanti, A.: Assembly automation with evolutionary nanorobots and sensor-based control applied to nanomedicine. IEEE Trans. Nanotechnol. **2**, 82–87 (2003)
17. Freitas, R.: What is nanomedicine? Nanomed **1**, 2–9 (2005)
18. Cavalcanti, A., Freitas, R.: Autonomous multi-robot sensor-based cooperation for nanomedicine. Int. J. Nonlinear Sci. Numer. Simul. **3**, 743–746 (2002)
19. Lewis, M., Bekey, G.: The behavioral self-organization of nanorobots using local rules. In: Proceedings of IEEE International Conference on Intelligent Robots and Systems, pp. 1333–1338 (1992)
20. Chandrasekaran, S., Hougen, D.: Swarm intelligence for cooperation of bio-nano robots using quorum sensing. In: Bio Micro and Nanosystems Conference, San Francisco, p. 104 (2006)
21. Zhang, S., Li, S., Guo, Y.: Cooperative control design for nano-robots in drug delivery. In: Guo, Y. (ed.) Selected Topics in Micro/Nano-Robotics for Biomedical Applications, pp. 101–123. Springer, New York (2013)
22. Lambe, A.: Reinforcement learning for optimal path length of nanobots using dynamic programming. In: Proceedings of IEEE International Conference on Industrial and Information Systems, Peradeniya, Sri Lanka (2017)
23. Ezzat, D., Amin, S., Shedeed, H.A., Tolba, M.F.: A new nano-robots control strategy for killing cancer cells using quorum sensing technique and directed particle swarm optimization algorithm. In: The International Conference on Advanced Machine Learning Technologies and Applications (AMLTA 2019), pp. 218–226. Springer, Cairo (2019)

Feature Selection Method Based on Chaotic Maps and Butterfly Optimization Algorithm

Asmaa Ahmed Awad[1], Ahmed Fouad Ali[1], and Tarek Gaber[1,2,3(✉)]

[1] Department of Computer Science, Faculty of Computers and Informatics, Suez Canal University, Ismailia, Egypt

[2] School of Science, Engineering and Environment, University of Salford, Salford, UK
t.m.a.gaber@salford.ac.uk

[3] Scientific Research Group in Egypt, (SRGE), Cairo, Egypt

Abstract. Feature selection (FS) is a challenging problem that attracted the attention of many researchers. FS can be considered as an NP hard problem, If dataset contains N features then 2^N solutions are generated with each additional feature, the complexity doubles. To solve this problem, we reduce the dimensionality of the feature by extracting the most important features. In this paper we integrate the chaotic maps in the standard butterfly optimization algorithm to increase the diversity and avoid trapping in local minima in this algorithm. The proposed algorithm is called Chaotic Butterfly Optimization Algorithm (CBOA).The performance of the proposed CBOA is investigated by applying it on 16 benchmark datasets and comparing it against six meta-heuristics algorithms. The results show that invoking the chaotic maps in the standard BOA can improve its performance with accuracy more than 95%.

Keywords: Butterfly optimization algorithm · Chaotic maps · Feature selection · Dimensionality reduction

1 Introduction

Nowadays, Dimensionality reduction becomes most common pre-processing step to prepare the dataset for machine learning algorithms. Dimensionality reduction helps machine learning and data mining algorithms to be more faster and efficient. Because of high-dimensional data and large number of features, the construction of a suitable machine learning model can be extremely demanding and often almost fruitless [4]. To reduce high dimensionality of data there are two approaches. The first one is feature extraction, which includes creation a new subset of features from the original features with low dimensionality [18]. The second one is feature selection which is used as pre-processing step to eliminate the irrelevant, redundant and noisy data and find set of informative features [19].

Feature selection (FS) is a challenging problem that attracted the attention of many researchers, its research dates back to the 1960s [10]. FS is a process

A.-E. Hassanien et al. (Eds.): AICV 2020, AISC 1153, pp. 159–169, 2020.
https://doi.org/10.1007/978-3-030-44289-7_16

for finding set of M features from original set N where $M < N$ without loss any information lead to decrease performance of learning algorithm. If dataset contains N features then 2^N solutions are generated with each additional feature, the complexity doubles.

FS algorithms are classified into two main categories: filter-based algorithms and wrapper-based algorithms [11]. Filter-based algorithms depend on statistical method for calculation relation between features and finding optimal parameters, in another hand, Wrapper-based algorithms based on machine learning algorithms for finding near optimal features. Despite Wrapper-based algorithms are computationally expensive, they obtain better results than filter-based algorithms in selecting features as show in the results in [13]. Meta-heuristic algorithms act as a source of concepts, mechanisms and principles for designing of artificial computing systems to deal with complex computational problems [7]. They divided into four categories (e.g. evolutionary algorithms, Physics-based algorithms, Swarm-based algorithms, Human-based algorithms). Swarm intelligence (SI) is a collection of intelligent systems inspired by the collective intelligence of a group. This collective intelligence is achieved through the direct or indirect interactions of agents that are homogeneous in nature, yet co-operate with each other in their local environment without being aware of global context or pattern [5].

There are many SI algorithms have been applied to solve feature selection problem such as: Grasshopper optimization algorithm [12], salp swarm algorithm [8], Whale optimization algorithm [15] and Dragon optimization algorithm [13]. Many Researchers add new features to swarm algorithms to get better results in feature selection such as using chaotic maps with Crow search algorithm in [16], adding binary version to Butterfly algorithms to select optimal features and achieve maximum accuracy [3], using sigmoid and v-shaped functions to design binary Grasshopper optimization algorithms [12], introducing chaotic version of salp algorithms by using four different chaotic maps to control the balance between exploration and exploitation [1], considering various chaotic maps are considered in whale optimization algorithms for tunning it's main parameters [9], proposing binary version of grey wolf optimization algorithm for maximization accuracy [6], implementing chaotic version of particle swarm algorithm for feature selection [17].

Butterfly Optimization Algorithm (BOA) is novel natural inspired SI optimization algorithm that mimics the food foraging behavior of butterflies [2].BOA encounters two problems similar to other meta-heuristic algorithms; (1) entrapment in local optima and (2) slow convergence speed. In order to increase the diversity of the BOA and avoid trapping in local minima, we combine it with ten chaotic maps. The proposed algorithm is called Chaotic Butterfly Optimization Algorithm (CBOA). CBOA is tested on different dataset with different feature dimensions. CBOA is compared with the standard BOA and six meta-heurestics algorithms. The results show that CBOA outperforms the other algorithms and it can reduce the feature dimensionality with high accuracy.

The rest of the paper is organized as follow. An overview of the BOA and chaotic maps are given in Sect. 2. The proposed algorithm, CBOA is described in Sect. 3. The experimental results and discussion of the CBOA are reported in Sect. 4. The conclusion of the paper is shown in Sect. 5.

2 Background

In this section, we give an overview of the algorithms/techniques used in the proposed CBOA.

2.1 Butterfly Optimization Algorithm (BOA)

To explain the main idea of BOA algorithm [2], firstly we highlight the biological and natural behaviors of the butterfly insect. Then we describe the main steps of the algorithm and how it mimics the social life of the butterflies.

Biological and Natural Behaviors. Butterflies are insects belong to Lepidoptera. They have five senses (smell, sight, taste, touch and hearing). They used these senses for finding foods, searching for mating parcener, immigration from one place to another and escaping from enemies. Although these senses are very important for butterflies, smell sense considers the most important sense which help them for finding food. In the mating process, male butterfly can identify the female butterfly through her pheromone. When butterfly moves from one place to another, it generates a fragrance with intensity which is propagate over the distance. The other butterflies can sense this fragrance and attracted to the butterfly according to the intensity of its fragrance. When a butterfly senses the best butterfly's fragrance it moves toward it. This process is called global search, while when it fails to sense the fragrance of any butterfly, it moves randomly to a new position in the search space. This process is called local search [2].

Magnitude of Fragrance. Butterfly emits a fragrance with intensity when it moves. The other butterflies attracted to the butterfly according to its magnitude of fragrance. The fragrance of each butterfly can be defined as in Eq. (1).

$$pf_i = cI^a \tag{1}$$

where pf_i represents the perceived magnitude of fragrance, I is fragrance intensity. The parameters c, a are a power exponent which represents the degree of the fragrance absorbtion and the sensor modality, respectively.

Butterflies Movement. The movement of butterflies are based on three phases as follow.

- **Global search phase.** Each butterfly emits fragrance when it moves and the other butterflies attracted to it according to its magnitude of fragrance. This process is called a global search and can be defined as follow

$$\boldsymbol{x}_i^{t+1} = \boldsymbol{x}_i^t + (r^2 \times \boldsymbol{g}^* - \boldsymbol{x}_i^t) \times f_i \tag{2}$$

  Where $\boldsymbol{x}_i^t$ is a vector which represent the butterfly (solution) at iteration t, $\boldsymbol{g}^*$ is the overall best solution, r is a random number in $[0, 1]$ and f_i is a fragrance of $i_t h$ butterfly. In this phase, g initial value is the position of the minimum fitness of all solutions and it calculated by assigning a fitness value for each solution and define the minimum fitness then update the g (position) according to the minimum fitness. Also, r value is not used to calculate fitness, it is controlled by p (switch probability) value and the intial value of p = 0.8. The r value is compared with p value to control the butterfly while moving to the best solution with minimum filness in local search or global search.
- **Local search phase.** When the butterflies fail to sense the fragrance of the other butterflies, they move randomly in the search space. The process is called local search and it can be defined as follow.

$$\boldsymbol{x}_i^{t+1} = \boldsymbol{x}_i^t + (r^2 \times \boldsymbol{x_j}^t - \boldsymbol{x}_k^t) \times f_i \tag{3}$$

  Where $\boldsymbol{x}_j^t, \boldsymbol{x_k}^t$ are two vectors that represent two different butterflies in the same population.
- **Solution evaluation.** The fragrance intensity of the butterfly represents its objective function. The butterfly attracts the other butterflies according to its magnitude of fragrance.

The BOA Algorithm. In this subsection, we present the main steps of the BOA as follow.

- **Parameter setting.** At the beginning, we initialize the algorithm's parameter values such as the population size n, parameters a (power exponent), c sensory modality, switch probability ρ, and the maximum number of iterations Max_{itr}.
- **Iteration initialization.** Set the initial value of the iteration counter t.
- **Initial population.** The initial population n is generated randomly $\boldsymbol{x}_i^t$.
- **Solutions evaluation.** Each butterfly $\boldsymbol{x}_i^t$ in the population is evaluated by calculating its fitness function $f(\boldsymbol{x}_i^t)$.
- **Global best solution.** Assign the overall best butterfly (solution) $\boldsymbol{g}^*$ in the population.
- **The main loop.** The following steps are repeated until the termination criterion satisfied.
  - **iteration increasing.** The iteration counter is increasing, $t = t + 1$.
  - **Random number generation.** We generate random number r, where $r \in [0, 1]$.
  - **Balancing between global and local search processes.** The global and local search processes are applied according to the parameters value of ρ and r.

Algorithm 1. Butterfly optimization algorithm (BOA)

Set the initial values of the population size n (butterflies), parameters a (power exponent), c sensory modality, switch probability ρ, and the maximum number of iterations Max_{itr}.
Set $t := 0$. {Counter initialization}.
for $(i = 1 : i \leq n)$ **do**
  Generate an initial population (butterflies) $\boldsymbol{x}_i^t$ randomly.
  Evaluate the fitness function of each butterfly (solution) $f(\boldsymbol{x}_i^t)$.
  Calculate the fragrance for $\boldsymbol{x}_i^t$ as shown in Eq. 1.
  Assign the overall best butterfly (solution) $\boldsymbol{g}^*$.
end for
repeat
  Set $t = t + 1$.
  for $(i = 1 : i \leq n)$ **do**
    Generate random number r, $r \in [0, 1]$.
    if $(r < \rho)$ **then**
      Move butterflies towards the best butterfly $\boldsymbol{g}^*$ as shown in Eq. 2. {Global search}.
    else
      Move butterflies randomly as shown in Eq. 3. {Local search}.
    end if
    Evaluate the fitness function of each butterfly (solution) $f(\boldsymbol{x}_i^t)$.
    Assign the overall best solution $\boldsymbol{g}^*$.
  end for
  Update the value of parameters $c = [0.01, 0.25]$.
until $(t > Max_{itr})$. {Termination criteria satisfied}.
Produce the best solution $\boldsymbol{g}^*$.

  - **Global search process.** The butterflies update their position according to the position of the overall best solution $\boldsymbol{g}^*$ as shown in Eq. 2.
  - **Local search process.** If the butterflies fail to sense the fragrance of any butterfly in the population, they move randomly as shown in Eq. 3.
- **Solutions evaluation.** Each butterfly $\boldsymbol{x}_i^t$ in the population is evaluated by calculating its fitness function $f(\boldsymbol{x}_i^t)$.
- **Global best solution.** Assign the overall best butterfly (solution) in the population $\boldsymbol{g}^*$.
- **Termination criteria satisfied.** The overall processes are repeated until termination criteria satisfied, which is reaching to the maximum number of iterations Max_{itr} in our case.
- **Produce the best solution.** Produce the best obtained butterfly (solution) so far $\boldsymbol{g}^*$.

2.2 Chaotic Maps

We investigate the effect of integrating ten chaotic maps in the proposed CBOA to increase the random behavior (diversity) of it to arrive at a global minima and avoid getting stuck at a local minima which lead to faster convergence. The initial point for the ten chaotic maps is random number between [0, 1]. We used the same chaotic maps and their parameters which are reported in [14].

3 The Proposed Chaotic Butterfly Optimization Algorithm (CBOA)

The proposed CBOA algorithm depends on the integration of chaotic maps in the standard BOA. The main steps of the proposed CBOA are as follows.

1. invoking the chaotic maps for updating butterfly positions instead of using random variables so this would improving the accuracy of CBOA. The Eqs. 2 and 3 will be modified by replacing r^2 by C_j as following:

$$\boldsymbol{x}_i^{t+1} = \boldsymbol{x}_i^t + (C_j \times \boldsymbol{g}^* - \boldsymbol{x}_i^t) \times f_i \tag{4}$$

$$\boldsymbol{x}_i^{t+1} = \boldsymbol{x}_i^t + (C_j \times \boldsymbol{x_j}^t - \boldsymbol{x}_k^t) \times f_i \tag{5}$$

   where C_j is the chaotic map and $j = 1, 2, \ldots, 10$. Note that the C_j values are chaotic values, generated using 10 chaotic maps, are replaced with the r value to get better results in accuracy and minimum fitness than original algorithm use random values.
2. transferring continues CBOA to binary CBOA: apply the binary CBOA will represent the search space in binary values [0, 1], so the binary CBOA can adaptively search the feature space for best feature combination and is expected to be much simpler than continuous version.binary CBOA represented in the following equations:

$$x_i^{t+1} = \begin{cases} 1 & if(s(x_i^{t+1})) \geq rand() \\ 0 & otherwise \end{cases} \tag{6}$$

   where s is a transfer function, $rand()$ is a random number generated from uniform distribution [0,1] and x_i^{t+1} is the updated solution.

$$s(x_i^{t+1}) = \frac{1}{1 + \exp^{10(x_i^{t+1} - 0.5)}} \tag{7}$$

4 Results and Discussion

In this section, we present the parameter setting and the obtained results of the CBOA as shown in Table 1.

4.1 Parameter Setting

The initial values of c, a are obtained from [2]: a is a constant and c is in the range [0.01, 0.25].The other parameters values are chosen based on a number of experiments where their values were found to be the most efficient ones.

Table 1. Parameter setting

Parameters	Definitions	Values
p	Switch probability	0.8
a	Power exponent	0.1
c	Sensory modality (c)	Min = 0.01, Max = 0.25
n	Search Agents no	7
Max_{itr}	Maximum number of iteration	30
M	Number of runs	20
K	Cross validation	5

4.2 Fitness Function

Each solution is evaluated based on its fitness function, which employs the k-nearest neighbors (KNN) classifier as an evaluator and considers the number of selected features in the solution. The fitness function in Eq. 8 is used to minimize the classification error in our proposed CBOA as follow.

$$Fitness = Minimize(\alpha \ \ Errrate + \beta \ \frac{|n|}{|N|}) \tag{8}$$

where *Errrate* represents the classification error rate, $|n|$ and $|N|$ are the number of selected features and the number of original features in the dataset respectively, α and β are the weights of the classification error rate and selection ratio, $\alpha \in [0, 1]$ and $\beta = (1 - \alpha)$.

4.3 Dataset Description

Sixteen dataset from UCI machine learning repository are used to prove the efficiency of our proposed algorithm. The dataset is reported in [3]. Two datasets (Breast-cancer, Congress) contain missing values. In this work, all the missing values are replaced by the median values of all known values of a feature given class as calculated in [16].

4.4 Performance Analysis

In this subsection we analyze the performance of CBOA algorithm as follows.

The Effect of Integrating the Chaotic Maps in CBOA. To verify the efficiency of invoking the chaotic maps in the proposed CBOA. We present the convergence curves for eight random datasets to show the performance of the standard BOA and the other ten modified CBOA as shown in Fig. 1.

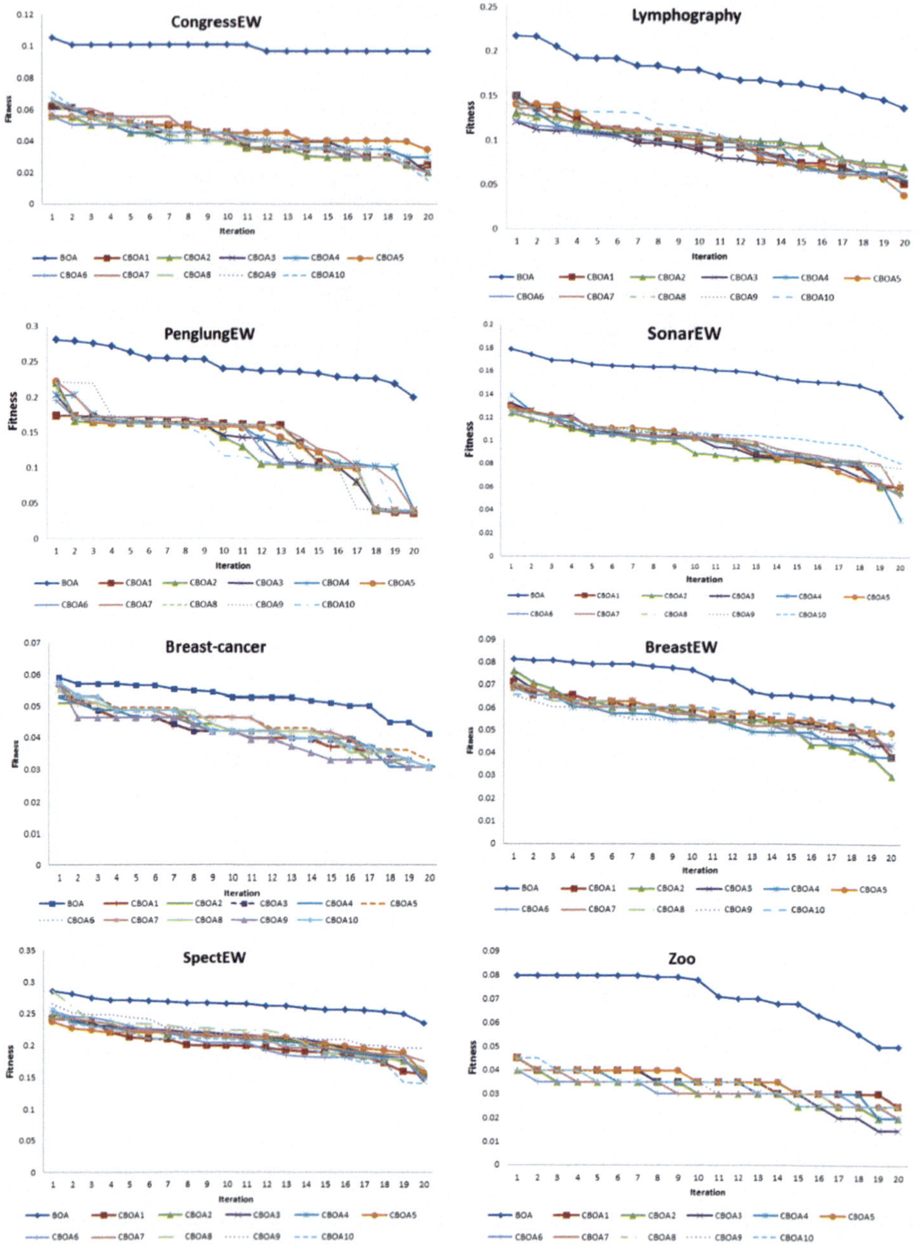

Fig. 1. The convergence curves of the standard BOA and the other modified CBOA

Comparison Between CBOA and the State-of-Art Algorithms. We investigate the performance of the proposed CBOA by comparing it against six algorithm. These algorithms are Antloin (ALO), brain storm optimizer (BSO), genetic algorithm (GA), greywolf (GWO), particle swarm optimization (PSO), binary butterfly optimizer (BBOA). The results of these algorithm are reported in [3]. The results in Table 2 show that CBOA achieves better results in maximize classification accuracy than the other six algorithms, while the results in Table 3 show that the mean fitness value for CBOA obtain the best results in most cases.

Table 2. Comparison between CBOA and other algorithms based on Average Accuracy:

Dataset	ALO	BSO	GA	GWO	PSO	BBOA	CBOA
Breast-cancer	0.959	0.920	0.959	0.960	0.960	0.969	0.997
Bresat EW	0.939	0.902	0.949	0.938	0.938	0.970	0.985
Clean1	0.847	0.826	0.870	0.858	0.855	0.883	0.955
Congress	0.937	0.855	0.941	0.933	0.924	0.959	0.997
HeartEW	0.777	0.695	0.787	0.777	0.779	0.824	0.879
IonosphereEW	0.860	0.854	0.894	0.868	0.880	0.907	0.977
KrVsKp	0.901	0.760	0.922	0.914	0.920	0.966	0.957
Lymphography	0.786	0.693	0.816	0.763	0.791	0.868	0.960
PenglungEW	0.807	0.768	0.672	0.834	0.814	0.878	0.912
Semeion	0.971	0.960	0.976	0.967	0.968	0.982	0.979
SonarEW	0.849	0.794	0.875	0.862	0.784	0.846	0.942
SpectEW	0.788	0.751	0.810	0.785	0.784	0.846	0.829
Tic-Tac-Toe	0.759	0.660	0.761	0.754	0.751	0.798	0.837
WaveformEW	0.707	0.615	0.692	0.709	0.719	0.743	0.803
WineEW	0.954	0.865	0.954	0.948	0.952	0.984	0.989
Zoo	0.922	0.813	0.929	0.953	0.945	0.978	1

From the results showed above, we can draw the following remarks. Firstly, it can be noticed that the results of BBOA are similar to CBOA ones. This is because some of dataset, such as breast EW, have large difference between its numerical values while other datasets have string values which give less fitness values than other datasets when using KNN classifier. Secondly, the results shows that the use of Chaotic maps increased the random behavior (diversity) that helped to reach to a global minima and avoid getting stuck at a local minima which lead to faster convergence. Thirdly, the results of the proposed method proved that irrelevant and redundant features are ignored while selecting the important features only so the dimension of dataset is reduced.

Table 3. Comparision between CBOA and other algorithms based on mean of fitness:

Dataset	ALO	BSO	GA	GWO	PSO	BBOA	CBOA
Breast-cancer	0.048	0.084	0.046	0.047	0.045	0.040	0.040
Breast EW	0.068	0.102	0.055	0.068	0.068	0.042	0.054
Clean1	0.160	0.177	0.134	0.147	0.150	0.113	0.082
Congress	0.069	0.149	0.063	0.073	0.082	0.045	0.039
HeartEW	0.228	0.307	0.216	0.228	0.226	0.180	0.150
IonosphereEW	0.145	0.149	0.109	0.136	0.124	0.096	0.059
KrvskpEW	0.108	0.242	0.083	0.094	0.086	0.054	0.511
Lymphography	0.219	0.309	0.187	0.241	0.214	0.139	0.087
PenglungEW	0.196	0.235	0.129	0.169	0.190	0.118	0.125
Semeion	0.035	0.044	0.029	0.040	0.039	0.026	0.126
SonarEW	0.158	0.209	0.128	0.143	0.138	0.086	0.086
SpectEW	0.216	0.252	0.192	0.219	0.219	0.160	0.199
Tic-Tac-Toe	0.249	0.342	0.243	0.252	0.253	0.205	0.223
WaveformEW	0.300	0.386	0.310	0.297	0.287	0.265	0.234
WineEW	0.054	0.139	0.051	0.060	0.055	0.023	0.045
Zoo	0.085	0.190	0.073	0.055	0.062	0.034	0.031

5 Conclusion

In this paper, the feature selection problem is discussed and how chaotic maps and bio-inspired optimization algorithms can be used to improve a feature selection algorithm. A new chaotic-based butterfly optimization algorithm, CBOA, is proposed. In this algorithm 10 chaotic maps are investigated to improve the performance of the standard butterfly optimization algorithm (BOA). Sixteen dataset from UCI machine learning repository are used to evaluate the proposed CBOA algorithm. Also its results are compared with the BOA and other optimization and feature selection algorithms. The results showed that the CBOA be used to maximize classification accuracy while minimizing the classification error. The results also showed that integrating chaotic maps into the standard BOA can reduce the dimension of a dataset.

References

1. Ahmed, S., Mafarja, M., Faris, H., Aljarah, I.: Feature selection using salp swarm algorithm with chaos. In: Proceedings of the 2nd International Conference on Intelligent Systems, Metaheuristics and Swarm Intelligence, pp. 65-69 (2018)
2. Arora, S., Singh, S.: Butterfly optimization algorithm: a novel approach for global optimization. Soft. Comput. **23**(3), 715–734 (2019)
3. Arora, S., Anand, P.: Binary butterfly optimization approaches for feature selection. Expert Syst. Appl. **116**, 147–160 (2019)

4. Brezocnik, L., Fister, I., Podgorelec, V.: Swarm intelligence algorithms for feature selection: a review. Appl. Sci. **8**(9), 1521 (2018)
5. Deshpande, A., Kumar, M.: Artificial Intelligence for Big Data: Complete Guide to Automating Big Data Solutions Using Artificial Intelligence Techniques. Packt Publishing Ltd., Birmingham (2018)
6. Emary, E., Zawbaa, H.M., Hassanien, A.E.: Binary grey wolf optimization approaches for feature selection. Neurocomputing **172**, 371–381 (2016)
7. Harifi, S., Khalilian, M., Mohammadzadeh, J., Ebrahimnejad, S.: Emperor Penguins colony: a new metaheuristic algorithm for optimization. Evol. Intel. **12**(2), 211–226 (2019)
8. Hegazy, A.E., Makhlouf, M.A., El-Tawel, G.S.: Improved salp swarm algorithm for feature selection. J. King Saud Univ.-Comput. Inform. Sci. (2018, in press)
9. Kaur, G., Arora, S.: Chaotic whale optimization algorithm. J. Comput. Design Eng. **5**(3), 275–284 (2018)
10. Li, Y., Li, T., Liu, H.: Recent advances in feature selection and its applications. Knowl. Inf. Syst. **53**(3), 551–577 (2017)
11. Liu, H., Yu, L.: Toward integrating feature selection algorithms for classification and clustering. IEEE Trans. Knowl. Data Eng. **4**, 491–502 (2005)
12. Mafarja, M., Aljarah, I., Faris, H., Hammouri, A.I., Ala'M, A.Z., Mirjalili, S.: Binary grasshopper optimisation algorithm approaches for feature selection problems. Expert Syst. Appl. **117**, 267–286 (2019)
13. Mafarja, M.M., Eleyan, D., Jaber, I., Hammouri, A., Mirjalili, S.: Binary dragonfly algorithm for feature selection. In: 2017 International Conference on New Trends in Computing Sciences (ICTCS), pp. 12–17 (2017)
14. Mirjalili, S., Gandomi, A.H.: Chaotic gravitational constants for the gravitational search algorithm. Appl. Soft Comput. **53**, 407–419 (2017)
15. Nematzadeh, H., Enayatifar, R., Mahmud, M., Akbari, E.: Frequency based feature selection method using whale algorithm. Genomics **111**, 1946–1955 (2019)
16. Sayed, G.I., Hassanien, A.E., Azar, A.T.: Feature selection via a novel chaotic crow search algorithm. Neural Comput. Appl. **31**(1), 171–188 (2019)
17. Xu, X., Rong, H., Trovati, M., Liptrott, M., Bessis, N.: CS-PSO: chaotic particle swarm optimization algorithm for solving combinatorial optimization problems. Soft. Comput. **22**(3), 783–795 (2018)
18. Tharwat, A., Gaber, T., Hassanien, A.E.: One-dimensional vs. two-dimensional based features: plant identification approach. J. Appl. Log. **24**, 15–31 (2017)
19. Tharwat, A., Gaber, T., Ibrahim, A., Hassanien, A.E.: Linear discriminant analysis: a detailed tutorial. AI Commun. **30**(2), 169–190 (2017)

Dynamic Cost Ant Colony Algorithm for Optimize Distributed Database Query

Sayed A. Mohsin[1(✉)], Saad M. Darwish[1(✉)], and Ahmed Younes[2(✉)]

[1] Institute of Graduate Studies and Research, Alexandria University, Alexandria, Egypt
sayed.abdelmohsin@gmail.com, saad.darwish@alexu.edu.eg

[2] Faculty of Science, Alexandria University, Alexandria, Egypt
ayounes@alexu.edu.eg

Abstract. Optimizing query in distributed database is considered as the most important part of a database system. The optimizer tries to find an optimal join order which reduces the query execution cost. Many factors may affect the execution cost of a query, including communication costs, resources, and access to large distributed data sets. When the number of relations and number of joins in a query increases, the complexity of the optimizer also increases. The success of query execution heavily influenced by the search method which is performed using the query optimizer. Processing of queries is considered as NP-hard problem and many researchers are focused on this problem in recent years. Researches are trying to build an appropriate algorithm to seek an optimal solution especially when the size of the database increases. In this paper, an ant colony algorithm as one of the hybrid strategy of evolutionary algorithms is utilized to find a solution for join query optimization problem in the distributed database systems. Unlike traditional ant colony-based query optimization techniques that based on static cost, the suggested model relies on dynamic cost which calculates the cost while the execution plan is built. Using this strategy, the algorithm aims to find an optimal join order which minimizes the total execution time. Experimental results show that the proposed model can handle different number of join entities. Also, the algorithm is affected by the number of ants used. Better results are obtained in case of large joined if the number of used ants increased.

Keywords: Distributed database system · Join query · Query optimization · Ant Colony Optimization

1 Introduction

Distributed database system is a collection of physically dispersed but logically centralized database system over a combination of computer network and Distributed Database Management System (DDBMS) [1]. A major problem with DDBMSs is the querying of data which is geographically distributed onto multiple sites [2]. The key success of the query processing tool mostly depends on their query optimization technologies [3, 4].

A.-E. Hassanien et al. (Eds.): AICV 2020, AISC 1153, pp. 170–181, 2020.
https://doi.org/10.1007/978-3-030-44289-7_17

Optimizing query in distributed database is considered as the most important part of a database system. The main task of the optimizer is to find a detailed plan, called a Query Execution Plan (QEP), for a user query. This plan explains to the executor how the query will be executed. When the query converted to a standard relational algebra form, the optimizer tries to find the best query execution plan [5]. Optimizing query in large-scale distributed databases is a combinatorial optimization problem and is classified as NP-hard problem which is difficult to solve by exploring all the query plans in this large search space [6, 7].

Many scalable methods have been developed to address the query optimization problem. System for Distributed Databases (SDD-1) was considered as one of the earliest algorithms used in query optimization in distributed database system. This algorithm built based on semi joins to reduce the communication cost on slow network by creating static unchangeable query plans without considering the vertical or horizontal data fragments of distributed database [8].

Recently, significant amount of work has been carried out on the way of generating optimal optimization for distributed database query. Randomized based models and algorithms have been applied successfully to many optimization problems. Simulated Annealing (SA), Particle Swarm Optimization (PSO), Iterative Improvement (II), and Genetic Algorithm (GA) have been suggested to optimize large scale recursive queries [9]. In [10] both SA and II applied on optimization of select-project-join queries.

Ant Colony Optimization Algorithm (ACO) is appropriate to use for query optimization in distributed database as a result of its features such as intelligent search mechanism, distributed computing, global optimization and ability to merge with other heuristics [3]. To solve the query optimization problem using ACO algorithm, the problem must be first represented as a graph G. The Objective of optimizer in this case is to find the Hamiltonian path for G which gives the minimum length. Query graph is represented as $G = (N, A)$ where N is the set of entities in the query graph and A is the set of relation between them. The join between entities represented as a connected edge on the graph G. No fixed cost exists over the edge, but it is dynamically calculated as an intermediate result while apply the join between quires. The main contribution of this paper is utilizing a dynamic version (dynamic cost) of ACO algorithm to tackle the problem of distributed join query optimization in a search space where entities are not replicated. The rest of this paper is organized as follows: Sect. 2 explains preliminaries that formally define the distributed database component and explains some general concepts of ACO algorithms which our proposed algorithm is based on. Section 3 covers related previous work. Section 4 introduces the proposed technique to optimize the distributed database query. Section 5 reports the accuracy and performance evaluation of the proposed technique and gives experimental results. Finally, Sect. 6 gives the conclusion and future work.

2 Preliminaries

In the relational database systems, the query optimization process consists of: (a) deciding the convenient join method for each join operator by considering the size of the relations, the physical structure of the data, and access paths, and (b) generating the joins order which is performed with respect to a cost model [10, 12]. The query optimizer that pursue this approach contains three components: A search space, a search strategy and a cost model.

Definition 1 Search Space: This refers to the generation of sets of alternative and equivalent QEPs of an input query by applying transformational rules such that they differ in the execution order of the operators. The QEPs are usually referred to as join trees whose operators are multiple types of joins or Cartesian products. Figure 1 shows two types of join tree, linear and bushy. The leaf nodes represent file materializations resulting from local processing and it is a reduced file. The root node represents the final step where the query result is obtained. Each parent node uses the result entities generated by its children as its inputs. In a database query which includes many relations and joins, the number of QEP grows exponentially as a result of the associative and commutative property of the join operation (O ($N!$) where N is the number of relations). Figure 2 shows the possible binary tree structures for a query involving 3 relations.

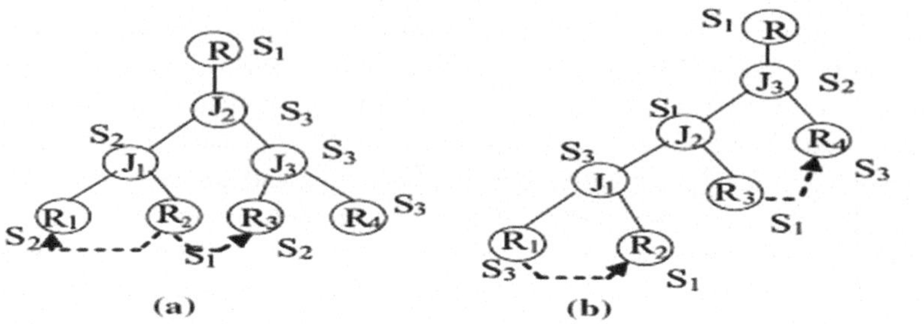

Fig. 1. Join tree structures for query with 3 joins: (a) bushy tree, (b) linear tree [21].

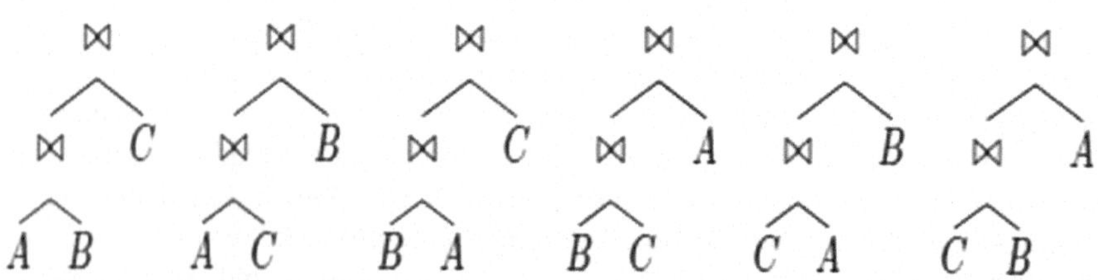

Fig. 2. Possible binary tree structures for query involved 3 relations [1].

Definition 2 Search Strategy: This refers to the algorithms applied to explore the search space and locate the best QEP based on join selectivity and join sites so as to minimize the cost of query optimization. There are essentially two classes of strategies that solve the problem of join operation [9] for optimizing the query. The first approach is the deterministic strategy that proceeds by building plans, starting from base relations, joining entities sequentially until complete plans are obtained. To minimize the optimization cost, the plans that do not lead to optimal solutions are neglected [1]. The other approach is the randomized strategies that look for the best solution around some particular points. These algorithms do not ensure optimal plan but they avoid the high cost of optimization [11, 12].

Definition 3 Cost Model: The purpose of query optimization in a distributed database environment is to minimize the total cost of computer resources. An optimizer cost model contains cost functions which estimate the cost of the operations used to evaluate the size of the results [13]. The cost function can be described as a form of either total time or response time. The total time consists of local processing cost (CPU Time + I/O Cost) and communication cost. Minimizing the total time means that the utilization of resources increases thus increasing the system throughput. The Response time is calculated as the time elapsed between initiation and completion of a query including parallelism [12].

2.1 Ant Colony Optimization

Ant colony optimization belongs to the large field of swarm intelligence in which researchers study the manner patterns of social insects in order to simulate processes. Initially, the algorithm was used to solve the traveling salesman problem and the quadratic assignment problem [14]. To solve the optimization problem with ACO algorithms, the problem should be abstracted as a graph. For example, the traveling salesman problem (TSP) can be represented by a graph $G = (N, E)$ where a fixed set of vertices N represents the cities and a fixed set of edges E shows the connection between the cities; the objective is to find Hamiltonian path for G which gives the minimal length [15]. Ants build their solutions by moving on the problem graph and laying pheromone so as to guide other ants. Pheromone trails, which provide a positive feedback mechanism, permit ants to cooperate and exploit each other's experiences. A negative feedback mechanism is also necessary to avoid stagnation which is implemented through pheromone evaporation.

The first ant colony optimization algorithm is known as Ant System (AS) and was proposed for solve TSP [14]. Since then, several other ACO algorithms have been proposed. Its main characteristics is that, at every iteration, the pheromone values are changed by all the m ants that construct a solution in the iteration itself. The pheromone τ_{ij}, exist on the edge joining cities i and j, is modified as follows [14]:

$$\tau_{ij} = (1 - \rho).\tau_{ij} + \sum\nolimits_{k=1}^{m} \Delta\tau_{ij}^{k} \tag{1}$$

where ρ is the evaporation rate, m is the number of ants, and $\Delta\tau_{ij}^{k}$ is the quantity of pheromone placed on edge (i, j) by ant k:

$$\Delta\tau_{ii}^{k} = \begin{cases} \frac{Q}{L_k} & \textit{if ant k used edge } (i,j) \textit{ in its tour,} \\ 0 & \textit{otherwise} \end{cases} \tag{2}$$

where Q is a constant, and L_k is the length of the tour constructed by ant k. In the construction of a solution, ants choice the following city to be visited through a stochastic technique. When ant k is in city i and has so far built the partial solution s^P, the probability of going to city j is given by:

$$p_{ij}^{k} = \begin{cases} \frac{\tau_{ij}^{\alpha} \cdot \eta_{ij}^{\beta}}{\sum_{c_{il} \in N(S^P)} \tau_{ij}^{\alpha} \cdot \eta_{ij}^{\beta}} & if\ c_{ij} \in N(S^P), \\ 0 & otherwise \end{cases} \tag{3}$$

where $N(s^P)$ is the set of suitable components; that is, edges (i, l) where l is a city not yet visited by the ant k. The parameters α and β control the relative importance of the pheromone versus the heuristic information η_{ij}, which is given by:

$$\eta_{ij} = \frac{1}{d_{ij}} \tag{4}$$

where d_{ij} is the distance between cities i and j.

2.2 Problem Definition

In distributed databases, data are distributed based on a distribution allocation scheme and may also be replicated at different sites. A query optimizer aims to find an execution strategy to execute a query that can minimize an objective cost function which could be based on either the total time or the response time. The optimizer solution space grows exponentially with large number of relations and distributed data storage sites included in the query. Therefore, the ability of the optimizer to find the best strategy for executing a query is computationally intractable and the real goal of the optimizer is to find a near optimal strategy and more importantly, to avoid bad strategies.

In order to build an algorithm for query optimization, three components should be addressed: (1) search space as a set of equivalent QEPs to abstract execution strategies, (2) cost model to give a cost to each QEP in order to show the performance of each plan and (3) search strategy to explore the search space in order to find the best QEP. In this paper, an ant colony algorithm is examined as a search strategy based on total dynamic time model to determine the best QEP.

3 Literature Survey

The most important part of a query optimization is its search technique or enumeration algorithm. Due to its importance, there is a rich literature on query optimization search technique. There are basically two major approaches used to find the best QEP. The first one is the deterministic strategies and the second is the randomized strategies. Dynamic programming belongs to deterministic strategies and it is used in most commercial database. The initial dynamic programming for query optimization is presented in [9, 16]. It works in a bottom-up way by repetitive building complex sub-plans from simpler sub-plans until construction of the complete plan. In this algorithm, costly expensive plans are discarded as early as possible. This approach is called pruning and plan will prune if there exist another equivalent plan that works at lower cost. Although this algorithm performs better than a randomize strategy for queries with small number of relations, but this situation is inverted when the query has large number of relations.

The authors in [17] presented and evaluated a new class of deterministic enumeration algorithms called Iterative Dynamic Programming (IDP). IDP works by apply dynamic programming many times in the process of query optimization. Many experiments have been made to show that IDP produce better plans than dynamic programming in situations in which dynamic programming is not viable because of its high complexity. When the query is simple, IDP produces a best plan with approximately the same time produced by dynamic programming. In case of high complex query, IDP produces plan which significantly better than other plans generated by other algorithms that are applicable under the same conditions but time in this case will take place.

One of the superior randomized algorithms is Iterative Improvement (II) [18]. II starts at random point and improves the solution by repeatedly accepting random descent moves until it arrives to a local minimum. This process repeats until a stopping condition (a pre-decided number of starting points are processed or a time limit is override) is met, at such point it returns the local minimum with lowest cost. The major drawback of II is that, in some cases, although high number of starting points may be used, the final result is still unacceptable. When the solution space contains a large number of high cost local minima, the algorithm gets easily trapped in one of them.

In [19], another randomized algorithm, Genetic Algorithm (GA), is presented. In this work, GA as a solution to query optimization is evaluated versus other models, but this work does not take into account customized crossover and mutation operators. Rho and March [20] presented a comprehensive distributed query optimization model that integrates copy identification, join order, join site selection and reduction by semi-joins into a one model. Here, the search space contains execution plans that reflects an environment where relations not only can be replicated, but also reduction with semi-joins are considered. Also, the joins can be executed at different sites rather the result site.

ACO algorithm is one of the swarm intelligence algorithms and used in this work as a search strategy for distributed join query optimization. This algorithm was successfully applied in centralized query optimization [21]. In [22], ACO algorithm was presented to solve the distributed database query optimization in replicated and not fragmented entities. Here, local and global ant colony was applied. Local optimization takes place at each site – with the same algorithms used in centralized database system – and results sent to a global optimization step which selects an optimum strategy to execute the query. Our proposed algorithm employs the distributed query optimization model that introduced in [22] and applied for non-replicated entities with total time as a cost model. Since the entities not replicated, local ant colony cannot be applied and global colony is utilized to handle overall optimization process. Moreover, the processing and communication cost for the joined query is dynamically calculated and depend on the ants walk path and the entitles location site.

4 Proposed Technique

In this section, the proposed model will be described based on the representation of search space, the method of cost calculation and the search strategy is used to find the best QEP.

4.1 Search Space

The search space can be represented as a query graph (annotated tree) denoted as $G = (N, A)$ where N is the set of nodes (vertices) in the query graph and A is the set of arcs (edges) [4]. Each node represents an entity in the join specification of the query. Two nodes are connected by an arc if the query joins the two corresponding entities. Each node in the query graph has an associated site.

4.2 Cost Model

The cost model of the suggested model is based on total time (some of all cost component). In this model, the I/O cost is computed for each join operation and the communication cost is considered only when it is needed to transfer entities between sites.

$$Total\ Cost = IO_{join} + COM_{Ri} \tag{5}$$

where IO_{join} is join cost and COM_{Ri} is cost of transfer entity R_i between sites. The number of page accesses required to perform the join operation is between two entities R_i and R_j

$$P_{join} = P_{Ri} * P_{Rj} \tag{6}$$

P_{Ri} is the number of pages in entity R_i and P_{Rj} is the number of pages in entity R_j .The number of pages needed to store the results is computed as [22]:

$$P_{writ} = \frac{card(R_i\ join\ R_i) * len(R_i\ join\ R_i)}{ps} \tag{7}$$

where $card(R_i)$ is the cardinality of relation R_i (the number of tuples in R_i), $len(R_i)$ is the average length of the tuple in R_i and ps is the page size. The local processing cost to perform a join at site S_k is computed as:

$$IO_{join} = \left(P_{join} + P_{write}\right) * IO(S_k) \tag{8}$$

where $IO(S_k)$ is the disk I/O time in site $S_{k.}$ In order to perform the joins, relations should be transmitted to the join sites if necessary. The cost of transferring relation R_i from site S_k to site S_p calculated as:

$$COM_{Ri} = card(R_i) * len(R_i) * COM\left(S_k, S_p\right) \quad (9)$$

where $COM(S_k, S_p)$ is the time to transfer one byte from site S_k to site $S_{p.}$

4.3 Search Strategy

In this work, the search strategy for the proposed algorithm is based on ACO meta-heuristic. Each entity in the join query represents as a node in a graph $G = (N, A)$. The cost of each ant journey to find the minimum spanning graph will dynamically calculate while building of this graph. An extra virtual node will be added to the graph to hold the intermediate result which helps in the cost calculation. The pseudo-code for the suggested model is as follow:

```
(1) Initialize all variables and set algorithm parameters.
(2) Create a graph for the joined entities.
(3) Initialize ants and Initialize pheromone.
(4) While (iteration < Max-Iteration)
(5) Randomly select the start-up node
(6) Move start-up node to Result Node
(7) For (ant = 1 to Ant-Numbers)
Choose next node.
Apply join entities.
Calculate the join cost
Increment the ant-journey cost
Move the join result to Result Node
(8) End For
(9) Calculate ants best trail
(10) Update pheromones.
(11) End While
(12) Calculate solution best trail
```

Initialization. In this stage, all the model parameters are initialized experimentally (in this version of work, the model' parameters), number of ants are determined, and pheromone trails are initialized. The graph that joins the entities is created so that every entity is connected to every other entity. Table 1 shows the description and the values of used parameters.

Table 1. ACO parameters [23]

Parameter	Description	Value
α	Pheromone influence factor	3
β	The influence of heuristic	2
ρ	Pheromone evaporation rate	0.02
Q	Constant used while pheromone update.	2
Max-Iteration	Varies with the number of entities	Experimentally determined
Ant-Numbers	Varies with the number of entities	Experimentally determined

Solution Construction. To find the solution for each ant, the model starts by randomly selecting an entity and set it as the start-up entity for the ant journey. This entity moves to the virtual node and wait for selecting the next entity for apply the join operation. To select the next entity, the algorithm will use Eq. (3). The equation, in this case, calculates the probability that ant k which currently in entity i will chooses to go to entity j. Here, τ_{ij} is the pheromone trail on the edge connecting entity i with entity j (join relation). Also, η_{ij} is the heuristic desirability and calculated as the inverse of the intermediate join cost (total cost) between entity i and entity j. If entity j is selected as the next entity, the cost of join is calculated and the result of join process is moved to the virtual node and then is used as the start-up entity for the next iteration. The process of entities selections and entities join cost calculation continues until all entities are processed and the ant journey cost is calculated.

Pheromone Update. When the ants construct their solutions, pheromone update is performed. Each pheromone value is decreased, simulating evaporation, and increased, simulating the deposit of pheromones by ants on the trail. The pheromone on all edges of the graph are updated as in Eq. (1). The update of pheromone also depend on L_k in Eq. (2) which represent the total cost of the of best ant trail tour constructed by ant k. After the iteration complete, the bets trail among the ants is selected.

5 Experimental Results

The proposed model was implemented in C# programming language and all experiments are running on a PC of Intel Core i5 2.4.0 GHz processor with DDR 8 GB main memory. Synthetic datasets are randomly generated for our experiments. The experiments run with different number of entities starts by 5 and increased to 10,15 and 20 with number of ants varies from one to five. All experiments are running with fixed parameters $\alpha = 3$, $\beta = 2$ and $\rho = 0.02$ as explained in Table 1. In the first experiment, as in Figs. 3, 4, 5 and 6, the relation between number of ants and the minimum cost for each ant is examined. This experiment runs for number of entities 5, 10, 15 and 20.

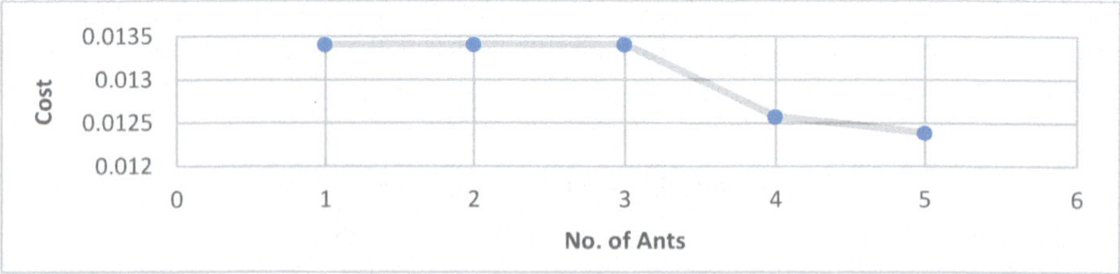

Fig. 3. Relation between the number of ants and cost (5 entities).

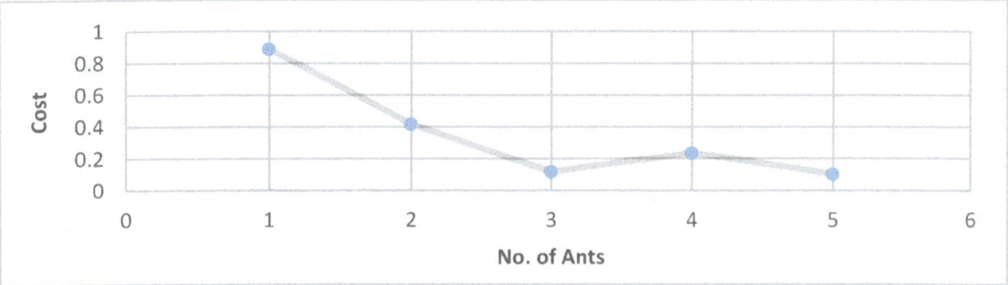

Fig. 4. Relation between the number of ants and cost (10 entities).

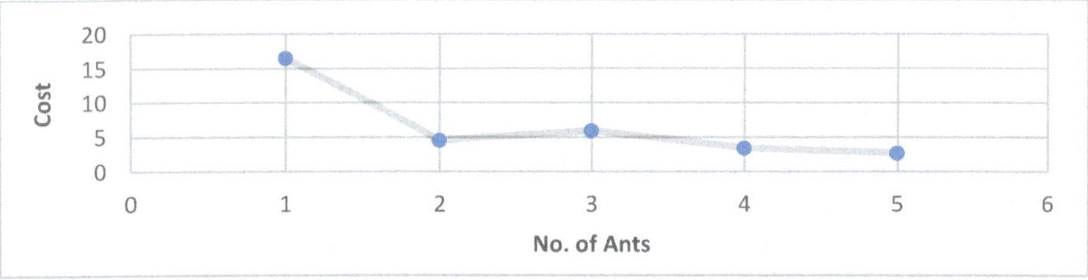

Fig. 5. Relation between the number of ants and cost (15 entities).

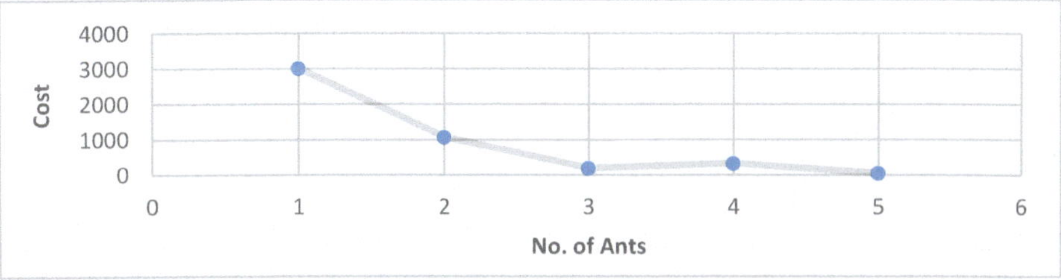

Fig. 6. Relation between the number of ants and cost (20 entities).

The experiments show that better results are obtained in case of large number of join entities exist in the query when number of ants increased. If the number of ants are extremely small as compared to the problem size, incomplete search takes place and the algorithm falls into local optimum leading to premature convergence. If large number of ants are involved in generating the optimal solutions, the randomness in the exploration increases leading to enhanced results.

In the second experiment, and for fixed number of ants, the relation between number of iterations and the minimum cost for each iteration is examined. This experiment runs for number of entities 5, 10, 15 and 20. As shown in Fig. 7, to gain minimal cost with the increase of the number of entities, more iterations are required. This is due to the increase in the complexity of search process. Using more ants helps to go faster for better cost which reduces the number of iterations.

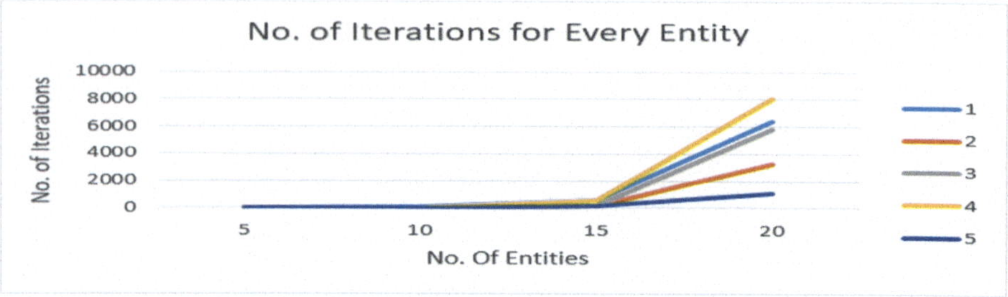

Fig. 7. Relation between the number of iterations and number of entities.

6 Conclusion and Future Work

This paper utilizes ant colony algorithm based on dynamic cost for optimizing the distributed database queries. The query cost involved both I/O cost and transfer cost in case of the joined entities lies in different sites. The model was tested with different number of entities and different number of ants. The results show that a better result will be obtained in large number of entities when the number of ants increased. There are several major areas for future studies. Using ACO with query after separating the query cost (I/O and data transfer) into two objective functions may enhance the query performance. Also, using the new paradigms such as quantum computing with the classical ACO algorithm may accelerate the process of finding the best path for entities join order. Furthermore, the model' parameters will be obtained using optimization algorithm.

References

1. Ramakrishnan, R.: Databases Management Systems, 3rd edn. McGraw-Hill Inc., New York (2003)
2. Tiwari, M.P., Chande, S.V.: Query optimization strategies in distributed databases. Int. J. Adv. Eng. Sci. **3**(3), 23–29 (2013)
3. Dökeroğlu, T., Coşar, A.: Dynamic programming with ant colony optimization metaheuristic for optimization of distributed database queries. In: Proceedings of 26th International Symposium on Computer and Information, pp. 107–113. Springer, London (2011)
4. Sharma, M., Singh, G., Singh, R.: A review of different cost-based distributed query optimizers. Progr. Artif. Intell. **8**(1), 45–62 (2019)
5. Hameurlain, A., Morvan, F.: Evolution of query optimization methods. Lect. Note Comput. Sci. **5740**, 211–242 (2009)
6. Chen, M., Yu, P.: Using join operations as reducers in distributed query processing. In: Proceedings of 2nd International Symposium on Databases in Parallel and Distributed System, July 1990
7. Pramanik, S., Vineyard, D.: Optimizing join queries in distributed database. IEEE Trans. Softw. Eng. **14**, 1391–1426 (1988)
8. Rothnie, J.B., Bernstein, P.A., Fox, S.: Introduction to a system for distributed database. ACM Trans. Database Syst. **5**(1), 1–17 (1980)

9. Aljanaby, A., Abuelrub, E., Odeh, M.: A Survey of distributed query optimization. Int. Arab J. Inform. Technol. **2**(1), 48–57 (2005)
10. Yannis, Y.C.K., Ioannidis, E.: Randomized algorithms for optimizing large join queries. ACM Sigmod Rec. **19**(2), 312–321 (1990)
11. Horng, J.T., Kao, C.Y., Jhiune, B.: A Genetic algorithm for database query optimization. In: Proceedings of the First IEEE Conference on Evolutionary Computation, Orlando, FL, USA, pp. 432–444 (1994)
12. Sevinc, E., Cosar, A.: An evolutionary genetic algorithm for optimization of distributed database queries. Comput. J. **54**(5), 717–725 (2010)
13. Sukheja, D., Singh, U.: A Novel approach of query optimization for distributed database system. Int. J. Comput. Sci. **8**(4), 307–312 (2011). No. 1
14. Dorigo, M., Birattari, M., Stützle, T.: Ant colony optimization. IEEE Comput. Intell. Mag. **1** (4), 28–39 (2006)
15. Dorigo, M., Stuzle, T.: Ant Colony Optimization. MIT Press, Cambridge (2004)
16. Kossmann, D.: The state of art in distributed query optimization. ACM Comput. Surv. **32**, 422–469 (2000)
17. Kossmann, D., Stocker, K.: Iterative dynamic programming: a new class of query optimization algorithm. ACM Trans. Database Syst. **25**, 43–82 (2000)
18. Steinbrunn, M., Moerkotte, G., Kemper, A.: Heuristic and randomized optimization for the join-ordering problem. Int. J. Very Large Data Bases **6**(3), 191–208 (1997)
19. Zhou, Z.: Using heuristics and genetic algorithms for largescale database query optimization. J. Inform. Comput. Sci. **2**(4), 261–280 (2007)
20. Rho, S., March, S.T.: Optimizing distributed join queries: a genetic algorithm approach. Ann. Oper. Res. **71**, 199–228 (1997)
21. Li, N., Liu, Y., Dong, Y., et al.: Application of ant colony optimization algorithm to multi-join query optimization. In: Proceedings of the 3rd International Symposium on Advances in Computation and Intelligence. Springer, Wuhan (2008)
22. Golshanara, L., Rankoohi, S.M.T.R., Shah-Hosseini, H.: A multi-colony ant algorithm for optimizing join queries in distributed database systems. Knowl. Inform. Syst. **39**(1), 175–206 (2014)
23. Tiwari, P., Chande, S.: Optimal ant and join cardinality for distributed query optimization using ant colony optimization algorithm. In: Proceedings of the 2nd International Symposium on Emerging Trends in Expert Applications and Security, Singapore, February 2019

Clustering Analysis Based on Coyote Search Technique

Asmaa Mohamed[1(✉)], Walaa Saber[1], Ibrahim Elnahry[1], and Aboul Ella Hassanien[2]

[1] Electrical Engineering Department, Port Said University, Port Said, Egypt
asmaa.khaled40@yahoo.com, walaa_saber@eng.psu.edu.eg, Fbsu.elnahry@gmail.com
[2] Faculty of Computers and Information, Cairo University, Cairo, Egypt
aboitcairo@gmail.com

Abstract. The clustering technique, which is widely employed for data analysis and data mining, represents the way of gathering similar objects or data with each other according to similar criteria in several clusters. Recently, meta-heuristic optimization techniques have become one of the most common approaches for researchers to solve clustering problems. We propose in this paper a novel algorithm for solving the data clustering problem based on a new optimization technique namely as Coyote Clustering Technique (CCT) inspired by the coyotes' behaviors. We have studied the proposed method by applying a famous and widely set of data used in this field. The outputs of the proposed Technique proved their efficiency by recording good results in speed, accuracy, and stability.

Keywords: Clustering · Meta-heuristic · Coyote Search Technique

1 Introduction

Data mining is the methodology of determining suitable and hidden items in a large collection of data. Supervised methods such as classification and unsupervised methods such as clustering are popularly utilized to determine suitable information from the set of data [1]. The clustering technique, which represents the way of gathering similar items with each other according to similar standards in many groups, is widely used as a means of data extraction and analysis. Accordingly, clustering algorithms are applied to different types of problems such as analysis of data and data mining [2], image processing [3], bioinformatics [4], wireless sensor networks [5], and mathematical programming [6].

Generally, algorithms of clustering can be defined in two major groups, namely partitioning and hierarchical. Among these two groups, of the greatest well-known partitioning algorithm is the k-means [7]. However, the K-means clustering algorithm has two foremost difficulties: first, it is highly sensitive to the initialization stage; second, it simply falls into local optimal solutions. To resolve these troubles, numerous researchers tried to resolve the clustering troubles using various swarm intelligence and meta-heuristic algorithms, such as Particle Swarm Optimization (PSO) [8], Ant Colony

A.-E. Hassanien et al. (Eds.): AICV 2020, AISC 1153, pp. 182–192, 2020.
https://doi.org/10.1007/978-3-030-44289-7_18

Optimization (ACO) [9], Tabu Search (TS) [10], and Simulated Annealing (SA) [11] in addition to the Gravitational Search Algorithm (GSA) [12], Genetic Algorithm (GA) [13], Cuckoo Search Algorithm (CSA) [14], Differential Evolution (DE) algorithm [15] and Black Hole (BH) algorithm [16]. Nevertheless, these algorithms possess limitations such as boor quality in output and slow speed. Therefore, we suggested a clustering algorithm based on the Coyote Search Technique (CST). The proposed technique offers advances in convergence speed, avoiding the local optima trouble.

The sections in the research are displayed as follows. In Sect. 2, an overview of the literature related to cluster analysis is provided, the principle of the clustering trouble, and demonstrates the basic principle of the Coyote Search Technique (CST). In Sect. 3, the execution of the proposed Coyote Clustering Technique (CCT) is illustrated. In Sect. 4, the results are explained and evaluated. Lastly, Sect. 5 Clarification of the conclusion and recommendations to be made in future research.

2 Background and Related Work

2.1 Related Clustering Algorithms

The clustering procedure is the approach of gathering a set of data into clusters such that the data in each cluster share a high degree of similarity. Although, the k-means method is the greatest widespread and fastest clustering algorithms, it suffers from two major shortcomings. The first it's extremely sensitivity to the initialization stage, whereas the second stage is its susceptibility to falling into local optimal solutions.

Several researchers have started to resolve the clustering troubles by applying meta-heuristic techniques. For example, Zhang et al. in 2010 were offered an ABC clustering algorithm [17]; the offered technique was compared with other well-known sets of data and different common meta-heuristic clustering algorithms. In 2012, a novel PSO method to the clustering data has implemented by Tunchan [18], irrespective of whether the number of clusters is known. In 2014 [19] Karaboga et al. has suggested a hybrid technique between the PSO and ABC clustering algorithms to develop the results of data clustering and compared them with other algorithms. In 2016, Wang et al. [20] proposed a hybrid technique composed of the Flower Pollination Algorithm (FPA) and the Bee Pollinator Algorithm (BA) to dealing with the weaknesses of the standard FPA. The Moth Flame Swarm Algorithm (MSA), which seeks to achieve a higher accuracy and stability, was applied to cluster analysis by Yang et al. in 2017 [21]. Most recently, in 2018, Nasiri and Khiyabani [22] offered a new clustering technique based on the Whale Optimization Algorithm (WOA) that simulates the behaviour of whales for searching about their food to reach a progress for the data clustering result.

2.2 Coyote Search Technique (CST)

CST is a new technique that is mimicked the social behaviour of the coyotes [23]. Coyotes mimic predators in the process of hunting a prey, but the difference here is that they hunt their prey in packs and use the nature of infiltration in this process. Coyotes have the distinctive characteristics, where they prefer to cooperate in moving in groups to hunt prey, but have the propensity to drop down their prey individually. The way a coyote searches for food and survives by avoiding enemies is used to formulation the search algorithm, where the coyote has an individual search ability and autonomous movement. The coyotes' behaviors have more advantages than other algorithms such as PSO [24] and Firefly [25], where the coyote acts as multiple leaders spreading to multiple directions for the best solution, rather than searching as a single herd. The coyote jumps out of the hunter's visual range to avoid falling into a trap, the previous acts are used in the algorithmic design, thus the coyote will jump away from the local optimal solution.

Coyotes have the best memory capability as they can hide foods, they also feel and track prey from far distances of miles. Coyotes hold a serious feature, they own highly powerful sense of smell that enables them to find their prey and locate an exact location. They mark their way in different behaviors and store previously visited sites. The direction and distance the coyote moves away from any random threat like as the mutation and crossover in the genetic algorithm. Each coyote is a possible solution for the optimization problem. Also, each coyote has a sensed distance that created a visual distance, which is used to notice the best solution, move to an improved position, and out of the local optimal search.

CST starts by assigning the coyotes randomly to the packs. Then the coyotes move between packs based on the number of packs and the number of coyotes per pack. In the CST algorithm, all these advantages taken into account, to get the best solution and move to it easily. Thus CST avoids reaching to local optima. In CST, the number of coyotes per pack and a number of packs defining the population size (PS) numbers, that are signified potential solutions numbers to the optimization problem. Finally, the aim of the CST is to discover the PS position that result in the best evaluation of a given objective function.

3 The Coyote Search-Based Clustering Technique

The main idea of the proposed CCT system is to partitioning the problem space into several clusters in a balancing way. Suppose that the PS denotes to cluster centroids U. Every population $D_{i,}$ is generated as follows [20]:

$$D_{i,} = (U_{i1}, U_{i2,} \ldots, U_{ik}) \tag{1}$$

where U_{ij} indicates the centroids of the cluster. Therefore, the number of population can denote to a number of clustering candidates' solution for the dataset. The primary purpose of the objective function is reducing the distance between these items and the centroids of the cluster by using the following Eqs. 2 and 3 [20].

$$d(D_i, U_j) = \sqrt{\sum_{k=1}^{m} (D_{ik}, U_{jk})^2} \quad (2)$$

where D_i is a given set of items while, m is the dimensions, and U_j is cluster centroids for the cluster k_{th}. Ultimately, the fitness function of the offered CCT technique is to minimize the value of the objective function that can be expressed as the next equation

$$f(D, U) = \sum_{i=1}^{n} \min\{\|D_i\_U_k\| | k = 1, 2, \ldots..k\} \quad (3)$$

Theoretically, the CCT clustering technique can be considered a global optimizer because it includes the ability to explore and exploit simultaneously. CST tries to find the position of the population that produces the best evaluation of a given objective function. In CST the new coyotes are updated through the following equation by using the effects of a fitness function as follow [23]:

$$soc_N^{p,t} = soc_N^{p,t} + r_1\delta_1 + r_2\delta_2 \quad (4)$$

where r_1 and r_2 are effected by an alpha and a pack, respectively. The initial values of r_1 and r_2 are randomization values into the selection [0, 1] through probability. Then, the novel situation of the coyotes is evaluated by the fitness for all the coyotes in the current pack based on the new coyotes' position by using the following equation [23]:

$$new\_fit_c^{p,t} = f(new\_soc_c^{p,t}) \quad (5)$$

After that, the information about packs and coyotes are updated. Lastly, the best coyote will be chosen as the optimal solution. The flowchart and the algorithm of the proposed CCT clustering algorithm are shown in Figs. 1 and 2.

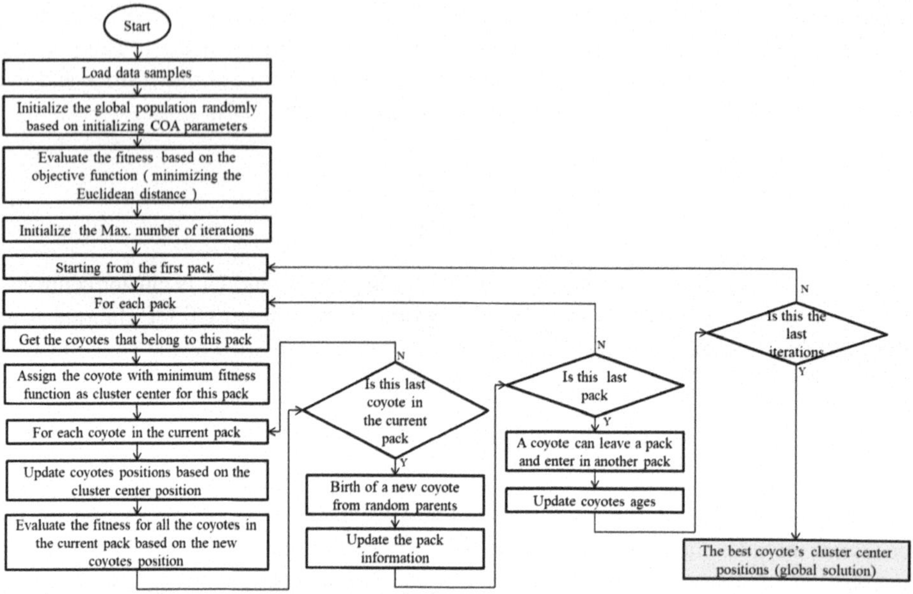

Fig. 1. The flow chart of the CCT clustering algorithm.

Algorithm 1 The Coyote Search-based Clustering Algorithm

```
Load data samples.
Initialize N_p packs with N_c coyotes.
Initialize the global population size randomly to contain U_k cluster centers based on N_p * N_c .
Defined the objective function, to minimize the distance between data object D_i and the center of cluster U_k (Eq. (3) ).
Evaluate the objective function.
while stopping condition is not attained
    for each pack p
        Get the coyotes that belong to this pack.
        Assign the coyote with minimum objective function as the cluster center of this pack.
        for each coyotes of the pack
            Update the coyote position (based on the cluster center) (Eq. (4) ).
            Evaluate the objective function based on the new position of coyotes (Eq. (5) ).
        end for
        Birth and death the coyotes.
        Update the pack information.
    end for
    Transition the coyotes between the packs.
    Update the coyotes' ages.
    t = t + 1
end while
Return the Best coyotes (Best solution).
```

Fig. 2. Pseudo code of the CCT clustering algorithm

4 Experimental Result

Several experiments were verified using the suggested technique and run it in MATLAB R2017a. For examined the influence of the proposed technique, 11 sets of data are used, including 2 artificial sets of data and 9 real life sets of data chosen from UCI [26]. These sets of data namely, Art1, Art2, Iris, Seeds, Ripley's glass (indicated as Glass), Heart, Balance Scale, Wisconsin-Breast-Cancer (indicated as Cancer) Contraceptive-Method-Choice (indicated as CMC), Wine and Vowel.

In Table 1 all datasets properties have been documented. The results obtained from running the program 20 times via an initial random solution for all sets of data. The objective function values for best, worst, average and standard values all registered in this table. Bold results show the technique that has been used achieves the best value. Examination of the outputs indicated that the better value is the lowest value recorded by the proposed technique, whereas the highest value is the worst value reached.

The data sets used in the algorithms are described as follows: Artificial data set one (Art1) contains 250 objects and 5 classes with 3 featured problem, while artificial data set two (Art2) contains 600 objects and 4 unique classes with 2 featured problems [27]. The Iris data set contains a 4 featured problem with 3 classes and 150 random samples of flowers from three iris species. The Wisconsin _breast _cancer (Cancer) data set contains 683 patterns with 9 attributes and 2 clusters. The Heart data set is a heart disease database containing 270 objects and 13 featured problems with 2 unique classes. The Balance Scale data set includes 4 features, 3 classes and 625 examples. The Seeds dataset contain 210 patterns belonging to 3 different types of wheat with 7 features and 3 classes. The Wine data set contains, 178 patterns and 13 vowel input features and 3 classes. The Vowel data set includes 871 patterns and 3 vowel classes with 6 input features. The settings of the aforementioned datasets are shown as following.

Table 1. Properties of the eleven datasets.

Data set	Art1	Art2	Iris	Seeds	Glass	Heart	Balance scale	Cancer	CMC	Wine	Vowel
Clusters number (M)	5	4	3	3	6	2	3	2	3	3	3
Features number (N)	3	2	4	7	9	13	4	9	10	13	6
Data size (O)	250	600	150	210	214	270	625	683	1473	178	871

Table 2 demonstrations the setting of parameters used for the suggested technique. The parameters are set after many trials to make the proposed algorithm work in an efficient way with different data sets. All algorithms are run 20 times, as shown in Table 3 all the results were recorded and that demonstrates the optimal values found by the CCT clustering algorithm are smaller than other values found by the other algorithms among the 11 data sets. Among these algorithms, the values found by the GWO algorithm are the largest, indicating that the GWO is the worst at solving the clustering problem among the five investigated algorithms.

The results of the CCT clustering operation in many cases such as Art, Iris, Seeds, Glass, Heart, and Balance scale data sets as shown in Table 2 shows that the best, worst and mean approximately the same value, which means that the CCT convergence to the optimal solution in a faster and stable way.

Table 2. Setting parameters for the suggested techniques

Algorithm	*CCT*	GA	GWO	PSO	DE
Parameters values	Nc = 5	pc = 0.8	PS = 30	w1 = 1	LSF = 0.2
	Np = 20	pm = 0.3	T = 200	w2 = 0.99	USF = 0.9
	T = 10000	gamma = 0.2		c1 = 2.0	PCR = 0.2
		beta = 8		c2 = 2.0	PS = 30
		mu = 0.02		PS = 30	T = 200
		PS = 30		T = 200	
		T = 200			

The convergence curves for the CCT clustering technique on the Art1, Iris, Cancer, and CMC data sets are provided in Figs. 3, 4, 5 and 6, respectively.

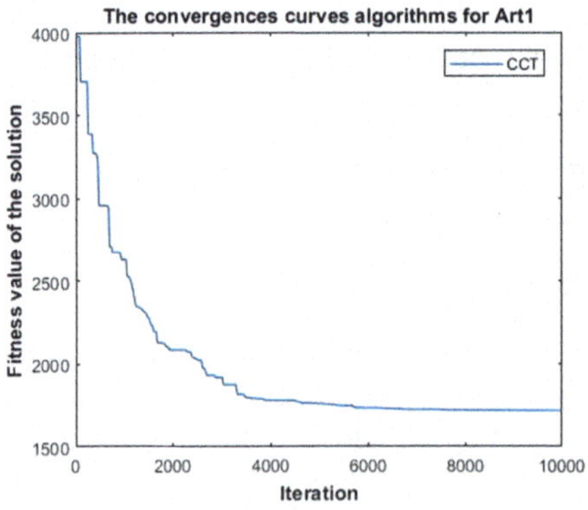

Fig. 3. The CCT convergence curve on the Art1 data set

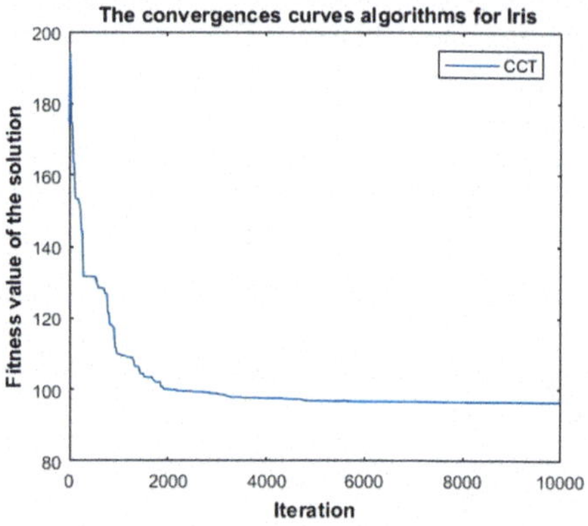

Fig. 4. The CCT convergence curve on the Iris data set

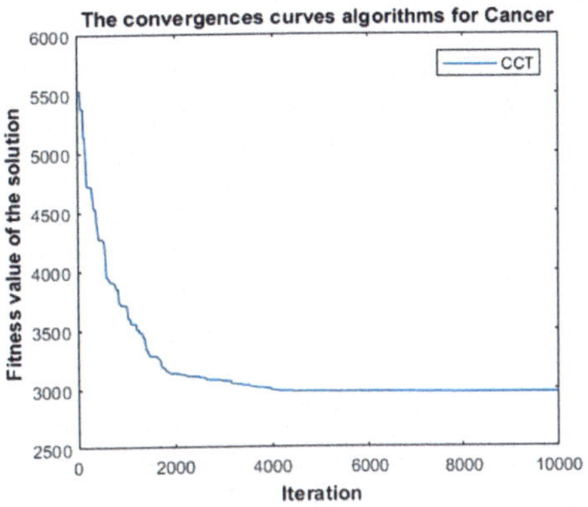

Fig. 5. The CCT convergence curve on the cancer data set

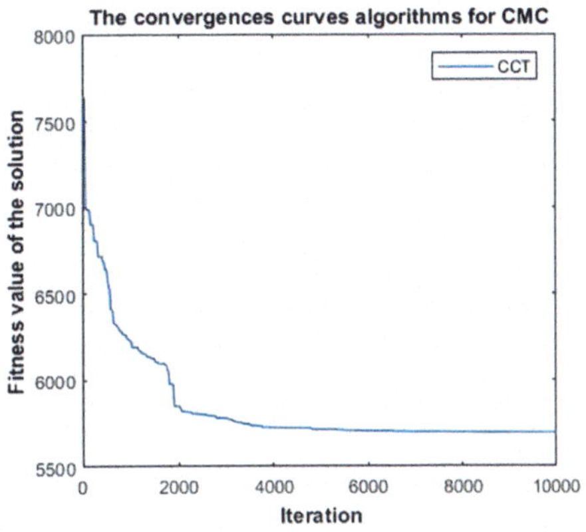

Fig. 6. The CCT convergence curve on the CMC data set

Table 3 shows the effects of the *CCT* technique on the different data sets.

Table 3. Comparison among the five clustering algorithms applied to the eleven datasets

Dataset	Criterion	PSO	GA	DE	GWO	CCT
Art1	Best	1808.4	1808.8	1981.0	3838.4	**1718.255**
	Worst	2491.0	3170.1	2711.5	5724.8	**1718.379**
	Mean	1948.4	2426.9	2311.9	4674.3	**1718.282**
	Std	229.5	365.7	99.9	481.7	**0.036**
Art2	Best	523.1	515.6	515.2	1102.6	**513.903**
	Worst	546.8	909.4	594.8	1531.1	**513.903**
	Mean	535.2	594.3	550.5	1288.2	**513.903**
	Std	7.1	145.5	19.9	129.3	**0.000**
Iris	Best	97.9	97.2	97.3	162.8	**96.655**
	Worst	100.0	128.4	103.5	226.8	**96.655**
	Mean	98.9	104.2	99.9	202.7	**96.655**
	Std	0.5	10.2	1.9	15.7	**0.000**
Seeds	Best	316.1	320.0	317.9	455.6	**311.798**
	Worst	324.8	406.4	342.4	567.2	**311.829**
	Mean	319.9	346.0	330.2	526.6	**311.807**
	Std	2.1	26.5	5.7	32.7	**0.009**
Glass	Best	281.0	286.6	249.1	490.7	**211.872**
	Worst	391.1	356.8	299.8	720.2	**234.982**
	Mean	303.6	325.2	279.2	603.1	**218.623**
	Std	23.2	18.1	13.6	59.1	**6.624**

(continued)

Table 3. (*continued*)

Dataset	Criterion	PSO	GA	DE	GWO	CCT
Heart	Best	10658.2	10644.0	10623.2	13171.0	**10622.985**
	Worst	10717.7	11159.4	10631.5	17872.5	**10623.032**
	Mean	10690.4	10763.1	10627.1	14938.3	**10622.993**
	Std	15.5	135.5	2.5	1165.5	**0.013**
Balance scale	Best	1425.5	1424.8	1427.9	1456.0	**1423.820**
	Worst	1427.8	1433.0	1432.3	1531.5	**1424.189**
	Mean	1426.6	1427.8	1430.2	1490.1	**1423.857**
	Std	0.7	2.4	1.2	20.2	**0.091**
Cancer	Best	2999.6	3061.8	2964.7	5582.2	**2964.388**
	Worst	3033.7	3495.8	2996.8	7233.2	**2964.421**
	Mean	3018.4	3304.1	2980.1	6496.4	**2964.393**
	Std	9.4	109.2	10.1	468.0	**0.008**
CMC	Best	5808.0	5971.1	5720.9	7252.2	**5693.725**
	Worst	5934.8	6577.7	5789.3	9126.1	**5694.491**
	Mean	5859.5	6220.5	5754.2	8202.3	**5693.940**
	Std	36.7	175.8	17.4	582.0	**0.259**
Wine	Best	16337.0	16334.5	16298.6	17692.5	**16292.190**
	Worst	16390.6	16903.0	16329.0	21647.2	**16293.949**
	Mean	16365.4	16469.4	16311.9	19730.9	**16292.770**
	Std	17.2	147.9	7.9	1202.2	**0.500**
Vowel	Best	154490.4	152544.7	159658.1	227948.1	**148968.679**
	Worst	166765.7	174429.6	170224.9	277922.6	**149468.3866**
	Mean	157646.5	161115.4	165787.2	253602.7	**149084.8812**
	Std	3370.8	6676.8	3053.6	14402.7	**111.497**

5 Conclusion and Future Work

In our paper, a novel technique CCT is used to resolve the clustering trouble. This new technique has a few parameters to adjust, it also offers novel technicality for balancing the exploration and exploitation for search space, in the way a coyote searches for food and survives by avoiding enemies in the search algorithm. Also, the ability of coyote search algorithm to discover the optimal solution, move to an improved position, and out of the local optimal search. All of these reasons making it much suitable to use than other algorithms.

The result of the CCT clustering technique is evaluated under various data sets with different features. Through the application of some comparisons between the technique that used in this paper and four stochastic techniques, namely, PSO algorithm, DE algorithm, GA and GWO algorithm, showed the superiority of our algorithm. The results obtained through experiments, statistical results and graphs proved the superiority of our technique more than the other techniques that used in resolve the clustering trouble. Some directions remain for additional research to develop the performance of the suggested algorithm by execution a combination between the CCT and other clustering methods to give an enhanced result in terms of speed in convergence, time and stability.

References

1. Witten, I.H., Frank, E.: Data Mining Practical Machine Learning Tools and Techniques. Morgan Kaufman, Burlington (2005, 2018)
2. Ng, R., Han, J.: Efficient and effective clustering method for spatial data mining. In: Proceedings of 1994 International Conference on Very Large Database, Santiago, Chile, pp. 144–155 (1994)
3. Imene, K., Yamina, T.: Colour image segmentation using spatial probabilistic clustering. Signal Imaging Syst. Eng. **7**(3), 173–179 (2014)
4. Iván, G., Grolmusz, V.: On dimension reduction of clustering results in structural bioinformatics. Biochim. et Biophys. Acta (BBA) – Proteins Proteom. **1844**(12), 2277–2283 (2014)
5. Zhu, J., Lung, C.-H., Srivastava, V.: A hybrid clustering technique using quantitative and qualitative data for wireless sensor networks. Ad Hoc Netw. **25**(Part A), 38–53 (2015)
6. Rao, M.R.: Cluster analysis and mathematical programming. J. Am. Stat. Assoc. **22**, 622–626 (1971)
7. Zalik, K.R.: An efficient k-means clustering algorithm. Pattern Recognit. Lett. **29**(9), 1385–1391 (2008)
8. Chen, C.-Y., Fun, Y.: Particle swarm optimization algorithm and its application to clustering analysis. In: 2004 IEEE International Conference on Networking, Sensing and Control (2004)
9. Shelokar, P.S., Jayaraman, V.K., Kulkarni, B.D.: An ant colony approach for clustering. Anal. Chim. Acta **509**(2), 187–195 (2004)
10. Güngör, Z., Ünler, A.: K-Harmonic means data clustering with tabu-search method. Appl. Math. Modell. **32**(6), 1115–1125 (2008)
11. Güngör, Z., Ünler, A.: K-harmonic means data clustering with simulated annealing heuristic. Appl. Math. Comput. **184**(2), 199–209 (2007)
12. Hatamlou, A., Abdullah, S., Nezamabadi-pour, H.: A combined approach for clustering based on K-means and gravitational search algorithms. Swarm Evol. Comput. **6**, 47–52 (2012)
13. Bouyer, A., Hatamlou, A., Abdullah, A.H.: An optimized clustering algorithm using genetic algorithm and rough set theory based on kohonen self-organizing map. Int. J. Comput. Sci. Inform. Secur. **8**(4), 39–44 (2010)
14. Bouyer, A.: An optimized K-Harmonic means algorithm combined with modified particle swarm optimization and cuckoo search algorithm. Found. Comput. Decis. Sci. **41**(2), 99 (2016)
15. Tian, Y., Liu, D., Qi, H.: K-harmonic means data clustering with differential evolution. In: International Conference on Future BioMedical Information Engineering, FBIE 2009. IEEE (2009)
16. Hatamlou, A.: Black hole: a new heuristic optimization approach for data clustering. Inform. Sci. **222**, 175–184 (2013)
17. Zhang, C., Ouyang, D., Ning, J.: An artificial bee colony approach for clustering. Exp. Syst. Appl. Sci. **37**, 4761–4767 (2010)

18. Tunchan, C.: A particle swarm optimization approach to clustering. Exp. Syst. Appl. Sci **39**, 1582–1588 (2012)
19. Karthikeyan, S., Christopher, T.: A hybrid clustering approach using artificial bee colony (ABC) and particle swarm optimization. Int. Comput. Appl. Sci. (100), 15 (2014)
20. Wang, R., et al.: Flower pollination algorithm with bee pollinator for cluster analysis. Inf. Process. Lett. **116**(1), 1–14 (2016)
21. Yang, X., et al.: Moth swarm algorithm for clustering analysis. In: International Conference on Intelligence Computational, pp. 503–514. Cham (2017)
22. Nasiri, J., Khiyabani, F.M.: A whale optimization algorithm (WOA) approach for clustering. Cogent Math. Stat. **5**, 1483565 (2018)
23. Pierezan, J., Coelho, L.D.S.: A new metaheuristic for global optimization problems. In: 2018 IEEE Congress on Evolutionary Computation (CEC). IEEE (2018)
24. Pitt, W.C., Box, P.W., Knowlton, F.F.: An individual-based model of canid populations. Modelling territoriality and social structure. Ecol. Model. **166**(1–2), 109–121 (2003)
25. Blake, C.L., Merz, C.J.: UCI repository of machine learning databases. http://archive.ics.uci.edu/ml/datasets.html
26. Niknama, T., Amiri, B.: An efficient hybrid approach based on PSO ACO and k-means for cluster analysis. Appl. Soft Comput. **10**(1), 183–197 (2010)

Brain Tumor Segmentation in 3D-MRI Based on Artificial Bee Colony and Level Set

Yasmine Mahmoud Ibrahim[1(✉)], Saad Darwish[1], and Walaa Sheta[2]

[1] Department of Information Technology, Institute of Graduate Studies and Research, Alexandria University, 163 Horreya Avenue, El-Shatby, P.O. Box 832, 21526 Alexandria, Egypt
{igsr.yasmine, saad.darwish}@alexu.edu.eg

[2] Department of Information Technology, Informatics Research Institute City of Scientific Research and Technology Applications, Borg EL Arab, Alexandria, Egypt
wsheta@mucsat.sci.eg

Abstract. Medical imaging technologies developed to satisfy the significant need for information on medical imaging by visualizing internal organs like tumors help the radiologists to extract critical clinical data accurately. Brain tumor segmentation is very needed for a contemporaneous planning system for brain surgery. This paper offers a quick and accurate level set segmentation using a Modified Artificial Bee Colony (ABC) clustering technique to extract the tumor. In ABC, the food source initial position is identified using *k*-means rather than random initialization. The suggested model calculates the centroids of clusters and utilizes level set segmentation to handle topological changes of contours as the brain tumors vary in their form, structure, and size. Our model consists of an initial, pre-processing step to extract the brain from the head and enhances contrast stretching. Second, two-step ABC is employed to extract tumor edges, which will be used as an initial contour of the Magnetic Resonance Images (MRI) sequence. In the last step, the level set segment is employed to extract the tumor region from all volume slices with a fewer number of iterations. The experimental results using the benchmark BraTs'2017 dataset confirm the accuracy of the suggested model.

Keywords: Artificial Bee Colony (ABC) · Clustering · Level set segmentation · 3D-MRI · Initial contour · Two-step

1 Introduction

From the American national brain tumor society, 700,000 humans suffered from brain tumors in 2017, and 79,000 humans could be ranging from 2018. All these numbers of individuals may die from a brain tumor [1]. Brain tumor means an abandoned growth of cancerous cells inside or around the brain. MRI is able to distinguish between soft tissue and the tumor tissue [2]. The extraction of brain tumor has to divide brain images into two segments. The primary is the tumor cells, whereas the second is healthy cells. Segmentation is dividing the image into meaningful parts [3]. So, the analysis of brain images became the most significant challenge job for MRI specialists. It depends on

A.-E. Hassanien et al. (Eds.): AICV 2020, AISC 1153, pp. 193–202, 2020.
https://doi.org/10.1007/978-3-030-44289-7_19

various factors like [4]: (1) brain tumors vary in their form, structure, and size, as there are different types of brain tumors. (2) tumor boundaries are usually unclear and irregular. (3) Doctors time-consuming per patient diagnosis in hospitals and clinics. (4) separate many tumor tissues like a solid or active tumor from healthy tissues.

The clustering method is to search out the similarity between objects in an image and divide it into multiple groups based on distance, intensity. ABC that mimics the behavior of honeybees [5, 6] has been applied to many improvement issues as it is a swarm intelligence algorithm. ABC is suitable for exploration, however weak in exploitation, that outcomes within the poor convergence. Two-step ABC speeds up the convergence and maintains the balance between exploration and exploitation. In this procedure, the initial position of bees and food sources is identified randomly, which is time-consuming for optimizing in clustering problems [7]. Nevertheless, using the *k*-means for ABC initial positions of food sources will make it more accurate.

The automatic segmentation of 3D MRI is a hard problem that has been solved by model-based segmentation techniques as parametric models and geometric models or level sets [6], Level set segmentation is keen to deal with split and merge as it easy to follow the topological difference in the object. The matter of this segmentation technique is high complexity, needs to specify initial curves properly that must be near to object boundary. Thus, it provides sensible results. The main contribution of this research is to create a new brain tumor segmentation model that deals with 3D-MRI images by adopting both two-step ABC clustering and level set segmentation. Clustering is utilized to take the clustering process output as an initial contour for the segmentation step of the MRI sequence. The steps repeated until all slices from entire MRI sequences obtained. So that to reduce the number of iterations in the level set and to increase accuracy.

The rest of the paper is organized as follows: Sect. 2 describes some of the recent related works. The detailed description of the proposed system has been made in Sect. 3. In Sect. 4, the results and discussions on the dataset are given. Finally, conclusions are drawn in Sect. 5.

2 Related Works

Many researches have been presented by researchers regarding MRI brain tumor segmentation techniques. A brief review of some of the recent researches is presented here. The authors in [8] introduced a convolutional neural network model that uses BRATS 2017 data sets for the segmentation of brain tumor and sub-tumor parts. The work segmented efficiently and automatically. The limitations of this method are that the training stage is time-consuming and not accurate for segmenting complete tumors.

The authors in [9] applied Fast Boundary Box (FBB) technique on BRATS 2017 dataset, which uses brain symmetry analysis to find the tumor. The limitations of this study are that brain imaging volumes with tiny tumor size and multiple tumors have not tested.

The authors in [10] applied Rough-Fuzzy C-Means (RFCM) to BRATS 2017 and BRATS 2013 to automatically predict and segment the brain tumor regions. The method's limitation includes that RFCM takes time longer than Hard C-Means (HCM).

The authors in [11] suggested an approach based on watershed algorithms applied to segment BRATS 2017 brain dataset. This technique is reliable and makes tumor selection accurate and solves the traditional over-segmentation problem. It is good compared to state-of-the-art-works. The authors in [12] offered a Convolutional Neural Networks (CNNs) model to segment benchmark BRATS 2017 & BRATS 2015 datasets. This model revealed accurate segmentation; however, it exhibited more time and space complexity.

The authors in [13] suggested a system that uses an adaptive *k*-means algorithm for segmenting MRI. It overcomes the limitation of *k*-means clustering algorithm. The system is very strong and superior for initial clusters optimization. The authors in [14] have designed a modified level set segmentation. They used BRATS 2017 & MICCAI 2015 datasets for evaluation. The technique shows good performance even if the object boundary is weak, and it is able to extract shapes with arbitrary are object shapes. Time-consuming represents the main limitation.

The authors in [15] introduced a hybrid approach based on *k*-means clustering and level sets to segment brain tumor, to track the traditional level set problems of poor convergence towards the concavities of the tumor boundary and attraction towards wrong image features. The accuracy of this algorithm depends mainly on initializing contours that are determined randomly in most cases. Despite the diversity of brain tumor segmentation techniques, we see that there are still many approaches to be explored within the field of tumor segmentation. Our model uses the same concept of [15] to segment brain tumors with less time and accurately with a small number of iterations by utilizing two-step ABC clustering to initialize the contour and to overcome the problem of sensitivity to the initial clustering center selection, the poor global search ability and falling into the local optimal solution.

3 Proposed Methodology

3.1 Preprocessing Phase

For MRI quality improvement, pre-processing operations are performed to enhance the input images. This step is applied to convert 3D volume into 2D slices, remove the skull from the head and enhance contrast.

3.2 Two-Step Artificial Bee Colony Clustering Phase

Artificial Bee Colony (ABC) is an optimization algorithm, which is based on the foraging behavior of honey bees for numerical optimization problems, is applied to classification and clustering benchmark problems [16]. In ABC algorithm, the colony of artificial bees contains three groups of bees: employed bees, onlookers, and scouts. A bee waiting on the dance area for making a decision to choose a food source is called onlooker, and one going to the food source visited by it before is named employed bee. The other kind of bee is a scout bee that carries out a random search for discovering new sources. The position of a food source represents a possible solution to the optimization problem, and the nectar amount of a food source corresponds to the quality (fitness) of the associated solution. ABC algorithm performance is better than different population algorithms [17]. Yet, there are three drawbacks with ABC algorithm that can be listed as follows: (a) It is good at exploration but bad at exploitation, which makes the lousy convergence [18]. (b) For solving a complex problem, it can be trapped in local optima [19], and to overcome this problem, a limit function used. (c) It takes a long time [20]. To handle these drawbacks, we use a multi-step *k*-means algorithm to determine the initial positions of food sources [21]. Figure 1 illustrates the main diagram of the proposed model.

3.3 Level Set Segmentation Phase

The zero-Level Set (LS) function for a given domain Ω can represent the boundary C as proceeds [9]:

$$\forall(x,y) \in \Omega \begin{cases} C = & \{(x,y) : \emptyset(x,y) = 0\} \\ inside(C) & \{(x,y) : \emptyset(x,y) > 0\} \\ outside(C) & \{(x,y) : \emptyset(x,y) < 0\} \end{cases} \quad (1)$$

The zero LS function separates the image into two regions: one inside (foreground), signified as inside *C*, and the region outside (background) indicated as outside *C*. The length of the contour *C* sketched as [22]:

$$Length(C) = \int_{\Omega} |\nabla H(\emptyset(x,y))|dxdy = \int_{\Omega} \delta(\emptyset(x,y))|\nabla \emptyset(x,y)|dxdy \quad (2)$$

and the area inside the contour *C* is defined as

$$Area(C) = \int_{\Omega} H(\emptyset(x,y))dxdy \quad (3)$$

The methods begin with an initial level set and a given image *I*. The method updating process is done by gradient descent by minimizing an energy function. Herein ABC clustering algorithm is utilized to find indexes (initial contours) for a level set method to accurately fragment tumors from images. If the initial contour is far from the actual contour, therefore it will need a more significant number of iterations, and a massive quantity of computation is needed. So, once using ABC clustering, the initial contour set near the actual tumor contour, it will observe good convergence and costs less time and no. of iterations. To conclude, a two-step ABC algorithm is utilized here to deal with the three critical issues of the ABC algorithm, such as initial positions of food sources, solution search equation, and abandoned food location. To resolve these issues, an improvement is made in the original ABC algorithm; the *k*-means algorithm is used to generate the initial food source position for the ABC algorithm.

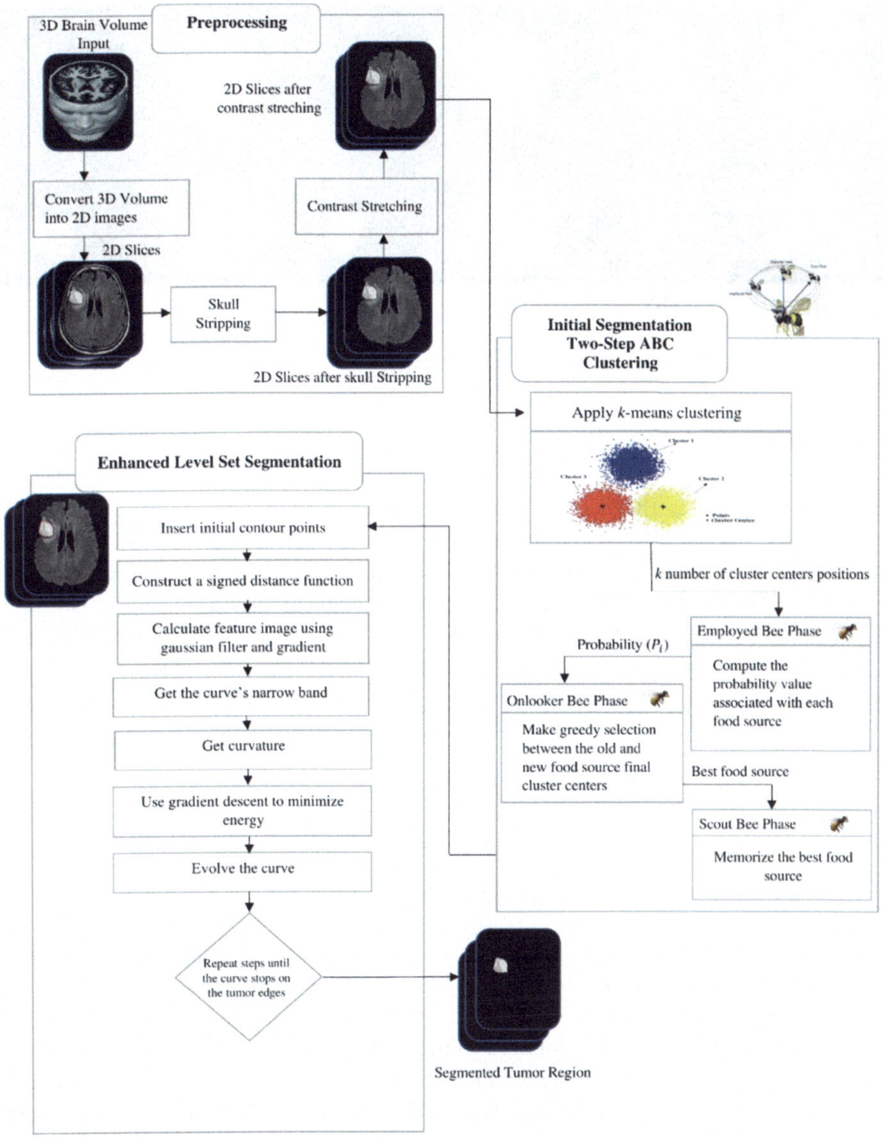

Fig. 1. The proposed diagram for 3D-MRI brain tumor segmentation

Figure 2 shows the segmentation results which represent one slice from one volume, and each column shows: (a) original MRI slice (b) skull stripping; (c) contrast stretching; (d) Two-step ABC Clustering; (e) using Two-step ABC clustering output indexes as initial contour; (f) tumor segmentation using level set segmentation method.

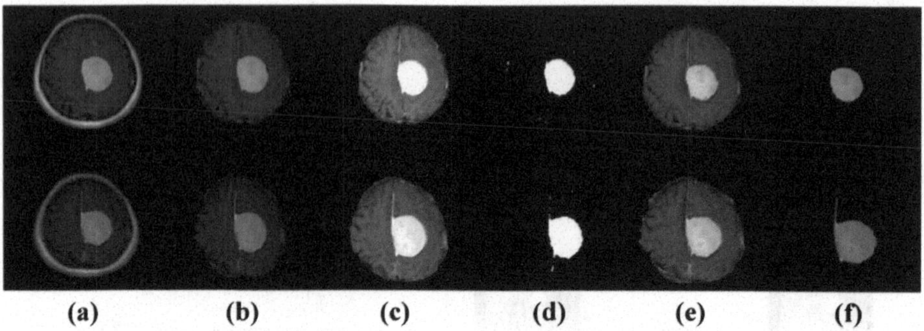

Fig. 2. 3D MRI slices segmentation with the proposed model

4 Experimental Results

This section evaluates the accuracy of the suggested model thorough conduction a series of experiments using benchmark brain datasets that include BRATS 2017 and Nanfang hospital dataset [23, 24]. BRATS 2017 dataset consists of 285 MRI Nanfang hospital datasets that have 233 patients. The experiments are done on core i3 HP laptop with eight GB RAM equipped with windows ten home edition platform. MATLAB language was used to implement the proposed system. Different metrics are used for assessment in brain tumor segmentation that includes Accuracy, Recall [25].

$$\boldsymbol{Accuracy} = \left(\frac{\boldsymbol{TP}+\boldsymbol{TN}}{\boldsymbol{TP}+\boldsymbol{TN}+\boldsymbol{FN}}\right) * \mathbf{100\%} \tag{4}$$

$$\boldsymbol{Recall} = \left(\frac{\boldsymbol{TP}}{\boldsymbol{TP}+\boldsymbol{FN}}\right) * \mathbf{100\%} \tag{5}$$

The first experiments were done to assess the accuracy of the proposed model in terms of evaluation matrices as compared to the state of the artworks for brain tumor segmentation. Tables 1 and 2 reveal the superiority of the suggested model as it increases the accuracy by 0.2 compared to the work presented in [14]. Despite the low of this percentage, it gives accurate results in medical images. One possible explanation of this result is that utilizing a two-step ABC classifier enhance the segmentation accuracy as they detect an accurate initial contour instead of choosing it randomly as the comparative methods work.

Table 1. Segmentation results for the BRATS 2017 dataset

Methods	Accuracy
Proposed system	99.5
Hasan et al. [11]	98.5
Virupakshappa et al. [14]	99.3

Table 2. Segmentation results for the BRATS 2017 dataset

Methods	Recall
Proposed system	98
Kermi et al. [9]	89
Chen et al. [12]	86.29
Bal et al. [10]	90

The second set of experiments was running to test the effect of ABC parameters in the segmentation accuracy. The two-step ABC algorithm has three main parameters that affect the accuracy and speed of execution. These parameters are considered essential; they are population size, number of cycles, and clusters number. Four clusters were chosen to efficiently partition the brain region into the Background (RB), Gray Matter (RG), White Matter (RW), and Tumor Cluster (RTC) [26]. Population size is the total number of bees used in searching the centroids of the clusters. Figure 3 shows the results after applying the two-step ABC clustering algorithm for different population sizes. If the population size is small, the problem will soon be convergent and will not close to a globally optimized solution. Nevertheless, if the population size is large, so a long time is needed for the algorithm to be convergent. The figure shows that the best result is 99.34% for the size of 10, 50, and 100. To save computational cost, population size = 10 is utilized.

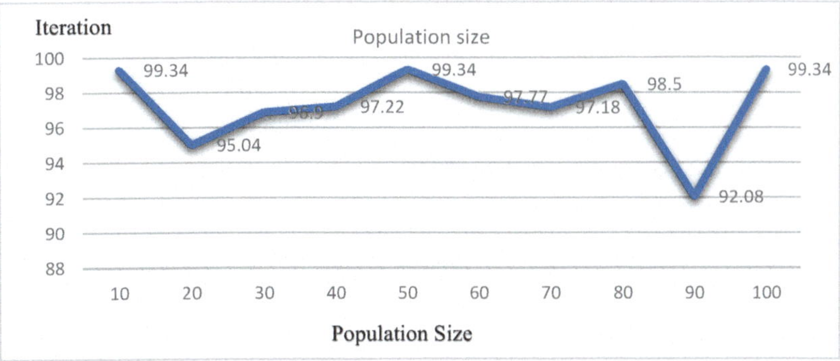

Fig. 3. The population size different numbers

Regarding the number of cycles parameter that usually measured by the number of loops. Figure 4 shows the results of using various numbers of cycles for testing one volume. The figure shows that the optimal result is 99.34% for 25 iterations. As a result of changing the number of iterations, the algorithm will strike some optimal balance between exploitation and exploration. Simultaneously, the two-step ABC iteration can speed up the convergence and guarantee continuous optimization. The optimal value of the cycle number is 25.

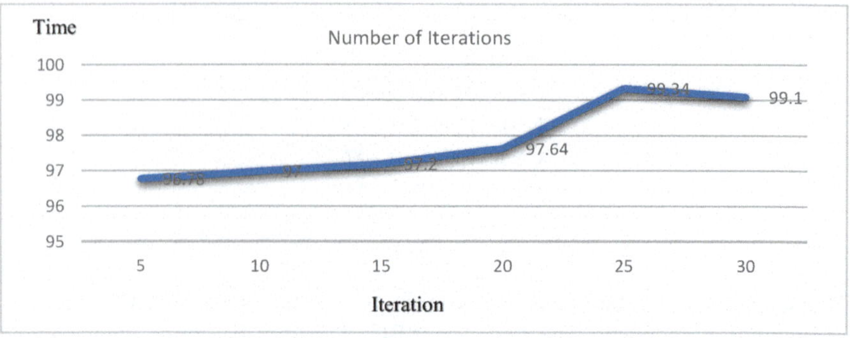

Fig. 4. The change of the iterations number

Using the best configured ABC parameters, the contour was figured very near to the tumor region. A variety of iterations numbers in level set segmentation affects the time. The obtained number of iterations for one volume is 15 with time 32.5 s. The experiments were repeated 30 times to guarantee the robustness of the proposed model against random factors, which may affect the algorithm behavior.

Figure 5 shows the time values of different iteration numbers for one volume.

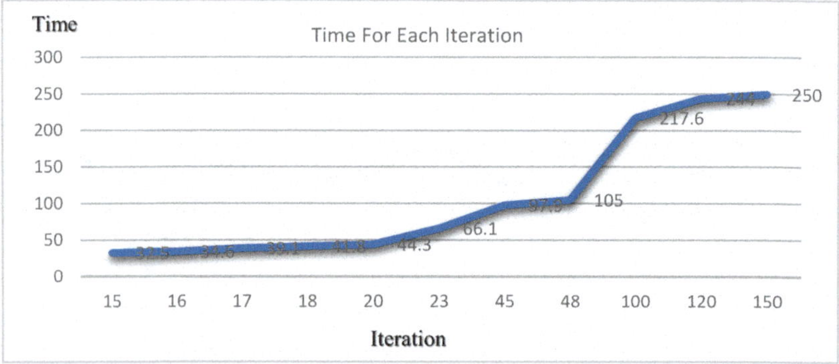

Fig. 5. The time values of different iteration numbers for the tested volume.

The level set has to be initialized well to achieve accurate output. Our ABC based level set segmentation algorithm decreases the number of iterations required; its time is reduced than the classical level set method.

5 Conclusion and Future Work

In this paper, an accurate model has been suggested to segment tumors from 3D MRI medical images. The proposed model uses a level set segmentation technique and ABC-based clustering for tumor region extracting. Using ABC, the select initial clusters based on k-means will help to extract accurate tumors instead of the commonly used trial and error contour detection methods. Also, will help to extract the tumor's edge in a fast way by reducing the number of iterations in segmentation. Experimental results confirmed the accuracy of the proposed model in terms of recall. Also, the suggested model achieves a small increase in accuracy as compared with [14]; however, it decreases the complexity in terms of the employed time by an average 7%. One limitation of the suggested model is that it has not been tested on volumes contains more than one tumor per one slice. Future work includes utilizing the parallel segmentation approach to further decrease the complexity of the suggested model.

References

1. El-Melegy, M.T., Abo El-Magd, K.M., Ali, S.A., Hussain, K.F., Mahdy, Y.B.: Ensemble of multiple classifiers for automatic multimodal brain tumor segmentation. In: International Conference on Innovative Trends in Computer Engineering (ITCE), Egypt (2019)
2. Aparna, R.M., Shanmugavadivu, P.: A survey of medical imaging, storage and transfer techniques. In: Proceedings of the International Conference on ISMAC in Computational Vision and Bio-Engineering 2018, pp. 17–29 (2019)
3. Popoola, J.J., Godson, T.E., Olasoji, Y.O., Adu, M.R.: Study on capabilities of different segmentation algorithms in detecting and reducing brain tumor size in magnetic resonance imaging for effective telemedicine services. Eur. J. Eng. Res. Sci. **4**(2), 23–29 (2019)
4. Işın, A., Direkoğlu, C., Şah, M.: Review of MRI-based brain tumor image segmentation using deep learning methods. Proc. Comput. Sci. **102**, 317–324 (2016)
5. Angulakshmi, M., Priya, G.L.: Automated brain tumour segmentation techniques- a review. Int. J. Imaging Syst. Technol. **27**(1), 66–77 (2017)
6. Mostafa, A.: CT liver segmentation using artificial bee colony optimisation. Proc. Comput. Sci. **60**, 1622–1630 (2015)
7. Akay, B., Karaboga, D.: A survey on the applications of artificial bee colony in signal, image, and video processing. SIViP **9**(4), 967–990 (2015)
8. Alqazzaz, S., Sun, X., Yang, X., Nokes, L.: Automated brain tumor segmentation on multimodal MR image using SegNet. Comput. Vis. Media **5**(2), 209–219 (2019)
9. Kermi, A., Andjouh, K., Zidane, F.: Fully automated brain tumor segmentation system in 3D-MRI using symmetry analysis of brain and level-sets. IET Image Proc. **12**(11), 1964–1971 (2018)
10. Bal, A., Banerjee, M., Chakrabarti, A., Sharma, P.: MRI brain tumor segmentation and analysis using rough-fuzzy C-means and shape based properties. J. King Saud Univ. – Comput. Inform. Sci. (2018)
11. Hasan, S., Ahmad, M.: Correction to: two-step verification of brain tumor segmentation using watershed-matching algorithm. Brain Inform. **5**(2), 8 (2018)
12. Chen, S., Ding, C., Liu, M.: Dual-force convolutional neural networks for accurate brain tumor segmentation. Pattern Recogn. **88**, 90–100 (2019)
13. Anitha, V., Murugavalli, S.: Brain tumour classification using two-tier classifier with adaptive segmentation technique. IET Comput. Vis. **10**(10), 9–17 (2018)
14. Virupakshappa, Amarapur, B.: Cognition-based MRI brain tumor segmentation technique using modified level set method. Cogn. Technol. Work **21**(3), 357–369 (2018)

15. Mahalakshmi, Krishnappa, H., Jayadevappa, D.: A hybrid approach for the segmentation of brain tumor using k-means clustering and variational level set. J. Adv. Res. Dyn. Control Syst. **10**(5), 258–264 (2018)
16. Hancer, E., Ozturk, C., Karaboga, D.: Artificial bee colony based image clustering method. In: IEEE Congress on Evolutionary Computation, Brisbane, Australia, pp. 10–15 (2012)
17. Kumar, Y., Sahoo, G.: A two-step artificial bee colony algorithm for clustering. Neural Comput. Appl. **28**(3), 537–551 (2015)
18. Gao, W., Liu, S., Huang, L.: A novel artificial bee colony algorithm based on modified search equation and orthogonal learning. IEEE Trans. Cybern. **43**(3), 1011–1024 (2013)
19. Chang, D., Zhang, X., Zheng, C.: A genetic algorithm with gene rearrangement for K-means clustering. Pattern Recogn. **42**(7), 1210–1222 (2009)
20. Kang, F., Li, J., Li, H.: Artificial bee colony algorithm and pattern search hybridized for global optimization. Appl. Soft Comput. **13**(4), 1781–1791 (2013)
21. Yan, X., Zhu, Y., Zou, W., Wang, L.: A new approach for data clustering using hybrid artificial bee colony algorithm. Neurocomputing **97**, 241–250 (2012)
22. Le, T.H.N., Gummadi, R., Savvides, M.: Deep recurrent level set for segmenting brain tumors. In: Frangi, A.F., Schnabel, J.A., Davatzikos, C., Alberola-López, C., Fichtinger, G. (eds.) MICCAI. LNCS, vol. 11072, pp. 646–653. Springer, Cham (2018)
23. Menze, B.H., et al.: The multimodal brain tumor image segmentation benchmark (BRATS). IEEE Trans. Med. Imaging **10**, 1993–2024 (2015)
24. James, J.A., Dasarathy, B.: Medical image fusion: a survey of the state of the art. Int. J. Inform. Fusion **19**, 4–19 (2014)
25. Hagwood, C., Bernal, J., Halter, M., Elliott, J.: Evaluation of segmentation algorithms on cell populations using CDF curves. IEEE Trans. Med. Imaging **31**(2), 380–390 (2012)
26. Bal, A., Banerjee, M., Sharma, P., Maitra, M.: Brain tumor segmentation on MR image using K-means and fuzzy-possibilistic clustering. In: 2nd International Conference on Electronics, Materials Engineering and Nano-Technology (IEMENTech) (2018)

Self-adaptive Parameters Optimization for Incremental Classification in Big Data Using Swarm Intelligence

Saad M. Darwish and Akmal I. Saber(✉)

Department of Information Technology,
Institute of Graduate Studies and Research,
Alexandria University, Alexandria, Egypt
{saad.darwish, igsr.Akmalibrahim}@alexu.edu.eg

Abstract. Nowadays, big data is one of the most technical challenges confront researchers and companies. The main challenge lies in the fact that big data sources usually formed in a continuous data stream. Thus, many previous researches present incremental data mining approaches to deal with the challenges of the data streams by adapting traditional machine learning algorithms. Artificial Neural Network (ANN) is a common technique in this field. The main challenge is how to optimize the neural network parameters to deal with a huge data arrive over time. These parameters, which are vital for the performance of a neural network, are called hyperparameters. Most earlier optimization approaches have dealt with big data containers instead of big data streams or handled big data streams with time consumed. This paper proposes an incremental learning process for ANN hyperparameters optimization over data stream by utilizing Grasshopper Algorithm (GOA) as a swarm intelligence technique. GOA is utilized to make a balance between exploration and exploitation to find the best set of ANN hyperparameters suitable for data stream. The experimental results illustrate that the proposed optimization model yields better accuracy results with appropriate CPU time.

Keywords: Big data · Incremental classification · Hyperparameters optimization · Grasshopper algorithm

1 Introduction

Since digital data sources are pervasively accessible nowadays, big data encompasses datasets of which size exceeds any typical database software tool to capture, accumulate, manage, or analyze [1]. Big data consists of increasingly complex yet non-structural sets of data. Such digitized data is further processed in a never-ending speed [2]. However, examining the big data is a challenge which researchers encounter at handling big data through adopting current analytical methods [3]. Those challenges are categorized into three groups: data challenges, which caused by the characteristics of the data itself, process challenges, which related to techniques, management challenges that related to privacy, security, authority [4]. The most cited data challenges are volume, variety, and velocity [1–5]. The volume indicates the huge amounts of data

A.-E. Hassanien et al. (Eds.): AICV 2020, AISC 1153, pp. 203–212, 2020.
https://doi.org/10.1007/978-3-030-44289-7_20

that must be handled. The variety indicates the variance of the data forms. The velocity indicates the speed at which the data arrives. These difficulties sore when the system is forced to manage it in real time [3].

Incremental learning gained more attention, especially in the context of big data and achieved learning via data streams [6]. This type of learning indicates the situation of the continuous model adaptation based on a regularly arriving data stream [7]. Training a classifier with data stream requires incremental training and modifications of the existing structure or parameters of the classifier to adapt to the changes, which arise from concurrent samples [8]. In addition, incremental learning is considered a train-and-forget strategy whereby only a segment of the data gets learned, one at a time. The algorithm needs to adapt rapidly to the recent contexts of the data stream. The contexts of the data stream may change in time, a phenomenon called concept drift. Concept drift is essential in incremental learning to determine how frequent a machine-learning model should refresh while handling data streams [9].

The main challenge in incremental learning techniques is the online model parameters adaptation. This is realized by incremental learning techniques, which use training samples one by one, without knowing all samples in advance. Thus, incremental learning can easily achieve by stochastic optimization approaches [7]. The incremental neural network learning problems solving grouped in three categories: (a) incremental construction algorithms on expanding the structure of the neural network, (b) evolving neural networks so the whole structure is optimized, and (c) hyperparameters optimization by tuning the values of hyperparameters to update the existing weights [9], which is the scope of this work.

The incremental learning process could be combined with hyperparameters optimization module, which searches for the best combination of hyperparameters values based on a given segment of the data stream [9]. To search a problem domain, a swarm intelligence algorithm processes a population of individuals. Each individual represents a potential solution to the problem being optimized. These solutions are initialized arbitrarily in the search space and are guided to more better areas. There are two types of techniques that apply swarm intelligence on big data. First, the swarm intelligence is directly applied on the big data itself. Second, it is applied to optimize the classification approach such as parameters tuning [10], which is the scope of this work.

In big data mining, specifically classification, a supervised learning model is trained by the values of attributes [11]. The swarm intelligence has arisen as a fast and reasonably accurate technique in solving complex search problems, such as classifying high-dimensionality data [12]. The grasshopper algorithm is presented recently as a swarm intelligence approach. A comparative study with genetic algorithm, particle swarm optimizer, artificial bee colony, flower pollination algorithm, bat algorithm, firefly algorithm, monarch butterfly optimization and biogeography-based optimization was showed the superiority of the grasshopper algorithm and ability of applied it to large-scale datasets [13]. This paper proposed a modified hyperparameters optimization model for Multilayer Perceptron neural network (MLP) in data stream mining without changing the network topology and structure. The suggested model words as a self-optimized hyperparameters for supervised incremental learning over the data stream through utilizing grasshopper algorithm. The grasshopper algorithm searches for the best combination of hyperparameters values based on a given segment of the data stream. The rest of the paper is organized as follows: Sect. 2 describes some of the recent related works. The detailed description of the proposed model has been shown in Sect. 3. In Sect. 4, the results and discussions are given. Finally, conclusions are drawn in Sect. 5.

2 Related Work

The field of ANN hyperparameters optimization has attracted more concern from the research community. The current state-of-the-art can be grouped into four categories. First, there are exhaustive methods that search exhaustively in the hyperparameters space such as grid search. Secondly, the model-specific methods that optimize hyperparameters for a certain model. Thirdly, the hyperparameters optimization methods based on the sequential model-based optimization model. Fourth, hyperparameters optimization based on evolutionary algorithms [14], which is the scope of this work. In [15] the authors presented an evolutionary algorithm-based model for automatically optimize the convolutional neural network structure by means of hyper-parameters. This approach based on optimization scheme and a new fitness function. The experiments on the MNIST dataset demonstrate enhancements to the state-of-the-art. However, they compare their results with a subset of representative state-of-the-art approaches without using augmentation on the training data.

The authors in [16] suggested a novel method for optimization of the hyperparameters of a support vector regression model. The method used a novel modified search pattern for genetic algorithms. The suggested method showed statistically improvement comparing to the traditional genetic algorithms, particle swarm optimization and firefly algorithm optimizers. On the downside, the method illustrated a slower convergence, due to the randomized regeneration, and the complexity was increased. The presented model in [17] is an automated parameter tuning for stream-processing systems. It supports standard black-box optimization algorithms as well as a novel gray-box optimization algorithm. The results proved that their approach outperformed the basic hill-climbing algorithm in three of the four test scenarios. A limitation of the approach is that it moves in a unidirectional way to change the parameters. Furthermore, the model is for offline parameter tuning and assumes that the scheduling is static.

The researchers in [18] offered a particle swarm optimization based parameter value selection approach to optimize the performance of deep learning models. The optimization approach is employed to select the number of hidden layers and the number of neurons in each one of them. The results are decreased the training times by 77%–85% with a better classification accuracy compared to the grid search. However, this approach is compared only with the grid search method without any indicates to other swarm intelligent algorithms. In [13] the authors employed grasshopper optimization algorithm (GOA) to achieve the optimal weights and biases for the multilayer perceptron neural networks. Their model applied to breast cancer and orthopedic patient's datasets. The results showed that this approach is beneficial for accurately classifying datasets of different sizes and complexities. However, it is based on the batch learning approach and should be tested with large-scale datasets. The work proposed in our paper, to the best of our knowledge based on google search, is the first work that utilizes swarm intelligence technique to optimize ANN hyperparameters on data streams.

3 The Proposed Model

A proposed model auto-adapts the hyperparameters of multilayer perceptron neural network (MLP) using grasshopper algorithm (GOA) for data streams that deal with concept drift. The MLP neural network learns incrementally by recognizing the relations between the input values and the output target class. The optimization is carried out under control by a mean-square error (MSE). The suggested model as shown in Fig. 1 uses two hyperparameters: Learning Rate and Momentum of neural network input weights. Learning rate [0, 1] determines the amount of weights to be updated; with 1 means all the weights are to be updates, 0 means none. Momentum is an intensity control parameter [0, 1] that is applied to the weights during updating. Default values of these hyperparameters not guarantee best results in training a neural network. Hence, the motivation for optimizing these values in the order to build an accurate classifier with shorten the optimization time along the process of incremental classification over the data stream. The optimization process using GOA can be described as follows:

1. The GOA starts by creating a random set of grasshoppers.
2. Perform the GOA internal phases.
3. The GOA assigned the Learning Rate and Momentum values for MLP.
4. Fitness function evaluates the quality of the MLP results using the MSE.
5. The process should find the lowest MSE value or the positions of grasshoppers updated to find new values for the hyperparameters of MLP.

In details, first, the samples of data are arrived to the memory from the data stream. When the amount of samples had been cached, the MLP neural network initialize with hyperparameters that are optimized by GOA. The inputs, weights and bias of MLP are summed using the summation function, as described in Eq. (1):

$$Q_j = \sum\nolimits_{i=1}^{n} w_{ij} I_i + B_j \tag{1}$$

where n represents the number of inputs. I_i indicates the input variable i. B_j is a bias. w_{ij} represents the connection weight. An activation function should be initiated using Q_j. Several forms of activation functions can be employed. By default, the sigmoid function is used:

$$f_j(x) = \frac{1}{1 + e^{-Q_j}} \tag{2}$$

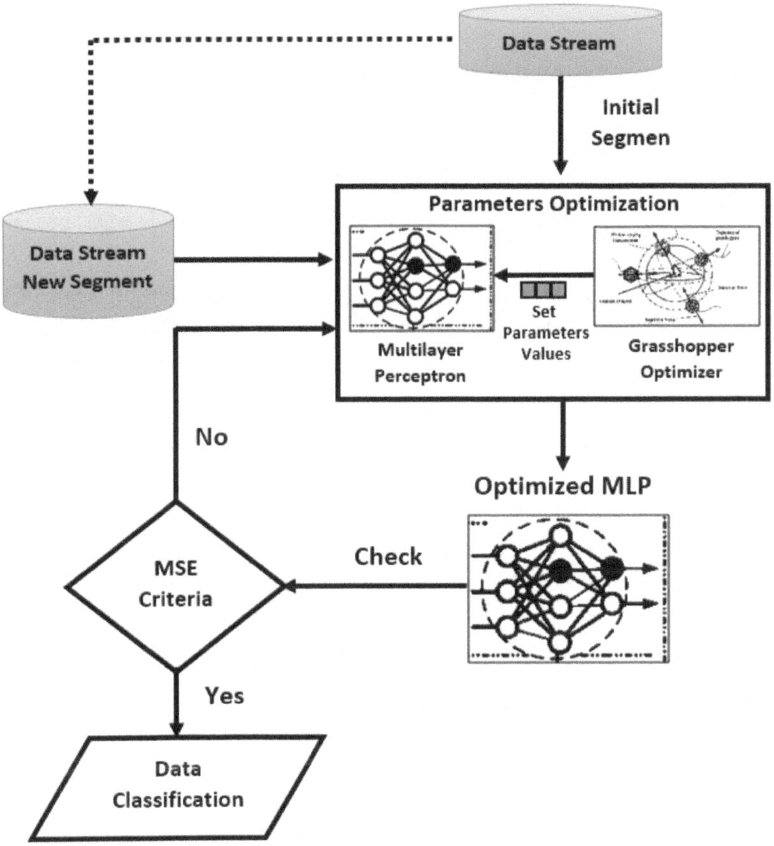

Fig. 1. The proposed hyperparameters optimization model

The output of the neuron_j can be obtained using:

$$y_i = f_j\left(\sum\nolimits_{i=1}^{n} w_{ij} I_i + B_j\right) \tag{3}$$

In [13, 19] to simulate the Grasshopper, a model was designed as:

$$x_i = S_i + G_i + A_i \tag{4}$$

where x_i the location of i insect, S_i shows the social communications, G_i indicates the gravity strength, and A_i the wind advection. These parameters can be obtained as:

$$S_i = \sum\nolimits_{j=1, j \neq i}^{N} s\left(d_{ij}\right)\hat{d}_{ij}, d_{ij} = \left|x_j - x_i\right|, \hat{d}_{ij} = \left(x_j - x_i\right)/d_{ij} \tag{5}$$

$$G_i = -g\hat{e}_g \tag{6}$$

$$A_i = -u\hat{e}_w \tag{7}$$

where d_{ij} is the distance of two grasshoppers $\hat{d}_{ij}$ is a unit vector, g indicates the gravitational constant, $\hat{e}_g$ is the unity vector of gravity, u indicates a constant drift, and $\hat{e}_w$ is the unity vector of wind. The s function computes either, social attraction or repulsion forces and can be attained by:

$$s(r) = fe^{-r/l} - e^{-r} \tag{8}$$

where f denotes the amplitude of attraction and l shows the length scale. Changing these parameters results in different social behaviors in artificial grasshoppers. In this work, the values of f and l parameters set to 0.5, and 1.6, respectively as stated in [13, 19]. According to Eqs. (5)–(7), the main rule in Eq. (4) can be reformulated as:

$$X_i = \sum\nolimits_{j=1, j \neq i}^{N} s\left(\left|x_j - x_i\right|\right) \frac{(x_j - x_i)}{d_{ij}} - g\hat{e}_g - u\hat{e}_w \tag{9}$$

where N is the size of swarm. According to Eq. (9), the population cannot be converged to the specific target, because the insects will rapidly reach to the comfort region. Therefore, a modified rule can be utilized as:

$$X_i^d = c\left(\sum\nolimits_{j=1, j \neq i}^{N} c\frac{UB_d - LB_d}{2} s\left(\left|x_j^d - x_i^d\right|\right) \frac{\left(x_j^d - x_i^d\right)}{dij}\right) + \hat{T}_d \tag{10}$$

where d shows dimension, UB_d and LB_d are the upper and lower boundaries, $\hat{T}_d$ is the best solution attained so far, and c is a decreasing factor. Note that the internal c assists GOA for decreasing of repulsion/attraction powers among grasshoppers, while the external c decreases the search tendency nearby the best location by more iteration. The c factor is attained as follows:

$$c = c_{max} - l\frac{c_{max} - c_{min}}{L} \tag{11}$$

where c_{max} and c_{min} are, respectively, the maximum and minimum values, l is the in progress iteration, and L denotes the upper bound of iterations. In this work, the c_{max} and c_{min} values are 1 and 0.00003, respectively [].

The second step, the optimized MLP is make a prediction and the mean-squared error (MSE) is counted as shown in Eq. (12):

$$MSE = \frac{1}{n}\sum\nolimits_{i=1}^{n} (y_i - \hat{y}_i)^2 \tag{12}$$

where y indicates the actual value, $\hat{y}$ shows the predicted one, and n indicates the total number of samples. In the proposed model, the MSE is used as a fitness function. If the MSE criteria is true, the optimized MLP is used and the incremental classification is happened. If not the MLP is readapted with new hyperparameters values using the GOA.

4 Experimental Results

All the experiments were conducted using MATLAB R2015a on a 64-bit Windows 10 with Intel core 2 duo T5670 processor at 1.80 GHz and a RAM of 3 GB. For training, testing and validation of the model, the dataset is partitioned into three parts: 70% of the dataset is utilized for training, 15% of the data is employed for testing and 15% for validation. The experimental dataset originally collected from the electricity market of New South Wales, Australia[1]. The electricity prices are determined by the dynamics of the demand and supply of the consumer market. The data are recorded every 5 min and are consisted of date/time, day of the week, present demand and present supply in both states of NSW and Victoria. Some attributes values change their ranges over time and leading to concept-drifts.

The dataset is sequentially divided into 10 portions of equal length. Each portion is 10% of the whole dataset. Having a large length than 10% may lead to long computation time for optimization and it may lead to bad prediction [9]. Default values of hyperparameters are set for all experimentation runs except the Learning Rate and Momentum of the neural network that are auto-optimized. For a fair comparison, both approaches are run 30 times to evaluate the means of values to obtain stable values. The number of swarms is 100 with 10 iterations for GOA. The results are tabulated in Table 1.

The results showed that the proposed model enhances the accuracy of incremental classification for whole dataset by 6.26%, and for the segments of data by 5.2–13.2% as shown in Fig. 2. Moreover, the CPU time of the optimization process is acceptable and in some segments less than the CPU time of MLP before optimization. However, the proposed model became time-consuming when attempting to decrease the MSE value below 0.1. Second limitation of the suggested model is that the increasing number of iterations leads the hyperparameters values assigned by GOA dropped to zero which causes the decline. The same results are obtained if the GOA amplitude of attraction and its length scale ware altered.

[1] http://sourceforge.net/projects/moaatastream/files/Datasets/Classification/elecNormNew.arff.zip/download.

Table 1. The incremental classification accuracy, MSE and CPU time.

Data	Default MLP (before optimization)			The optimized MLP		
	Accuracy	MSE	CPU time	Accuracy	MSE	CPU time
All	73.67%	0.1849	212.9323	79.93%	0.1473	233.9687
Segment1	79.13%	0.1698	28.0313	88.63%	0.1035	21.7813
Segment2	79.17%	0.1700	22.5729	85.37%	0.1245	34.2292
Segment3	69.27%	0.2370	22.0886	82.47%	0.1373	34.0990
Segment4	75.30%	0.1916	22.2604	82.37%	0.1401	53.9479
Segment5	77.03%	0.1775	21.8021	82.97%	0.1366	20.7083
Segment6	74.03%	0.1921	22.1406	82.93%	0.1355	27.3906
Segment7	69.30%	0.2361	21.8646	79.23%	0.1540	80.6510
Segment8	70.90%	0.2124	21.9688	79.43%	0.1520	60.5990
Segment9	71.37%	0.2093	22.0696	80.37%	0.1487	33.9219
Segment10	76.20%	0.1795	22.0729	81.40%	0.1400	33.8802
Average	74.17%	0.1975	22.6872	82.52%	0.1372	40.1208
Minimum	69.27%	0.1698	21.8021	79.23%	0.1035	20.7083
Maximum	79.17%	0.2370	28.0313	88.63%	0.1540	80.6510

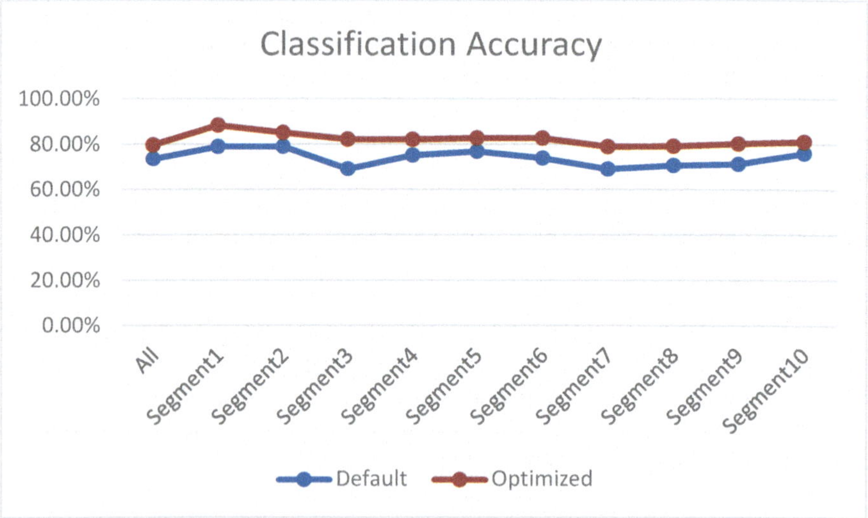

Fig. 2. The accuracy of MLP under the two incremental classification approaches

5 Conclusion

The streaming nature of big data presents huge computational challenges in the data mining field. Therefore, it requires a novel computation approach to dynamically deal with a large scale of data that change over time. This paper investigated the possibility of using incremental learning for MLP neural network combined with adaptive hyperparameters-optimization technique. The literature exposed that GOA can be used in optimization level to obtain a high classification accuracy. Thus, this work employed GOA to identify optimal values for the hyperparameters of MLP aiming to create fast and accurate classifier over data streams. This work is based on data streams taken from the Australian electricity market as a case study. The proposed optimization model showed statistically improvements comparing to the MLP before optimization. For future works, the effects of GOA parameters itself and the optimization process for more than two hyperparameters need more investigation.

References

1. Emani, C.K., Cullot, N., Nicolle, C.: Understandable big data: a survey. Comput. Sci. Rev. **17**, 70–81 (2015)
2. Bayera, H., Aksogana, M., Celikb, E., Kondilogluc, A.: Big data mining and business intelligence trends. J. Asian Bus. Strategy **7**(1), 23–33 (2017)
3. Ruzgas, T., Jakubėlienė, K.: Big data mining and knowledge discovery. J. Commun. Technol. Electron. Comput. Sci. **9**(1), 5–9 (2016)
4. Sivarajah, U., Kamal, M.M., Irani, Z., Weerakkody, V.: Critical analysis of big data challenges and analytical methods. J. Bus. Res. **70**, 263–286 (2017)
5. Mnich, M.: Big data algorithms beyond machine learning. KI-Künstliche Intell. **32**(1), 9–17 (2018)
6. Losing, V., Hammer, B., Wersing, H.: Incremental on-line learning: a review and comparison of state of the art algorithms. Neurocomputing **275**, 1261–1274 (2018)
7. Gepperth, A., Hammer, B.: Incremental learning algorithms and applications. In: Proceedings on European Symposium on Artificial Neural Networks, Computational Intelligence and Machine Learning, Belgium, pp. 357–368 (2017)
8. Bose, S., Huber, M.: Incremental learning of neural network classifiers using reinforcement learning. In: Proceedings on IEEE International Conference on Systems, Man, and Cybernetics (SMC), Budapest, pp. 97–103 (2016)
9. Fong, S., Fang, C., Tian, N., Wong, R., Yap, B.W: Self-adaptive parameters optimization for incremental classification in big data using neural network. In: Big Data Applications and Use Cases, pp. 175–196. Springer (2016)
10. Cheng, S., Shi, Y., Qin, Q., Bai, R.: Swarm intelligence in big data analytics. In: Intelligent Data Engineering and Automated Learning, pp. 417–426. Springer (2013)
11. Fong, S., Yang, X., Deb, S.: Swarm search for feature selection in classification. In: Proceedings of IEEE International Conference on Computational Science and Engineering, Sydney, pp. 902–909 (2013)
12. Fong, S., Deb, S., Yang, X., Li, J.: Feature selection in life science classification: metaheuristic swarm search. IT Prof. **16**(4), 24–29 (2014)
13. Heidari, A.A., Faris, H., Aljarah, I., Mirjalili, S.: An efficient hybrid multilayer perceptron neural network with grasshopper optimization. Soft Comput. 1–18 (2018)
14. Schilling, N., Wistuba, M., Drumond, L., Schmidt-Thieme, L.: Hyperparameter optimization with factorized multilayer perceptrons. In: Joint European Conference on Machine Learning and Knowledge Discovery in Databases, pp. 87–103. Springer, Cham (2015)

15. Bochinski, E., Senst, T., Sikora, T.: Hyper-parameter optimization for convolutional neural network committees based on evolutionary algorithms. In: Proceedings of IEEE International Conference on Image Processing (ICIP), Beijing, pp. 3924–3928 (2017)
16. Tsirikoglou, P., Abraham, S., Contino, F., Lacor, C., Ghorbaniasl, G.: A hyperparameters selection technique for support vector regression models. Appl. Soft Comput. **61**, 139–148 (2017)
17. Bilal, M., Canini, M.: Towards automatic parameter tuning of stream processing systems. In: Proceedings of the 2017 Symposium on Cloud Computing, pp. 189–200. ACM (2017)
18. Qolomany, B., Maabreh, M., Al-Fuqaha, A., Gupta, A., Benhaddou, D.: Parameters optimization of deep learning models using Particle swarm optimization. In: Proceedings of 13th International Wireless Communications and Mobile Computing Conference (IWCMC), pp. 1285–1290. IEEE (2017)
19. Saremi, S., Mirjalili, S., Lewis, A.: Grasshopper optimisation algorithm: theory and application. Adv. Eng. Softw. **105**, 30–47 (2017)

A Modified Sunflower Optimization Algorithm for Wireless Sensor Networks

Aliaa F. Raslan[1(✉)], Ahmed Fouad Ali[2], and Ashraf Darwish[3]

[1] Higher Institute for Management Information Systems, Suez, Egypt
aliaa.raslan2012@gmail.com
[2] Department of Computer Science, Faculty of Computers and Informatics, Suez Canal University, Ismailia, Egypt
[3] Faculty of Science, Helwan University, Cairo, Egypt

Abstract. Maximizing the lifetime of the wireless sensor networks (WSNs) is one of the biggest challenges due to the difficulty of changing their batteries when they run out of energy. Low Energy Adaptive Clustering Hierarchy (LEACH) is one of the most famous protocols which have applied to solve this problem. The main drawback of LEACH is that it may choose a cluster head that has less energy. Therefore, it will die in a short time and the network lifetime will finish rapidly. Many researchers have applied swarm intelligence algorithm to solve this problem however most of these algorithms trapped in local minima and suffer from premature convergence. In this paper, we combine the sunflower optimization algorithm (SFO) with the lèvy flight to maximize the WSNs lifetime. Such a combination can help the SFO algorithm to avoid trapping in local minima due to the random walk of the lèvy flight. The proposed algorithm is called a modified sunflower optimization algorithm (MSFO). To verify the superiority of the MSFO we compare it with five algorithms in literature for different numbers of nodes and cluster heads. The results show that the lifetime of the WSNs which is using the proposed MSFO is longer than their lifetime when they applied the other algorithms.

Keywords: Sunflower optimization algorithm · Lèvy flight · Wireless sensor network · Cluster head selection

1 Introduction

One of the biggest challenges for wireless sensor networks (WSNs) is maximizing the network lifetime. Wireless sensor network is comprised of a major amount of sensor nodes characterized by delimited energy (mostly work on batteries). On another hand sensor nodes are usually deployed in serious and risky places, it will be difficult to exchange it if their batteries run out of energy and this causes the network to die. So, it is needed to use the sensor node's energy in a more efficient way to keep the network work for a longer time. Due to the significance of WSNs and its diverse applications, there are many researchers

A.-E. Hassanien et al. (Eds.): AICV 2020, AISC 1153, pp. 213–222, 2020.
https://doi.org/10.1007/978-3-030-44289-7_21

concerned to apply their algorithms to resolve the problems confronting wireless sensor networks.

The clustering approach is an effective way for decreasing the energy consumption of sensor nodes. Consequently, the life period of the network will be increased. Many various algorithms and protocols have been produced for clustering in WSNs. Some of them depend on heuristic methods such as: LEACH protocol [4], which choose the cluster heads (CHs) in a random way without taking into consideration the residual energy for each node. The main weakness of LEACH is that may choose a cluster head that has less energy. Therefore, it will die in a short time and the network lifetime will finish rapidly. E-LEACH protocol has been proposed in [11], It takes the remaining energy of nodes into consideration when electing the cluster heads and thus increase the network life. Least distance clustering (LDC) protocol has been proposed in [2] by Bari et al. for enhancing the WSN lifetime by taking into account the distance between nodes and CHs when assigning nodes to clusters. In PEGASIS [7], the nodes have been arranged in a chain and every node can send or receive data from its neighbors only and a random node from the chain is selected to be a cluster head. Other clustering algorithms depend on nature-inspired techniques such as: LEACH-Centralized (LEACH-C) [5] in which the base station is responsible for choosing the CHs depending on the energy and the distance of nodes. Rao et al. [9] proposed algorithm using particle swarm optimization [6] named (PSO-ECHS) for selecting the cluster heads more efficiently and thus reducing the consumption of energy and maximizing the network lifespan. The authors in [1] proposed a novel algorithm that depends on decreasing the intra-cluster distance among nodes to increase the network lifetime by using a cuckoo search (CS) algorithm [12]. The authors in [10] proposed a number of algorithms to resolve the WSN lifetime problem by using the Bat algorithm (BA) [13], Flower pollination algorithm (FPA) [14] and Antlion algorithm (AL) [8].

This paper proposes a modified sunflower optimization algorithm (MSFO) by combining the lèvy flight with the standard sunflower optimization algorithm [3] for WSN efficient cluster head selection. The algorithm includes two significant phases: the election of cluster head phase and the cluster formation phase. The algorithm takes into consideration the distance between nodes and CHs, the distance between CHs and the base station, the remaining energy and the degree of nodes. The proposed algorithm evaluated and proved its superiority to some existing algorithms.

The paper is organized as follow. The objective function and the problem formulation of the WSNs lifetime is presented in Sect. 2. In Sect. 3, we highlight the steps of the standard sunflower optimization algorithm and explain our modified algorithm. In Sect. 4, we present the performance analysis and the results of the MSFO algorithm. In the end, we conclude the paper in Sect. 5.

2 Problem Formulation

In order to reduce the energy consumption in the WSNs, we have to minimize the distance between the CHs and the base station as shown in the following function.

$$Minimize\ \ f_1 = \sum_{j=1}^{m} \frac{1}{l_j} \Big(\sum_{i=1}^{l_j} dist(s_i, CH_j) + dist(CH_j, BS) \Big) \tag{1}$$

Where l_j is the number of sensor nodes in cluster j, dist (s_i, CH_j) is the distance between the sensors and the cluster head, dist (CH_j, BS) is the distance between the cluster head and the base station.

Also, we have to maximize the energy of the selected cluster heads as follow.

$$Maximize\ \ f_2 = \frac{1}{\sum_{j=1}^{m}(E_{CH_i})} \tag{2}$$

Where E_{CH_i} is the total energy for the selected cluster heads.

The final objective function can be formulated as follow.

$$\begin{aligned} Minimize \quad & F = \alpha \times f_1 + (1-\alpha) \times f_2 \\ S.t \quad & dist(s_i, CH_i) \leq d_{max}\ \ \forall\ \ s_i \in S,\ \ CH_j \in C \\ & dist(CH_j, BS) \leq R_{max}\ \ \forall\ \ CH_j \in C \\ & E_{CH_i} > T_H,\ \ 1 \leq j \leq m \\ & 0 < \alpha < 1 \\ & 0 < f_1, f_2 < 1 \end{aligned} \tag{3}$$

Where α is the control parameter between the distance and the energy, S, C are the set of sensor nodes and cluster heads respectively, R_{max} is the maximum communication range of a selected cluster heads, T_H is the threshold energy for a cluster head, E_{CH_i} is the energy of the selected cluster head.

3 A Modified Sunflower Optimization Algorithm (MSFO)

The structure of the proposed MSFO algorithm and its main steps will be described in the following subsections.

3.1 Sunflower Optimization Algorithm (SFO)

In the following subsections, we will present the steps of the standard Sunflower optimization algorithm (SFO) and the modifications which we have applied to improve its performance.

Biological and Natural Behaviors. The Sunflower optimization algorithm (SFO) is a nature inspired algorithm that mimics the movement of the sunflowers toward the sun. During the movement, fertilization may be happen between the closest neighboring sunflowers. The aggregate absorbed radiation for each sunflower is based on the distance between it and the sun. Increasing the distance between the sunflowers and the sun will reduce the absorbed radiation for them as shown in the following equation.

$$I = \frac{P_s}{4\pi d^2} \tag{4}$$

Where I is the intensity of the solar radiation, P_s is the power of the sun and d is the distance between the sun and each sunflower.

Sunflower Direction Adjustment. Each sunflower adjusts its direction to the sun as shown in Eq. 5

$$\boldsymbol{s}_i = \frac{X^* - X_i}{\|X^* - X_i\|} \quad i = 1, 2, \ldots, n \tag{5}$$

Where X^* is the overall best solution, X_i is the current solution and n is the population size.

The step of individuals (sunflowers) towards the sun is represented as follow.

$$d_i = \gamma \times P_i(\|X_i + X_{i-1}\| \times \|X_i + X_{i-1}\| \tag{6}$$

Where γ is the sunflower's inertial displacement, $P_i(\|X_i + X_{i-1}\|$ is the probability of the pollination that each sunflower can fertilize the closest neighboring sunflower to produce a new sunflowers in a new position based on the distance among the sunflowers. Each sunflower step is restricted to avoid exceeding the following value

$$d_{max} = \frac{\|X_{max} - X_{min}\|}{2 \times n} \tag{7}$$

Where X_{max}, X_{min} are upper and lower bounds and n is the population size.

Sunflower Position Update. Each sunflower (individual) i updates its position to produce a new generation of sunflowers based on the direction and the step of sunflowers towards the sun as follow.

$$X_{i+1} = X_i + d_i \times \boldsymbol{s}_i \tag{8}$$

Where X_{i+1} is the position of the new generated sunflower (individual).

Algorithm 1. Sunflowers optimization algorithm

Set the initial values for the mortality rate m, the population size n, pollination rate P and the maximum iteration number max_{itr}.
Set $t := 0$.
Generate the population $X_i^{(t)} \in [L, U]$ randomly, $i = 1, \dots, n$.
Evaluate the fitness function of the individuals (sunflowers) in the population $f(X_i^{(t)})$.
Assign the overall best solution in the population X^*.
Adjust all the individuals (sunflowers) toward the best solution (sun) as shown in Equation 5.
repeat
  Calculate the direction vector (fitness) for each individual.
  Remove the worst $m\%$ individuals.
  Compute the step of the individuals toward the sun as shown in Equation 6.
  Fertilize the best sunflowers around the sun.
  Check the maximum step of individuals as shown in Equation 7.
  Update the individuals as show in Equation 8.
  Evaluate the new individuals.
  Accept the new individuals if their fitness are better than the current values.
  Set $t = t + 1$.
until $(t > max_{itr})$.
Produce the best individual.

The SFO Algorithm. We present the main steps of the SFO algorithms as shown in Algorithm 1 as follow

We can summarize the steps of Algorithm 1 as follow.

- **Parameter initialization.** The parameters of the SFO algorithm are initialized such as mortality rate m, the population size n, pollination rate P and the maximum iteration number max_{itr}.
- **Counter initialization.** The iteration counter is initialized $t := 0$.
- **Individuals generation.** The population contains individuals which are generated randomly $X_i^{(t)}$.
- **Individuals evaluation.** The individuals in the population are evaluated by calculating their fitness value.
- **Assign the global best individual.** The overall best individual is assigned as the global best solution.
- **Individuals adjustment.** Each individual adjust its position towards the sun as show in Eq. 5.

- **Individuals mortality.** The direction vector for each individual is calculated and the worst $m\%$ individuals are removed.
- **Individuals movements.** The step of each individual towards the best individual is calculated as shown in Eq. 6 and the maximum step for each individual is calculated as shown in Eq. 7.
- **Individuals update.** The individuals are update their positions as show in Eq. 8.
- **Assign the new global best individual.** The individuals are evaluated and the overall best individual is assigned.
- **Counter increment.** The iteration number is increased $t = t + 1$.
- **Termination criteria satisfaction.** The overall processes are repeated until termination criteria are satisfied which are in our algorithm reaching to the maximum number of iterations max_{itr}.
- **The global best individual production.** The overall best individual is produced.

3.2 The Modified Sunflower Optimization Algorithm (MSFO)

The steps of the MSFO algorithm are the same steps of the SFO algorithm except that we have invoked the lèvy flight in the MSFO algorithm to escape from trapping in local minima. The step size of the lèvy flight is random which can increase the diversity (exploration) of the MSFO algorithm, However in the standard SFO algorithm, the step size is based on the location of the best solution (sun) as shown in Eq. 6. Such way of solution's update can cause a premature convergence and let the algorithm trapping in local minima.

So, In the MSFO, each solution can update itself as follow.

$$X_{i+1} = X_i + levy(\nu) \times \boldsymbol{s}_i \tag{9}$$

Where Lèvy (ν) is the Lèvy distribution.

4 Numerical Experiments

The MSFO algorithm was investigated using MATLAB R2018a and on Core i5 processor 2.30 GHz, RAM 8 GB running on Windows 7 64bit. The energy and the network models in this work are used on [4].

4.1 Parameter Setting

We summarize the parameter values for the network and the MSFO algorithm in Tables 1 and 2 respectively as follow.

Table 1. The parameters of the network.

Parameters	Definitions	Values
NS	Network size	$200 \times 200\,\text{m}^2$
BS	Base station location	(200, 200)
n	Number of nodes	[300–500]
R	Communication range	100
CHs	Percentage of cluster heads	10%
E_0	Initial energy of sensor node	2 J
E_{elec}	Dissipated energy per bit	50 nJ/bit
ε_{mp}	Amplification energy for multi-path model	$100\,\text{PJ/bit/m}^2$
K	Packet length	4000 bits
EDA	Data aggregation energy	5 nJ/bit

Table 2. The parameter of the MSFO algorithm.

Parameters	Definitions	Values
Pl	Number of plants	30
P	Pollination rate	0.05
m	Mortality rate	0.1
α	Control parameter between the distance and the energy	0.3
max_{itr}	Number of iterations	5000

4.2 Performance Analysis

We present in this subsection, the effect of combining the lèvy flight in the proposed MSFO algorithm and the results of it as follow.

4.3 The Effect of Invoking the Lèvy flight in the MSFO algorithm

To investigate the effect of combining the lèvy flight with the standard SFO algorithm in the proposed MSFO, we present the convergence curves of the standard SFO, the LEACH protocol and the proposed MSFO algorithm at 300, 400 and 500 nodes as shown in Fig. 1. As sown in Fig. 1, the number of alive nodes for the LEACH protocol and the SFO algorithm is decreased rapidly and faster than their number in the proposed MSFO. We can conclude from Fig. 1 that combining the lèvy flight with the SFO algorithm can improve its performance and increase the network lifetime.

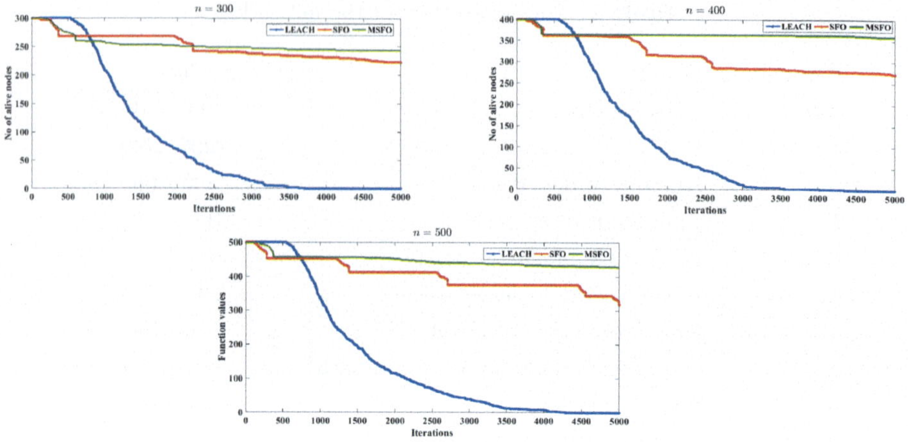

Fig. 1. The effect of invoking the lèvy flight in the MSFO algorithm

Comparison Between the MSFO Algorithm and Other State of Art Algorithms. The performance of the proposed MSFO algorithm is tested by comparing it with five algorithms. The total energy consumption of the LEACH, E-LEACH, LEACH-C and LDC are reported from their original paper [9]. The results of all algorithms are reported at 5000 round for 30, 40 and 50 CHs as shown in Table 3 and Fig. 2.

Table 3. Total energy consumption e_{total} at (CHs, n).

Algorithm	(30, 300)	(40, 400)	(50, 500)
	e_{total}	e_{total}	e_{total}
LEACH	601.13	801.48	1002.19
E-LEACH	600	800	1000
L EACH-C	599.79	796.97	991.07
LDC	536.53	721.58	898.56
SFO	407.16	544.42	690.98
MSFO	**377.17**	**477.31**	**601.09**

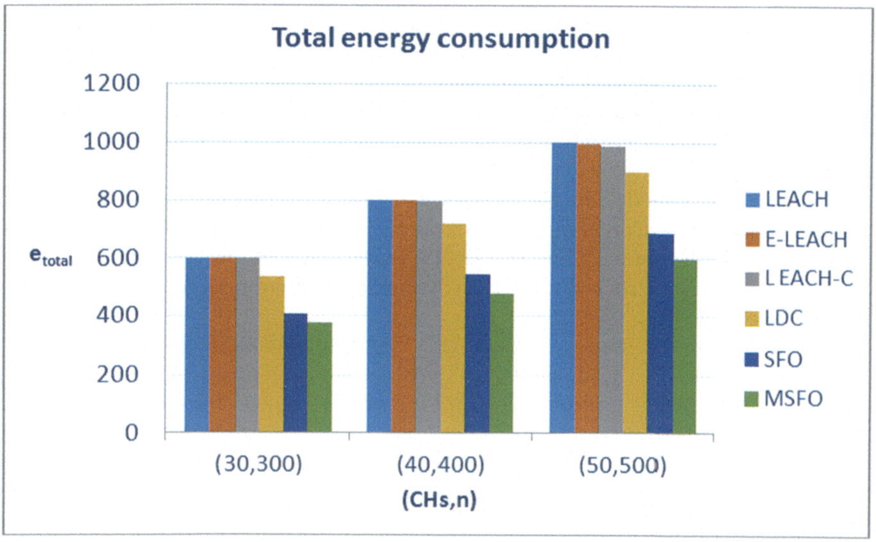

Fig. 2. Total energy consumption e_{total} at (CHs, n)

5 Conclusion

In this paper, we proposed a new algorithm to reduce the consumption energy in WSNs. The proposed algorithm is called a modified sunflower optimization (MSFO). In MSFO, The CHS have selected in a dynamic way to aggregate the data and transferring it to the base station with low energy consumption. Combining the lèvy flight operator in the standard sunflower optimization algorithm (SFO) can accelerate the convergence of the MSFO algorithm by escaping from the local minima. The MSFO algorithm was investigated by comparing it against five algorithms in the literature by forming different clusters size with different numbers of CHS and nodes. The results of the MSFO show that it is a promising algorithm and it can reduce the consumption energy of the WSNs.

References

1. Adnan, M.A., Razzaque, M.A., Abedin, M.A., Reza, S.S., Hussein, M.R.: A novel cuckoo search based clustering algorithm for wireless sensor networks. In: Advanced Computer and Communication Engineering Technology, pp. 621–634. Springer, Cham (2016)
2. Bari, A., Jaekel, A., Bandyopadhyay, S.: Clustering strategies for improving the lifetime of two-tiered sensor networks. Comput. Commun. **31**(14), 3451–3459 (2008)
3. Gomes, G.F., da Cunha, S.S., Ancelotti, A.C.: A sunflower optimization (SFO) algorithm applied to damage identification on laminated composite plates. Eng. Comput. **35**(2), 619–626 (2019)

4. Heinzelman, W.R., Chandrakasan, A., Balakrishnan, H.: Energy-efficient communication protocol for wireless microsensor networks. In: Proceedings of the 33rd Annual Hawaii International Conference on System Sciences, p. 10. IEEE (2000)
5. Heinzelman, W.B., Chandrakasan, A.P., Balakrishnan, H.: An application-specific protocol architecture for wireless microsensor networks. IEEE Trans. Wirel. Commun. **1**(4), 660–670 (2002)
6. Eberhart, R., Kennedy, J.: Particle swarm optimization. In: Proceedings of the IEEE International Conference on Neural Networks, vol. 4, pp. 1942–1948 (1995)
7. Lindsey, S., Raghavendra, C.S.: PEGASIS: power-efficient gathering in sensor information systems. In: Proceedings, IEEE Aerospace Conference, vol. 3, pp. 3–3. IEEE (2002)
8. Mirjalili, S.: The ant lion optimizer. Adv. Eng. Softw. **83**, 80–98 (2015)
9. Rao, P.S., Jana, P.K., Banka, H.: A particle swarm optimization based energy efficient cluster head selection algorithm for wireless sensor networks. Wirel. Netw. **23**(7), 2005–2020 (2017)
10. Sharawi, M., Emary, E.: Clustering optimization for WSN based on nature-inspired algorithms. In: Nature-Inspired Computation in Engineering, pp. 111–132. Springer, Cham (2016)
11. Xiangning, F., Yulin, S.: Improvement on LEACH protocol of wireless sensor network. In: 2007 International Conference on Sensor Technologies and Applications (SENSORCOMM 2007), pp. 260–264. IEEE (2007)
12. Yang, X.S., Deb, S.: Cuckoo search via Lévy flights. In: 2009 World Congress on Nature & Biologically Inspired Computing (NaBIC), pp. 210–214. IEEE (2009)
13. Yang, X.S.: A new metaheuristic bat-inspired algorithm. In: Nature Inspired Cooperative Strategies for Optimization (NICSO 2010), pp. 65–74. Springer, Heidelberg (2010)
14. Yang, X.S., Karamanoglu, M., He, X.: Multi-objective flower algorithm for optimization. Procedia Comput. Sci. **18**, 861–868 (2013)

Deep Learning and Applications

Hybrid Approach for Improving Intrusion Detection Based on Deep Learning and Machine Learning Techniques

Merna Gamal[1(✉)], Hala Abbas[2(✉)], and Rowayda Sadek[1(✉)]

[1] Department of Information Technology, Helwan University, Cairo, Egypt
merna_gamal43@yahoo.com, rowayda_sadek@yahoo.com
[2] Department of Computer Science, Helwan University, Cairo, Egypt
ass.hala@gmail.com

Abstract. An intrusion detection system works to recognize the attacks using either the signature or signature-less method. The signature-less method suffers from a lot of false alarms that affect accuracy and recall. Commonly used IDS (intrusion detection system) Dataset experiences imbalance which causes a high false alarms rate. Nowadays CNN (convolution neural network) excels in image and computer vision. Using CNN in IDS is promising. The paper proposes a hybrid approach between CNN and ML (SVM, KNN). CNN is efficiently utilized to get important features from the dataset. Then ML used to classify the data. Using the hybrid approaches to benefit from the advantage of machine learning (high accuracy, Low false alarms) and Deep learning which deal with a large amount of data and reduce the number of feature of the dataset (feature extraction). In this paper we used 10% of KDDcup1999 dataset. The experimental results showed enhancement in the detection accuracy to 99.3 and reduction in losses to 0.03.

Keywords: CNN (Convolution Neural Network) · SVM (Support Vector Machine) · KNN (K-Nearest Neighbor) · ML (Machine learning)

1 Introduction

The intrusion detection systems can be defined as software products or as hardware products [1], which detects and identifies the episode caused by the attackers. IDSs monitor the information that related to the intrusions, disconnect them, and make reports for security administrators to check [2] in the real-time environment. The intrusion detection system can be set as a security tool that protects the network from Variety attacks, e.g., firewalls [3]. It also provides the prevention from different intrusions that occurred by the attackers. Because of the variation of network configuration, several types of IDS systems exist to deal with this variation. Each type of IDS has benefits and disadvantages in detection, cost, and configuration. There are two types of IDS techniques which are Host-based intrusion detection system (HIDS) and Network-based intrusion detection system (NIDS). Host-based Intrusion Detection System monitors the events that happen on the host side and collects the data on the host computer for the analysis. Network-based intrusion detection systems monitor the

A.-E. Hassanien et al. (Eds.): AICV 2020, AISC 1153, pp. 225–236, 2020.
https://doi.org/10.1007/978-3-030-44289-7_22

traffic and analyze the collected data to detect suspicious packets. This system is very good at recognizing the users that do not have authorization before accessing the computer. SNORT is a popular example of NIDS to detect attacks in the network. There are two types of detection which are anomaly detection and misuse detection. Anomaly Detection detects the intruders based on anomalous behavior in network usage and admits that intruders highly associated with abnormal actions. Misuse detection tries to model attacks as specific patterns and then scan any traffic based on these patterns systematically. Misuse-based detection techniques depend on the signature of the attack [4] and used for known attacks and results in a small number of false alarms. The main abuse of signature-less (Anomaly-based) technique is the increasing number of false alarms. This paper presents machine learning (ML) and deep learning (DL) techniques for solving high false alarm rates of signature-less techniques. Machine learning is a portion of artificial intelligence (AI) and related to the statistical computer, that concentrates on alert the computer [5]. ML thinks as human and tries to learn the action of the attack to get the figure print of the attack [6]. ML is the study of scientifically and statistical methods that computers need to learn and do the task without direct instructions, depending on figure print and deduction. ML models establish mathematical ways depending on the data samples so that it considering part of artificial intelligence (AI). DL is a Wide family of machine learning that depends on artificial neural networks. The method of DL might be Unsupervised, supervised or semi-supervised [7]. DL approach is making feature extraction in the system. The distinction between deep learning (DL) and machine learning (ML) the amount of the data defines the execution of the system. Deep learning algorithms need a huge amount of data to understand clearly and perform well but machine learning works well with the little data [8]. A deep learning act similar to a human using a logical framework that can continually learn and resolve the data, draw illation. Deep learning implementation uses different layers to achieve good results. DL model getting high Performance with GPUs than the machine-learning model [9]. Feature processing is to convert variables to significative meaning. We do that after having your data summaries and in a visualized way. Feature processing is taking much time for getting specific knowledge. The accuracy of the feature extractor controls the performance of the machine learning model. The main difference between DL and ML models based on how to get a high-level feature of the data [10]. DL reduces the feature extractor design of the problem. Machine learning divided the problem into small problems and solves the small problems to achieve the result. Deep learning optimal end-to-end to solve the problem. It takes much time in the training phase because of the large number of the parameter in the DL models; so, the training step takes much time. The most advanced DL algorithm, such as ResNet, takes many periods to training part but ML training takes little time a few seconds. The test phase is the opposite in time. The rest of the paper works as follows: Sect. 2 provides related work and background. Section 3 provides the proposed hybrid approach for intrusion detection system. Section 4 Experimental result. Section 5 provides conclusion of the paper.

2 Related Work and Background

2.1 Background

2.1.1 Convolution Neural Network (CNN)

CNN is a recent type of neural network that working on to learn and reach suitable features for representing the input data. There are two differences with MLPs which are weight sharing and pooling. CNN has many layers and each layer consists of many convolution kernels which are used to create different feature map. Each region of the neuron of a feature map is connected to the next layer. All the spatial locations of the input share the kernel for generating the feature map. One or multiple fully connected layers are used for the classification [11] after some convolution and pooling layers. Because of the using of shared weights in a convolution neural network, the model learns the same pattern happening at different positions of inputs without asking to learn separate detectors for each position. For that, the model can power to the translation of inputs [12]. The pooling layers minimize the computational encumbrance because it decreases the number of connections between convolutional layers. However, pooling layers increasing the properties of translation and enhancing the receptive field of convolution layers. One or multiple fully connected layers can be added at the end of the stream of the network. For measuring the errors in the training part, loss function can be used for counting the errors [13].

2.1.2 Support Vector Machine (SVM)

The SVM is one of the important algorithms in the classification process, especially the binary classification process. The support vector machine making the classification for patterns based on some statistical techniques to use them in many processes such as regression and classifications with different kernel functions. SVM used recently to detect intrusions for accurately making a secure System. SVM has many features such as good generalization and powerful to deal with complex dimensions so SVM considered as one of the most popular methods for anomaly intrusion detection [14]. SVM can deal with the little training dataset. SVM can choose an optimal setup parameter than traditional techniques. SVM has many benefits such as high-speed, real-time, and can learn large data to classify in a good way. SVM always upgrade the training dataset when new patterns come [15].

2.1.3 K - Nearest Neighbor

K- Nearest Neighbor (KNN) a data mining algorithm that used for classification based on the examples which train in the feature space. KNN can be considered a lazy learning algorithm and it is one of the simplest techniques of machine learning. The classification of an object made by its neighbors voting, with the object that is specified to the class most common between its K nearest neighbors where k is a positive integer usually small. If $K = 1$ so the object will be specified to the class of its nearest neighbor [16]. The nearest neighbor in some cases refers to a single feature and in another case, it refers to a multidimensional feature vector that both works to describe the sample on the closest, and the closest one could be the Euclidean distance of feature vector [17].

2.2 Related Work

Vishwakarma et al. [18] propose a hybrid model between SVM and KNN_ACO for intrusion detection system using data set KDDCUP 1999. This approach achieves good results compared with other model's accuracy 94.17% and false alarm rate 5.32%. E.G. Dada [19] proposed a new approach that hybridizes different classifiers for better accuracy in the detection of intrusion attacks. The result of experiments conducted shows that the fusion of Support Vector Machine, k-Nearest Neighbor, and Primal-Dual Particle Swarm Optimization produced better classification accuracy than each of the singular classifiers on the KDD99 dataset the classification accuracy is 93.55. Tang et al. [20] proposed a new model for anomaly detection systems in SDN ambiance using the NSL-KDD dataset. This approach using a deep neural network for reducing the features from forty-one to six feature. This model resulted in test accuracy = 75.75%. Kokila et al. [21] Proposed a model for DDoS attack detection system by using signature method with (EMD) Earth Mover's Distance the simulation results show that accuracy increase with comparing EMD values by the case of no attacks and this model increase the true positive rate of the model the accuracy of SVM 95.11% and false alarms rate 0.3% Chowdhury [22] Proposed a hybrid model between support vector machine(SVM) and 1-nearest neighbor (1-NN) algorithms using KDD CUP1999, NSL_KDD datasets. Extract feature by using the deep convolutional neural network (CNN). The result of the test accuracy of 94.62%. Liu [23] studies a more automated and intelligent approach to build an intrusion detection (CNN) model based on behavior detection with self-adaptively and dynamic expansibility. And compared CNN with SVM and KNN algorithms efficiency for CNN = 93.11. SVM = 90.4, KNN = 90.74. Meena et al. [24] Apply Hybrid model PSO-SVM using KDDcup1999 trying to reduce attributes from forty-one to eighteen. The accuracy of the classification was DOS 99.4%, Probe 99.3%, R2L 98.7% and U2R 98.5% this approach have a better detection rate but little precision Probe84.2% and U2R89.4%. Researchers try to get high accuracy and low losses on the KDDcup 1999 dataset working on 41 features, and 5 attacks and overcome the problem of the imbalanced dataset but the result not good enough, so we try to make a new approach to enhance these result.

3 The Proposed Hybrid Approach for Intrusion Detection System

This paper, trying to make a hybrid approach between DL" Deep Learning" and ML" Machine Learning" to extract features for datasets and then classify the record. Figure 1 shows the steps of the proposed technique step by step in detail.

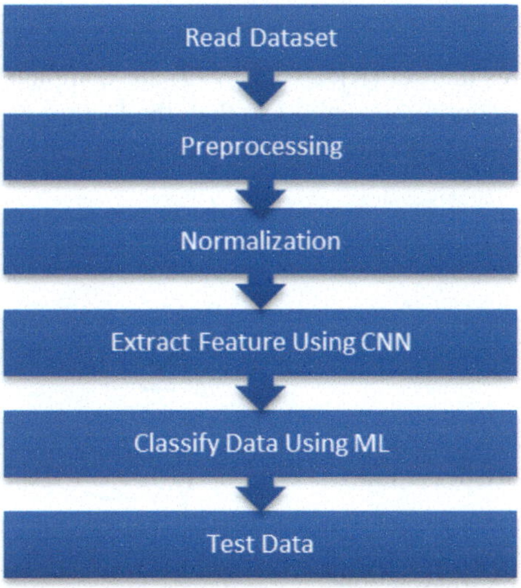

Fig. 1. The proposed approach

3.1 Preprocessing

Firstly, we try to extract the labels there are different types of fifteen attacks in the testing data that not included in the training data. The thirty-seven attack in the dataset clustered into four attack types:

- Denial of service attacks (DOS).
- Remote to Local attacks (R2L).
- User to Root (U2R).
- Probe attacks (Probe).

Secondly, In the pre-processing phase trying to convert non-numerical features to numerical values because the convolution neural network process numbers only assign protocol in the dataset numerical values such as (TCP = 1, UDP = 2, ICMP = 3). Convert attack types (non-categories) to numeric categories. Category 1 assigned to data is normal, Category2 assigned to DoS, Category 3 assigned to Probe, Category 4 assigned to R2L and Category 5 assigned U2R attack. To overcome the imbalance in the dataset, the train and test data will be collected in the same data frame and then divided that to 20% testing data and 80% training data.

3.2 Normalization

The values of the features dataset (KDDCUP1999) are completely different because of the data have discrete or continuous values so that, the feature was normalized the same range of values using StandardScaler () in the sci-kit-learn library which Calculated in math as:

$$\text{Standardization:} \quad z = \frac{x - \mu}{\sigma} \tag{1}$$

$$\text{Mean :} \quad \mu = \frac{1}{N}\sum\nolimits_{i=1}^{N} x_i \tag{2}$$

$$\text{Standard deviation:} \quad \sigma = \sqrt{\frac{1}{N}\sum\nolimits_{i=1}^{N} (x_i - \mu)^2} \tag{3}$$

$$\text{Min - Max scaling:} \quad X_{norm} = \frac{X - X_{min}}{X_{max} - X_{min}} \tag{4}$$

3.3 Extract Feature Using CNN

In this section, extracting the features using a convolution neural network (CNN) will be explained. Table 1 shows the architecture of CNN that used in the proposed techniques. Based on the architecture[], we use to get the feature to extract important and most useful features to reduce the number of features that make the model well understand the data set and achieve good results using that with the library sci-kit-learn. (Feature extraction layer) feeding this function with the first three layers from [0] to [3] layer and using get Prediction (classification layer) feed this function with the remaining layer from [4] to [5] layer. The number of features used in this experiment is 41 features. This experiment tries 10 epochs passing the output to the SVM and KNN to begin to classify and calculate the accuracy and confusion matrix for each algorithm. Figure 2 shows the Architecture of CNN.

Table 1. CNN_SVM results in details

Measurement	Normal	Dos	Probe	U2r	R2l
Precision	0.999	0.97	0.96	0.89	0.92
Recall	0.999	0.87	0.31	0.68	0.99
F1score	0.999	0.46	0.24	0.39	0.48
FPR	0.0039	0.0002	0.00029	0.000012	0.021
FNR	0.0007	0.126	0.68	0.32	0.005

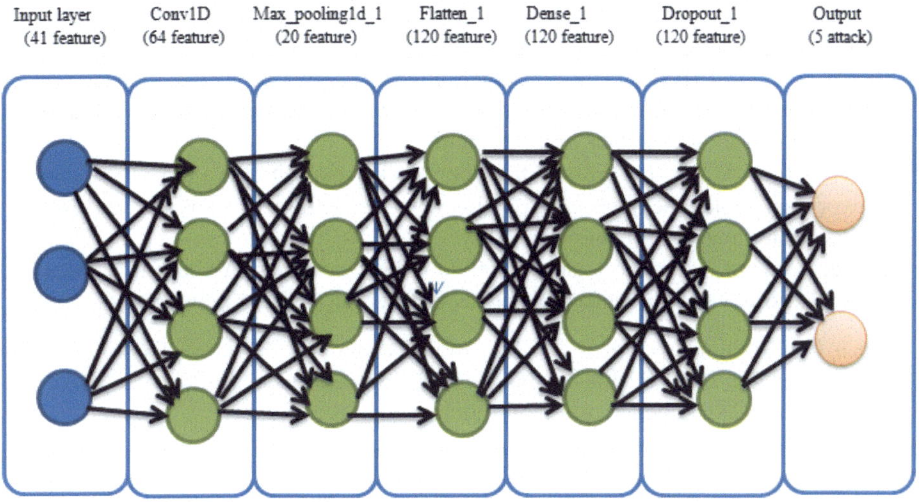

Fig. 2. The Architecture of CNN

3.4 Classify Data Using ML

Ml (machine learning) is part of AI (artificial intelligence). It is a field that interested in resolve and evolves the algorithms based on datasets. Machine learning concentrate on how to learn and be intelligent to learning and realize composed samples and get good result relying on datasets it works like a human brain. Machine learning has different fields of implementation such as classification techniques. The classification of the dataset can be made using a variety of techniques such as k-mean, support vector machine (SVM), K nearest neighbor (KNN), and naïve Bayes. In this paper, we will work on SVM and KNN. The proposed approach works on SVM and KNN using skit learn library SVC (C = 1.0, cache size = 200), K-Neighbors Classifier (leaf size = 30, neighbors = 5, p = 2).

4 Experimental Results

We used Google open-source data flow engine Tensor Flow using the Python Keras package. Keras is the most important and famous library in deep convolution network study because it has prototype backup use it easily and quickly. Runs on CPU and GPU.10% of KDDCUP 1999 were used in this experiment.

4.1 Dataset

The KDDCup 1999 is a famous benchmarking dataset established for an intrusion detection system to assess various methods. It is an extension of DARPA1998 [25]. KDDCup has about five million records. It has twenty–two attack and forty-one feature of traffic in any record. The KDDcup1999 features can be classified as Basic features, Traffic features and Content features the packets in traffic network using TCP format the emulation run in nine-week in local network. Experts grouped the dataset into four families [23].

- Denial of Service (DOS): It occurs when memory resources so occupied and full to deal with the requests this action prevents the user to access to the machine.
- Remote to Local (R2L): It occurs when attacker begins to access such as normal user and try to exploit the vulnerability such as sniff password, thesaurus attack or public engine to access root to the system.
- User to Root (U2R): It Occurs when the attacker exploits the vulnerability of the network and sends the message over this network without has any account to have access to the machine or this network.
- Probe: It occurs when the attacker tries to collect information about machines and networks to control and surround the security control of the networks.

Each flow training weight in the IDS model cannot be balanced. Compared with other kinds of flows, DOS attacks, and normal flow are easier to detect to solve this problem; we needed to generate a new training dataset averagely.

4.2 Result and Analysis

4.2.1 Summary of Test Results for CNN _SVM in Multi-class Classification

In the implementation of the Convolution neural network (CNN) and support vector machine (SVM) to detect the attacks and its types, the accuracy, precision, and recall were calculated and get a detailed result. Figure 3 shows the performance of CNN-SVM in detecting different attacks such as DOS, R2L, U2R, and Probe with his accuracy, precision and recall comparison.

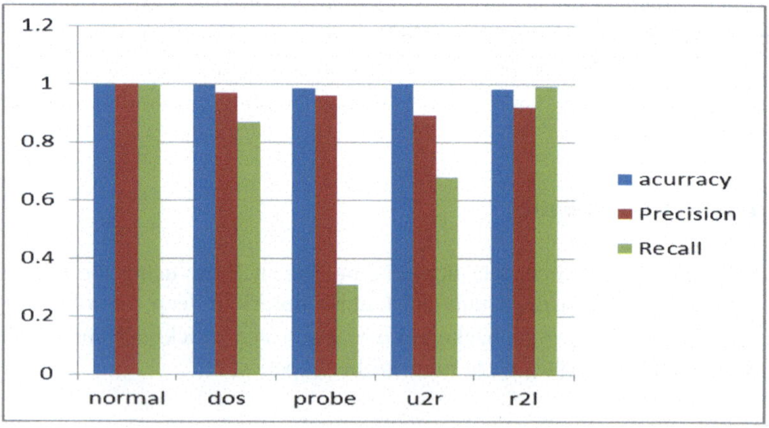

Fig. 3. The performance of CNN_SVM

Figure 4 shows the CNN_SVM Confusion Matrix. Table 2 shows the precision, recall, F1 score, FPR (false positive rate), FNR (False Negative Rate) of attacks in detail when using CNN-SVM.

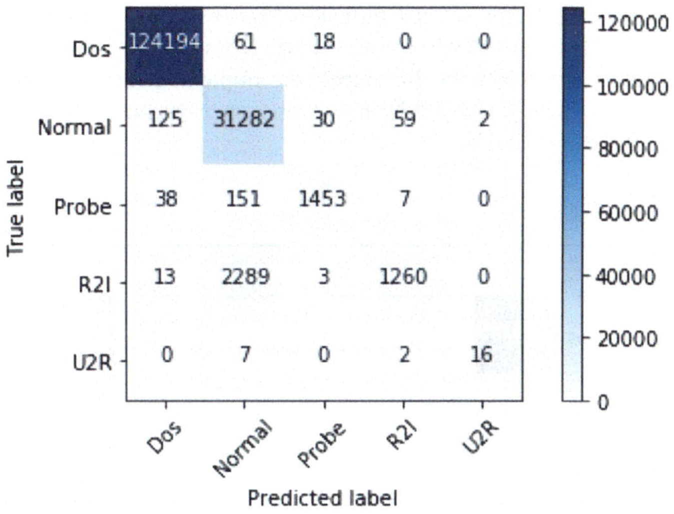

Fig. 4. CNN_SVM confusion matrix

4.2.2 Summary of Test Results for CNN_KNN in Multi-class Classification

In the implementation of the Convolution neural network (CNN) and K-Nearest Neighbor (KNN) to detect the attacks and its types, the accuracy, precision, and recall were calculated and get a detailed result. Figure 5 shows the performance of CNN-KNN in detecting different attacks such as DOS, R2L, U2R, and Probe with his accuracy, precision and recall comparison.

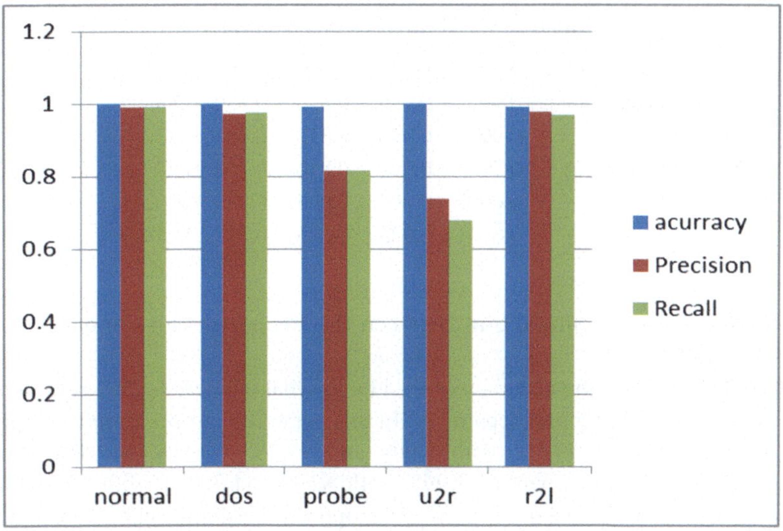

Fig. 5. The performance of CNN_KNN

Figure 6 shows the CNN_KNN Confusion Matrix. Table 2 shows the precision, recall, F1 score, FPR (False Positive Rate), FNR (False Negative Rate) of attacks in detail when using CNN-KNN. In this paper, the implementation of CNN_SVM and CNN_KNN proved that CNN_KNN has higher performance and accuracy than CNN_SVM in detecting attacks.

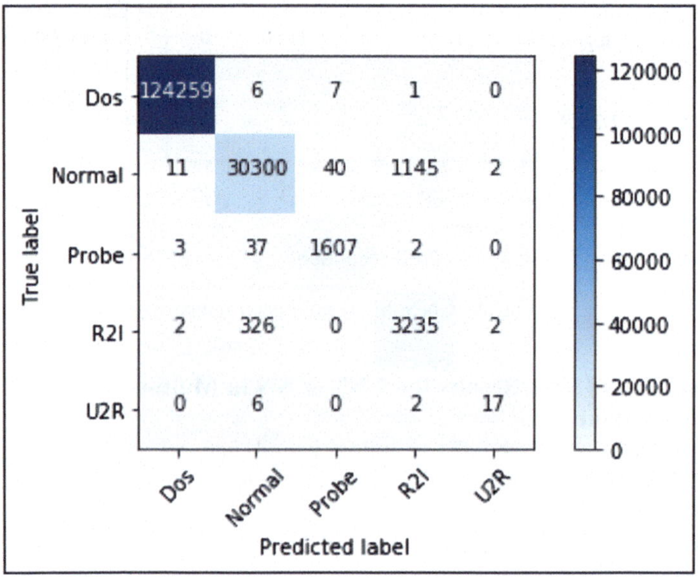

Fig. 6. CNN_KNN confusion matrix

Table 2. CNN_KNN results in details

Measurement	Normal	Dos	Probe	U2r	R2l
Precision	0.99	0.973	0.815	0.739	0.98
Recall	0.99	0.976	0.817	0.68	0.97
F1score	0.99	0.974	0.816	0.708	0.97
FPR	0.0005	0.0003	0.0041	0.036	0.005
FNR	0.0001	0.023	0.183	0.32	0.158

4.2.3 Test Accuracy Comparison Between the Proposed and Other Techniques

Chowdhury [16] reaches the accuracy of 95.11% by making new feature representation and the training and testing part separately. In this paper, the proposed hybrid approach reaches the accuracy of 99.1% with KNN and 98.1% with SVM. The proposed approach makes an improvement of 3.99% with KNN and 2.99% with SVM. Table 3 shows an accuracy comparison between the proposed work and other techniques.

Table 3. Test accuracy comparison

Algorithms	Test accuracy
Vishwakarma et al. [12]	94.17%
Chowdhury [16]	95.11%
Kokila et al. [15]	94.62%.
The proposed hybrid approach	KNN 99.2% SVM 98.3%

5 Conclusion and Future Work

In this research, we make a hybrid approach for the intrusion detection system between deep learning (DL) and machine learning (ML) firstly we resample that data to overcome imbalanced of the dataset and then make use of DL for extract feature and train SVM or KNN classifier for intrusion detection. Our method obtained a good result on the KDDcup99 dataset. Further work of our work try other techniques for ML and make CNN more complex and use real-time connections.

References

1. Milan, H.S., Singh, K.: Reducing false alarms in intrusion detection systems – a survey. Int. Res. J. Eng. Technol. (IRJET) **05**(02), 9–12 (2018)
2. Abdullah, B., Abd-Alghafar, I., Salama, G.I., Abd-Alhafez, A.: Performance evaluation of a genetic algorithm based approach to network intrusion detection system. In: 13th International Conference on Aerospace Sciences and Aviation Technology (ASAT), 26–28 May 2009 (2009)
3. Ashoor, A.S., Gore, S.: Importance of intrusion detection system (IDS). Int. J. Sci. Eng. Res. **2**(1), 1–4 (2011)
4. Modi, C.N., Acha, K.: Virtualization layer security challenges and intrusion detection/prevention systems in cloud computing: a comprehensive review. J. Supercomput. **73**(3), 1–43 (2016)
5. Louridas, P., Ebert, C.: Machine learning. IEEE Softw. **33**(5), 110–115 (2016)
6. Jordan, M.I., Mitchell, T.M.: Machine learning: trends, perspectives, and prospects. Science **349**(6245), 255–260 (2015)
7. Lecun, Y., Bengio, Y., Hinton, G.: Deep learning. Nature **521**(7553), 436–444 (2015)
8. Xin, Y., Kong, L., Liu, Z.: Machine learning and deep learning methods for cyber security. IEEE Access 1–9 (2017)
9. Coelho, I.M., Coelho, V.N., Luz, E.J.D.: A GPU deep learning metaheuristic based model for time series forecasting. Elsevier **201**(1), 412–418 (2017)
10. Deng, L., Yu, D.: Deep learning: methods and applications. Found Trends® Signal Process **7**(3), 197–387 (2014)
11. Vinayakumar, R., Soman, K.P.: Applying convolutional neural network for network intrusion detection. In: International Conference on Advances in Computing, Communications and Informatics (ICACCI), pp. 1222–1228 (2017)
12. Kapoor, A.J., Fan, H.: Intelligent detection using convolutional neural network (ID-CNN). In: Earth and Environmental Science, pp. 1–10 (2019)

13. Liu, H., Lang, B.: Machine learning and deep learning methods for intrusion detection systems: a survey. In: Applied Sciences, PP. 1–28 (2019)
14. Patgiri, R., Akutota, T.: An investigation on intrusion detection system using machine learning. In: IEEE Symposium Series on Computational Intelligence SSCI, pp. 1684–1691 (2018)
15. Shon, T., Kim, Y., Lee, C., Moon, J.: A machine learning framework for network anomaly detection using SVM and GA. In: Proceedings of the IEEE, pp. 176–183 (2005)
16. Liao, Y., Vemuri, R.V.: Use of k-nearest neighbor classifier for intrusion detection. In: ICACCI, pp. 1–10 (2016)
17. Shirazi, H.M.: Anomaly intrusion detection using information theory, k-NN and KMC algorithms. Aust. J. Basic Appl. Sci. **3**(3), 2581–2597 (2009)
18. Vishwakarma, S., Sharma, V., Tiwari, A.: An intrusion detection system using KNN-ACO algorithm. Int. J. Comput. Appl. **171**(10), 13–23 (2017)
19. Dada, E.G.: A hybridized SVM-KNN-pdAPSO approach to intrusion detection system. Fac. Semin. Ser. **8**, 1–8 (2017)
20. Tang, T.A., Mhamdi, L., McLernon, D., Zaidi, S.A.R., Ghogho, M.: Deep learning approach for network intrusion detection in software-defined networking. In: International Conference on Wireless Networks and Mobile Communications (WINCOM), pp. 1–6, October 2016
21. Kokila, R.T., Selvi, S.T., Govindarajan, K.: DDoS detection and analysis in SDN-based environment using support vector machine classifier. In: Sixth International Conference on Advanced Computing, pp. 205–210 (2015)
22. Chowdhury, M.M.U., Hammond, F., Konowicz, G.: A few-shot deep learning approach for improved intrusion detection. In: IEEE, pp. 456–462 (2017)
23. Liu, Y., Liu, S.: Intrusion detection algorithm based on convolutional neural network. In: International Conference on Engineering Technology and Application, pp. 9–13 (2017)
24. Meena, G., Choudhary, R.R.: A review paper on IDS classification using KDD 99 and NSL KDD dataset in WEKA. In: International Conference on Computer, Communications, and Electronics, pp. 553–558 (2017)
25. http://kdd.ics.uci.edu/databases/kddcup99/kddcup99.html

Prediction of the Electrical Load for Egyptian Energy Management Systems: Deep Learning Approach

Mohammed E. El-Telbany(✉)

Faculty of Information Technology and Computers Science,
University of Sinai, North Sinai, Egypt
mohammed.elsaid@su.edu.eg

Abstract. In the era of the Internet of Things (IoT) and machine learning, energy utilities have now the possibility of providing many benefits to the customers. Using *advanced metering infrastructure* is an important step to optimize and manage energy consumption. Smart meters are deployed in millions of households worldwide, representing an opportunity for monitoring systems and data analytics. Egypt has started its path into smart metering and needs to use big data analytics to endow the Egyptian grid with intelligence. In this paper, we investigate the role of big data analytics and available machine learning technologies that mine the data to generate insights into smart grid management process. We investigate a deep learning-based methodology such as long short-term memory (LSTM) network model to perform prediction of the electrical load for energy management systems. The LSTM model presents a decrease of 14% in forecasting error compared to deep forward neural networks.

Keywords: Big data analytics · Machine learning · Smart meters · Deep learning · LSTM · Predictive analytics · Electrical load

1 Introduction

The widespread deployment of smart meters rises the promise of real-time data analytics [1,2]. The smart meters' ability to communicate high-resolution consumption data on both the *spatial scale* and the *temporal scale* in enables novel data analytics applications, such as energy consumption analysis, theft detection, outage management, pervasive demand response at the residential level and personalized energy feedback. Smart electricity meters are a fundamental component of the smart grids which act as an intelligent system that allows information sharing among producers and consumers. Using data analytics aims to help utilities and consumers understand electricity consumption patterns and enables dynamic pricing strategies. The smart meter analytics is seen so important to the Egyptian market which has been growing rapidly. The government's energy efficiency plan include the roll-over *thirty million* smart electricity meters

A.-E. Hassanien et al. (Eds.): AICV 2020, AISC 1153, pp. 237–246, 2020.
https://doi.org/10.1007/978-3-030-44289-7_23

to customers within the next ten years period with a LE 60 billion budget [3]. The first locations of roll-over will be 6th October City and Sheikh Zayed districts. Currently, the electrical utilities have turned to big data analytics for management the flow of smart meters data to detect patterns useful for prediction and constantly self-learn from previous responses using advanced cognitive capabilities [4]. On a larger scale, there is a vast amount of data available, using big data technologies coupled with the IoT, all of this data can be gathered, structured and analyzed to yield valuable knowledge on utility management. There are many analytics tools available in the market such as SAP [5], Oracle/Data Raker [6] and IBM's event-driven data management system for the energy sector [7]. There is also open-source software which is provided through cloud computing infrastructures such as Spark [8] to provide a complete ecosystem for handling Big Data and generating useful insights. Most of the tools are provided by Apache and released under the Apache License. The analytical tools technologies have been used to manage the big data obtained from a smart grid. The major applications of such analytical tools are operations, asset management, demand-side management and customer satisfaction [9]. The demand side management is an important task for the smart grid to avoid the failure of electricity distribution networks, the suppliers rely on generation and demand balancing which can be implemented by an intelligent system that is able to provide appropriate responses to changes in load dynamics. For balancing demand generation and filling the demand response gap, the utilities have to estimate the energy demand patterns of different consumers to improved grid reliability. Many data analytics algorithms for smart meter have been proposed in demand side management such as electricity consumption prediction, consumption profile extractions, clustering consumers, and recommendations feedback to consumers to adjust their uses and reduce their bills. For electricity consumption prediction, there are several prediction methods where the data cleaning algorithms such as dimensionality reduction have been applied to keep only the useful predictors and machine learning algorithms like deep learning or decision trees have been used to create model customer load curves. This has been followed by consumption forecasting techniques to provide values for price and demand forecasts [10]. The *main contributions* of this paper are twofold: (1) it provide a model for big data analytics platform for endow the Egyptian smart grid with intelligence [11]. The using of big data analytics gives intelligence for smart grid operations support, and also provide real-time knowledge to increase grid reliability, balance loads, reduce outages, and detect fraud. The proposed solution is based on open source tools. (2) it investigated deep learning algorithms using LSTM based network to forecast the Egyptian electrical daily peak load. In the remainder of this paper, we describe a brief description of the structure of smart meter and Egyptian effort to in this domain in Sect. 2. Section 3 presents the big data analytics architecture. Section 4 presents the big data analytics applications in smart metering data. Section 5 gives the problem formulation and describes the deep learning for daily load forecasting case study. Finally, Sect. 6 concludes the paper.

2 Smart Meters

The smart meter is an electronic device of recording and processing customer energy consumption and providing real-time feedback regarding it. The smart meter has non-volatile data storage, tamper detection, power quality monitoring and two-way communication using a Power Line Carrier (PLC), a wireless modem (GSM) or an existing Internet connection [12]. The smartness of a smart meter lies in its communications functions. Figure 1 shows a typical smart meter structure where Egyptian researches have developed a prototype of single-phase smart meter [13].

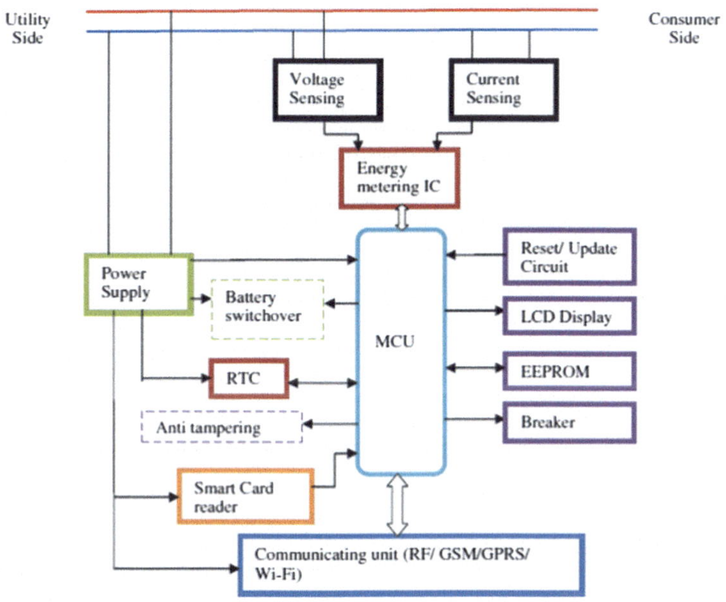

Fig. 1. Modern smart meter structure [14].

Smart meters use a secure communication network to automatically and wirelessly send your actual energy usage to utility companies for monitoring energy usage and management as well as with customers for observing their energy consumption. The smart meters technologies act as main components of a smart grid, which acts as an intelligent system that monitors generation, transmissions and consumption in the complete electric grid and performs dynamic energy management.

3 System Overview

This section presents a high-level overview of the proposed system's core components, as well as our data pipeline as streams, are ingested into our in-memory store and subsequently interacted with and analyzed. The proposed architecture is shown in Fig. 2. The focus of this paper is on the analytics layer using Spark [15,16], which consists of Spark SQL database and MADLib as the (in-database) machine learning library, and Spark programming tools. It provides high-level APIs in Java, Scala and Python, and an optimized engine that supports general execution graphs. This system aims to implement the whole process of smart meter data analytics, including data ingestion, data transformation, loading, analyzing, and visualization. The analytics layer applies many mining algorithms, including load time-series analytics, load disaggregation, consumption pattern discovery, segmentation, forecasting, and consumer feedback.

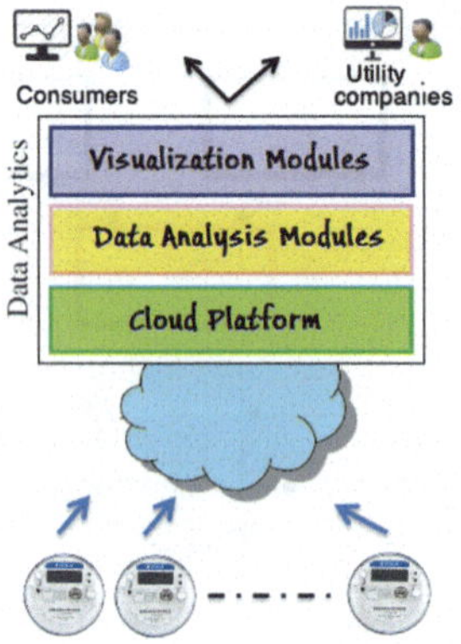

Fig. 2. Big data analytics architecture on cloud system.

4 Smart Meter Data Analytics

Recently, there are an increasing number of researchers have applied machine learning and data mining techniques to model, analyze, and understand residential electricity consumption and realize many different applications based on smart meter data analytics [17]. Most of the applications focus on real-time demand-response optimization, electricity consumption prediction, consumption profile extractions, clustering consumers, and personalized feedback to consumers. Actually, the *prediction*, *classification* and *clustering* are the most important data analytics applications based on smart metering data [12]. Forecasting the electricity demand has been important for helping utilities to perform short-term, mid-term, and long-term supply and infrastructure planning [18–20]. The customer segmentation approaches mostly focus on the socio-economic status of the household, the geographic location of the residence, the amount of energy attributed to heating and cooling [21,22], or inferring activities performed by the occupants [23]. Another application of smart meter data analytics is the detection of anomalies in the electricity network [24]. The reasons for anomalies could be meter malfunctioning, power leakage, reading errors, or intentional customer fraud.

5 Load Forecasting: A Deep Neural Networks Approach

There are many algorithms are suggested in the literature concerning time-series based prediction, regression-based models and statistical learning algorithms. These methods were widely applied to improve the accuracy of load forecasting. These traditional statistical algorithms are inadequate to fully model the complex nature of electricity demand and often result in lower accuracy [25]. Most of the studies that use support vector machines and artificial neural networks apply a single hidden-layer neural network which are considered among the most successful machine learning algorithms for this purpose [20,26,27]. However, with a limited number of hidden nodes considered, the capability of exploring non-linearity are usually constrained. The overfitting is the major disadvantage of traditional artificial neural networks with a large number of hidden units [28]. Recently, *deep learning* is becoming extremely appealing thanks to the high predictive power that it has empirically shown on real-world problems to forecast the electrical loads [28–30]. This section presents the results obtained from applying deep neural networks of the big data in electrical load forecasting of the unified the Egyptian grid.

5.1 Problem Formulation

In this work, we are interested in developing an predictive model that trained on labeled data for electrical peak load forecasting. Formally, the problem we investigate in this work can be stated as: Given a set of input-output pairs $\mathcal{D} = \{\boldsymbol{x}_i, y_i\}_{i=1}^n = (X, \boldsymbol{y})$, supervised machine learning algorithms aim at finding a function of the inputs $f(\boldsymbol{x})$ that approximates the output y. This translates into the minimization problem defined in Eq. (1).

$$\underset{f \in \mathcal{F}}{\operatorname{argmin}} \frac{1}{n} \sum_{i=1}^{n} L(f(\boldsymbol{x}_i), y_i) \tag{1}$$

The loss function $L(f(\boldsymbol{x}), y)$ can be seen as a measure of *adherence* to the available training data.

5.2 Deep Neural Networks

The Deep Neural Network (DNN) architecture is based on multiple hidden layers of neurons stacked together such as recurrent neural networks which learn representations of data [31]. In general, a neuron is a non-linear transformation of the linear sum of its inputs. The first layer models directly the data. A hidden layer in the neural network architecture is built as an array of neurons taking the inputs from the previous layer. The activation function of a neuron on top of k stacked layers in the architecture is using composite functions, such as $x \otimes h_1 \otimes h_2 \otimes \cdots \otimes h_k$. The supervised training of a very deep neural network with hard non-linearities is faster if the hidden layers are composed of Rectified Linear Units (ReLU) [32]. Recently, the logistic sigmoid and the hyperbolic tangent activation are outperformed by ReLU. Formally, ReLU is defined as a function $f(x_i) = \max(0, x_i)$, where x_i is its input. The resulting output (y'_i) is fed into a loss function L to calculate the error for this training example. A typical loss function for this purpose is the sum of squared errors:

$$L(\boldsymbol{x}, \boldsymbol{y}, \boldsymbol{w}, \boldsymbol{b}) = \sum_{i=1}^{n} (y_i - y'_i)^2 \tag{2}$$

where $\boldsymbol{w}$ and $\boldsymbol{b}$ specify the weights and biases of the neural network respectively, $\boldsymbol{x}$ is the set of training examples, $\boldsymbol{y}$ is the set of training labels, n is the number of training examples, and y_i and y'_i are the training label and output value for training example i, respectively. The network adjusts its weights and biases to minimize loss function L, using *gradient descent*. Different building blocks including convolutional neural networks (CNN) [33] and long short-term memory (LSTM) [34] have allowed deep neural networks to be highly flexible and effective. A comprehensive overview of deep learning methodologies can be found in [31]. The LSTM networks are a special kind of Recurrent Neural Networks (RNN) with capability of handling long-term dependencies. The LSTM network structure is a good candidate for tackling the learning problem of electrical load prediction due to its capability to learn complex non-linear patterns, temporal relationship catching and automatic feature extraction abilities. The LSTM uses a series of gates and feedback loops to control the flow of information that are themselves trained on the input data. The LSTM cell takes a new input x_i and the hidden vector h_{i-1} from the last time step, then produces an estimate output y'_i for the target output y_i given all the previous input sequence $x_1, x_2, \ldots, x_i$ also with a new hidden vector h_i and a new memory vector c_i. Figure 3 shows the structure of a LSTM cell.

5.3 Dataset and Experimental Results

Next, we present the experimental results obtained through forecasting algorithms and the parameters setup. The data being used for the future prediction is hourly interval data obtained from Egyptian electricity demand measured in Megawatt (MW) for daily peak load. The readings are for the period between January 2, 2012, to the November 30, 2013. The data is shown in Fig. 4. To evaluate the goodness of fit of algorithms we use the root mean square error (RMSE) as evaluation metric. The preprocessing consists in scaling and de-trending the dataset to prepare it for the learning algorithms. The LSTM model was implemented using the Tensorflow library [35]. The entire dataset is divided into train and test parts in $4:1$ ratio. The model is trained on training data and validated on the test dataset. Our architecture consists of two hidden LSTM layers with 60,20 and 20 LSTM units respectively, a dense output layer with 1 neuron, and a dropout of 0.05. The prediction model is trained with Adam optimizer using a learning rate of 0.05, a decay rate of 0.99, and a batch size of 1. Training was done for 50 epochs with early stopping. We compare the results of the LSTM algorithm to multilayer perceptron (MLP) with $20-10-1$ architecture.

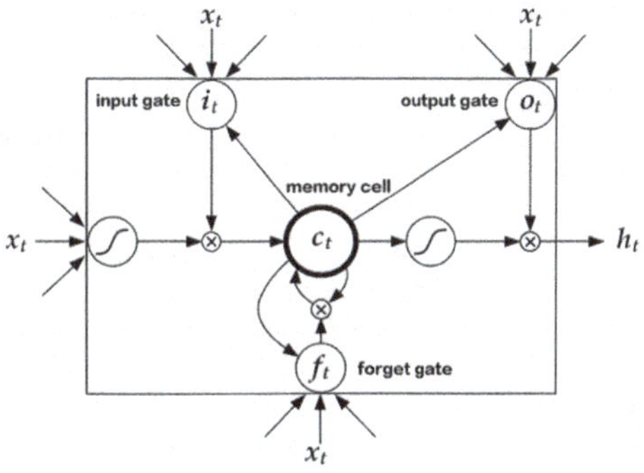

Fig. 3. LSTM cell structure

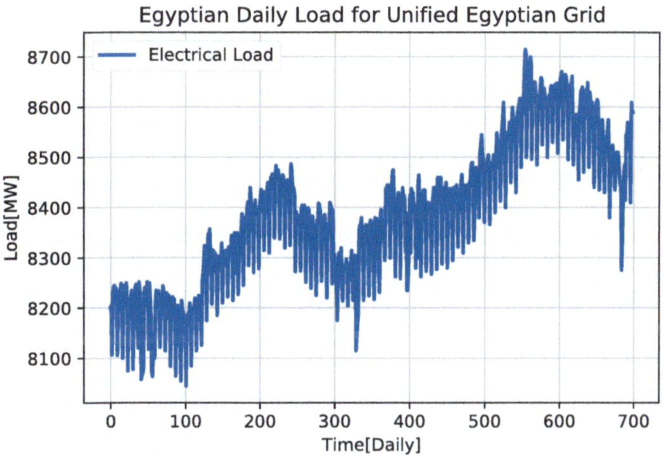

Fig. 4. Egyptian electricity demand dataset.

The performance of both LSTM model and MLP is listed in Table 1. Figure 5 shows the prediction of the LSTM model. As expected, the deep learning methods outperform the MLP forecast which performs poorly. Based on these experimental results we conclude that LSTM model is effective time series predictor due to utilizing hidden states using information from the previous value of a given series.

Table 1. Comparison of RMSE (MW/day) of MLP and LSTM models for Egyptian load prediction

Modes	Training	Validation
MLP	0.120	0.641
LSTM	0.111	0.501

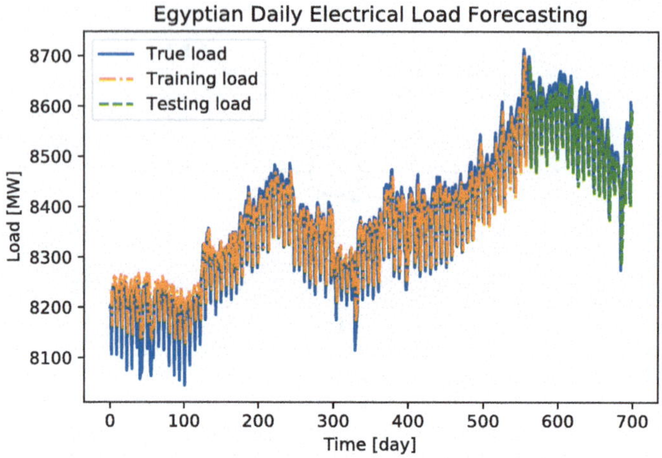

Fig. 5. LSTM predictions for Egyptian daily load.

6 Conclusions

Egypt has starting its path into smart metering and needs to use big data analytics for endow the Egyptian grid with intelligence. In Egypt, smart metering is still a developing industry and required to explore the possibilities of data management using data analytics technologies. We have strong evidence that big data analytic based algorithms have the ability to endow the Egyptian grid with the intelligence. Moreover, we have achieved high prediction accuracy for the load forecasting problem by using LSTM deep learning technique. In future, we hope to incorporates load data with weather and calendar features. Another possible improvement would be by using reinforcement learning for doing shorter term predictions.

References

1. Leskovec, J., Rajaraman, A., Ullman, J.: Mining of Massive Datasets, 2nd edn. Cambridge University Press, Cambridge (2014). ISBN 978-1107015357
2. Tsai, C.W., Lai, C.F., Chao, H.C., Vasilakos, A.V.: Big data analytics: a survey. J. Big Data **2**(21), 1–32 (2015). https://doi.org/10.1186/s40537-015-0030-3
3. Ministry of Electricity: Egyptian Electricity Holding Company. Annual Report (2017)
4. Alahakoon, D., Yu, X.: Smart electricity meter data intelligence for future energy systems: a survey. IEEE Trans. Ind. Inf. (2015). https://doi.org/10.1109/TII.2015.2414355
5. Ardakanian, O., Koochakzadeh, N., Singh, R., Golab, L., Keshav, S.: Computing electricity consumption profiles from household smart meter data. In: Proceedings of the EDBT Workshop on Energy Data Management (EnDM), pp. 140–147 (2014)

6. Birt, B., Newsham, G., Beausoleil-Morrison, I., Armstrong, M., Saldanha, N., Rowlands, I.: Disaggregating categories of electrical energy end-use from whole-house hourly data. Energy Build. **50**, 93–102 (2012). https://doi.org/10.1016/j.enbuild.2012.03.025
7. Downey, P.: Deliver more intelligence to intelligent energy systems. IBM (2017)
8. Aven, J.: Data Analytics with Spark Using Python. Addison-Wesley, Boston (2019). ISBN 978-0134846019
9. Gellings, C., Parmenter, K.: Demand-side management. In: Kreith, F., Pepper, D. (eds.) Energy Efficiency and Renewable Energy Handbook, 2nd edn, pp. 399–420. CRC Press, Boca Raton (2015)
10. Diamantoulakis, P., Kapinas, V., Karagiannidis, G.: Big data analytics for dynamic energy management in smart grids. Big Data Res. **2**(3), 94–101 (2015). https://doi.org/10.1016/j.bdr.2015.03.003
11. Alahakoon, D., Yu, X.: Smart electricity meter data intelligence for future energy systems: a survey. IEEE Trans. Ind. Inf. **12**(1), 425–436 (2016). https://doi.org/10.1109/TII.2015.2414355
12. Uribe-Pérez, N., Hernández, L., de la Vega, D., Angulo, I.: State of the art and trends review of smart metering in electricity grids. Appl. Sci. **6**(68), 1–24 (2016). https://doi.org/10.3390/app6030068
13. Riad, M.: Smart meters project. DLMEI Alliance (2019). http://www.dlmei.sci.eg/?q=en/smart%20meter
14. Weranga, K., Kumarawadu, S., Chandima, D.: Smart metering design and applications. Springer, Singapore (2014). ISBN 978-981-4451-81-9
15. Wills, J., Ryza, S., Owen, S., Laserson, U.: Advanced Analytics with Spark. Cambridge University Press, Cambridge (2015)
16. Guller, M.: Big Data Analytics with Spark. Apress, New York (2015). ISBN 978-1484209653
17. Eichinger, F., Pathmaperuma, D., Vogt, H., Muller, E.: Data analysis challenges in the future energy domain. In: Yu, T., Chawla, N., Simoff, S. (eds.) Computational Intelligent Data Analysis for Sustainable Development, pp. 1–55. Taylor & Francis, Didcot (2013)
18. Lee, K., Cha, Y., Park, J.: Short-term load forecasting using an artificial neural network. IEEE Trans. Power Syst. **7**(1), 124–132. https://doi.org/10.1109/59.141695
19. Metaxiotis, K., Kagiannas, A., Askounis, D., Psarras, J.: Artificial intelligence in short term electric load forecasting: a state-of-the-art survey for the researcher. Energy Conv. Manag. **44**(9), 1525–1534 (2003). https://doi.org/10.1016/S0196-8904(02)00148-6
20. El-Telbany, M., Fawwaz El-Karmi, F.: Short-term forecasting of Jordanian electricity demand using particle swarm optimization. Electr. Power Syst. Res. **78**(3), 425–433 (2008). https://doi.org/10.1016/j.epsr.2007.03.011
21. Chicco, G.: Overview and performance assessment of the clustering methods for electrical load pattern grouping. Energy **42**(1), 68–80 (2012). https://doi.org/10.1016/j.energy.2011.12.031
22. Wijaya, T., Ganu, T., Chakraborty, D., Aberer, K., Seetharam, D.: Consumer segmentation and knowledge extraction from smart meter and survey data. In: 14th SIAM International Conference on Data Mining (SDM), pp. 226–234 (2014). https://doi.org/10.1137/1.9781611973440.26

23. Chen, F., Dai, J., Wang, B., Sahu, S., Naphade, M., Lu, C.: Activity analysis based on low sample rate smart meters. In: 17th International Conference on Knowledge Discovery and Data Mining (KDD), pp. 240–248. ACM (2011). https://doi.org/10.1145/2020408.2020450
24. Mashima, D., Cardenas, A.: Evaluating electricity theft detectors in smart grid networks. In: Balzarotti, D., Stolfo, S., Cova, M. (eds.) Research in Attacks, Intrusions, and Defenses, LNCS, vol. 7462 , pp. 210–229. Springer, Heidelberg (2012)
25. Marin, F., Garcia-Lagos, F., Joya, G., Sandoval, F.: Global model for short-term load forecasting using artificial neural networks. IEE Proc. Gener. Transm. Distrib. **149**(2), 121 (2002)
26. Ceperic, E., Ceperic, V., Baric, A.: A strategy for short-term load forecasting by support vector regression machines. IEEE Trans. Power Syst. **28**(4), 4356–4364 (2013). https://doi.org/10.1109/TPWRS.2013.2269803
27. Rodrigues, F., Cardeira, C., Calado, J.: The daily and hourly energy consumption and load forecasting using artificial neural network method: a case study using a set of 93 households in Portugal. Energy Procedia **62**, 220–229 (2014). https://doi.org/10.1016/j.egypro.2014.12.383
28. Ouyang, T., He, Y., Li, H., Sun, Z., Baek, S.: A deep learning framework for short-term power load forecasting. CoRR (2017). https://doi.org/10.1109/TETCI.2018.2880511
29. He, W.: Deep neural network based load forecast. Comput. Model. New Technol. **18**(3), 258–262 (2014). http://www.cmnt.lv/upload-files/ns_8art43_CMNT1803-34_N289_He.pdf
30. Tian, C., Ma, J., Zhang, C., Zhan, P.: A deep neural network model for short-term load forecast based on long short-term memory network and convolutional neural network. Energies **11**(3493), 1–13 (2018). https://doi.org/10.3390/en11123493
31. Goodfellow, I., Bengio, Y., Courville, A.: Deep Learning. MIT Press, Cambridge (2016). ISBN 9780262035613
32. Glorot, X., Bordes, A., Bengio, Y.: Deep sparse rectifier neural networks. In: Proceedings of the Fourteenth International Conference on Artificial Intelligence and Statistics (AISTATS 2011), pp. 315–323 (2011)
33. Krizhevsky, A., Sutskever, I., Hinton, G.: Imagenet classification with deep convolutional neural networks. In: Advances in Neural Information Processing Systems, pp. 1097–1105 (2012). https://doi.org/10.1145/3065386
34. Hochreiter, S., Schmidhuber, J.: Long short-term memory. Neural Comput. **9**(8), 1735–1780 (1997). https://doi.org/10.1162/neco.1997.9.8.1735
35. Abadi, M., et al.: Tensorflow: large-scale machine learning on heterogeneous systems Software, vol. 1 (2015). http://tensorflow.org

Comparative Performance of Machine Learning and Deep Learning Algorithms for Arabic Hate Speech Detection in OSNs

Ahmed Omar[1(✉)], Tarek M. Mahmoud[1], and Tarek Abd-El-Hafeez[1,2]

[1] Computer Science Department, Faculty of Science, Minia University, EL-Minia, Egypt
{Ahmed.omar, d.tarek, tarek}@mu.edu.eg
[2] Deraya University, EL-Minia, Egypt

Abstract. Nowadays, Online Social Networks (OSNs) are the most popular and interactive media that used to express feelings, communicate and share information between people. However, along with useful and interesting content, sometimes unsuitable or abusive content can be published on these networks, such as hate speech and insults. Hate speech includes any type of online abuse concepts like cyberbullying, discrimination, abusive language, profanity, flaming, toxicity, and harassment. Most of the Hate speech detection attempts have concentrated on the English text, while work on the Arabic text is sparse. In this paper, we constructed a standard Arabic dataset that can be used for hate speech and abuse detection. In contrast to most previous work the datasets were collected from one platform, the proposed dataset is collected from more social network platforms (Facebook, Twitter, Instagram, and YouTube). To validate the effectiveness of the proposed datasets twelve machine learning algorithms and two deep learning architecture were used. Recurrent Neural Network (RNN) outperformed other classifiers with an accuracy of 98.7%.

Keywords: Arabic hate speech · Hate speech detection · Arabic text classification · OSN

1 Introduction

The rapid development of online social networks (OSNs) led to the widening of the scope of the internet. Nowadays OSN is the most popular and interactive media that used to express feelings, communicate and share information between people. The communication includes all data types such as text, image, audio, and video. Text is the most exchanged content type on social media via posts, tweets, comments, replies, and messages.

Facebook is the top social network site with 2.414 billion active users. YouTube, as a video streaming site, ranked second with 2 billion active users, Instagram, as a photo-sharing site, ranked sixth with 1 billion active users. Despite his widespread reputation, Twitter ranked eleventh with 300,000 active users [6]. Every minute, 510,000 comments, and 293,000 status updates are posted on Facebook, 4.15 Million videos are

A.-E. Hassanien et al. (Eds.): AICV 2020, AISC 1153, pp. 247–257, 2020.
https://doi.org/10.1007/978-3-030-44289-7_24

watched on YouTube, 456,000 tweets are written on Twitter, and 46,740 photos are uploaded on Instagram [14, 17].

With that huge amount of data published every minute, the user's page will be full of different types of topics and contents. The OSN's users are not given the ability to prevent the content before they appear on their timeline [5]. The shared content may be Hate speech, offensive language, harassing, racism, insults, and other types of online abusive behavior. Unfortunately, OSNs are a perfect platform for publishing hurtful content such as cyberbullying and hate speech [9].

This type of behavior is harmful both socially and psychologically on the users, reducing the tendency trust of users to the online social media. The research community has also made attempts for detecting online hate speech and abusive behavior, Therefore, a solution must be found before this growing hates gets control and destroys the lives of individuals, families, and communities [9].

Most of the efforts in the field of abuse and hate speech detection have focused on the English language, while research on the Arabic language is scant. The Arabic language is one of the widely spoken languages in the world, It is the fifth most used language on the internet [12]. In the last years, online Arabic content increased quickly, surpassing 3% of the entire online content, and In the eighth position in the whole internet content [1], However, there are a few research efforts and scarcity of resources for the Arabic language datasets [7, 19].

Because of the increasing prevalence of the Arabic language on the Internet and the challenges of analysis the Arabic language, in addition to the scarcity of studies in the field of Arabic hate speech detecting, we contribute in this paper the following points:

1. We constructed a standard Arabic dataset for hate speech and abusive detection in online social networks.
2. We train machine learning and deep learning models for hate speech detection in OSNs.

The remaining sections of this paper are organized as follows: Related work is presented in Sect. 2. Section 3 describes the Hate speech problem. The methodology, which includes data collection, data filtering, and annotation, is presented in Sect. 4. The experimental results are presented in Sect. 5. Finally, Sect. 6 includes the conclusion and future work.

2 Related Work

Abuse detection has been an increasing attention research topic over the past few years. Some studies have been published trying to address this problem in online social networks, there is very little focus on the Arabic language among them. Several of the online abuse concepts found in the literature were treated as hate speech, like cyberbullying, discrimination, abusive language, profanity, flaming, toxicity, and harassment. Some literature slightly differentiates between these concepts and hate speech, but confirm that they are related to it, and it contains them in one way or another [8]. The following paragraphs present some of these efforts and are summarized in Table 1.

Hatem et al. [11] built an Arabic dataset in the Tunisian dialect contains 6,039 comments classified as hate, abusive, and normal. They mentioned that the comments were collected from different social media platforms but they did not specify them. Naive Bayes and SVM were used to evaluate the classification performance and Naive Bayes gave the best performance, which achieved precision, recall, F1_score, and accuracy of 93.5%, 91.5%, 92.3%, and 92.9%, respectively.

Hanane et al. [15] use the dataset proposed by [2] and apply different deep learning technique on it, CNN gave the best performance which achieved precision, F1_score, and accuracy of 86.10%, 84.5%, and 87.84%, respectively.

Table 1. Summary of the related work in Arabic hate speech detection and illustrating for each study: dataset size, collection source(s), classes names, best classifier, and the performance metrics values.

Study	Dataset size	Source	Classes	Best classifier	Performance
Hatem et al. [11]	6,039	Social media platforms	Hate, abusive, and normal	Naive Bayes	Prec = 93.5%, Rec = 91.5%, F1 = 92.3%, Acc = 92.9%
Mohaouchane et al. [15]	16,000	YouTube	Offensive, inoffensive, neutral	CNN	Prec = 86.10% F1 = 84.5%, Acc = 87.84%
Albadi et al. [4]	6,000	Twitter	Hate, not hate	Deep learning	Prec = 76%, Rec = 78%, F1 = 77%, Acc = 79%
Ghosh et al. [10]	6,000	Twitter	Hate, not hate	Word Embedding, Social Network Graphs	Prec = 74%, Rec = 89%, F1 = 78%, Acc = 86%
Mubarak et al. [16]	1,100 32,000	Twitter Aljazeera.net	Obscene, offensive, clean	None	None
Mulki et al. [18]	6,000	Twitter	Hate, Abusive, Normal	Naïve Bayes	Prec = 90.5%, Rec = 89%, F1 = 89.6%, Acc = 90.3%

Albadi et al. [14] built an Arabic dataset for Religious hate speech detection, the dataset contains 6,000 Arabic tweets manually annotated to hate and not hate. They evaluate the dataset using Lexicon-based, SVM and deep learning, deep learning gave the best performance, which achieved precision, recall, F1_score, and accuracy of 76%, 78%, 77%, and 79%, respectively.

Ghosh et al. [10] used Word Embedding and Social Network Graphs to evaluate the dataset proposed in [4], the performance in terms of precision, recall, F1_score, and accuracy of 74%, 89%, 78%, and 86%, respectively.

Mulki et al. [18] built a dataset for Arabic hate speech and abusive detection, the dataset contains 6,000 tweets, manually classified as Hate, Abusive, and Normal. They applied two classifiers Naïve Bayes and SVM, the best classifier was Naïve Bayes, which achieved precision, recall, F1_score, and accuracy of 90.5%, 89%, 89.6%, and 90.3%, respectively.

It is noticeable that the previous studies in Arabic hate speech detection are still scarce, and most of them were collected from one source, which is Twitter, and the algorithms used are mostly limited to Naïve Bayes and SVM and some studies have used a dataset from other studies instead of building their own dataset.

3 Hate Speech

Hate speech can occur in different linguistic styles and through different acts, such as insulting, abusing, provocation, and aggression. Anis [23] lists nine fundamental meanings in the vocabulary for making hate speech in Arabic, we have added a tenth meaning in addition to examples for each type as shown in Table 2, with an English translation for each example.

Table 2. Hate speech meaning with words examples and its English translation

Hate Speech Meaning	Arabic Words	English translation
low	حقير، وضيع	Despicable, mean
dirty	نجس، معفن، وسخه، قذر	Dirty, moldy, filthy
double faced	منافق، عميل، خاين	Hypocrite, traitor
irrational and unreasonable	مجنون، متخلف، أهبل، اهطل، معتوه	Crazy, subnormal, asshole, fool
Sick	مريض، يشفى الكلاب ويضرك، تعبان فدامغه	Sick, God heals the dogs and harm you, Confused
immoral and irreligious	كافر، مرتد، عاهره، شرموطة، ديوث	Infidel, apostate, prostitute, cuckold
Shameless	وقح، سافل، منحط	Shameless, raffish, decadent
ethnic labels	فلاح، صعيدي، يهودي	A farmer, From Upper Egypt, a Jew
Animal	خنزير، حمار، لبوة كلب، خروف، نعجة،	Dog, sheep, pig, donkey, lioness
jobs	عربجي، زبال، بواب	Caterer, scavenger, concierge

4 Methodology

The point of the research, as well as the availability of datasets, will determine whether a ready and published dataset will be used, or a new dataset will be created. The dataset building process involves three stages: data collection, data filtering, and data annotation. A data source, or multiple sources, must be specified before starting the collection process, because each source can differ in terms of text length, writing style, accessibility, and privacy. The collected dataset will include samples of text relevant to the research problem [19]. The process of building datasets involves three stages: data collection, data filtering, and data annotation. Figure 1 depicts the dataset building process.

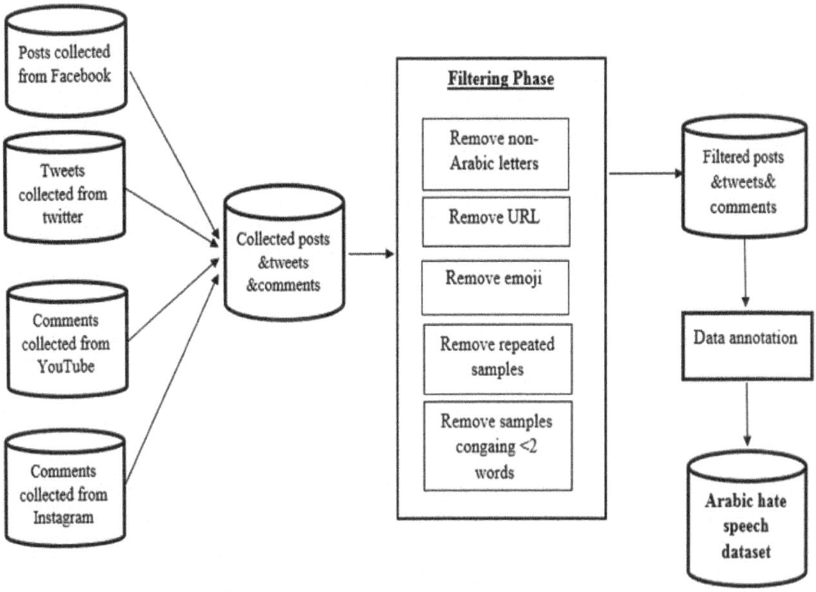

Fig. 1. The process of building the proposed dataset

4.1 Data Collection

The dataset acquisition step is fundamental and the most time-consuming part of the text classification process [21]. We selected the most four popular online social network platforms Facebook, Twitter, YouTube, and Instagram [6]. Arabic pages and accounts containing controversial issues and opinions, whether religious, political or sports were chosen, as were the pages of some of the controversial personalities. To collect the text data from these platforms we developed a web crawler that automatically scrolls down on each page then gathers all the posts, tweets, and comments and saves them in a text file.

4.2 Data Filtering

The collected data may contain unnecessary text that adversely affects the classification performance such as non-Arabic characters, emoji, or URLs. Therefore, the type of text must be filtered and cleaned, also text contains less than two words are deleted from the dataset because one word often will not make sense and its presence will increase the total number of posts. Table 3 shows a sample of posts before filtering and filtering decisions. The algorithm used for collecting and filtering the dataset can be summarized as:

Input: *L: A list of Facebook pages, twitter account,*
Instagram account, YouTube videos
Output: *D: array of strings //collected samples (posts, tweets, comments)*
Output: *P: array of strings//Filtered Samples*
Input: *j: =0, k: =0*

I. *Scroll Through the social network platforms page (L) and save texts(D)*
II. *Foreach Di in D do*
    a. *If not((*D_i *:= URL) or (* D_i*:= non-Arabic) or(*D_i*:= emoji)or (P contains* D_i*)) then*
        i. *P[j]:=*D_i
        ii. *j:=j+1*
    b. *End if*
    c. *If not((*D_i *contains URL) or (* D_i *contains non-Arabic) or (*D_i *contains emoji) then*
        i. *Remove URL or non-Arabic or emoji*
III. *End for*
IV. *Save P to text file*
V. *End.*

Table 3. A sample of text before the filtering and the filtering action.

Text before	Action
Hate you سافل ابن سافل اقسم بالله هذا المذيع اكرهه من خلقته مبين عليه	Remove non arabic characters
انت عار على الإسلام والمسلمين وخسارة فيكى الحجاب ☹☹☹	Remove emoji
http://www.youm7.com/4588539 ازاي وهي كافرة عروس الجنة	Remove URL

4.3 Data Annotation

The dataset was labeled by three native Arabic speakers represent varied cultures, ages, genders, and jobs, to facilitate their task, we built an online webpage to help the annotators access the dataset remotely, from any device, and at any time. Table 4 shows a sample of the labeled text and their translation into English.

Table 4. A sample of the labeled text and their translation into English.

Translation	text
she will enter hell because she is an infidel	هتدخل النار دي كافرة
You are cattle, retarded, disabled, and disgusted, and do not understand anything	انتوا بهايم ومتخلفين ومعاقين .. ومقرفين ولا تفهموا حاجة خالص
The most filthy society we can live in and see as it	اوسخ واقذر مجتمع بنعيشه وبنشوفه
I will kill you, Mohammed Nasser,	هقتلك يامحمد يابن ناصر
she looks so ugly	شكلها وحش جدا

5 Results and Analysis

5.1 Dataset

After completing the three dataset building steps, the data acquisition, data filtering, and data annotation, we formulated a balanced dataset of 20,000 posts, tweets, and comments collected from the most popular Facebook pages, Twitter accounts, Instagram accounts, and YouTube videos. All samples in the dataset were manually annotated to hate or not hate.

5.2 Model Evaluation

Model evaluation is a two main steps process: the training phase, and the test phase, so, the dataset must be portioned into two parts training data and test data, the training part contains the larger portion of the dataset and is used for training the classification algorithm. The testing part is used to test the classification algorithm performance. We split our dataset to 80% for training and 20% for testing.

Several metrics are used to evaluate the performance of the classification algorithm, such as precision (Prec), and recall (Rec), F1-score, and accuracy (Acc), and they are defined as follows:

$$Precision = \frac{TP}{(TP + FP)} \quad (1)$$

$$Recall = \frac{TP}{(TP + FN)} \quad (2)$$

$$F1\_\text{Score} = \frac{(2 \times Precision \times Recall)}{(Precision + Recall)} \quad (3)$$

$$Accuracy = \frac{(TP + TN)}{(TP + TN + FP + FN)} \quad (4)$$

where TP, TN, FN, and FP, refer to True Positive, True Negative, False Negative, and False Positive, respectively [1].

5.2.1 Machine Learning

For model evaluation, we use python and select twelve Machine Learning algorithms, they are MultinomialNB, Complement NB, BernoulliNB, SVC, NuSVC, LinearSVC, LogisticRegression, Decision Tree, SGD, Ridge, Perceptron, and Nearest Centroid, [20]. Accuracy, F1_score, Recall, and Precision are the performance evaluation metrics for comparison. The results of the trained models on the test data are shown in Table 5.

It can be readily inferred from Table 5 and Fig. 3 that Complement NB performing better compared to other classifiers in terms of Accuracy, F1_score, Recall, and Precision equal to 97.59%, 97.59%, 97.59%, and 97.63%, respectively. The lowest performance metrics are 49.92%, 33.25%, 49.92%, and 24.92%, respectively when using SVC. The results of the trained models on the test data are shown in Table 5 and Fig. 2.

Table 5. Models evaluation performances comparison

	Classifier	Accuracy	F1_score	Recall	Precision
1	MultinomialNB	97.52%	97.52%	97.52%	97.55%
2	ComplementNB	**97.59%**	**97.59%**	**97.59%**	**97.63%**
3	BernoulliNB	96.19%	96.18%	96.19%	96.38%
4	SVC	49.92%	33.25%	49.92%	24.92%
5	NuSVC	92.73%	92.70%	92.73%	93.38%
6	LinearSVC	97.49%	97.49%	97.49%	97.49%
7	LogisticRegression	97.54%	97.54%	97.54%	97.55%
8	DecisionTree	93.41%	93.40%	93.41%	93.50%
9	SGD	97.42%	97.42%	97.42%	97.42%
10	Ridge	97.07%	97.07%	97.07%	97.07%
11	Perceptron	96.69%	96.69%	96.69%	96.69%
12	NearestCentroid	89.39%	89.38%	89.39%	89.58%

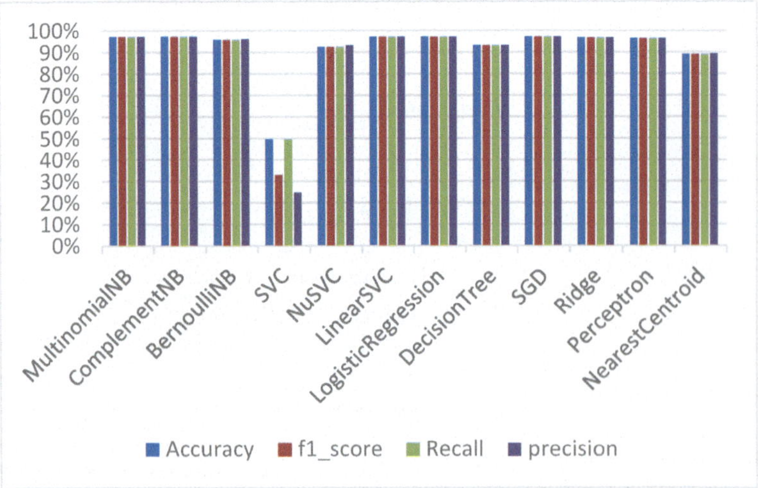

Fig. 2. Models evaluation performances comparison

5.2.2 Deep Learning

We select the two most effective architecture in text classification Convolutional Neural Networks (CNN) and Recurrent Neural Network (RNN) [13]. Table 6 shows a comparison of the performance of the two architecture. Figure 3 shows the accuracy and loss of the RNN in the 10 epochs.

Table 6. Deep learning architectures performance comparison

	Accuracy	F1_score	Recall	Precision
RNN	**98.70%**	**98.70%**	**98.70%**	**98.70%**
CNN	98.35%	98.35%	98.35%	98.36%

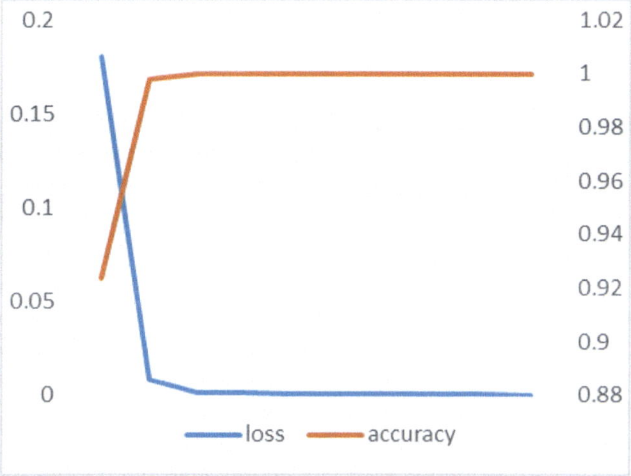

Fig. 3. RNN accuracy and loss for each epoch

It appears from Table 6 that deep learning gives a better performance than machine learning, and RNN is the best architecture in classification Arabic hate speech in online social networks with Accuracy, F1_score, Recall, and Precision equals to 98.70%, 98.70%, 98.70%, and 98.70%, respectively.

Table 7 shows the performance comparison between the mentioned evaluation algorithms on our proposed Arabic hate speech dataset, and other available datasets mentioned in the related work section.

Table 7. Arabic hate speech dataset performances comparison

Evaluation metric	Proposed dataset	Hatem et al. [11]	Alakrot et al. [3]	Mohaouchane et al. [15]	Albadi et al. [4]	Ghosh et al. [10]	Mulki et al. [18]
Accuracy	98.70	92.9	–	87.84	79	86	90.3
F1_score	98.70	92.3	82	84.5	77	78	89.6
Recall	98.70	91.5	80	–	78	89	89
Precision	98.70	93.5	88	86.1	76	74	90.5

It can be readily inferred that our proposed Arabic hate speech gave the best performance comparing with the other dataset. It is also distinguished by being collected from four social media platforms (Facebook, Twitter, Instagram, and YouTube), in contrast, most of the previous datasets collected from one platform.

6 Conclusions

This paper presented a dataset for Arabic hate speech detection in OSNs. The dataset containing 20,000 posts, comments, and tweets collected from Facebook, Twitter, Instagram, and YouTube and manually labeled by three Arabic annotators into two balanced classes: Hate, and not hate. To the best of our knowledge, this is the first Arabic hate speech collected from more than one platform. Twelve machine learning algorithms (e.g. are MultinomialNB, Complement NB, BernoulliNB, SVC, NuSVC, LinearSVC, LogisticRegression, Decision Tree, SGD, Ridge, Perceptron, and Nearest Centroid) and two deep learning architectures (e.g. CNN, and RNN) were applied to evaluate the performance of the dataset. In machine learning algorithms, Complement NB yielded the best performance achieving an accuracy score of 97.59%. While in deep learning architectures, RNN gave the highest performance achieving an accuracy score of 98.70. In future work, we will extend the dataset by adding more features extracted from user behavior, number of likes, emotions types, number of replies. Also, collect data from different Arabic regions to cover different dialects and cultures, in addition to detect Hate speech on disability.

References

1. Al-Tahrawi, M.M., Al-Khatib, S.N.: Arabic text classification using polynomial networks. J. King Saud Univ. Comput. Inf. Sci. **27**(4), 437–449 (2015)
2. Alakrot, A., Murray, L., Nikolov, N.S.: Dataset construction for the detection of anti-social behaviour in online communication in Arabic. Procedia Comput. Sci. **142**, 174–181 (2018)
3. Alakrot, A., Murray, L., Nikolov, N.S.: Towards accurate detection of offensive language in online communication in Arabic. Procedia Comput. Sci. **142**, 315–320 (2018)
4. Albadi, N., Kurdi, M., Mishra, S.: Are they our brothers? Analysis and detection of religious hate speech in the Arabic Twittersphere. In: Proceedings of the 2018 IEEE/ACM International Conference on Advances in Social Networks Analysis and Mining. ASONAM 2018, pp. 69–76 (2018)

5. Bodkhe, R., Ghorpade, T., Jethani, V.: A novel methodology to filter out unwanted messages from OSN user's wall using trust value calculation. In: Satapathy, S.C., Raju, K.S., Mandal, J.K., Bhateja, V. (eds.) Proceedings of the Second International Conference on Computer and Communication Technologies, pp. 755–764. Springer, New Delhi (2016)
6. Clement, J.: Most popular social networks worldwide as of October 2019, ranked by number of active users (2019). https://www.statista.com/statistics/272014/global-social-networks-ranked-by-number-of-users/. Accessed 01 Jan 2020
7. Elayeb, B.: Arabic word sense disambiguation: a review. Artif. Intell. Rev., 1–58 (2018)
8. Fortuna, P., Nunes, S.: A survey on automatic detection of hate speech in text. ACM Comput. Surv. **51**(4), 1–30 (2018)
9. Founta, A.M., Chatzakou, D., Kourtellis, N., Blackburn, J., Vakali, A., Leontiadis, I.: A unified deep learning architecture for abuse detection. In: WebSci 2019 – Proceedings of the 11th ACM Conference on Web Science, pp. 105–114 (2019)
10. Ghosh Chowdhury, A., Didolkar, A., Sawhney, R., Shah, R.R.: ARHNet - leveraging community interaction for detection of religious hate speech in Arabic, pp. 273–280 (2019)
11. Haddad, H., Mulki, H., Oueslati, A.: T-HSAB: a tunisian hate speech and abusive dataset. Springer International Publishing (2019)
12. Internet World Stats: INTERNET WORLD USERS BY LANGUAGE (2019). https://www.internetworldstats.com/stats7.htm. Accessed 30 July 2019
13. Lecun, Y., Bengio, Y., Hinton, G.: Deep learning. Nature **521**(7553), 436–444 (2015)
14. Mar, B.: How Much Data Do We Create Every Day? The Mind-Blowing Stats Everyone Should Read, Forbes (2018). https://www.forbes.com/sites/bernardmarr/2018/05/21/how-much-data-do-we-create-every-day-the-mind-blowing-stats-everyone-should-read/. Accessed 12 Dec 2019
15. Mohaouchane, H., Mourhir, A., Nikolov, N.S.: Detecting offensive language on Arabic social media using deep learning. In: 2019 Sixth International Conference on Social Networks Analysis, Management and Security, pp. 466–471 (2019)
16. Mubarak, H., Darwish, K., Magdy, W.: Abusive language detection on Arabic social media. In: Proceedings of the First Workshop on Abusive Language Online, pp. 52–56. Association for Computational Linguistics, Stroudsburg (2017)
17. Mulani, J., Heda, S., Tumdi, K., Patel, J., Chhinkaniwala, H., Patel, J.: Deep Learning Techniques for Biomedical and Health Informatics. Springer, Cham (2020)
18. Mulki, H., Haddad, H., Bechikh Ali, C., Alshabani, H.: L-HSAB: a levantine Twitter dataset for hate speech and abusive language, pp. 111–118 (2019)
19. Omar, A., Mahmoud, T.M., Abd-El-Hafeez, T.: Building online social network dataset for Arabic text classification. In: The International Conference on Advanced Machine Learning Technologies and Applications (AMLTA2018). Advances in Intelligent Systems and Computing, pp. 486–495 (2018)
20. Pedregosa, F., Grisel, O., Weiss, R., Passos, A., Brucher, M.: Scikit-learn: machine learning in Python **12**, 2825–2830 (2011)
21. Stieglitz, S., Mirbabaie, M., Ross, B., Neuberger, C.: Social media analytics – challenges in topic discovery, data collection, and data preparation. Int. J. Inf. Manag. **39**(October 2017), 156–168 (2018)

Abnormal Events and Behavior Detection in Crowd Scenes Based on Deep Learning and Neighborhood Component Analysis Feature Selection

Alaa Atallah Almazroey[1(✉)] and Salma Kammoun Jarraya[1,2]

[1] Computer Science Department, Faculty of Computing and Information Technology, King Abdulaziz University, Jeddah, Kingdom of Saudi Arabia
aalmazroey0001@stu.kau.edu.sa, smohamad1@kau.edu.sa
[2] MIR@CL Laboratory, Sfax University, Sfax, Tunisia

Abstract. In the last few years, surveillance cameras have been massively distributed both indoors and outdoors in public, due to security concerns creating a need to monitor unexpected actions or activities in a scene. An intelligent automated approach is highly required to detect anomalies from the scene as well, to save the time and cost required by laborers to detect the anomalies manually from monitor screens. In this research, we propose a deep learning-based approach to detect abnormal events and behavior from surveillance videos in crowd scenes. Thus, by using the keyframe extractor to localize the keyframes that contain important information from video frames. The selected keyframes are used to compute the optical flow values of magnitudes, orientations, and velocities of each keyframe to generate several 2D templates. Then, the obtained 2D templates are supplied to a pre-trained model 'AlexNet' to extract high-level features. The Neighborhood Component Analysis (NCA) feature selection method is applied to select the appropriate features, then use these features to generate a classification model via Support Vector Machine (SVM) classifier. Results are evaluated on several public datasets, along with a new dataset that we built it that contains different videos covering abnormal events and behaviors. The obtained results proved that the proposed method outperforms other methods.

Keywords: Abnormal events detection · Keyframes selection · Neighborhood Component Analysis · Video surveillance

1 Introduction

Anomaly detection from a video is one of the fundamental tasks in a video surveillance system. Surveillance cameras are increasingly being distributed and used for different purposes, such as security reasons, avoiding traffic congestion, and crowd control. An abnormal event is an event that affected by external causes; for example, an escape that may be caused by a natural disaster i.e. earthquake, or by other abnormal behavior, i.e. explosion. While abnormal behavior is an attitude related to conducting individual or group activities, such as walking or driving in the wrong direction. However, it is still

A.-E. Hassanien et al. (Eds.): AICV 2020, AISC 1153, pp. 258–267, 2020.
https://doi.org/10.1007/978-3-030-44289-7_25

quite challenging to design a general framework for detecting all abnormal events and behaviors from a scene. This is because there is no exact definition or dictionary term that is currently used to explain what the abnormal events or abnormal behaviors are, since the boundary between normal and abnormal is often unclear and might vary depending on the situation, i.e. day/night time, peak hours/off-peak hours, etc. [1]. Therefore, activities related to each other by space and time form the context. The early technique to detect anomalies from a surveillance camera was a non-intelligent practice that monitored multiple screens continuously and were checked primarily by a human, which was considered as a critical task that needs high levels of attention, since the anomalies rarely occur when compared to normal activities. Therefore, to alleviate this issue, developing an intelligent system is much desired to detect abnormal events and behaviors in crowd scenes automatically. Although, detecting abnormal events is still a challenging task due to different scenarios, video quality, crowded and complex scenes, occlusions, and other factors. Most of the existing models are proposed based on a hand-crafted approach that trained and designed for a particular dataset to detect a specific kind of abnormal events or behavior from a particular scene under a specific condition. These approaches are facing a problem when using them with new datasets since each dataset has its own bias. In reality, any scene may contain complicated and different types of abnormal events and behaviors under different conditions that all need to be detected. In view of this, we propose a deep learning-based approach to extract deep features and detect abnormal events and behavior in crowd scenes. Firstly, by applying a keyframe extractor to extract keyframes, which are video frames that summaries the video and eliminate adjacent redundant frames. Thus, by using the Cosine Similarity algorithm (CS) [2] to only select the frames that contain new information. Then the selected keyframes are processed to extract the optical flow components magnitude, orientation, and spatial components via the Lucas–Kanade method [3] before being combined and fed to a pre-trained convolutional neural network AlexNet [4] to obtain deep features. Eventually, the Neighborhood Component Analysis (NCA) method [5] is performed to obtain the most effective features before submitting the selected features to a binary SVM classifier. The proposed method is tested with UCSD Ped1and UCSD Ped2 [6], Avenue [7], Hockey dataset [8], including our dataset that consists of collected videos that cover various abnormal events, such as panic, smoke and fire, and abnormal behaviors such as fighting, throwing an object, etc.

The rest of this paper is structured as follows: In Sect. 2, we present the literature review for some of the existing methods. In Sect. 3, we illustrate the proposed methodology; the experiment's performance, quantitative results are discussed in Sect. 4; finally, we conclude the paper in Sect. 5.

2 Literature Review

Abnormal events and behavior detection refer to detecting and responding to the abnormal changes in videos. Recently, many researchers have been exploring how to design an effective model in order to accurately detect an anomaly. One of the early attempts at anomaly detection approaches is a hand-crafted representation. In these approaches, features are extracted from the input videos, which then requires an expert to design a model based on the specific extracted features. Such an approach can be classified into two categories: object trajectory-based approaches, as in [9, 10], and non-object trajectory-based approaches [11, 12]. In [9, 10], the authors proposed a trajectory-based approach that tracks objects and defines the global anomaly behavior. Such methods can provide an acceptable result when a scene contains few objects, but in crowded scenes, a trajectory-based approach becomes unreliable, as it is limited by occlusion. As alternatives, non- trajectory based approaches do not attempt to track objects, and propose to extract low-level features from pixel-level, such as optical flow (OF) [13], and Histogram of Oriented Gradients (HOG) [14], to detect both local and global anomaly in crowd scenes. However, the hand-crafted approach is suboptimal, as it is limited by the characteristics obtained. In recent years, deep learning has become gradually popular and commonly used in intelligent video surveillance systems to overcome the limitations of hand-crafted approaches, as no specific extraction procedures are needed. Deep learning approaches demonstrated significant performance on different domains like image classification [15], object detection [16], and activity recognition [17]. The generative approach is one of the deep learning approaches that used with unsupervised learning model that can understand and describe how the data is generated, in terms of a probabilistic model. One of the generative approach is Auto-Encoders model (AE), which has the ability to extract features efficiently without needing prior knowledge [18]. Sabokrou et al. [19] learned and serves two independent AE models, one to extract local features for local anomalies, and the second model to extract global features. Then, combine results from both models and trained Gaussian models to detect anomalies. Sparse Auto-Encoder (SAE) [20], and stack denoising auto-encoder (SDAE) [21, 22] are AE techniques that applied to speed up the model generating and reduce the dimensionality required for training. In [20], (SAE) is used to extract features from video frames. Then, the recurrent neural network (RNN) is used to generate a model. Feng et al. [21] applied two stacked denoising auto-encoder (SDAE), one for extracting appearance, texture, and short-term motion features, and the other for fusing the extracted features. Likewise, Xu et al. [22] applied stacked denoising autoencoders (SDAs) to extract both motion and appearance information. However, the networks used in their work are shallow and based on small image patches. Recently, fusion LSTM and Convolutional AE were proposed in [23] authors applied RNN model to learn appearance and motion features of long-term video. While in [24], the authors applied the Epic-Flow method to extract optical flow from the input data, then feeding the weighted convolutional Auto-Encoder long short-term memory (WCAE-LSTM) networks with these optical flows and raw input. Another deep learning approach is a discriminative approach that is a supervised learning method that uses labelled data for classification. Convolution neural networks (CNN) is an example of the discriminative model, which has a powerful tool to extract deeper and more discriminate hidden features. However, CNN requires a large number of training videos to avoid overfitting. To alleviate the limitations of CNN, fully convolutional neural networks (FCN) have been used for the first time by Sabokrou et al. [25]. Thus, by employing a pre-trained CNN model which reduces the computational cost and achieved three times faster detection of abnormal events than only using a traditional CNN. As in [26], proposed by computing the OF values and used these values to generate several templates that used as input into a pre-trained CNN to extract deep features. Wei et al. [27] proposed two-stream FCN network. The first stream of FCN for original frame input is to extract appearance features, and the second stream uses OF to extract motion features between the video frames. Then, the combination of these features provides convolutional features. Lately, a hybrid approach was applied by combining the discriminative and generative architectures together. As in [28], the authors used (3D-GANs) to detect a temporal anomaly. They trained the model with only regular videos and detected irregularities according to the deviation estimated. In [29] proposed a combination of pre-trained CNN and LSTM to extract spatial-temporal features. However, the work in [30] enhanced the method used in [29] by proposing a Bidirectional Convolutional LSTM (BiConvLSTM) network. Thus, by applying a pre-trained network to extract appearance features, and then passing these features to the BiConvLSTM to encode temporal information in both directions.

In summary, there are many successful approaches in the related field of anomaly detection, as represented above. However, all these approaches are designed for detecting either abnormal events or abnormal behavior from video, but not both. In our work, we adopt a discriminative approach, as we provided labelled data. In addition, we tend to use a pre-trained CNN, since it does not require a large dataset for training.

3 Proposed Method Overview

The proposed method is based on a supervised deep learning-based approach for detecting abnormal events and behaviors from a crowd scene. The process consists through two stages: The first stage is the pre-processing step as illustrated in Fig. 1. Initially, we converted the input videos to a set of frames. After that, a keyframe selection method is applied to these frames using the Cosine Similarity (CS) algorithm, to calculate the difference between two frames (i.e. the current frame and the previous keyframe). The procedure starts by checking if the current frame is a first frame in the video frames, if it is then we save that frame in a buffer as the first keyframe; otherwise, (is not the first frame) then the CS algorithm is used to calculate the differences

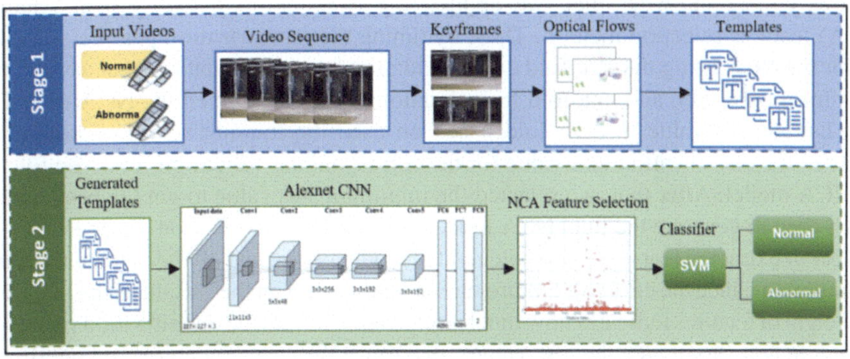

Fig. 1. The two stages of the proposed method to generate abnormal events and behavior detection model.

between the current frame and the previously extracted keyframe as in (1) [2]. As A refers to the previous selected keyframe and B refers to the current frame and n states number of frames.

$$\text{Cosine Similarity (CS)} = \frac{\sum_{i=1}^{n} A_i B_i}{\sqrt{\sum_{i=1}^{n} A_i^2}\sqrt{\sum_{i=1}^{n} B_i^2}} \quad (1)$$

When the obtained CS result does not exceed the similarity threshold value, then it means that the two frames are different, and the current frame will be considered as a new keyframe. Then we saved it in a buffer and used it to extract the next keyframe.

The similarity threshold value was selected by calculating the CS between video frames for all input videos in order to determine the appropriate threshold value. The closer CS value to 1, the smaller changes between the two frames [2]. Figure 2 represents the average CS values for the abnormal and normal videos from Hockey, UCSD ped1, UCSD Ped2, and Avenue datasets. Moreover, we noticed that most of the CS values are ranged between 0.90 and 1. After several empirical experiments, we set a similarity threshold by 0.995. Secondly, in this stage, we used the nominated keyframes to estimate the optical flow of vertical and horizontal velocity, magnitude and orientation for each adjacent keyframes via the Lucas–Kanade method [3]. So, each video generates four different 2D templates [26]. The first template represents the orientation of motion by accumulating orientation values of each pixel. The second template calculates the dot product of magnitude and orientation values to obtain the speed of motion for an object. Moreover, the last two templates contain the horizontal and vertical velocity components along with magnitude information. Thus, the use of these templates makes the feature extraction step focus on motion regions instead of the all spatiotemporal features. In the second stage (Fig. 1), the generated templates from the previous stage are used as an input to a pre-trained CNN, Alexnet [4] to extract high-level features from each template. The network consists of five convolutional layers and three fully connected layers (FC). A dropout regularization [31] (50%) is placed between FC layers, to prevent an overfitting problem. Then we merged each template's extracted features to get the final big features collection. After that, we applied the Neighborhood Component Analysis (NCA) features selection method. Thus, by tuning the regularization parameter lambda (λ) for the NCA using the five-fold cross-validation for training dataset that assigns four folds for training set, and one-fold for validation. Then, we trained the NCA model for each lambda (λ) value in each fold based on used the training sets. In addition, we recorded the classification loss value for the corresponding validation set in the fold via the NCA model. After that, we selected the minimum loss value to get the best lambda, which used to generate the final NCA model used for feature selection. Furthermore, the number of extracted features in each generated model varies, and all the extracted features are important since the effective features are selected based on the weight of the feature, as the weight of features indicates how much each feature affects the classification issue. This procedure helps to reduce the loss values, maximize the accuracy and avoid overfitting. After that, we used the selected features to test the model through different classifiers, where the linear SVM is the one that produced better accuracy results.

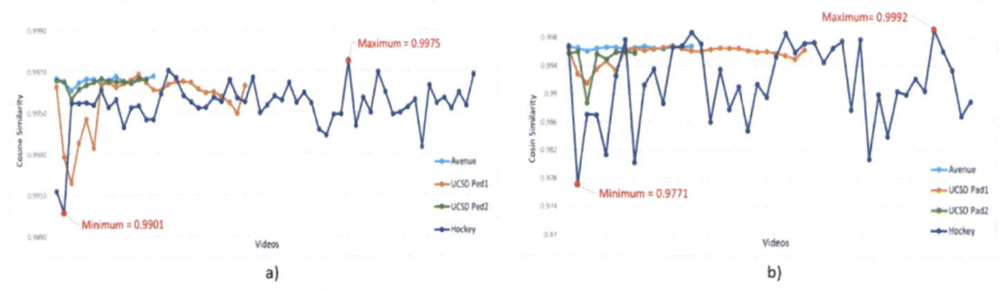

Fig. 2. The average of Cosine Similarity values (a) normal videos; (b) abnormal videos for all datasets.

4 Experiments and Results

4.1 Datasets Description

Our proposed model is trained and tested with four different public datasets, which are UCSD Ped1 [6], UCSD Ped2 [6], Avenue [7], and Hockey dataset [8]. Including our dataset that consists of other collected videos that contain abnormal events and behaviors. We divided the dataset by splitting 70% of the videos for the training set and 30% for the testing set. A combination of normal and abnormal videos is used for training.

Public Datasets. **UCSD Ped1** has 70 videos for training and testing. Normal behavior is people moving towards and away from the camera. **UCSD Ped2** contains 28 videos. The normal behavior here is people moving parallel to the camera. The anomalies for both datasets include people walking on grass, skaters, bikers, and people in wheelchairs. **Avenue dataset** consists of 37 videos for training and testing. The anomalies are persons walking toward the camera, running, and throwing an object. **Hockey dataset** is used for fighting detection, it consists of 500 clips containing fighting and another 500 clips for non-fighting. Overall, all the previous public datasets contain only abnormal behaviors. However, in real-life, both abnormal events and behavior may occur simultaneously in the scene and need to be detected.

Collected Videos. Due to the limitation in the previous public datasets (as all these datasets contain only abnormal behaviors), we recorded videos containing numerous abnormal events, such as panic from a natural disaster (earthquake), fire smoke, and flames from cars and motorcycles in a petrol station. These videos were collected from YouTube and we only selected videos captured by CCTV cameras. Then, we combined these videos with the previous public dataset videos to obtain a large dataset that contains both abnormal events and behaviors.

We applied the keyframes selection method to only work with the important frames from each dataset. Table 1 represents the average number of the selected keyframes from each video in these public datasets. As represented in Table 1, there is a significant reduction in the number of frames to be processed. Therefore, keyframe selection improves the efficiency of computation.

Table 1. Dataset description: public anomaly datasets and our training and testing dataset

Dataset name	No. of videos	Average frames	Average keyframes	Anomaly type
UCSD Ped1	70	200	93	Behavior
UCSD Ped2	28	163	57	Behavior
Avenue	37	180	76	Behavior
Hockey	1000	41	33	Behavior
Collected videos	73	395	156	Events
Our training and testing dataset	**1,208**	**195**	**83**	**Events and behavior**

4.2 Experimental Results Discussion

We used the Area Under ROC Curve (AUC) as the measurement to evaluate our proposed model. All experiments are conducted with processor Intel® Core™ i7–8750H and a graphics processor NVIDIA GeForce GTX 1050 Ti.

The keyframe selection method improved the required processing time for classification. Figure 3 shows the required time for classification of one video from the Hockey dataset with and without keyframe selection are 0.72 s and 0.89 s, respectively. The measurement includes the generation of templates, features extraction, and prediction.

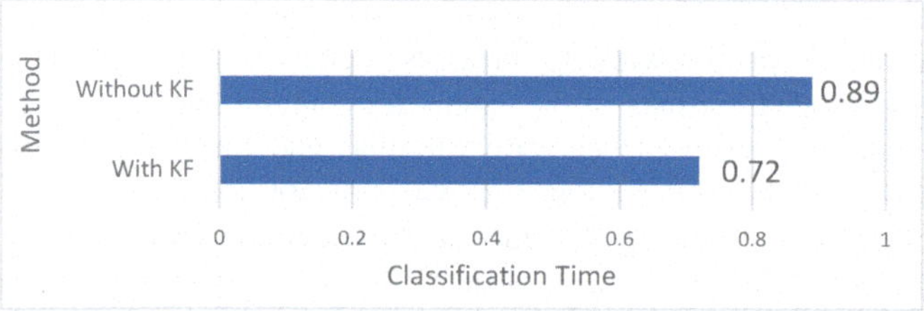

Fig. 3. Required time for classification with and without keyframe selection method for one video of Hockey dataset of 2 s.

We compared the results of our model with other existing methods on all the four datasets for anomaly detection as presented in Table 2. The obtained results showed that our model performs better than state-of-the-art methods on UCSD ped1, Avenue, and Hockey dataset. We record 93.75%, 87.5%, and 98.14%, respectively. Thus, the obtained model is sufficiently robust to detect anomalies from the crowd scene.

Table 2. AUC comparison of our model against other state-of-the-art methods.

Methods	AUC (%)			
	UCSD Ped1	UCSD Ped2	Avenue	Hockey
Stack Denoising AE (SDAs) [22]	92.1	90.8	N/A	N/A
Convolutional AE + LSTM [23]	89.9	87.4	80.3	N/A
Convolutional AE [24]	89.1	**94.8**	87.2	N/A
3D_GAN [28]	N/A	N/A	79.68	N/A
Optical flow + CNN [26]	N/A	N/A	N/A	94.40
CNN + BiConvLSTM [30]	N/A	N/A	N/A	96.54
CNN + LSTM [29]	N/A	N/A	N/A	97.1
Proposed model on testing Datasets	**93.75**	83.3	**87.50**	**98.14**

In addition, we compared the results of our model with Keçeli et al. [26]. The authors in [26], applied the Relieff features selection method [31] to select the best 10% of features from all video frames. However, in our method we used the NCA feature selection method with k-fold cross-validation from only keyframes. The set of features selected by the NCA method gives better results in terms of accuracy and generalization. Table 3 shows the obtained accuracy for the Hockey dataset using keyframes is 98.14%, while in [26] it records an accuracy of 94.4% using all frames. As for our dataset, we tested it with both feature selection methods, with and without the keyframes method. The features selected by the NCA method that applied on keyframes provide higher accuracy than those selected by Relieff method for all frames. This improvement in accuracy is due to the NCA method tunes the regularization parameter lambda (λ) through k-fold cross-validation by updating the (λ) value for each fold to find the best lambda that produces the lowest classification loss.

Table 3. AUC percentage obtained from the Hockey and Our dataset, by using the Relieff method [26] with all frames in comparison with the NCA feature selection method with keyframes only.

Features selection methods	AUC (%)	
	Hockey dataset	Our testing dataset
Relieff feature selection from all video frames [26]	94.4	97.0
Proposed model based on the NCA feature selection from Keyframes	**98.14**	**98.3**

5 Conclusions

In this paper, we proposed a deep-learning approach to detect real-world anomalies from surveillance videos. This approach achieves fast anomaly detection from the videos since it is based on a keyframe selection method to process only with important frames, then we calculated the optical flow values to generate multiple templates, which represent the continuity of the motion between frames in the video. After that, we fed the pre-trained CNN with these templates to extract deep features. We used the NCA feature selection method to select the appropriate features and trained the model using linear SVM. To validate the obtained model for abnormal events and behavior detection, we tested the model on different well-known public datasets such as UCSD Ped1, UCSD Ped2, Avenue, and Hockey dataset. Also, we have collected several videos from YouTube containing other anomalies like panic, fire smoke, and fire on cars and motorbike. The experiment results proved that our proposed abnormal events and behavior detection model performs significantly better than state-of-the-art methods.

References

1. Yuan, Y., Feng, Y., Lu, X.: Statistical hypothesis detector for abnormal event detection in crowded scenes. IEEE Trans. Cybern. **47**(11), 3597–3608 (2016)
2. Li, Y., et al.: Key frames extraction of human motion capture data based on cosine similarity. Vectors **11**(12), 1 (2017)
3. Barron, J.L., Fleet, D.J., Beauchemin, S.S.: Performance of optical flow techniques. Int. J. Comput. Vis. **12**(1), 43–77 (1994)
4. Krizhevsky, A., Sutskever, I., Hinton, G.E.: Imagenet classification with deep convolutional neural networks. In: Advances in Neural Information Processing Systems (2012)
5. Yang, W., Wang, K., Zuo, W.: Neighborhood component feature selection for high-dimensional data. JCP **7**(1), 161–168 (2012)
6. Mahadevan, V., et al.: Anomaly detection in crowded scenes. In: 2010 IEEE Computer Society Conference on Computer Vision and Pattern Recognition. IEEE (2010)
7. Lu, C., Shi, J., Jia, J.: Abnormal event detection at 150 FPS in MATLAB. In: Proceedings of the IEEE International Conference on Computer Vision (2013)
8. Nievas, E.B., et al.: Violence detection in video using computer vision techniques. In: International Conference on Computer Analysis of Images and Patterns. Springer, Heidelberg (2011)
9. Ghrab, N.B., Fendri, E., Hammami, M.: Abnormal events detection based on trajectory clustering. In: 2016 13th International Conference on Computer Graphics, Imaging and Visualization (CGiV). IEEE (2016)
10. Coşar, S., et al.: Toward abnormal trajectory and event detection in video surveillance. IEEE Trans. Circ. Syst. Video Technol. **27**(3), 683–695 (2016)
11. Li, N., et al.: Spatio-temporal context analysis within video volumes for anomalous-event detection and localization. Neurocomputing **155**, 309–319 (2015)
12. Cho, S.-H., Kang, H.-B.: Abnormal behavior detection using hybrid agents in crowded scenes. Pattern Recogn. Lett. **44**, 64–70 (2014)
13. Liu, W., et al.: Future frame prediction for anomaly detection–a new baseline. In: Proceedings of the IEEE Conference on Computer Vision and Pattern Recognition (2018)
14. Li, F., Yang, W., Liao, Q.: An efficient anomaly detection approach in surveillance video based on oriented GMM. In: 2016 IEEE International Conference on Acoustics, Speech and Signal Processing (ICASSP). IEEE (2016)

15. Deecke, L., et al.: Image anomaly detection with generative adversarial networks. In: Joint European Conference on Machine Learning and Knowledge Discovery in Databases. Springer, Heidelberg (2018)
16. Kumaran, S.K., Dogra, D.P., Roy, P.P.: Anomaly Detection in Road Traffic Using Visual Surveillance: A Survey. arXiv preprint arXiv:1901.08292 (2019)
17. Zhu, Y., Nayak, N.M., Roy-Chowdhury, A.K.: Context-aware activity recognition and anomaly detection in video. IEEE J. Sel. Top. Sig. Process. **7**(1), 91–101 (2012)
18. Liou, C.-Y., et al.: Autoencoder for words. Neurocomputing **139**, 84–96 (2014)
19. Sabokrou, M., et al.: Real-time anomaly detection and localization in crowded scenes. In: Proceedings of the IEEE Conference on Computer Vision and Pattern Recognition Workshops (2015)
20. Zhou, X.-G., Zhang, L.-Q.: Abnormal event detection using recurrent neural network. In: 2015 International Conference on Computer Science and Applications (CSA). IEEE (2015)
21. Feng, Y., Yuan, Y., Lu, X.: Deep representation for abnormal event detection in crowded scenes. In: Proceedings of the 24th ACM International Conference on Multimedia. ACM (2016)
22. Xu, D., et al.: Detecting anomalous events in videos by learning deep representations of appearance and motion. Comput. Vis. Image Underst. **156**, 117–127 (2017)
23. Chong, Y.S., Tay, Y.H.: Abnormal event detection in videos using spatiotemporal autoencoder. In: International Symposium on Neural Networks. Springer, Heidelberg (2017)
24. Yang, B., et al.: Anomalous behaviors detection in moving crowds based on a weighted convolutional autoencoder-long short-term memory network. IEEE Trans. Cogn. Dev. Syst. **11**, 473–482 (2018)
25. Sabokrou, M., et al.: Deep-anomaly: fully convolutional neural network for fast anomaly detection in crowded scenes. Comput. Vis. Image Underst. **172**, 88–97 (2018)
26. Keçeli, A., Kaya, A.: Violent activity detection with transfer learning method. Electron. Lett. **53**(15), 1047–1048 (2017)
27. Wei, H., et al.: Crowd abnormal detection using two-stream Fully Convolutional Neural Networks. In: 2018 10th International Conference on Measuring Technology and Mechatronics Automation (ICMTMA). IEEE (2018)
28. Yan, M., Jiang, X., Yuan, J.: 3D convolutional generative adversarial networks for detecting temporal irregularities in videos. In: 2018 24th International Conference on Pattern Recognition (ICPR). IEEE (2018)
29. Sudhakaran, S., Lanz, O.: Learning to detect violent videos using convolutional long short-term memory. In: 2017 14th IEEE International Conference on Advanced Video and Signal Based Surveillance (AVSS). IEEE (2017)
30. Hanson, A., et al.: Bidirectional convolutional LSTM for the detection of violence in videos. In: Proceedings of the European Conference on Computer Vision (ECCV) (2018)
31. Robnik-Šikonja, M., Kononenko, I.: Theoretical and empirical analysis of ReliefF and RReliefF. Mach. Learn. **53**(1–2), 23–69 (2003)

Classifying Upper Limb Activities Using Deep Neural Networks

Hassan Ashraf Elkholy[3,4], Ahmad Taher Azar[1,2](✉), Ahmed Magd[3,4], Hagar Marzouk[3,4], and Hossam Hassan Ammar[3,4]

[1] Robotics and Internet of Things Lab, Prince Sultan University, Riyadh, Saudi Arabia
aazar@psu.edu.sa

[2] Faculty of Computers and Artificial Intelligence, Benha University, Banha, Egypt
ahmad.azar@fci.bu.edu.eg

[3] Smart Engineering Systems Research Center (SESC), Nile University, Sheikh Zayed City 12588, Giza, Egypt
{h.ashraf,a.magd,h.marzouk,hhassan}@nu.edu.eg

[4] School of Engineering and Applied Sciences, Nile University Campus, Sheikh Zayed District, Juhayna Square, 6th of October City 12588, Giza, Egypt

Abstract. This paper presents a classification method using Inertial Measurement Unit (IMU) in order to classify six human upper limb activities. The study was also carried out to investigate whether theses activities are being performed normally or abnormally using two different neural networks: Artificial neural network (ANN) and convolutional neural network (CNN). Human activities that were included in the study: arm flexion and extension, arm pronation and supination, shoulder internal and external rotations. Before activities categorization, training data was obtained by the means of an IMU sensor fixed on an armband worn around the forearm. The training data obtained were positions, velocities, accelerations and jerks around x, y and z axes. Training samples of 264 have been collected from 10 participants, 2 women and 8 men from ages 19 to 23. Then, 204 features were extracted from IMU data, nonetheless, 15 features only have been used as inputs to the proposed neural networks because they were the most distinguished ones. After all, the networks classify the data into one of 6 classes and their results were compared. Furthermore, these proposed methods of classification have been validated by real experiments showing that ANN network gives the best performance.

Keywords: Upper limb activities · Classification · Convolutional Neural Network (CNN) · Artificial Neural Network (ANN)

1 Introduction and Literature Survey

Wearable sensors that measure human's daily activities have become widely available with a noticeable raising interest in many scientific fields [25]. These

A.-E. Hassanien et al. (Eds.): AICV 2020, AISC 1153, pp. 268–282, 2020.
https://doi.org/10.1007/978-3-030-44289-7_26

sensors are not only important because they are able to measure users' activities, body capabilities and rehabilitation progress in the clinic or in a structured environment but also they provide daily tracking feature for people over long periods with very low cost [2]. Most of these sensors generally capture major body motions which include static postures, such as sitting, standing, etc., and dynamic motions such as walking, running, jogging, however, algorithms with more specific data can underestimate overall expenditure on activities [20]. Many types of sensors are available to satisfy the purpose of activity classification [6,7]. They can be electromagnetic, mechanical or optical sensors and there are inertial sensors as well [22]. Activity classification can be obtained using single or multi-sensor configuration, but it is found that single systems simplify the process and more costly wise when multi-sensor systems are practically accurate [20]. Smartphones are one of the best wearable computing devices as they have many powerful sensors embedded in their systems enabling them to collect, process and transmit users' data easily [3,26]. Other effective sensors in this area are accelerometer-based ones, they consume low power and are perfectly suitable for wearable devices due to their small dimension and light weight in addition to be not expensive and easy to use [9,20,30]. They are capable of quantifying human motion as they are not only measuring the motion angels and movement directions but also they outputs the acceleration amplitudes range and frequencies [4,16]. After data collection is done through wearable sensors and in order to make the recognition of human activities, the problem is treated as a pattern recognition and a classification problem to be specific [26]. Many authors have presented different logical-influences algorithms using artificial neural networks, fuzzy systems, genetic algorithms and neuro-fuzzy systems [14,27]. Many studies presented deep algorithms of convolutional neural network (CNN) [8], Recurrent Neural Network (RNN), decision tree [31], KNN [18], etc. Others used Restricted Boltzmann Machine (RBM) [21], Stacked Auto-encoder (SAE), Deeply-Connected Network (DFN), and Deep Belief Network (DBN) [26]. The purpose of this paper is to classify six human upper limb activities for normal and abnormal people using an IMU sensor with 6 Degree of Freedom (DOF). This user-friendly device has been chosen in order to achieve our goal of presenting a light weight, low-cost and a functional sensor with the maximum accuracy. To do the classification activity of human elbow after data collection step, two different neural networks were applied and compared to determine which movement has been performed and also for checking whether this is a normal or abnormal movement to be able to present a treatment for this abnormal action later. CNN and ANN methods were chosen due to their ability to recognize patterns across time and space respectively and they provide state-of-the-art results on challenging activity recognition tasks in similar work in [11,13,19]. This paper is organized as follow: In Sect. 2, a brief description of the upper limb activities which are being classified is presented. Data collection and features extraction are described in Sect. 3. Neural Network architectures are presented in Sect. 4. Results with discussions are shown in Sect. 5. Finally, the paper is concluded in Sect. 6.

2 Upper Limb Movements and Experiments

2.1 Kinematics of the Upper Limb

In this paper, we are addressing six movements: elbow's four basic motions and two shoulder movements. First, human elbow is a hinged joint that consists of three bones, Humerus bone, Radius bone and the Ulna bone and this joint is linking between the upper arm and lower arm portions [5]. The four basic motions for the human elbow joint: arm flexion, arm extension, pronation and supination are shown from Figs. 1a and 2a [23]. On the other hand, shoulder structure is away complicated. In this paper, two of its basic motions are only used: shoulder internal (medial) rotation and external (lateral) rotation as shown in Fig. 3a. Figure 1b shows arm flexion and extension movements during experiments. In normal arm movement, position A1 is the start position and position A3 is the end position. While in abnormal arm movement, position A1 is the start and position A2 is the end. Figure 3 shows shoulder internal and external rotations where positions B1 and B3 are the start and end position in normal shoulder activity. Figure 2 shows Arm pronation and supination movements where positions C1 and C3 are define the normal range of motion in this joint.

Table 1. Upper limb movements and their assigned classes

Movement	Class
Normal arm flexion/extension	C1
Abnormal arm flexion/extension	C2
Normal arm pronation/supination	C3
Abnormal arm pronation/supination	C4
Normal shoulder internal/external rotation	C5
Abnormal shoulder internal/external rotation	C6

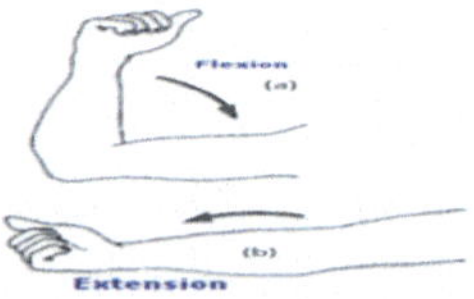

(a) Arm flexion and extension movements schematics

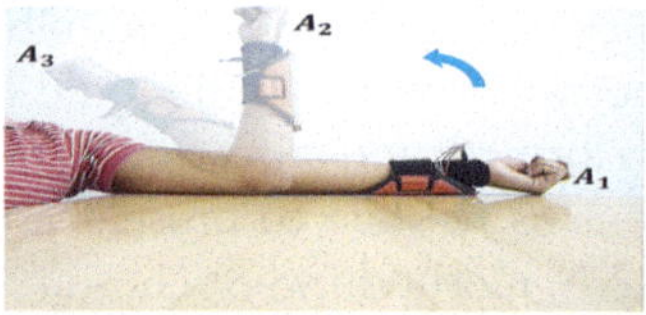

(b) Arm flexion and extension movements during experiments

Fig. 1. Arm flexion and extension movements

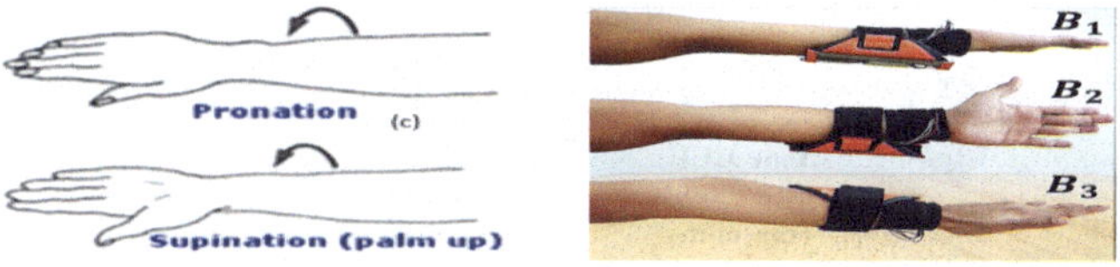

(a) Arm pronation and c movements schematics

(b) Arm pronation and supination during experiments

Fig. 2. Arm pronation and supination movements

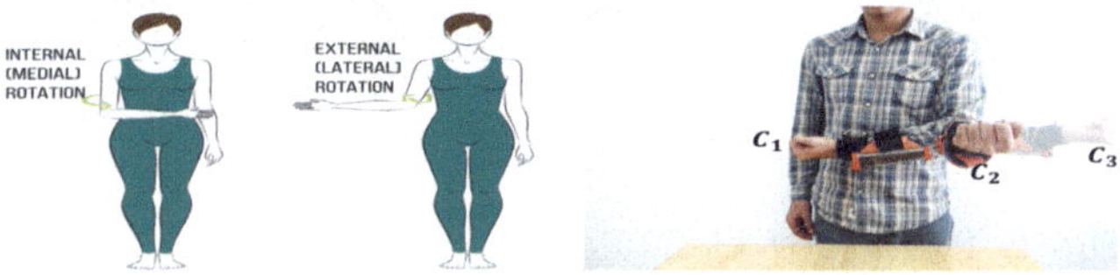

(a) Shoulder internal and external rotations schematics

(b) shoulder internal and external rotations during experiments

Fig. 3. Shoulder internal and external rotations

2.2 Hardware Configuration and Data Acquisition

Physical activity monitoring and analyzing of the upper limb are done using IMU 6050 sensor. The IMU sensor was attached to an armband. This armband was placed on participants' forearms as shown in Fig. 4. The sensor readings are sent to a NodeMCU that transmits the signals using wifi to a receiver NodeMCU that is connected to Arduino Uno. The NodeMCU is also attached to the armband and powered by a regulated power of 5 V.

2.3 Participants and Protocol

In order to provide enough training data for training the neural networks, 264 samples have been collected. The 264 samples are divided into 44 samples for each activity class as shown in Table 1. To generalize the training process and prevent over fitting, ten subjects, two female and eight male students, with age 21 ± 2 years have participated in this study. They were asked to perform six different types of physical activities. Seven subjects were right handed and the other three were left handed. The physical activities included were the extension and flexion of the arm back and forth, arm pronation and supination movements. The final activity was shoulder internal and external rotations. The participants were asked to perform the previous activities for one minute at a normal speed. Afterwards, the participants were asked to perform the same activities for the same amount of time but with smaller range of motion as illustrated in Positions A2, B2 and C2 in Figs. 1b, 3b and 2b, respectively.

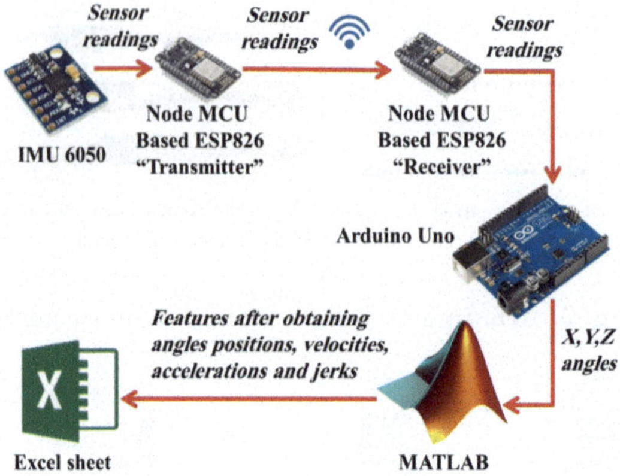

Fig. 4. The process of data acquisition during experiments

3 Data Collection and Features Extraction

3.1 Obtaining General Features

The IMU 6050 is capable of sending senor readings to the arduino which in turn calculates the angles around the x, y and z axes. Afterwards these values are fed into MATLAB software to be differentiated and to obtain velocities, accelerations and jerks values around the three axes. Therefore, there are 12 different signals for each training sample. For each signal, 17 different general features have been calculated which means there are 204 calculated features for each training sample. Equations (1) to (17) show the mathematical formulas for calculating the features [28,29]. Mean Absolute Value (MAV) represents the average of the absolute values in the signal. Modified Mean Absolute Value (MMAV) is similar to MAV but with assigning weight function to the signal values depending on their location among the signal. Moreover, MMAV2 is the same but with assigning a continuous weights function. On the other hand, Enhanced Mean Absolute Value (EMAV) is like MAV but the signal values are raised to a power value depending on their location among the signal. Wavelength (WL) can be obtained by calculating the cumulative length of the waveform and then simplifying it. Similarly, Enhanced Wavelength (EWL) is similar to WL but with raising the signal values to an exponent depending on their location among the signal. On the other hand, value of the Zero Crossing (ZC) represents the frequency information. Moreover, the Slope Sign Change (SSC) expresses how many times did the waveform change its sign. Average amplitude change measures the average of changing the amplitude values. Log Detector (LD) can be used to calculate the exerted force. Then, Root Mean Square (RMS) is used to evaluate the average of squaring the signal. It is used to calculate the direct current in alternative electric current and it is useful in expressing an average value for an alternating signal. Difference Absolute Standard Deviation Value (DASDV) is used to evaluate the absolute standard deviation. Moreover, Myopulse percentage rate (MYOP) can be used to calculate the mean of how many times did the signal values exceeded a predefined threshold. On the other hand, Willison Amplitude (WA) is similar to MYOP but without calculating the mean. Furthermore, Simple Square Integral (SSI) represents the summation of squaring the signal values. In addition, Variance of Signal (VAR) is used to calculate the signal power.

$$EMAV = \frac{\sum_{i=1}^{n} |x_i|^b}{n}, \begin{Bmatrix} b = 0.75, 0.2n \leq i \leq 0.8n \\ b = 0.5, 0.2n > i > 0.8n \end{Bmatrix} \tag{1}$$

$$EWL = \frac{\sum_{i=2}^{n} |x_i - x_{i-1}|^b}{n}, \begin{Bmatrix} b = 0.75, 0.2n \leq i \leq 0.8n \\ b = 0.5, 0.2n > i > 0.8n \end{Bmatrix} \tag{2}$$

$$MAV = \frac{\sum_{i=1}^{n} |x_i|}{n} \tag{3}$$

$$SSC = \sum_{i=2}^{n-1} b, \begin{Bmatrix} & x_{i-1} < x_i > x_{i+1} \\ b = 1, & x_{i+1} > x_i < x_{i-1} \\ & |x_i - x_{i+1}| \geq 0.016 \\ & |x_i - x_{i-1}| \geq 0.016 \\ b = 0, & else \end{Bmatrix} \tag{4}$$

$$ZC = \sum_{i=2}^{n-1} b, \begin{Bmatrix} & x_{i-1} < x_i > x_{i+1} \\ b = 1, & x_{i+1} > x_i < x_{i-1} \\ & |x_i - x_{i+1}| \geq 0.016 \\ & |x_i - x_{i-1}| \geq 0.016 \\ b = 0, & else \end{Bmatrix} \tag{5}$$

$$WL = \sum_{i=2}^{n} |x_i - x_{i-1}| \tag{6}$$

$$RMS = \sqrt{\frac{\sum_{i=1}^{n} x_i^2}{n}} \tag{7}$$

$$AAC = \frac{\sum_{i=1}^{n} (x_{i+1} - x_i)}{n} \tag{8}$$

$$DASDV = \sqrt{\frac{\sum_{i=1}^{n} (x_{i+1} - x_i)^2}{n-1}} \tag{9}$$

$$LD = e^{\frac{\sum_{i=1}^{n} \log|x_i|}{n}} \tag{10}$$

$$MMAV = \frac{\sum_{i=1}^{n} b\left|x_i\right|}{n}, \begin{Bmatrix} b = 1, & 0.25n \le i \le 0.75n \\ b = 0.5, & 0.25n > i > 0.75n \end{Bmatrix} \tag{11}$$

$$MMAV2 = \frac{\sum_{i=1}^{n} b\left|x_i\right|}{n}, \begin{Bmatrix} b = 1, & 0.25n \le i \le 0.75n \\ b = \frac{4i}{n}, & 0.25n > i \\ b = \frac{4(i-n)}{n}, & i > 0.75n \end{Bmatrix} \tag{12}$$

$$MYOP = \frac{\sum_{i=1}^{n} b}{n}, \begin{Bmatrix} b = 1, \left|x_i\right| > 0.016 \\ b = 0, \left|x_i\right| < 0.016 \end{Bmatrix} \tag{13}$$

$$SSI = \sum_{i=1}^{n} x_i^2 \tag{14}$$

$$VAR = \frac{1}{n-1}\sum_{i=1}^{n} x_i^2 \tag{15}$$

$$WA = \sum_{i=1}^{n-1} b, \begin{Bmatrix} b = 1, \left|x_i - x_{i+1}\right| \ge 0.016 \\ b = 0, \left|x_i - x_{i+1}\right| < 0.016 \end{Bmatrix} \tag{16}$$

$$MFL = \log\sqrt{\sum_{i=1}^{n-1} \left(x_{i+1} - x_i\right)^2} \tag{17}$$

3.2 Features Reductions

Reduced feature selection increases the exploration of information and models [1,10,12,15,17,24]. Since it is difficult to determine which features are the best to describe the behaviour or the activity, all the 204 features have been calculated at the beginning using Eqs. (1) to (17). Afterwards, a feature reduction algorithm has been used to provide the networks or the classification algorithms with the most distinguished features. The idea of this algorithm is to calculate the standard deviation of each feature of the 204 features separately. Equation (18) shows the mathematical formula of calculating standard deviation. Afterwards, the values will be sorted and the highest 15 features will be the selected features. This means that these features are the most changing features and their values differ the most from their mean. Table 2 shows the selected features and their calculating standard deviations.

$$Standard\ deviation = \frac{\sqrt{\sum_{i=1}^{N} \left(x_i - \bar{x}\right)^2}}{N-1} \tag{18}$$

Table 2. The most distinguished features and their standard deviations

Feature name	Standard deviation
$SSI_{z-position}$	2.7e+06
$SSI_{x-position}$	1.7e+06
$SSI_{y-position}$	1.3e+06
$SSI_{x-velocity}$	8.3e+03
$VAR_{z_position}$	2.7e+03
$SSI_{y-velocity}$	2.4e+03
$SSI_{z-velocity}$	1.8e+03
$VAR_{x-position}$	1.7e+03
$VAR_{y-position}$	1.3e+03
$WL_{x-position}$	1.2e+03
$WL_{y-position}$	708
$WL_{z-position}$	654
$EWL_{x-position}$	647
$EWL_{y-position}$	451
$EWL_{z-position}$	449

3.3 Data Scaling

Data scaling is usually used to minimize the numerical differences between the training data. The normalized data range will be from 0 to 1. This technique helps the neural network in optimizing the weight values much better and faster. Equation (19) shows the mathematical formula used to normalize the training data.

$$Normalized\ data = \frac{Data - Min}{Max - Min} \tag{19}$$

4 Neural Network Architectures

Two Different neural network architecture have been used in the upper limb activities classification: artificial neural network and convolutional neural Network. The architecture of the artificial neural networks consists of 1 hidden layers with 15 neurons and the activation function used is sigmoid. Figure 5 illustrates the architecture of the artificial neural network. Moving to the convolutional neural network, Fig. 6 illustrates the CNN architecture. The loss function used was cross-entropy.

5 Results and Discussion

In order to classify hand activities into three classes as mentioned before, 2 neural architecture of ANN and CNN were used. In the training process of ANN's model, the error was decreasing with a high rate till it reached its minimum at the 31st epoch as shown in Fig. 7. On other hand, CNN's model also started out with a low accuracy and climbed all the way up during the training phase. It started with an initial learning rate of 0.1 and finished with 50 epochs with a validation accuracy of 94.87% as it can be seen in Fig. 8.

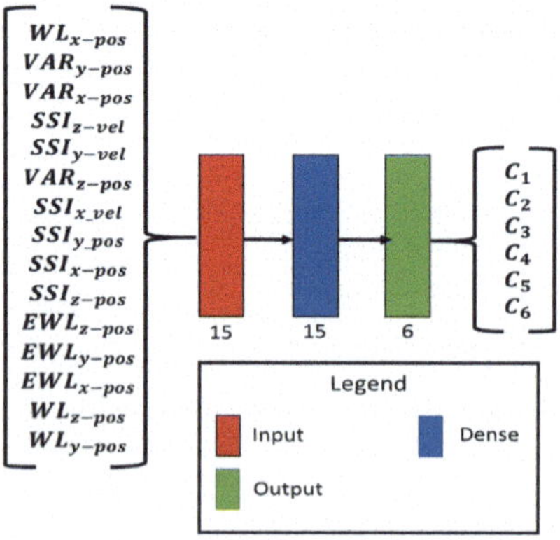

Fig. 5. Architecture of the ANN

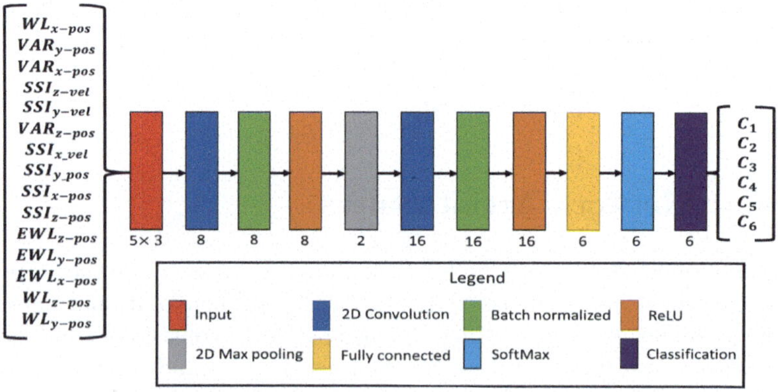

Fig. 6. Architecture of the CNN

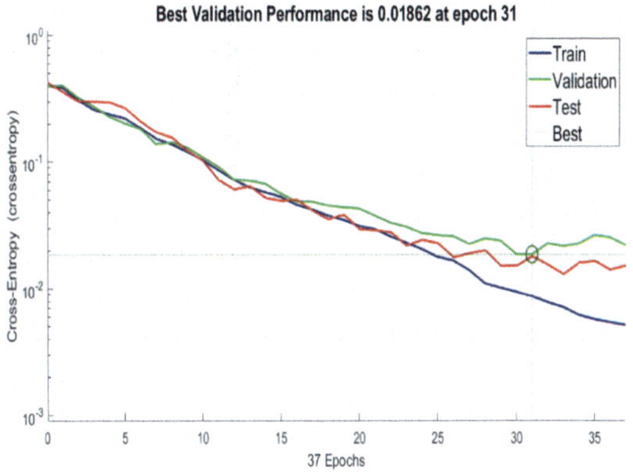

Fig. 7. Loss in ANN.

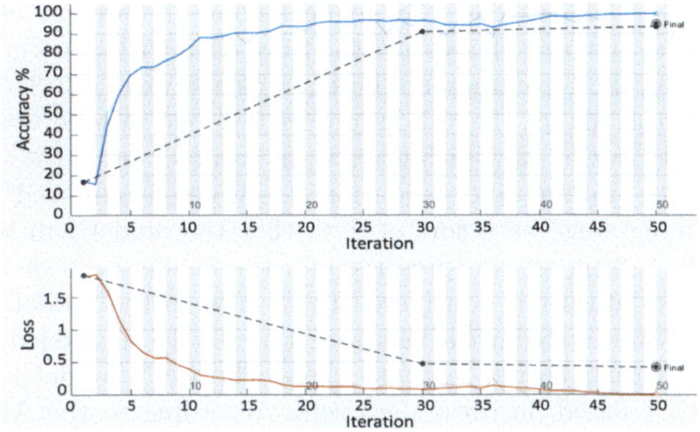

Fig. 8. Accuracy and loss in CNN.

Unlike CNN, the ANN's model does not estimate the class at once, it rather gives an estimation of how the input signal is close to each class. For example: a given sample belongs to class 1 by 60% and to class 2 by 20% and so on till the final class. Therefore, the error in this case for each sample does not have to be either one or zero, it could be a fraction between both. The error histogram in Fig. 9, shows that error in each sample of the data set was used to train, test and validate the ANN's model. Contrarily, error histogram for the CNN's model showing that error is either 0 or 1 which is 100%, because this model splits out the final class prediction without additional information about how much this sample is close to that class.

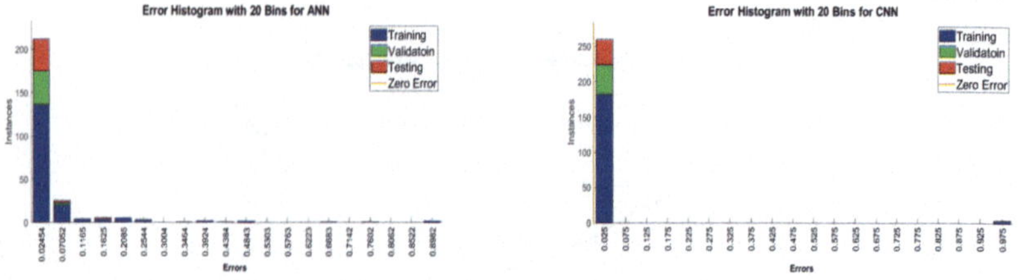

(a) Histogram of errors in ANN (b) Histogram of errors in CNN

Fig. 9. Errors histograms in ANN and CNN

It is a common practice in classification problems to use confusion matrix to measure the performance of the classification model. In this case it was used twice, once for the CNN model and the other was for the ANN model. As it can be seen from Fig. 10, when the model was fed with the training set, it identified four of the activity classes correctly while having the second class with two elements identified in wrong classes and the third class with one sample identified incorrectly. Hence, final accuracy of the training set was (187-3)/187, which is 98.4%. On the other hand, the validation set had one sample was classified in the wrong activity out of 42, which gives it an accuracy of 97.6%. It is worth to mention that both training and validation sets were encountered by the model during the training stage as a sort of data that the model will learn from. In the contrary, the testing set was preserved for the final evaluation of the trained model; thus, this data set is considered to be new to the model. Surprisingly, when the model was tested with this set, it perfectly succeeded to match all the samples with their correct activity class. The CNN model had an overall accuracy of 98.5% based on these data sets. According to the ANN confusion matrix in Fig. 10, each data set had some elements that were identified in the incorrect activity class. It has an accuracy of 99.5%, 95.0% and 97.5% for the training, validation and testing sets respectively. This implies that this model succeeded to predict the correct class most of the given samples with an overall accuracy of 98.5%.

Both models had the same final accuracy, which is 98.5%, with the 264 samples. Thus, in order to decide which model actually predicts better than the other, a random set was generated and used to judge if these models actually work as expected. This set consists of 24 samples with 4 samples in each class, and to make it more difficult, half of the data set was generated with left hand and the other half with the right one.

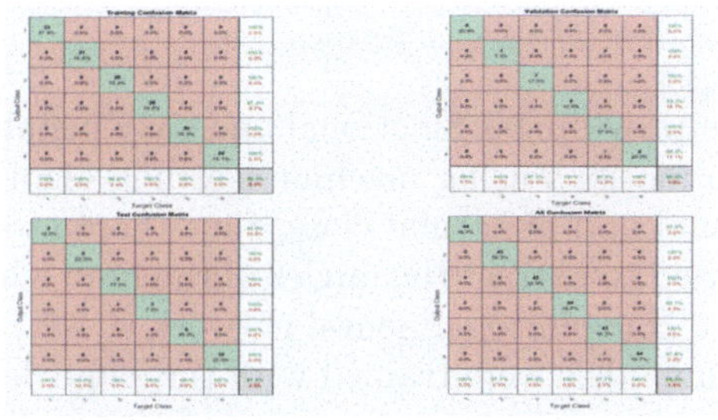

(a) Confusion matrix of ANN

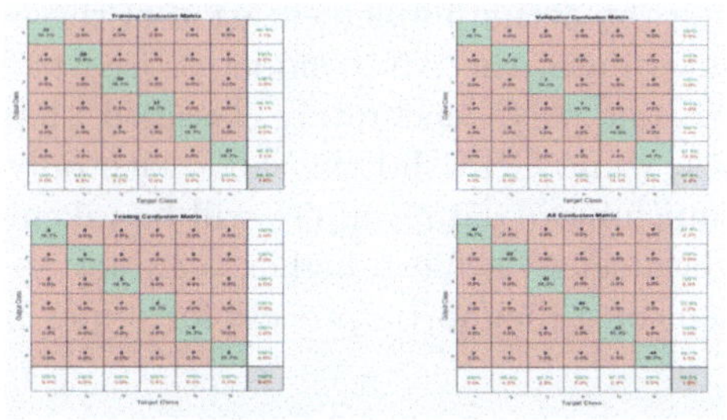

(b) Confusion matrix of CNN

Fig. 10. Confusion matrices of ANN and CNN

According to that study, the accuracy of the ANN was 87.5%, whilst the CNN is 62.5%. All the three wrong samples in the ANN model were not correct, only if they were able to classify the behavior as normal or abnormal. Table 3 provides a comparison between ANN and CNN after testing 24 new samples based on some errors and accuracy calculations methods. It is clear that ANN model performed better when being tested on completely new data.

Table 3. Comparison between ANN and CNN after validation on new samples

Method	ANN	CNN
Mean of accuracy	0.8750	0.6250
Standard deviation of accuracy	0.3378	0.4945
Mean square error	0.1250	0.3750
Root mean square error	0.3536	0.6124
Normalized root mean square error	0.3536	0.6124

6 Conclusion

This paper focuses on providing a functional and easy installed sensor that classify three upper limb motions using machine learning algorithms. Each motion is divided into a normal and abnormal class; thus there are a total of 6 classes. An IMU sensor was used to extract the angles of the limb while in motion. Then, the data is supplied to convolutional neural network (CNN) and artificial neural network (ANN). Both models were trained with the same data sets, so that they can classify any data they encounter into one of the 6 classes. After training both models, it turned out that they produced the same overall accuracy of 98.5%. However, during the actual implementation of these two models with an actual new experimental data, the ANN model proved to be more robust and accurate than the CNN one. ANN accurately classified 21 samples out of the 24 new experimental data, while CNN did the same with only 15 correct outputs. For the future work, much more data samples will be taken from participants to furnish the training data set to achieve higher accuracy so that it can be used as a part of a treatment or a diagnostic process for upper limb rehabilitation.

References

1. Anter, A.M., Azar, A.T., Fouad, K.M.: Intelligent hybrid approach for feature selection. In: Hassanien, A.E., Azar, A.T., Gaber, T., Bhatnagar, R., Tolba, M.F. (eds.) The International Conference on Advanced Machine Learning Technologies and Applications (AMLTA 2019), pp. 71–79. Springer, Cham (2020)
2. Arif, M., Kattan, A.: Physical activities monitoring using wearable acceleration sensors attached to the body. PLoS ONE **10**(7), e0130851 (2015)
3. Bayat, A., Pomplun, M., Tran, D.A.: A study on human activity recognition using accelerometer data from smartphones. Procedia Comput. Sci. **34**, 450–457 (2014)
4. Bhattacharya, A., McCutcheon, E., Shvartz, E., Greenleaf, J.: Body acceleration distribution and O2 uptake in humans during running and jumping. J. Appl. Physiol. **49**(5), 881–887 (1980)
5. Card, R.K., Lowe, J.B.: Anatomy, shoulder and upper limb, elbow joint. In: StatPearls [Internet]. StatPearls Publishing (2018)
6. Chen, C., Jafari, R., Kehtarnavaz, N.: UTD-MHAD: a multimodal dataset for human action recognition utilizing a depth camera and a wearable inertial sensor. In: 2015 IEEE International Conference on Image Processing (ICIP), pp. 168–172. IEEE (2015)
7. Chen, C., Jafari, R., Kehtarnavaz, N.: A survey of depth and inertial sensor fusion for human action recognition. Multimedia Tools Appl. **76**(3), 4405–4425 (2017)
8. Chen, Y., Xue, Y.: A deep learning approach to human activity recognition based on single accelerometer. In: 2015 IEEE International Conference on Systems, Man, and Cybernetics, pp. 1488–1492. IEEE (2015)
9. Dehzangi, O., Taherisadr, M., ChangalVala, R.: IMU-based gait recognition using convolutional neural networks and multi-sensor fusion. Sensors **17**(12), 2735 (2017)
10. Ganesan, J., Inbarani, H.H., Azar, A.T., Polat, K.: Tolerance rough set firefly-based quick reduct. Neural Comput. Appl. **28**(10), 2995–3008 (2017)

11. Hammerla, N.Y., Halloran, S., Plötz, T.: Deep, convolutional, and recurrent models for human activity recognition using wearables. arXiv preprint arXiv:160408880 (2016)
12. Hannah Inbarani, H., Nizar Banu, P.K., Azar, A.T.: Feature selection using swarm-based relative reduct technique for fetal heart rate. Neural Comput. Appl. **25**(3), 793–806 (2014)
13. Ignatov, A.: Real-time human activity recognition from accelerometer data using Convolutional Neural Networks. Appl. Soft Comput. **62**, 915–922 (2018)
14. Ijjina, E.P., Chalavadi, K.M.: Human action recognition using genetic algorithms and convolutional neural networks. Pattern Recogn. **59**, 199–212 (2016)
15. Inbarani, H.H., Bagyamathi, M., Azar, A.T.: A novel hybrid feature selection method based on rough set and improved harmony search. Neural Comput. Appl. **26**(8), 1859–1880 (2015)
16. Iosa, M., Picerno, P., Paolucci, S., Morone, G.: Wearable inertial sensors for human movement analysis. Expert Rev. Med. Dev. **13**(7), 641–659 (2016)
17. Jothi, G., Inbarani, H.H., Azar, A.T., Devi, K.R.: Rough set theory with Jaya optimization for acute Lymphoblastic Leukemia classification. Neural Comput. Appl. **31**(9), 5175–5194 (2019)
18. Kaghyan, S., Sarukhanyan, H.: Activity recognition using k-nearest neighbor algorithm on smartphone with tri-axial accelerometer. Int. J. Inform. Models Anal. (IJIMA) **1**, 146–156 (2012)
19. Lee, S.M., Yoon, S.M., Cho, H.: Human activity recognition from accelerometer data using convolutional neural network. In: 2017 IEEE International Conference on Big Data and Smart Computing (BigComp), pp. 131–134. IEEE (2017)
20. Mannini, A., Intille, S.S., Rosenberger, M., Sabatini, A.M., Haskell, W.: Activity recognition using a single accelerometer placed at the wrist or ankle. Med. Sci. Sports Exerc. **45**(11), 2193 (2013)
21. Mocanu, D.C., Ammar, H.B., Lowet, D., Driessens, K., Liotta, A., Weiss, G., Tuyls, K.: Factored four way conditional restricted Boltzmann machines for activity recognition. Pattern Recogn. Lett. **66**, 100–108 (2015)
22. Nam, H.S., Lee, W.H., Seo, H.G., Kim, Y.J., Bang, M.S., Kim, S.: Inertial measurement unit based upper extremity motion characterization for action research arm test and activities of daily living. Sensors **19**(8), 1782 (2019)
23. Norkin, C.C., White, D.J.: Measurement of Joint Motion: A Guide to Goniometry. FA Davis, Philadelphia (2016)
24. Sayed, G.I., Hassanien, A.E., Azar, A.T.: Feature selection via a novel chaotic crow search algorithm. Neural Comput. Appl. **31**(1), 171–188 (2019)
25. Seshadri, D.R., Li, R.T., Voos, J.E., Rowbottom, J.R., Alfes, C.M., Zorman, C.A., Drummond, C.K.: Wearable sensors for monitoring the internal and external workload of the athlete. NPJ Digit. Med. **2**(1), 1–18 (2019)
26. Sousa Lima, W., Souto, E., El-Khatib, K., Jalali, R., Gama, J.: Human activity recognition using inertial sensors in a smartphone: an overview. Sensors **19**(14), 3213 (2019)
27. Subramanian, K., Suresh, S.: Human action recognition using meta-cognitive neuro-fuzzy inference system. Int. J. Neural Syst. **22**(06), 1250028 (2012)
28. Too, J., Abdullah, A.R., Mohd Saad, N., Tee, W.: EMG feature selection and classification using a Pbest-guide binary particle swarm optimization. Computation **7**(1), 12 (2019)
29. Too, J., Abdullah, A.R., Saad, N.M.: Classification of hand movements based on discrete wavelet transform and enhanced feature extraction. Int. J. Adv. Comput. Sci. Appl. **10**(6), 83–89 (2019)

30. Yu, H., Cang, S., Wang, Y.: A review of sensor selection, sensor devices and sensor deployment for wearable sensor-based human activity recognition systems. In: 2016 10th International Conference on Software, Knowledge, Information Management & Applications (SKIMA), pp. 250–257. IEEE (2016)
31. Zhao, H., Liu, Z.: Human action recognition based on non-linear SVM decision tree. J. Comput. Inf. Syst. **7**(7), 2461–2468 (2011)

Experimental Kinematic Modeling of 6-DOF Serial Manipulator Using Hybrid Deep Learning

Nada Ali Mohamed[1], Ahmad Taher Azar[2,3](✉), Nada Elsayed Abbas[1], Mamdouh Ahmed Ezzeldin[1], and Hossam Hassan Ammar[1]

[1] Smart Engineering Systems Research Center (SESC), Nile University, Sheikh Zayed City, Giza 12588, Egypt
{n.ali,n.Elsayed,m.ezz,hhassan}@nu.edu.eg

[2] Robotics and Internet of Things Lab, Prince Sultan University, Riyadh, Saudi Arabia
aazar@psu.edu.sa

[3] Faculty of Computers and Artificial Intelligence, Benha University, Banha, Egypt
ahmad.azar@fci.bu.edu.eg

Abstract. According to its significance, robotics is always an area of interest for research and further development. While robots have varying types, design and sizes, the six degrees of freedom (DOF) serial manipulator is a famous robotic arm that has a vast areas of applications, not only in industrial application, but also in other fields such as medical and exploration applications. Accordingly, control and optimization of such robotic arm is crucial and needed. In this paper, different analyses are done on the chosen design of robotic arm. Forward kinematics are calculated and validated, then simulation using MSC ADAMS is done, followed by experimentation and tracking using Microsoft Kinect. Two approaches are used in this study: adaprive neuro fuzzy (ANF) system optimized by simulated annealing (SA) algorithm and convolutional neural networks (CNNs) optimized by adaptive moment estimation (Adam). The same inputs are given to both models and their results are compared in order to determine the best fit algorithm for higher precision in the given robotic model. The findings have shown that the accuracy of CNNs is higher. Furthermore, this advantage has a higher cost for the time of computation than for NFs with SA.

Keywords: Serial manipulator · Neuro-Fuzzy Control (NFC) · Convolutional Neural Network (CNN) · Simulated annealing (SA) · Adaptive moment estimation (Adm)

1 Introduction

Robots have already shaped the industry during the last decade, and this influence is still increasing with new technological advanced and the need for higher precision robot navigation [2,16]. Because advancing in the field of robotic control is

A.-E. Hassanien et al. (Eds.): AICV 2020, AISC 1153, pp. 283–295, 2020.
https://doi.org/10.1007/978-3-030-44289-7_27

a promising approach, robotic issues such nonlinear control systems, path planning and obstacles avoidance had been the subjects for analysis and development. For these tasks, a smooth trajectory between the start and end configurations is needed. This type of control uses inverse kinematics, in which positions and orientation are given and the corresponding joints angles are needed to achieve the given configuration. Despite that the inverse kinematics are complex and need a long process and considerable time to be solved, it becomes quickly and accurately solvable by the artificial intelligence. From a control point of view, Artificial intelligence (AI) is one of the most used control technologies nowadays because it enhances the performance of the robot to execute the motion in less time and more accuracy [20]. Moreover, Artificial intelligence is used to solve the issues happening because of complex mathematical problems. Under the umbrella of artificial intelligence, come the artificial neural networks and machine learning: the two technological advances that shape the future of the robotics field. The idea of training and learning from data is what distinguish the neural networks from many other control and optimization technique. This feature boosts the accuracy of the navigation problem, and thus it has been used in many applications [4,5]. Since each field of robot design and control system is an extensive area with ongoing technological developments, the intersection between these two fields is an ever-changing, rich science environment. For example, in [3], a fuzzy logic controlled mobile robot was tracked using machine vision. In [9], CNNs were used to drive a robot with accuracy using sensors readings as inputs. This research was concerned with applying two different algorithms of artificial neural networks to the data obtained from experimentation and tracking of a serial manipulator. To reach the step of safe actuation of a real robotic arm, a simulation was first done to study the dynamics and kinematics of the system and to analyze the feasible ranges of motion of each joint from the six revolute joints. The design used in our analysis is a compact one that takes a smaller space in its home position that the workspace of it. After simulation and testing, results from both optimization algorithms are compared to each other and to the actual data obtained from machine learning. The objective was to provide a complete valid easily-repeatable process of optimization of a commonly used serial manipulator that can be of great use in many applications in different fields due to its greater capability to achieve specific position and orientation within the workspace, compared to other robots with lower degrees of freedom. This paper is organized as follow: In Sect. 2, forward kinematics are presented. ADAMS Simulation is described in Sect. 3. Neural Network architectures are presented in Sect. 4. Experimental Analysis with machine vision are described in Sect. 5. Results with discussions are shown in Sect. 6. Finally, the paper is concluded in Sect. 7.

2 Forward Kinematics

To study the forward kinematics of a serial manipulator, full understanding of its joints locations and axis of rotation is essential. In this research, a 6-DOF robotic arm was used in the study; its design is shown in Fig. 1. In order to link between the global coordinates and the coordinates in any frame attached to the joints, forward kinematics is used. Denavit–Hartenberg parameters are to derive the relations between different frame of references and facilitate transformations between them, as used in other research works [14,15].

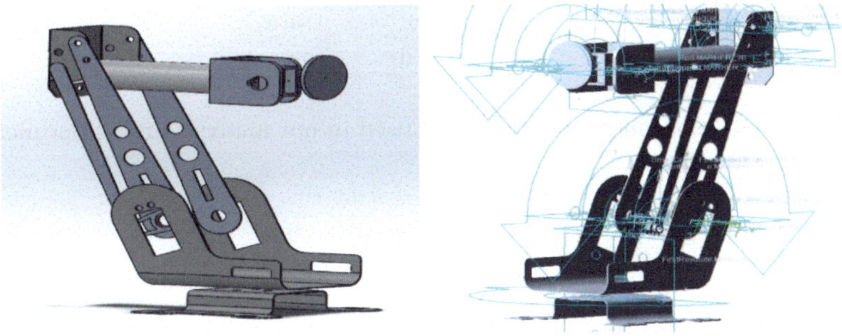

Fig. 1. SolidWorks Design of the 6 DOF robotic arm

Assuming the base as a major reference with performed coordinators θ_0, x_0, y_0, z_0. Using the homogeneous transformation matrix, the position and orientation of end effector in the workspace are found in relation to the base coordinates; the general homogeneous transformation matrix t_6^0 for the whole system is shown in Eq. 1

$$t_6^0 = \begin{bmatrix} R_6^0 & O_6^0 \\ 0 & 1 \end{bmatrix} \tag{1}$$

Where R_6^0 represents the rotational matrix and O_6^0 represents the translational vector for the end effector in the workspace. In this method, each homogeneous transformation matrix A_i is represented as a product of four basic transformations matrices, as in Eqs. 2 and 3. Table 1 illustrates the symbols references. Then, Table 2 shows the DH parameters for our robot; q_n refers to the input angle for joint n. These parameters can be used to calculate a separate transformation matrix for each joint using Eq. 3, and the resulting total transformation between the end effector and base frame is the multiplication of all the 4 by 4 transformation matrices.

$$A_i = Rot_{z,\theta_i} * Trans_{z,d_i} * Trans_{x,a_i} * Rot_{x,\alpha_i} \tag{2}$$

$$A_i = \begin{bmatrix} b & -ad & ac & a_i b \\ a & bd & -bc & a_i a \\ 0 & c & d & d_i \\ 0 & 0 & 0 & 1 \end{bmatrix} \tag{3}$$

Forward kinematics study was essential before making the simulation to have a complete understanding of kinematics and constrains, and use that knowledge in controlling the model inputs in the simulation.

Table 1. Symbols meanings for the DH parameters and equations

Symbol	Reference	Symbol	Reference
θ_i	angle between x_i and x_{i-1} about z_{i-1}	α_i	angle between z_i and z_{i-1} about x_i
d_i	angle between x_i and x_{i-1} along z_{i-1}	a_i	angle between z_i and z_{i-1} along x_i
a	$sin(\theta_i)$	c	$sin(\alpha_i)$
b	$cos(\theta_i)$	d	$cos(\alpha_i)$

Table 2. The DH parameters for the robot used in our analysis and experimentation

$Link_i$	θ_i	d_i	a_i	α_i
1	q_1	α_i	d_i	−90
2	q_2	0	0	−90
3	q_3	d_3	0	90
4	q_4+90	0	0	90
5	q_5	d_5	0	90

3 ADAMS Simulation

MSC Adams software is a widely used tool in the robotics problems as it simulates the motion of joints in different robot types, and provide data about the dynamics and kinematics of the links, as in [18] where a serial robot is used in medical application. In order to simulate the response of our serial manipulator to different motor inputs, SolidWorks and MSC ADAMS have been used to provide data about the kinematics and dynamics of the robotic arm. This data was further used as inputs for the control problem in which neural networks were applied. First, the robot was designed and assembled in SolidWorks with all its features. After checking that the robot moves in all the six DOF, motors and bolts were removed from the design before importing the model into MSC ADAMS workspace, as shown in Fig. 1. Second, the assembly file as saved as an (IGS) format to enable importing in ADAMS. At this step, joints were applied between different bodies; these joints were fixed and revolute joints, as shown in Fig. 1. And, motions are applied to the six revolute joints that control the robotic arm motion according to the design. At that point, motion is applied to every revolution joint by STEP function which takes the initial time, reference angle, the end time and final angle for each motion segment. The input motion takes 80 s; the first 65 s simulate the movement of each joint alone, and the last 15 s simulate motion under all joints, as shown in Fig. 2. Third, motion is simulated and results are graphed using the Adams Post Processor functionality. The robot has six DOF, and data for all them are obtained from MSC ADAMS. Data for linear positions is shown in Fig. 3, and they are exported as Excel files for the next step. Using MATLAB in optimization is a well-known approach and is implemented in other researches as in [19].

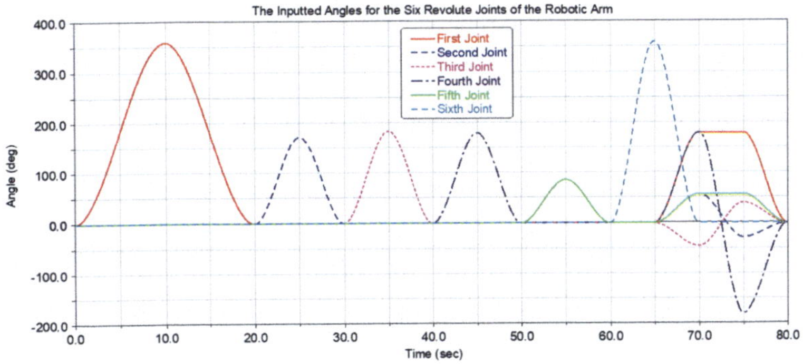

Fig. 2. The inputs for the six revolute joints given in degrees

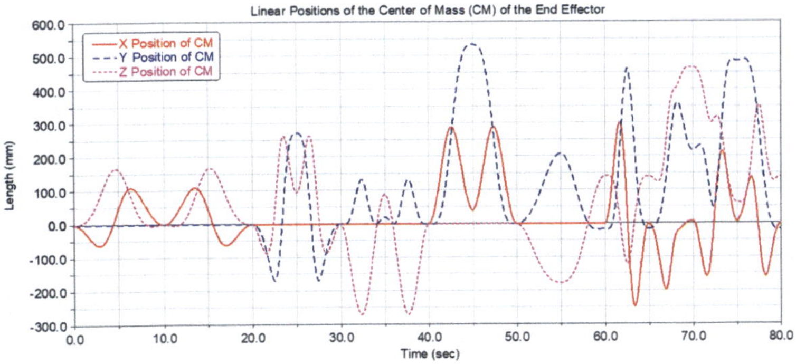

Fig. 3. The linear positions of the end effector under the inputted motions

The simulation was essential for the control and optimization of dynamics using neural network. Data from simulation was extracted and inputted to the control models.

4 Neural Network Architectures

The accuracy in estimation and controlling of neural networks in inverse kinematics and navigation has been widely explored and applied to many mobile robots as well as unmanned aerial vehicle [6,11]. Accordingly, neural networks technique was our choice for controlling the 6-DOF robotic arm, for which machine vision is used to track the end effector location and orientation. Two algorithms are applied and compared: the neuro fuzzy system optimized with SA and the convolutional neural network optimized with adaptive moment estimation.

4.1 Neuro Fuzzy System Optimized by SAA

The first algorithm that we applied was neuro fizzy system (NF) optimized by simulated annealing. This approach is very effective as it adds the advantages of both the fuzzy and neural control as well as optimizes the behavior using the simulated annealing algorithm (SAA). Neuro fuzzy logic is well recognized for mimicking the brain neurons interconnections and transforming crisp data into fuzzy linguistic domains using membership functions [17]. Accordingly, each new case is connected to the fuzzy sets through degrees of memberships [8]. Then, SAA is applied as an iterative process that aims to find the global optimum solution using the difference in an objective function to decide whether to accept the solution completely or with a small probability [10]. This objective function is called the energy function, and it is evaluated on every new set of data. This technique has a low possibility to stuck in local maxima or minima, which makes it efficient and suitable for different problems and for navigation between obstacles. Simulated annealing algorithm works quickly at first and slows down as it approaches the goal. This technique helps in finding the optimum solution for the inverse kinematic problem and decreases the computational time. The pseudo code for the SAA that we implemented is shown Algorithm 1

In the previous algorithm, it is obvious that at greater temperature, we have a high probability to accept even bad configuration because we are exploring the available options. However, at low temperature, we become more focused on good configurations only. Additionally, the scheme is shown in Fig. 4.

4.2 CNN Optimized by Adaptive Moment Estimation

The second applied algorithm is convolutional neural network (CNN), which is widely known for applications in image processing. This technique has been not only applied to many problems in robotics and computer vision, but also developed by many researches in order to decrease the computational time and give

Algorithm 1. Simulated Annealing Algorithm

Choose a start random configuration for the process and call it C **Define** two variables to simulate the temperature of the annealing; they are T_{min} and T_{max}
for $T = T_{min}$ *to* T_{max} **do**
  Find the energy of the given configuration C and call it E_c
  By using configuration updating technique, get a next configuration and call it N
  Find the energy of the given configuration N and call it E_n
  Evaluate ΔE which is the difference between En and E_c
  if ΔE *is positive* **then**
    Assign the current configuration C to be the next configuration N
  else
    if *the exponential function of* ΔE *over* T *is bigger than a random number between 0 and 1* **then**
      Assign the current configuration C to be the next configuration N
    end
  end
end

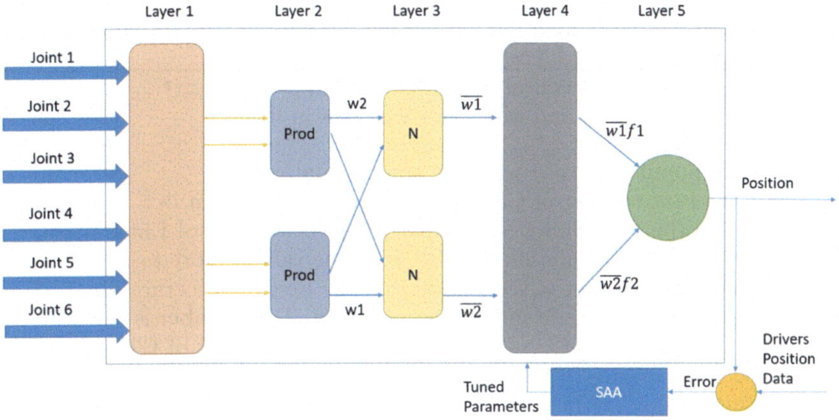

Fig. 4. The working scheme for NF optimized by SAA for obtaining optimization results

a real time result for critical applications [7]. This technique is a special class of neural network technique in which layers and nodes are used, and weights are evaluated for data transferring between different nodes in order to find the optimal solution. In the CNN algorithm, each node has a filter or more that is used to transform the data into a new format to evaluate it. This approach is what distinguishes the CNN and what makes it convolutional. CNNs is ultimately accurate in finding a path for the robotic arm with high credibility using the images from the camera system. Moreover, it can also work with precision using a single uncalibrated image from the camera as mentioned in [13]. The pseudo code for the CNNs algorithm that we implemented is shown in Algorithm 2 [1]:

The applied CNNs algorithm is optimized by adaptive moment estimation (Adam) to give accurate results. The scheme used in the analysis is shown graphically in Fig. 5.

Such optimization method was applied on simulation data to optimize behavior and control. The adaptive moment estimation and simulated annealing were the optimization methods used for CNNs and Neural Fuzzy algorithms, respectively. These optimization methods enhance the training capabilities of neural networks and the results accuracy. Taking a step forward towards implementation, experimental analysis was done on the robot as shown in the next section and the two neural networks used on simulation were reapplied on data from actual experiment.

Algorithm 2. Convolutional Neural Network Algorithm

Generate p convolutional neural networks (CNNs) structures (C_1, ..., C_p)
Create a population (pp) and save the structures in it
for *i = 1 to n* **do**
  Train the population (pp) with the training dataset
  Select the best number of CNN structures and save them in a new population (pp_{new}). This best number is called Number of Elites (noe).
  Validate the new population (pp_{new}) with the validation dataset
  Find and save the best CNN structure in the best architecture file
  Calculate the number of random parents (ran). This number is the difference between the number of CNN structures (p) and Number of Elites (noe).
  for *j = 1 to ran* **do**
    From (pp_{new}), select two CNN structures Ca and Cb according to validation loss criteria
    Using a crossover rate R, generate offspring Ca' and Cb' by one-point cross over with the selected two structures
    Save the two offspring to a new population (pp_{offs})
    Select CNN structure Cm
    Under a mutation rate (M), mute Cm and generates a mutated structure Cm'.
    while *the mutated structure Cm' is not unique* **do**
      Update and repair Cm' to be unique
    end
    Save Cm' to (pp_{offs})
  end
  Update the original population (pp) to be the addition of (pp_{new}) and (pp_{offs})
end
From the best CNN architecture file, select a group of CNN architectures
Apply majority voting ensemble using the test dataset

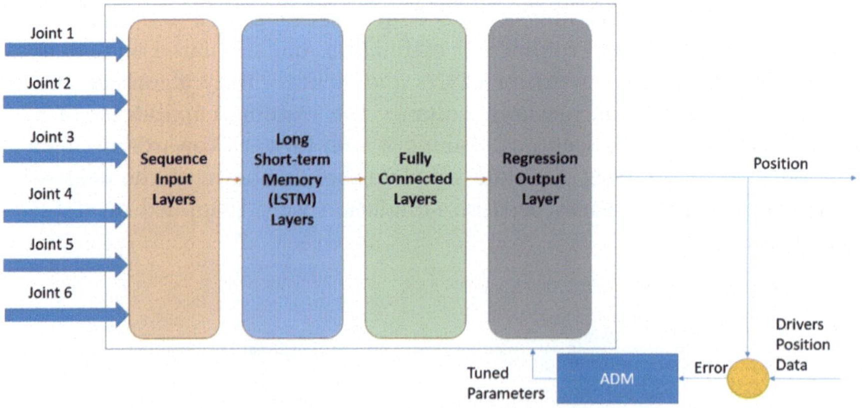

Fig. 5. The working scheme for CNNs for obtaining optimization results

5 Experimental Analysis with Machine Vision

Two main steps characterize our experimental analysis: (1) actuating the robot and tracking the position; and (2) orientation of the robot using machine vision. Similar approach was used by researchers such as in [12]. For the first step, we used a robot with the same design and dimensions as the one used in our SolidWorks design and MSC ADAMS analysis. We utilized the data gained from ADAMS in understanding the joints limits and avoid points of singularity. For the second part, we used Kinect for determining with accuracy the position of the end effector. At this point, we focused only on positions, not orientations. A red marker was inserted into the middle of the circular weight of the end effector. These data were further used in the CNNs and neuro fuzzy based simulated annealing to obtain the results. The setup of our experimentation is shown in Fig. 6.

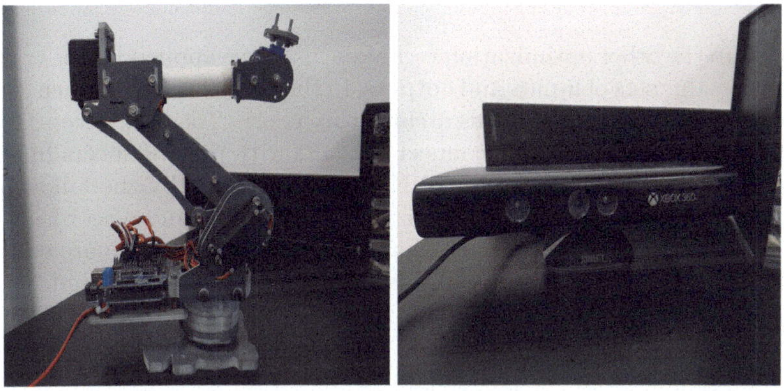

Fig. 6. The experimental setup of the actuation of 6 DOF serial manipulator. First, it shows the serial manipulator with its links and joint at a given configuration. Second, it shows the XBOX 360 Kinect which has been used to track the end effector location during actuation

6 Results and Discussions

In order to run the chosen optimization techniques on the dataset obtained from experimentation, MATLAB Deep Learning toolbox was used. Performance were based on the algorithms as well as the mean square error (MSE). Six sets of angles represent the inputs of the system, while the outputs are represented by three sets of linear positions. To start with neuro fuzzy system trained by SA, Fig. 7 shows the membership functions of the six revolute joints. The figure clearly points out the differences in the allowable range for each joint based on the design used. The used scheme of NF-SA is characterized by the following: 3 membership functions for each variable, Gaussian type of membership functions and linear type of outputs of membership functions.

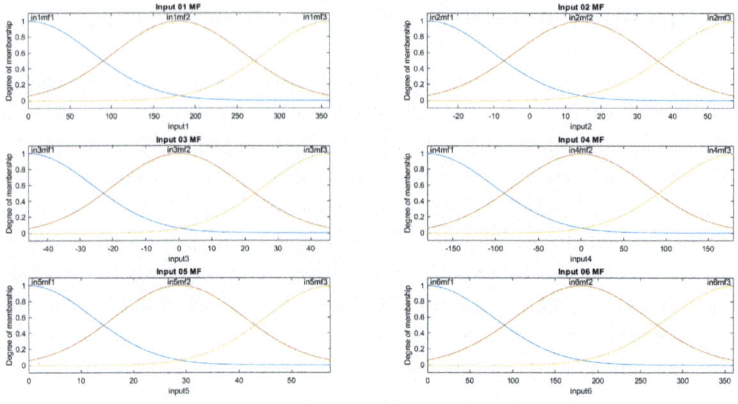

Fig. 7. The membership functions for the six revolute joints inputs in serial robotic arm

Moving to the other optimization technique used, we applied CNNs trained by Adam to the same sets of inputs and outputs. In the used CNNs approach, AlexNet architecture was implemented; this architecture consists of eight layers: five convolutional layers and three fully connected layers with pooling layers in between for activation. The used scheme of CNNs is characterized by the following: 500 epochs, 0.05 initial learning rate, 0.0004 final learning rate and piecewise learning rate schedule. During simulation, loss function and root mean square error function (RMSE) are graphed with each iteration, as shown in Figs. 8 and 9. Graphs demonstrate the fact that both the loss function and the RMSE function starts with a very large initial value, 35000 and 261, respectively. However, they end up with much lower values as the number of iterations increase. Another observation about the results is that the rate of change in each function decreases as the iterations increase.

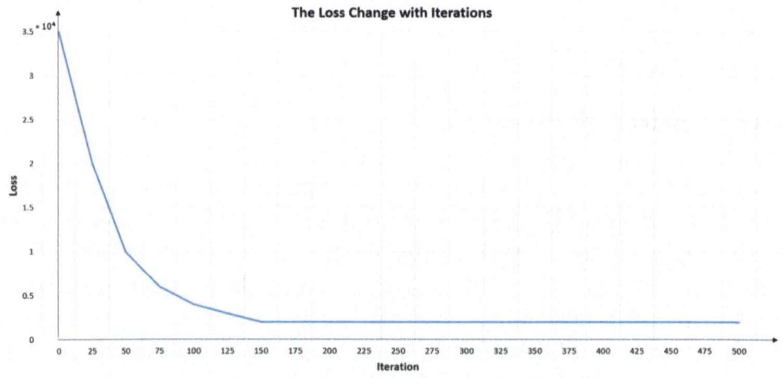

Fig. 8. Performance measures for CNNs analysis of robotic arm data loss function

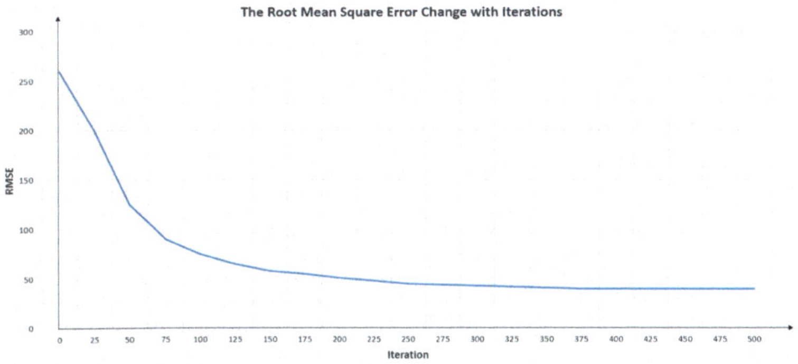

Fig. 9. Performance measures for CNNs analysis of robotic arm data root mean square error

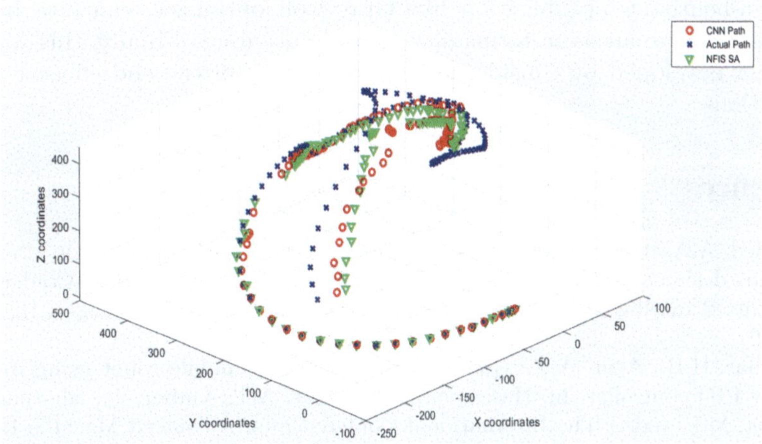

Fig. 10. Three dimensional representations of actual, CNNs and NF SAA paths

A necessary step after using different techniques in optimizing the results is to compare these results with the actual data. Visual comparison is shown in Fig. 10, where the three dimensional representations of the outputs is displayed for the actual data, CNNs data and NF SAA data. From the figure, it appears that at lower values of y and z, the results of all the three data sets is very similar; however, it differs greater with larger values. Another important observation is that CNNs is better than NF-SA in mimicking the actual results of the robotic arm. Although CNNs appears as a great choice for optimization, the larger computational time for this method should be in consideration in implementation.

7 Conclusion

In this research, simulation, experimentation and optimization are applied to the design of a serial 6-DOF robot arm. Simulation results were of great use in providing us with the limits of angles and the workspace to avoid crashes in real experimentation. Then, work was concerned with applying neural networks, either CNNs or NF optimized by SA, in optimizing the control and actuation. Such development is useful for applications in many field. Experimentation was done using Kinect to track the linear positions of the end effectors, and data are extracted for optimization algorithms. After running CNNs and NF, CNNs proved to give more accurate results with lower error. However, the computational time for CNNs is considerably greater than NF-SA. This research can be further extended by using multiple other control and optimization methods and compare between them. Moreover, advanced techniques, such as reinforcement and meta-heuristics optimization like Gray wolf optimizer, can also be great addition to our analysis in terms of increasing accuracy. Finally, this work can be further developed by considering the orientation of the end effector besides the positions.

References

1. Ahmed, A.A., Darwish, S.M.S., El-Sherbiny, M.M.: A novel automatic CNN architecture design approach based on genetic algorithm. In: International Conference on Advanced Intelligent Systems and Informatics, pp. 473–482. Springer, Heidelberg (2019)
2. Ammar, H.H., Azar, A.T.: Robust path tracking of mobile robot using fractional order PID controller. In: Hassanien, A.E., Azar, A.T., Gaber, T., Bhatnagar, R., Tolba, M.F. (eds.) The International Conference on Advanced Machine Learning Technologies and Applications (AMLTA 2019). pp. 370–381. Springer, Cham (2020)
3. Azar, A.T., Ammar, H.H., Mliki, H.: Fuzzy logic controller with color vision system tracking for mobile manipulator robot. In: International Conference on Advanced Machine Learning Technologies and Applications, pp. 138–146. Springer, Heidelberg (2018)
4. Azar, A.T., Aly, A.M., Sayed, A.S., Radwan, M.E., Ammar, H.H.: Neuro-fuzzy system for 3-DOF parallel robot manipulator. In: 2019 Novel Intelligent and Leading Emerging Sciences Conference (NILES), vol. 1, pp. 1–5. IEEE (2019)
5. Azar, A.T., Ammar, H.H., Ibrahim, Z.F., Ibrahim, H.A., Mohamed, N.A., Taha, M.A.: Implementation of PID controller with PSO tuning for autonomous vehicle. In: International Conference on Advanced Intelligent Systems and Informatics, pp. 288–299. Springer, Heidelberg (2019)
6. Braga, J.R., Velho, H.F., Conte, G., Doherty, P., Shiguemori, É.H.: An image matching system for autonomous UAV navigation based on neural network. In: 2016 14th International Conference on Control, Automation, Robotics and Vision (ICARCV), pp. 1–6. IEEE (2016)
7. Howard, A.G., Zhu, M., Chen, B., Kalenichenko, D., Wang, W., Weyand, T., Andreetto, M., Adam, H.: Mobilenets: Efficient convolutional neural networks for mobile vision applications. arXiv preprint arXiv:170404861 (2017)

8. Mamlook, R., Ahamed, T.I., Maqbool, S.D., Al-Ammar, E.A., Malik, N.: A fuzzy simulated annealing algorithm for minimizing consumer electricity bill under demand response. Int. J. Comput. Sci. Inf. Secur. **14**(11), 144 (2016)
9. Moeys, D.P., Corradi, F., Kerr, E., Vance, P., Das, G., Neil, D., Kerr, D., Delbrück, T.: Steering a predator robot using a mixed frame/event-driven convolutional neural network. In: 2016 Second International Conference on Event-based Control, Communication, and Signal Processing (EBCCSP), pp. 1–8. IEEE (2016)
10. Pandey, A., Parhi, D.R.: Autonomous mobile robot navigation in cluttered environment using hybrid takagi-sugeno fuzzy model and simulated annealing algorithm controller. World J. Eng. **13**(5), 431–440 (2016)
11. Panigrahi, P.K., Ghosh, S., Parhi, D.R.: Navigation of autonomous mobile robot using different activation functions of wavelet neural network. Arch. Control Sci. **25**(1), 21–34 (2015)
12. Pérez, L., Rodríguez, Í., Rodríguez, N., Usamentiaga, R., García, D.F.: Robot guidance using machine vision techniques in industrial environments: a comparative review. Sensors **16**(3), 335 (2016)
13. Ran, L., Zhang, Y., Zhang, Q., Yang, T.: Convolutional neural network-based robot navigation using uncalibrated spherical images. Sensors **17**(6), 1341 (2017)
14. Sawaqed, L., Al-Ali, A.I., Hatamleh, K.S., Jaradat, M.A.: Modeling and simulation of a moving robotic arm mounted on wheelchair. In: 2017 7th International Conference on Modeling, Simulation, and Applied Optimization (ICMSAO), pp. 1–5. IEEE (2017)
15. Serrezuela, R.R., Chavarro, A.F.C., Cardozo, M.A.T., Toquica, A.L., Martinez, L.F.O.: Kinematic modelling of a robotic arm manipulator using MATLAB. ARPN J. Eng. Appl. Sci. **12**(7), 2037–2045 (2017)
16. Soliman, M., Azar, A.T., Saleh, M.A., Ammar, H.H.: Path planning control for 3-Omni fighting robot using PID and fuzzy logic controller. In: Hassanien, A.E., Azar, A.T., Gaber, T., Bhatnagar, R., Tolba, F.M. (eds.) The International Conference on Advanced Machine Learning Technologies and Applications (AMLTA 2019), pp. 442–452. Springer, Cham (2020)
17. Li, W.: Neuro-fuzzy systems for intelligent robot navigation and control under uncertainty. In: Proceedings of 1995 IEEE International Conference on Fuzzy Systems, vol. 4, pp. 1747–1754 (1995). https://doi.org/10.1109/FUZZY.1995.409918
18. Yang, J., Yu, L., Wang, L., Wang, Z., Zhuang, Z.: Dynamic characteristics analysis based on ADAMS for general robotic arm of minimally invasive surgical robot. In: 2015 IEEE International Conference on Mechatronics and Automation (ICMA), pp. 854–859. IEEE (2015)
19. Zhou, L., Bai, S., Hansen, M.R.: Design optimization on the drive train of a light-weight robotic arm. Mechatronics **21**(3), 560–569 (2011)
20. Zhu, Q., Azar, A.T.: Complex System Modelling and Control Through Intelligent Soft Computations, Studies in Fuzziness and Soft Computing, vol. 319, Berlin, Germany (2015)

A Hybrid Deep Learning Based Autonomous Vehicle Navigation and Obstacles Avoidance

Habiba A. Ibrahim[1,2], Ahmad Taher Azar[3,4](✉), Zahra Fathy Ibrahim[1,2], and Hossam Hassan Ammar[1,2]

[1] Smart Engineering Systems Research Center (SESC), Nile University, Shaikh Zayed City, Giza 12588, Egypt
{h.ibrahim,z.fathy,hhassan}@nu.edu.eg
[2] School of Engineering and Applied Sciences, Nile University Campus, Juhayna Square, Sheikh Zayed District, 6th of October City, Giza 12588, Egypt
[3] Robotics and Internet-of-Things Lab (RIOTU), Prince Sultan University, Riyadh, Saudi Arabia
aazar@psu.edu.sa, ahmad_t_azar@ieee.org
[4] Faculty of Computers and Artificial Intelligence, Benha University, Banha, Egypt
ahmad.azar@fci.bu.edu.eg

Abstract. Technological revolution has reached all life activities starting from day planning reaching communication, entertainment, industry, and transportation. Each of previously mentioned categories get improved in a way making human life easier and safer. In the use of automatic control, several researches focused on automating vehicles' systems to make driving easier and safer. The availability of autonomous vehicles will avoid accidents caused by taking a late decision or lack of driving experience in such situation. Approaching autonomous driving, an autonomous vehicle must be able to respond to the state of objects in the surrounding, be it stationary or in motion. This paper outlines the techniques which enable the car to become conscious of its immediate environment while it moves independently and to decide its next course of action to avoid obstacles. It investigates two approaches which are Neuro-Fuzzy System tuned by Particle Swarm Optimization (PSO) and Convolutional Neural Network (CNN) tuned by Adaptive Moment estimation (Adam). Such control can allow cars on roads to operate smoothly and, according to trained data, take quick accurate decisions. Results showed high performance of deep learning algorithms specially CNN with Adam; however, it needs more computational time than Neuro-Fuzzy system tuned with PSO.

Keywords: Autonomous vehicle · Neuro-fuzzy control · Convolutional Neural Network (CNN) · Particle Swarm Optimization (PSO) · Adaptive moment estimation (Adam)

A.-E. Hassanien et al. (Eds.): AICV 2020, AISC 1153, pp. 296–307, 2020.
https://doi.org/10.1007/978-3-030-44289-7_28

1 Introduction and Related Work

Keeping up with technology revolution and changes nowadays, people are turning everything in their lives to be smart. Clearly, they enhancing performance of any used device/equipment to reduce the man power or stress required. Accordingly, autonomous cars have been one of brilliant developed upcoming ideas [11]. They rely mainly on teaching logic to cars control units to be able to drive safely according to algorithm derived by a driving expert. Basically, driving skills can be easily taught to a human; however, the real problem occurs when facing unavoidable crashing scenarios [12]. In such situations, the best-case scenario is to hurt someone; as a result, a man in the situation would panic from overthinking of what should be done to get the least damage or not to hit innocent ones. Therefore, people thought about teaching cars controllers code of ethics and morality according to an expert which will help the car behaving better than a human as it will only collect data required for the algorithm to take a decision without panicking or being overwhelmed with emotions [12]. Actually, different companies, such as Nissan and Mercedes Benz, started to develop their own ethics-based controllers to save their rights in such a development. In this context, A scenario is discussed based on fuzzy logic in addition to Neural Network that mimics human mind in its technique of thinking. Trying to keep up with Industry 4.0 and automation revolution, autonomous vehicles got the priority to be implemented than the manual car. They are able to take fast right decision in situations where the best case scenario is to hurt someone. Such decision is taken according data that was previously entered and taught to vehicle controller. Therefore, many companies have started their journeys for releasing their own autonomous vehicles as soon as possible. For example, Volvo started its journey of implementing such vehicles in 2006. Then, it releases their first full autonomous test in 2017 intending to bring the first car to market in 2021. Another model was that of Tesla; all Tesla cars nowadays are equipped with self-driving capabilities. Predicted in 2020 for rest of competitors such as Audi, BMW, Mercedes-Benz, and Nissan to release their own AVs [9]. Over the past few years, Convolutional Neural Networks (CNNs) have achieved state-of-the-art performances in solving problems related to computer vision especially in image classification. The development of autonomous motion took a new path following the breakthrough in deep learning for image recognition [20]. Depth learning is able to detect and classify various moving entities: cars, people or obstacles along the road. In the field of vision and an inertial measurement unit with differential global positioning system (DGPS) to set up the vehicle's position [4], the neuro-fuzzy-based controller is expected to be supplied from a vision sensor source. Neuro-fuzzy navigation logic applies on various road conditions for overcoming obstacles. The researcher warned, however, that the approach performed well on the experimental scenario and that certain constraints remain, such as the catering for the unpredictable movements of the other drivers, tuning for diversions and road blocks. There has also been work in creating hybrids between the Neuro-Fuzzy and particle swarm optimization (PSO) [4]. A hybrid algorithm is proposed in [13], where randomly selected particles from the Neuro-fuzzy inference system (NFIS) are evolved using PSO and returned to the NFIS for further optimization. The authors in [20] proposed randomly partitioning the particle population into sections where PSO and NFIS are combined then applied to achieve the optimal model then re-partitioned after each iteration [15]. Recent improvements considerably over CNNs, which can only recognize simple object functions based on supervised learning with templates that are manually entered [16]. This makes the profound learning perfect for classifying objects as traffic scenes in arbitrary scenes. The deep learning on the basis of CNN represents an essential step in the development of artificial perceptions [7]

Lastly, this paper is concerned with most common and important system in autonomous vehicle navigation such as object avoidance. Scenario and rule base model is discussed and the generated data is then compared to actual driving data. The key contribution of this paper is in designing an optimal navigation system to mimic human decision to overtake an obstacle based on its proximity and direction and complement with deep neural network and Neuro-Fuzzy based PSO to navigate the vehicle so that it is adaptive to various obstacles sizes. The design feasibility was demonstrated with a scaled down physical vehicle fitted with a low-cost ultrasonic sensors to ascertain the obstacle proximity and direction.

This paper is organized as follow: In Sect. 2, mathematical modeling of the vehicle dynamics is presented. Hardware and data generation is described in Sect. 3. The Proposed deep learning techniques are presented in Sect. 4. Results with discussions are shown in Sect. 5. Finally, the paper is concluded in Sect. 6.

2 Mathematical Modeling of the Vehicle Dynamics

For a four-wheel vehicle to steer ideally, it follows one of the steering mechanisms. Ackermann's mathematical model calculate the actual steering angel transmitted from the steering wheel by the driver. Ackermann's model requires specific geometry conditions in the four-wheel vehicle. Figure 1 shows the required geometry for a vehicle to turn right. Taking l as the car length, b the width of the vehicle from the wheels' center, θ_l and θ_2 the turning angles for the front wheels, Eq. 1 presents the ideal Ackermann's steering criteria. Ackermann's model requires also all the four wheels to be rolling around a specific and common center [4,8].

$$\cot(\theta 1) - \cot(\theta 2) = \frac{b}{l} \tag{1}$$

In the simplified form of Lagrange where m is the total mass of the vehicle, J is the moment of inertia, J_k is wheels' moment of inertia, and coordinates were represented by x, y [4].

$$L = \frac{1}{2}m\left(\dot{x}^2 + \dot{y}^2\right) + J\dot{\theta}^2 + J\dot{\theta}\dot{\Phi} + \frac{1}{2}J\dot{\Phi}^2 + J_k\dot{\Psi}^2 \tag{2}$$

$$\dot{x}\ sin\theta - \ \dot{y}cos\theta = 0 \tag{3}$$

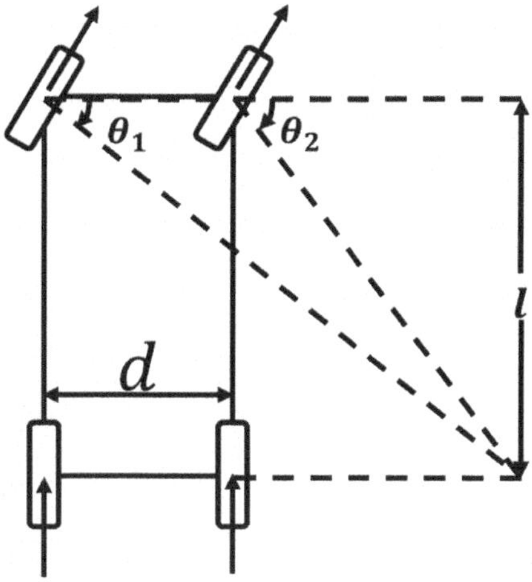

Fig. 1. Schematic representation for Ackermann's steering model

$$\dot{x}\ \sin(\theta+\Phi)-\dot{y}\cos(\theta+\Phi)-l\dot{\theta}cos\Phi=0 \tag{4}$$

$$\dot{x}\ cos\theta+\ \dot{y}sin\theta-R\dot{\psi}=0 \tag{5}$$

The previous equations helped concluding the equation of motion for the system as follows:

$$m\ddot{x}+\lambda_1 sin\theta+\lambda_2\sin(\theta+\Phi)+\lambda_3 cos\theta=0 \tag{6}$$

$$m\ddot{y}-\lambda_1 cos\theta-\lambda_2\cos(\theta+\Phi)+\lambda_3 sin\theta=0 \tag{7}$$

$$2J\ddot{\theta}+J\ddot{\Phi}-\lambda_2 l\cos\theta=0 \tag{8}$$

$$J\ddot{\Phi}+J\ddot{\theta}=\tau_1 \tag{9}$$

$$J_k\ddot{\psi}-\lambda_3 R=\tau_2 \tag{10}$$

Where (x,y,θ) are the three degrees of freedom of Ackermann's model and (ψ,ϕ) are the control variables. Based on the control variables, Ackermann's model can be described as non-holonomic mechanical system that has three kinematic equations of motion (Eq. 3) [2,6,17,19].

3 Hardware Description and Data Generation

The basic architecture for self-driving cars has two main systems that are responsible for the behavior of the car. The first system is the perception system which is responsible for returning the current state of the vehicle and capturing the surrounding environment using sensors attached to the vehicle outer body. In this model, Inertial Measurement Unit (IMU) sensor, two ultrasonic sensors, and camera module were used with prior data about the surrounding environment traffic rules as shown in Fig. 2. The decision-making system is responsible for navigating the car based on the collected data from the sensors and the camera module. In order to decide the optimum action for the vehicle, the data about the surrounding objects was collected from the ultrasonic sensors and the position, acceleration, and velocity of the car were generated from the IMU sensor. To generate the data from the hardware, a wireless module and a controller were used to navigate the car in a created environment, then the data was used in the training models of neural network and fuzzy logic.

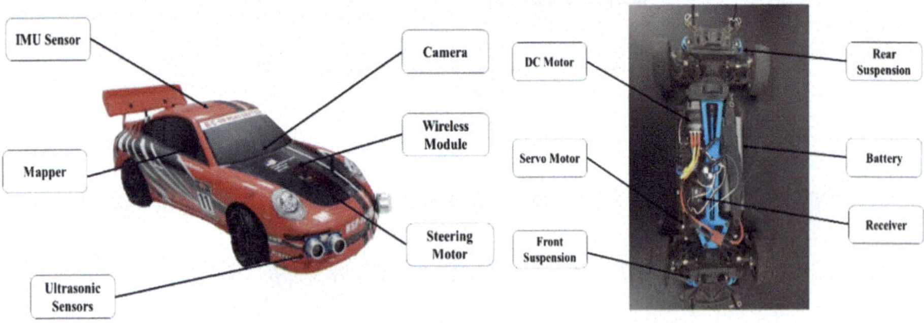

Fig. 2. Hardware architecture

Real-time datasets were generated through creating a real-life environment with various obstacles on using a real car model as shown in Fig. 2. Two ultrasonic sensors were positioned in front of the model and their detection data was used in the training model. The car was controlled and navigated to follow the traffic rules and avoid the different obstacles in the environment and the generated datasets were exported to MATLAB to be trained using the Deep Learning Toolbox.

The navigation of the car is based on number of inputs and outputs where the inputs are the objects relative speed and distance while the outputs are the angular and the linear velocity of the car. The relative speeds of the objects were calculated from the distance detected by the ultrasonic sensors and given the real time of the paths as shown in Fig. 3.

4 The Proposed Deep Learning Techniques

The field is based upon the presumption that human intelligence "can be described so precisely that it can be simulated by a machine." This gives rise to philosophical discussion of the nature of the mind and the ethics of man-like intelligence-creation. In this section Neuro-Fuzzy Inference System (NFIS) optimized with PSO and Convolutional Neural Network (CNN) optimized using Adaptive moment estimation (Adam) have been used to make the car behaves the same way as a human driver.

4.1 Neuro-Fuzzy Based PSO Optimization

The neuro-fuzzy method merges the advantages of neural networks and fuzzy theory to design a model that uses a fuzzy theory to represent knowledge in an interpretable manner and the learning ability of a neural network [3,10]. Neuro-fuzzy system includes input layer, hidden layers, output layer. Input layer is represented with a vector of inputs. The sampled data for each input vector is passed on to the hidden layer. Hidden layer contains two networks: the premise network, and the consequent network. Fuzzification interface is first performed with its three stages; fuzzification, Decision-making unit, and defuzzification. First stage, fuzzification, assigns each input a value that ranges from 0 to 1 and represents its membership value, or its involvement percentage to a specific category [5]. Membership functions can be triangular, sigmoidal, polynomial, or Gaussian distribution and their number in each input is a compromise between complexity and accuracy. Membership functions are given by:

$$\lambda_{in} = exp\left(-\frac{(x_i - \mu_{in})^2}{(\sigma_{in})^2}\right) \tag{11}$$

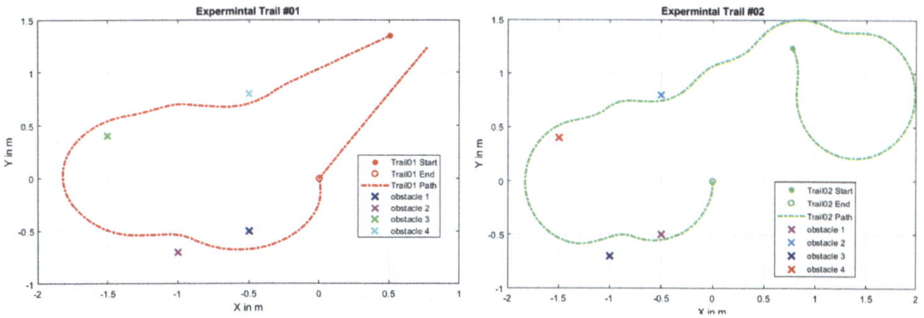

Fig. 3. Experimental paths for data generation results

Where λ_{in} defines the corresponding mean value and σ_{in} the spread factor of the n-th Gaussian function.

This stage generates the single output with set of fuzzy values/membership values that allocates it accurately. Premise network aims to separate inputs area into partitions using Mamdani approach which gives a number of membership functions for each input variable to represent its linguistic partitions. After decision-making, defuzzification is performed to get the final output crisp value again [1]. In consequent network, a set of linear equations is computed given the input dataset and outputs, then the result is transferred to output layer. In this final layer, output data generated from premise and consequent networks are multiplied in order to predict the model output such as:

$$\rho = [w_1G \quad w_1T \quad w_2G \quad w_2T \quad w_{64}] \tag{12}$$

Output parameters are finally tuned using PSO which utilizes the swarm intelligence generated by the cooperation and competition between the particles in a swarm. Such technique has emerged as a useful tool for engineering optimization of control parameters [7]. The proposed Neuro-Fuzzy based PSO is shown in Fig. 4.

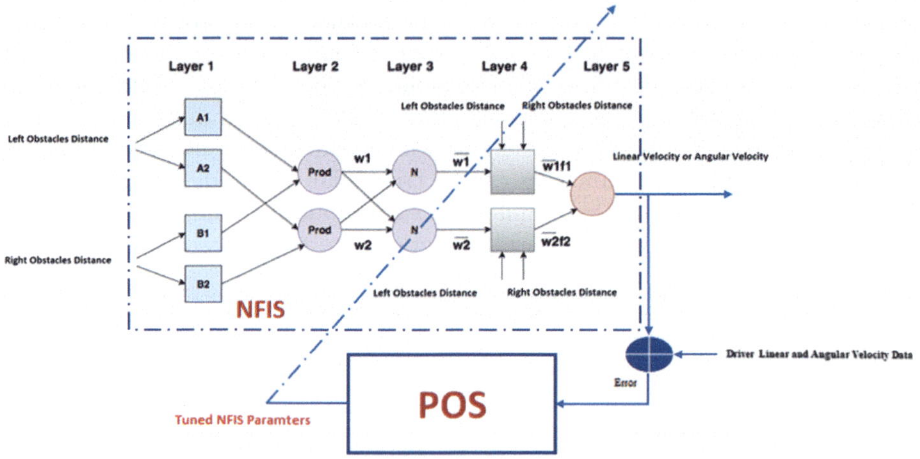

Fig. 4. The proposed Neuro-Fuzzy Tuned with PSO

4.2 Convolutional Neural Network (CNN)

In deep learning, CNN is a class of neural network that is used for visual imagery. It takes the entire image as an input and start to extract its features through CNN layers; Convolutional layer, RELU layer, pooling layer, and Fully-connected (FC) layer. The most important layer is the convolutional due to the number of filters included inside both forward and backward passes. Other layers are responsible for down sampling, preventing over-fitting, running activation function, or connecting neurons to solve non-linear problems [16]. Activation function has several types; for instance, it can be exponential or linear unit. Second part of this method is the classifier where data is categorized according features extracted in the convolutional part [14]. This part contains fully connected layers, each is separated with activation function layer; the last FC layer will have number of neurons as same as number of categories. Batch Normalization (BN) layers aim to accelerate the training of the network by normalizing each scalar feature inside a mini batch independently. CNN is trained using back-propagation algorithm where the loss function and optimization function play key roles. Loss function is mathematically calculated as shown in Eq. 13 to minimize it [18].

$$L_{crossentropy} = -\sum_{i=0}^{N} y_i log(\hat{y_i}) \tag{13}$$

The proposed CNN optimized by Adam is shown in Fig. 5.

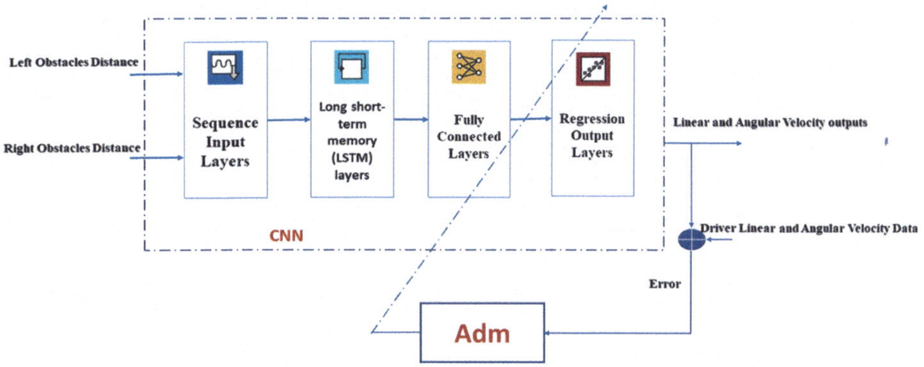

Fig. 5. The proposed CNN optimized by Adam

5 Results and Discussion

The obtained dataset is used later in training the CNN and Neuro-Fuzzy model on MATLAB Deep Learning Toolbox using the proposed learning algorithms and Mean Square Error (MSE) as the performance function. The inputs were represented in two sets of obstacles distances from the ultrasonic sensors and the outputs were the angular and linear velocity of the car. The CNN model using Adam and Stochastic Gradient Descent (GD) tested on the system features. AlexNet architecture was implemented for CNN. This architecture basically consists of eight layers; five convolutional layers and three FC layers with pooling layers in between for activation. The neuro-fuzzy model structure is illustrated in Table 1 and the inputs membership functions are shown in Fig. 6. Comparative analysis has been performed using another two testing cases. The models were not trained to evaluate the performance of CNN and NFIS models. Table 2 illustrates the analysis results using different performance indices which show that the better results was obtained using CNN with Adam learning algorithm, but they need longer computational time and higher processing power than NFIS trained by PSO.

The accuracy in the autonomous car navigation is achieved by implementing CNN optimized with Adam and Neuro-Fuzzy based on PSO. The Neuro-Fuzzy based PSO training performance is shown in Fig. 7.

CNN training performance using Adam for 500 iterations is shown in Fig. 8. As for the Fuzzy Logic model, the inputs were read from the two sensors and then the best action for the car is selected.

Table 1. Structure of neuro-fuzzy model

Fuzzy inferance Type	Sugeno
Number of MF for each variable	3
Input MF Type	Gaussian
Output MF Type	Linear
Number of nodes	20
Number of rules	36

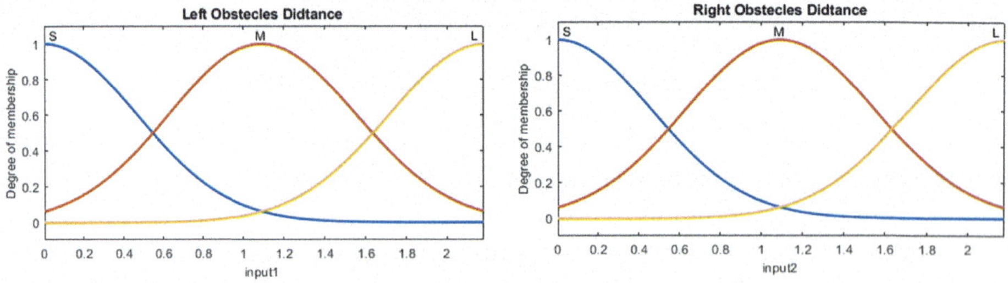

Fig. 6. Neuro-fuzzy inputs' memberships functions MFs

Table 2. Statistical analysis of CNN optimized with Adam and NFIS-PSO Testing Results

	CNN-Adam	NFIS-PSO
Mean Square Error (MSE)	7.8351	9.0655
Error Standard Deviation SD	4.7793	31.2125
Mean Absolute Error (MAE)	0.6463	7.8241
Root means squared error (RMSE)	1.3053	23.7914
Normalized root means squared error (NRMSE)	0.1633	0.3015
Pearson_Rp	0.9622	0.7560
Pearson_R2p	0.9258	0.6716
Spearman_Rs	0.9486	0.9419

CNN optimized with Adam has shown better results when compared to NFIS Tunied with PSO. The comparison was performed using statistical analysis and calculation of the correlation factors. It shows mean absolute error of 0.6463 and 7.8241 for linear and angular velocities respectively; additionally, Normalized Root Squared Error (NRMSE) was 0.1633 and −0.3015. Such similarity between the actual data and the generated optimized data encourages the idea of autonomous car production lowering fear of smart technologies.

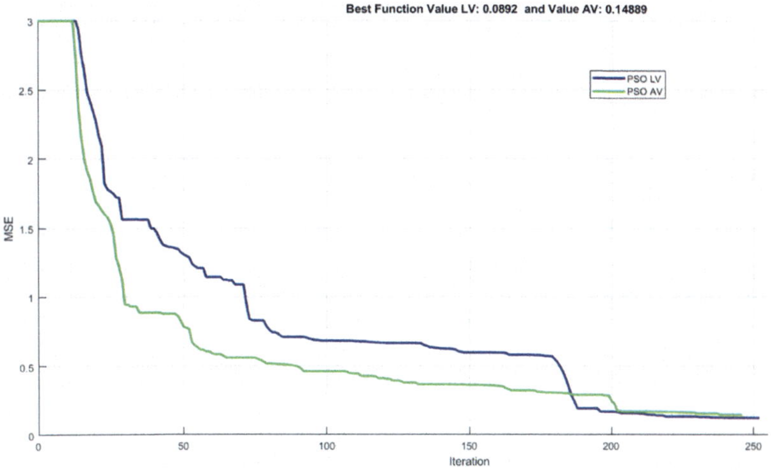

Fig. 7. Neuro-Fuzzy optimization using PSO performance MSE

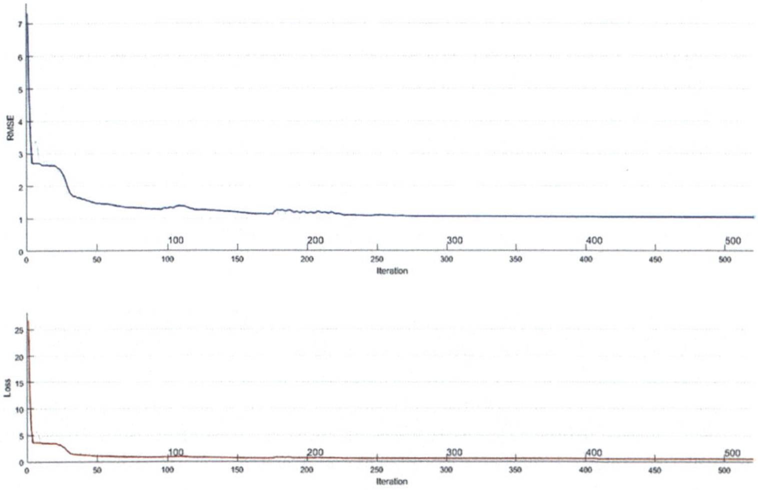

Fig. 8. CNN Neural Network training performance using Adm

6 Conclusion and the Future Work

This research is concerned about autonomous car scenarios on road to approach higher level of safety and ease. Neuro-Fuzzy system tuned with PSO have been applied to system data to get the results of autonomous navigation. The findings have shown that the optimized results for CNN with Adam are higher than NFIS Tunied with PSO. This similarity between actual data and the optimized data generated promotes the idea of autonomous car production, which reduces the fear of intelligent technologies. This research could be extended using more deep learning algorithms like reinforcement and metaheuristics optimization such as Gray wolf optimizer.

References

1. Abdel-Hamid, O., Deng, L., Yu, D.: Exploring convolutional neural network structures and optimization techniques for speech recognition. In: Interspeech, vol. 11, pp. 73–75 (2013)
2. Ammar, H.H., Azar, A.T.: Robust path tracking of mobile robot using fractional order PID controller. In: Hassanien, A., Azar, A., Gaber, T., Bhatnagar, R., F. Tolba, M. (eds.) The International Conference on Advanced Machine Learning Technologies and Applications (AMLTA2019). AMLTA 2019. Advances in Intelligent Systems and Computing, vol. 921. Springer, Cham (2020)
3. Azar, A.T., Aly, A.M., Sayed, A.S., Radwan, M.E., Ammar, H.H.: Neuro-fuzzy system for 3-DOF parallel robot manipulator. In: 2019 Novel Intelligent and Leading Emerging Sciences Conference (NILES), vol. 1, pp. 1–5 (2019). https://doi.org/10.1109/NILES.2019.8909333
4. Azar, A.T., Ammar, H.H., Ibrahim, Z.F., Ibrahim, H.A., Mohamed, N.A., Taha, M.A.: Implementation of PID controller with PSO tuning for autonomous vehicle. In: International Conference on Advanced Intelligent Systems and Informatics, pp. 288–299. Springer, Heidelberg (2019)
5. Azar, A.T., Sayed, A.S., Shahin, A.S., Elkholy, H.A., Ammar, H.H.: PID controller for 2-DOFs twin rotor MIMO system tuned with particle swarm optimization. In: International Conference on Advanced Intelligent Systems and Informatics, pp. 229–242. Springer, Heidelberg (2019)
6. Babu, V.C., Rao, P.G., Santa Rao, K., Krishna, B.M.: Design of accurate steering gear mechanism. Mech. Mech. Eng. **22**(1), 93–103 (2018)
7. Bingül, Z., Karahan, O.: A fuzzy logic controller tuned with PSO for 2 DOF robot trajectory control. Expert Syst. Appl. **38**(1), 1017–1031 (2011)
8. Choi, M.W., Park, J.S., Lee, B.S., Lee, M.H.: The performance of independent wheels steering vehicle (4WS) applied Ackerman geometry. 2008 International Conference on Control, Automation and Systems, pp. 197–202. IEEE (2008)
9. Faisal, A., Yigitcanlar, T., Kamruzzaman, M., Currie, G.: Understanding autonomous vehicles: a systematic literature review on capability, impact, planning and policy. J. Transp. Land Use **12**(1), 45–72 (2019)
10. Jang, J.S.R., Sun, C.T., Mizutani, E.: Neuro-fuzzy and soft computing-a computational approach to learning and machine intelligence [book review]. IEEE Trans. Autom. Control **42**(10), 1482–1484 (1997)
11. Martinez-Diaz, M., Soriguera, F.: Autonomous vehicles: theoretical and practical challenges. Transp. Res. Procedia **33**, 275–282 (2018). XIII Conference on Transport Engineering, CIT 2018
12. Mordue, G., Yeung, A., Wu, F.: The looming challenges of regulating high level autonomous vehicles. Transp. Res. Part A: Policy Pract. **132**, 174–187 (2020)
13. Nauck, D., Klawonn, F., Kruse, R.: Foundations of Neuro-Fuzzy Systems. Wiley, Hoboken (1997)
14. Özsert Yiğit, G., Özyildirim, B.M.: Comparison of convolutional neural network models for food image classification. J. Inf. Telecommun. **2**(3), 347–357 (2018)

15. Sallam, O.K., Azar, A.T., Guaily, A., Ammar, H.H.: Tuning of PID controller using particle swarm optimization for cross flow heat exchanger based on CFD system identification. In: International Conference on Advanced Intelligent Systems and Informatics, pp. 300–312. Springer, Heidelberg (2019)
16. Shalaby, M.A.W., Ortiz, N.R., Ammar, H.H.: A neuro-fuzzy based approach for energy consumption and profit operation forecasting. In: International Conference on Advanced Intelligent Systems and Informatics, pp. 58–69. Springer, Heidelberg (2019)
17. Soliman, M., Azar, A.T., Saleh, M.A., Ammar, H.H.: Path planning control for 3-omni fighting robot using PID and fuzzy logic controller. In: Hassanien, A., Azar, A., Gaber, T., Bhatnagar, R., F. Tolba, M. (eds.) The International Conference on Advanced Machine Learning Technologies and Applications (AMLTA2019). AMLTA 2019. Advances in Intelligent Systems and Computing, vol. 921. Springer, Cham (2020). https://doi.org/10.1007/978-3-030-14118-9_45
18. Wang, J.S., Lee, C.G.: Self-adaptive recurrent neuro-fuzzy control of an autonomous underwater vehicle. IEEE Trans. Robot. Autom. **19**(2), 283–295 (2003)
19. Zhao, J.S., Liu, X., Feng, Z.J., Dai, J.S.: Design of an Ackermann-type steering mechanism. Proc. Inst. Mech. Eng. Part C: J. Mech. Eng. Sci. **227**(11), 2549–2562 (2013)
20. Zhou, T., Ruan, S., Canu, S.: A review: deep learning for medical image segmentation using multi-modality fusion. Array **3–4**(100), 004 (2019)

Deep Learning Based Kinematic Modeling of 3-RRR Parallel Manipulator

Abdelrahman Sayed Sayed[1,2], Ahmad Taher Azar[3,4(✉)], Zahra Fathy Ibrahim[1,2], Habiba A. Ibrahim[1,2], Nada Ali Mohamed[1,2], and Hossam Hassan Ammar[1,2]

[1] Smart Engineering Systems Research Center (SESC), Nile University, Sheikh Zayed City 12588, Giza, Egypt
{ab.sayed,z.fathy,h.ibrahim,n.ali,hhassan}@nu.edu.eg

[2] School of Engineering and Applied Sciences, Nile University Campus, Juhayna Square, Sheikh Zayed District, 6th of October City 12588, Giza, Egypt

[3] Robotics and Internet-of-Things Lab (RIOTU), Prince Sultan University, Riyadh, Saudi Arabia
aazar@psu.edu.sa, ahmad_t_azar@ieee.org

[4] Faculty of Computers and Artificial Intelligence, Benha University, Banha, Egypt
ahmad.azar@fci.bu.edu.eg

Abstract. This paper presents a novel low cost design for a 3-RRR Planar Parallel Manipulator (PPM). These manipulators proved their superiority over serial manipulators due to their speed, precision and smaller work space where the work space area is accounted for in the design to ensure that the robot is performing its task in a smooth and simple way without getting into any singularity points. The challenge with PPM is to obtain the kinematic constraint equations of the manipulator due to their complex non-linear behavior. Screw theory is a new approach that is used to compute the direct and inverse kinematics based on the relation between each link and its' predecessor. The design is then inserted into ADAMS to study its dynamical behavior and to obtain a data set that would be used in analyzing the system in MATLAB. A Neuro-Fuzzy Inference System (NFIS) model was constructed in order to predict the end-effector position inside the work space and it is tuned with Particle swarm optimization (PSO) and Genetic algorithm (GA).

Keywords: Planar Parallel Manipulator (PPM) · Kinematic analysis · ADAMS · Neural networks · Neuro-Fuzzy Inference System · Particle Swarm Optimization (PSO) · Genetic Algorithm (GA)

1 Introduction

Parallel robots have attracted the attention of the industrial and research communities due to their notable characteristics over the serial ones. They are considered to be a form of closed-loop mechanisms where there is a moving end-effector connected to the base by independently actuated serial kinematic chains [3,15].

A.-E. Hassanien et al. (Eds.): AICV 2020, AISC 1153, pp. 308–321, 2020.
https://doi.org/10.1007/978-3-030-44289-7_29

Compared to serial manipulators, parallel manipulators have high stiffness, high load to weight ratio and higher flexibility. Indeed, some parallel industrial robots such as the Delta [18,23] and the Stewart [9] have been used in very high speed pick and place systems and in flight simulators. The high stiffness of parallel manipulators allows to design high compact structures where all the actuators are placed on the robot base [20,21]. However, parallel manipulators have limited work spaces, large number of kinematic singularities and complex kinematic models [10]. A solution to the previously mentioned drawbacks is to use a redundant actuation and sensing systems that can be used to reduce singularities [17], thus improving the kinematic manipulability and dexterity [14]. Unlike serial manipulators the forward kinematic solution of a parallel mechanism is difficult to obtain than the inverse kinematic solution due to the large number of equations which describe the relation between different chains in the robot [19]. The combination of mobile robots and machine learning was a favored research area as developing kinematic model of a parallel or serial robot is the most complex challenge [5,6,12]. Especially for parallel robots, it requires many computations of independent equations that describe each chain of the robot and how it affects the robot final linear velocity or angular velocity [4]. Nevertheless, researchers and engineers performed forward and inverse kinematics along the way to control robots and understand their motion. With the combination with machine learning and training algorithms, it has been proved that the headache of doing manual forward/inverse kinematic analyses manually can be exchanged with a machine learning model using Neural Network (NN). Accordingly, based on training data, NN can be used easily to detect behavior of parallel robot and describe it. In other words, machine learning allowed the idea of developing forward and inverse kinematic models of a robot in an easier way. A research developed by Xu and Li in [24] discussed a 3-PRS (Prismatic - Revolute - Spherical) parallel manipulator Control which is based of NN. They used NN model and trained their data for Stewart platform parallel robot and get a perfect ordinary control of the robot. Also, safety analysis for delta robot has been developed by Liu et al. in [13] where a back- propagation NN has been designed to implement forward kinematics of the robot. Results show that NN model allows the endpoint of the delta robot system to achieve 97.75% accurate positioning and is overcomes weakness points of manual time-consuming calculations [13]. In this context, a new step has been taken of not only applying machine learning models to get kinematic analyses, but it also uses several metaheuristic optimization techniques such as Particle Swarm Optimization (PSO) and Genetic Algorithm (GA) to get the best results among models. So, two different optimization methods were developed and used to train the NFIS using a generated data for the PPM extracted from ADAMS. Results are then compared to choose the most suitable one among the two applied optimization algorithms.

This paper is organized as follow: In Sect. 2, kinematic model of PPM is presented. ADAMS simulation and analysis is described in Sect. 3. Neuro-Fuzzy Inference System structure is presented in Sect. 4. Results with discussions are shown in Sect. 5. Finally, the paper is concluded in Sect. 6.

2 Kinematic Model of PPM

2.1 Inverse Kinematics of PPM

The three DOF PPM analyzed in this work is shown in Fig. 1. It consists of an equilateral triangular end-effector ($C_1C_2C_3$) linked to a circular plate using three legs and a fixed base ($A_1A_2A_3$) joined together with three kinematic chains. Each of these chains consists of three revolute joints and two links. These links are either an active type A_iB_i or a passive type B_iC_i and the revolute joints are located at A_i, B_i and C_i, where in this paper $i = 1, 2, 3$. The length of the active link l_1 is the same of the passive link l_2. O is the center of the triangular end-effector $\triangle A_1A_2A_3.l_3 = C_1P$ and $l_4 = A_1O$. The base frame is a coordinate system that is defined and set with respect to the fixed base of the manipulator, with center O and X and Y axes. The position and orientation of the triangular end-effector are denoted by (x, y) and q_c. The active joint vector is $q_a = [q_{a1}q_{a2}q_{a3}]^T$, the passive joint vector is $q_b = [q_{b1}q_{b2}q_{b3}]^T$ and the vector $q_y = [xyq_c]^T$ is the task coordinate vector of the outputs of the 3-RRR PPM. PPM kinematics are usually described using geometrical methods with a close chain vectors. A set of non-redundant, non-linear and implicit equations are obtained. These three set of vector equations can be expresses as:

$$OA_i + A_iB_i + B_iC_i + C_iP = OP \tag{1}$$

Equation (1) is then divided into x-axis component and y-axis component, then the closed loop constraint equations can be obtained as:

$$x = x_{ai} + l_1cosq_{ai} + l_2cosq_{bi} + l_3cos[q_c + \frac{\pi}{6} + \frac{2\pi}{3}(i-1)] \tag{2}$$

$$y = y_{ai} + l_1sinq_{ai} + l_2sinq_{bi} + l_3sin[q_c + \frac{\pi}{6} + \frac{2\pi}{3}(i-1)] \tag{3}$$

After obtaining Eqs. (2) and (3), the input values can be obtained as:

$$q_{ai} = 2 \times tan^{-1}\left(\frac{-e_{1i} \pm \sqrt{e_{1i}^2 + e_{2i}^2 - e_{3i}^2}}{e_{3i} - e_{2i}}\right) \tag{4}$$

where

$$e_{1i} = -2l_1 \cdot \left[y + l_3 \cdot sin\Big(q_c + \frac{\pi}{6} + \frac{2\pi}{3}(i-1)\Big) - l_4 \cdot sin\Big(\frac{7\pi}{6} + \frac{2\pi}{3}(i-1)\Big)\right] \tag{5}$$

$$e_{2i} = -2l_1 \cdot \left[x + l_3 \cdot cos\Big(q_c + \frac{\pi}{6} + \frac{2\pi}{3}(i-1)\Big) - l_4 \cdot cos\Big(\frac{7\pi}{6} + \frac{2\pi}{3}(i-1)\Big)\right] \tag{6}$$

$$e_{3i} = \frac{e_{2i}}{-2l_1} + l_1^2 + \frac{e_{1i}}{-2l_1} - l_2^2 \tag{7}$$

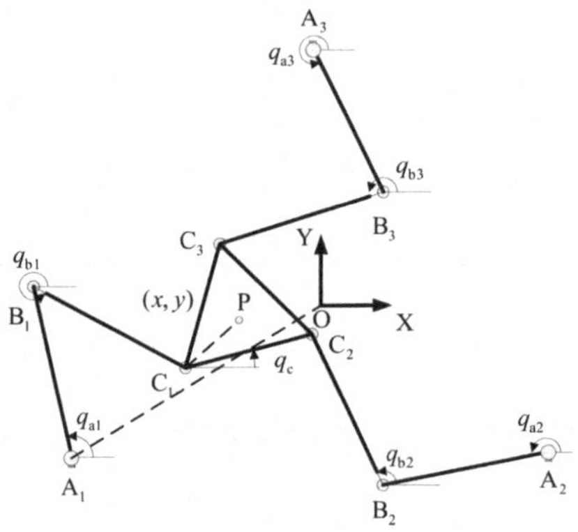

Fig. 1. A general case 3RRR PPM

2.2 PPM Working Modes

The resulting solutions from Eq. (4) correspond to the two configurations of each kinematic chain in terms of x-axis and y-axis for a certain pose position and orientation of the end-effector. Due to the fact that there are eight different solutions to the inverse kinematic problem, each solution refers to a different working mode of the manipulator identified by sign triple $\triangle_j = [\delta_1, \delta_2, \delta_3]$, where δ_i gives the orientation of the triangular end-effector $A_iB_iC_i$ and $\triangle_j$ is working mode index, where $j = 1, 2, ..., 8$. For each working mode, the work space covers a different portion of the forward singularities and the boundary of these portions can be computed by calculating the sub-regions corresponding to the inverse and forward singularities [7]. To change the manipulator working mode, it is compulsory for the end-effector to pass over an inverse singular configuration. For that reason the work space size should be taken into consideration in order to allow the working modes changes [26].

3 ADAMS Simulation and Analysis

In order to simulate the design and study its behavior, MSC ADAMS software was used. In this simulation environment, dynamical and kinematical behaviors of the systems can be studied with great accuracy [11]. Our methodology in using MSC ADAMS is divided into three main stages. In the first stage, the SolidWorks design is converted into IGS file format to enable manipulation using ADAMS. Then the design is inserted into the work space, and then the design constrains were applied to the different components of the design. In the second stage, the joints were formed which are the positions where the PPM robot is actuated by motors. Thus, three rotational joints were applied between the end-effector and the passive joints, as shown in Fig. 2. Then, rotational motions were applied to the joints. In order to have a time-variant motion of the joints and to test different cases of motors speeds, step function is used to describe the rotational motion. This function is entered in each motion segment with the start- and end-time and the angle of rotation. The whole simulation time was 1 min and it was divided into 12 motion segments. Then, in the third stage, the results were exported in form of graphs and data-sheets. The graphs would help us to visualize the kinematic behavior of the system, while the data-sheets are used to store data for analysis in MATLAB. The idea of using for further in detail analysis and interpreting of data is well known in robotics research [1]. In this process, six quantities were extracted which are the three input angles describing motors inputs, the x and y qualities for the center of gravity of the end-effector and the angle of the end-effector in the x-y plane. By finishing the third step, we succeeded in simulating the system and get useful results from it using MSC ADAMS to move to the next step of data analysis.

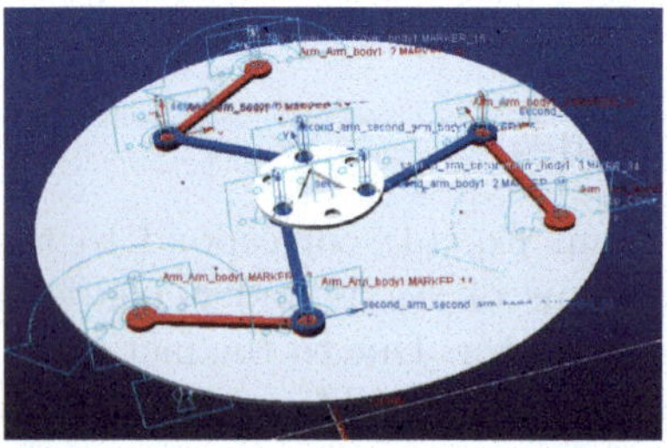

Fig. 2. Defining the constrains and joints before running the simulation in MSC ADAMS

While the MSC Adams is a useful simulation tool that measures and displays all kinematical and dynamical quantities, only some data were of great use in our analysis. In Fig. 3, the inputs of the three rotational joints are shown. During the first thirty seconds, each joint was allowed ten seconds to reach the 90° and return back to its normal state, while all the other joints are kept stationary. All joints are reached in the next ten seconds with the same motion input range. Because all joints are moving simultaneously, the maximum range that each joint could reach was slightly above 25°.

Then, random different motions were assigned to the joints to further explore the behavior. Then, in Fig. 4, the motion of the end-effector is displayed because it is the main interest. Because this robot has three degrees of freedom, quantities for x, y and theta are displayed. It is unsurprising that in the period between thirty and forty seconds, where all joints have the same inputs, the x and y positions of the end effector are zeros; however, the angular position is nonzero. That happened because the end effector just rotates around the center of mass.

4 Neuro-Fuzzy Inference System Structure

Neuro-Fuzzy Inference System (NFIS) is used to predict the end-effector position and orientation of the 3-RRR PPM. Neuro-Fuzzy can take multiple inputs

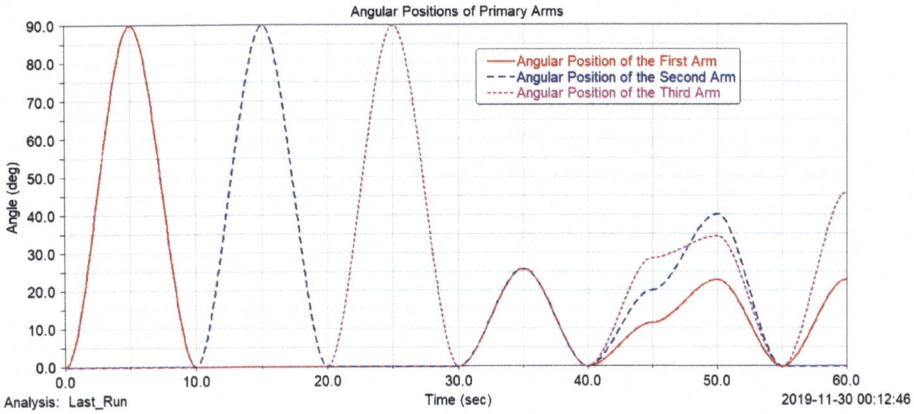

Fig. 3. The input angles for the three joints during the simulation time

at once, but can only generate one output each time. Therefore, three models were used to obtain the three required outputs. The NFIS is a combination of both the Neural Network and the Fuzzy system. It needs a training set to learn as required in the neural systems. It also uses the advantage of fuzzy systems, where the output of the membership functions lies between 0 and 1, which makes the computation easier for the PC. Moreover, two metaheuristic techniques were used to tune the NFIS which are Particle Swarm Optimization (PSO) and Genetic Optimization Algorithm (GA). They will be discussed in the two following sub-sections.

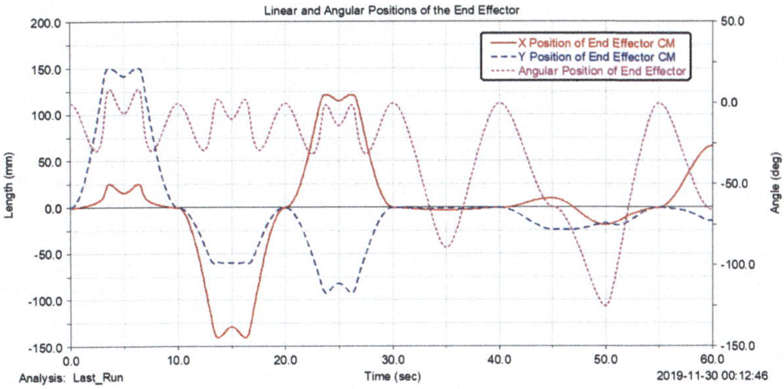

Fig. 4. The linear and angular positions of the center of mass of the end effector

4.1 Particle Swarm Optimization (PSO)

PSO is a stochastic search algorithm, as it uses a number of particles that represents a swarm moving around searching for the best solution and each particle is always adjusting its condition according to its own experience as well as of the other particles [6]. Based on that, the position of the particle is changed by the change of velocity compared to the current state as shown in Eq. 8, where $x_i(t)$ is position of particle i, $v_i(t)$ is the added velocity, and t is the discrete time step [16].

$$x_i(t+1) = x_i(t) + v_i(t+1) \tag{8}$$

4.2 Genetic Optimization Algorithm (GA)

Genetic Algorithms are a class of algorithms based on abstraction of Darwin's evolution of biological systems. It simulates the idea of getting better generations of individuals; however, population in this context is data set, and individuals are solutions for a specific problem. Such technique of optimization is used to find the fittest solution to the challenge exist. Genetic Optimization define individual solution with characterizing variables, called Genes. Genes are, then, set as a string named Chromosome which is represented as String and is unique for each solution individual [2,22]. Equation 9 shows the simplified theorem that calculates the number of chromosomes that carries the schema s, where $u(s,t)$ is the average fitness of the chromosomes, $N(s,t)$ is the number of chromosomes in generation t, and $N(s,t+1)$ is the number in the next generation [8]. Such Algorithm considers five different phases in its process. First of all, it has an initial population of solutions that all able to solve the problem with different capabilities. Fitness Function, the second phase, determines how each individual is capable to compete with others in the population. Individuals' capabilities are computed according variables/genes giving each individual a fitness score. Moving to Selection phase, individuals of highest fitness scores are chosen to pass their genes to the next generation. It goes to crossover stage where a crossover point is selected; then, parents exchange genes before this point to get new fitter children. Genetic Algorithm has several stopping conditions depending on which factor is the most effective in the optimization process. The process can be conditioned by a time limit for the process, it can be limited by a fitness limit that stops the process if reached by any of the children, and it even can be stopped giving the fitness function a tolerance to change within. Whatever the stopping condition is, the further you stop the process, the better results will be [25].

$$N(s,t+1) = u(s,t)[1-e]N(s,t) \tag{9}$$

5 Results and Discussion

By using PSO as a metaheuristic approach to tune the training of the NFIS model, each input was designed to have three membership function varying between Gaussian and triangular. Figure 5 shows the memberships functions for the three inputs and the resulted rule base surface for the optimized first output. The process is repeated for each output and the data generated was used in calculating the cost functions. Table 1 shows the parameters used to train the NFIS model with PSO algorithm. The same process is repeated using the GA to tune the same model of NFIS with number of generations was set to 1000.

Table 1. NFIS tuned with PSO parameter

Parameter	Value
Fuzzy structure	Sugeno-type
Type of MFs	Gaussian, Triangular
Output MF	Linear
No. of inputs	3
No. of outputs	1
No. of fuzzy rules	24
Epoch number	1000
No. of samples	600
No. of population	50

PSO and Genetic tuning algorithms are developed on NFIS with number of iterations equal to 1000. The algorithms are applied on the data collected from the mentioned model using the three angles of the robot arms as inputs and the coordinates and end effector twist angle as outputs. The data used to train the NFIS model contained 600 samples after filtering. The cost functions regarding the two algorithms with each output is almost the same, however genetic algorithm showed faster stability with less computational time. Figures 6, 7 and 8 show the cost function of the two mentioned algorithms for X coordinate, Y coordinate, and the end effector twist angle, respectively.

The trained data was tested and then used in a comparison process between the two optimization algorithms. The actual output data for the model was plotted and compared to the optimized output generated from the two tuning algorithms. Figures 9, 10, and 11 show the three mentioned outputs before and after applying the optimization techniques.

In order to compare the results of the two models, the error of each output was calculated. Tables 2 and 3 show the RMSEs of each output in each model, in addition to other statistical parameters used to show which model was better. The outputs from both models were close to the actual outputs from the PPM. But, PSO performed slightly better than GA as it was exactly aligned with actual data from the PPM compared to GA which showed some deviations from the actual results. This is because of the stochastic behavior of the PSO algorithm.

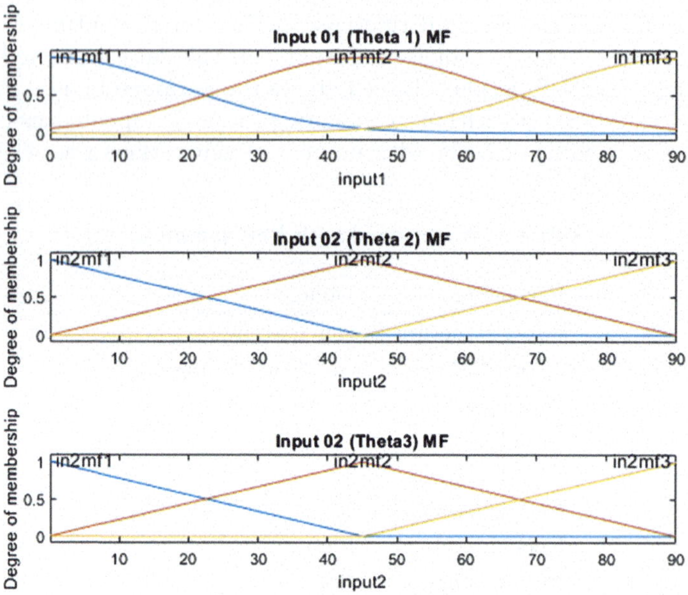

Fig. 5. Input membership functions (MFS)

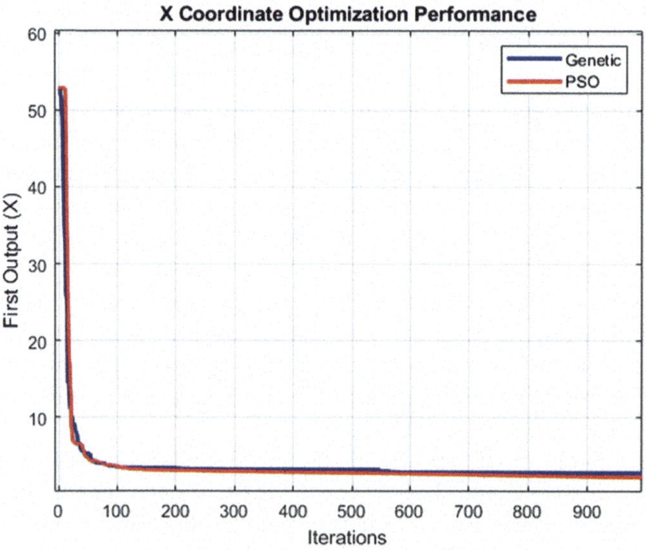

Fig. 6. Cost function convergence for X-coordinate

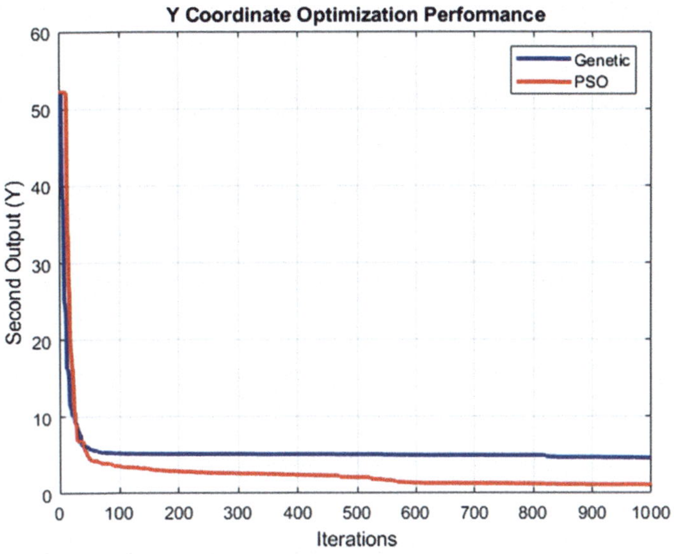

Fig. 7. Cost function convergence for Y-coordinate

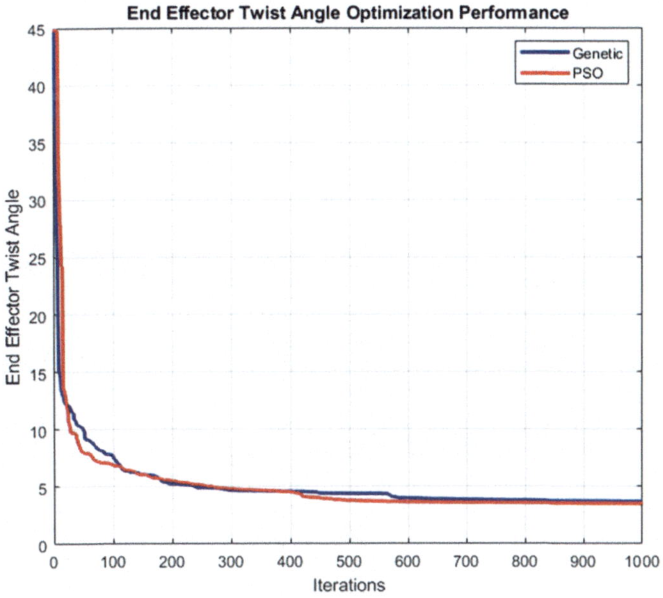

Fig. 8. Cost function convergence for end-effector twist angle

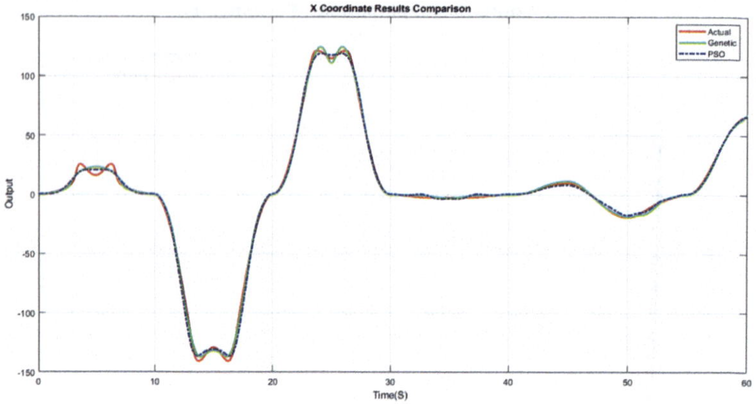

Fig. 9. X-axis output compared results

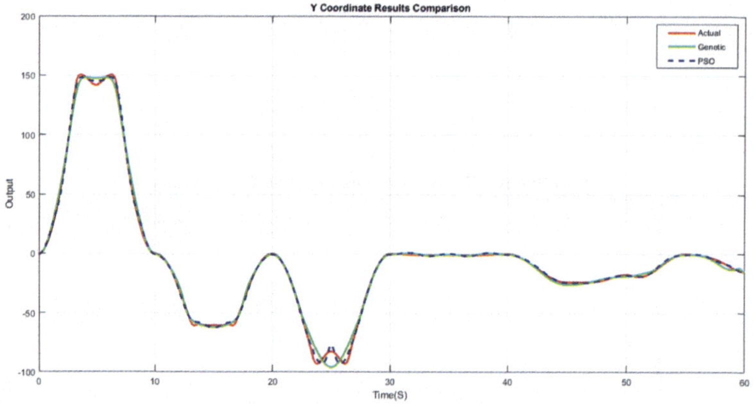

Fig. 10. Y-axis output compared results

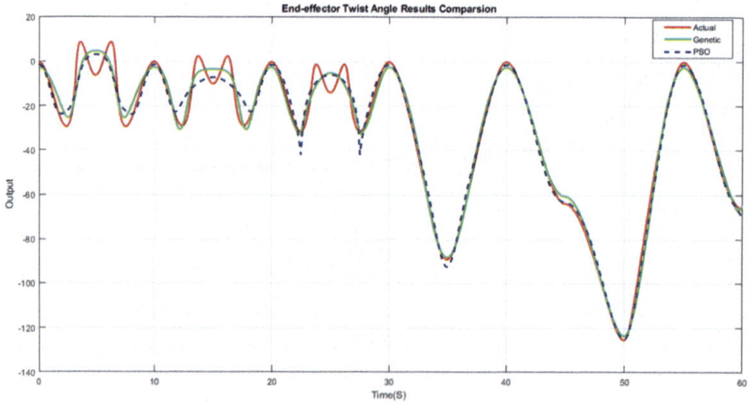

Fig. 11. End-effector output compared results

Table 2. Statistical analysis for NFIS Tuned with PSO model

	X	Y	θ
Mean	0.2437	−0.0175	0.1353
SD	2.5353	1.5153	4.3608
MAE	1.6873	1.0448	3.0220
RMSE	2.5448	1.5142	4.3592
NRMSE	0.0097	0.0062	0.0325
Pearson Rp	0.9989	0.9996	0.9909
Pearson R^2p	0.9977	0.9992	0.9818
S-Rs	0.9964	0.9914	0.9616

Table 3. Statistical analysis for NFIS Tuned with GA model

	X	Y	θ
Mean	0.0089	−0.0577	0.0276
SD	1.9949	2.9479	3.7604
MAE	1.3446	1.6860	2.7084
RMSE	1.9932	2.9460	3.7574
NRMSE	0.0076	0.0121	0.0280
Pearson Rp	0.9993	0.9984	0.9932
Pearson R^2p	0.9986	0.9968	0.9865
S-Rs	0.9985	0.9938	0.9700

6 Conclusion

In this paper, a new low-cost PPM structure with 3-DOF is proposed. The robot structure was investigated and its kinematic model was developed and for the proof of concept, a prototype was tested in ADAMS environment and then fabricated. The NFIS algorithm was implemented and tuned with Particle Swarm Optimization (PSO) and Genetic Algorithm (GA) in order to simplify the complex mathematical model of the PPM. Further developments on the robot will focus on the singularity analysis of the proposed structure, design optimization and control exploiting actuation with sensing redundancy, which will enhance the robot capabilities and performances.

References

1. Angel, L., Pérez, M., Díaz-Quintero, C., Mendoza, C.: Adams/matlab co-simulation: dynamic systems analysis and control tool. In: Applied Mechanics and Materials, vol. 232, pp. 527–531 (2012)

2. Arora, J.: Genetic algorithms for optimum design. In: Introduction to Optimum Design, pp. 643–655 (2004)
3. Azar, A.T., Zhu, Q., Khamis, A., Zhao, D.: Control design approaches for parallel robot manipulators: a review. Int. J. Model. Identif. Control **28**(3), 199–211 (2017)
4. Azar, A.T., Aly, A.M., Sayed, A.S., Radwan, M.E., Ammar, H.H.: Neuro-fuzzy system for 3-DOF parallel robot manipulator. In: 2019 Novel Intelligent and Leading Emerging Sciences Conference (NILES), vol. 1, pp. 1–5 (2019)
5. Azar, A.T., Ali, N., Makarem, S., Diab, M.K., Ammar, H.H.: Design and implementation of a ball and beam PID control system based on metaheuristic techniques. In: Hassanien, A.E., Shaalan, K., Tolba, M.F. (eds.) Proceedings of the International Conference on Advanced Intelligent Systems and Informatics, pp. 313–325. Springer, Cham (2020)
6. Azar, A.T., Sayed, A.S., Shahin, A.S., Elkholy, H.A., Ammar, H.H.: PID controller for 2-DOFS twin rotor MIMO system tuned with particle swarm optimization. In: Hassanien, A.E., Shaalan, K., Tolba, M.F. (eds.) Proceedings of the International Conference on Advanced Intelligent Systems and Informatics, pp. 229–242. Springer, Cham (2020)
7. Boudreau, R., Nokleby, S.: Force optimization of kinematically-redundant planar parallel manipulators following a desired trajectory. Mech. Mach. Theory **56**, 138–155 (2012)
8. Goldberg, D.E.: Genetic Algorithms in Search, Optimization and Machine Learning, 1st edn. Addison-Wesley/Longman Publishing Co., Inc., Upper Saddle River (1989)
9. Gough, V.: Contribution to discussion of papers on research in automobile stability, control and tyre performance. Proc. Auto Div. Inst. Mech. Eng. **171**, 392–395 (1957)
10. Hamdoun, O., El Bakkali, L., Baghli, F.Z.: Analysis and optimum kinematic design of a parallel robot. Procedia Eng. **181**, 214–220 (2017)
11. Kubela, T., Pochyly, A., Singule, V.: Advanced tools for multi-body simulation and design of control structures applied in robotic system development. Solid State Phenom. **164**, 387–391 (2010)
12. Kucuk, S.: Dexterous workspace optimization for a new parallel robot manipulator. J. Mech. Robot. **10**(6), 8 p. (2018). Article no. 064503
13. Liu, C., Cao, G., Qu, Y.: Safety analysis via forward kinematics of delta parallel robot using machine learning. Saf. Sci. **117**, 243–249 (2019)
14. Luces, M., Mills, J.K., Benhabib, B.: A review of redundant parallel kinematic mechanisms. J. Intell. Robot. Syst. **86**(2), 175–198 (2017)
15. Merlet, J.: Parallel Robots, vol. 74. Springer, Sophia-Antipolis (2012)
16. Mitchell, M., Toroczkai, Z.: Complexity: a guided tour. Phys. Today **63**, 47 (2010)
17. Müller, A.: Redundant actuation of parallel manipulators. In: Parallel Manipulators, Towards New Applications. IntechOpen (2008)
18. Pierrot, F., Reynaud, C., Fournier, A.: Delta: a simple and efficient parallel robot. Robotica **8**(2), 105–109 (1990)
19. Seibel, A., Schulz, S., Schlattmann, J.: On the direct kinematics problem of parallel mechanisms. J. Robot. **2018**, 1–9 (2018). Article ID 2412608. https://doi.org/10.1155/2018/2412608
20. Shi, H., Su, H.J.: An analytical model for calculating the workspace of a flexure hexapod nanopositioner. J. Mech. Robot. **5**(4), 041,009 (2013)
21. Shi, H., Su, H.J., Dagalakis, N.: A stiffness model for control and analysis of a mems hexapod nanopositioner. Mech. Mach. Theory **80**, 246–264 (2014)

22. Talaslioglu, T.: Optimal design of steel skeletal structures using the enhanced genetic algorithm methodology. Front. Struct. Civ. Eng. **13**(4), 863–889 (2019)
23. Xie, F., Liu, X.J.: Design and development of a high-speed and high-rotation robot with four identical arms and a single platform. J. Mech. Robot. **7**(4), 041,015 (2015)
24. Xu, Q., Li, Y.: A 3-PRS parallel manipulator control based on neural network. In: Liu, D., Fei, S., Hou, Z.G., Zhang, H., Sun, C. (eds.) Advances in Neural Networks - ISNN 2007, pp. 757–766. Springer, Heidelberg (2007)
25. Yang, X.S.: Optimization and Metaheuristic Algorithms in Engineering, pp 1–23 (2013)
26. Zhang, Z., Shao, Z., Wang, L.: Improving the kinematic performance of a planar 3-RRR parallel manipulator through actuation mode conversion. Mech. Mach. Theory **130**, 86–108 (2018)

Deep Learning in Breast Cancer Detection and Classification

Ghada Hamed(✉), Mohammed Abd El-Rahman Marey, Safaa El-Sayed Amin, and Mohamed Fahmy Tolba

Faculty of Computer and Information Sciences, Ain Shams University, Cairo, Egypt
ghadahamed@cis.asu.edu.eg

Abstract. Breast cancer is considered one of the primary causes of mortality among women aged 20–59 worldwide. Early detection and treatment can allow patients to have proper treatment and consequently reduce rate of morbidity of breast cancer. Research indicates that most experienced physicians can diagnose cancer with 79% accuracy while 91% correct diagnosis is achieved using machine learning techniques. In this paper, we present the most recent breast cancer detection and classification models that are machine learning based models by analyzing them in the form of comparative study. Also, in this paper, the datasets that are public for use and popular as well are listed in the recent work to facilitate any new experiments and comparisons. The comparative analysis shows that the recent highest accuracy models based on simple detection and the classification architectures are You Only Look Once (YOLO) and RetinaNet.

Keywords: Breast cancer · Mortality · Detection · Classification · YOLO · RetinaNet

1 Introduction

The cancer tumor is developed through the abnormal growth of cells, which invades the surrounding tissues in the human body. There are two classes of tumors, benign and malignant, and if there is no tumor in the breast it is considered normal. The benign tumor cells are non–cancerous cells and grow only locally and cannot spread by invasion. While malignant tumors are cancerous cells and they have the ability to multiple uncontrollably, to spread to various parts of the body and invade surrounding tissue. Breast cancer is one of the most common cancers and a severe public health issue for women all over the world. According to the report of International Agency for Research on Cancer (IARC) presented by the World Health Organization (WHO), in the year 2012, approximately 8.2 million deaths are caused by cancer, which further states that in 2030 the death ratio from cancer is expected to increase up to 27 million [1]. So, on-time and accurate detection, early diagnosis, and active prevention are critical requirements for reducing mortality rate among women.

A.-E. Hassanien et al. (Eds.): AICV 2020, AISC 1153, pp. 322–333, 2020.
https://doi.org/10.1007/978-3-030-44289-7_30

Abnormalities such as masses, micro-calcifications, and areas of asymmetry and distortion within the breast may indicate the existence of breast cancer. Among these abnormalities, masses are the most representative and common lesion type. However, masses can be easily hidden by overlapping breast tissues, making it difficult to detect them. Moreover, some breast tissues are morphologically similar to masses, and thus misidentified as masses. An undetected mass is a false negative, which delays a patient diagnosis until the next screening. A misidentified mass is a false positive, which leads to additional tests including re-screening and biopsy, causing unnecessary anxiety and pain to patients. These problems limit the effectiveness and utility of mammography. So, this task is considered a daily challenge for radiologists since there is a huge number of mammograms that consumes time and effort to examine each view of a mammogram [2–4]. Consequently, there is a tradeoff between sensitivity and specificity that has been realized during the diagnosis process of the existing masses. Through the use of a second reading, either by other experts or by a Computer-Aided Diagnosis (CAD) systems, the overall accuracy and specificity of the mass detection, segmentation, and classification could be improved [5,6] and false positive and negative cases reduced.

There are few studies involving a completely integrated system that include all phases of detection, segmentation, and classification. In fact, the variation of the masses within the surrounding tissues in terms of texture, shape, size, as well as the location in mammograms, makes the detection task challenging [7,8]. Besides this, improving overall accuracy in addition to reducing false positive and negative rates by mass segmentation is a big challenge due to the strong association between the presence of masses and their irregularities in shape, size, and location with low contrast and ambiguous boundaries [9–12]. Unfortunately, these methods still lack performance in handling mass segmentation automatically, because the simple hand-crafted or semi-automatic features based on prior knowledge cannot deal with complex shape variations, as well as the different density distribution of the masses and their surrounding tissues [9]. Besides that the manual feature extraction and selection are extremely time consuming. The manual feature extraction and selection make it difficult to define as well meaningful features. Recently, a few studies based on deep learning models have offered a good alternative to other conventional segmentation methods, by automatically extracting deep high-level hierarchy features for mass segmentation directly from input raw data to avoid the problems of hand-crafting features [13].

A Convolutional Neural Network (CNNs) consists of a number of convolutional layers which can extract features that represent the various contexts of images without feature engineering. That is why CNN has become the most widely used method for image interpretation tasks in many domains such as the breast cancer detection and classification. After the success of CNNs in standard object detection tasks [14], several studies have exploited the advantages of deep CNNs to overcome the drawbacks of conventional mass detection models [15–22].

The main purpose of mammography is to detect early signs of cancer and to diagnose breast masses from the images [23]. Mammograms are considered as the most important tool doctors have not only to screen for breast cancer, but also to diagnose, evaluate, and follow people who have had breast cancer. Screening with mammography is performed in two steps. First, the breast is compressed between two small flat plates. Then, a low X-ray dose is applied directly through the breast and acquired by a two-dimensional (2D) panel detector.

The rest of the paper is organized as follows. In the dataset section, the most common used and public datasets are given and the main differences between each are discussed. Then, the dataset section is followed by the comparative study and analysis section which begins with the detection and classification accuracy metrics used to detect how accurate the proposed model is. The metrics are listed with their definitions and how they are calculated as well. Then, the recent published work in the breast cancer detection and classification from 2017 to 2019 are compared with respective to different aspects including the used datasets, data preprocessing and data augmentation in section three followed by the main building blocks they build on the cancer classification and detection. Finally the conclusion section is presented in the last section.

2 Dataset

This section lists the most common datasets containing mammograms used for the breast cancer detection and classification. There are five datasets presented which are: DDSM, CBIS-DDSM, MIAS, INbreast and BCDR datasets.

DDSM Dataset: The Digital Database for Screening Mammography (DDSM) is an available online dataset [24] that is used for the research purposes in the breast cancer detection and classification systems. This dataset is collected by South Florida University [25]. It is collected to represent real breast data with an average size of 3000×4800 pixels, resolution of 42 microns, and 16 bits.

The DDSM database consists of 2,620 scanned film mammography studies that are categorized into 43 volumes. Each case involves four breast images, two of them are Mediolateral Oblique (MLO) views and the others are Cranio-Caudal (CC) views of each breast. Benign and malignant masses in all mammograms are recognized and annotated by expert radiologists.

CBIS-DDSM Dataset: Curated Breast Imaging Subset of DDSM (CBIS-DDSM) is an updated and standardized version of the Digital Database for Screening Mammography (DDSM) [26]. The CBIS-DDSM collection is available as well [27] which includes a subset of the DDSM data selected and curated by a trained mammographer. The CBIS-DDSM contains 6775 studies. The images have been decompressed and converted to DICOM format. Updated ROI segmentation and bounding boxes, and pathologic diagnosis for training data are also included.

MIAS Dataset: The Mammographic Image Analysis Society (MIAS) is an organisation of UK research groups interested in the understanding of mammograms and has generated a database of digital mammograms [28]. Mammographic images are available via the Pilot European Image Processing Archive (PEIPA) at the University of Essex. The database contains 322 digitized films and is available on 2.3 GB 8 mm (ExaByte) tape. The database has been reduced to a 200 micron pixel edge and padded/clipped so that all the images are 1024×1024.

INbreast Dataset: The INbreast was obtained from the S. João Hospital Centre in Porto [29]. It consists of 410 full digital mammograms. All lesions including masses were assigned a standardized Breast Imaging-Reporting and Data System (BI-RADS) category [30] by a radiologist after interpreting a mammogram. INbreast does not exist publicly on the web but it can be obtained by a request from [31].

BCDR Dataset: The Breast Cancer Digital Repository (BCDR) is a compilation of Breast Cancer anonymized patients' cases annotated by expert radiologists containing clinical data (detected anomalies, breast density, BIRADS classification, etc.), lesions outlines, and image-based features computed from CC and MLO mammography image views.

Currently, two repositories (examples) are available for public domain: one containing digitalized Film mammography (FM) and other one containing Full Field Digital (DM) mammography and related ultrasound images. Also, four benchmarking datasets (two masses-based and two microcalcifications/calcifications-based) representatives of benign and malignant lesions (biopsy-proven) comprising instances of clinical and image-based features are available for free download to registered users [32].

Finally, Table 1 gives the main characteristics of the mentioned common used and public datasets for breast mammograms.

Table 1. The common breast mammographic datasets (N: Normal cases, B: Benign cases, M: Malignant cases)

Dataset	Number of cases	Number of images	Available classes	Image format	Publicly available
DDSM	2620	10,480	N, B & M	JPEG	Yes
CBIS-DDSM	6775	10,239	N, B & M	DICOM	Yes
MIAS	–	322	N, B & M	PGM	Yes
INbreast	115	410	N, B & M	DICOM	No
BCDR	1734	3703 FM - 3612 DM	N, B & M	TIFF	Yes

3 Comparative Study for the Used Evaluation Metrics and Datasets

There have been active developments in the research in each specific area of abnormalities detection, segmentation, and classification. The Mass detection step is considered an important preprocessing stage to detect potential cancerous regions for further analysis by a Computer-Aided Diagnosis system (CAD) system. Breast mass segmentation also plays a crucial role in accurately extracting discriminative shape features of specific mass regions, while excluding surrounding tissues.

This section presents first the metrics used to evaluate a proposed breast mass detection and classification system and how to deduce each metric with its intent. Then, a comparison between the selected recent proposed approaches in the breast cancer detection and classification from the perspective of the data preparation done in each. Finally, a comparative analysis is proposed to show the detection and classification models used and their performance.

3.1 Detection and Classification Accuracy Metrics

This section presents the accuracy metrics used to evaluate the breast cancer detection and classification work. They are used as well to indicate the accuracy measurement process. They are:

- True Positives (TP): They are the cases when the actual class of the data point was 1 (True) and the predicted is also 1 (True).
- True Negatives (TN): They are the cases when the actual class of the data point was 0 (False) and the predicted is also 0 (False).
- False Positives (FP): They are the cases when the actual class of the data point was 0 (False) and the predicted is 1 (True).
- False Negatives (FN): They are the cases when the actual class of the data point was 1 (True) and the predicted is 0 (False).
- Confusion Matrix: It is used for finding the correctness and accuracy of the model. It derives TP, TN, FP and FN in one matrix.
- Accuracy: It is the number of correct predictions made by the model over all kind of predictions made as follows:

$$Accuracy = \frac{TP + TN}{TP + TN + FP + FN} \tag{1}$$

- Precision: It is a measure that tells us the proportion of patients that we diagnosed as having cancer or actually had cancer. The predicted positives (People predicted as cancerous are TP and FP) and the people actually having a cancer are TP as follows:

$$Precision = \frac{TP}{TP + FP} \tag{2}$$

- Recall (Sensitivity): It is a measure that tells us the proportion of patients that actually had cancer and were diagnosed by the model as having cancer. The actual positives (People having cancer are TP and FN) and the people diagnosed by the model having a cancer are TP as follows:

$$Sensitivity = \frac{TP}{TP + FN} \tag{3}$$

- Specificity: It is a measure that tells us what proportion of patients that did NOT have cancer and were predicted by the model as non-cancerous. The actual negatives and the people diagnosed by us not having cancer are TN as follows:

$$Specificity = \frac{TN}{TN + FP} \tag{4}$$

- Dice similarity coefficient (F1-score): It represents a harmonic average of the precision and sensitivity and measures the similarity rate between predicted and ground-truth regions as follows:

$$F1 - score = \frac{2 * Precision * Recall}{Precision + Recall} \tag{5}$$

- Matthews Correlation Coefficient (MCC): It measures the classification quality even if the classes are in different sizes, counting true and false positives and negatives as follows:

$$MCC = \frac{TP * TN - FP * FN}{[(TP + FP) * (FN + TN) * (FP + TN) * (TP + FN)]^{1/2}} \tag{6}$$

- Intersection over union (IoU): IoU measures the overlap between 2 boundaries to measure how much the predicted boundary overlaps with the ground truth (the real object boundary). In some datasets, an IoU threshold is predefined (say 0.5) in classifying whether the prediction is a true positive or a false positive.

$$IOU = \frac{AreaofOverlap}{AreaofUnion} \tag{7}$$

- mean Average Precision (mAP): AP (Average precision) is a popular metric in measuring the accuracy of object detectors which quantifies how good our model is at performing a specific query. Given Q which is the number of queries in the set and AveP(q) which is the average precision (AP) for a given query, q, the mAP is calculated as follows:

$$mAP = \frac{\sum_{q=1}^{Q} AveP(q)}{Q} \tag{8}$$

3.2 Datasets Preparation

Seven recent proposed ideas for breast cancer detection and classification are selected to be compared in this section which are recent and considered end to

Table 2. Comparative study for the dataset preparation for the recent work in the breast cancer detection and classification

Ref.	Dataset	Size	Views	Augmentation	Preprocessing
[33]	INbreast	107 M	CC/MLO	Augment mammograms eight times by rotating them with the angles of 45 = ° (i.e., 0°, 45°, 90°, 135°, 180°, 225°, 270° and 315°)	A contrast-limited adaptive histogram equalization (CLAHE) method is utilized as a preprocessing step for all detected masses
[4]	DDSM	600 M	CC/MLO	Rotating mammograms with 90, 180, 270°	Otsu thresholding technique to generate breast region. Then, multi-threshold peripheral equalization enhanced peripherals thickness and remove irrelevant information
[2]	DDSM	150 M	CC/MLO	No data augmentation	Multi-threshold peripheral equalization to enhance peripherals thickness and remove irrelevant information
[34]	BCDR-F03	736 M	CC/MLO	To each refined breast lesion, another seven new samples are generated using a combination of flipping and rotation ($\pi/2$, Π and $3\pi/2$) transformations	No preprocessing
[35]	DDSM + NP	2242 M	CC/MLO	Each ROI was flipped and rotated four times to obtain eight augmented samples	No preprocessing
[36]	DDSM	1820 M	CC/MLO	The selected mammograms augmented by 5 random rotations and sampled 5 random crops for each rotation, thus effectively multiplying the training set size by 25	First extract the mass from the full mammogram by taking a bounding box around the pixel-level mask applied to the original image. Followed by a fixed padding of 50 pixels all around the mass
[37]	INbreast + NP	329 M	CC/MLO	Mammograms were flipped, randomly cropped, and rotated up to 90°, 180°, and 270°	Original mammograms are divided into small sub-sections which do not require the resizing method

end systems as well. They are studied with respect to different aspects including the dataset they used, if they made preprocessing on this data or not, or data augmentation exist and the mammograms views they trained with.

In Table 2: Ref. column is used as the abbreviation for the paper Reference, Dataset column indicates the used dataset, size column is the used dataset size, views column is the views included in the training process which may be Mediolateral Oblique (MLO) views or Cranio-Caudal (CC) views or both, augmentation columns determines if augmentation is done by the proposed method and what was done to augment data and finally preprocessing column is the steps used for preprocessing. Inside the table, M is used instead of mammograms and NP is used instead of Non Public Dataset.

Table 3. Comparative study for the recent proposed work in the breast cancer detection and classification

Ref.	Detection & segmentation	Classification	Classes	Fine tuned	Accuracy
[33]	YOLO is used for mass detection. Then, they proposed a new full resolution convolutional network (FrCN) deep learning model for pixel-to-pixel mass segmentation	AlexNet	B/M	They used the transfer learning method to initialize the parameters of all deep learning models	– Detection accuracy = 98.96% – Classification accuracy = 95.64%
[4]	YOLO is used to detect the mass	YOLO	B/M	Fine-tuned with the pre-trained weights with a large computer vision ImageNet dataset	– Detection accuracy = 99.7% – Classification accuracy = 97%
[2]	Adaptive threshold & morphological operations	Deep Belief Network (DBN)	B/M/N	Not fine tuned	– Overall accuracy = 92.86%
[34]	ROIs are first extracted and then Otsu segmentation algorithm followed by morphological operation to refine the mass	AlexNet, GoogLeNet and Shallow CNN	B/M	Pre-trained on a large-scale visual database	GoogLeNet AUC = 88% & AlexNet AUC = 83%
[35]	No detection and segmentation	DCNN	B/M	Fine-tuned with the pre-trained weights with a large computer vision ImageNet dataset	The multi-task transfer learning DCNN was found to have significantly (p = 0.007) higher performance generalization compared to the single-task transfer learning DCNN
[36]	No detection and segmentation	AlexNet, GoogLeNet and Shallow CNN	B/M	Fine-tuned with the pre-trained weights with a large computer vision ImageNet dataset	The GoogleNet with Deeper Training outperforms the other models with a recall of 0.934 compared to at most 0.901 for the other models
[37]	RetinaNet	No classification	No classification	They used ResNet50 that is pre-trained on the ImageNet dataset for the backbone netw	They did a lot of experiments and they achieved higher true positive rate than the complex model which is average 97%

4 Comparative Study for the Applied Detection and Classification Models

The same seven approaches are also compared based on the used detection model if exist, the used classification model if exist, the classes the model trained on if benign (B), malignant (M) or normal (N) and the achieved accuracy as shown in Table 3. In this table: Ref. column is the paper Reference, detection and classification column shows the applied model for mass detection or segmentation or both, classification columns shows the mass classification model followed by the classes column then the fine tuned columns to check if the model is fine tuned or not and finally the performance in the last accuracy column.

From this study, it is found that the recent and common proposed models for the mass detection and classification are based on AlexNet [37], GoogleNet [38], YOLO [4] and RetinaNet [39]. The most recent and best achieved accuracy work in the breast detection and classification area is [4] and [39].

4.1 Limitation of the Used Detection and Classification Models

Besides the high accuracy achieved using the above models, they have some limitations. AlexNet model causes duplication of data due to the overlapping blocks of pixels which leads to more memory to be used. While GoogleNet used sparse connections (not all the output feature maps are connected to all the input feature maps), to overcome the problem of redundant information and reduced cost by omitting feature maps that were not relevant. However, the main drawback of the GoogleNet was its heterogeneous topology that needs to be customized from one module to another. Besides that its representation bottleneck that drastically reduces the feature space in the next layer and thus sometimes may lead to loss of useful information. On the other-hand the YOLO model is customizable but it does have some limitations on how close objects can be, i.e. it couldn't get good accuracy with the small sized objects and tends to do a few mistakes in localization. This results in introducing the RetinaNet to fill in for the imbalances and inconsistencies of the single shot object detectors like YOLO in a fast and efficient manner.

5 Conclusion

In this paper, we compare between the recent proposed models in the breast cancer detection and classification in order to reduce the death rate caused by the breast cancer. We integrated from the recent published work, the various datasets that are available public such as DDSM, CBIS-DDSM and INbreast. The differences are discussed between these datasets from many points as the size, the views included, the classes contained with the images format as well. Then, an overall comparative study is held between various machine learning based detection and classification methods that are recently published. It is deduced that the common database used and considered large is the CBIS-DDSM which is an enhanced version from the DDSM. It is deduced that YOLO and RetinaNet is the new models that are recently used and considered more simpler than the conventional CNN networks and achieve better results and accurate performance for the mass detection and malignancy classification. Also, the models that use balanced number for the classes the model shall detect, are more fair and achieve better overall accuracy than the unbalanced sized data models.

References

1. Boyle, P., Levin, B., et al.: World cancer report 2008. IARC Press, International Agency for Research on Cancer (2008)
2. Al-antari, M.A., Al-masni, M.A., Park, S.U., Park, J.H., Metwally, M.K., Kadah, Y.M., Han, S.M., Kim, T.-S.: An automatic computer-aided diagnosis system for breast cancer in digital mammograms via deep belief network. J. Med. Biol. Eng. **38**(3), 443–456 (2017)
3. Al-masni, M., Al-antari, M., Park, J., Gi, G., Kim, T., Rivera, P., Valarezo, E., Han, S.-M., Kim, T.-S.: Detection and classification of the breast abnormalities in digital mammograms via regional convolutional neural network. In: 39th Annual International Conference of the IEEE Engineering in Medicine and Biology Society (EMBC 2017), Jeju Island, South Korea, pp. 1230–1236(2017)
4. Al-masni, M.A., Al-antari, M., Park, J.-M.P., Gi, G., Kim, T.-Y.K., Rivera, P., Valarezo, E., Choi, M.-T., Han, S.-M., Kim, T.-S.: Simultaneous detection and classification of breast masses in digital mammograms via a deep learning YOLO-based CAD system. Comput. Methods Programs Biomed. **157**, 85–94 (2018)
5. Al-antari, M.A., Al-masni, M.A., Park, S.U., Park, J.H., Kadah, Y.M., Han, S.M., Kim, T.S.: Automatic computer-aided diagnosis of breast cancer in digital mammograms via deep belief network. In: Global Conference on Engineering and Applied Science (GCEAS), Japan, pp. 1306–1314 (2016)
6. Al-antari, M.A., Al-masni, M.A., Kadah, Y.M.: Hybrid model of computer-aided breast cancer diagnosis from digital mammograms. J. Sci. Eng. **04**(2), 114–126 (2017)
7. Dromain, C., Boyer, B., Ferré, R., Canale, S., Delaloge, S., Balleyguier, C.: Computed aided diagnosis (CAD) in the detection of breast cancer. Eur. J. Radiol. **82**(3), 417–423 (2013)
8. Dhungel, N., Carneiro, G., Bradley, A.P.: Automated mass detection in mammograms using cascaded deep learning and random forests. In: 2015 International Conference on Digital Image Computing: Techniques and Applications (DICTA), Australia (2015)
9. Wang, Y., Tao, D., Gao, X., Li, X., Wang, B.: Mammographic mass segmentation: embedding multiple features in vector-valued level set in ambiguous regions. Pattern Recognit. **44**(9), 1903–1915 (2011)
10. Rahmati, P., Adler, A., Hamarneh, G.: Mammography segmentation with maximum likelihood active contours. Med. Image Anal. **16**(9), 1167–1186 (2012)
11. Domínguez, A.R., Nandi, A.: Toward breast cancer diagnosis based on automated segmentation of masses in mammograms. Pattern Recognit. **42**(6), 1138–1148 (2009)
12. Qiu, Y., Yan, S., Gundreddy, R.R., Wang, Y., Cheng, S., Liu, H., Zheng, B.: A new approach to develop computer-aided diagnosis scheme of breast mass classification using deep learning technology. J. X-Ray Sci. Technol. **25**(5), 751–763 (2017)
13. Dhungel, N., Carneiro, G., Bradley, A.P.: A deep learning approach for the analysis of masses in mammograms with minimal user intervention. Med. Image Anal. **37**(1), 114–128 (2017)
14. LeCun, Y., Boser, B.E., Denker, J.S., Henderson, D., Howard, R.E., Hubbard, W.E., et al.: Handwritten digit recognition with a back-propagation network. In: Advances in Neural Information Processing Systems, pp. 396–404 (1990)
15. Becker, A.S., Marcon, M., Ghafoor, S., Wurnig, M.C., Frauenfelder, T., Boss, A.: Deep learning in mammography: diagnostic accuracy of a multipurpose image analysis software in the detection of breast cancer. Invest. Radiol. **52**(7), 434–440 (2017)

16. Kooi, T., Litjens, G., van Ginneken, B., Gubern-Mérida, A., Sánchez, C.I., Mann, R., et al.: Large scale deep learning for computer aided detection of mammographic lesions. Med. Image Anal. **35**, 303–312 (2017)
17. Dhungel, N., Carneiro, G., Bradley, A.P.: Automated mass detection in mammograms using cascaded deep learning and random forests. In: 2015 International Conference on Digital Image Computing: Techniques and Applications (DICTA), pp. 1–8. IEEE (2015)
18. Dhungel, N., Carneiro, G., Bradley, A.P.: A deep learning approach for the analysis of masses in mammograms with minimal user intervention. Med. Image Anal. **37**, 114–128 (2017)
19. Akselrod-Ballin, A., Karlinsky, L., Alpert, S., Hasoul, S., Ben-Ari, R., Barkan, E.: A region based convolutional network for tumor detection and classification in breast mammography. In: Deep Learning and Data Labeling for Medical Applications, pp. 197–205. Springer (2016)
20. Akselrod-Ballin, A., Karlinsky, L., Hazan, A., Bakalo, R., Horesh, A.B., Shoshan, Y., et al.: Deep learning for automatic detection of abnormal findings in breast mammography. In: Deep Learning in Medical Image Analysis and Multimodal Learning for Clinical Decision Support, pp. 321–329. Springer (2017)
21. Ribli, D., Horváth, A., Unger, Z., Pollner, P., Csabai, I.: Detecting and classifying lesions in mammograms with deep learning. Sci. Rep. **8**(1), 4165 (2018)
22. Choukroun, Y., Bakalo, R., Ben-Ari, R., Akselrod-Ballin, A., Barkan, E., Kisilev, P.: Mammogram classification and abnormality detection from nonlocal labels using deep multiple instance neural network (2017)
23. Jalalian, A., Mashohor, S., Mahmud, H., Saripan, M., Rahman, A., Ramli, B., et al.: Computer-aided detection/diagnosis of breast cancer in mammography and ultrasound: a review. Clin. Imaging **37**(3), 420–426 (2013)
24. Digital Database for Screening Mammography. http://www.eng.usf.edu/cvprg/Mammography/Database.html
25. Heath, M., Bowyer, K., Kopans, D., Moore, R., Kegelmeyer, W.P.: The digital database for screening mammography. In: Yaffe, M.J. (ed.) Proceedings of the Fifth International Workshop on Digital Mammography, pp. 212–218. Medical Physics Publishing (2001). ISBN 1–930524–00–5
26. Lee, R.S., Gimenez, F., Hoogi, A., Rubin, D.: Curated breast imaging subset of DDSM. Cancer Imaging Arch. **8** (2016)
27. Curated Breast Imaging Subset of DDSM. https://wiki.cancerimagingarchive.net/display/Public/CBIS-DDSM
28. The Mini-MIAS Database of Mammograms. http://peipa.essex.ac.uk/info/mias.html
29. Moreira, I.C., Amaral, I., Domingues, I., Cardoso, A., Cardoso, M.J., Cardoso, J.S.: INbreast: toward a full-field digital mammographic database. Acad. Radiol. **19**(2), 236–248 (2012)
30. Orel, S.G., Kay, N., Reynolds, C., Sullivan, D.C.: BI-RADS categorization as a predictor of malignancy. Radiology **211**(3), 845–850 (1999)
31. The INbreast Dataset. http://medicalresearch.inescporto.pt/breastresearch/index.php/Get_INbreast_Database
32. The Breast Cancer Digital Repository. https://bcdr.eu/
33. Al-antari, M.A., Al-masni, M.A., Choi, M.-T., Han, S.-M., Kim, T.-S.: A fully integrated computer-aided diagnosis system for digital X-ray mammograms via deep learning detection, segmentation, and classification. Int. J. Med. Inform. **117**, 44–54 (2018)

34. Jiang, F., Liu, H., Yu, S., Xie, Y.: Breast mass lesion classification in mammograms by transfer learning. In: Proceedings of the 5th International Conference on Bioinformatics and Computational Biology, pp. 59–62. ACM (2017)
35. Samala, R.K., Chan, H.-P., Hadjiiski, L.M., Helvie, M.A., Cha, K.H., Richter, C.D.: Multi-task transfer learning deep convolutional neural network: application to computer-aided diagnosis of breast cancer on mammograms. Phys. Med. Biol. **62**(23), 8894 (2017)
36. Levy, D., Jain, A.: Breast mass classification from mammograms using deep convolutional neural networks. arXiv:1612.00542 (2016)
37. Yuan, Z.-W., Jun, Z.: Feature extraction and image retrieval based on AlexNet. In: Eighth International Conference on Digital Image Processing (ICDIP 2016), vol. 10033, p. 100330E. International Society for Optics and Photonics (2016)
38. Wu, C., Wen, W., Afzal, T., Zhang, Y., Chen, Y.: A compact DNN: approaching GoogleNet-level accuracy of classification and domain adaptation. In: Proceedings of the IEEE Conference on Computer Vision and Pattern Recognition, pp. 5668–5677 (2017)
39. Jung, H., Kim, B., Lee, I., Yoo, M., Lee, J., Ham, S., Woo, O., Kang, J.: Detection of masses in mammograms using a one-stage object detector based on a deep convolutional neural network. PloS One **13**(9), e0203355 (2018)

Hyperspectral Image Classification Using Deep Learning Technique

Mayar A. Shafaey[1,4(✉)], Mohammed A.-M. Salem[1,2], Maryam N. Al-Berry[1], Hala M. Ebied[1], Elsayed A. El-Dahshan[3,4], and Mohammed F. Tolba[1]

[1] Faculty of Computers and Information Sciences, Ain Shams University, Cairo, Egypt
mayaraly22@gmail.com, {salem,maryam_nabil,halam}@cis.asu.edu.eg, fahmytolba@gmail.com

[2] Faculty of Media Engineering and Technology, German University in Cairo, Cairo, Egypt

[3] Faculty of Science, Department of Physics, Ain Shams University, Cairo, Egypt

[4] National Egyptian E-Learning University (EELU), Giza, Egypt
seldahshan@eelu.edu.eg

Abstract. Today, the classification process is demanded for modern city planning, agriculture and environmental monitoring, and many other applications. The optimum classification degree is still insufficient so far. The present classification methods for remote-sensing images are grouped according to the features they use into: manual feature-based methods, unsupervised feature learning methods, and supervised feature learning methods. In recent times, the supervised deep learning approaches are extensively introduced in various remote-sensing applications, such as object detection and land use scene classification. In this article, an experiment is conducted using one of the widespread deep learning models, Convolution Neural Networks (CNNs), specifically, *AlexNet* architecture on a standard sounded hyper spectral dataset, *Pavia University (PaviaU)*. The model achieved an overall accuracy of 91% ± 0.01. A comparison with other different techniques is also introduced.

Keywords: Deep learning · Convolution Neural Networks (CNNs) · Remote-sensing · Satellite images · Hyperspectral images

1 Introduction

For the time being, there are different types of satellite images acquired by artificial satellites [1]. These image types are: high spatial, spectral, and temporal resolution images which cover the earth within a day [2]. Spectral imaging is imaging that consists of different bands over the electromagnetic range [3]. In fact, the traditional camera captures light over three wavelength bands through the visible range of red, green, and blue (RGB). The acquisition of spectral imaging data may use the infrared,

A.-E. Hassanien et al. (Eds.): AICV 2020, AISC 1153, pp. 334–342, 2020.
https://doi.org/10.1007/978-3-030-44289-7_31

the visible spectrum, the ultraviolet, x-rays, or some combination in visible and non-visible bands simultaneously. It is possible to capture hundreds of wavelength bands for each pixel in an image [4].

Several remote sensing systems [5] record energy over multiple separate wavelength ranges at various spectral resolutions. These are stated as multi-spectral sensors [6]. Innovative multi-spectral sensors are called hyper-spectral sensors [7]. The analysis and classification of large-scale nature of the data acquired using these sensors introduce new challenges in the image analysis field.

The detailed spectral information provided by hyper-spectral sensors increases the power of accurately differentiating materials of interest with increased classification accuracy. Moreover, with respect to advances in hyper-spectral technology, the fine spatial resolution of recently operated sensors makes the analysis of small spatial structures in images possible [4].

The above-mentioned advances make the hyper-spectral data a useful tool for an extensive variety of applications. Thus, the main objective of this paper is to address the problem of pixel wise classification of hyper-spectral imagery and understanding the shape of objects, which is required to aid the classification process (semantic analysis).

In the next section, a quick review of common and free hyper-spectral images is presented. A review on recent classification approaches applied on one or some of these datasets is presented in Sect. 3. The experimental work followed by results and evaluations are presented in Sect. 4. Finally, the conclusions are highlighted in Sect. 5.

2 Related Work

The hyper-spectral sensors are usually composed of several hundreds of spectral channels of the same scene as illustrated in Fig. 1). Recently, several hyper-spectral remote-sensing image datasets have been announced by different groups to enable machine-learning based research for scene classification and the evaluation of different methods in this field. A review of some publicly available data sets is presented in this section, as given in Table 1. The table shows a comparison between different data sets with respect to the number of scene classes, labeled pixels, total pixels, number of bands, and spatial resolution (Ground Sample Distance - GSD). The lowest value of GSD points to the high quality of the image and vice versa.

The popular and sounded *PaviaU* hyper-spectral dataset [9] has been acquired by the Reflective Optics System Imaging Spectrometer (ROSIS) over Pavia University, northern Italy. It consists of 103 spectral bands within the wavelength range 430–770 nm. Each band's size is a 610 $\times$ 340 pixel image. A color composite image of the scene is illustrated in Fig. 2(left) where spectral bands number 28, 19 and 10 are used for the RGB channels respectively [8]. It has nine ground-truth classes of representative urban ground objects. Its details are shown in Fig. 2(right).

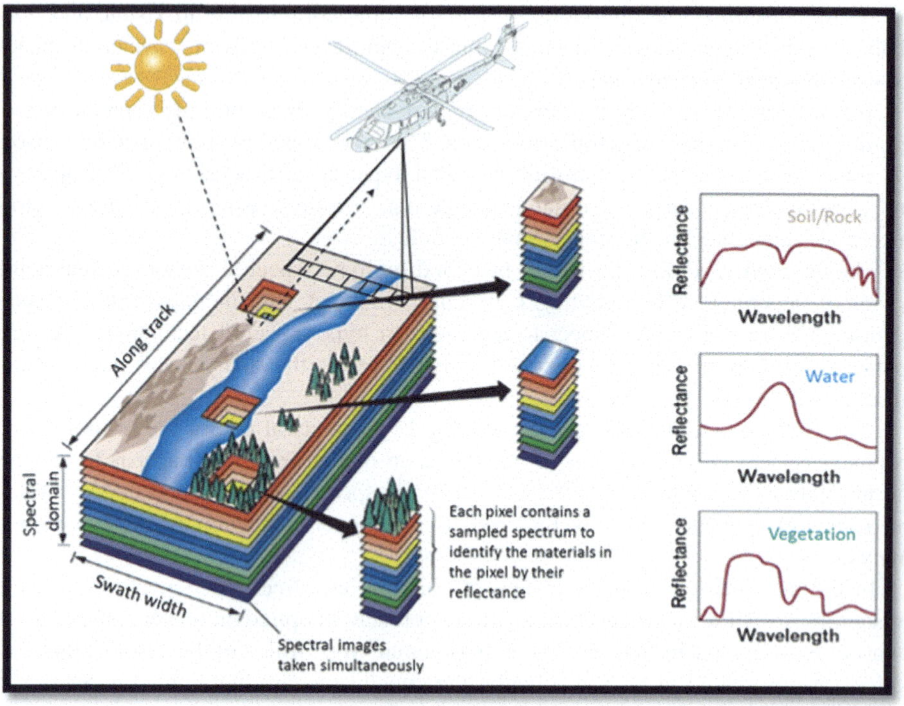

Fig. 1. Hyper-spectral sensors [7]

Table 1. Comparison between the several hyper-spectral remote-sensing datasets [9]

Data set	Pixels	Bands	Spatial resolution	Labels	Classes	Mode
PaviaU	991,040	103	1.3 m	50,232	9	Aerial
Indian Pines	21,025	224	20 m	10,249	16	Aerial
Salinas	111,104	227	3.7 m	54,129	16	Aerial
KSC	314,368	176	18 m	5,211	13	Aerial
Botswana	377,856	145	30 m	3,248	14	Satellite

According to *Zhou et el.* [10], the most important bands of *paviaU* are 8, 31, and 53 whereas they are containing more discriminatory information. Hence, the main goal of our work is to classify the hyper-spectral image based on these bands instead of RGB bands (28, 19, and 10) [8].

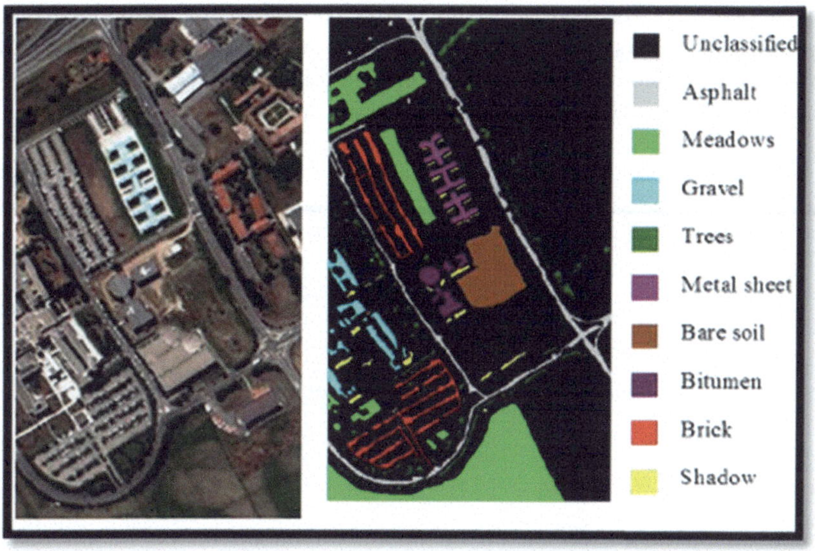

Fig. 2. *PaviaU* dataset. (left) True-color composite. (middle) Ground reference map. (right) Ground-truth classes [9]

3 Remote Sensing Image Classification Methods

With respect to the vast publications of remote-sensing images for the task of scene classification, the existing scene classification methods could be categorized into three main categories, as shown in Table 2. The three categories are: manual feature based methods, unsupervised classification methods and supervised learning methods.

Table 2. The existing image scene classification methods

Handcrafted feature based method	Unsupervised feature learning based methods	Deep feature learning based methods
Color histogram [11]	Principle Component Analysis (PCA) [15]	Stacked Auto-Encoder (SAE) [19]
Texture descriptor [12]	K-mean [16]	Convolutional Neural Networks (CNNs) [20]
GIST [13]	Spare coding [17]	Deep Belief Nets (DBN) [21]
SIFT [14]	Auto-encoder [18]	Deep Belief Machines (DBM) [22]

In this article, the authors mainly use deep learning method, especially CNNs, to train huge multi-layer networks for classifying large scale input images. The CNN itself has different models like: *AlexNet, GoogleNet, ResNet, VGGNet, CaffeNet*… etc. [20]. Limited by the space, a short description of *AlexNet* architecture is given. The network consists of 25 layers: 5 convolution layers, 5 max-pooling layers, 5 dropout layers, and 3 fully connected layers, as shown in Fig. 3. It is trained on *ImageNet* data, which contains over 15 million annotated images from a total of over 22,000 categories and uses the rectified layer (ReLU) for the nonlinearity functions [23].

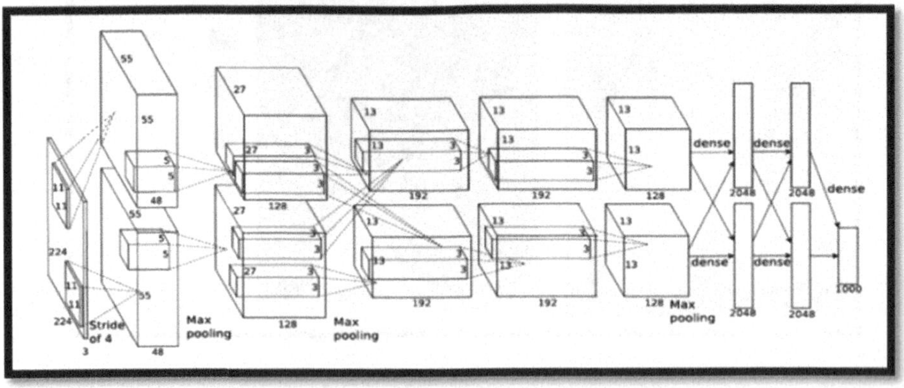

Fig. 3. *ImageNet* classification with *AlexNet* CNN [23]

Table 3 represents the results published by authors who used the CNN models in their experiments for large scale image scene classification and gave the outstanding accuracy values which are clues for the power of CNN learning model.

Table 3. Some of recent publications that applied CNNs in experiments on large scale remote-sensing (RS) images – UC Merced dataset

Ref.	Year	Application	Method	Accuracy
[24]	2015	Multi-spectral land use classification	Deep CNN	93.48%
[25]	2015	Land use RS classification	GoogleNet, and CaffeNet	97%, and 95.48%
[26]	2016	RS scene classification	Large patch CNN	Effective results
[27]	2016	Large scale RS image classification	CNN	92.4%
[28]	2019	Remote sensing scene classification	AlexNet	94%

4 Experiments and Results

The *AlextNet* CNN approach was applied for large scale image classification. In this section, the experiment's steps are described; i.e., software, hardware specification, results, comments, and comparisons.

4.1 Experimental Procedure

The experiments were conducted on a PC with a processor: Intel® Core™ i7-2670QM CPU @ 2.20 GHz – 8 GB RAM and using Matlab® software using *alexnet* built-in function which is trained on a subset of the *ImageNet* database – ~1.2 million images - and can classify images into 1000 object categories. This function requires Neural Network Toolbox™ Model for *AlexNet* Network.

Initially, the input image dimension was resized from 610 × 340 to 227 × 227 as a required input size for the *alexnet* function. Then, the features were extracted using the CNN, and passed to the trained multiclass SVM classifier to get the labels. Finally, the classification results were given by computing the summation of main diagonal of the confusion matrix divided by the diagonal elements number.

The experiment was carried out twice. The first trial was for bands 8, 31, and 53 of *PaviaU* dataset. The second one was for bands number 25, 50, and 75 for the same dataset. Figure 4 shows the bands 25, 50 and 75 of a sample from *PaviaU* dataset. Table 4 gives a comparison of classification accuracy for *PaviaU* scene based on the selected bands.

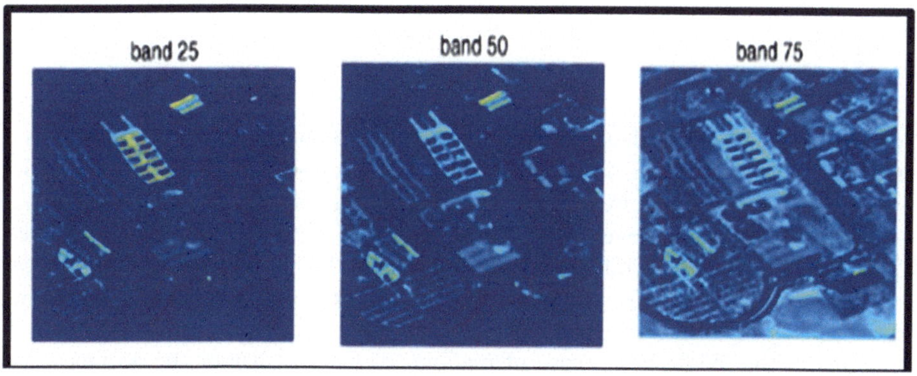

Fig. 4. Bands number 25, 50, and 75 of *PaviaU* scene [29]

Table 4. Comparison of classification accuracy for *PaviaU* scene

Method	Bands number	Selection criteria	Feature extractor	Accuracy
Our Exp. #1	8, 31, and 53	More discriminatory	CNN	91% ±0.01
Our Exp. #2	25, 50, and 75	Random	CNN	77% ±0.04
Zhou *et al.* [10]	RGB	Basic & common bands	PCA	85.82%

4.2 Evaluation and Discussions

By comparing our work with other published results, such as [10], the authors reached 85.82% using SVM classifier based on 3 labeled bands (RGB), while, our experiment reached 91% ± 0.01, in experiment 1, using the aforementioned discriminatory bands. Nevertheless, experiment 2 recorded 77% ± 0.04 of accuracy value. The interpretation of that high accuracy is the power of CNN as a feature extractor model, while the low value of the accuracy percentage refers to the randomly selection of image bands. It may contain no useful information for the classification process.

On one hand, by comparing the traditional handcrafted features versus deep features learning based method, the conventional methods consume a high mental thinking and skills, while the deep learning features are learned automatically from data through deep layers of neural networks architecture. This is the key advantage of deep features learning based methods. On the other hand, in comparison with unsupervised feature learning methods, i.e. auto-encoder, the deep learning models are consisted of multiple processing layers. It can learn more powerfully and it is more applicable for the large scale image scene classification. The deep learning architectures performance is similar to the human brain, in which each layer uses the information from the previous layer to learn intensely and precisely [30].

There is no doubt that increasing the number of bands will affect the classification accuracy positively. This can be noted for future work. Another comparison is between the achieved results, in this article, versus the results which are mentioned in Table 3. The reader may note that the accuracy of our experiment is slightly less than others. This is due to the difference of the used dataset. The presented results didn't use the *PaviaU* hyper-spectral dataset.

5 Conclusions

Nowadays, there are two hot research topics of the automated target detection, and large scale remotely sensed image classification. Therefore, this paper gave a quick review of the common and free hyper-spectral remote-sensing datasets, i.e. *PaviaU*, to enable the community to develop the large scale image scene classification task. Then, it gave a short summary of different methods used for this task. Finally, the CNN deep learning method (AlexNet) is applied on only 3 bands of this dataset, through two experiments. One of the experiments has gained a high accuracy for the CNN use as feature extractor. While in the case of random selection of bands, the accuracy was lower. The results were reported and compared to state-of-the-art work and as a baseline for future research.

Deep learning methods can guarantee better feature representations for the remote-sensing related works, and there is a bright prospect of sight for more and more researchers dedicated to learning better features for the target detection and scene classification tasks.

References

1. NASA: What is a satellite? NASA Knows! (Grades 5–8) series (2014)
2. Zhang, L., Xia, G., Wu, T., Lin, L., Tai, X.: Deep learning for remote sensing image understanding. J. Sens. **2016**, 1–2 (2016)
3. Yaniv, O., Isaac, A., Vladimir, F., Daniel, G., Adrian, S.: Compressive sensing hyperspectral imaging by spectral multiplexing with liquid crystal. J. Imaging **5**(1), 3 (2019)
4. Roy, B., Lawrence, T., Mahdi, N.: Spectral imaging using a commercial colour-filter array digital camera. In: The Fourteenth Triennial ICOM-CC Meeting, pp. 743–750 (2005)
5. Adam, C.W., Vincent, G.A., Everett, A.H.: Unmanned aircraft systems in remote sensing and scientific research: classification and considerations of use. Remote Sens. **4**(6), 1671–1692 (2012)
6. Jose, A.J.B., Pablo, J.Z., Lola, S., Elias, F.: Thermal and narrowband multispectral remote sensing for vegetation monitoring from an unmanned aerial vehicle. IEEE Trans. Geosci. Remote Sens. **47**(3), 722–738 (2009)
7. Elhadi, A., Onisimo, M., Denis, R.: Multispectral and hyperspectral remote sensing for identification and mapping of wetland vegetation: a review. Wetlands Ecol. Manag. **18**(3), 281–296 (2010)
8. Shaheera, R., Nicolas, D.: A split-and-merge approach for hyperspectral band selection. IEEE Geosci. Remote Sens. Lett. **14**(8), 1378–1382 (2017)
9. https://rslab.ut.ac.ir/data Accessed 25 May 2019
10. Zhou, Z., Edoardo, P., Melba, M.C., James, C.T.: An active learning framework for hyperspectral image classification using hierarchical segmentation. IEEE J. Sel. Top. Appl. Earth Obs. Remote Sens. **9**(2), 640–654 (2016)
11. Swain, M.J., Ballard, D.H.: Color indexing. Int. J. Comput. Vis. **7**(1), 11–32 (1991)
12. Shalaby, M.M., Salem, M.A.-M., Khamis, A., Melgani, F.: Geometric model for vision-based door detection. In: 2014 9th International Conference on Computer Engineering & Systems (ICCES), Cairo, pp. 41–46 (2014). https://doi.org/10.1109/icces.2014.7030925
13. Salem, M.A.-M., Appel, M., Winkler, F., Meffert, B.: FPGA-based smart camera for 3D wavelet-based image segmentation. In: 2nd ACM/IEEE International Conference on Distributed Smart Cameras, ICDSC, pp. 1–8 (2008). https://doi.org/10.1109/icdsc.2008.4635720
14. Salem, M.A.-M., Alaa, A., Alaa, S., Marwa, S.: Recent survey on medical image segmentation. In: Computer Vision: Concepts, Methodologies, Tools, and Applications, pp. 129–169. IGI Global (2018). https://doi.org/10.4018/978-1-5225-5204-8.ch006
15. Jolliffe, I.: Principal Component Analysis. Springer, New York (2002)
16. Zhao, B., Zhong, Y., Zhang, L.: A spectral–structural bag-of-features scene classifier for very high spatial resolution remote sensing imagery. ISPRS J. Photogramm. Remote Sens. **116**, 73–85 (2016)
17. Olshausen, B.A., Field, D.J.: Sparse coding with an over complete basis set: a strategy employed by V1? Vis. Res. **37**(23), 3311–3325 (1997)
18. Hinton, G.E., Salakhutdinov, R.R.: Reducing the dimensionality of data with neural networks. Science **313**(5786), 504–507 (2006)
19. Vincent, P., Larochelle, H., Lajoie, I., Bengio, Y., Manzagol, P.-A.: Stacked denoising autoencoders: learning useful representations in a deep network with a local denoising criterion. J. Mach. Learn. Res. **11**, 3371–3408 (2010)
20. Simonyan, K., Zisserman, A.: Very deep convolutional networks for large-scale image recognition. In: Proceedings of International Conference on Learning Representations, pp. 1–13 (2015)

21. Hinton, G.E., Osindero, S., Teh, Y.-W.: A fast learning algorithm for deep belief nets. Neural Comput. **18**(7), 1527–1554 (2006)
22. Salakhutdinov, R., Hinton, G.: An efficient learning procedure for deep Boltzmann machines. Neural Comput. **24**(8), 1967–2006 (2012)
23. Krizhevsky, A., Sutskever, I., Hinton, G.: ImageNet classification with deep convolutional neural networks. In: Proceedings Conference on Advances in Neural Information Processing Systems, pp. 1097–1105 (2012)
24. Luus, F., Salmon, B., Van Den Bergh, F., Maharaj, B.: Multiview deep learning for land-use classification. IEEE Geosci. Remote Sens. Lett. **12**(12), 2448–2452 (2015)
25. Castelluccio, M., Poggi, G., Sansone, C., Verdoliva, L.: Land Use Classification in Remote Sensing Images by Convolutional Neural Networks. Cornell University, Ithaca (2015)
26. Zhong, Y., Fei, F., Zhang, L.: Large patch convolutional neural networks for the scene classification of high spatial resolution imagery. Appl. Remote Sens. **10**(2), 025006 (2016)
27. Marmanis, D., Datcu, M., Esch, T., Stilla, U.: Deep learning earth observation classification using ImageNet pretrained networks. IEEE Geosci. Remote Sens. Lett. **13**(1), 105–109 (2015)
28. Shafaey, M.A., Salem, M.A.-M., Ebied, H.M., Al-Berry, M., Tolba, M.F.: Deep learning for satellite image classification. In: Proceedings of the International Conference on Advanced Intelligent Systems and Informatics, vol. 4, pp. 383–391 (2019)
29. Baofeng, G., Robert, I.D., Steve, R.G., James, D.B.: Improving hyperspectral band selection by constructing an estimated reference map. Appl. Remote Sens. **8**, 083692 (2014)
30. Hu, F., Xia, G., Wang, Z., Huang, X., Zhang, L., Sun, H.: Unsupervised feature learning via spectral clustering of multidimensional patches for remotely sensed scene classification. IEEE J. Sel. Top. Appl. Earth Obs. Remote Sens. **8**(5), 2015–2030 (2015)

Solving Inverse Kinematics of a 7-DOF Manipulator Using Convolutional Neural Network

Hassan Ashraf Elkholy[1,2(✉)], Abdalla Saber Shahin[1,2(✉)], Abdelaziz Wasfy Shaarawy[2(✉)], Hagar Marzouk[1,2(✉)], and Mahmoud Elsamanty[1,2,3(✉)]

[1] Smart Engineering Systems Research Center (SESC), Nile University, Shaikh Zayed City 12588, Giza, Egypt
{h.ashraf,a.saber,h.marzouk,melsamanty}@nu.edu.eg

[2] School of Engineering and Applied Sciences, Nile University Campus, Sheikh Zayed District, Juhayna Square, 6th of October City 12588, Giza, Egypt
a.wasfy@nu.edu.eg

[3] Faculty of Engineering at Shoubra, Benha University, Cairo, Egypt
mahmoud.alsamanty@feng.bu.edu.eg

Abstract. This paper presents a way to solve inverse kinematics of a 7-DOF manipulator using artificial neural networks. The manipulator consists of a 6-DOF articulated arm installed on a linear guide system to increase the workspace of the robot. The purpose of this paper is to provide an alternative to the traditional and complicated way to solve inverse kinematics by using artificial neural networks. The training data is generated from MATLAB after obtaining the DH parameters and workspace of the manipulator. Then, it was fed to the convolutional neural architecture to obtain a model for the manipulator. The input of the CNN is the end effector desired pose, and the outputs are the position angles of each joint. Two different architectures of artificial neural networks are compared to decide the most efficient architecture that produces the most accurate descriptive model of the manipulator.

Keyword: 7-DOF manipulator, Inverse kinematics, Convolutional Neural Network (CNN), Artificial Neural Network (ANN)

1 Introduction and Literature Survey

Nowadays, serial robotic manipulators are being used in many applications especially in the industrial sector. In general, they consist of several rigid links that are connected together by the means of rotational or prismatic joints. Increasing the number of degrees of freedom (DOF) results in more motion flexibility and more complex inverse kinematics problem (IK) [3]. The traditional method of controlling serial robotic manipulators is done by sending the desired coordinates of the end effector to the controller unit after solving its IK problem. For higher

A.-E. Hassanien et al. (Eds.): AICV 2020, AISC 1153, pp. 343–352, 2020.
https://doi.org/10.1007/978-3-030-44289-7_32

order of DOF, the Jacobean inverse kinematics algorithm can be used [13,14]. However, IK problem is difficult to be solved analytically because the solution will have high redundancy and many solutions [7]. Nowadays, Artificial Neural Network (ANN) is commonly used to solve the IK problems with higher number of DOF because of its advantages such as the ability of self-learning as well as parallel processing [16]. It is also able to fit non-linear relationships between inputs and outputs with a very good accuracy [5]. Neural networks depend on the input-output relationship specified by the training data, unlike traditional methods of control [3]. This paper is related to previous work in the aspect of using ANN to solve the IK problem of the manipulator [5–9,12]. According to Almusawi et al. [1], a new artificial neural network was implemented on a 6-DOF Denso robotic manipulator to increase the ability of the neural network to estimate the output joint angles [1]. Moreover, a study proposed multi-layered perceptron neural network architecture that has six sub-neural networks and back-propagation algorithm applied on 2-DOF manipulator [4]. In this study, a neural network was used to solve the problem of matrix inversion by iterating on the joint position directly [4]. Furthermore, according to Toshani et al. [13], a neural network based on a numerical method and one based on ANN were used to solve the inverse kinematics problem of a 7-DOF robot manipulator. The aim of this paper is to provides an ANN to solve the IK of a proposed 7-DOF manipulator. Figure 1 shows the whole stages that this paper went through. First, the 3D model of the proposed robotic arm was designed on SolidWorks software, then the DH parameters were taken and fed into the kinematics equations of this robotic arm to get the workspace which was validated using Moveit environment and SolidWorks. Afterwards, these obtained data of the workspace was fed into two different types of ANN on matlab. The paper is organized as the following: kinematic model of the roboticator, data obtaining and processing, artificial neural network architecture, simulation results and discussion, CNN validation, and conclusion.

2 Kinematics Model of the 7-DOF Manipulator

The system used in the paper is a 7-DOF manipulator consisting of six revolute joints mounted on a prismatic joint as shown in Fig. 2. The DH parameters table relates each joint with its previous joint to finally come up with the relation between the world frame and the end effector frame [11]. This is done by multiplying the 4×4 transformation matrices of the seven joints which yields very complex equations. One single transformation matrix of these equations is shown in Eq. (1).

$$T_i = \begin{bmatrix} cos(\Theta_i) & -sin(\Theta_i) \times cos(\alpha_i) & sin(\Theta_i) \times sin(\alpha_i) & ai \times cos(\Theta_i) \\ sin(\Theta_i) & cos(\Theta_i) \times cos(\alpha_i) & -cos(\Theta_i) \times sin(\alpha_i) & ai \times sin(\Theta_i) \\ 0 & sin(\alpha_i) & cos(\alpha_i) & di \\ 0 & 0 & 0 & 1 \end{bmatrix} \tag{1}$$

where i ranges between 1 to 7. In order to obtain the DH parameters shown in Table 1 accurately, joint limits were defined and assigned according to its own hardware and mechanical limitations. To validate these DH parameters, a model of the robot was created in MATLAB using the robotics toolbox as shown in Fig. 3(a). The workspace was validated by applying random angle joints of the workspace in the CAD model in SolidWorks and comparing the pose of the manipulator in the two models. This comparison resulted in a 0.01 mm error in the manipulator's pose between the two models.

Robot Operating System (ROS) is a widely used framework for robotics motion planning and inverse kinematics solving [10]. Moveit is a navigation software in ROS to enhance the process of trajectory planning and help creating a collision-free environment [2]. The 7-DOF manipulator is parsed in moveit environment to further validate the manipulator's workspace generated by MATLAB as shown in Fig. 3(b). In order to make the robot ready to be used in moveit environment, a Unified Robot Description Format (URDF) of the robot in its home position was created [10] as shown in Fig. 2. Figure 4 shows the real hardware of the 7-DOF serial manipulator. Figure 5 illustrates the front and side views of the workspace of the robot.

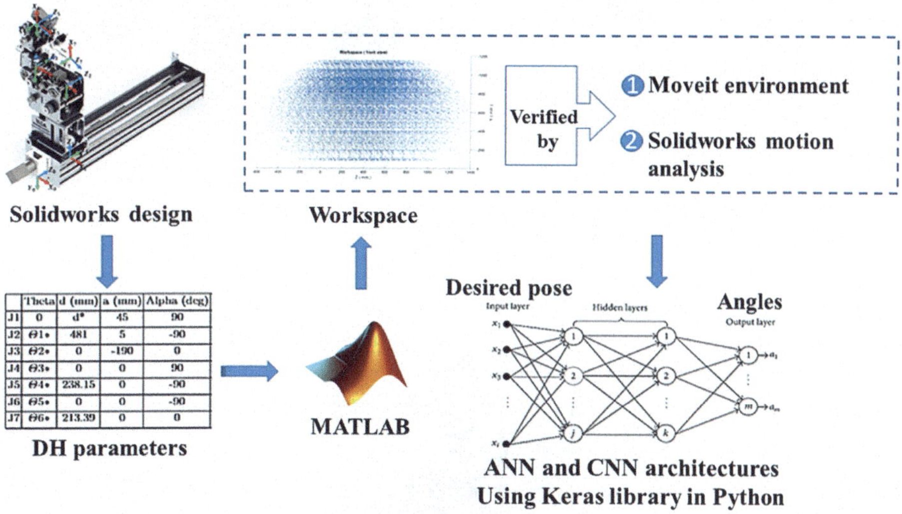

Fig. 1. Process of collecting training data

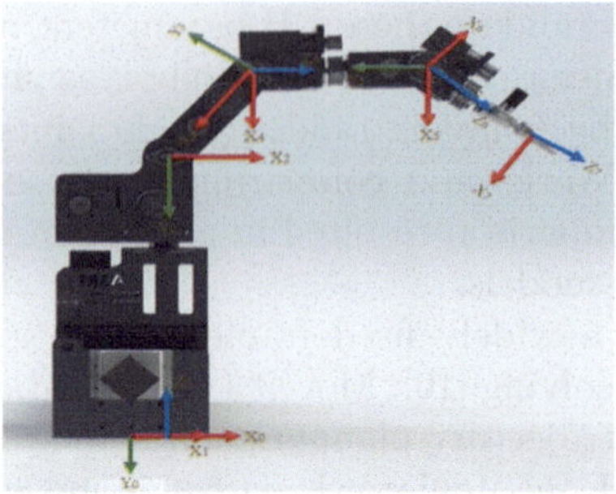

(a) Side view of the manipulator

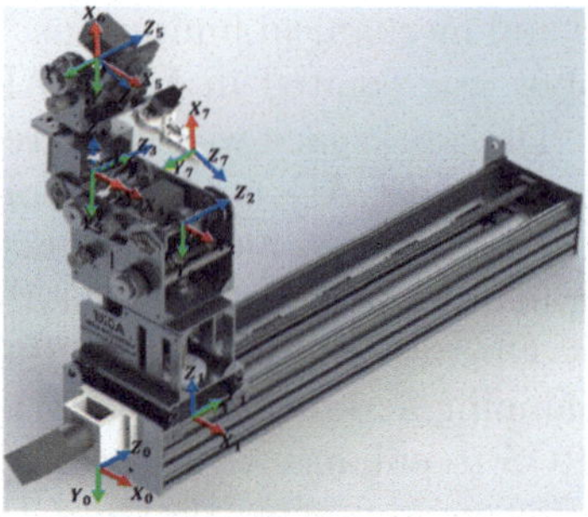

(b) Isometric view of the manipulator in home position

Fig. 2. 7-DOF serial manipulator with joints frames

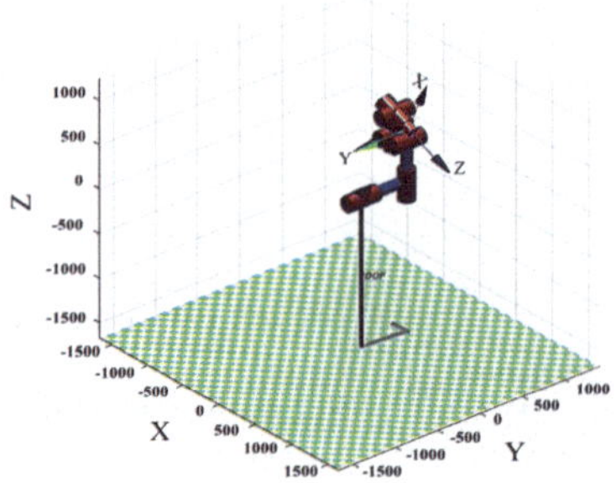

(a) MATLAB model describing the direction of rotation of each joint

(b) Motion Analysis of the manipulator on moveit

Fig. 3. Matlab model and moveit analysis

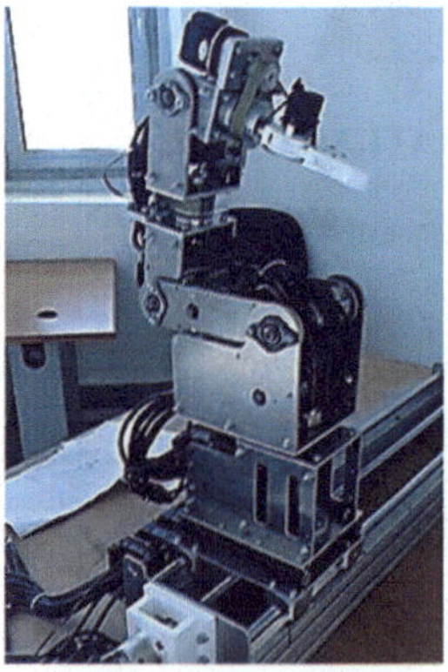

Fig. 4. The real hardware

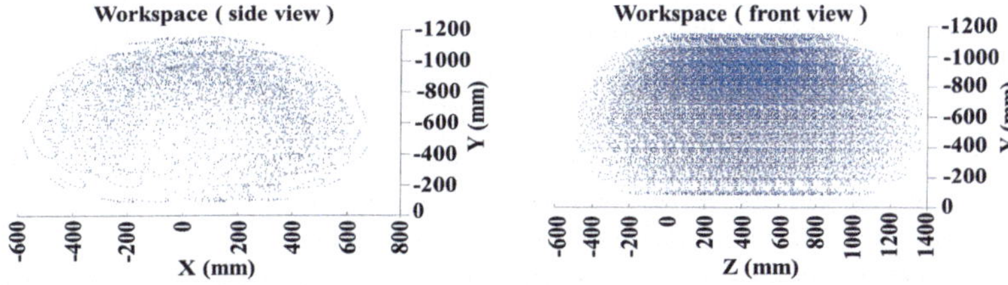

(a) Workspace side view (b) Workspace front view

Fig. 5. Workspace of the manipulator

Table 1. The DH parameters of the manipulator

	Theta	d (mm)	a (mm)	Alpha (deg)	Joints in home position
J1	0	d*	45	90	97.5 mm
J2	$\Theta 1*$	481	5	−90	0°
J3	$\Theta 2*$	0	−190	0	−0.4°
J4	$\Theta 3*$	0	0	90	4.36°
J5	$\Theta 4*$	238.15	0	−90	0°
J6	$\Theta 5*$	0	0	−90	−70.44°
J7	$\Theta 6*$	213.39	0	0	0°

3 Data Obtaining and Processing

After obtaining and validating the data from Matlab, it was fed into the neural network to start training the model. Range of data values was large; thus, it was scaled down to produce data values that are close to each other, so the weights of the artificial neural network are adjusted quickly, and the model converges fast. Two types of scaling were used: normalization and standardization. Sometimes normalization or standardization of data is required to make the training process much easier [15]. Standardization of the data produced more accurate model than normalization shown in Eq. (2), and so was preferred upon normalization. Only the output data was standardized as shown in Eq. (3):

$$Normalized\ Data = \frac{Data - Min}{Max - Min} \tag{2}$$

$$Standardized\ Data = \frac{Data - Mean}{Standard\ deviation} \tag{3}$$

4 Artificial Neural Network Architecture

Two different types of ANN architectures are used in this paper. The first architecture is a Multi-Layer Perceptron (MLP) ANN with 7 hidden layers and "linear" activation function. The second one consists of the same previous architecture with replacing the first two layers with two 1D CNN layers. Each ANN architecture is evaluated in two aspects: loss and accuracy. The architecture that reaches minimum loss and maximum accuracy is chosen to be the best neural architecture that is capable of generating the most descriptive model for the manipulator. The two architectures were constructed using keras a neural network library in python. About 70% of the data obtained was fed to each neural net, while 15% were used as validation data, and other 15% were used as testing data. Figure 6 illustrates the architecture of the neural network with CNN and the number of neurons in each layer. The activation functions in all layers were "Tanh", the optimizer function was 'Adam' and the loss function was mean square error. The learning rate during the training was 1×10^{-3}.

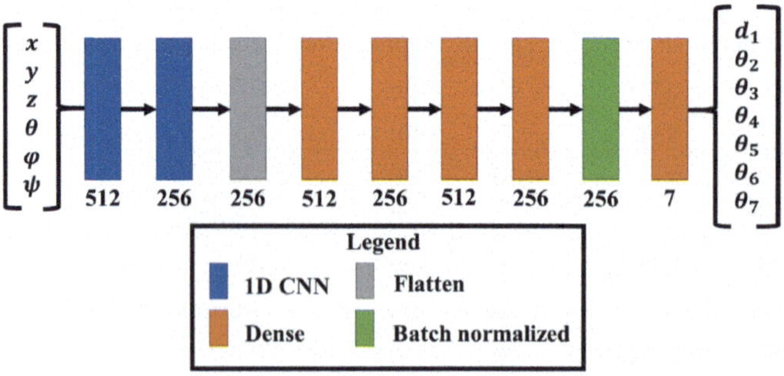

Fig. 6. CNN architecture

5 Simulation Results and Discussion

As mentioned before, the accuracy of the CNN with and without data scaling is illustrated in Fig. 7 where two methods of data scaling were used: normalization and standardization. The accuracy of the standardized training data was found to be 92.16% and that of the normalized training data was 91.68%. Whereas the accuracy of non-scaled training data was only 85%. The network that produces the best model for the serial manipulator is shown in Fig. 8. The results of this ANN is compared with results of another ANN that consists only of MLP 2NN with 7 hidden layers. The accuracy and loss of the training data in the CNN-based network are 92.16% and 0.346 respectively. However, they were 91.6% and 0.027 respectively in the MLP-based network as shown in Fig. 9. Both models were generated after 100 epochs with a batch size of 10000 and a learning rate of 1×10^{-3}.

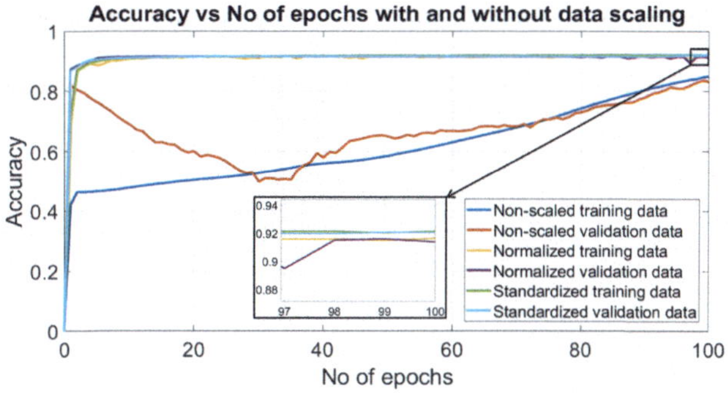

Fig. 7. Comparison between results of data with and without scaling

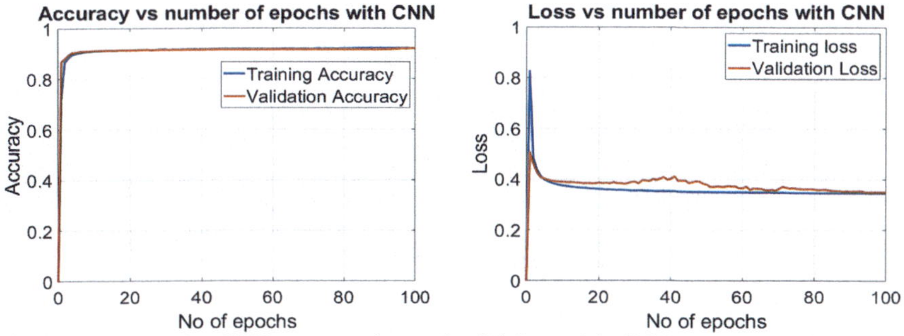

(a) Accuracy with CNN vs number of epochs

(b) loss with CNN vs number of epochs

Fig. 8. Accuracy and loss of the neural network with CNN

6 CNN Validation

To validate the accuracy of the predicted data from the CNN, a path of 4 points was fed to the CNN as the desired poses to obtain the required angles. The predicted angles were fed to the forward kinematics model in MATLAB to visualize the position of the end effector. Afterwards, a comparison between the generated positions and the desired positions is done as shown in Fig. 10. According to Fig. 10, the two paths are compared in four-point path, and the error between each two points in the x-axis and y-axis is shown in Fig. 11. The values of the error range between 14.8 mm to 89.3 mm in the x-axis, and 7 mm to 75 mm in the y-axis proving the 92.16% accuracy of the network.

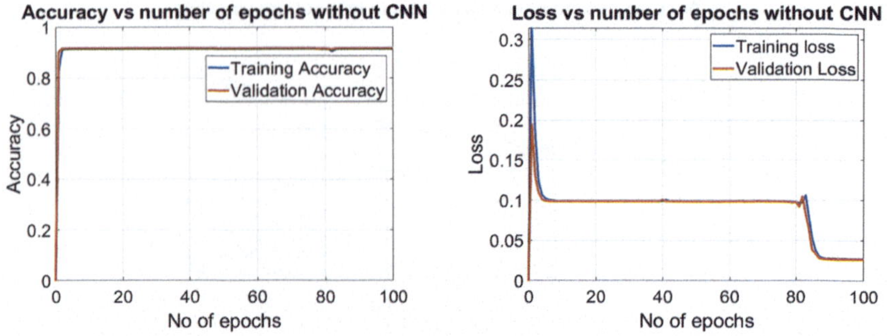

(a) Accuracy without CNN vs number of epochs (b) loss without CNN vs number of epochs

Fig. 9. Accuracy and loss of the neural network without CNN

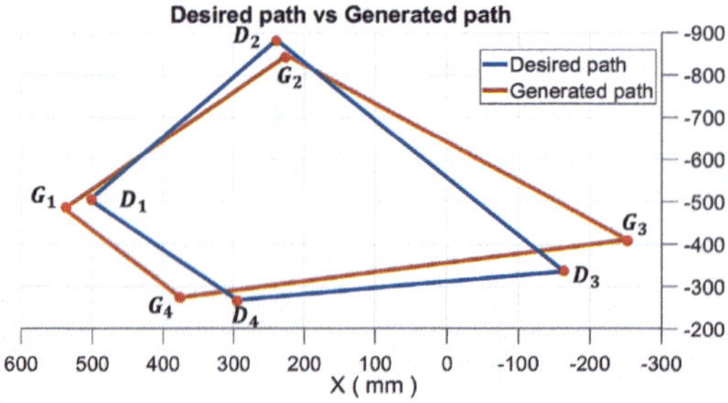

Fig. 10. Desired path vs generated path from the CNN

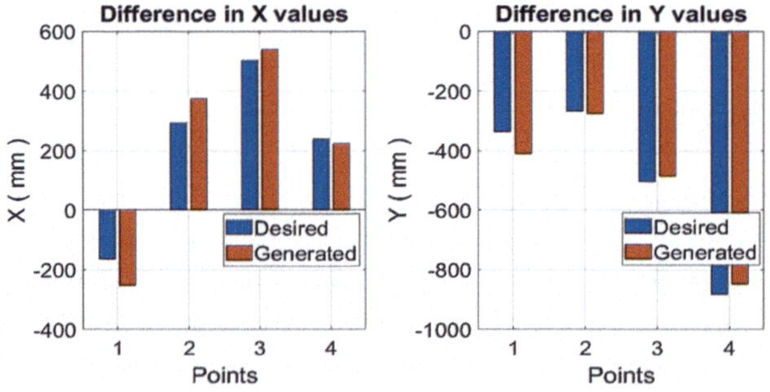

Fig. 11. Error values between the desired path and generated path

7 Conclusion

In this paper, the kinematic analysis of a 7-DOF manipulator, consisting of one prismatic and six revolute joints, using artificial neural networks is presented. The manipulator is designed on SolidWorks and simulated in moveit environment to extract the data of its workspace. The data is reformed and manipulated to be fed into the neural network architecture to be trained to produce a strong model to be later used in order to control the manipulator. Therefore, various combinations of neural network architectures were evaluated and compared. The most accurate architecture combined CNN connected to traditional ANN with dense layers, and it produced an accuracy of 92.16%. For future work, more complex ANN, that combines different neural architectures, will be used with much more layers and neurons. After obtaining the model from the artificial neural network, it will be used to control the real hardware system.

Acknowledgment. The authors would like to thank Eng. Bahaa Ashraf, Eng. Ahmed Tarek, Eng. Mahmoud Bakr, Eng. Lotfy Monir from Nile University for their design of the 7-DOF serial manipulator, and Eng. MennaAllah Soliman, teaching assistant in Mechatronics Engineering department at Nile University, for her great help and support in this work.

References

1. Almusawi, A.R.J., Dülger, L.C., Kapucu, S.: A new artificial neural network approach in solving inverse kinematics of robotic arm (denso vp6242). Comput. Intell. Neurosci. **2016**, 1–10 (2016). https://doi.org/10.1155/2016/5720163
2. Chitta, S., Sucan, I., Cousins, S.: Moveit! [ros topics]. IEEE Robot. Autom. Mag. **19**(1), 18–19 (2012). https://doi.org/10.1109/mra.2011.2181749
3. Dash, K., Choudhury, B., Khuntia, A., Biswal, B.: A neural network based inverse kinematic problem. In: 2011 IEEE Recent Advances in Intelligent Computational Systems (2011). https://doi.org/10.1109/raics.2011.6069357
4. Daya, B., Khawandi, S., Akoum, M.: Applying neural network architecture for inverse kinematics problem in robotics. J. Softw. Eng. Appl. **03**(03), 230–239 (2010). https://doi.org/10.4236/jsea.2010.33028
5. Duka, A.V.: Neural network based inverse kinematics solution for trajectory tracking of a robotic arm. Procedia Technol. **12**, 20–27 (2014). https://doi.org/10.1016/j.protcy.2013.12.451
6. Farzan, S., DeSouza, G.N.: From D-H to inverse kinematics: a fast numerical solution for general robotic manipulators using parallel processing. In: 2013 IEEE/RSJ International Conference on Intelligent Robots and Systems (2013). https://doi.org/10.1109/iros.2013.6696709
7. Hasan, A.T., Hamouda, A., Ismail, N., Al-Assadi, H.: An adaptive-learning algorithm to solve the inverse kinematics problem of a 6 D.O.F serial robot manipulator. Adv. Eng. Softw. **37**(7), 432–438 (2006). https://doi.org/10.1016/j.advengsoft.2005.09.010
8. Köker, R.: A genetic algorithm approach to a neural-network-based inverse kinematics solution of robotic manipulators based on error minimization. Inf. Sci. **222**, 528–543 (2013). https://doi.org/10.1016/j.ins.2012.07.051
9. Nagata, F., Kishimoto, S., Kurita, S., Otsuka, A., Watanabe, K.: Neural network-based inverse kinematics for an industrial robot and its learning method. In: The Proceedings of the 4th International Conference on Industrial Application Engineering 2016 (2016). https://doi.org/10.12792/iciae2016.015
10. Qian, W., Xia, Z., Xiong, J., Gan, Y., Guo, Y., Weng, S., Deng, H., Hu, Y., Zhang, J.: Manipulation task simulation using ROS and Gazebo. In: 2014 IEEE International Conference on Robotics and Biomimetics (ROBIO 2014) (2014). https://doi.org/10.1109/robio.2014.7090732

11. Singh, A., Singla, A., Soni, S.: D-H parameters augmented with dummy frames for serial manipulators containing spatial links. In: The 23rd IEEE International Symposium on Robot and Human Interactive Communication (2014). https://doi.org/10.1109/roman.2014.6926379
12. Srisuk, P., Sento, A., Kitjaidure, Y.: Inverse kinematics solution using neural networks from forward kinematics equations. In: 2017 9th International Conference on Knowledge and Smart Technology (KST) (2017). https://doi.org/10.1109/kst.2017.7886084
13. Toshani, H., Farrokhi, M.: Kinematic control of a seven DOF robot manipulator with joint limits and obstacle avoidance using neural networks. In: The 2nd International Conference on Control, Instrumentation and Automation (2011). https://doi.org/10.1109/icciautom.2011.6356794
14. Wang, J., Li, Y., Zhao, X.: Inverse kinematics and control of a 7-DOF redundant manipulator based on the closed-loop algorithm. Int. J. Adv. Robot. Syst. **7**(4), 37 (2010). https://doi.org/10.5772/10495
15. Wilamowski, B.: Neural network architectures and learning algorithms. IEEE Ind. Electron. Mag. **3**(4), 56–63 (2009). https://doi.org/10.1109/mie.2009.934790
16. Zhou, Y., Tang, W., Zhang, J.: Algorithm for multi-joint redundant robot inverse kinematics based on the Bayesian - BP neural network. In: 2008 International Conference on Intelligent Computation Technology and Automation (ICICTA) (2008). https://doi.org/10.1109/icicta.2008.406

Remote Sensing Image Classification Based on Convolutional Neural Networks

Mayar A. Shafaey[1,2(✉)], Mohammed A.-M. Salem[1,2], Maryam N. Al-Berry[1], Hala M. Ebied[1], and Mohammed F. Tolba[1]

[1] Faculty of Computers and Information Sciences, Ain Shams University, Cairo, Egypt
mayaraly22@gmail.com, fahmytolba@gmail.com, {salem,maryam_nabil,halam}@cis.asu.edu.eg

[2] Faculty of Media Engineering and Technology, German University in Cairo, Cairo, Egypt

Abstract. Nowadays, large amounts of high resolution remote-sensing images are acquired daily. However, the satellite image classification is requested for many applications such as modern city planning, agriculture and environmental monitoring. Many researchers introduce and discuss this domain but still, the sufficient and optimum degree has not been reached yet. Hence, this article focuses on evaluating the available and public remote-sensing datasets and common different techniques used for satellite image classification. In recent years, there has been an extensive popularity of supervised deep learning methods in various remote-sensing applications, such as geospatial object detection and land use scene classification. Thus, the experiments, in this article, are carried out based on HSV Color Space using one of the popular deep learning models, Convolution Neural Networks (CNNs), precisely, *AlexNet* architecture with SVM classifier on 7 different standard datasets. It has reached about $99.7 \pm 0.02\%$ of HSV color space for the high resolution dataset, *PatternNet*. Finally, a comparison with other different techniques is highlighted.

Keywords: HSV color model · Deep learning · Convolution Neural Networks (CNNs) · Remote-sensing · Satellite images

1 Introduction

The artificial satellites can scan the whole earth in less than a day. These satellites produce high spatial resolution images which are considered as large-scale ones. The analysis and classification of these remote-sensing images have been considered hot topics recently. It is very important to analyze and classify remote-sensing images in everyday applications, including urban planning, earth observation, military monitoring, and natural hazards detection [1]. For these reasons, many researchers carried out their experiments on the satellite images for the scene classification mission.

In fact, the traditional camera captures light over three wavelength bands through the visible range of red, green, and blue (RGB). In general, RGB color space is a general color space widely used in digital image display and optical instruments.

A.-E. Hassanien et al. (Eds.): AICV 2020, AISC 1153, pp. 353–361, 2020.
https://doi.org/10.1007/978-3-030-44289-7_33

However, RGB color space is not sensitive to human visual perception or statistical analysis. HSV color space is a non-linear transform from the RGB color space that can describe perceptual color relationship more accurately than the RGB color space. In this paper, HSV color space is adopted in following procedures [2, 3].

HSV color space is made by the hue (H), saturation (S) and value (V) components. Hue means the property of color such as blue, green, red, and so on. Saturation means the perceived intensity of a specific color. Value indicates brightness perception of a specific color. Thus, it can be seen that HSV color space is different from RGB color space in color variations [5]. When a color pixel-value in RGB color space is adjusted, intensities of red channel, green channel, and blue channel of this color pixel are modified. This means that color, intensity, and saturation of a pixel are involved in color variations. It is difficult to observe the color variation in complex color environment or content. However, HSV color space separates the color into hue, saturation, and value which means that the observation of color variation can be individually discriminated. According to the above descriptions of HSV color space, we can conclude that the HSV color space can describe the color in a more detailed way than RGB color space in terms of color, intensity and brightness [6].

Thus, the main objective of this paper is to present a literature review on the recent deep-learning based techniques for satellite image classification and the available training datasets. Moreover, testing results will present on multiple datasets based on HSV color space via the *AlexNet* model of the Convolution Neural Networks (CNNs).

In the next section, a quick review of common and free remote sensing images is presented. A review on recent classification approaches applied on one or some of these datasets is presented in Sect. 3. The experimental work followed by results and evaluations are presented in Sect. 4. Finally, the conclusions are highlighted in Sect. 5.

2 Remote Sensing Image Datasets

During the recent years, different research groups introduced various high-resolution remote sensing image datasets to enable machine learning based research for scene classification [7]. For evaluating different techniques in this field, the authors will firstly review some available datasets in this sub-section. Table 1, shows the number of scene classes, images per each class, total images, size of images, and spatial resolution value for each dataset.

Table 1. Comparison between the different remote-sensing datasets proposed

Data set	Scene classes	Images/class	Total images	Spatial resolution	Image sizes
AID [8]	30	200–400	10000	High	600 × 600
Patter Net [9]	38	800	30400	Up to 0.8	256 × 256
UC-Merced Land Use [10]	21	100	2100	0.3	256 × 256
WHU-RS19 [11]	19	~50	1005	Up to 0.5	600 × 600
SIRI-WHU [12]	12	200	2400	2	200 × 200
RSSCN7 [13]	7	400	2800	–	400 × 400
NWPU-RESISC45 [14]	45	700	31500	~30–0.2	256 × 256

These datasets are mostly produced by Google® Earth Engine. They cover a widespread of areas, such as, parking, runway, agricultural, airplane, baseball diamond, beach, buildings, chaparral, overpass, forest, freeway, harbor, sparse residential, intersection, river, mobile home park, storage tanks, and so on, as in Fig. 1.

Fig. 1. 21 representative classes of the UC-Merced Land-Use dataset [10]

Indeed, remote-sensing image quality is affected by the Spatial Resolution (SR) measure. SR describes how much details in a photographic image are visible to the human eye. It is usually expressed by Ground Sample Distance (GSD). The GSD is defined as the distance between pixel centers measured from air or space on the ground. For instance, in an image with a 1 m GSD, adjacent pixels image positions are 1 m apart on the ground. Figure 2 shows the degradation of the image resolution from left to right. The lowest value points to the best quality and vice versa. So, the classification accuracy is affected by the image quality [15].

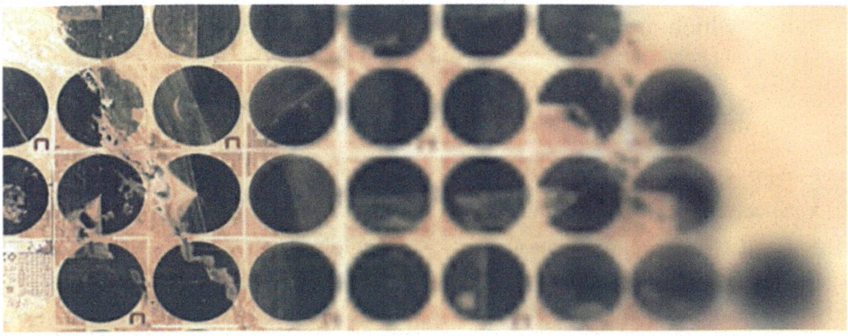

Fig. 2. Image resolution in meters (5–10–15–20–30–60–100–120–250) [15]

3 CNNs Applied for Remote Sensing Image Classification

With respect to the vast publications of remote-sensing images for the task of scene classification, the existing scene classification methods can be categorized into three main categories: manual feature based methods, unsupervised classification methods and supervised learning methods, especially, a deep learning method which made by *Hinton and Salakhutdinov* [24]. There exists different numbers of deep learning models, such as deep belief nets (DBN) [21,25,26], deep Boltzmann machines (DBM) [21], stacked auto-encoder (SAE) [26], Convolutional Neural Networks (CNNs) [27].

The CNN itself has different models like: *AlexNet, GoogleNet, ResNet, VGGNet, CaffeNet*… etc. [16]. Limited by the space, a short description of *AlexNet* architecture is given. The network consists of 25 layers: 5 convolution layers, 5 max-pooling layers, 5 dropout layers, and 3 fully connected layers, as shown in Fig. 3. It is trained on *ImageNet* data [10], which contains over 15 million annotated images from a total of over 22,000 categories and uses the rectified layer (ReLU) for the nonlinearity functions [17].

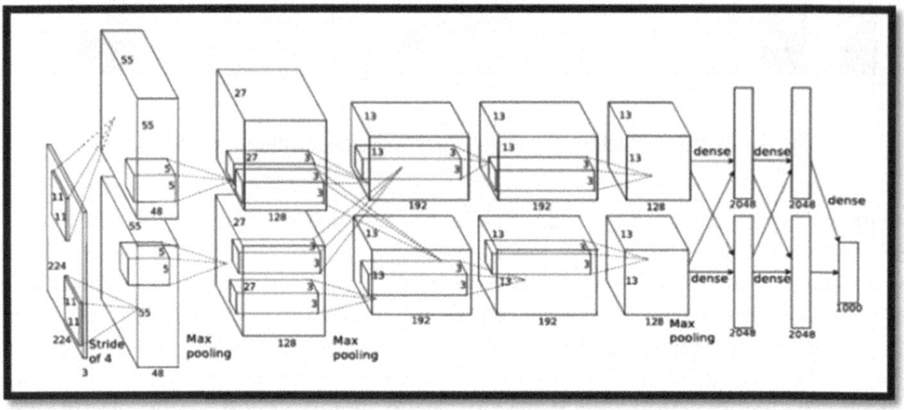

Fig. 3. *ImageNet* classification with *AlexNet* CNN [17]

Table 2 represents the results published by authors who used the CNN models in their experiments for large scale image scene classification and gave the outstanding accuracy values which are clues for the power of CNN learning model.

Table 2. Recent work on CNN for remote sensing classification

Ref.	Year	Application	Method	Accuracy
[18]	2015	Spectral land use classification	Deep CNN	93.48%
[19]	2015	Land Use RS classification	GoogleNet, and CaffeNet	97%, and 95.48%
[20]	2016	RS scene classification	Large patch CNN	Effective results
[21]	2016	Large scale RS image classification	CNN	92.4%
[22]	2019	RS scene classification	AlexNet	94%

In this article, the authors mainly train the huge multi-layer networks for classifying large scale input images via deep learning method, especially CNNs. As mentioned in the table above, the CNNs have proven their power in the classification task. Therefore, the *alexNet* was built from scratch with its 25 layers using ~80,000 images. Table 3 shows the layers of the network architecture with the chosen parameters.

Table 3. *AlexNet* architecture

Name	Description	Parameters
data	Image input	227 × 227 × 3
conv1	Convolution	11 × 11 × 3 window, stride [4 4], padding [0 0 0 0]
norm1	Cross normalization	5 channels per element
pool1	Max pooling	3 × 3, stride [2 2], pad [0 0 0 0]
conv2	Convolution	5 × 5 × 48 window, stride [1 1], padding [2 2 2 2]
norm2	Cross normalization	5 channels per element
pool2	Max pooling	3 × 3, stride [2 2], pad [0 0 0 0]
conv3	Convolution	3 × 3 × 256 window, stride [1 1] and padding [1 1 1 1]
conv4	Convolution	3 × 3 × 192 window, stride [1 1] and padding [1 1 1 1]
conv5	Convolution	3 × 3 × 192 window, stride [1 1] and padding [1 1 1 1]
pool5	Max pooling	3 × 3, stride [2 2], pad [0 0 0 0]
fc6	Fully connected	4096 fully connected layer
drop6	Dropout	50% dropout
fc7	Fully connected	4096 fully connected layer
drop7	Dropout	50% dropout
fc8	Fully connected	50 fully connected layer
prob	Softmax	Softmax
output	Classification output	50 classes

4 Experimental Results

The *AlextNet* CNN model was applied for large scale image classification. In this section, the experiment's steps are described; i.e., software, hardware specification, results, comments, and comparisons.

4.1 Experiments Setting

The experiments were conducted on a PC with a processor: Intel® Core™ i7-2670QM CPU @ 2.20 GHz – 8 GB RAM and using Matlab® software. The *alexnet* network was trained on different remote sensing datasets, which mentioned in Table 2, – ~80,000 images - and was able to classify images into 50 object categories. These experiments required the use of the Neural Network Toolbox™ Model for *convNet*.

Initially, the input image dimension was resized to 227 × 227 as a required input size for the *alexnet* model. Then, the features were extracted using the CNN, and passed to a trained multiclass SVM classifier to get the labels. The ratio of training data was 0.8, while the rest was used for testing. Finally, the classification results were computed by computing the summation of the main diagonal of the confusion matrix divided by the diagonal elements number.

The experiment was carried out on 7 different datasets, mentioned in Table 1. The first trial was for RGB color space. The second one was for HSV space on the same dataset. Table 4 gives a comparison between the RGB and HSV spaces with respect to the classification accuracy. The results were obtained using 7 datasets.

Table 4. Comparison between classification accuracies based on SVM for different datasets in RGB vs HSV color spaces

Model	Dataset	Classification accuracy %
AlexNet using RGB	WHU-RS19	95.7 ± 0.05
	UC-Merced Land Use	94.2 ± 0.04
	SIRI-WHU	90.2 ± 0.07
	RSSCN7	85.6 ± 0.11
	AID	85.2 ± 0.02
	PatterNet	98 ± 0.02
	NWPU-RESISC45	79.2 ± 0.01
AlexNet using HSV	WHU-RS19	98.4 ± 0.05
	UC-Merced Land Use	96.6 ± 0.04
	SIRI-WHU	94 ± 0.07
	RSSCN7	90.1 ± 0.12
	AID	89.3 ± 0.02
	PatterNet	**99.97 ± 0.02**
	NWPU-RESISC45	86.5 ± 0.01

4.2 Evaluation and Discussions

By comparing our work with other published results, such as [10], the authors reached 94.2% using SVM classifier based on 3 labeled bands (RGB), while, in our experiments we reached 96.6% using HSV color space. The interpretation of this high accuracy is the power of CNN as a feature extractor model

The dark black number has the max accuracy value of *patternNet* dataset which indicates to the *SVM* classifier. As mentioned in before, the low value of spatial resolution points to better quality. So, if we are comparing the accuracy of *patternNet* and *NWPU-RESISC45* datasets, they have approximately the same number of images and the same image size, but obviously, the accuracy of all networks on *patternNet* exceeds by far the accuracy of *NWPU-RESISC45*. The reason is the spatial resolution of each one. For *patternNet*, the max spatial resolution is 0.8 while for *NWPU-RESISC45* is up to 30 m.

On one hand, by comparing the traditional handcrafted features versus deep features learning based method, the conventional methods consume a high mental thinking and skills, while the deep learning features are learned automatically from data through deep layers of neural networks architecture. This is the key advantage of deep features learning based methods. On the other hand, in comparison with unsupervised feature learning methods, such as, Auto-encoder, the deep learning models consists of multiple processing layers. It can learn more powerfully and is more applicable for large scale image scene classification. The deep learning architectures performance is similar to the human brain, in which each layer uses the information from the previous layer to learn intensely and precisely [30].

There is no doubt that increasing the number of bands will affect the classification accuracy positively. This can be noted for future work. Another comparison is between the achieved results, in this article, versus the results which are mentioned in Table 2. The reader may note that the accuracy of our experiment is slightly less than others. This is due to the difference of the used dataset. The presented results didn't use the *PaviaU* hyper-spectral dataset.

5 Conclusions

Nowadays, there are two hot research topics that are the automated target detection, and large scale remotely sensed image classification. Therefore, this paper gives a quick review of the common and free remote-sensing datasets, to enable the community to develop the large scale image scene classification task. Then, a short summary of different methods used for this task is given. Finally, the CNN deep learning method *AlexNet* is applied on 7 different datasets, through two experiments. The first experiment is based on RGB color space, and the second is based on the HSV space, and achieved a high classification accuracy using the CNN as feature extractor. The results were reported and compared to state-of-the-art work and as a baseline for future research. Deep learning methods can guarantee better feature representations for the remote-sensing related works, and there is a bright prospect of sight for more and more researchers dedicated to learning better features for the target detection and scene classification tasks.

References

1. Zhang, L., Xia, G., Wu, T., Lin, L., Tai, X.: Deep learning for remote sensing image understanding. J. Sens. **2016**, 1–2 (2016)
2. Yaniv, O., Isaac, A., Vladimir, F., Daniel, G., Adrian, S.: Compressive sensing hyperspectral imaging by spectral multiplexing with liquid crystal. J. Imaging **5**(1), 3 (2019)
3. Roy, B., Lawrence, T., Mahdi, N.: Spectral imaging using a commercial colour-filter array digital camera. In: The Fourteenth Triennial ICOM-CC Meeting, pp. 743–750 (2005)
4. Adam, C.W., Vincent, G.A., Everett, A.H.: Unmanned aircraft systems in remote sensing and scientific research: classification and considerations of use. Remote Sens. **4**(6), 1671–1692 (2012)
5. Chang, J.-D., Yu, S.-S., Chen, H.-H., Chwei-Shyong, T.: HSV-based color texture image classification using wavelet transform and motif patterns. J. Comput. **20**(4), 63–69 (2010)

6. Elhadi, A., Onisimo, M., Denis, R.: Multispectral and hyperspectral remote sensing for identification and mapping of wetland vegetation: a review. Wetlands Ecol. Manag. **18**(3), 281–296 (2010)
7. Shaheera, R., Nicolas, D.: A split-and-merge approach for hyperspectral band selection. IEEE Geosci. Remote Sens. Lett. **14**(8), 1378–1382 (2017)
8. AID Dataset. http://www.lmars.whu.edu.cn/xia/AID-project.html. Accessed 16 Feb 2018
9. PatternNet Dataset. https://sites.google.com/view/zhouwx/dataset?authuser=0. Accessed 16 Feb 2018
10. UC-Merceed Land Use Dataset. http://weegee.vision.ucmerced.edu/datasets/landuse.html. Accessed 16 Feb 2018
11. WHU-RS19 Dataset. https://www.mediafire.com/file/o88sj9zbv7jau26/RS19.rar/file. Accessed 16 Mar 2020
12. SIRI-WHU Dataset. http://www.lmars.whu.edu.cn/prof_web/zhongyanfei/e-code.html. Accessed 16 Feb 2018
13. RSSCN7 Dataset. https://www.mediafire.com/file/bqkqee7bwecx96d/RSSCN7.rar/file. Accessed 16 Mar 2020
14. NWPU-RESISC45 Dataset. https://www.mediafire.com/file/5xtt5rkr64ie7nh/NWPU-RESISC45.rar/file. Accessed 16 Mar 2020
15. Zhou, Z., Edoardo, P., Melba, M.C., James, C.T.: An active learning framework for hyperspectral image classification using hierarchical segmentation. IEEE J. Sel. Top. Appl. Earth Obs. Remote Sens. **9**(2), 640–654 (2016)
16. Salem, M.A.-M., Appel, M., Winkler, F., Meffert, B.: FPGA-based smart camera for 3D wavelet-based image segmentation. In: 2nd ACM/IEEE International Conference on Distributed Smart Cameras, ICDSC, pp. 1–8 (2008). https://doi.org/10.1109/icdsc.2008.4635720
17. Salem, M.A.-M., Alaa, A., Alaa, S., Marwa, S.: Recent survey on medical image segmentation. In: Computer Vision: Concepts, Methodologies, Tools, and Applications, pp. 129–169. IGI Global (2018). https://doi.org/10.4018/978-1-5225-5204-8.ch006
18. Zhao, B., Zhong, Y., Zhang, L.: A spectral–structural bag-of-features scene classifier for very high spatial resolution remote sensing imagery. ISPRS J. Photogramm. Remote Sens. **116**, 73–85 (2016)
19. Olshausen, B.A., Field, D.J.: Sparse coding with an over complete basis set: a strategy employed by V1? Vis. Res. **37**(23), 3311–3325 (1997)
20. Simonyan, K., Zisserman, A.: Very deep convolutional networks for large-scale image recognition. In: Proceedings of International Conference on Learning Representations, pp. 1–13 (2015)
21. Salakhutdinov, R., Hinton, G.: An efficient learning procedure for deep Boltzmann machines. Neural Comput. **24**(8), 1967–2006 (2012)
22. Zhong, Y., Fei, F., Zhang, L.: Large patch convolutional neural networks for the scene classification of high spatial resolution imagery. Appl. Remote Sens. **10**(2), 025006 (2016)
23. Shafaey, M.A., Salem, M.A.-M., Ebied, H.M., Al-Berry, M., Tolba, M.F.: Deep learning for satellite image classification. In: Proceedings of the International Conference on Advanced Intelligent Systems and Informatics, vol. 4, pp. 383–391 (2019)
24. Hinton, G., Salakhutdinov, R.: Reducing the dimensionality of data with neural networks. Science **313**(5786), 504–507 (2006)
25. Hinton, G., Osindero, S., Teh, Y.-W.: A fast learning algorithm for deep belief nets. Neural Comput. **18**(7), 1527–1554 (2006)

26. Vincent, P., Larochelle, H., Lajoie, I., Bengio, Y., Manzagol, P.-A.: Stacked denoising autoencoders: learning useful representations in a deep network with a local denoising criterion. Mach. Learn. Res. **11**, 3371–3408 (2010)
27. Sermanet, P., Eigen, D., Zhang, X., Mathieu, M., Fergus, R., LeCun, Y.: OverFeat: integrated recognition, localization and detection using convolutional networks. In: Proceedings of the International Conference on Learning Representations, pp. 1–16 (2014)

Combination of Convolutional and Recurrent Neural Networks for Heartbeat Classification

Abdelrahman M. Shaker(✉), Manal Tantawi(✉), Howida A. Shedeed(✉), and Mohamed F. Tolba(✉)

Faculty of Computer and Information Sciences, Ain Shams University, Cairo, Egypt
{Abdelrahman.shaker,manalmt,dr_howida, fahmytolba}@cis.asu.edu.eg

Abstract. Electrocardiogram (ECG) plays an essential role in the medical field, it records the electrical activity of the heart over time and provides information about the heart condition. Hence, the cardiologist uses it to detect the abnormalities of the heart and to diagnose the heart diseases. Convolutional Neural Networks (CNNs) have proven their ability in extracting the most important features, Long Short-Term Memory (LSTM) has the capabilities of learning the temporal dependencies between the sequential data. In this paper, a novel method based on the combination of CNN and LSTM is proposed to classify 15 classes of the MIT-BIH dataset automatically without any hand-engineering feature extraction methods. The proposed method consists of data filtering, dynamic technique for heartbeat segmentation, and CNN-LSTM model consists of 12 layers.

Our experimental results of the proposed method achieved promising overall accuracy of 98.16% in classification between 15 classes of the MIT-BIH dataset, which outperforms several heartbeat classification methods.

Keywords: ECG classification · Heart diseases · Deep learning · Convolution Neural Networks (CNN) · Long Short-Term Memory (LSTM)

1 Introduction

Heart disease is one of the most common health problems in the world and the leading global cause of death. According to the World Health Organization (WHO), approximately 17.9 million die each year from Cardiovascular Diseases (CVDs), an estimated 31% of all deaths worldwide [1], and by 2030, almost 23.6 million people will die from CVDs [2]. There are three main categories of CVDs; electrical (arrhythmias and abnormal heartbeats due to cardiac disorders), structural (the diseases of the heart muscle), and circulatory (diseases of the blood vessels).

Arrhythmias can be identified by analyzing the abnormal (fast, slow, or irregular) heartbeats and can be grouped into two groups (life-threating and non-life-threating). The diagnosis of arrhythmias is based on identifying the abnormal heartbeats from the Electrocardiogram (ECG), which is the process of tracing the heart's electrical activity over a period of time by placing a collection of electrodes on the surface of the body

A.-E. Hassanien et al. (Eds.): AICV 2020, AISC 1153, pp. 362–371, 2020.
https://doi.org/10.1007/978-3-030-44289-7_34

such as the chest, neck, and arms. Hence, the placed electrodes are responsible for detecting the electrical changes in the heart. ECG signals consist of several features, including P waves, QRS complexes, and T waves. The atria contractions are shown in the P wave, while QRS complex demonstrates the ventricular contractions, and the electrical activity produced during the recharge of the ventricles for the next contractions is reflected by the T wave [3].

Arrhythmias diagnosis using ECG records requires the classification of heartbeats. Therefore, the heart rhythm category can be determined by identifying the classes of consecutive heartbeats [4]. The screening of cardiac arrhythmias requires detailed study of the ECG records by the cardiologists, this process is a time-consuming and too difficult. Hence, the automation process of arrhythmias identification and ECG analysis is crucial in the medical field, many Computer-aided diagnosis (CAD) models have been proposed in the last decade to automate this process.

In the literature, different conventional approaches—which require hand-engineering features extraction methods such as Wavelet and Fourier transforms—have been proposed for automatic heartbeats classification [5–7].

Recently, deep learning has got a lot of attention because its models have achieved remarkable success and promising results that were not possible before. It has demonstrated state-of-the-art performances in many fields, and sometimes exceeded the human-level performance. The techniques of deep learning, specially Convolutional Neural Networks (CNN), have been utilized in several fields such as medical diagnosis [8], bioinformatics [9], computer vision [10]. They also have been applied to several tasks such as brain tumor classification [11], breast cancer classification [12], MRI analysis [13] and image classification [14].

End-to-end learning is one of the main advantages of the CNN [15], where feature extraction, feature reduction, and classification stages are performed together with no need to hand-engineering features unlike the conventional approaches. Besides the CNN, Long Term-Short Memory (LSTM) [16]—which is a Recurrent Neural Network (RNN) architecture—is widely used in the field of deep learning for time series analysis. LSTM has been applied to several tasks such as Natural Language Processing (NLP) [17] and speech recognition [18].

In this paper, an end-to-end solution based on the combination of CNN and LSTM is proposed. Firstly, preprocessing step is needed to reduce the noise from the ECG records and dynamically segment the filtered records into heartbeats. The proposed model is able to extract the most relevant features automatically and classify them into 15 classes of the MIT-BIH arrhythmia dataset with superior performance than other studies.

The remainder of this paper is organized as follows: The related work is provided in section two, while section three proposes the architecture and methodology in details. Section four represents the achieved results. Finally, the conclusion and the future work are proposed in the fifth section.

2 Related Work

The studies in the literature are following two approaches; conventional approach and deep learning approach. The conventional approach of ECG heartbeat classification is to filter the input signal to reduce the noise and apply a segmentation technique to divide the ECG records into separate heartbeats. After that, develop an algorithm to extract the most important relevant features from the heartbeats and then choose the optimal features by using reduction techniques. Finally, choose an appropriate classifier to be used in the classification stage. Figure 1 shows the steps of this approach.

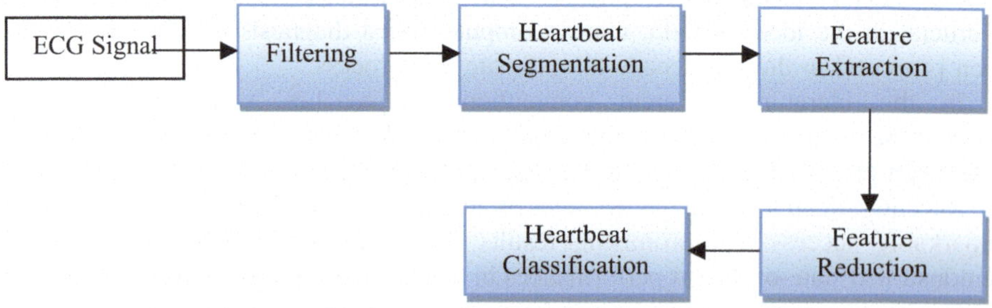

Fig. 1. The main steps for ECG heartbeat classification using conventional approach.

Undoubtedly, there are many studies have been done in the literature of such way for ECG heartbeat classification [19–25].

Sahoo et al. [19] used Discrete Wavelet Transform (DWT) to extract the features and used Support Vector Machines (SVM) to classify between four classes from the MIT-BIH arrhythmia dataset, they achieved an accuracy of 98.39%. While in [20], Independent Component Analysis (ICA) was used to extract the features and Neural Networks were utilized to classify different eight classes with an accuracy of 98.71%.

In [21], the authors combined the QRS wave with DWT to extract the features and ICA, PCA, and LDA were used to reduce the extracted features with the use of SVM and NN for classification between 15 classes mapped to five categories (N, S, V, F, Q). They achieved an accuracy of 99.28%.

Martis et al. [22] considered five different classes with comparison for different feature extraction methods such as DWT and PCA, they achieved an accuracy of 98.11% using PCA, while SVM was used in the classification stage. On the other hand, the same five classes have been considered by Yazdanian et al. [23] and they achieved an accuracy of 96.67% using wavelet transforms, morphological and time-domain features, SVM is utilized in the classification stage.

In [24], they have proposed a two-stage hierarchical classification approach to classify 15 classes from MIT-BIH arrythmia dataset. The morphological features are extracted using DWT and then reduced by using PCA algorithm. Therefore, the reduced features are concatenated with four RR features and fed into SVM classifier, they have achieved an overall accuracy of 94.94%.

Acharya et al. [26] have proposed a CNN consists of nine layers to classify the five main categories (N, S, V, F, Q) in the MIT-BIH arrhythmia dataset. They augmented the heartbeats to balance the dataset and achieved an accuracy of 94.03%, while Zhang [27] proposed six-layer CNN to classify between five classes of the same dataset, an overall accuracy of 97.50% is achieved.

In [7], they have explored the use of DWT layer with LSTM network to classify five types of heartbeats (Normal, Ventricular Premature Contraction, Paced Beat, Left Bundle Branch Block, and Right Bundle Branch Block) obtained from MIT-BIH arrhythmia dataset, an accuracy of 99.39% is obtained.

To summarize, the following points can be concluded from the studies in the literature: (1) The most commonly used dataset in the literature is the MIT-BIH. (2) In the conventional approach, PCA, ICA and DWT are the most widely used in the feature extraction and reduction stages, and SVM is the dominant method in the classification stage. (3) In the deep learning approach, CNN is the commonly used network with different architectures. LSTM—which has been preferred in sequential data analysis—is also used for ECG signal classification. (4) Most of the studies consider only few classes.

The aim of this paper is to propose an automated ECG heartbeat classification system using a combination of CNN and LSTM to consider both morphological and temporal features to achieve superior classification performance. 15 classes are considered and data from lead 1 only is utilized to reduce the computational time.

3 Methodology

This section discusses the techniques used for preprocessing and introduces the proposed method for the classification stage in details.

3.1 Preprocessing Stage

Data filtering is utilized to reduce the noise from the ECG records as well as improving the quality of the signals and increasing the Signal-to-Noise Ratio (SNR). Butterworth bandpass filter with bandwidth 0.5 Hz–45 Hz is applied because that range includes the most valuable information of the ECG [28]. Thereafter, heartbeats are dynamically segmented using the locations of the R peaks which are associated with the dataset. In the literature, there are many studies that applied fixed segmentation techniques due to the difficulty of detecting the beginning and the ending of the heartbeats. However, such assumption doesn't consider the heart rate variability, which is not reliable. Dynamic heartbeat segmentation technique is utilized to overcome the variability of the heartrate as proposed in [24]. The dynamic technique considers the number of samples before and after each R peak location based on the distance between the current R peak and the previous R peak (RR previous interval) as well as the distance between the current and the next R peaks (RR next interval). Subsequently, the number of samples of the larger interval is divided into two parts, the first part is considered before the R peak while the other part is considered after the R peak to ensure that the three main waves are considered in an invariant way to the heart rate variability. Finally, the amplitude of each heartbeat is normalized between [0-1] and the heartbeat is resampled to have 300 samples. Figure 2 shows the result of the preprocessing stage.

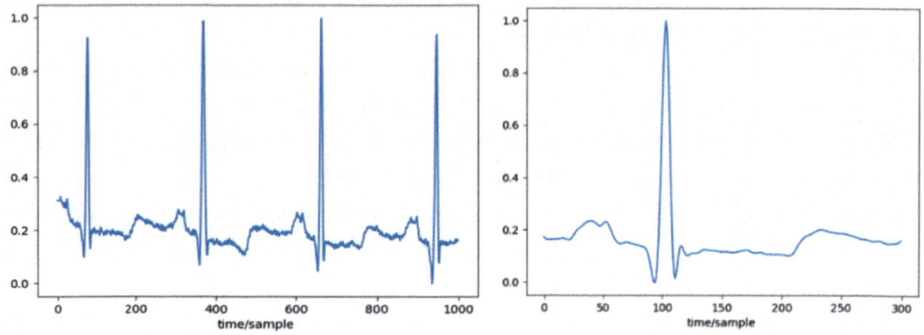

(a) ECG record (b) Filtered and segmented heartbeat

Fig. 2. The results of the preprocessing stage.

3.2 Classification Stage

Convolution Neural Networks (CNN) are known for their ability to extract the most important features from the input automatically and classify these features to one of the target classes. It combines feature extraction and classification stages into a single architecture. The first part of the CNN consists of convolutional and pooling layers, the convolutional layers are responsible to extract the useful features from the ECG signals by applying a convolution operation between the input and convolutional filters. Each filter is a matrix of trainable values, called weights, that are responsible to detect specific features and ReLU activation function [29] is utilized after the convolution operation. Pooling layers are optionally applied after the convolutional layers to speed up the training procedure and to reduce the number of trainable parameters in the network. In this study, the reduced features passed to LSTM layer before going to the classification stage.

LSTM has demonstrated their ability in learning the sequential data by considering the temporal dependencies between the data. The difference between the standard RNN and the LSTM is that the former consists of repeating modules contain a simple computation process while the latter consists of repeating modules contain interactive layers that control the information flow. Each LSTM module consists of a cell and three gates (forget, input, output). Firstly, the forget gate is responsible to pass the relevant information to the cell state and forget the irrelevant part using sigmoid layer. The input gate decides what is the information—the old information or the new information or fraction of them—that will be updated and stored in the cell state. Finally, the output gate determines which parts of the cell state are going to output after filtering it [16].

In this paper, a novel model based on the combination of CNN and LSTM is proposed to classify 15 arrhythmia classes of the MIT-BIH arrhythmia dataset. The architecture of the proposed model is shown is Fig. 3. The input is the heartbeat, which is a vector of 300x1, followed by two convolutional layers with 64 filters and kernel size of three for each one. Thereafter, max pooling layer is followed with pool size of three, followed by two convolutional layers with 128 filters and kernel size of three for each one, followed by another max pooling layer with pool size of three. LSTM layer is utilized after the pooling layer with 64 LSTM modules. After that, three fully connected layers are developed with number of neurons 128, 64, and 32 respectively. Finally, Softmax function is used in the output layer, which contains 15 neurons corresponding to the utilized 15 arrhythmia classes.

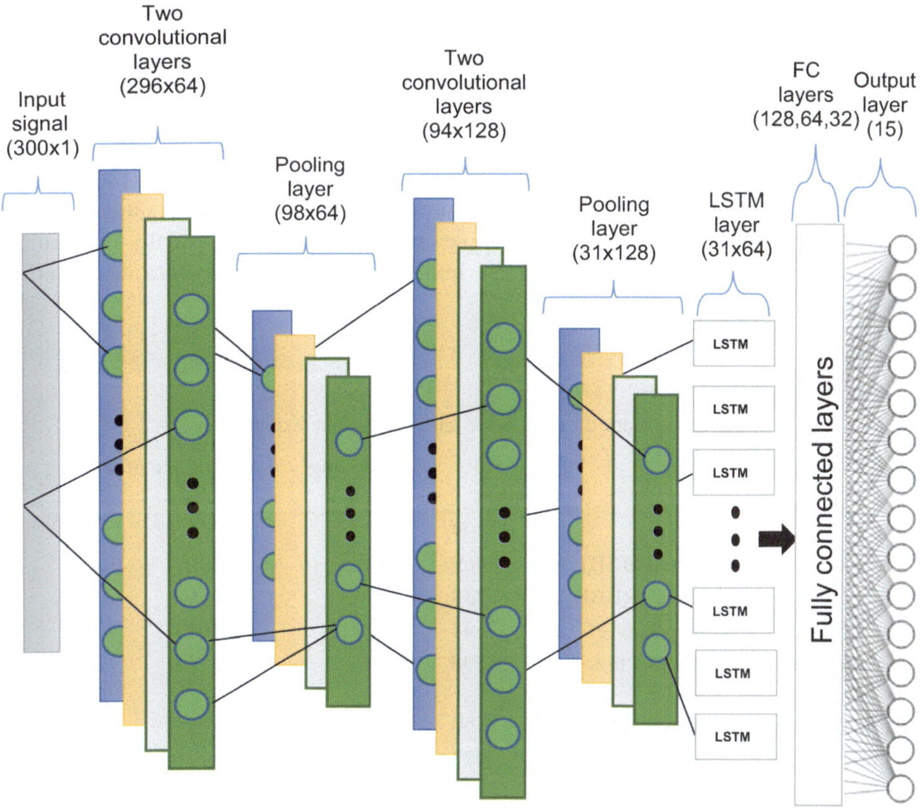

Fig. 3. Overview of the proposed CNN-LSTM architecture.

The convolutional layers are perfect in extracting features of interests from the input and produce the spatial features maps, while pooling layers are used to reduce the size of the feature maps and steed up the training. The LSTM layer helps the model to learn the temporal dependencies that present in these feature maps [16]. The final part of the proposed architecture is a series of fully connected layers, which used to predict the class label.

4 Experimental Results

4.1 Dataset

MIT-BIH arrhythmia dataset [30] is the most commonly used in the literature. It consists of 48 records obtained from 47 subjects; each record is a 30 min duration with a sampling frequency of 360 Hz.

According to ANSI/AAMI EC57: 1998 standard [31], only 44 out of 48 records are recommended and can be utilized because there are four paced records. Each record is associated with the annotations of the beats and the locations of the R peaks. These locations and annotations are used as the ground truth in the learning and in the evaluation steps. ECG data from lead 1 only is considered.

In this paper, the utilized records are segmented into heartbeats. After that, the heartbeats are divided into training and testing sets. The data division in [24, 25] has been followed for comparison sake. The training and testing percentages are not equal for the classes because the number of samples in the dataset is not equally distributed. The training ratio for each class and the number of training beats are described in Table 1.

Table 1. Data division for each class in our experiments.

Class	Total beats	Training ratio	Number of training beats
Normal beat (N)	74309	13%	9660
Left Bundle Branch block (LBBB)	8069	40%	3227
Right Bundle Branch block (RBBB)	7238	40%	2895
Atrial Premature Contraction (APC)	2544	40%	1017
Premature Ventricular Contraction (PVC)	6890	40%	2756
Aberrated Atrial Premature (AP)	150	50%	75
Ventricular Flutter Wave (VF)	472	50%	236
Fusion of Ventricular and Normal (VFN)	802	50%	401
Blocked Atrial Premature (BAP)	193	50%	96
Nodal (junctional) Escape (NE)	229	50%	114
Fusion of Paced and Normal (FPN)	982	50%	491
Ventricular Escape (VE)	106	50%	53
Nodal (junctional) Premature (NP)	83	50%	41
Atrial Escape (AE)	16	50%	8
Unclassifiable (UN)	15	50%	7
15 Classes	102098	20.6%	21077

4.2 Results

In this study, the training set is selected randomly according to the data division in Table 1 and the other 81021 beats were used as the testing set. Adam optimizer [32] is utilized to fine tune the parameters of the network, the weights were initialized with random values from uniform distribution.

The proposed model is applied to 15 classes of the MIT-BIH arrythmia dataset and data from lead 1 only is considered to reduce the computational time. The evaluation is done by measuring the achieved overall accuracy for the classes. An overall accuracy of 0.9816 (98.16%) is achieved after 40 epochs of training. The accuracy of the testing set and the loss of the network for each epoch are shown in Fig. 4.

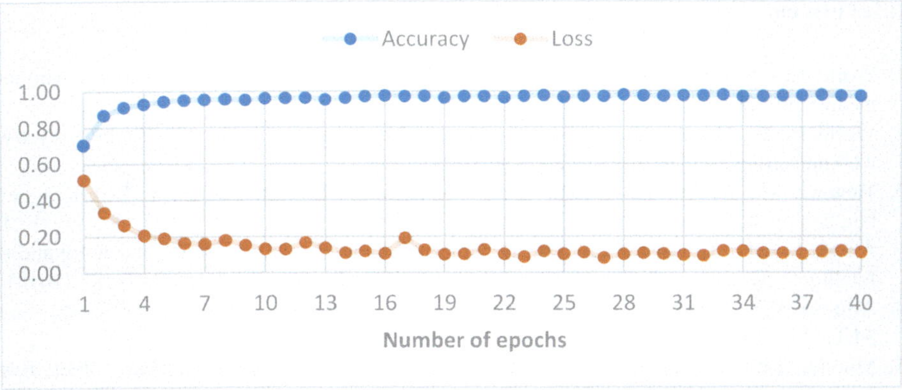

Fig. 4. The achieved accuracy and the loss of the network for each epoch.

The comparison between the existing studies and our work is given in Table 2. It demonstrates that the proposed method is applied to more classes than other studies. Moreover, superior overall accuracy is achieved compared to the accuracies of the published studies.

Table 2. Comparison of the existing studies and our proposed work.

Study	Number of classes	Feature set	Classifier	Accuracy
Sahoo et al. [19]	4	DWT	SVM	98.39%
Yu, Chou et al. [20]	8	ICA	NN	98.71%
Martis et al. [21]	5	DWT + ICA	PNN	99.28%
Martis et al. [22]	5	PCA	SVM	98.11%
Yazdanian et al. [23]	5	Wavelet	SVM	96.67%
Acharya et al. [26]	5	End-to-end	1D-CNN	94.03%
Zhang et al. [27]	5	End-to-end	1D-CNN	97.50%
Yildirim et al. [7]	5	End-to-end	LSTM	99.25%
El-saadawy et al. [24]	15	DWT	SVM	94.94%
Proposed method	15	End-to-end	CNN-LSTM	98.16%

5 Conclusion and Future Work

To conclude, dynamic technique for heartbeat segmentation is utilized to be invariant to the variability of the heart rate. An end-to-end method based on the combination of CNN and LSTM has been applied to classify the heartbeats into 15 classes of MIT-BIH arrythmia dataset. An overall accuracy of 98.16% is achieved, which outperforms other studies as well as more classes are considered. This means that the CNN succeeded to capture the most relevant features automatically without any hand-crafted methods. The future work is to test the performance of the proposed model using heartbeats from different arrhythmia datasets.

References

1. World Health Organization: Cardiovascular diseases (CVDs) (2017). http://www.who.int/mediacentre/factsheets/fs317/en/
2. World Health Organization: About cardiovascular diseases (CVDs) (2017). https://www.who.int/cardiovascular_diseases/about_cvd/en/
3. Tantawi, M., Revett, K., Salem, A.B., Tolba, M.F.: Electrocardiogram (ECG): a new burgeoning utility for biometric recognition. In: Hassanien, A., Kim, T.H., Kacprzyk, J., Awad, A. (eds.) Bio-inspiring Cyber Security and Cloud Services: Trends and Innovations. Intelligent Systems Reference Library. Springer, Heidelberg, vol 70, pp. 349–382 (2014)
4. Alfonso, V.X., Tompkins, J.: Detecting ventricular fibrillation. IEEE Trans. Biomed. Eng. **54**(1), 174–177 (2007)
5. Moody, G.B., Mark, R.G.: The impact of the MIT-BIH arrhythmia database. IEEE Eng. Med. Biol. Mag. **20**, 45–50 (2001)
6. Kastor, J.A.: Arrhythmias, 2nd edn. W.B. Saunders, London (1994)
7. Yildirim, Ö.: A novel wavelet sequence based on deep bidirectional LSTM network model for ECG signal classification. Comput. Biol. Med. **96**, 189–202 (2018). https://doi.org/10.1016/j.compbiomed.2018.03.016
8. Bakator, M., Radosav, D.: Deep learning and medical diagnosis: a review of literature. Multimodal Technol. Interact. **2**, 47 (2018). https://doi.org/10.3390/mti2030047
9. Min, S., Lee, B., Yoon, S.: Deep learning in bioinformatics. Brief. Bioinf. **18**, 851–869 (2017)
10. Zebardast, B., Ghaffari, A., Masdari, M.: A new generalized regression artificial neural networks approach for diagnosing heart disease. Int. J. Innov. Appl. Stud. **4**, 679 (2013)
11. Abiwinanda, N., Hanif, M., Hesaputra, S.T., Handayani, A., Mengko, T.R.: Brain tumor classification using convolutional neural network. In: Lhotska, L., Sukupova, L., Lacković, I., Ibbott, G. (eds.) World Congress on Medical Physics and Biomedical Engineering 2018. IFMBE Proceedings, vol. 68, no. 1. Springer, Singapore (2019)
12. Karthik, S., Srinivasa Perumal, R., Chandra Mouli, P.V.S.S.R.: Breast cancer classification using deep neural networks (2018). https://doi.org/10.1007/978-981-10-6680-1_12
13. Liu, J., Pan, Y., Li, M., Chen, Z., Tang, L., Lu, C., Wang, J.: Applications of deep learning to MRI images: a survey. Big Data Min. Anal. **1**, 1–18 (2018)
14. Krizhevsky, A., Sutskever, I., Hinton, G.E.: ImageNet classification with deep convolutional neural networks. In: Advances in Neural Information Processing Systems, pp. 1097–1105 (2012)

15. Bengio, Y.: Learning deep architectures for AI. Found. Trends Mach. Learn. **2**, 1–127 (2009)
16. Hochreiter, S., Schmidhuber, J.: Long short-term memory. Neural Comput. **9**(8), 1735–1780 (1997). https://doi.org/10.1162/neco.1997.9.8.1735
17. Wu, Y., et al.: Google's neural machine translation system: bridging the gap between human and machine translation (2016)
18. Kim, M., et al.: Speaker-independent silent speech recognition from flesh-point articulatory movements using an LSTM neural network. IEEE/ACM Trans. Audio Speech Lang. Process. **25**(12), 2323–2336 (2017)
19. Sahoo, S., Kanungo, B., Behera, S., Sabut, S.: Multiresolution wavelet transform based feature extraction and ECG classification to detect cardiac abnormalities. Measurement **108**, 55–66 (2017)
20. Yu, S.N., Chou, K.T.: Integration of independent component analysis and neural networks for ECG beat classification. Expert Syst. Appl. **34**, 2841–2846 (2008)
21. Martis, R.J., Acharya, U.R., Min, L.C.: ECG beat classification using PCA, LDA, ICA and discrete wavelet transform. Biomed. Signal Process. Control **8**, 437–448 (2013)
22. Martis, R.J., Acharya, U.R., Mandana, K., Ray, A.K., Chakraborty, C.: Application of principal component analysis to ECG signals for automated diagnosis of cardiac health. Expert Syst. Appl. **39**, 11792–11800 (2012)
23. Yazdanian, H., Nomani, A., Yazdchi, M.R.: Autonomous detection of heartbeats and categorizing them by using support vector machines. IEEE (2013)
24. El-Saadawy, H., Tantawi, M., Shedeed, H.A., Tolba, M.F.: Hybrid hierarchical method for electrocardiogram heartbeat classification. IET Signal Process. **12**(4), 506–513 (2018). https://doi.org/10.1049/iet-spr.2017.0108
25. Ye, C., Kumar, B.V.K.V., Coimbra, M.T.: Heartbeat classification using morphological and dynamic features of ECG signals. IEEE Trans. Biomed. Eng. **59**(10), 2930–2941 (2012)
26. Acharya, U.R., Oh, S.L., Hagiwara, Y., Tan, J.H., Adam, M., Gertych, A., San, T.R.: A deep convolutional neural network model to classify heartbeats. Comput. Biol. Med. (2017). https://doi.org/10.1016/j.compbiomed.2017.08.022
27. Li, D., Zhang, J., Zhang, Q., Wei, X.: Classification of ECG signals based on 1D convolution neural network. In: 2017 IEEE 19th International Conference on e-Health Networking, Applications and Services (Healthcom) (2017). https://doi.org/10.1109/healthcom.2017.8210784
28. Thakor, N.V., Webster, J.G., Tompkins, W.J.: Estimation of QRS complex power spectra for design of a QRS filter. IEEE Trans. Biomed. Eng. **BME-31**(11), 702–706 (1984). https://doi.org/10.1109/tbme.1984.325393
29. Rectifier (neural networks). https://en.wikipedia.org/wiki/Rectifier_(neural_networks)
30. MIT-BIH Arrhythmias Database. http://www.physionet.org/physiobank/database/mitdb/. Accessed 5 Sept 2019
31. Testing and reporting performance results of cardiac rhythm and ST segment measurement algorithms. ANSI/AAMI EC57:1998 standard. Association for the Advancement of Medical Instrumentation (1998)
32. Kingma, D.P., Ba, J.: Adam: a method for stochastic optimization. CoRR, abs/1412.6980 (2014)

A Two-Stage Method for Bone X-Rays Abnormality Detection Using MobileNet Network

Hadeer El-Saadawy(✉), Manal Tantawi, Howida A. Shedeed, and Mohamed F. Tolba

Faculty of Computer and Information Science, Ain Shams University, Cairo, Egypt
hadeer_ibrahim@cis.asu.edu.eg

Abstract. Abnormality in bones is considered to be from the critical medical cases, since the wrong diagnosis could lead to bad side effects. Moreover, exhausted and over loaded doctors may miss some cases. Hence, Computer aided diagnosis systems have a vital role nowadays. This paper presents a method for detecting the fractures in the seven extremity upper bones (shoulder, humerus, forearm, elbow, wrist, hand and finger) using X-ray images. A two-stage classification method based on MobileNet network is proposed. Enhanced X-ray image is fed into the first stage to detect bone type. Thereafter, the bone image is directed according to the result of the first stage to one of seven classifiers (one for each bone type) to detect the abnormality in the bone. MURA dataset is utilized as a performance dataset and average accuracy 73.42% has been achieved after merging the two classification stages.

Keywords: Computer aided diagnosis · Medical imaging · Deep learning

1 Introduction

Medical imaging techniques have been used in various medical domains. Among the oldest and frequently used domains is the X-Rays domain. Digital X-ray imaging devices are being widely used in many medical applications as a result of the combination of radiation and computer image processing technologies [1].

Computer aided bone abnormality detection is utilized to assist the doctors to diagnose the abnormality in human bones. However, bones abnormality detection is a very difficult task as the abnormality can occur in any part of human skeleton bones like head bones, trunk bone, upper and lower extremity bones. Hence, the variety in the bones is very large, so the need for computer aided bone abnormality detection for more than one bone type is essential nowadays [2].

X-ray medical imaging is the main tool for the diagnosis of the bone's abnormality, especially the fractures. X-ray images are utilized by the medical practitioners to be able to take the critical decisions in the diagnosis [2, 3]. Wrong diagnosis of the bone's abnormality can lead to ineffective patient management, increased dissatisfaction and

A.-E. Hassanien et al. (Eds.): AICV 2020, AISC 1153, pp. 372–380, 2020.
https://doi.org/10.1007/978-3-030-44289-7_35

expensive litigation. Hence, correct diagnosis and appropriate treatment of these abnormalities is considered a critical issue [1].

Recently, many research papers [4–11] have attempted to develop computer aided systems to detect fractures in different bones of the human body. However, most of the previous studies [4–11] consider only one bone due to the variety in the bone's abnormality cases and the huge differences in bones shape. Hence, to cover the whole bones in the human body, huge number of systems is required which is not realistic at all.

In this paper, the seven bones of upper extremity (shoulder, humerus, forearm, elbow, wrist, hand and finger) are considered instead of only one bone type. X-ray images are fed into two-stage classification method to classify and detect the abnormality of a bone. Bone X-ray image is fed first into a classifier to recognize its type. Subsequently, the abnormality of this bone is detected by a specific classifier in stage 2 based on its type defined already in stage 1. Hence, the proposed method includes eight classifiers: one for stage 1 and seven for stage 2. MobileNet Deep learning model is utilized for all classifiers in this study. A large public dataset for extremity upper bones called MURA dataset has been considered for training and testing purposes.

The remainder of this paper is organized as follows: Sect. 2 presents a summary of the related previous work. Section 3 discusses in detail the proposed method and dataset utilized. Section 4 provides the system specifications used and the achieved results and finally Sect. 5 presents the conclusion and future work.

2 Related Work

Recently, many studies have been emerged to detect bone abnormality, especially fractures in bones and few of them detect the fracture type as well. In this section, a brief survey on the previous studies is provided.

A computer aided diagnosis method is proposed by Al-ayyoub et al. [4] to classify the abnormalities of long bones. X-ray images are preprocessed to remove the noise using a set of filters to reduce darkness, brightness, blurring, Poisson and Gaussian noise. Subsequently, edge detection, corner detection, peak detection, texture features and fracture lines are utilized to extract the features. The extracted features are then fed into four different classifiers: (1) Support Vector Machine (SVM); (2) Decision Tree; (3) Nave Bayes; and (4) Neural Network (NN). 300 X-ray images were collected from hospitals and websites to train and test the proposed approach. The best accuracy has been achieved using SVM classifier, which is above 85% for both presence/ absence of the fracture(s) and the fracture(s) type.

A generalized Stacked Random Forests Feature Fusion method is proposed by Cao et al. [5] to detect multiple bone fracture types and multiple bone structures throughout the body. 145 X-ray images are utilized as a performance dataset to evaluate the proposed method. The proposed method achieved an accuracy of 81.2%.

Hmeidi et al. [6] proposes a hand fracture detection method. The images are filtered to remove the noise and the edges are detected using Sobel operator. Wavelet transform and Curvelet transform are applied to preprocessed images to extract features after performing edge detection. On the other hand, Gray Level Co-occurrence Matrix (GLCM) is used to extract features from only filtered images. The extracted features are then fed into four different classifiers: (1) Bayesian Networks; (2) Naive Bayes; (3) NN and (4) Decision Tree. 98 X-ray images are utilized as a performance dataset to achieve an accuracy of 98.1%.

Haar Wavelet transforms is utilized by Dimililer [7] to remove the noise from the X-ray images. Then, Scale-Invariant Feature Transform (SIFT) is considered to extract features. Thereafter, the extracted features are fed into Back Propagation Neural Networks (BPNN) with 1024 input neurons. Dataset of 100 X-rays images are utilized as a benchmark dataset to achieve an accuracy of 94.3%.

Mahendran et al. [8] proposes a method to detect fracture in long bones of legs called Tibia. X-ray images are preprocessed using Simultaneous Automatic Contrast adjustment, Edge Enhancement and Noise Removal (SACEN) algorithm. The preprocessed images are segmented using two algorithms to segment the bone image from the X-ray image and identify the diaphysis region of the bone image. Moreover, GLCM is utilized to extract the textural features. The extracted features are fed into three classifiers: (1) BPNN; (2) SVM and (3) Naïve Bayes. 1000 X-ray images are utilized as a performance dataset to achieve 90% accuracy after applying a fusion step to the classifiers output using highest majority.

Akter et al. [9] introduces a method to detect the fractures of hand fingers. X-ray images are preprocessed by removing the noise and changing them to binary images. 32 features are extracted using moment feature, GLCM features, entropy, major axis length, minor axis length, eccentricity, orientation, convex area, area, filled area, equivalent diameter, solidity, extent, perimeter, mean, standard deviation, correlation coefficient, median, variance, height, width, pixel count and Euclidian distance. Thereafter, the extracted features are then fed into SVM and NN classifiers. SVM provides the best result 92.24% accuracy for the presence/absence of fractures.

Normal and four proximal humerous fracture types (greater tuberosity, surgical neck, 3-part and 4-part) using plain antero-posterior shoulder radiographs are considered by Chung et al. [10]. The images are resized to be 256 × 256. Thereafter, the preprocessed images are then fed into Convolutional Neural Network (CNN) classifier. 1891 images are utilized as a performance dataset to achieve 96%, 99% and 97% for accuracy, sensitivity and specificity, respectively.

A 169-layer densely connected convolutional network is utilized by Rajpurkar et al. [11] to detect and localize abnormality in upper extremity bones which include shoulder, forearm, humerus, elbow, wrist, hand and finger. MURA dataset has been utilized for the training and testing purposes. The dataset contains more than 40,000 images and each bone has more than one view. The images are resized to be 320 × 320. Then, for each view, the 169-layer convolutional network predicts the abnormality of the bone. 70.5% is the best average accuracy achieved for the all seven bones.

Finally, the following observations can be drawn from the existing studies: (1) Generally speaking there are few studies that consider the detection of the fractures or their types; (2) Most of the studies consider only one bone due to the strong variations in bone and fracture types; (3) Lack of public published datasets can be justified by the difficulty in collecting medical images from hospitals, which can be considered another reason why none of the existing studies considers all body bones; 4) Most of the previous studies utilize traditional classification methods and very few ones utilize deep learning classification methods. In this study, seven upper extremity bones which include shoulder, forearm, humerus, elbow, wrist, hand and finger are considered using two-stage deep learning classification method.

3 Proposed Method

In this paper, a hierarchical two stages-based method for upper extremity bones classification is proposed as shown in Fig. 1. The first stage is responsible for the classification of the type of the bone into seven classes: shoulder, humerus, elbow, forearm, wrist, hand, and finger. Thereafter, the second stage is responsible for the detection of the bone abnormality (i.e. whether the bone is normal or abnormal). In the second stage, each bone has its own classifier to detect the abnormality of each bone separately. A detailed description of the utilized dataset and the steps of the proposed method are presented in the next subsections.

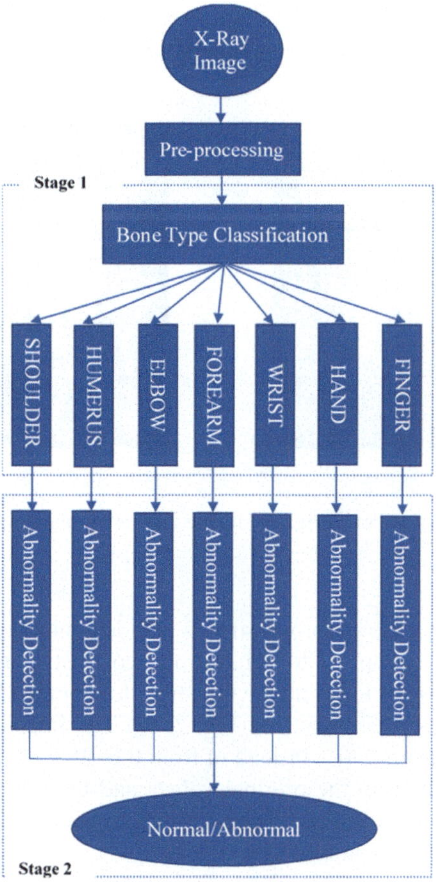

Fig. 1. Proposed method architecture

3.1 Dataset

MURA dataset [11] is utilized in this study. MURA dataset is considered one of the largest public radiographic image datasets in general and the largest public dataset for the bone's abnormality. MURA dataset consists of 9,067 normal and 5,915 abnormal musculoskeletal radiographic studies of the upper extremity with more than one view for each study including the shoulder, humerus, elbow, forearm, wrist, hand, and finger. MURA dataset consists of overall 40,561 multi-view X-ray images.

3.2 Preprocessing

The difference between the bone and its background and any noise in the x-ray image should be maximized as much as possible to enhance the prominence of the bone in the image. Thus, to improve contrast, Adaptive Histogram Equalization is considered. The main idea behind the Histogram equalization is to change the intensity of the image to make the distribution of its intensity uniform. Adaptive Histogram Equalization has the same idea, but it divides the image into different pars and applies histogram equalization on each region separately to enhance the local contrast as well [12]. Accordingly, the results after using adaptive histogram equalization are much better.

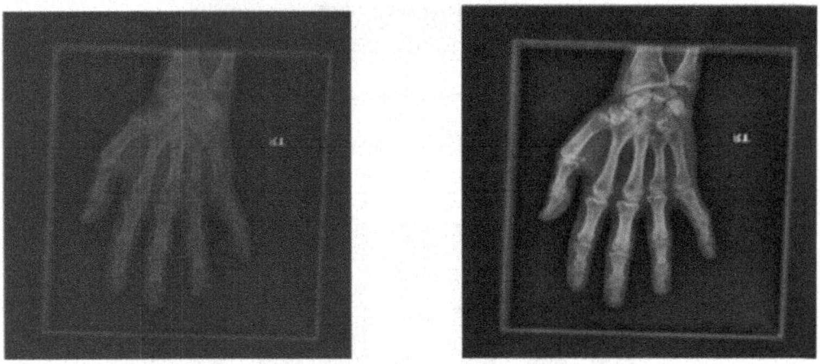

(a) (b)

Fig. 2. Image preprocessing: (a) original image and (b) image after adaptive histogram equalization

Hence in this study, the preprocessing is divided into four steps: (1) Adaptive histogram equalization is applied to enhance the contrast of the images where each image is divided into 8*8 regions and the threshold for contrast limiting is set to 4 after many trials to get the best contrast, see Fig. 2; (2) The intensity values of the RGB image are scaled to be from 0 to 1 instead of from 0 to 255; (3) The data is normalized by subtracting the mean from the original values and then dividing the result by the standard deviation as shown in Eqs. 1 and 2; (4) The data is randomly augmented by flipping some images horizontally on the x-axis.

$$Training\,Data = \frac{Training\,Data - Mean\,of\,Training\,Data}{Standard\,Deviation\,of\,Training\,Data} \tag{1}$$

$$Testing\,Data = \frac{Testing\,Data - Mean\,of\,Training\,Data}{Standard\,Deviation\,of\,Training\,Data} \tag{2}$$

3.3 Features Extraction and Classification

As mentioned before, a two stage-based classification method is proposed. Regarding the two stages, pre-trained MobileNet network is considered for features extraction and classification. The initial weights for the networks are those of the ImageNet dataset [13].

MobileNet Network [14] is a CNN based model that is built from depthwise separable convolution to build light weight deep neural networks. MobileNet consists of mainly 30 layers and its architecture is as follows: (1) Convolutional layer with stride 2; (2) Depthwise layer; (3) Pointwise layer that doubles the number of channels; (4) Depthwise layer with stride 2; (5) Pointwise layer that doubles the number of channels.

Depthwise convolution applies a single filter on each input channel separately. Then, the pointwise convolution applies a 1 * 1 convolution for the depthwise convolution to be combined. Inputs are combined and filtered into new set of outputs in one step by the means of standard convolution. After that, the depthwise separable convolution splits the output into two separable layers, one for filtering and the other for combining [14]. In this study, MobileNet is considered by freezing only the first 10 layers in the network and train the other 20 layers with the utilized dataset weights.

4 Experimental Results

As mentioned before, MURA dataset is utilized for training and testing phases. However, according to the limitations of the used hardware only part of the dataset is utilized. RGB images are preprocessed using Adaptive histogram equalization and normalized as mentioned before. The preprocessed images are then fed into pre-trained MobileNet network. The classification process is divided into two main stages, the first stage is to classify the bones type and the second stage is to detect the abnormality of the bone. All experiments were carried out using GPU (2496 CUDA cores and 12 GB (11.439 GB Usable) GDDR5 VRAM), disk 320 GB and RAM 12.6 GB. The next subsections provide the results of each stage separately and the results after merging the two stages.

4.1 Results of First Stage

In this stage, the x-ray image is fed into MobileNet pre-trained network to be classified to one of the seven upper extremity bones namely shoulder, humerus, elbow forearm, wrist, hand, and finger. Table 1 summarizes the classification accuracy for each bone separately.

Table 1. Results of bones type classification (stage 1)

Bone type	Bone classification accuracy
Shoulder	92.33%
Humerus	91.33%
Elbow	92.00%
Forearm	92.33%
Wrist	93.67%
Hand	97.33%
Finger	94.33%

As noticed, the accuracy for each bone separately is above 90%. Consequently, the average accuracy for this stage is 93.33%.

4.2 Results of Second Stage

In this stage, the x-ray image is fed into one of seven classifiers according to the result of stage 1. The seven classifiers in this stage are trained separately to detect the abnormality of specific type of bones. Each classifier is a MobileNet pre-trained network. Table 2 summarizes the abnormality detection results for each bone separately.

Table 2. Results of abnormality detection (stage 2)

Bone type	Abnormality detection accuracy
Shoulder	77.00%
Humerus	86.52%
Elbow	81.29%
Forearm	78.41%
Wrist	81.94%
Hand	75.43%
Finger	80.04%

As noticed, the accuracy for each bone separately is above 75%. Consequently, the average accuracy for this stage is 80.09%.

4.3 Results of Merging the Two Stages

In the previous subsections, the results of each stage are discussed separately. In this subsection, the results of the two stages are merged. X-Ray image is fed into the bone type classifier (stage 1) and then fed into one of the seven MobileNet classifiers of stage 2 to detect the abnormality (whether the bone is normal or abnormal). Table 3 summarizes the results after merging the two stages.

Table 3. Results of merging stage 1 and stage 2

Bone type	Final accuracy
Shoulder	69.33%
Humerus	77.85%
Elbow	73.29%
Forearm	70.74%
Wrist	75.61%
Hand	72.76%
Finger	74.37%

As noticed, the accuracy for each bone separately after the merging step is above 70%. Consequently, the average accuracy for this stage is 73.42%. To the best of our knowledge, the proposed method couldn't be compared with any of the previous work as the only paper [11] that used MURA dataset utilized private testing data.

5 Conclusion

To sum up, this paper presents a method to detect the abnormality of upper extremity bones including shoulder, humerus, forearm, elbow, wrist, hand and finger. X-ray images are normalized and enhanced to increase their contrast. A two-stage method based on MobileNet network is proposed and examined using on MURA dataset. X-ray images are fed into the first stage to classify the bone into one of the seven upper extremity bones. Then, the classified bone image is fed into a specific classifier in the second stage trained to detect the absence/presence of abnormality in this type of bones. Average accuracy of 73.42% has been achieved using MURA dataset as a benchmark dataset. Hence, the proposed method considers two problems: Bones type classification and Fractures detection. Seven upper extremity bones are considered, instead of one bone as in literature. In addition, the proposed method is scalable, since there is no need to repeat the training process of all classifiers if a new bone is added, only the classifier of the first stage.

In the future, we are looking forward to examining different deep learning classifiers and architectures to improve both the bones classification and the abnormalities detection accuracies.

References

1. Mahendran, S.K., Santhosh Baboo, S.: Automatic fracture detection using classifiers - a review. IJCSI Int. J. Comput. Sci. **8**(6), 340–345 (2011)
2. Kinnari, P., Dangar, Y.: A survey for fracture detection in hand bone X-Ray images. IJARIIE **3**(1), 1529–1533 (2017)
3. Swathika, B., Anandhanarayanan, K., Baskaran, B., Govindaraj, R.: Radius bone fracture detection using morphological gradient based image segmentation technique. IJCSIT Int. J. Comput. Sci. Inf. Technol. **6**(2), 1616–1619 (2015)

4. Al-Ayyoub, M., Al-Zghool, D.: Determining the type of long bone fractures in X-Ray images. Wseas Trans. Inf. Sci. Appl. **10**(8), 261–270 (2013)
5. Cao, Y., Wang, H., Moradi, M., Prasanna, P., Syeda-Mahmood, T.F.: Fracture detection in X-Ray images through stacked random forests feature fusion. In: IEEE 12th International Symposium on Biomedical Imaging, New York, NY, USA (2015)
6. Hmeidi, I., Al-Ayyoub, M., Rababah, H., Khatatbeh, Z.: Detecting hand bone fractures in X-Ray images. In: Association of Computer Electronics and Electrical Engineers (2013)
7. Dimililer, K.: IBFDS: intelligent bone fracture detection system. In: 9th International Conference on Theory and Application of Soft Computing, Computing with Words and Perception, ICSCCW 2017, Budapest, Hungary (2017)
8. Mahendran, S.K., Santhosh Baboo, S.: An enhanced tibia fracture detection tool using image processing and classification fusion techniques in X-Ray images. Glob. J. Comput. Sci. Technol. **11**(14), 22–28 (2011)
9. Miah, M.B.A., Akter, A.: Detecting and determining the types of hand bone fracture using k-means clustering. J. Comput. Sci. Appl. Inf. Technol. **3**(3), 1–10 (2018)
10. Chung, S.W., Han, S.S., Lee, J.W., Oh, K.S., Kim, N.R., Yoon, J.P., Noh, Y.M.: Automated detection and classification of the proximal humerus fracture by using deep learning algorithm. Acta Orthopaedica **89**(4), 468–473 (2018)
11. Rajpurkar, P., Irvin, J., Bagul, A., Ding, D., Duan, T., et al.: MURA dataset: towards radiologist-level abnormality detection in musculoskeletal radiographs. In: 1st Conference on Medical Imaging with Deep Learning, arXiv:1712.06957 [physics.med-ph] (2017). https://stanfordmlgroup.github.io/competitions/mura/. Accessed 3 Nov 2019
12. Gonzalez, R.C., Woods, R.E.: Digital Image Processing, 4th edn. Pearson, London (2017)
13. Russakovsky, O., Deng, J., Su, H., Krause, J., Satheesh, S., et al.: Image Net large scale visual recognition challenge. arXiv:abs/1409.0575 (2014)
14. Howard, A.G., Zhu, M., Chen, B., Kalenichenko, D., Wang, W., et al.: MobileNets: efficient convolutional neural networks for mobile vision. arXiv:1704.04861v1 [cs.CV] (2017). https://towardsdatascience.com/transfer-learning-using-mobilenet-and-keras-c75daf7ff299. Accessed 8 Sept 2019

Use of Deep Learning for Bird Detection to Reduction of Collateral Damage in Fruit Fields

Humberto Garcia[1], Alberto Ochoa-Zezzatti[2(✉)], Abraham Martínez-Retamoza[2], Gilberto Ochoa[3], Lina Aguilar[4], Diego Oliva[5], and José Mejía[2]

[1] Instituto Tecnológico de Ciudad Juárez, Ciudad Juárez, Mexico
[2] Universidad Autónoma de Ciudad Juárez, Ciudad Juárez, Mexico
alberto.ochoa@uacj.mx
[3] ITESM Campus Guadalajara, Zapopan, Mexico
[4] Universidad Autónoma de Guadalajara, Zapopan, Mexico
[5] CUCEI, UdG, Guadalajara, Mexico

Abstract. Since the beginning of agriculture there have been all kinds of pests that have reduced crop production. Throughout this time, different methods of scarring have been used to control pests and increase crop yields. In the same way today, pest control has other goals compared to before, one of the main objectives is the integrity of the animal to be treated, this is intended to preserve the ecosystem balance and the protection of the environment in which it develops. Pest detection today has become one of the most important factors to consider in this maintenance area in agricultural and fruit fields. This document is mainly focused on the detection of flocks of birds that are considered as pests in apple orchards, in Cuauhtémoc Chihuahua, although, the same method can be applied for different production areas.

Keywords: Flocks · Bird detection · Object detection · Convolutional Neural Network · CNN

1 Introduction

1.1 The Current Problem

Pest control in agricultural fields is a very important issue to deal with for people who develop this type of economic activities, that is why nowadays there are different methods of scaring off, which are used for specific pest control. Some of these types of technologies that are implemented in agricultural fields are fence crop protection, gas cannons or sound alarms, among others [1]. As we mentioned at the beginning of the document, the main purpose today is the efficiency of the scare method, but also to maintain the integrity of the environment without damaging the ecosystem. An example of a repelling method that is harmful to the ecosystem is the gas gun as it generates noise pollution in the environment, not to mention how uncomfortable it is for people living around the field where this type of integrated device is being used. In

A.-E. Hassanien et al. (Eds.): AICV 2020, AISC 1153, pp. 381–392, 2020.
https://doi.org/10.1007/978-3-030-44289-7_36

recent years the use of Machine Learning has made many advances and innovations in the aspect of artificial intelligence applied in problems and situations that the use of an autonomous and independent system would be the most appropriate. In the article entitled "A Study on CNN Transfer Learning for Image Classification," made by students of the Aston University School of Engineering and Applied Sciences, mention how they performed a neural network applied to the classification of objects, and check the effectiveness and potential of using a convolutional neural network for the problem applications they need could have better performance if they had the help of an automated and independent system. [2] One aspect to consider is, a human is performing the task, and is susceptible to corruption and to the poor expansion of his work, the use of a neural network applied by a computer or system, is incorruptible and can be improved. In the document written by IEEE, by the name of "A Machine Learning-Based Framework for Automatic Visual Inspection of Microdrill Bits in PCB Production," he talks about how to use a model implemented with Machine Learning, it is very effective when inspecting Micro-perforations in the production of PCB plates, and also improves the effectiveness in identifying failures in PCB plates, and reducing the cost of inspection tools. [3] This information was very useful to us because with this article you can check on a problem applied using an automated system, intelligent and independent, improving production, or improving quality inspection.

2 Previous Research

The purpose of our work is mainly in the recognition of birds for the control of bird pests in agricultural and fruit fields. This using convolutional neural networks (CNN), as well as being of help for the development of other neural networks in the future. In the literature we review and look for different investigations concerning this problem, of which we find the article entitled, "A Study on Detecting Drones Using Deep Convolutional Neural Networks", students of the University of Sydney in Australia [4] in 2007, which mentions the methodology they used to create a trained neural network to detect drones and differentiate them from birds in an image. During the practice they mention that they used a database already available, specifically a video of which they used as a sample to train the model. This information was useful to us at the time of making proposals for the architecture of the neural network and discussing the samples that we will use as training elements. The use of a video for the imaging of the object to be detected was a very tempting alternative, however, because we were going to create a complete neural network and segment the objects manually, and in the short term, it was impossible for us to use a database of a lot of input data. In the literature we find a document that was of great importance for the elaboration of this research since the purpose of this same document was also the application of deep learning in drones, although the same research was focused on the performance of GPU for data processing and the computational capacity they had for such tasks. The document is called the Deep Learning-Based Real-Time Multiple-Object Detection and Tracking from Aerial Imagery via a Flying Robot with GPU-Based Embedded Devices. [15] This is not the first research using neural networks for the detection of animals. In the research of name "Part-Based R-Cnns for Fine-Grained Category Detection," a research and implementation of a convolutional model for the categorized detection of certain objects for bird classification was conducted, this research was useful to us, because here we could find some examples of different types of convolutional neural networks and one of the applications it has, this for example is dedicated more, to the detection of certain patterns in specific to obtain much more accurate results, however, this neural network is focused for another type of activity. [5] Different types of neural network models have been developed for different purposes, for example in the article called "Object Recognition in Very Low Resolution Images using Deep Collaborative Learning", mention as the use of this type of neural network, which in this precise case is not a convolutional network but a collaborative neural network, used for the recognition of objects having a low image quality. The compilation of information obtained from the document was useful for us, the stage prior to the development of the convolutional neural network, when we were observing what kind of method for object recognition would be the most optimal for the type of work we were developing, although we ended up choosing to use a neural network instead of a collaborative neural network, it was useful for us to observe the methodology used and to make a small comparison between the two methods and to determine which could carry out the work more efficiently. [13] One aspect that we had to consider when planning a methodology for the recognition of objects in this case of birds was the fact of evaluating and considering the possibility of the recognition of objects in a three-dimensional plane, so that the same neural network could evaluate not only the position of height and width, which are the two dimensional planes, but also evaluate the depth of the same object, in the literature fortunately we find in the document of name [BV-Cnns]: Binary Volumetric Convolutional Networks for 3D Object Recognition, a lot of information that was useful to us. Among the information that we consider important was, within the methodology proposed by the authors of the document, is that they themselves used solid objects that were generated by three-dimensional design programs, that is one of the main aspects that represent something different since the purpose of us in this neural network is that this same one has a direct application in some specific problem. So one of the limitations that we had in making the neural network is the lack of computational capacity that we had, this was a very limiting aspect that we cannot process images of a very high resolution, because the same computer does not have the capacity to process so much information. [14] The same document was very helpful in deciding for a two-dimensional vision that this same one evaluates the width and the height in which the object to be analyzed is located. Another clear example of the use of convolutional networks for pattern recognition is in the text "Real-Time Detection of Apple Leaf Diseases Using Deep Learning Approach Based on Improved Convolutional Neural Networks" [9] where the recognition that the convolutional network has to do is to see and analyze the patterns that are generated in tree leaves, so that you can know if the trees have some disease that deprives them of health as a plant, this is a very big help since the computer is taught different types of patterns that correspond to different diseases, so on our side you could teach the computer to recognize different types of birds (depending on the area where the project is implemented) to know which ones pose a threat, in our case, to fruit fields.

3 Methodological Approach

What we propose to do is make use of the different images that were obtained from the web to use them as samples and train a convolutional neural network, so that in this way it is able to recognize in this specific case and for the resolution of this problem, birds.

4 Methodology Implemented

During the whole methodology we worked without a pre-established database if not, we used 100 specimens of bird images that we collected on the web. The language with which the neural network was built, is Python in its version 3.7. The main library that was used for the model of the neural network was keras, but this was not the only one, we also gave utility to a numpy-named library in processes that will be explained later. The implementation of Keras as the main library for the next convolutional neural network was because it is the library in which we have more knowledge and experience programming it, and to personal consideration is a little more efficient and less complicated to understand the commands and much more visible the behavior of the different convolution layers (Illustration 1).

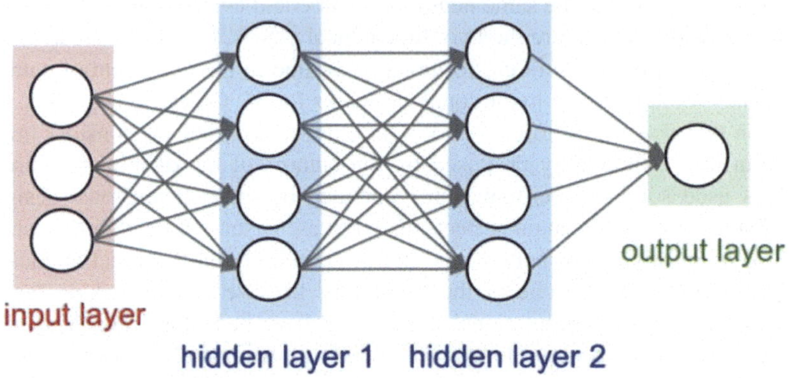

Illustration 1. Basic layer principle of a convolutional neural network.

In the article named "Development of Predictive Emissions Monitoring System Using Open Source Machine Learning Library", Keras: A Case Study on a Cogeneration Unit which the research consisted of an analysis of the use of an open source library and how it performed, and in the same way as it performed in the same way using a library already borrowed as Keras, in conclusion of the research, the use of Keras, R and/or Python for the elaboration of a convolutional neural network, it ends up being a very useful help and a facility for people dedicated or related to the field of Deep Learning, the fact that it is a free library also means a positive aspect to consider when using it. As we mentioned at the beginning of the methodological approach, the same way in the document is mentioned as working with Keras to develop projects related to deep learning, is very manageable and very easy to maintain control and maintain [15].

5 Architecture of the Neural Network

In the next diagram we will resume all this process to be easier to understand. And be clear in how we made the neural network (Illustration 2).

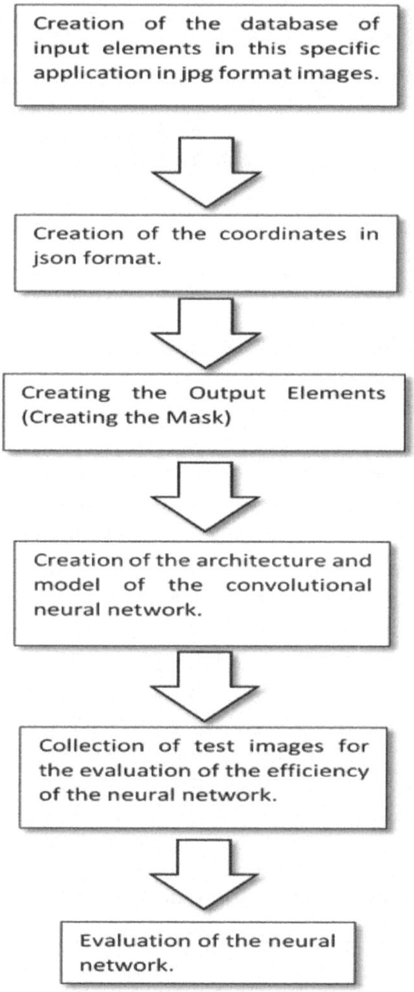

Illustration 2. Neural network diagram.

5.1 Creation of Training Data

For the realization of the neural network, before we started adding the layers that our neural network has, we had to create our own database of bird specimens, in which we use 100 images of birds in different types of locations, and different positions and profiles, this with the purpose of obtaining the maximum number of birds at different angles of view so that the model is able to recognize it at different angles. The images were saved with the format "jpg", this was mainly done for us to be able to work all the 100 images with the same format and be easier programming the code for processing images (Fig. 1).

Fig. 1. Datasets of training images

The compilation of these images was supported by Google's search engine, which we analyzed different copies of, which we considered to be the most convenient to use, so that one could clearly observe the objects that one wanted to identify that in this case are the birds. This is very important as; these are the input elements that the convolutional neural network is going to have then if the network is trained with specimens in which a bird is not clearly observed it will not relate any type of pattern between them. However, data collection should have used different positions and different contexts in which birds are found so that they could be observed with a background where trees or branches are found, this for the purpose that the neural network is trained for the recognition of birds even when in the background branches or some other type of interference are found. He was trained not only to recognize a single bird, but also to recognize a number of birds in flock.

5.2 Creation of Segmented Images for Our Research

After collecting the 100 color images in which birds or birds are visualized according to the type of image, we proceeded to identify the coordinates in which they are located, this procedure we did using the page. The page is essential for the segmentation of the image, in which it locates the precise region of the image and expresses them in coordinates in a format ".Json" (Fig. 2).

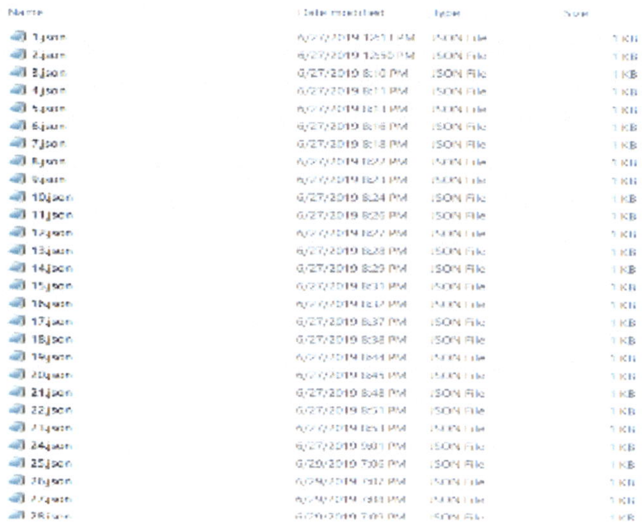

Fig. 2. Datasets of Json files of each training image

Each of these elements in contain within them a code similar to the following;

{"2.jpg57637":{"fileref":"""size":57637,"filename":"2.jpg","base64_img_data":""", "fileuattributes":{},"regions:{"0":{"shapeuattributes"{"{"{",897,897,897,917,847,917, 847,847,917,847,847,847,847,917,917,847,917,847,847,825,815,813,822,830,821,809, 799,805,817],"all_points_y":[425,393,354,369,377,371,364,332,328,330,322,299,250, 236,248,278,288,306,310,310,304,306,319,325,330,338,351,392,425,425]},"region_ attributes":{}}}}}"

Basically, the above-mentioned code are the specific coordinates of the input image No. 1, these coordinates are going to be used in the next step which is the creation of the mask.

5.3 Creation of the Mask

Once we have the two components, we need to create the segmented image [11], which is the image and the json of each image, we can continue the programming. This whole

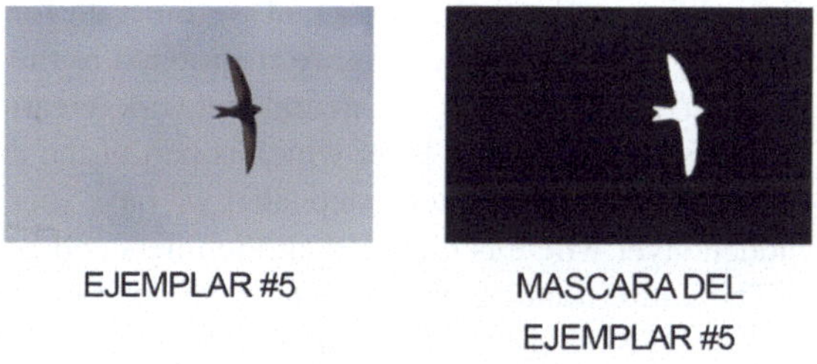

Fig. 3. Example of input and output of each training image.

process we decided to call a mask creation, which basically consists of making the original image be in black and white where the white color is where the bird is found so that they look like this, looking forward to an inverted silhouette (Fig. 3):

For the elaboration the methodology consisted of the creation of an iterative process for the opening of each one of the images contained in the datasets mentioned in the section "Creation of training data"., along with that the assignment in the json document the specific coordinates of "x" and "y", to place them in the image so that in the same way with the precise coordinates, it is possible to trace an irregular polygon which is the specific figure of the bird specific to each image. After all this process an image will be produced which we will call a mask which is the output of this whole process and is part of the output images that we will use in the training process of the neural network. As 100 training images were used and it was an iterative process that included each of these, 100 images were obtained with a specific mask.

For the elaboration of this whole process specific libraries were used as:

- Numpy
- Matplotlib
- Skimage
- Os
- Json

5.4 Neural Network Selected

To improve the optimization and reduce the computational work of the team we decided to make different codes for each stage, and save the input variable which is a four-dimensional numpy array, of [100,128,128,1], this in the same way to unify all the images in the same size large enough to be able to analyze the images efficiently [10], and not so large to optimize the performance of computational work, and we converted the image to grayscale in order to facilitate computational work [7]. Once we have already saved our input and output variables that we decide to name "X_test and Y_test", respectively, we work with the training input data and the training output data. We worked with a convolutional neural network, so during the training different filters were applied that is part of the convolution process, and pooling and max pooling the images. We used the 32 Bach size with 500 epochs for the first neural network model. However, in the same way we made another neural network this time trained with 1000 epochs hoping in this way to obtain greater results compared to the first model. All part of this process could be resume as we could tell that this this part of the process is part of the include of all the filters and convolutions that we must do for have information that could be processed by the final model. As we mentioned at the beginning of this paper, we could say that the convolutional neural network creation process has 3 important steps. The first step is the input layer which is part of the creation of datasets and the creation of the mask, all the previous steps that we have mentioned before, the second step is the hidden layer where is included all the filters and convolution that the neural network did. The entire training process was developed on a laptop with an i7 7500u processor, Nvidia GTX 950M, and 12 GB RAM.

6 Results and Discussion

When we finished the model of the neural network we used images of birds different from those we had used for the training of the network, so that in this way we could test the efficiency and the work that the neural network would do for images of which I had no knowledge of the location of the object to be identified. This process is the final step of all the process to made the neural network, being this one the third one; the output layer (Fig. 4).

Imagen de entrada al modelo.

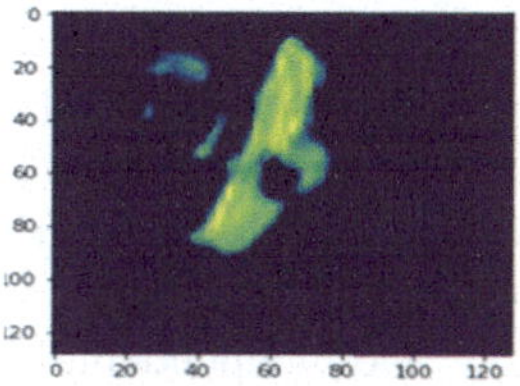

Elemento de salida del modelo, después del procesamiento de la imagen

Fig. 4. Final performance of the CNN.

The error rate which the model determined after the 1000 epochs was. 007 that this can be translated to the algorithm having a precision of 99.993% at the time of identifying bird images. The model can be improved by using more images to train the model however so that the model is completely accurate and has no interference with the background or other object, you would have to have a much larger database, where you would have many more bird images and also have the segmented image of each image you want to use for training the neural network.

The output data obtained when processing an image that was not within the training images managed to recognize the silhouette which includes the bird, this in a way we have two very important data, that is precisely the neural network can recognize with a good performance the presence of a bird and similarly the location in a two-dimensional plane. The research that we did and the neural network that we carried out, was mainly for the verification of the capacity of the intelligent systems to carry out different activities with an efficiency equal to that of the humans. Where while we realized that there is a minimal error, this algorithm could fail, but as mentioned before, to achieve an even smaller error would require more training to make the error basically negligible, and so you can say that the machine works just like a human being, as long as it's about recognition.

7 Conclusions

The implementation of this type of neural networks for the identification of birds to carry out jobs that include functions or capabilities of recognition of extensive lands during natural disasters [12], or even basic surveillance and recognition of aggressive behavior [8] may represent a saving in the economic sphere and in the expenditure made by people and/or companies engaged in the business of fruit and agriculture and see birds as a pest, as well as the insects that consume the product they grow. The model that was created from which we mention in this document this fact to be implemented mainly in fruit and agricultural fields, as mentioned in the beginning, the objective of the model creation is to reduce human interaction in jobs and activities that a system can perform in a more efficient way. However, it may present different problems, that is why it is advisable that the system be assisted by people so that the relationship between machine and human can benefit the automated system so that the same one grows in experience and capacities until reaching the degree that is completely autonomous and independent. [5] The development of this convolutional neural network can be used for many tasks involving the recognition of objects specifically in this case birds, and the neural network is to be performed in an efficient manner, with a percentage of assertiveness greater than 90%.

8 Future Direction

The training of this neural network was done for the implementation of the same in an autonomous vehicle capable of making decisions based on obtaining environmental information and carrying out the most appropriate actions according to the system considers it more appropriate depending on the situation. The future research that has been planned is after including it in the aforementioned vehicle, is to train in the same way a neural network for the identification of behaviors and patterns of violence [8] for control purposes and that safety is currently assisted by means of autonomous unmanned vehicles. Using a generative adversarial networking you can cast the shadow from the drone of an unknown predator bird, this will strongly influence the distance of the birds in question, as can be seen in Figs. 5 and 6.

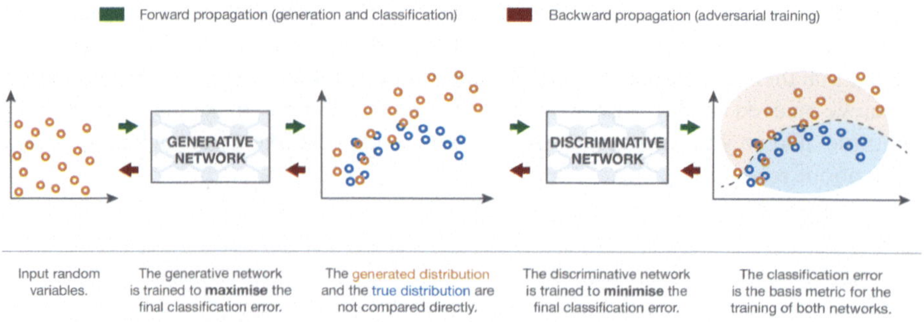

Fig. 5. Use of an adversarial networking generative to build shadows that may come from a cluster of shadows associated with predator bird.

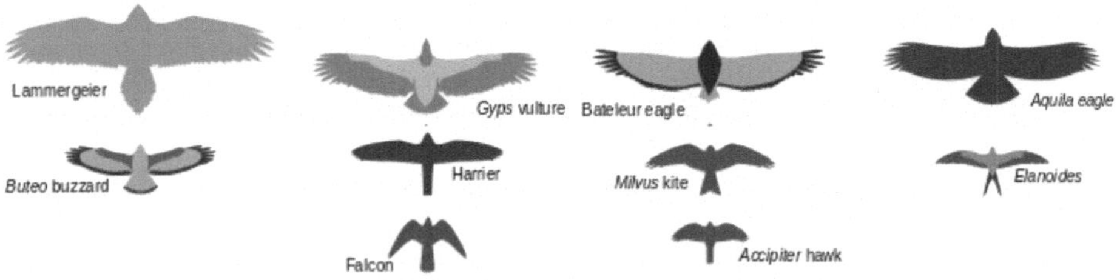

Fig. 6. Examples of some common bird silhouettes

It is worth mentioning that there are more than 11 thousand species of birds of different sizes and shapes so there is an inherent difficulty in correctly recognizing a flock depending on the scenario where their performance is contextualized and the way in which their collective behavior is decisive in determining whether they are a potential threat from shooting down a plane or using their beaks to affect other species in the stratification of tree species or in the fight for fruit food.

Flight missions will be programmed of points of interest according to the patterns detected in the neural network by adding them as specific navigation charts to the flight software library.

References

1. Bishop, J., McKay, H., Parrott, D., Allan, J.: Review of international research literature regarding the effectiveness of auditory bird scaring techniques and potential alternatives. Central Science Laboratories (2003)
2. Hussain, M., Bird, J.J., Faria, D.R.: A study on CNN transfer learning for image classification. University of Aston (2018)
3. Duan, G., Wang, H., Liu, Z., Chen, Y.: A machine learning-based framework for automatic visual inspection of microdrill bits in PCB production. IEEE Trans. Syst. Man Cybern. **42** (6), 1679–1689 (2012)
4. Saqib, M., Khan, S.D., Sharma, N., Blumenstein, M.: A study on detecting drones using deep convolutional neural networks. NSW Australia (2007)
5. Zhang, N., Donahue, J., Girshick, R., Darrell, T.: Part-based R-CNNs for fine-grained category detection. University of California, Berkeley, USA (2014)
6. Huber, A., Weiss, A.: Developing human-robot interaction for an Industry 4.0 Robot. Institute of Automation and Control Vienna University of Technology, Vienna, Austria (2017)
7. Krizhevsky, A., Sutskever, I., Hinton, G.E.: ImageNet classification with deep convolutional neural networks (2012)
8. Alom, Z., Bontupalli, V.R., Taha, T.M.: Intrusion detection using deep belief network and extreme learning machine, Department of Electrical and Computer Engineering, University of Dayton, OH, USA, Dayton (2016)
9. Jiang, P., Chen, Y., Liu, B.: Real-time detection of apple leaf diseases using deep learning approach based on improved convolutional neural networks (2019)

10. Xiao, Y., Xing, C., Zhang, T., Zhao, Z.: An intrusion detection model based on feature reduction and convolutional neural networks. In: College of Information and Communication Engineering, Harbin Engineering University, Harbin, P.R. China (2019)
11. Girshick, R., Donahue, J., Darrell, T., Malik, J.: Rich feature hierarchies for accurate object detection and semantic segmentation (2014)
12. Adams, S.M., Friedland, C.J.: A survey of unmanned aerial vehicle (UAV) usage for imagery collection in disaster research and management (2011)
13. Seo, J., Park, H.: Object recognition in very low resolution images using deep collaborative learning. School of Computer Science and Engineering, Kyungpook National University, Buk-gu, Korea (2019)
14. Ma, C., An, W., Lei, Y., Guo, Y.: BV-CNNs: binary volumetric convolutional networks for 3D object recognition. College of Electronic Science and Engineering, National University of Defense Technology, Changsha, China (2017)
15. Si, M., Tarnoczi, T.J., Wiens, B., Du, K.: Development of predictive emissions monitoring system using open source machine learning library – Keras: a case study on a cogeneration. Tetra Tech Canada Inc., Calgary, Canada (2019)
16. Hossain, S., Lee, D.J.: Deep learning-based real-time multiple-object detection and tracking from aerial imagery via a flying robot with GPU-based embedded devices. Sensors **19**(15), 3371 (2019)

Machine Learning and Applications

Machine Learning Model for Predicting Non-performing Agricultural Loans

Mohamed Ahmed Elnaggar(✉), Mostafa Abed EL Azeem, and Fahima A. Maghraby

College of Computing and Information Technology, Arab Academy for Science, Technology and Maritime Transport, Cairo, Egypt
mohamed.elnaggar@hotmail.com, melbaqary@gmail.com, fahima@aast.edu

Abstract. Extending agricultural loans to individuals is essential to support the agriculture sector and markets. One of the most important risks affecting the banking sector is the concept of credit risk. Predicting the probability of non-performing loans for an individual is a vital and beneficial role for banks to decrease credit risk and make the right decisions. These decisions are based on credit study and in accordance with generally accepted standards, loan payment history, and demographic data of the clients. The subject paper here is proposing an ensemble-based model, to enhance classification accuracy. For the building model, the dataset was gathered from an agricultural bank in Egypt. Egyptian credit dataset involves 112907 instances and 17 features that are used in the current study. Variable selections were used to select important features for the classification. Cross-classification has also been used with ten subsets. Classification methods have been applied with Logistics Regression (LR), k-nearest neighbors (KNN), Support Vector Machine (SVM), Decision Tree (DT) and Meta-classifier methods for training and testing toward the dataset. The outcome of the specified experiments showed that the accuracy of the ensemble method is the highly recommended one for individuals.

Keywords: Data pre-processing · Credit risk · Loan · Machine learning

1 Introduction

The agricultural bank in Egypt (AB) offers a range of agricultural loans in order to fulfill its national duty in the implementation of the country's agricultural policy and the service of national income to achieve comprehensive development that suits economic variables. The Bank offers a variety of agricultural loans to individuals.

The agricultural loan has different criteria from other loans in terms of interest rate, guarantees, and loan amount. The loan amount can be calculated based on different features like the Agricultural holding which is composed of (Area in Fadan, Area in Kerat, and Area in Sahm), the type of crop and the agricultural season which is divided into three agricultural seasons (the winter season, the summer season and the Nile season), area of the cultivated land, the region, etc.

A.-E. Hassanien et al. (Eds.): AICV 2020, AISC 1153, pp. 395–404, 2020.
https://doi.org/10.1007/978-3-030-44289-7_37

On the other hand, the bank is concerned about whether borrowers can meet their obligations or not before granting loans. How to reduce and prevent the risk of Non-Performing Loans (NPLs) has become an issue for banks [1], especially since the bank has gone through a difficult period with bad loans in the recent period, which forced the bank to announce the launch of an initiative to settle bad debts with borrowers in default. An agricultural loan can be considered ineffective after 90 days have elapsed without paying the outstanding loan. In case the loan exceeds 90 days without repaying, the bank has to allocate provision for those loans [2]. So, an increase in NPLs will affect the bank's capability to expand new loans to individuals and reduction of the bank's profit will result in failure in the long term, and the bank must keep NPLs at the minimum level [3].

Machine learning approaches are widely used to analyze a large amount of data and to understand client's transactions in the financial sector and considered the right tool in solving real problems. Because of the problem of credit risk assessment, different classification techniques can be used to classify credit risk. In 2015 Paireekreng et al. [3] proposed a classification model with feature selection by NN and weight of attributes by genetic algorithm. The author used different individual techniques for classification then ensemble vote algorithm which combine three best-perfumed classifiers to get higher performance. In 2016 Zhang et al. [4] presented multiple classifier fusion algorithms that integrate base classifiers (J48, IBK, SVM) to build NPLs prediction. In 2016 Goyal et al. [5] proposed a new ensemble method to predict the credit risk of customers to get optimum accuracy. The author used eleven models to predict the credit risk of applied customers depends on nine features. To make sure the model is robust the author used the k-fold method. In 2017 Okesola et al. [6] built an improved credit scoring model for high speed and effective memory usage by using a Bayesian algorithm to predict loan requests. In 2019 Soni and Paul [7] proved that random forest classifier with optimization of parameters results in better accuracy for the capability prediction of loan repayment by the customers.

Limitations or cons of [3–7], are the lack of the number of samples which may lead to increase the direct training difficulty and over-fitting, which leads to the decrease of classification accuracy and may manifest issues relating to generalization [8, 9].

This paper introduces an ensemble-based model on credit data for agricultural loans with 112907 instances. The rest of this paper is organized as follows: Sect. 2 presents an overview of classification methods. Section 3 introduces Feature selection techniques. Section 4 is allocated to the proposed model. The results of experiments and analysis are illustrated in Sect. 5. Finally, the conclusion is in Sect. 6.

2 Classification Methods

2.1 Overview of Individual Classifiers

K Nearest Neighbors (KNN) is one of the simplest machine learning techniques. It can be used for classification and also can be used for regression. KNN searches for the entire training data to see which one contributes more by finding nearest neighbors k or similarity [10].

Decision Tree (DT) works for both regression as well as classifications and is also considered one of the most popular methods [11]. It's a little bit like a flow diagram where trained data passed from the root node to the final nodes which called leaves or terminals. It works by Split based on set impurity criteria and stopping criteria such as how many levels to stop [12]. However, DT widely used as a base classifier in ensemble models.

Support Vector Machine (SVM) is widely used in classification as well as regression and has a feature that can help in outlier's detection and recommend in binary classification. Robustness in classification depends on finding optimal hyperplane with maximum margin while misclassification nearly comes from the low margin [13]. SVM has a kernel function that transforms to higher dimension space to let data be as linearly separable as possible from low dimensional input space also called kernel-based.

Logistic Regression (LR) is a form of regression analysis and widely used in classification usually dichotomous variables prediction. Fitting the features in the logistic curve depends on the sigmoid function and is also sensitive to correlations with the independent variables its target variable is a non-linear [14].

2.2 Overview of the Meta-Classifiers

Ensemble Vote Classifier is one of the easiest ensemble methods. It's easy to understand and implement. It's used for classification. It aggregates similar or conceptually classifiers for classification then ensemble them and predict the class that gets the majority (1) or plurality (2) voting as described in Fig. 1.

$$\sum_{t=1}^{T} d_t, J = \max_{j=1}^{c} \sum_{t=1}^{T} d_t, J \quad (1)$$

$$\sum_{t=1}^{T} d_t, J = \max_{j=1}^{c} \sum_{t=1}^{T} w_t d_t, J \quad (2)$$

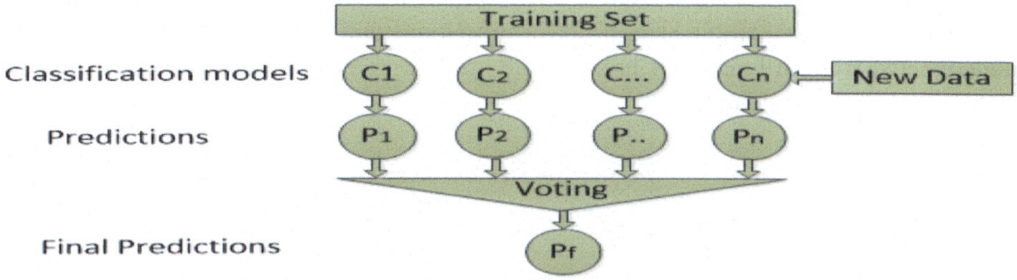

Fig. 1. Meta classifier via majority or plurality voting.

Bagging is one of the simplest powerful ensemble-based algorithms that can be used to improve the stability, accuracy and reduce the miss classification error of the base classifier. The name Bagging stands for the abbreviation of bootstrap aggregation [15]. It is based on the idea of randomly sampling data to fit base classifiers with the replacement of a given training sample where all of them have an equal probability of being selected and a classification, a rule can be derived by simple aggregation or a combination to form a final prediction.

3 Feature Selection Techniques

Our model uses one feature selection method which is Random forest (RF). RF can be used in different classification domains of applications as a supervised learning model [16–19]. Forests introduced by Breiman et al. [20], considered an ensemble of several individual trees that treat as weak classifiers (e.g., using C4.5 algorithm or CART) but with aggregating them it becomes robust learner [20]. RF becomes nowadays more and more one of the most powerful learning methods which implicitly performs feature selection. In the context of Variable importance, Breiman suggested measuring the importance of a variable Xi for predicting Y based on the GINI index of node impurity the Gini index is defined as (3) [20].

$$G(t) = 1 - \sum\nolimits_{k=1}^{\mathrm{Q}} (\mathrm{k}|\mathrm{t}) \tag{3}$$

4 The Proposed Model

In this paper, we propose a framework model as shown in Fig. 2 to optimize the classification, through the evaluation of the base classifiers with The Ensemble method. Figure 2 elaborated in the subsequent sections as follows.

4.1 Data Collection

The experiment data were collected from an agricultural bank in Egypt. The bank has different types of loans, but we focused only on agricultural loans, the data collected from January 2014 to 2018 collected in a random form. The data set has 22 features and 112907 instances, the sample size of performing loans and NPLs is 91485: 21422. The data collected from different AB branches and only agricultural data are retrieved after that the data is consolidated.

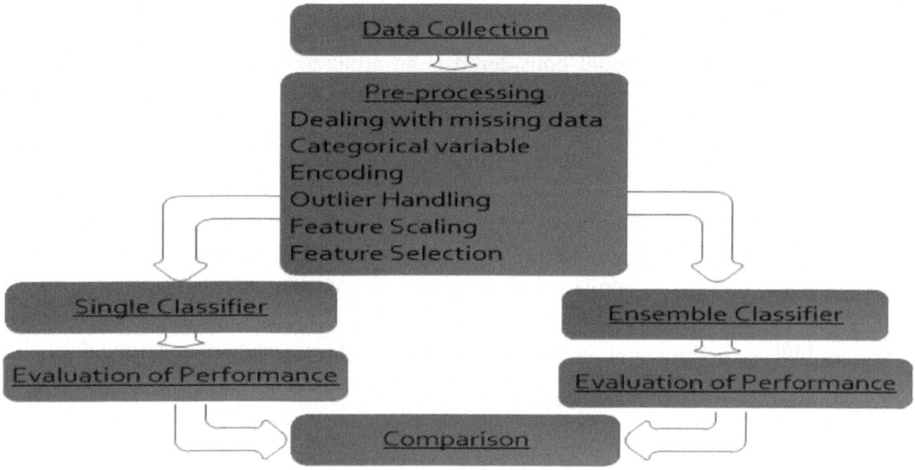

Fig. 2. A framework of the proposed methodology

4.2 Data Pre-processing

In the original dataset, there are a lot of features related to loans and customer information that will not affect the classification model and is not worth to consider and have to exclude. Some features did not exist in the original dataset created from existing features as the target feature. Target feature is created according to the criteria of AB bank, whether the loan non-performing or not that will be 1 or 0 respectively.

Many processes were run to find null values and columns with more than 80% null values to drop. This process includes an analysis of all variables and Handling non-numeric variables and categorical. After that, data normalization is used for rescaling variables to be in the form of a normal distribution form between zero and one as well as to confirm the suitability of each comparison technique [21].

4.3 Training Based Dataset

Once the pre-processing stage is completed, several base classification techniques with selected variables are applied to the data set. Then, we used a strategy to divide the data set into k parts and randomly split the training dataset into k equal parts without replacement (K-fold cross-validation). In our case, we employed 10-parts.

K-1 parts used for the model training where the remaining one subset part (with leave-one-out) is retained for testing and also for performance evaluation. Classification techniques as individuals applied in this research were Decision Tree, Support Vector Machine, Logistics Regression, and K Nearest Neighbors. First, experiments with each classification method get an accuracy value, which refers to the percentage of each classification method. The prediction values for each classification method, are then compared to the proposed Meta-classifier.

4.4 Ensemble Models

Bagging Classifier in our model considered is an adopted technique that got the best decision result and Bagging algorithm is summarized in Fig. 3. Supposed that training dataset, m containing m number of training examples, bagging is simply training T independent classifiers each of them trained by sampling, with replacement from the original training dataset.

With bagging, some examples may be sampled more than once several times, while others may be left out (not sampled, called out-of-bag OOB instances).

When the T process is applied multiple times, T samples are obtained from the M training examples, then, from each sample, a basic learner can be trained by applying a basic learning algorithm and to obtain better predictive performance the class that obtains the majority of the votes wins for classification and averaging in case of regression [22].

Input: Data set $D = \{(x_1, y_1), (x_2, y_2), \ldots, (x_m, y_m)\}$;

Base learning algorithm $\mathcal{L}$;

Number of base learners T.

Process:

1. **for** $t = 1, \ldots, T$:
2. $h_t = \mathcal{L}(D, \mathcal{D}_{bs})$% $\mathcal{D}_{bs}$ is the bootstrap distribution
3. **end**

Output: $H(x) = \underset{x \in y}{argmax} \sum_{t=1}^{T} \mathbb{I}(h_t(x) = y)$

Fig. 3. Bagging classifier

Followed by Majority Voting (Meta-classifier) which combines a set of different classifiers then it compares the predicted result in each classifier then takes the majority voting result of them to get a unique best decision result.

Figure 4 represents an example of the majority voting model. The techniques within DT, SVM, and LR classifiers were aggregated with the Ensemble model. Thereafter, the predicted outcome from those single classifiers was combined via majority voting.

For example, we can observe that the result at ID_1 of Support Vector Machine and Decision tree classifiers have the same predicted class label PL while the logistic regression classifier predicts the class label NPL. Finally, the majority voting classifier collects weighted votes from all classifiers and gives PL prediction where it has the highest vote.

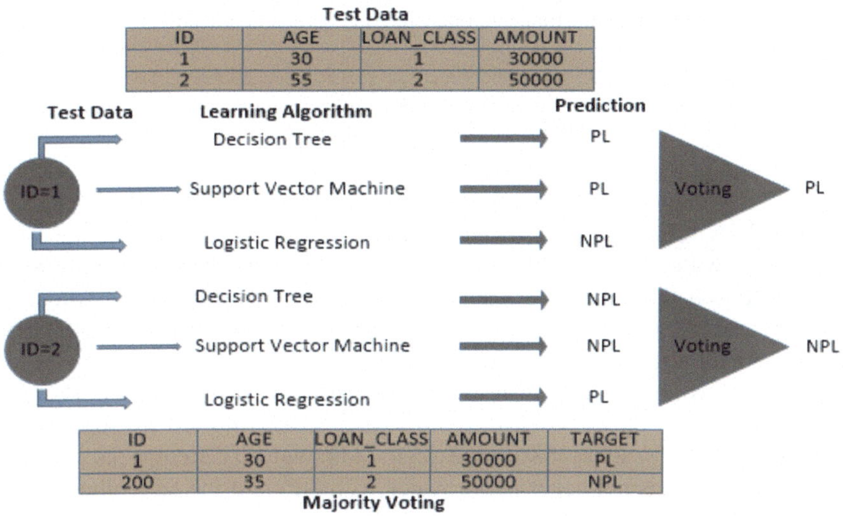

Fig. 4. Example of an ensemble of majority voting.

5 Results and Analysis

Feature selection includes analysis for all variables and also variable selection was applied to reduce the number of variables for the model.

To find which variable affects the model more and which one has more weight. We applied random forest and the least important variables were excluded. Consequently, the variables set include 17 variables instead of 22 features as shown in Fig. 5. For validating our model performance, 10-fold cross-validation used in the model stage to train the data. For the classification of non-performing agricultural loans, four single techniques (KNN, SVM, LR, and DT) and two ensemble techniques used in this experiment with reduced features. These four single techniques may not perform well as ensemble methods where the voting algorithm combines the four single classifiers to increase the performance of the four classifiers. On the other hand, we used a bagging ensemble method with the Decision tree as a base classifier to increase the classification accuracy.

To measure the performance of the models, recall and precision are complementary to each other and have been used to measure the number of examples classified as negative or positive in coordination with the actual examples, and F-score is considered a harmonic mean of precision and recall (4).

$$F\alpha = \left(\alpha\frac{1}{\text{Recall}} + (1-\alpha)\frac{1}{\text{Precesion}}\right)^{-1} \quad (4)$$

Another common measure that has been used to evaluate classification models was accuracy (5).

$$ACC = \frac{TP + TN}{TP + FP + FN + TN} \tag{5}$$

Where TP refers to True Positives, TN refers to True Negatives, FP refers to False Positives and FN refers to False Negatives [19].

In order to evaluate the results of classifiers' performance based on our approach. Table 1 and Fig. 6 show the experimental results of the comparative classifiers based on different evaluation measures such as Accuracy, Precession, Recall, and F-mean.

The observation from those results as follow:

For basic classifiers, The Logistic Regression has the best Accuracy, Precession, Recall, and F-mean with 87.91%, 89%, 88%, and 85% Followed by support vector machine with 87.90%, 89%, 88%, and 85%. Although DT known as one of the strongest classifiers has not better performance with NPLs of the agriculture bank on, the other hand, the classification accuracy of our meta classifier was the best performance with NPLs of the agriculture bank as follow: The Bagging has the best Accuracy, Precession, Recall, and F-mean with, 88.02% Accuracy, 89% Precession, 88% Recall, and 86% F-mean Followed by the Voting Algorithm with 87.94%, 89%, 88%, and 85%.

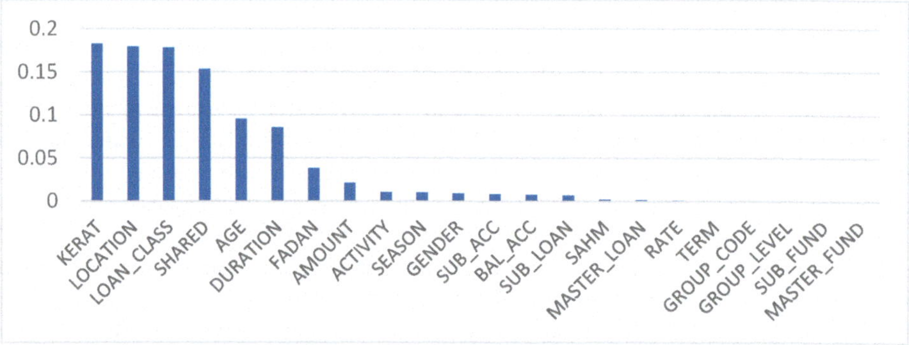

Fig. 5. Relative importance.

Table 1. Performance evaluation of the various algorithms.

Algorithm	Accuracy	Precession	Recall	F-mean
DT	0.8045	0.81	0.80	0.81
SVM	0.8790	0.89	0.88	0.85
LR	0.8791	0.89	0.88	0.85
KNN	0.8748	0.88	0.87	0.85
Voting	0.8794	0.89	0.88	0.85
Bagging	0.8802	0.89	0.88	0.86

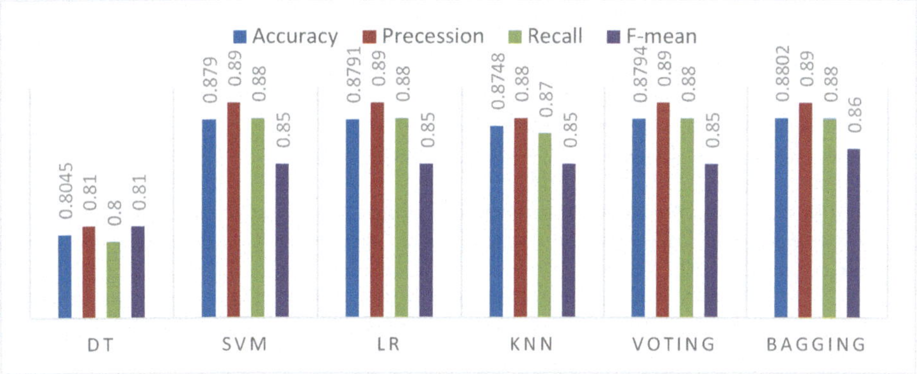

Fig. 6. Performance evaluation of all classifiers.

6 Conclusion

Predicting non-performing loans for the Agricultural Bank is essential because the agricultural activity is one of the most dangerous. It can be a reliable solution that can help both the credit officer and decision-makers avoid losses by making an appropriate decision. In this research, we applied machine learning algorithms to construct Non - performing agricultural loan prediction models.

We are just considered important variables of Non - performing agricultural loans by using the random forest technique. The cross-classification was performed using ten data partitions.

We applied four different base machine learning models to a dataset with significantly reduced features, consisting of KNN, SVM, LR, and DT classification methods. We compared the classifiers using several major measurement parameters such as Accuracy, Precision, Recall, and F-mean. We found that The Logistic Regression to achieve the highest accuracy Followed by Support Vector Machine, in contrast to the Decision tree, which achieved the worst accuracy in our experiments. To achieve higher classification accuracy, a bagging ensemble method used with the Decision tree as a base classifier which is a kind of voting algorithm, followed by ensemble voting learning techniques. The Rating results of the evaluation showed that Bagging performed well and had better performance than all classifiers. In future work, alternative methods of ensembles will be implemented to provide extra valid prediction results.

References

1. Chen, W.H., Shih, J.Y.: A study of Taiwan's issuer credit rating systems using support vector machines. Expert Syst. Appl. **30**, 427–435 (2006)
2. European Central Bank: What are non-performing loans? https://www.ecb.europa.eu/explainers/tell-me/html/npl.en.html. Accessed 11 Dec 2019

3. Paireekreng, W., Choensawat, W.: An ensemble learning based model for real estate project classification. In: 6th International Conference on Applied Human Factors and Ergonomics and the Affiliated Conferences, vol. 3, pp. 3852–3859 (2015)
4. Zhang, Y., et al.: Predicting non-performing loan of business bank by multiple classifier fusion algorithms. J. Interdisc. Math. **19**(4), 657–667 (2016)
5. Goyal, A., Kaur, R.: Loan prediction using ensemble technique. Int. J. Adv. Res. Comput. Commun. Eng. **5**(3), 523–526 (2016)
6. Okesola, O.J., et al.: An improved bank credit scoring model: a naïve Bayesian approach. In: 2017 International Conference on Computational Science and Computational Intelligence (CSCI), pp. 228–233 (2017)
7. Soni, P.M., Paul, V.: A novel optimized classifier for the loan repayment capability prediction system. In: 2019 3rd International Conference on Computing Methodologies and Communication (ICCMC), pp. 23–28 (2019)
8. Zhao, W.: Research on the deep learning of the small sample data based on transfer learning. In: AIP Conference Proceedings, vol. 1864, p. 020018 (2017)
9. Maheswari, J.P.: Breaking the curse of small datasets in Machine Learning: Part 1. https://towardsdatascience.com/breaking-the-curse-of-small-datasets-in-machine-learning-part-1-36f28b0c044d. Accessed 10 Dec 2019
10. Hand, D.J., Vinciotti, V.: Choosing k for two-class nearest neighbour classifiers with unbalanced classes. Pattern Recogn. Lett. **24**(9–10), 1555–1562 (2003)
11. Rish, I.: An empirical study of the naive Bayes classifier. In: IJCAI 2001 Workshop on Empirical Methods in Artificial Intelligence, vol. 3, pp. 41–46 (2001)
12. Patel, B., Rana, K.: A survey on decision tree algorithm for classification. Int. J. Eng. Dev. Res. (IJEDR) **2**(1), 1–5 (2014)
13. Gestel, V., et al.: A support vector machine approach to credit scoring. Bank en Financiewezen **2**, 73–82 (2003)
14. Wang, G., et al.: A comparative assessment of ensemble learning for credit scoring. Expert Syst. Appl. **38**(1), 223–230 (2011)
15. Breiman, L.: Bagging predictors. Mach. Learn. **24**, 123–140 (1996). https://doi.org/10.1023/A:1018054314350
16. Lin, W.-Z., et al.: iDNA-Prot: identification of DNA binding proteins using random forest with grey model. PLoS ONE **6**, 9 (2011)
17. Khalilia, M., et al.: Predicting disease risks from highly imbalanced data using random forest. BMC Med. Inf. Decis. Making **11**(1), 51 (2011)
18. Mohan, A., et al.: Automatic classification of protein structures using physicochemical parameters. Interdisc. Sci.: Comput. Life Sci. **6**(3), 176–186 (2014)
19. Seera, M., Lim, C.P.: A hybrid intelligent system for medical data classification. Expert Syst. Appl. **41**(5), 2239–2249 (2014)
20. Breiman, L.: Random forests. Mach. Learn. **45**, 5–32 (2001). https://doi.org/10.1023/A:1010933404324
21. Zajkowski, A., et al.: Data Normalization. U.S. Patent US20030110250 (2003)
22. Zhou, Z.-H.: Ensemble Methods: Foundations and Algorithms. Chapman & Hall/CRC, Boca Raton (2012)

Machine Learning-Based Sentiment Analysis for Analyzing the Travelers Reviews on Egyptian Hotels

Lamiaa Mostafa(✉)

Business Information System Department, Arab Academy for Science and Technology and Maritime Transport, Alexandria, Egypt
Lamiaa.mostafa31@aast.edu, Lamiaa.mostafa31@gmail.com

Abstract. Tourism affects the economy of any country; actually, it is the foundation of the country on the economic side. Egyptian Government is giving a big concern in developing the tourism sector. Hotel companies are using E-commerce technology for online booking and online reviewing. Travelers choose hotels based on their prices, facilities and other traveler's review. Sentiment analysis is a very important topic that can be used to analyze the opinion of online users. Different websites are classifying the traveler reviews such as Tripadvisor, Expedia. The research aims to propose a Traveler Review Sentiment Classifier that will analyze the traveler's reviews on Egyptian Hotels and provide a classification of each sentiment based on hotel features. Travelers Sentiment about five hotels located in Aswan in Egypt with a total of 11458 reviews were collected and analyzed. Sentiment model uses three classification techniques: Support Vector Machine, Naive Bayes and Decision Tree. Results had shown that Naïve Bayes has the highest accuracy level.

Keywords: Sentiment analysis · Tourism · Hotel · Online review · Trip advisor · Mining

1 Introduction

User-generated content (UGC) is used in different fields like the tourism and hospitality industry. Tourists share travel experiences through online travel agencies (OTAs) such as TripAdvisor, and Expedia [27]. Sentiment analysis is used to enhance the quality delivered by the customers. In the tourism industry, companies collect and analyze online travel reviews to obtain more information about travel experiences and customer feedback [28]. Online travel reviews can include hotel reviews, sightseeing, and restaurant reviews. Authors in [1] stated that 77% of traveler does not make the decision to travel unless they read online reviews. Text mining can collect these reviews easily while providing data analytics that helps companies manage their competitive advantage and traveler's loyalty.

Online reviews were collected previously by questionnaire surveys and mathematical models, however, text mining can collect these reviews easily while providing data analytics that helps companies manage their competitive advantage and travelers' loyalty [4].

A.-E. Hassanien et al. (Eds.): AICV 2020, AISC 1153, pp. 405–413, 2020.
https://doi.org/10.1007/978-3-030-44289-7_38

Authors in [2] examines the effect of personalized management responses on topic matching, he tested data of 500 hotels collected from Trip Advisor. Hoteliers must monitor the performance of their properties across the rating topics and develop specific strategies to conduct online reputation management. Authors in [5] studied the effect of hotel brand positioning and competitive by text-mining user-generated content (UGC). They combined content analysis and repertory grid analysis (RGA) to analyze 111,986 hotel reviews from the two biggest Chinese cities. They concluded that brand positioning helps developing companies' brand strategies. The paper is organized as follows: Sect. 2 provides a literature review including background about tourism, hotel review, and opinion mining. Section 3 provides details about the sentiment analysis model and its components. Section 4 conducts the experimental results and Sect. 5 discusses the conclusions, limitations and future directions.

2 Literature Review

This section will discuss the previous work in tourism, hotel review, and opinion mining.

2.1 Tourism and Hotel Review

Online travel reviews have a strong impact on tourist satisfaction and perceived quality. Researchers in [2] examines the effect of personalized management responses on topic matching; they tested data of 500 hotels collected from TripAdvisor. They revealed that hoteliers must monitor the performance of their properties across the rating topics and develop specific strategies to conduct online reputation management.

Researchers in [5] studied the effect of hotel brand positioning and competitive by text-mining user-generated content (UGC). They combined content analysis and repertory grid analysis (RGA) to analyze 111,986 hotel reviews from the two biggest Chinese cities. They concluded that brand positioning helps developing companies' brand strategies.

Consumers book rooms based on the online review features posted. Many web applications that provide booking online from these websites are Trip advisor, Booking, Expedia, and Trivage. Hotels have many features such as Location, room cleanliness, service and staff, value for money, comfort, facilities, free wifi, hotel condition, location, and Food. Table 1 shows some hotel attributes.

A semantic analytical approach was proposed in [3] outperforms baseline algorithms and achieves high precision (0.95) and recall (0.96). Based on the visual analytics results had shown that business traveler gives lower ratings and couples give higher ratings. Month December is the highest rate while July is the lowest rate, finally, business traveler expresses their opinions using words like dirty, horrible and broken. An aspect detection also is known as opinion target indicates the opinion topic. Customer reviews differ based on traveler nature such as family may focus on travel safety while another traveler may prefer room attributes.

Table 1. Hotel attributes

Feature	Trip advisor	Booking	Expedia	Triage
Location	✓	✓		✓
Room cleanliness	✓	✓	✓	✓
Service and staff	✓		✓	✓
Value for money	✓	✓		✓
Comfort	✓	✓		
Facilities	✓	✓		✓
Free Wifi		✓		
Hotel condition	✓		✓	✓
Food	✓			✓

Text mining techniques are used in [21] to understand the drivers of online hotel reviews (OHR), they concluded that content and title polarity lead to OHRs while negative emotion with low arousal makes OHR helpful. OHR can affect a company's economic level.

2.2 Opinion Mining

The aim of opinion mining is to understanding the objective expressions used for describing the opinion. Sentiment analysis defined as "a computational study that analyses opinionated texts (the texts that include opinions), which contains people's opinions, appraisals, emotions, and sentiments related to entities [6].

It summarized and classifies people's opinions, and emotions towards organizations [7]. Opinion mining is divided into six categories: sentiment analysis, opinion extraction, sentiment mining, subjection analysis, affect or emotion analysis, and review mining [22]. Opinion mining is lexicon-based approaches [8] and machine learning [9]. The lexicon includes a corpus and a dictionary-based approach. The machine learning approach divided into three groups: supervised learning, semi-supervised learning and unsupervised learning [7]. The machine learning approach extracts text features and uses a classification method [10].

Opinion mining or sentiment analysis is performed using aspect level sentiment analysis which includes different methods: frequency-based method, syntax-based method, supervised machine learning approach, unsupervised machine learning approach [11].

The frequency-based method depends on counting frequencies of single nouns and compound nouns. The syntax-based method defines the relationships between words in the form of an adjectival relationship [6]. Supervised machine learning approach depends on text classification algorithms like Support Vector Machine (SVM) [12–14],

NB classifier (NB) [13, 15], Decision Tree classifier [13], Neural Network classifier [14]. Aspect based sentiment analysis was used in [23] to extract aspects of hotels which are important by the traveler from the online hotel reviews, and then expect the traveler traveling Mode. The study concluded that social media content affects a traveler's preferences.

Authors in [16] analyze the sentiments of online reviews based on product categories of television and camera extracted from an e-commerce web site to create ranking lists that focus on a brand score of each product.

3 Sentiment Analysis Model

The sentiment analysis model is composed of three stages: data collection, processing, and classification. Sentiments are collected from four websites (TripAdvisor, Booking, Expedia, and Trivage) which are online websites for hotel reservations all over the world. Data set was specified in Table 2 into five hotels located in Aswan in Egypt with a total of 11458 reviews. A java tool is used to collect all sentiments while removing HTML tags. The sentiment model classified the reviews into (good and bad). The sentiment model classifies each sentiment into one of these six groups: Location, Staff, Hotel condition (Cleanliness, Comfort, and facilities), Value for money, Food and Free Wifi.

Table 2. List of hotels and reviews

Hotel name	TripAdvisor	Booking	Expedia	Triage
Pyramisa Isis Corniche Aswan Resort	542	1,002	129	666
Sofitel Legend Old Cataract Aswan	1824	491	210	597
Movenpick Resort Aswan	1790	527	269	747
Kato Dool Nubian House	213	51	25	57
Pyramisa Isis Island Aswan	895	769	135	519
Total	5264	2840	768	2586

The processing step is designed using Knime. Knime is an analytical tool that can be used in web mining [20]. Sentiment model is divided into three components: pre-processing, feature selection and classification as shown in Fig. 1, each component is described in the following sections.

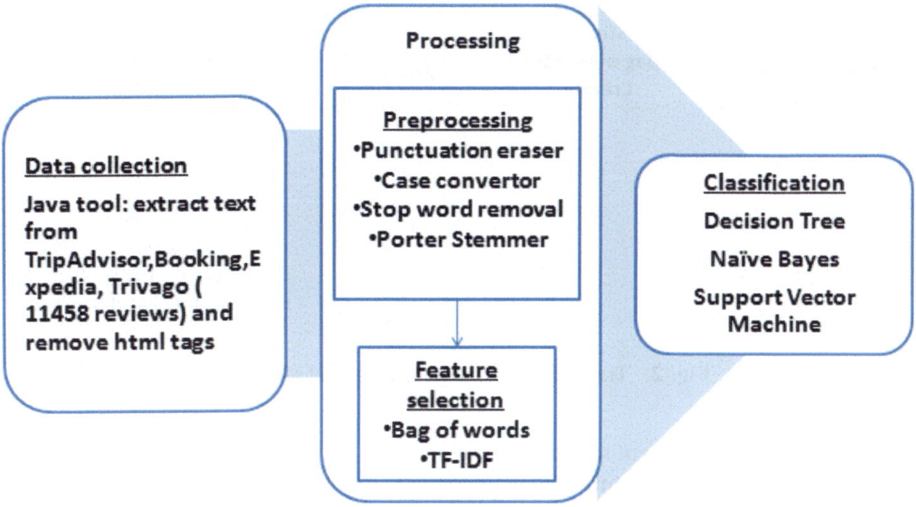

Fig. 1. Traveler review sentiment model

3.1 Pre-processing Step

Pre-processing is cleaning and removes unwanted words [17]. Different pre-processing techniques are used in mining such as remove stop words and stemming. Sentiment model uses Porterstemmer in processing.

3.2 Features Selection Step

Features are extracted from sentiments and considered to be the inputs of the classifiers. Bag of words (BOW) is a visual representation of words in the sentences [17, 18]. Term Frequency Inverse Document Frequency (TF-IDF) defines the most important words in the document [19]. TF-IDF "calculates values for each word in a document through an inverse proportion of the frequency of the word in a particular document to the percentage of documents the word appears in" [17]. Both BOW and TFIDF are used in the traveler review sentiment model.

3.3 Machine Learning Step

Different machine learning approaches are used in sentiment analysis such as Support Vector Machines [12], and Naive Bayes [19]. In the Sentiment analysis guilt model classifiers used are SVM, Naive Bayes and decision tree [17]. Decision tree induction chooses relevant features. Decision tree classifiers build tree structures to identify the feature that represents each property test. The classifier is validated using the following measures: accuracy, precision, and recall have been used. The following figure represents the phases of traveler review sentiment model in Knime implementation (Fig. 2):

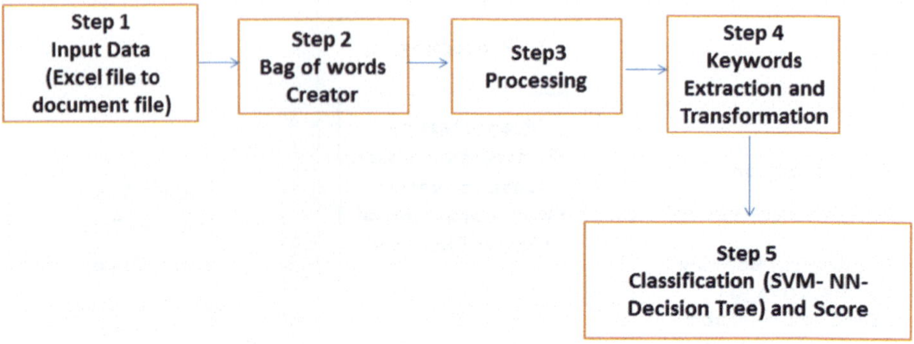

Fig. 2. Traveler review sentiment model (Knime)

Travel Review Sentiment model composes of five steps: The first step is processing in which the input data includes two documents for the training process. The two documents are converted from a string (excel reader) to document (text), step two creates the bag words. Step three includes punctuation eraser, case convertor, stop word removal, porter stemmer, and term grouper. Step four extracts the keyword that represents each class and the transformation process is used to present the keyword to each category. Step five is the classification and scoring that uses three classifiers (SVM, NN, and Decision Tree), the scorer calculates the accuracy level. Precision and recall are calculated based on the following equations:

$$\text{recall} = \frac{\textit{number of relevant items retrieved}}{\textit{a number of relevant items in the collection}} \tag{1}$$

$$\text{precion} = \frac{\textit{number of relevant items retrieved}}{\textit{total number of items retrieved}} \tag{2}$$

3.4 Classifiers Used in Traveler Review Sentiment Model

- A Naive Bayes classifier is probabilistic classifier based on Bayes' theorem, it focuses on the dimensionality of the inputs are high, NB works on the following equation [24]:

$$p_{NB}(c|d) = \frac{p(c)(\prod_{i=1}^{m} p(f_i|c)^{n_{i(d)}}}{p(d)} \tag{3}$$

  Where, c is the class, d is the document, m is the no of features and f_i is the feature vector.
- Support vector machines are supervised learning methods that find the best possible surface to separate positive and negative training samples, SVM uses the following equation [25]:

$$min_{w,b,\varepsilon}\frac{1}{2}w^{T}w+C\sum\nolimits_{i=1}^{l}\varepsilon_{i} \tag{4}$$

Given training vectors w^T text weight, x is the input vector, b is the bias, i represent class.

- Decision Tree is a K-array tree where each node examines some characteristics, decision tree uses the following equation [26]:

$$info(D)=\sum\nolimits_{i=1}^{m}p_{i}log_{2}(p) \tag{5}$$

where pi is the likelihood, that self-assertive vector in D fits in with class ci.

4 Experimental Results

Online reviews were classified into two classes: Good and Bad. Sentiment analysis classifier tests the machine learning classifiers based on 11,458 sentiments, sentiments are classified into good sentiment and bad sentiment. Naive Bayes, SVM, and Decision Tree were used for the classification of the sentiments. Table 3 shows the results of the sentiment model.

Table 3. Traveler review sentiment model results

Classifier	Accuracy	Precision	Recall
SVM	79%	0.61	0.83
NB	85%	0.76	0.67
Decision tree	71%	0.51	0.8

The results of the sentiment analysis classifier have shown that NB is the highest accuracy 85% while SVM 79% and the decision tree 71%. Sentiment analysis techniques is a very good reflector of traveler review, NB had shown the highest accuracy results which is agreed with [13, 14].

5 Conclusion and Future Work

The paper proposed a traveler review sentiment model that classified traveler's online reviews into Good reviews or bad reviews. Five hotels located in Aswan in Egypt with a total of 11458 reviews were collected and analyzed from four online travel agencies (Trip advisor, Booking, Expedia and Trivago). The sentiment analysis model includes three modules: text processing, feature selection, and machine learning classification. Three machine learning classifiers were used NB, SVM and decision tree. Experimental results had shown that NB is the most accurate classifier.

Sentiment model results had shown that NB outperforms SVM in precision with values 0.76 and 0.61, however, SVM outperforms NB in recall with the values 0.83 and 0.67. The accuracy level is accepted from the two classifiers SVM and NB and decision tree [29], the sentiment model can be applied in different areas. Decision tree induction chooses relevant features.

The traveler review sentiment model is limited to first: the dataset which includes 11458 with only 5 hotels, second: the other machine learning technique must be tested including clustering techniques and Neural Networks. Plans including testing the results in more hotels and different types of text like Arabic text.

References

1. Ye, Q., Law, R., Gu, B.: The impact of online user reviews on hotel room sales. Int. J. Hosp. Manag. **28**(1), 180–182 (2009)
2. Zhang, X., Qiaoa, S., Yang, Y., Ziqiong, Z.: Exploring the impact of personalized management responses on tourists' satisfaction: atopic matching perspective. Tour. Manag. **76**(2020), 103953 (2020)
3. Changa, Y., Kuc, C., Chend, C.: Social media analytics: extracting and visualizing Hilton hotel ratings and reviews from TripAdvisor. Int. J. Inf. Manag. **48**, 263–279 (2019)
4. Boo, S., Busser, J.A.: Meeting planners' online reviews of destination hotels: a twofold content analysis approach. Tour. Manag. **66**(6), 287–301 (2018)
5. Hua, F., Trivedi, R.: Mapping hotel brand positioning and competitive landscapes by text-mining user-generated content. Int. J. Hosp. Manag. **84**, 102317 (2020)
6. Calı, S., Balaban, S.: Improved decisions for marketing, supply, and purchasing: Mining big data through the integration of sentiment analysis and intuitionistic fuzzy multi-criteria assessment. Comput. Ind. Eng. **129**, 315–332 (2019)
7. Hermitian, F., Sohrabi, M.K.: A survey on classification techniques for opinion mining and sentiment analysis. Artif. Intell. Rev. **1**, 1–51 (2017)
8. Daud, A., Khan, W., Che, D.: Urdu Language Processing: A Survey. Kluwer Academic Publishers, Alphen aan denrijn (2017)
9. Weiler, B., Walker, K.: Enhancing the visitor experience: reconceptualising the tour guide's communicative role. J. Hosp. Tour. Manag. **21**(21), 90–99 (2014)
10. Raisi, H., Baggio, R., Barratt-Pugh, L., Willson, G.: Hyperlink network analysis fa tourism destination. J. Travel Res. **57**(2), 1–27 (2017)
11. Schouten, K., Frasincar, F.: Survey on aspect-level sentiment analysis. IEEE Trans. Knowl. Data Eng. **28**(3), 813–830 (2016)
12. Abdi, A., Shamsuddin, S.M., Hasan, S., Piran, J.: Machine learning-based multi documents sentiment-oriented summarization using linguistic treatment. Expert Syst. Appl. **109**, 66–85 (2018)
13. Raut, V.B., Londhe, D.D.: Opinion mining and summarization of hotel reviews. In: Proceedings - 2014 6th International Conference on Computational Intelligence and Communication Networks, pp. 556–559 (2014)
14. Dehkharghani, R., Yanikoglu, B., Tapucu, D., Saygin, Y.: Adaptation and use of subjectivity lexicons for domain-dependent sentiment classification. In: IEEE 12th International Conference on Data Mining Workshops Adaptation, pp. 669–673 (2012)

15. Smetana, M., Koncz, P., Smetana, P., Parali, J.: Active learning enhanced semiautomatic annotation tool for aspect-based sentiment analysis. In: IEEE 11th International Symposium on Intelligent Systems and Informatics, pp. 191–194 (2013)
16. Najmi, E., Hashmi, K., Malik, Z., Rezgui, A., Khan, H.U.: CAPRA: a comprehensive approach to product ranking using customer reviews. Computing **97**(8), 843–867 (2015)
17. Mostafa, L., Abd Elghany, M.: Investigating game developers' guilt emotions using sentiment analysis. Int. J. Softw. Eng. Appl. (IJSEA), **9**(6), (2018)
18. Mostafa, L., Farouk, M., Fakhry, M.: An automated approach for webpage classification. In: ICCTA 2009 Proceedings of the 19th International Conference on Computer Theory and Applications, Alexandria, Egypt (2009)
19. Abdelghany, M., Abdelghany, M., Mostafa, L.: The analysis of the perceptions of service facilities and their impact on student satisfaction. In: IJBR, vol. 19, no. 1 (2019)
20. Knime. http://www.knime.com/. Accessed 11 Sept 2019
21. Chatterjee, S.: Drivers of the helpfulness of online hotel reviews: A sentiment and emotion mining approach. Int. J. Hosp. Manag. (2019)
22. Kim, D., Park, B.J.: The moderating role of context in the effects of choice attributes on hotel choice: a discrete choice experiment. Tour. Manag. **63**, 439–451 (2017)
23. Yadav, M., Roychoudhury, B.: Effect of trip mode on opinion about hotel aspects: a social media analysis approach. Int. J. Hosp. Manag. **80**, 155–165 (2019)
24. Pang, B., Lee, L., Vaithyanathan, S.: Thumbs up?: sentiment classification using machine learning techniques. In: Proceedings of the ACL-02 Conference on Empirical Methods in Natural Language Processing, vol. 10, pp. 79–86 (2002)
25. Chang, C., Lin, C.: LIBSVM: a library for support vector machines. ACM Trans. Intell. Syst. Technol. (TIST) **2**(3), 27 (2011)
26. Suresh, A.: Sentiment classification using decision tree-based feature selection. Int. J. Control Theory Appl. **9**, 419–425 (2016)
27. Xiang, Z., Schwartz, Z., Gerdes Jr., J.H., Uysal, M.: What can big data and text analytics tell us about hotel guest experience and satisfaction? Int. J. Hosp. Manag. **44**(44), 120–130 (2015)
28. Zhang, F., Song, Y., Cai, W., Liu, S., Liu, S., Pujol, S., Feng, D.D.: Pairwise latent semantic association for similarity computation in medical imaging. IEEE (2016)
29. Korovkinas, K., Dennis, P., Garšva, G.: SVM and Naïve Bayes classification ensemble method for sentiment analysis. Baltic J. Mod. Comput. **5**(4), 398–409 (2017)

Machine Learning Techniques for Handwritten Digit Recognition

Ahmad Taher Azar[1,2(✉)], Alaa Khamis[3], Nashwa Ahmad Kamal[4], and Brian Galli[5]

[1] Robotics and Internet-of-Things Lab (RIOTU), Prince Sultan University, Riyadh, Saudi Arabia
aazar@psu.edu.sa

[2] Faculty of Computer and Artificial Intelligence, Benha University, Benha, Egypt
ahmad.azar@fci.bu.edu.eg

[3] Centre for Pattern Analysis and Machine Intelligence (CPAMI), University of Waterloo, Waterloo, Canada
alaakhamis@gmail.com

[4] Faculty of Engineering, Cairo University, Giza, Egypt
nashwa.ahmad.kamal@gmail.com

[5] Hofstra University, Hempstead, NY, USA
Brian.Galli@liu.edu

Abstract. The following study examines how various classification algorithms perform on the problem of handwritten digit recognition. The classifiers discussed are k-Nearest Neighbour (k-NN), Single Classification Decision Trees and Bagged Decision Trees. These algorithms were evaluated with the use of information from the United States Postal Service (USPS). This study's results show that the k-NN classifier had the fastest performance while the bagged decision trees were the slowest. In terms of classification performance, the bagged decision tree method was found to have the fewest misclassifications and outperformed k-NN and single classification trees in all of the considered metrics.

Keywords: Optical Character Recognition · k-Nearest Neighbor · Single Classification Decision Trees · Bagged Decision Tree

1 Introduction

The recognition of handwritten digits is an active research area and has many applications, such as recognizing postal codes in post offices and recognizing digits on mobile devices [1–3]. Furthermore, Optical Character Recognition (OCR), an application of machine learning, involves the conversion of printed or handwritten text images into machine-encoded text [4]. The use of early OCR techniques can be traced to 1914 where it was used in the development of technologies, such as reading products for blind individuals and telegraphs [3]. More recently, OCR has become more common, and several object-oriented OCR systems have been developed for applications

A.-E. Hassanien et al. (Eds.): AICV 2020, AISC 1153, pp. 414–426, 2020.
https://doi.org/10.1007/978-3-030-44289-7_39

involving invoices, sales receipts, checks, template matching, and legal billing documents.

In general, the character recognition pipeline is made up of four stages: image analysis, image enhancement, character recognition and contextual processing. Image analysis involves aspects such as image quality assessment, text line detection and word/character extraction. From here, the image is enhanced to improve its quality by filling holes and removing noise. Character recognition is performed to identify words and characters based on shapes within the image and other features. Contextual processing is the last part of the pipeline and aims to limit the feature search space [5]. The focus of this paper is the classification problem associated with the character recognition aspect of the OCR pipelines, more specifically, the problem of handwritten digit recognition. Three algorithms will be considered for this problem: k-Nearest Neighbor (k-NN), Single Classification Decision Tree and Bagged Decision Trees. For the purpose of this study, the data set for the United States Postal Service (USPS) Data Set will be utilized. This data set consists of several samples of each digit and will be used when benchmarking the performance of the three algorithms.

The remaining of this paper is organized as follows. Section 2 of this paper provides preliminaries of the algorithms that were examined as part of this study. Section 3 defines the problem statement and corresponding model. The proposed approaches to solve the digit classification problem are covered within Sect. 4, followed by the results and discussion in Sect. 5. Finally, Sect. 6 gives a concise conclusion and prospective for further work.

2 Preliminaries

Formally speaking, the learning algorithm seeks a scoring function $f : X \times C \to \mathbb{R}$, where X represents pixel/feature space, C represents handwritten digit space. This policy can be learned using N training examples of the form $\{(x_{11}, x_{21}, \ldots., x_{k1}, c_1), (x_{12}, x_{22}, \ldots., x_{k2}, c_2), \ldots, (x_{1N}, x_{2N}, \ldots., x_{kN}, c_N)\}$, where x_i is the feature vector of the i-th example, k is number of samples and c_i is the handwritten digit class. The learned mode returns the recognized digit that gives the highest score of f, i.e., $c(x) = arg\ \max_c f(x, c)$.

As can be seen, the handwritten digit recognition problem is handled as a multi-classification problem. Different learning algorithms can be used to learn the classification policy that allows predicting the handwritten digit given an image. The following subsections introduce K-NN, decision trees and bagged decision trees techniques.

(a) **k-Nearest Neighbor**

The k-NN technique is one of the simplest approaches for machine learning. It performs classification from the closest training samples within the feature space by calculating the space between the particular feature vector to be classified and every other feature vector within the data set. The given feature vector is then assigned the class label corresponding to its k nearest neighbors (based on a majority vote) [6].

Furthermore, a set of pairs is provided: (x_1, c_1), (x_2, c_2), …, (x_n, c_n). The variables, x_i, assume values in a metric space X, where a metric, d, is clarified. The variables, c_i, take on values from the set [1, 2, …, M], as M indicates how many classes. Each c_i reflects the sample of the ith class, as x_i is also the resulting measurement made on that data point. When classifying a new pair, (x, c) the goal of this classification is to estimate the value of c [7]. The value of $x_n' \in [x_1, x_2, \ldots, x_n]$ is considered the closest neighbor to x when:

$$\min d(x_i, x) = d(x_n', x)\ i = 1, 2, \ldots, n \tag{1}$$

A commonly used distance metric used for k-NN is the Euclidean distance shown in (2). Other metrics include Cityblock and Mahalanobis distance.

$$(xi, x) = \|xi - x\| = \sqrt{}(xi - x) \cdot (xi - x) \tag{2}$$

An advantage of the k-NN method is that it works well with multi-modal classes. Since the algorithm bases its decision on a small neighborhood of objects it can result in good accuracy despite dealing with a multi-modal target class. However, k-NN considers all features equally when measuring similarities. This can result in lower accuracy for cases where only a small number of features are useful for classification [7].

(b) **Single Classification Decision Trees**

Decision tree learning is a popular method for inductive inference [8]. Its tree structure is similar to a flow chart, as the non-leaf nodes indicates an analysis on a feature, and the branches signify the test results with every leaf node holding a label of class [9]. The decision tree recursively splits training data into subsets based on the value of a single attribute. Each split corresponds to a node in the tree. Splitting stops when every subset is pure (all elements belong to a single coordination mode). Different algorithms can be used to achieve this procedure such as ID3, C4.5, CART. The most popular of these algorithms are ID3 (Iterative Dichotomiser 3), CART (Classification and Regression Tree) algorithms, and C4.5. While the ID3 algorithm is simple, it builds trees by testing the values of the properties to maximize information gain and to split the data set. C4.5 is an extension of ID3, as it constructs a decision tree by recursively splitting the data and using the gain ratio as the splitting criteria [10]. Finally, The CART algorithm is used to generate binary decision trees (where each non-leaf node has two children). This algorithm recursively partitions the data into two subsets, which makes the data within each subset more homogeneous. Also, it maximizes the decrease in impurity. The CART algorithm is popular because it can handle both numerical and categorical data, and can easily handle outliers. It can even identify the most significant variables while eliminating the non-significant ones; however, this can result in an unstable tree [10]. In general, decision trees are very popular because they perform well with large data sets and can manage categorical and numerical data. Decision trees also utilize statistical tests to validate models, as well as a white-box model that is easily explained using Boolean logic. Additionally, decision trees are robust, as well as easy to interpret. However, this technique has its weaknesses as well. One of the main disadvantages of an optimal decision tree is that it is "NP-complete" [11]. Furthermore, the algorithms used with decision trees are greedy algorithms. At every node, they tend to make locally-optimal decisions, but they cannot ensure a globally-optimal decision tree. Decision tree methods can also result in "over-complex trees" that overfit the training data and do not generalize well. When dealing with data involving categorical variables with multiple levels, information gain in the tree can be biased towards attributes with more levels [12].

(c) **Bagged Decision Trees**

The decision tree technique can be extended using ensemble learning [9]. One popular ensemble learning method is bootstrap aggregation, or bagging [13, 14]. In this method, training data is repeatedly re-sampled in order to improve prediction performance. By re-sampling the training data, the prediction error is decreased by reducing the variance of the results. This is especially useful when working with decision trees, a relatively unstable algorithm whose results can change dramatically if the training data is modified slightly.

In order to construct bagged decision trees for a data set that is the size of n, M bootstrap samples of size n are generated and consist of n examples from the original data set, with replacement. The probability of selecting each sample is shown in (3) and on average, each bootstrap sample will consist of 63.2% of unique training data while the rest are duplicates. From here, a decision tree is grown for each bootstrap sample and the final prediction is obtained by averaging the results of each tree.

$$Probability = (1 - 1/n)^n \tag{3}$$

The main advantage of using a bagging approach is that it uses averaging to reduce the variance of the model without affecting the bias. This is very helpful when dealing with overfitted models where there is a high dependence on the training data. However, bagging does not provide much improvement in cases with high bias where changes in training data have minimal influence on the base model [14]. The use of bagged decision trees is also more complex than using a single tree, resulting in reduced interpretability.

3 Formulating and Modeling a Problem

In this section, an OCR problem will be defined and used to illustrate potential solutions. Thus, the data set of The United States Postal Service (USPS) will be considered [15]. Furthermore, The USPS data set is commonly used for experimentation with handwriting recognition applications. It consists of 16-bit grayscale images of the digits from 0–9 with 1100 examples of each character. These handwritten digits were obtained from envelopes by the USPS and were de-slanted and size normalized to facilitate classification [15].

The images contain 16 × 16 pixels with grayscale values of 0 to 255. The 256-pixel values will be used as the feature vector X and will provide input for the classifier to assign one of the ten-digit classes to the image based on its features. The digit class is denoted by the 10-level categorical variable, C. In this work, 770 samples of every digit were utilized for training, while the final 330 were utilized for testing. As a result, the total training dataset, D, consisted of 7700 images and the test dataset consisted of 3300 images.

4 Proposed Method

Template matching will be used for classification. The templates, consisting of character image prototypes for each digit, will be collected using the training data. Digit classification will be performed by comparing the input image against the templates to obtain a similarity measure. The input digit will be assigned a class label based on the most similar template. The three classification techniques discussed previously were examined as potential solutions for this problem. They are the k-NN, single classification tree, and bagged decision tree algorithms.

(a) **k-Nearest Neighbor**
The k-NN algorithm is non-parametric and is used for classification and regression [16]. In k-NN classification, both the distance matrix and number of neighbors can be altered. The pseudocode for k-NN classification is provided in Fig. 1. The k-NN model was implemented in Matlab using built-in functions. First, the model was created by acquiring the randomly partitioned trained data sample containing predictor and classification values. This was done using the fitcknn function. Based on this model, the predict function was used to generate a vector of predicted class labels to minimize the expected cost. A value of $k = 5$ was used for this study; thus, the five nearest neighbours will be used for classification.

Algorithm 1: k-NN

Input : Image as feature set, X
Handwritten digits as a set of classes, C
Training data, $D = (X, C)$
n_D is the number of training samples
Output: Recognized handwritten digit class

begin
    for $i = 1 \ldots n_D$ **do**
        Calculate Euclidean distance $d(x_i, x)$
        Order $d(x_i, x)$ from lowest to highest
        Select k nearest neighbours to x
        Assign class label to x based on most frequent classs within these k neighbours

---

Single Classification Decision Trees

The single classification tree is a predictive model that performs classification by observing an item's target value [17]. The pseudocode for constructing a decision tree is provided in Algorithm 2.

Algorithm 2: Build Decision Tree

Input : Image as feature set, X
Handwritten digits as a set of classes, C
Training data, $D = X, C$
n_D is the number of training samples

Output: Decision tree

Split: $(node, D)$
begin
    $A \leftarrow$ the best attribute for splitting the D
    Decision attribute for this node $\leftarrow A$
    for *each value of* A **do**
        Create new $child_{node}$
        Split training D to $child_{nodes}$
        for *each* $child_{node}/subsect$ **do**
            if $\{$*subset is pure* **then**
                Stop
            else
                Split $(child_{node}, subsect)$

This technique was implemented using built-in Matlab functions. The single classification tree was generated using the fitctree function. Then, based on the model generated, the predict function was used to generate a vector of predicted class labels to minimized the expected cost.

(b) **Bagged Decision Trees**

When using bagged decision trees training data is repeatedly re-sampled in order to improve prediction performance. The pseudocode for constructing the ensemble trees is presented in Algorithms 3 and 4.

Algorithm 3: Build Bagged Decision Tree

Input : A training data, D and a test set E
M is the number of bootstrap samples

Output: training tree

begin
  for $i = 1...M$ **do**
    Uniformly obtain bootstrap sample of size n from D with replacement
    Learn training tree, T_i using D

Algorithm 4: Bagged Decision Tree

Input : A sample, x and a test set E

Output: Recognized handwritten digit class c

begin
  for $i = 1...M$ **do**
    Obtain a class label c_i from each of the M decision trees
  Find the final class label c as average of the M results
  Return c

When implementing the bagged decision tree model, several built-in functions from Matlab were used. An ensemble of 200 decision trees was generated using the fitensemble function since it is an extension of ensemble learning. Then, based on the model generated, a vector of predicted class labels was generated using the predict function to minimize the expected cost. To facilitate comparison, a confusion matrix was generated using the results of each algorithm [18–22]. The confusion matrix is known as a contingency table or an error matrix. It shows the predicted and actual classifications performed by a model. The rows of the matrix show the actual values, while the predicted values are represented by the columns. Thus, the elements on the diagonal are the correct predictions, while off-diagonal elements represent misclassifications [19, 23–28]. The confusion matrix was generated using the built-in Matlab function, confusionmat.

5 Results and Performance Assessment

The algorithms of the k-NN, the single classification tree, and the bagged decision tree had classification times of 33.7, 39.8, and 133.6 ms respectively. This measure reflects the time required by each classifier to learn and test the model. Furthermore, the k-NN classifier had fastest performance, and the bagged decision tree classifier was the slowest. Figure 1 utilizes the k-NN algorithm to show the confusion matrix. Figure 2 shows the confusion matrix using a single classification tree. Also, Fig. 3 illustrates the confusion matrix with an ensemble of bagged classification trees. Customizable heat maps were used when generating these confusion matrices in order to better visualize the results. In this case, the magnitude of the values is indicated by the intensity of the color red with low values shown in a light red and high values shown in bright red.

From simply observing Figs. 1, 2 and 3, it can be seen that the diagonal elements are shown in bright red, indicating that the majority of predictions are correct for all three classifiers. These results also show that the bagged decision trees had the fewest misclassifications and therefore the best performance. The k-NN algorithm had similar performance, although it had more misclassifications than the bagged decision trees. The single classification decision tree had the most misclassifications. Figures 1, 2 and 3 also show that all three classifiers had the weakest performance when classifying the digit 8. In all three cases, this digit was frequently confused with other digits, resulting in more misclassifications.

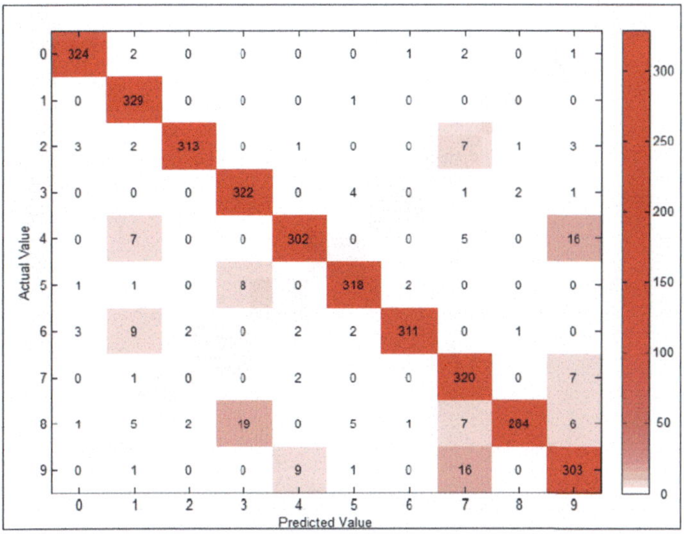

Fig. 1. Confusion matrix using k-NN

Further analysis was performed by converting each of the confusion matrices above into confusion matrices for binary classifiers. This facilitated the calculation of metrics such as accuracy, precision and misclassification rate.

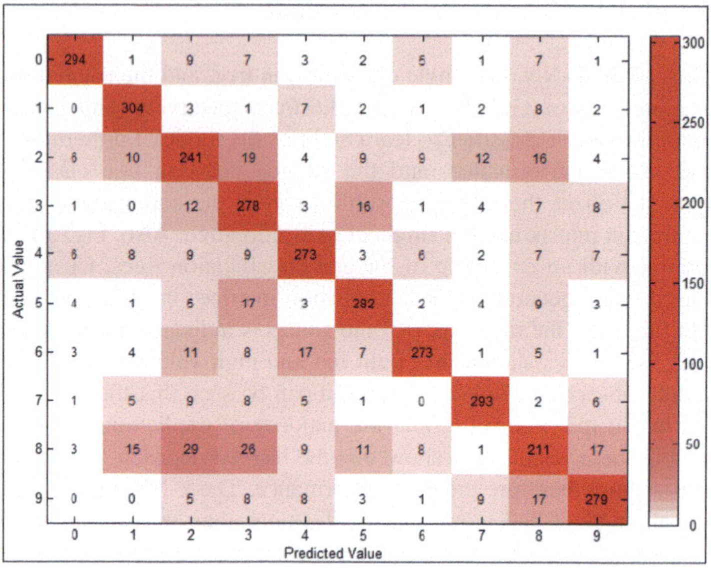

Fig. 2. Confusion matrix using single classification tree

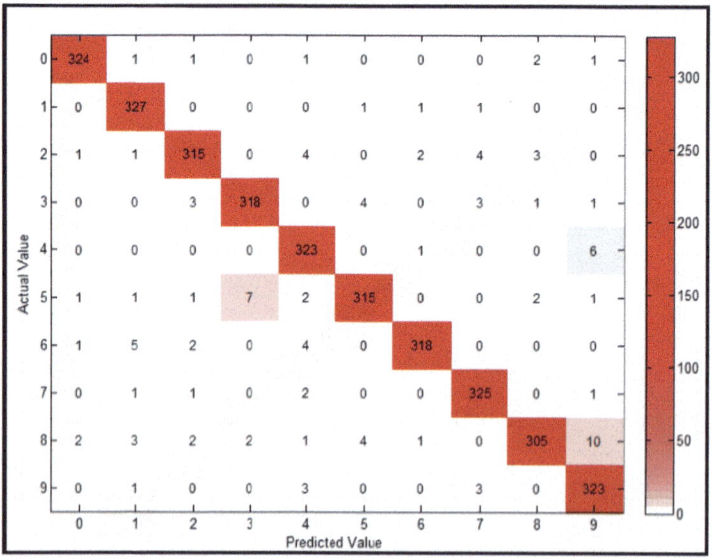

Fig. 3. Confusion matrix using ensemble of bagged classification trees

Transforming the confusion matrix into a binary classifier requires that the classification performance for each digit is considered. For example, when analyzing the number 8, True positive (TP) occurs when 8 is classified correctly, False positive (FP) occurs when other digits are classified as 8, False negative (FN) occurs when 8 is classified as other digits, and True negative (TN) occurs when other digits are classified as other digits (not the digit 8). The values for TP, TN, FP, and FN were calculated and summarized in Table 1. These tables present the performance metrics for the digit 8. This digit was chosen as the focus for this example because it was found that all three classifiers had the weakest performance when classifying the number 5.

Table 1. Confusion matrix using KNN, single classification tree and Bagged Classification Trees for detecting number 8

KNN classifier		
	Predicted positive	*Predicted negative*
Actual positive	284	46
Actual negative	4	2966
Single classification tree		
	Predicted positive	*Predicted negative*
Actual positive	211	119
Actual negative	78	2892
Bagged classification trees		
	Predicted positive	*Predicted negative*
Actual positive	305	25
Actual negative	8	2962

Next, a set of evaluation metrics was used to assess the performance from the confusion matrix. The overall accuracy divides the correct predictions by the total number of predictions. The confusion matrix's accuracy rate using k-NN algorithm is found to be 0.947. Also, the confusion matrix's overall accuracy rate using single classification tree is found to be 0.827. Finally, the confusion matrix's overall accuracy rate using an ensemble of bagged classification trees is found to be 0.968.

Table 2 shows the evaluation metrics for the k-NN, single classification tree and bagged decision tree algorithms, but only for the number 8. From the table, it is clear that the bagged decision trees had the highest accuracy and sensitivity, and the lowest miss rate and misclassification rate. The k-NN method was found to perform best in terms of fall-out rate, specificity and precision. The single classification tree method did not outperform the other algorithms but still performed well with an accuracy rate of 0.94.

Table 2. The results of K-NN, the single classification tree, and the bagged classification trees for the number 8

Evaluation metrics	KNN	Single classification tree	Ensemble bagged classification trees
Accuracy	0.985	0.940	0.990
Sensitivity/Recall	0.861	0.639	0.924
Fall-out	0.001	0.026	0.003
Specificity	0.999	0.974	0.997
Miss rate	0.139	0.361	0.076
Precision	0.986	0.730	0.974
Misclassification	0.015	0.060	0.010

The accuracy, sensitivity, specificity and precision of the classifiers for each digit are summarized in Fig. 4.

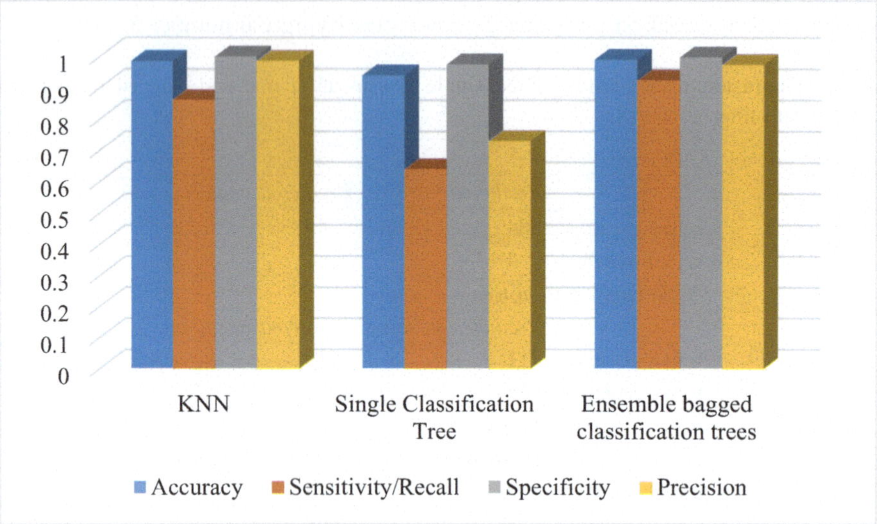

Fig. 4. The accuracy, sensitivity, specificity and precision of k-NN, single classification tree and the bagged decision tree

Overall, the bagged decision tree classifier outperformed the other algorithms. This result was expected since the use of ensemble learning has been shown to have better performance than the use of a single classifier. By performing classification based on the average of 200 decision trees, the number of misclassifications is greatly decreased in comparison to the single decision tree performance. Similarly, the results of the bagged decision trees are more accurate than the results of the k-NN algorithm. Since it uses the average of 200 decision trees, this algorithm is less affected by outliers and diminishes the variance within the results. Though, the k-NN method had similar performance to the bagged decision trees, it was found to be much faster. The bagged decision tree technique is more computationally expensive as it must train and store 200 decision trees; thus, the difference in classification time is expected. However, in this case, the classification accuracy between the two algorithms is not significant. This indicates that the performance increase presented by the bagged decision trees may not be worth the additional time required for training and prediction.

6 Conclusion

As part of this study, three techniques for handwritten digit recognition were examined: k-NN, single classification decision tree and bagged decision trees. These techniques were evaluated using the USPS data set to compare their performance. A confusion matrix was generated using the results of each algorithm and additional metrics were used to facilitate comparison. From here it was found that the bagged decision tree method had the best classification performance; however, it required much more time for training and classification. Since the performance of the k-NN classifier was comparable to that of the bagged decision trees, it is recommended that future studies consider the use of ensemble learning with the k-NN algorithm. In this study it was found that the use of ensemble learning with decision trees greatly improved the classification performance. Therefore, it can be expected that similar improvements can be obtained for the kNN classifier when implemented using ensemble learning. Similarly, the use of hybrid techniques should also be considered for this application. These techniques seek to reduce the weaknesses of intelligent methods by using complementary techniques.

Further testing could be done to evaluate the performance of the three algorithms when used with different features of the data set. While this study performed template matching using pixel values as features, other information such as character shapes or other attributes such as concavities and corners could be used. The use of such features has the potential to improve classification performance.

References

1. Arica, N., Vural, F.T.Y.: An overview of character recognition focused on offline handwriting. IEEE Trans. Syst. Man Cybern. Part C: Appl. Rev. **31**(2), 216–233 (2001)
2. Cheriet, M., Kharma, N., Liu, C.L., Suen, C.Y.: Character Recognition Systems: A Guide for Students and Practitioners. Wiley, Hoboken (2007)
3. Dholakia, K.: A survey on handwritten character recognition techniques for various Indian languages. Int. J. Comput. Appl. **115**(1), 17–21 (2015)
4. Chaudhuri, A., Mandaviya, K., Badelia, P., Ghosh, S.K.: Soft computing techniques for optical character recognition systems. In: Optical Character Recognition Systems for Different Languages with Soft Computing. Studies in Fuzziness and Soft Computing, vol. 352. Springer, Cham (2017)
5. Yu, F.T.S., Jutamulia, S.: Optical Pattern Recognition. Cambridge University Press, Cambridge (1998)
6. Mittal, K., Aggarwal, G., Mahajan, P.: Performance study of K-nearest neighbor classifier and K-means clustering for predicting the diagnostic accuracy. Int. J. Inf. Technol. **11**(3), 535–540 (2019)
7. Li, Y., Lu, B.L.: Feature selection for identifying critical variables of principal components based on k-nearest neighbor rule. In: Qiu, G., Leung, C., Xue, X., Laurini, R. (eds.) Advances in Visual Information Systems. VISUAL 2007. Lecture Notes in Computer Science, vol. 4781. Springer, Heidelberg (2007)
8. Azar, A.T., El-Metwally, S.M.: Decision tree classifiers for automated medical diagnosis. Neural Comput. Appl. **23**(7–8), 2387–2403 (2013)
9. Dong, X., Yu, Z., Cao, W., Shi, Y., Ma, O.: A survey on ensemble learning. Frontiers Comput. Sci. **14**(2), 241–258 (2020)
10. Batra, M., Agrawal, R.: Comparative analysis of decision tree algorithms. In: Panigrahi, B., Hoda, M., Sharma, V., Goel, S. (eds.) Nature Inspired Computing. Advances in Intelligent Systems and Computing, vol. 652, pp 31–36. Springer, Singapore (2018)

11. Hyafil, L., Rivest, R.L.: Constructing optimal binary decision trees is NP-complete. Inf. Process. Lett. **5**(1), 15–17 (1976)
12. Hothorn, T., Hornik, A.: Unbiased recursive partitioning: a conditional inference framework. J. Comput. Graph. Stat. **15**(3), 651–674 (2006)
13. Yildiz, O.T., Irsoy, O., Alpaydin, E.: Bagging soft decision trees. In: Holzinger, A. (ed.) Machine Learning for Health Informatics. Lecture Notes in Computer Science, vol. 9605, pp. 25–36. Springer, Cham (2016)
14. Khiari, J., Moreira-Matias, L., Shaker, A., Ženko, B., Džeroski, S.: MetaBags: bagged meta-decision trees for regression. In: Berlingerio, M., Bonchi, F., Gärtner, T., Hurley, N., Ifrim, G. (eds.) Machine Learning and Knowledge Discovery in Databases. ECML PKDD 2018. Lecture Notes in Computer Science, vol. 11051, pp. 637–652. Springer, Cham (2019)
15. Roweis, S.: USPS Handwritten Digit Dataset. http://www.cs.nyu.edu/~roweis/data.html. Accessed 11 Nov 2019
16. Nielsen, F.: Supervised learning: practice and theory of classification with the k-NN rule. In: Introduction to HPC with MPI for Data Science. Undergraduate Topics in Computer Science, pp. 213–229. Springer, Cham (2016)
17. Zhu, Q., Azar, A.T.L Complex System Modelling and Control Through Intelligent Soft Computations. Studies in Fuzziness and Soft Computing, vol. 319, Springer, Heidelberg (2015). ISBN 978-3-319-12882-5
18. Azar, A.T., El-Said, S.A.: Probabilistic neural network for breast cancer classification. Neural Comput. Appl. **23**(6), 1737–1751 (2013)
19. Inbarani, H.H., Banu, P.K.N., Azar, A.T.: Feature selection using swarm-based relative reduct technique for fetal heart rate. Neural Comput. Appl. **25**(3–4), 793–806 (2014)
20. Azar, A.T., El-Said, S.A.: Superior neuro-fuzzy classification systems. Neural Comput. Appl. **23**(1), 55–72 (2013). https://doi.org/10.1007/s00521-012-1231-8
21. Kumar, S.S., Inbarani, H.H., Azar, A.T., Polat, K.: Covering-based rough set classification system. Neural Comput. Appl. **28**(10), 2879–2888 (2017)
22. Jothi, G., Inbarani, H.H., Azar, A.T., Polat, K.: Tolerance rough set firefly-based quick reduct. Neural Comput. Appl. **28**(10), 2995–3008 (2017)
23. Azar, A.T., Inbarani, H.H., Devi, K.R.: Improved dominance rough set-based classification system. Neural Comput. Appl. **28**(8), 2231–2246 (2017)
24. Inbarani, H.H., Kumar, S.U., Azar, A.T., Hassanien, A.E.: Hybrid rough-bijective soft set classification system. Neural Comput. Appl. **29**(8), 67–78 (2018)
25. Jothi, G., Inbarani, H.H., Azar, A.T., Devi, K.R.: Rough set theory with Jaya optimization for acute lymphoblastic leukemia classification. Neural Comput. Appl. **31**(9), 5175–5194 (2019)
26. Banu, P.K.N., Azar, A.T., Inbarani, H.H.: Fuzzy firefly clustering for tumor and cancer analysis. Int. J. Model. Ident. Control (IJMIC) **27**(2), 92–103 (2017)
27. Inbarani, H.H., Azar, A.T., Jothi, G.: Supervised hybrid feature selection based on PSO and rough sets for medical diagnosis. Comput. Methods Programs Biomed. **113**(1), 175–185 (2014)
28. Jothi, G., Inbarani, H.H., Azar, A.T.: Hybrid tolerance-PSO based supervised feature selection for digital mammogram images. Int. J. Fuzzy Syst. Appl. (IJFSA) **3**(4), 15–30 (2013)

Short Term Electricity Load Forecasting Through Machine Learning

Ahmad Taher Azar[1,2](✉), Alaa Khamis[3], Nashwa Ahmad Kamal[4], and Brian Galli[5]

[1] Robotics and Internet-of-Things Lab (RIOTU), Prince Sultan University, Riyadh, Saudi Arabia
aazar@psu.edu.sa
[2] Faculty of Computer and Artificial Intelligence, Benha University, Benha, Egypt
ahmad.azar@fci.bu.edu.eg
[3] Center for Pattern Analysis and Machine Intelligence (CPAMI), University of Waterloo, Waterloo, Canada
alaakhamis@gmail.com
[4] Faculty of Engineering, Cairo University, Giza, Egypt
nashwa.ahmad.kamal@gmail.com
[5] Hofstra University, Hempstead, NY, USA
Brian.Galli@liu.edu

Abstract. Essentially, Electricity Load Forecasting is an approximation of upcoming active loads from a variety of load buses before the active loads occur. Also, it is an important factor for power system energy management. Accurate and precise load forecasting can help to reduce the capacity of the power system, to make unit commitment decisions, and to increase the dependability of power systems. Hence, this paper presents a generalized regression Neural Network (GRNN) based approach for Short Term Load Forecasting (STLF). The results showed that the performance of GRNN with 30 neurons is better of short-term load forecasting in comparison with 10 neurons. For 10 neurons, the Mean Absolute Percentage Error (MAPE) was 2.10% and Mean Absolute Error (MAE) was 306.21 MWh. However, for 30 neurons it was observed that MAPE is 1.81% and MAE 268.48 MWh.

Keywords: Electricity Load Forecasting · Energy management · Power system · Short Term Load Forecasting (STLF)

1 Introduction

Machine learning has become, for the past couple of decades, one of the cornerstones of information systems and a rather central part of our life used for solving real-world problems [1–7]. Through rising information accessibility, it is worth believing that the use of intelligent data analysis as a necessary component of technological development will become even more prevalent [3, 8–14]. The focus of Short-Term Load Forecasting (STLF) is to predict system loads over the course of one hour to one week for sufficiently operating and scheduling the power systems [15, 16]. Furthermore, STLF

A.-E. Hassanien et al. (Eds.): AICV 2020, AISC 1153, pp. 427–437, 2020.
https://doi.org/10.1007/978-3-030-44289-7_40

conventionally has been the most important element of energy management systems because its input data is vital for load flow analysis and contingency. Any electrical utility operation aims to serve dependable power to its consumers. In electronic distribution, customer load demands may change all the time, as human activities vary over the cycles of days, weeks, and months. Also, there are different levels of load consumption during holidays. Load demands are usually high throughout the day than in the evening, as lights are used during the day more so than at night when most people sleep [17]. So, approximating loads at any expected time becomes an important factor. If it is done incorrectly, then there would be a direct impact on the planning of future loads which eventually affects in overall planning process of the power system. Hence, load forecasting is crucial in operating, preparing, and controlling electric systems.

Load forecasting is difficult because the load on the consumer side is highly complex. Load at any given hour depends on several elements (i.e., the load from an hour prior, the load during the same hour from the day prior, and the load from the same hour a week prior) [18]. Furthermore, the other factors affecting the load consumption are external components, especially weather [19]. There are wide number of solutions available for the load forecasting problem. Traditional approaches have been time series modeling, regression and interpolation methods. Modern day approaches for STLF utilize machine learning techniques, including Artificial Neural Networks (ANN) based methods [20, 21], Fuzzy Logic approaches [22, 23], Support Vector Machine (SVM) based techniques [24], Wavelet Neural Network with Particle Swarm Optimization [25], etc.

The time series method approaches load pattern as a time series signal to then predict future loads with multiple time series analysis methods [26]. The most common issues with this approach are inaccurate predictions and numerical instability [27]. Additionally, the time series approach does not account for weather variables, which leads to unreliable results. The weather variables always have a strong correlation, such as humidity, temperature etc. The time series approach uses cumbersome matrix-oriented algorithms that could be unstable in some instances. Regression approaches for STLF try to reveal a functional relationship between load demands and weather variables. Furthermore, conventional regression techniques utilize linear or piecewise linear to represent the forecasting functions [27]. Though linear combinations regression approaches can attempt to establish a functional correlation, there is no linear relationship between weather and loads. Hence, there is a need for more adaptive techniques.

In this paper, a machine learning approach, such as Artificial Neural Network (ANN), is used for STLF. Similar to the time series method, ANN observes preceding load patterns to predict or extrapolate a load pattern by utilizing the current load model. Also, it uses weather and holiday data to model the system, can carry out non-linear modeling, and has adaptive capabilities. However, ANN does not require any prior assumption of functional relationships between weather variables and load. One of the greatest advantages of ANN is that although it might take hours to train the model when the training data is big, results are given in micro seconds for the new test data.

This study is organized in the following sections: section two features a brief literature assessment of some of pre-existing algorithms for STLF. Section three describes problem formulation and modeling of ANN. Finally, conclusions and future work are presented in section five.

2 Preliminaries

A. *Time Series Modeling*

The time series method is rooted in the intuition that load pattern is only a time series signal with established hourly, weekly, and monthly periodicities that roughly estimate future load [28]. Any difference among the estimated load and the actual is the stochastic process, which results in a random signal. The assessment of the random signal is carried out with certain techniques, such as the Box-Jenkins method [29], Kalman Filtering [30], spectral expansion techniques [31], and the auto-regressive moving average (ARMA) techniques. Estimation of covariance matrix is required in case of Kalman Filtering method. High levels of non-stationary load pattern make it hard to give an accurate estimate of the load pattern. The Box-Jenkins method uses autocorrelation to identify ARMA models [29]. This is accomplished using pattern recognition techniques, but a disadvantage is slow performance. Also, the ARMA model explains the stochastic pattern for hourly loads. ARMA models make use of linear combinations of previous hour load patterns to estimate the current load pattern. It usually requires larger data set for accurate results. For parameter identification, there needs to be a longer computational time. In general, time series techniques perform optimally if there are no significant environmental changes or sociological variables that directly affect the load pattern. Any changes in those parameters, time series fails to produce precise results. Also, time series uses many complex relationships that can take a large computational time, which can result in numerical instabilities [29].

B. *Regression*

The regression method involves mainly three steps. First, it selects the appropriate or accessible weather variables. Second, basic useful elements are assumed. Third, appropriate coefficients are found for the assumed functional elements. As temperature information is the most important factor it is used commonly in regression-based approaches [16, 17]. Most approaches use simple linear or piecewise linear functions for basic functional elements. Equation (1) shows the commonly utilized functional relationship between load (L) and temperature (T).

$$L = \sum_{i=1}^{N} a_i T\{U(T - T_{i1}) - U(T - T_{i2})\} + C \quad (1)$$

Where

$$U(t) = \begin{cases} 1 & \text{if } T \geq 0 \\ 0 & \text{otherwise} \end{cases}$$

ai, T_{i1}, T_{i2}, and *C*, are constants and $T_{i1} > T_{i2}$ for all *i*.

Clearly, the variables (a_i, T_{i1}, T_{i2}, C, and L) are all temporally varying. Once the basic practical forms of the subclasses of temperature are obtained, then the appropriate coefficients of the functional forms are established to represent linear combinations of functions. There are additional approaches for finding functional coefficients that have been proposed in the literature and they include expert systems-based approaches [32].

C. *Fuzzy Logic*
Fuzzy Logic is a machine learning based approach used for STLF. It is one of the most commonly used approaches for STLF besides ANN. First step involved in fuzzy logic is fuzzification of the input data. It is a procedure for converting crisp numeric values into the degrees of memberships that correspond with the fuzzy sets. Also, a membership function accepts the crisp values as its argument and returns the degree to those values that belong to the fuzzy set [33]. Different types of membership functions can be used for this application. Subsequently, the fuzzy logic method forms the fuzzy rule base, which is the core of fuzzy system. Furthermore, the heuristic information of the forecast is stored based on the "IF-THEN" rules that send necessary information to the fuzzy inference system to find the forecast output [33]. The last step of fuzzy logic is defuzzification, which changes the fuzzified output back to crisp values as required for the outer world. Unfortunately, the fuzzy method requires a huge knowledge base when the data is big. This might consume a lot of memory of the system which may slow down the process.

3 Proposed Method

The data set used is a table of historical hourly loads and temperature observation from the New England ISO for the year 2004–2009 [34]. A separate list of New England holidays that span the historical data range is also used as one of the input parameters. The data set is categorized by training and test sets. The data set from 01-01-2004 to 31-12-2007 is used as training set. Also, the data from 01-01-2008 to 31-12-2009 is used as the test set. A predictor matrix was generated with historical weather, holidays, hourly load, dry bulb temperatures, etc as an input to the neural network. There are eight total input values to the network. All the experiments were carried out in MATLAB.

A general regression neural network (GRNN) based solution in this study is proposed for STLF as shown in Fig. 1 [35].

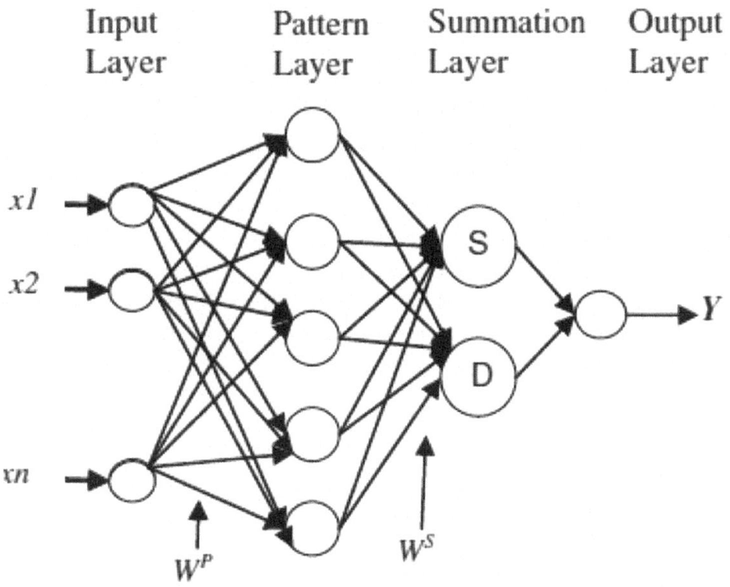

Fig. 1. Generalized Regression Neural Network (GRNN)

The GRNN structure is identical to the radial base network, but the second layer is slightly different. The first layer includes as many neurons as input/objective vectors. The distance between the input vector and its weight vector (W^P) is the weighted entry of each neuron. The net input of each neuron is the outcome of its weighted input. The output of each neuron is the net input passed by radial layer of the base. The two neurons in a summing layer are connected to each pattern layer component, the summation layer weights (W^S): the neuron of the S-summation and the neuron of the D-summation. The S-sum neuron calculates the sum of the weighted results of the surface of the model while the D-summing neuron calculates the unweighted results of the pattern neurons. The output layer separates only the output of each S neuron by that of each D neuron, giving the forecasted value of the input x as follows:

$$Y(x) = \frac{\sum_{K=1}^{N} y_k K(x, x_k)}{\sum_{K=1}^{N} K(x, x_k)} \tag{2}$$

Where Y(x) is the prediction value of input, y(k) is the activation weight for the pattern layer neuron at k, N is the number of training patterns and K(x, x_k) is the Gaussian kernel and expressed as:

$$K(x, x_k) = e^{-d_k/2\sigma^2},\ d_k = (x - x_k)^T (x - x_k) \tag{3}$$

Where d_k is the squared Euclidean distance between the training samples $x_{k,}$ the input x, and σ, or the size, is considered to assess the output of the GRNN generalization. A larger value of σ will generalize better, its optimum value is determined by test and error or tuning optimization techniques. It should be noted that all units in the pattern layer have a single spread in conventional GRNN applications [35].

GRNN is pretty simple learning. The input weights are the transposed learning outputs and the output weights are the training goals. Because GRNN is an associative memory, the number of hidden neurons after learning is the same as the number of exercise samples. However, when there are many training samples this training process is not effective, thus, a data reduction technique such as clustering or main component analysis (PCA) is used as one of the suggested solutions. A GRNN growth algorithm is one of the new solutions for data reduction [36, 37]. It's summarized in Algorithm 1.

Algorithm 1: GRNN Growing Algorithm [50]

```
1: Train GRNN with 10% of the training data
2: for i ≤ i_final do
3:     Find the output of GRNN y_i of input_i
4:     Find the MSE: 1/2 (y_i − Target_i)^2
5:     if MSE>Threshold then
6:         Train GRNN with the input_i
           Do not train GRNN with the input_i
7:     end if
8: end if
```

The algorithm will check to see if a training input needs to be included based on a prediction error before GRNN is trained. When the prediction error exceeds the given level, GRNN should be trained with the error without including that input. In this paper, the number of neurons chosen is 10, 20, and 30. The number of neurons beyond 30 was not used due to memory issues of the system. GRNN training determines the network parameters, such as weights, for accomplishing the desired goal derived from the available training sets. The back-propagation method is utilized for the training. The algorithm used for back-propagation in this paper is Levenberg-Marquardt method.

4 Results and Performance Assessment

Both Mean Absolute Error (MAE) and Mean Absolute Percentage Error (MAPE) are used as performance evaluation metrics. They are given by the following equations:

$$MAE = |X_t - X_f| \tag{4}$$

$$MAPE = \frac{1}{N}\sum_{i=1}^{N}\frac{|X_t - X_f|}{X_t} * 100 \tag{5}$$

where X_t is the actual load and X_f is the forecasted load.

Empirically, it was observed that when neural network was trained with 20 neurons within the hidden layer, then MAPE is 1.86%, while MAE is 275.15 MWh.

For 10 neurons, the MAPE is given by 2.10% and MAE is 306.21 MWh. And, for 30 neurons it was observed that MAPE is 1.81% and MAE 268.48 MWh. Hence, performance with 30 neurons is better. However, there is not much of a performance improvement when there are 30 neurons compared to 20 neurons.

Figure 2 illustrates both the actual and the predicted model for the test data between 2008 and 2010. It also shows the residual for the test data. Figure 3 shows the forecast errors calculated on the test data with 20 neurons. It can be observed that error distribution follows a Gaussian model i.e., it increases initially and gradually decreases.

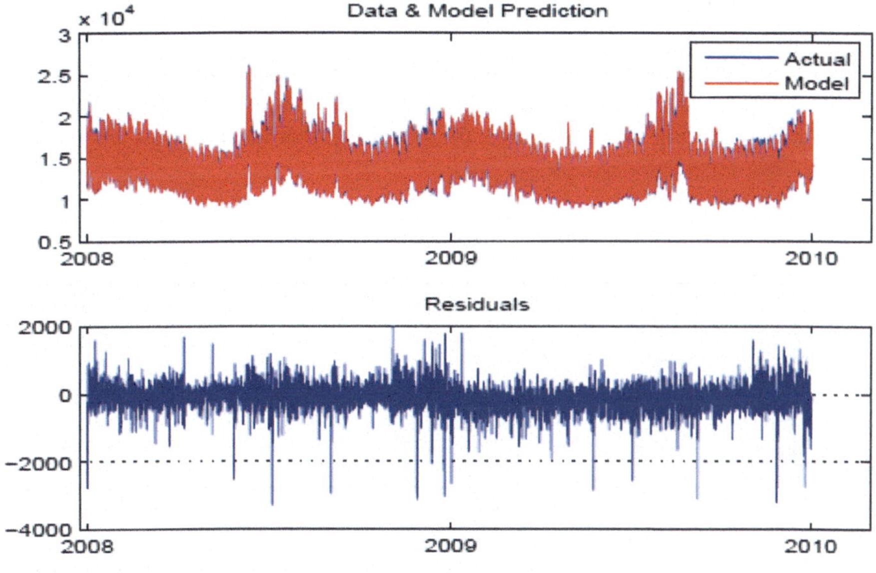

Fig. 2. Predicted model and residual curves on the test data from 2008 to 2010.

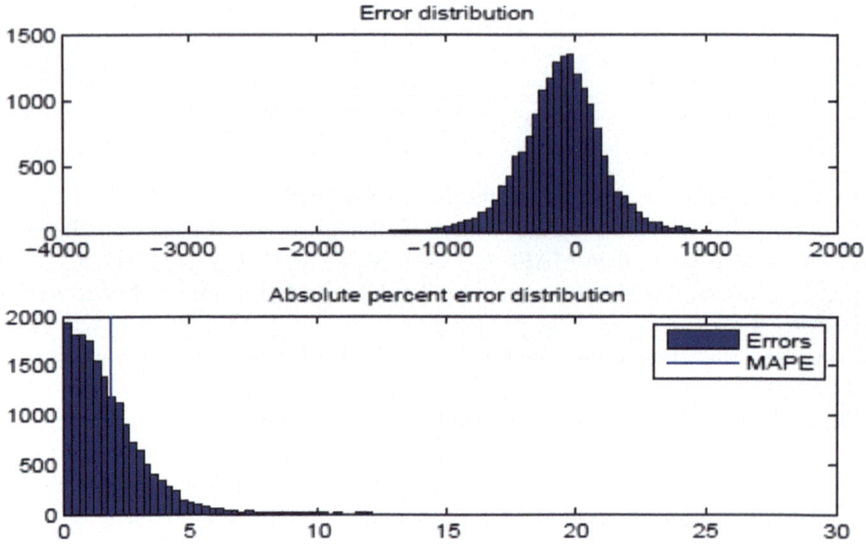

Fig. 3. Calculated errors on the test data with 20 neurons.

Figures 4 and 5 show the actual and predicted data model curves for the first two weeks of 2008 and last two weeks of 2009 respectively with the corresponding MAPE and residual. According to these figures, the actual and predicted models are almost the same.

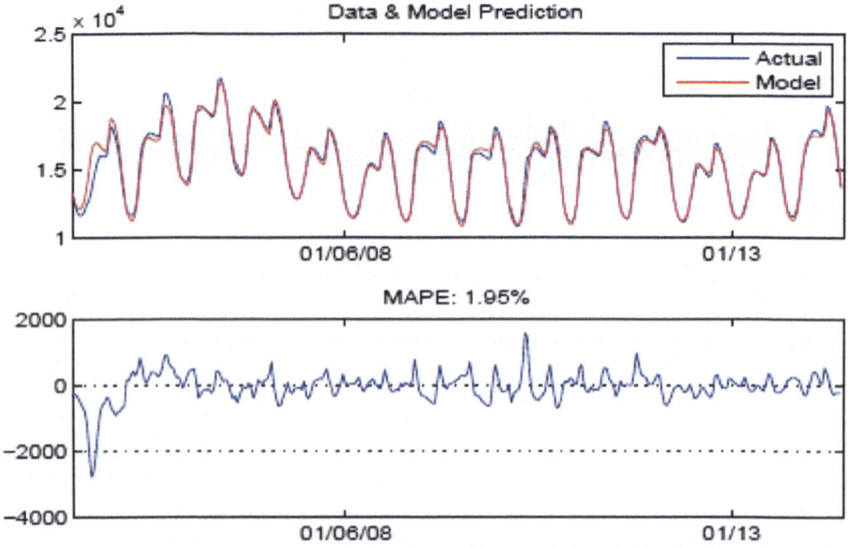

Fig. 4. Data model for the first two weeks of 2008 with the corresponding MAPE

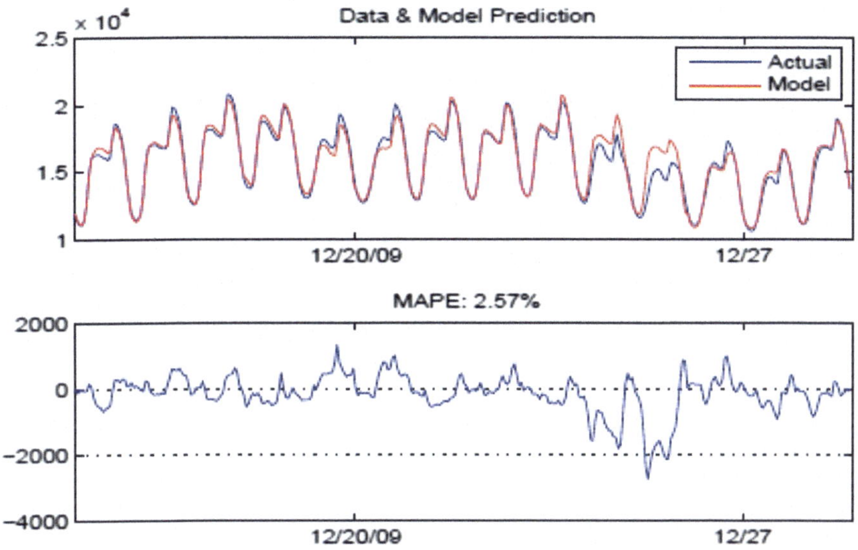

Fig. 5. Data model for the last two weeks of 2009 with the corresponding MAPE

5 Conclusion

This paper utilizes GRNN method for STLF with noise-free data. Thus, in the future, noisy data can be considered, and experiments can be carried out with the same setup. Other back-propagation algorithms, such as resilient back-propagation and gradient descent back propagation, can become methods of training in the future for the neural network. Finally, GRNN methods for price prediction on electricity loads can be modeled and tested as a part of future research.

References

1. Sayed, G.I., Hassanien, A.E., Azar, A.T.: Feature selection via a novel chaotic crow search algorithm. Neural Comput. Appl. **31**(1), 171–188 (2019)
2. Azar, A.T., El-Metwally, S.M.: Decision tree classifiers for automated medical diagnosis. Neural Comput. Appl. **23**(7–8), 2387–2403 (2013)
3. Inbarani, H.H., Banu, P.K.N., Azar, A.T.: Feature selection using swarm-based relative reduct technique for fetal heart rate. Neural Comput. Appl. **25**(3–4), 793–806 (2014)
4. Kumar, S.S., Inbarani, H.H., Azar, A.T., Polat, K.: Covering-based rough set classification system. Neural Comput. Appl. **28**(10), 2879–2888 (2017)
5. Jothi, G., Inbarani, H.H., Azar, A.T., Polat, K.: Tolerance rough set firefly-based quick reduct. Neural Comput. Appl. **28**(10), 2995–3008 (2017)
6. Azar, A.T., Inbarani, H.H., Devi, K.R.: Improved dominance rough set-based classification system. Neural Comput. Appl. **28**(8), 2231–2246 (2017)
7. Inbarani, H.H., Kumar, S.U., Azar, A.T., Hassanien, A.E.: Hybrid rough-bijective soft set classification system. Neural Comput. Appl. **29**(8), 67–78 (2018)

8. Jothi, G., Inbarani, H.H., Azar, A.T., Devi, K.R.: Rough set theory with Jaya optimization for acute lymphoblastic leukemia classification. Neural Comput. Appl. **31**(9), 5175–5194 (2019)
9. Banu, P.K.N., Azar, A.T., Inbarani, H.H.: Fuzzy firefly clustering for tumor and cancer analysis. Int. J. Model. Ident. Control (IJMIC) **27**(2), 92–103 (2017)
10. Inbarani, H.H., Azar, A.T., Jothi, G.: Supervised hybrid feature selection based on PSO and rough sets for medical diagnosis. Comput. Methods Programs Biomed. **113**(1), 175–185 (2014)
11. Jothi, G., Inbarani, H.H., Azar, A.T.: Hybrid tolerance-PSO based supervised feature selection for digital mammogram images. Int. J. Fuzzy Syst. Appl. (IJFSA) **3**(4), 15–30 (2013)
12. Moftah, H.M., Azar, A.T., Al-Shammari, E.T., Ghali, N.I., Hassanien, A.E., Shoman, M.: Adaptive K-means clustering algorithm for MR breast image segmentation. Neural Comput. Appl. **24**(7–8), 1917–1928 (2014)
13. Azar, A.T., Elshazly, H.I., Hassanien, A.E., Elkorany, A.M.: A random forest classifier for lymph diseases. Comput. Methods Programs Biomed. **113**(2), 465–473 (2014)
14. Hassanien, A.E., Moftah, H.M., Azar, A.T., Shoman, M.: MRI breast cancer diagnosis hybrid approach using adaptive ant-based segmentation and multilayer perceptron neural networks classifier. Appl. Soft Comput. **14**(Part A), 62–71 (2014)
15. Jacob, M., Neves, C., Vukadinovic Greetham, D.: Short term load forecasting. In: Forecasting and Assessing Risk of Individual Electricity Peaks. Mathematics of Planet Earth, pp. 15–37. Springer, Cham (2020)
16. Walther, J., Spanier, D., Panten, N., Abele, E.: Very short-term load forecasting on factory level – a machine learning approach. Proc. CIRP **80**, 705–710 (2019)
17. Fan, G.F., Peng, L.L., Hong, W.C.: Short term load forecasting based on phase space reconstruction algorithm and bi-square kernel regression model. Appl. Energy **224**, 13–33 (2018)
18. Zhao, J., Liu, X.: A hybrid method of dynamic cooling and heating load forecasting for office buildings based on artificial intelligence and regression analysis. Energy Build. **174**, 293–308 (2018)
19. Raza, M.Q., Khosravi, A.: A review on artificial intelligence-based load demand forecasting techniques for smart grid and buildings. Renew. Sustain. Energy Rev. **50**, 1352–1372 (2015)
20. Hsu, Y.Y., Tung, T.T., Yeh, H.C., Lu, C.N.: Two-stage artificial neural network model for short-term load forecasting. IFAC-PapersOnLine **51**(28), 678–683 (2018)
21. Singh, P., Dwivedi, P., Kant, V.: A hybrid method based on neural network and improved environmental adaptation method using Controlled Gaussian Mutation with real parameter for short-term load forecasting. Energy **174**, 460–477 (2019)
22. Sideratos, G., Ikonomopoulos, A., Hatziargyriou, N.D.: A novel fuzzy-based ensemble model for load forecasting using hybrid deep neural networks. Electr. Power Syst. Res. **178**, 106025 (2020). https://doi.org/10.1016/j.epsr.2019.106025
23. Ali, D., Yohanna, M., Ijasini, P.M., Garkida, M.B.: Application of fuzzy – neuro to model weather parameter variability impacts on electrical load based on long-term forecasting. Alexandria Eng. J. **57**(1), 223–233 (2018)
24. Yang, A., Li, W., Yang, X.: Short-term electricity load forecasting based on feature selection and Least Squares Support Vector Machines. Knowl.-Based Syst. **163**, 159–173 (2019)
25. López, C., Zhong, W., Zheng, M.: Short-term electric load forecasting based on wavelet neural network, particle swarm optimization and ensemble empirical mode decomposition. Energy Proc. **105**, 3677–3682 (2017)

26. Sadaei, H.J., Enayatifar, R., Abdullah, A.H., Gani, A.: Short-term load forecasting using a hybrid model with a refined exponentially weighted fuzzy time series and an improved harmony search. Int. J. Electr. Power Energy Syst. **62**, 118–129 (2014)
27. Lindberg, K.B., Seljom, P., Madsen, H., Fischer, D., Korpås, M.: Long-term electricity load forecasting: current and future trends. Utilities Policy **58**, 102–119 (2019)
28. Hagan, M.T., Behr, S.M.: The time series approach to short term load forecasting. IEEE Trans. Power Syst. **2**(3), 785–791 (1987)
29. Box, G.E.P., Jenkins, G.M., Reinsel, G.C., Greta, M., Ljung, G.M.: Time Series Analysis: Forecasting and Control, 5th edn. Wiley, San Francisco (2015)
30. Al-Hamadi, H.M., Soliman, S.A.: Short-term electric load forecasting based on Kalman filtering algorithm with moving window weather and load model. Electr. Power Syst. Res. **68**(1), 47–59 (2004)
31. Christiaanse, W.: Short-term load forecasting using general exponential smoothing. IEEE Trans. Power Apparatus Syst. **90**(2), 900–910 (1971)
32. Chiu, C.C., Cook, D.F., Kao, J.L., Chou, Y.C.: Combining a neural network and a rule-based expert system for short-term load forecasting. Comput. Ind. Eng. **32**(4), 787–797 (1997)
33. Ranaweera, D.K., Hubele, N.F., Karady, G.G.: Fuzzy logic for short term load forecasting. Int. J. Electr. Power Energy Syst. **18**(4), 215–222 (1996)
34. Pricing Reports - ISO New England. http://iso-ne.com/markets/hstdata. Accessed 22 Nov 2019
35. Specht, D.F.: A general regression neural network. IEEE Trans. Neural Netw. **2**(6), 568–576 (2002)
36. Al-Mahasneh, A.J., Anavatti, S.G., Garratt, M.A.: Altitude identification and intelligent control of a flapping wing micro aerial vehicle using modified generalized regression neural networks. In: 2017 IEEE Symposium Series on Computational Intelligence (SSCI), pp. 2302–2307. IEEE (2017)
37. Al-Mahasneh, A.J., Anavatti, S.G., Garratt, S., Pratama, M.: Applications of General Regression Neural Networks in Dynamic Systems, Digital Systems, Vahid Asadpour, IntechOpen (2018). https://doi.org/10.5772/intechopen.80258

A Modified Query Processing Algorithm Based on Dynamic Clustering for Big Data Applications

Saad M. Darwish[1], Ramy D. Elsaidy[1(✉)], and Saleh Mesbah[2]

[1] Department of Information Technology, Institute of Graduate Studies and Research, Alexandria University, 163 Horreya Avenue, El-Shatby, P.O. Box 832, Alexandria 21526, Egypt
Saad.darwish@alexu.edu.eg, ramy.elsaidy@dmu.edu.eg
[2] College of Computing and Information Technology, Arab Academy for Science, Technology and Maritime Transport, Abo-Keer, Alexandria, Egypt
saleh.mesbah@aast.edu

Abstract. Existing approaches for query processing are insufficient to provide accurate results quickly in a big data environment due to the needing to identify approximation measures and execution time for the queries are increasing. The practical results are obtained through the clustering process. However, how to scale up and speed up clustering algorithms with minimum sacrifice to the clustering quality remain a big problem. The main contribution of this work is to propose a modified dynamic distributed clustering approach applicable in spatial query processing for big data. The dynamic nature of the approach comes from the fact that it does not need to give the number of correct clusters in advance. The suggested model works through two main phases: the first phase builds local clusters based on its portion of the entire dataset; this stage takes full advantage of task parallelism paradigm based on Spark framework; whereas the second phase aggregate the local clusters in such a way that the final clusters are compact and accurate. Experimental results show that the approach is scalable and produces results of high quality for spatial query processing.

Keywords: Big data clustering · Dynamic clustering · Query processing · Apache Spark

1 Introduction

Big data analytics is a method for storing, collecting, processing, and analyzing a massive volume of structured and unstructured data that is hard to process using a conventional database [1]. Query processing is a huge challenge in the field of big data. Managing data for optimized query performance is mainly a challenging task. This is due to the growing complexity of analytical data computations involved in querying data, where traditional systems failed to expand beneficially. Clustering is a popular data mining approach to divide data points into uniform groups so that the data points in the same group are as similar and as different as possible. However, due to their high difficulty and computing costs, conventional clustering strategies cannot deal with this large amount of data.

A.-E. Hassanien et al. (Eds.): AICV 2020, AISC 1153, pp. 438–447, 2020.
https://doi.org/10.1007/978-3-030-44289-7_41

The main issue for big data clustering is how to scale up and speed up clustering algorithms with minimal sacrifice to the quality of clustering. In recent years, parallel and distributed algorithms that use multiple machines are employed to increase the scalability of current techniques of big data clustering [2]. To handle the massive amount of data, many tools that are parts of Hadoop and HDFS framework for storing and managing big data are used. Recently, Apache Spark is more efficient than MapReduce, specifically when dealing with iterative algorithms [3].

1.1 Problem Statement

Generally, for the large data, the execution time for the queries is increased. For that, the split-up method is evolved in research studies, which slightly reduces the execution time of query processing. The contemporary research community has suggested several clustering approaches. For finishing the clusters, these approaches have either curse of the dimensionality or require multiple iterations. The contemporary methods of clustering are, therefore, not capable of analyzing big data. Parallel-distributed clustering methods using Apache Spark suffer from noisy data that make it challenging to cluster an object, thereby affecting the results of clustering.

1.2 Motivation and Aim of the Work

Clustering needs to improvement nowadays more than before to help data analysts in extracting knowledge from terabytes and petabytes of data. While traditional sampling and dimension reduction algorithms still are useful, but they do not have enough strength to deal with a huge amount of data; consequently, Distributed computing is linked to the future of clustering. Parallel clustering is very useful for clustering, but it is a challenge to implement such algorithms. The observation motivates this work that dynamic clustering decides the best cluster head by changing the cluster heads that is best fitting without setting the cluster head as static and thus playing a role in improving the query processing for big data. This study is aimed to propose a dynamic clustering model to facilitate spatial query processing for big data. One of the key outputs of this distributed clustering technique is that the number of global clusters is dynamic; no need to be pre-fixed.

1.3 Contribution and Methodology

The work presented in this paper address the issue of how to improve the scalability and performance of big spatial data queries using adopting a distributed dynamic clustering model on top of Apache Spark. The proposed distributed dynamic clustering approach has two main phases: a completely parallel phase in which each system node calculates its local clusters based on its entire dataset portion. There is no interaction during this step; the parallelism model takes full advantage of it. The second phase is dispersed as well, but there is some communication between the nodes. However, the overhead due these communications has minimized by using a new concept of cluster representatives. In addition, for spatial data sets, the distributed clustering approach is dynamic because the number of correct clusters need not be given in advance (as an input parameters). This resolves one of *K*-Medoids' main weaknesses.

The rest of this paper is organized as follows: Sect. 2 presents a review of the recent related works. Section 3 introduces the proposed dynamic distributed clustering model. Section 4 exhibits the experimental results to evaluate the performance study of the proposed mechanism. Finally, we conclude the paper and give future directions in Sect. 5.

2 Related Work

There has been a lot of research recently conducted to get a more efficient and effective result related to clustering in big data algorithms. Fahad et al. [4] presented the survey of current clustering in different categories. In their work, they compared the results of five categories with their most representative algorithm; their goal was to identify the best performing for big data.

It can be concluded from the results of this survey that no clustering algorithm can solve all the big data issues. The complexity of the implementation of these algorithms remains a great challenge though the parallel classification is potentially very useful for big data clustering. However, the MapReduce or Spark framework can form an excellent basis for the implementation of such parallel algorithms. In general, we need to develop clustering algorithms in order to manage a large amount of data while preserving appropriate resource needs by decreasing their memory and time complexity.

In [5] the authors suggested a new *K*-Medoids++ spatial clustering algorithm based on MapReduce for clustering massive spatial data. With the same idea, Jiang and Zhang [6] suggested the parallel *K*-Medoids clustering algorithm based on Hadoop for big data clustering called H*K*-Medoids. Every job submitted has a lot of iterative procedures for MapReduce.

In 2017, Martino et al. [7] suggested a novel implementation for the large-scale *k*-medoids algorithm using the Apache Spark framework as parallel and distributed computing environment. In 2018, the same author's [8] discussed techniques for potential speedups in *k*-medoids clustering using the Apache Spark framework. In another work, in 2017, Kusuma et al. [9] designed intelligent *k*-means based on Spark for big data clustering. Recently, Song et al. [10] suggested a parallel *k*-Medoids clustering technique with high accuracy and efficiency called (PAMAE) on Spark as well as Hadoop. They identified two factors: "global search" and "entire data" that are essential to achieving high accuracy but are also very time consuming if considered simultaneously.

In 2016, Bendechache et al. [11] suggested a new distributed clustering approach that utilizes both local results and generation of global models by aggregation. The same author's [12] enhanced the above dynamic distributed clustering techniques by incorporating an efficient data reduction phase which reduces significantly the size of the data ex-changed, therefore, it deals with the problem of communication overhead and developed this technique for spatial datasets.

According to the review mentioned above, it can be found that past studies were primarily devoted to (1) No clustering algorithm can be used to solve all the big data issues. (2) Parallel classification can be useful for the clustering of large data, but the difficulty of these algorithms implementation remains an important challenge. (3) The MapReduce framework can provide a very good basis for the implementation of such parallel algorithms but not useful when dealing with iterative algorithms. (4) None of the existing algorithms satisfies both high accuracy and high efficiency in the *k*-medoids algorithm for a big data environment. The effective results are obtained through the clustering process.

However, to the best of our knowledge, little attention has been paid to devising a new dynamic distributed *K*-medoids clustering approach applicable in spatial query processing for big data. The next chapter discusses in detail the suggested model that integrates the Spark framework to handle big data clustering with a dynamic distributed clustering technique.

3 The Proposed Dynamic Distributed Clustering Model

This section introduces in details the proposed dynamic distributed *K*-medoids clustering model for improving efficiency and scalability for the study of large datasets. This model operates on two phases; generation of local clustering and generation of global clustering by aggregation. This approach belongs to the category of hierarchical clustering. The model operates above the Spark as a distributed computing framework, which is based on Hadoop MapReduce algorithms. Figure 1 shows the main components of the suggested model and how these components are linked together and the following subsections discuss its steps in details.

Step 1: Input Dataset. The data set provides signs of movement for taxis in San Francisco, USA. This comprises GPS coordinates for about 500 cabs obtained over 30 days in San Francisco Bay.

Step 2: Distributed Dataset in spark. Given the spatial dataset, these massive datasets are too large and complex for humans to extract useful information without the aid of computational tools effectively. Technologies like Spark have new and exciting ways to process and turn Big Data into meaningful information, described as complicated, unstructured or large volumes of data. On these data, operations (transformations) are carried out which generate RDDs or dataset fragment [13]. In the next step, after the dataset is divided into multiple trunks and stored in a set of machine nodes, the preprocessing phase is done in multiple parallel operations. Preprocessing phase is the opportunity to impute missing value, to treat noise, normalize, transformation, integration, mitigate inconsistencies and enhance the reliability of data significantly, reduction and discretization [14].

Step 3: (Phase 1) b. Reduction Stage. The data reduction methods can be used to obtain a simplified, much smaller volume representation of the data set, while maintaining the quality of the original data tightly. The suggested model applies the same reduction technique as in [15] in which the boundary of clusters as well as their density information will construct the reduction set that can be seen as the local model M_i at the site i in the system. This local model is sent to the server in which global models will be built.

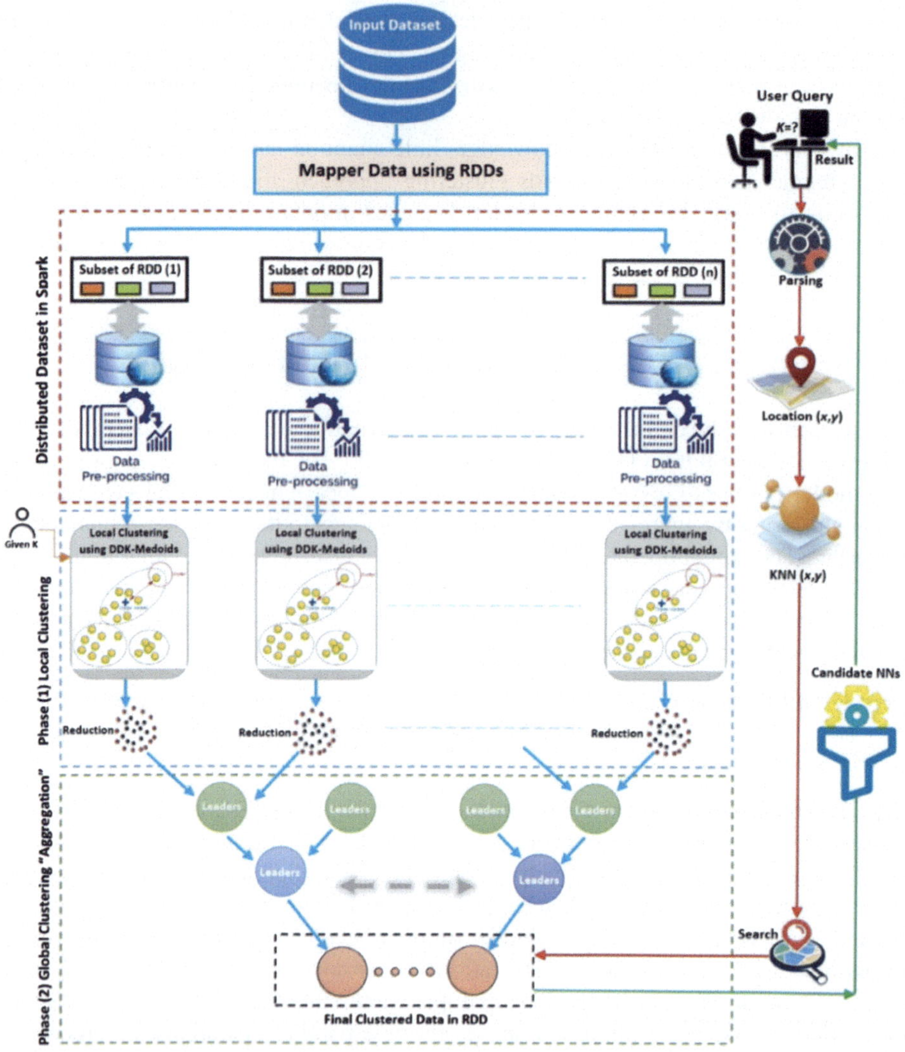

Fig. 1. The proposed dynamic distributed clustering model

Step 4: (Phase 2) Global Clustering "Aggregation". There are two main steps in this phase, that are repeated until all the global clusters were generated. First, each leader collects the local clusters of its neighbors. Second, the leaders will merge the local clusters using the overlay technique. The root node will contain the global clusters, which is reached by the continuous process of merging clusters. The second phase consists of exchanging the reduced version of the cluster (medoid plus the boundary points of the cluster) located in each node with its neighborhood nodes. This will allow us to identify overlapping representatives (clusters). Every leader merge overlapping representatives of it's group. There are some characteristics that are needed in the leaders to be selected like capacity, processing power, connectivity, etc.

Step 5: User Querying Processing. Given the input's user query; lexical analysis is performed to check grammar and extract important tokens from the query. A lexical analyzer is used to construct tokens from an input text string flow divided into components that form coherent phrases. Given the extracted tokens, the two most common spatial queries are nearest neighbor (*k*NN) queries and window queries [16]. A *k*NN query retrieves the *k* nearest objects with respect to the location of the querying user, while a window query displays all the objects within the requested window centered at the query location. *K*-nearest neighbor search queries are used to find *K* nearest spatial objects from a given query point. A Spark data frame API *K*NN_query (spark_object, query_point, *K*) is implemented to find *K* nearest neighbors. It is implemented by using selection and merge method.

4 Experimental Results

This section provides performance analysis and comparative analysis of the proposed and existing models. This set of experiments evaluates the performance results of both existing and proposed techniques for proving the superiority of the clustering proposed system. Here, the results are analyzed and evaluated in terms of clustering error, convergence time, clustering accuracy, and execution time. The suggested model has been running over Hadoop 2.7.1 and Spark 2.2.0 for distributed parallel processing. Ten of 12 instances were used as nodes for the worker, and the other two were used as master nodes. All algorithms have been written and run on JDK 1.7.0_101 in Java programming language. The system has been implemented using HP ProLiant DL180 Gen9 Server with the following specifications: Processor: Intel Xeon E5-2620v3 @ 2.4 GHz (12 CPUs), RAM: 32 GB, SSD Hard Disk: 3 * 1T.B (Raid 5), System type: 64-bit operating system and Microsoft Server 2008 R2 Enterprise 64 bit. Different datasets of variable size (5 million to 30 million records; 800 MB to 5 GB) for the benchmark taxi dataset are generated to evaluate the performance of the model.

The first set experiments is conducted to evaluate the performance of different queries in terms of average latency against the variable size of datasets. Pickup location is considered as a 2D spatial object. Different test cases of *K*NN search query are designed for evaluation. Ten test queries are performed to find *K* (K = 3 to 15) nearest pickup locations from a given query point. Different query points are selected from high-density areas using the sampling method. The performance of *K*NN search query is shown in Fig. 2. As expected, the average latency increases as both of the number of neighbors for *k*NN query retrieval engine and the number of dataset rows increase. Form $k = 3$ to $k = 15$ the average latency increases by approximately 4% for the dataset with 5 million dataset rows. This average increases by roughly about 3.7% from 5 million dataset rows to 30 million dataset rows. In general, using distributed clustering (spark framework) is considered one of the important features of the suggested model to obtain the final clustered which in turn decreases the run time. This helps to increase the query processing for spatial query inside a big data environment. Furthermore, the method used to aggregate spatial local clusters into global clusters allows only to exchange a small amount of the original datasets, which is highly efficient.

The second set of experiments is running to compare the query performance in terms of average latency of the suggested model with related technique in [17] that is built on top of Spark framework NoSQL database i.e. Cassandra. The same dataset that is employed in the first experiment is used. Ten test queries are performed to find K (K = 5 to 25) nearest pickup locations from a given query point. The performance of KNN search query for both models is shown in Fig. 3. The suggested model achieves 23% reduction on average latency. The superiority of the suggested model relies on the fact that it built based on two levels of clustering local and global clustering. Unlike the comparative model, it only depends on one level of clustering. So the total number of generated clusters for the suggested model is much fewer as compared with the alternative method. In general, searching within a few numbers of clusters takes less time.

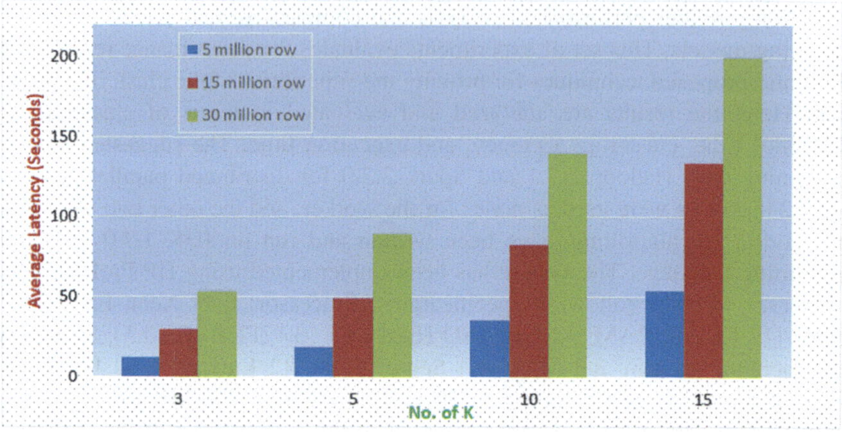

Fig. 2. KNN query performance (effect of data size and number of K)

The third set of experiments evaluates the performance results of both existing and proposed techniques for proving the superiority of the clustering proposed model. Here, the results are analyzed and evaluated in terms of clustering error. The existing algorithms considered in this analysis are K-means, K-prototype, object clustering iterative learning (OCIL), and similarity-based k-medoids clustering [18]. Figure 4 shows the cluster error of both existing and proposed clustering techniques using adult dataset. From the results, it is observed that the proposed dynamic clustering model reduced the error rate to 0.36% on average when compared to the other techniques. The suggested model decreases the clustering error by 16%, 25%, 51.6%, and 51.7% as compared with similarity-based *k*-Medoids clustering, K-prototype, OCIL, and *K*-means respectively. Multiphase *K*-Medoids for clustering is better in clustering error but with the drawback, that the complexity is high as compared to *K*-Means. This drawback was handled by means of distributed computing. In the suggested model, the compact clusters contain the main features of the local clusters. The local cluster of the cluster significantly affects the quality of the final models.

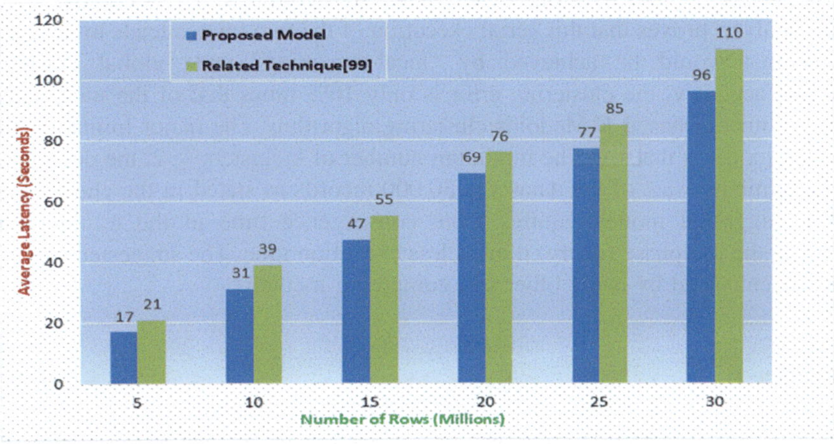

Fig. 3. Comparative study for *K*NN query performance

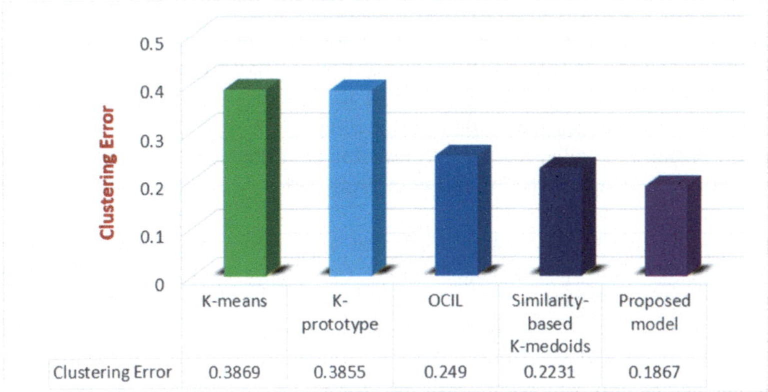

Fig. 4. The clustering error versus different clustering algorithms

5 Conclusion

The work in this paper offers a modified approach that alleviates the performance problems associated with the state of the query processing methods. The essence of the suggested approach, which differentiates it from related research, rests on adopting a dynamic distributed clustering-based query processing instead of the static clustering-based system. The distributed clustering approach is moreover dynamic, for spatial datasets, as it does not need to give the number of correct clusters in advance (as an input parameter). This solves one of the major shortcomings of K-Medoids. The theoretical analysis proves that this serial execution of the two phases leads to an accurate solution that would be achieved by "local-clustering" and "global clustering". Regarding accuracy, the clustering error is only 16% times that of the most accurate existing similarity-based K-Medoids clustering algorithm. The major limitation of the suggested model is that: (1) The minimum number of k clusters ≥ 2, the default is 10. (2) The minimum size of the dataset is 30,000 records as stated in the chess data set. (3) The suggested model requires more convergence time in the training phase, whereas in the use phase (query) it takes less execution time. The suggested model can be further enhanced by using other soft computing methods.

References

1. Zulkarnain, N., Anshari, M.: Big data: concept, applications, & challenges. In: International Conference on Information Management and Technology, Indonesia, pp. 307–310 (2016)
2. Suganya, R., Pavithra, M., Nandhini, P.: Algorithms and challenges in big data clustering. Int. J. Eng. Tech. **4**(4), 40–47 (2018)
3. Tang, S., He, B., Yu, C., Li, Y., Li, K.: A Survey on Spark Ecosystem for Big Data Processing, pp. 1–21 (2018). arXiv:1811.08834v1
4. Fahad, A., Alshatri, N., Tari, Z., Alamri, A.: A survey of clustering algorithms for big data: taxonomy and empirical analysis. IEEE Trans. Emerg. Top. Comput. **2**(3), 267–279 (2014)
5. Yue, X., Man, W., Yue, J., Liu, G.: Parallel K-Medoids++ Spatial Clustering Algorithm based on MapReduce, pp. 1–8 (2016). arXiv:1608.06861
6. Jiang, Y., Zhang, J.: Parallel K-medoids clustering algorithm based on Hadoop. In: IEEE 5th International Conference on Software Engineering and Service Science, China, pp. 649–652 (2014)
7. Martino, A., Rizzi, A., Mascioli, F.: Efficient approaches for solving the large-scale k-medoids problem. In: 9th International Joint Conference on Computational Intelligence, Portugal, pp. 338–347 (2017)
8. Martino, A., Rizzi, A., Mascioli, F.: Distance matrix pre-caching and distributed computation of internal validation indices in k-medoids clustering. In: International Joint Conference on Neural Networks, Brazil, pp. 1–8 (2018)
9. Kusuma, I., Ma'sum, M., Habibie, N., Jatmiko, W., Suhartanto, H.: Design of intelligent K-means based on spark for big data clustering. In: International Workshop on Big Data and Information Security, Indonesia, pp. 89–96 (2016)
10. Song, H., Lee, J., Han, W.: PAMAE: parallel k-medoids clustering with high accuracy and efficiency. In: 23rd ACM SIGKDD International Conference on Knowledge Discovery and Data Mining, Canada, pp. 1087–1096 (2017)

11. Bendechache, M., Le-Khac, N., Kechadi, M.: Efficient large scale clustering based on data partitioning. In: International Conference on Data Science and Advanced Analytics, Canada, pp. 612–621 (2016)
12. Bendechache, M., Le-Khac, N., Kechadi, M.: Performance evaluation of a distributed clustering approach for spatial datasets. In: 15th Australasian Data Mining Conference, Canada, pp. 38–56 (2018)
13. Kamaruddin, S., Ravi, V., Mayank, P.: Parallel evolving clustering method for big data analytics using apache spark: applications to banking and physics. In: International Conference on Big Data Analytics, Switzerland, pp. 278–292 (2017)
14. Shah, J.: News clustering using spark. Int. J. Latest Technol. Eng. Manag. Appl. Sci. **5**(9), 58–61 (2016)
15. Laloux, J.-F., Le-Khac, N.-A., and Kechadi, M.-T.: Efficient distributed approach for density-based clustering. In: IEEE 20th International Workshops on Enabling Technologies: Infrastructure for Collaborative Enterprises, France, pp. 145–150 (2011)
16. Bai, J.-W., Wang, J.-Z., Huang, J.-L.: Spatial query processing on distributed databases. Adv. Intell. Syst. Appl. **1**, 251–260 (2013)
17. Shah, P., Chaudhary, S.: Big data analytics framework for spatial data. In: Lecture Notes in Computer Science, vol. 11297, pp. 250–265 (2018)
18. Narayana, G., Vasumathi, D.: An attributes similarity-based k-medoids clustering technique in data mining. Arab. J. Sci. Eng. **43**(8), 3979–3992 (2018)

Metadata Extraction for Low-Quality Semi-structured Spreadsheets

Arwa Awad[1(✉)], Rania Elgohary[1(✉)], Ibrahim Moawad[1(✉)], and Mohamed Roushdy[2(✉)]

[1] Faculty of Computer and Information Sciences, Ain Shams University, Cairo, Egypt
arwaawad91@yahoo.com, {rania.elgohary, ibrahim_moawad}@cis.asu.edu.eg

[2] Faculty of Computer and Information Technology, Future University in Egypt, Cairo, Egypt
Mohamed.Roushdy@fue.edu.eg

Abstract. The approach in which data is stored in Excel spreadsheets is not the most ideal approach to sort out and access it. Despite using a spreadsheet is in a growing exponentially in the last years, the increases in volume and complexity of these data have led to increased requirements to preserve these data and reuse it. Most of the applications focused on extracting structured data from semi-structured data on the web by HTML and XML formats. Nowadays, there is an explosion of semi-structured documents that generate outside the web. Metadata is important to extract the spreadsheet data. This paper produced an automated relation extractor tool that lets ordinary users extract structured relational tables from spreadsheets without previous experience. To define a structure for the spreadsheet table, this paper introduces a framework that automatically extracts relational data (tables) from spreadsheets and converts from Low-Quality data to High-Quality data. The paper considers some perspective approaches for the clustering-based table detection, heuristic-based table metadata extraction and rule-based table analysis for attribute names to automate insertion into a structured database to facilitate, integrate and reuse the data stored in spreadsheets.

Keywords: Spreadsheets low-quality · Metadata extractor · Relational table extractor · Spreadsheet conversion

1 Introduction

An Excel spreadsheet is low-quality as Excel documents may further contain some items. Firstly, multiple worksheets that might be autonomous or interrelated. Secondly, visual impacts that imply to outskirts, concealing, text style styling and so forth. Such organizing might be contingent on cell esteems. Thirdly, hidden content: concealed sections, lines, worksheets, and cell remarks, which are not shown until the client effectively uncovers them inside the spreadsheet. Besides of filters that changed perspectives that non-damagingly reorder as well as conceal the basic information and charts: perceptions (diagrams, plots, and so on.) that are figured from the basic

A.-E. Hassanien et al. (Eds.): AICV 2020, AISC 1153, pp. 448–457, 2020.
https://doi.org/10.1007/978-3-030-44289-7_42

information. Fourthly, references to information esteem: a worksheet may reference cell esteems in another worksheet in the equivalent Excel document. Fifthly, multiple tables may exist on the same sheet.

So the change of spreadsheet information is a significant and basic issue. Accordingly, the interest to unravel this issue being nonstop till now. Numerous kinds of research work at this issue utilizing various calculations and various rules and improve their application itself. This paper proposes a new framework architecture and discusses the semantic perspective on their metadata extraction explained how the conversion process baseline did from spreadsheets to relational database tables. The application experience of this paper confirms that such arrangements are requested to populate databases and create ontologies with tabular data extracted from semi-structured documents. This solution allows the refactoring of spreadsheets to prevent duplication data, migration of spreadsheet applications to database applications, and advanced querying of spreadsheets with SQL queries.

Metadata is a collection of key information about particular content, which can be used to facilitate the understanding, use, and management of data [1, 2]. The paper focuses only on how to extract the table stores in a spreadsheet, not on graphs or text or anything else. As a result, in this paper, the metadata extractor acts in extract the table name, attributes and values to each table in the spreadsheet.

Along these lines, it is necessary to extract metadata from semi-structured data to convert a spreadsheet document into a relational database table [3]. The tables of a relational database have some significant attributes as each line or row contains one and only one incentive for every column. As a result, a user can benefit from the integration model, extraction knowledge and decision making.

This paper is organized as follows. Section 2 reviews the related work of the information extraction from spreadsheets. Whereas Sect. 3 introduces the proposed framework architecture, both Sects. 4 and 5 explain in details the proposed architecture phases and the implemented algorithms. Finally, the experimental results are presented in Sect. 6 and the paper is concluded in Sect. 7.

2 Related Work

The work presented in [4] discusses the problem of no available electronic metadata for the recognition tables. For this kind of issue, three tasks are studied: The detection of tables/forms in heterogeneous documents, the classification of forms into models and the recognition of tables. While the work presented in [5] focuses on table extraction using a conditional random field. Concluded from the previous related work, the table extraction might be broken down into six overlapping sub-issues: 1. Locate the table position. 2. Identify the table borders. 3. Segment the table into cells. 4. Identify the positions of the attributes row or column. 5. Tag the cells as attributes or values or table name. 6. Associate data cells with their corresponding labels. Most of the research focuses on the first and second issues [6] and others focus on another issue. Anyway, this paper covered all these issues. The paper proposes a novel algorithm for detecting and extracting tables with its metadata in spreadsheets and automatic insertion or conversion to a relational database. The paper uses a hierarchal clustering for table detection; rules-based a heuristic approach for metadata table extraction that represented in a different functional component. For example, table name, table attributes and table values. Also, rule-based table analysis for attribute names to automatic insertion to a database. A checklist evaluation comparison is introduced in Table 1 to evaluate the previous works. The table shows that the proposed approach solved the issue of table recognition similar to the work presented in [7, 12] while the work presented in [8–11, 13] does not discuss this issue. Also, the proposed approach does not need any programing language for the user to convert a spreadsheet into a relational database. However, the work presented in [9–11, 13] needs the user to know a programming language. Also, the proposed approach supports semantic. For example, it implies matching different formats of the same data type. However, the work presented in [7, 8, 10] and [12] does not achieve in their conversion process. The proposed approach supports a generic framework. For example, a flow of control is not dictated by the user, but by the framework. While the work presented in [7–10, 13] has not achieved. Like the work presented in [7, 13], the proposed approach exploits the user interface component to convert the spreadsheets to a relational database, while the work presented in [8–12] does not include user interface components.[1]

Table 1 shows checklist evaluation comparison of reference [7–13] the citation of paper. Where Table recognition means frame finder or table boundaries. Programming required deals with the required user to learn a programming language. Semantic deals with the question "is coercion of data type supported and implies matching of different formats of the same data type". Generic framework is the overall program's flow of control is not dictated by the user, but by the framework. UI is the User interface. In the entire table, the symbols Y and N implies to yes and No Respectively.

Table 1. Checklist evaluation comparison

Reference	Table recognition	Programming required	Semantic	Generic framework	UI
[7]	Y	N	N	N	Y
[8]	N	N	N	N	N
[9]	N	Y	Y	N	N
[10]	N	Y	N	N	N
[11]	N	Y	Y	Y	N
[12]	Y	N	N	Y	N
[13]	N	Y	Y	N	Y
Proposed approach	Y	N	Y	Y	Y

[1] * Y and N imply Yes and No respectively, and UI refers to User Interface.

3 The Proposed Framework Architecture

As shown in Fig. 1, the framework architecture detects tables from a spreadsheet and extracts their metadata, then converts it into a database table.

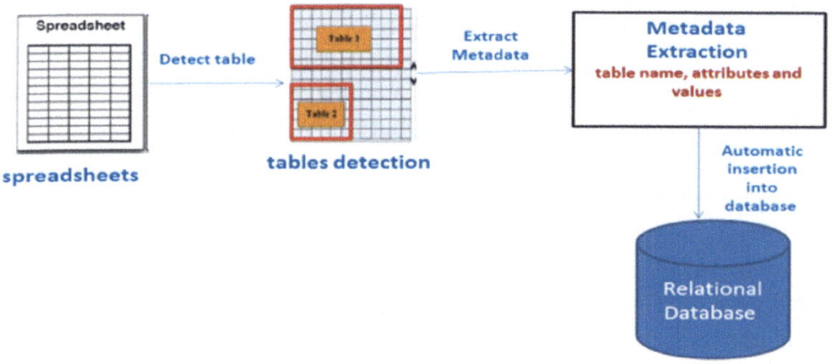

Fig. 1. The proposed framework architecture.

The proposed approach consists of three phases. The first phase is to detect tables from a spreadsheet. The second phase is to extract its metadata, which acts in a functional component of the table and consists of the table name, its attributes, and values. The third phase is to automate insertion into a relational database. The paper will present a concise gauge discussion of each phase with the algorithms utilized in the following segment.

4 A Spreadsheet Conversion Phases

4.1 Table Detection Based on a Clustering Approach

Data clustering refers to the strategy of grouping data into various groups depending on their characteristics. This grouping gets to request the data and consequently, further handling of this data is made simpler. The paper uses a clustering approach to specify the location of data cells. Each set of cells in a set of rows or columns are grouped side by side, be a cluster of cells called table. Through the detection of the table data cells [14], the algorithm starts from the start row, end row, start column, and end column of the table to identify the table borders.

Apportioning calculations depend on determining an underlying number of gathering, and iteratively reallocating items among gatherings to the combination. Interestingly, progressive calculations consolidate or partition existing gatherings, making a various leveled structure that reflects the request wherein bunches are blended or partitioned. When combined, Linkage strategies utilize the maximum, average, and minimum separations between the individuals from two groups, respectively. This procedure repeats until all items are in a solitary gathering. Various variations of agglomerative progressive grouping calculations may utilize distinctive cost functions.

1. Register the nearness framework containing the separation between each pair of examples. 2. Locate the most comparable pair of groups utilizing the closeness framework. 3. On the off chance that all examples are in one group, stop.

4.2 Metadata Extraction Based on a Heuristic Approach

The spreadsheet tables are a mix of attributes, values and other elements such as title and footnotes. The first challenge is which cell carries values versus attributes. Thus, the need to define the structure/template of a spreadsheet table is a critical issue.

The potential geometric configurations that allow for a set of domain values with labels (features) are called templates. A notation for specifying simple domain templates is defined as follows. It consists of a triple (row width (X); column depth (Y); type (T)). X specifies the extent of an area of cells that match the template, given by T followed by either an integer or string, which matches any width of cells. Y specify the extent of area cells that match the template, given by T followed by either an integer string, which matches any depth of cells. T specifies whether the template is to be counted as a value (Val) or an attribute (Att), given by T followed by either Val or Att. For example, if a set of cells value is numbers then the cells are classified into the data region or values of the table. Rules can be defined on the template shown in Table 2.

Table 2. Heuristic rules.

Rules	Area must
Top	Top row
Left	Left column

The attributes may be top or left of the table. The first row or first column (Top or Left) of each table is used as labels if they are encodings, like color, bold or typeface formatting and the remaining elements are assumed to be data. The system algorithm of this paper identifies the functional component of the table which represented in a table name, attributes [15], and values found in a data region [16].

The metadata extraction is based on a heuristics-based approach to classify spreadsheet table cells. The algorithm starts reading row by row. It starts reading from the first row with data which means the row is not empty and stopped when it reads an empty row to detect the frame to a table extracted. The features used to specify the data cells depending on layout features and content features. A layout that means the data stored in a box or has a frame and content or textual features that has not a frame. The heuristics approach for cell features can be defined as shown in Table 3.

Table 3. Cell features.

Cell	Features
Attributes	Bold, Italic, Underline, Starts with letters or text, Contains keyword like: "total", " Name", "ID", Contains colon, Capitalized, Color, Font size, Adjacent
Values	Numbers, Font Size, Color, and Style

The conditions attribute features for example attribute is bold, italic or underline. Attribute starts with (letters or text) not a number. Attribute contains keyword like: "total", "Name", "ID", etc. The attribute contains a colon, centered or aligned. The attribute letters are all capitalized. Attributes are written with a different color other than values, for example, red. Attributes font size is bigger than the values font size. The attribute is adjacent.

The conditions value features for example numbers, values font size is smaller than the attributes font size or values are written with the same color and style, for example, black.

4.3 Automatic Insertion Based on a Table Analysis for Attributes Name

The transformation approach is done automatically into a relational database [17, 18] by taking every one of the attributes extracted from a spreadsheet with its type as a column name and sends it to make a new table inside the database consequently. Each table should have a key which means each attribute may be a key. Thus, we conclude the rules to analyze the attribute name and achieve integration with a database. Primary Key is a column in the database should store a unique value that means no duplicate. Also, the attribute name should be not null which means no missing in attributes name of the table created in a database. Now the generated source code is ready for compilation and building of executable programs for domain-specific spreadsheet data extraction and conversion.

5 The Proposed Approach Algorithms

5.1 Algorithm 1: Table Detection

The algorithm is designed to detect spreadsheets tables. It considers only filled data sheets and each sheet looks for filled rows. Empty rows and columns are the main separators between tables if one data sheet contains more than one table. For example, if a data sheet contains two tables placed vertically, undoubtedly they are separated by one or more empty rows. Similarly, if a data sheet contains two tables placed horizontally, undoubtedly they are separated by one or more empty columns. To detect a table in the spreadsheet the following steps are pursued as shown in Algorithm 1.

- Input: Excel workbook with multiple sheets A = {S1,S2,…,Sn}
- Output: Separate tables detected

1. Start from the rightmost, base-most, non-empty cell, and add it to the table cells set (T).
2. Read the cells of the next row and cluster these cells with the previous row.
3. Add these cells to the table cells set T.
4. Stop if there are several empty rows.

Algorithm 1. Table detection.

5.2 Algorithm 2: Excel Metadata Extraction

The algorithm is designed to extract excel spreadsheets metadata for tables as shown in Algorithm 2.

- Input: Excel workbook with multiple sheets A = {S1,S2,…,Sn}
- Output: Separate tables each includes:Table name as a variable TblNm and Attributes as an array listAtt ={att1, att2,…..,attn}, values as Val[m,n] = {valm1n1,valm1n2,…….valm,n}

1. Initiate A
2. For each sheet in A
3. Initiate new TBL
4. For each filled row
5. If cell have Attributes format and no other data in this row
6. TblNm<- cell data
7. Else cell have Attributes format and there is other data in this row
8. Att <- cell data
9. If the cell does not have attributes format
10. Val[m,n]<- cell data
11. Else empty row go to 3
12. End if
13. End if
14. End for
15. End for

Algorithm 2. Metadata extraction.

5.3 Algorithm 3: Automatic Insertion into a Database

After the extraction process and data values have been analyzed, now the data is ready to be inserted into a database engine. The table name used to create a new table in the database, the attributes used to create the table columns considering the column type with its interconnected data types as shown in Algorithm 3. The user of the system can choose any field of the table column, new identity field or without primary key. In the case of the existing column should be checked to be sure that the column is computable with the primary key constraints, otherwise, the system will ask the user to choose another field or fix the data.

- Input: Separate tables each includeTBL:Table name as a variable TblNm Attributes as an array listAtt ={att1, att2,.....,attn}, column types as an array listClmTyp = { Typ1, Typ2, }, Table primary key name TBLPKattributesdata Att[m,n] = {attm1n1,attm1n2, attm,n} and existing Database DB
- Output: Database table with its primary key.

1. For each table TBL
2. Create a new Table TblNm with Att and ClmTyp
3. Create new constraint TBLPK
4. For each Att
5. For each row
6. Insert data
7. End for
8. End For

Algorithm 2. Automatic insertion into a database.

6 Experiments Results and Analysis

We obtained 17 real public datasets (Water Quality, Libraries Board of South Australia Annual Report data, DHS Annual Report Data – Contractors) from **data.sa.ga.au,** (Australian Domestic Airline Activity (Link), Australian Government ICT Expenditure 2008–09–2011–12 Report, Rental Bond and Rental Data Tasmania June 2016 to July 2017, Taxation statistics - postcode data, Births Deaths and Marriages Applications by Year 2011 to 2016, Financial Advisers Dataset – Current, Ad-hoc requested data, City of Casey Transport Data, Intellectual Property Government Open Data 2017 (GovHack), Public trust and confidence in Australian charities 2013) from **data.gov.au,** (datasets, Fuel poverty, Improving the local area) from **Data North Yorkshire,** and (fish-benthos-1999–2014) from **eatlas.org.au.**

We searched over Excel spreadsheet style with '.xls' or '.xlsx' formats. All these datasets contain 43 Excel workbooks to examine. Every workbook may contain multiple sheets. The examination performs on spreadsheets contains a simple table structure. The evaluation based on table detected and extraction without a table name presented in Table 4.

Table 4. Test results.

Number of Excel workbook	43
Total number of simple tables	80
Number of successfully detected tables	78
Percentage of successfully detected tables	97.5%

As shown in Table 4, the percentage of correct table detection is 97.5%, where the algorithm fails to detect only 2 tables. In summary, to extract a large corrected number of tables with relational form, you have to identify a variety of tables cells feature.

The application is written in JAVA. To test it, an executable jar file is created. The hardware specifications for the system are, Intel® Core™ 2 Duo CPU with 4 GB of RAM. The version of Microsoft Excel installed on the machine is 2013 however the application works with '.xlsx' format. The H2 database version used is 1.3.172.

7 Conclusion and Future Work

This paper introduced novel algorithms to extract the metadata of semi-structured spreadsheets tables and transform them into a relational database. The paper used clustering approach for detecting each table inside the spreadsheet. A heuristic approach is used to identify table name, attributes, and values for each detected table. A rule-based table analysis is used for attribute names to automate insertion into a database. The experimental results showed that the accuracy of table detection is 97.5%.

In this paper, the focus was on the low-quality of the spreadsheet structure itself, not on the low-quality of the table content such as missing data value and duplicated rows. Besides, the extraction algorithm works on a simple spreadsheet table. As future work, it is worthy to work on these issues and extract complex data like hierarchal attributes or merged cells.

References

1. Horry, Y.: Financial information description language and visualization/analysis tools. Comput. Lang. Syst. Struct. **50**, 31–52 (2017)
2. Roy, S., Hermans, F., Aivaloglou, E., Winter, J., van Deursen, A.: Evaluating automatic spreadsheet metadata extraction on a large set of responses from MOOC participants. In: 2016 IEEE 23rd International Conference on Software Analysis, Evolution, and Reengineering (SANER), vol. 1, pp. 135–145. IEEE, March 2016
3. Bhowan, U., Sacristan, P., O'Malley, L., Miranda, A.A., Corcoran, M.: US Patent Application No. 15/810,885 (2019)
4. Hu, J., Kashi, R.S., Lopresti, D., Wilfong, G.T.: Evaluating the performance of table processing algorithms. Int. J. Doc. Anal. Recogn. **4**(3), 140–153 (2002)
5. Coüasnon, B., Lemaitre, A.: Recognition of Tables and Forms. Handbook of Document Image Processing and Recognition, pp. 647–677 (2014)
6. Pinto, D., McCallum, A., Wei, X., Croft, W.B.: Table extraction using conditional random fields. In: Proceedings of the 26th Annual International ACM SIGIR Conference on Research and Development in Information Retrieval, pp. 235–242. ACM, July 2003
7. Chen, Z., Cafarella, M.: Integrating spreadsheet data via accurate and low-effort extraction. In: Proceedings of the 20th ACM SIGKDD International Conference on Knowledge Discovery and Data Mining, pp. 1126–1135. ACM, August 2014
8. Barowy, D.W., Gulwani, S., Hart, T., Zorn, B.: FlashRelate: extracting relational data from semi-structured spreadsheets using examples. ACM SIGPLAN Not. **50**(6), 218–228 (2015)

9. Goto, K., Ohta, Y., Inakoshi, H., Yugami, N.: Extraction algorithms for hierarchical header structures from spreadsheets. In: EDBT/ICDT Workshops (2016)
10. Cunha, J., Saraiva, J., Visser, J.: From spreadsheets to relational databases and back. In: Proceedings of the 2009 ACM SIGPLAN Workshop on Partial Evaluation and Program Manipulation, pp. 179–188. ACM, January 2009
11. Le, V., Gulwani, S.: FlashExtract: a framework for data extraction by examples. In: ACM SIGPLAN Notices, vol. 49, no. 6, pp. 542–553. ACM, June 2014
12. Chen, Z., Cafarella, M.: Automatic web spreadsheet data extraction. In: Proceedings of the 3rd International Workshop on Semantic Search over the Web, p. 1. ACM, August 2013
13. Gulwani, S., Harris, W.R., Singh, R.: Spreadsheet data manipulation using examples. Commun. ACM **55**(8), 97–105 (2012)
14. Schreiber, S., Agne, S., Wolf, I., Dengel, A., Ahmed, S.: DeepDeSRT: deep learning for detection and structure recognition of tables in document images. In: the 2017 14th IAPR International Conference on Document Analysis and Recognition (ICDAR), vol. 1, pp. 1162–1167. IEEE, November 2017
15. Bendre, M., Venkataraman, V., Zhou, X., Chang, K., Parameswaran, A.: Towards a holistic integration of spreadsheets with databases: a scalable storage engine for presentational data management. In: 2018 IEEE 34th International Conference on Data Engineering (ICDE), pp. 113–124. IEEE, April 2018
16. Chen, Z., Cafarella, M.: Integrating spreadsheet data via accurate and low-effort extraction. In: Proceedings of the 20th ACM SIGKDD International Conference on Knowledge Discovery and Data Mining, pp. 1126–1135. ACM, August 2014
17. Spoth, W., Arab, B.S., Chan, E.S., Gawlick, D., Ghoneimy, A., Glavic, B., Hammerschmidt, B., Kennedy, O., Lee, S., Liu, Z.H., Niu, X.: Adaptive schema databases. In: 8th Biennial Conference on Innovative Data Systems Research, CIDR 2017, Online Proceedings, Chaminade, CA, USA, 8–11 January 2017, January 2017
18. Santos, M.Y., e Sá, J.O., Costa, C., Galvão, J., Andrade, C., Martinho, B., Lima, F.V., Costa, E.: A big data analytics architecture for industry 4.0. In: World Conference on Information Systems and Technologies, pp. 175–184. Springer, Cham, April 2017

Classification of Imbalanced Data Using Decision Tree and Bayesian Classifier

Ajay Malik, Abhishek Singh, and Maroti Deshmukh(✉)

Department of Computer Science and Engineering, National Institute of Technology, Uttarakhand, Srinagar, India
ajaymalik.cse16@nituk.ac.in, abhishek.cse16@nituk.ac.in,
marotideshmukh@nituk.ac.in

Abstract. The Bayesian classification is a method based on the Bayes theorem which gives best result when attributes are independent of each other and data is normalized. In this paper, a two-step approach is proposed to classify the data attributes which are locally normalized but not globally. The first step involves finding value for each attribute where the gain ratio is maximum. The classification occurs in the second step on the two separate parts of data using the Bayesian classifier. The experimental results show that the accuracy of the proposed method is better than the Bayesian classification and Decision tree.

Keywords: Gain ratio · Decision tree · Bayesian classification · Imbalanced data

1 Introduction

The classification problem is the most general problem in data science which has wide range of applications across various fields in computer science. As things move forward in time the data collected from various sources grows larger and larger. On one hand, greater the amount of data [12], the better results the classification strategies can achieve, while at the same time it makes classification an expensive and time consuming process. The Naive Bayes classifier [1] technique comes handy when the size of dataset is large as it can achieve very fast in classification of the data.

1.1 Bayesian Classifier

Bayes theorem is used to find the probability of an event C occurring, while X has occurred, when it is known that the conditional probability of X, given X has occurred. It can be stated as follows:

$$P(C_k|X) = \frac{P(C_k)P(X|C_k)}{P(X)} \quad (1)$$

A.-E. Hassanien et al. (Eds.): AICV 2020, AISC 1153, pp. 458–466, 2020.
https://doi.org/10.1007/978-3-030-44289-7_43

where, $P(X)$ is the total class probability, $P(C_k|X)$ is the posterior probability or probability of class C_k for given sample X, $P(X|C_k)$ is the conditional probability or probability of sample X for given class C_k, and $P(C_k)$ is the class probability. Here, X is the set of attributes given as input

$$X = (x_1, x_2, x_3, \cdots, x_n)$$

and based on the given probabilistic model the Naive Bayes classifier classifies the set of given attributes [7,11] into a class y based on the following formula:

$$y = argmax_{k \in \{1,2,3,\cdots,K\}} P(C_k) \prod_{i=0}^{n} P(X_i|C_k) \tag{2}$$

In the proposed model have utilised Gaussian Naive Bayes classifier as it works best for values which are continuous. The probability of an event using Gaussian Naive Bayes can be calculated as

$$P(x = v|C_k) = \frac{1}{\sqrt{2\pi\sigma_k^2}} e^{-\frac{(v-\mu_k)^2}{2\sigma_k^2}} \tag{3}$$

where, $P(x = v|C_k)$ is the conditional probability, σ_k^2 variance of class C_k and μ_k is the mean of class C_k, v is set of possible value of that unknown sample X.

1.2 Decision Tree

A decision tree [5] is a flowchart-like tree structure, where each internal node denotes a test on an attribute, each branch represents an outcome of the test, and each leaf node holds a class label. The topmost node in a tree is the root node. Given a sample X, for which the associated class label is unknown, the attribute values of the sample are tested against the decision tree [10]. A path is traced from the root to a leaf node, which holds the class prediction for that sample.

At every step, gain ratio for each branch is calculated and the attribute with the maximum gain ratio is selected as the splitting attribute. The gain ratio is calculated as given below:

$$Gain\ Ratio(D, X) = \frac{Gain(D, X)}{Split\ Information(D, X)} \tag{4}$$

$$Gain(D, X) = Entropy(D) - Gain\ Information(D, X) \tag{5}$$

$$Entropy(D) = \sum_{i=1}^{k} -p_i \log_2 p_i \tag{6}$$

$$Gain\ Information(D, X) = \sum_{i=1}^{v} \frac{|D_j|}{|D|}\ Entropy(D_j) \tag{7}$$

$$Split\ Information(D, X) = -\sum_{j=1}^{v} \frac{|D_j|}{|D|}\ log_2 \frac{|D_j|}{|D|} \tag{8}$$

where, a data set is D, the set of possible values of an attribute X is v, k is the number of classes. $|D|$ is the number of samples present in the dataset D, $|D_j|$ number of samples present in the dataset D for attribute possible value is j.

The performance of the model is analyzed with the help of a confusion matrix as shown in Table 1. TP stands for True Positives, it means the number of positive records correctly classified. FP stands for False Positives, the number of negative records wrongly classified. FN stands for False Negatives, the number of positive records wrongly classified. TN stands for True Negatives, the number of negative records correctly classified. Based on the confusion matrix following measures calculated.

Table 1. Confusion matrix

	Predicted	
Actual	TP	FN
	FP	TN

$$Accuracy = \frac{TP + TN}{TP + FP + FN + TN} \tag{9}$$

$$Recall = \frac{TP}{TP + FN} \tag{10}$$

One downside of the Naive Bayes classifier is that it works best when the values of attributes are globally normalized as it assumes that the mean value if the one with most frequency. Our proposed model eliminates this drawback [9] by using a two step approach. In the first step, the model, with the help of decision tree parameter finds the split value at which maximum gain ratio is achievable. The attributes is then divided into two parts based on the optimum value. In the next step a Bayesian classification approach is used for getting the output class.

The rest of this paper is organized in the following way. Section 3 provides the method itself and the inner functioning of the model in detail. Section 3 provides a description of the datasets utilised for experimentation purposes and also shows the various experiments and analysis performed on the datasets in order to provide a comparison between the different models. Section 4 concludes the paper.

2 Methodology

Naive Bayesian classifier works best for normalized datasets and fails for non-normalized datasets. In this paper, the model proposes a two step approach that eliminates this drawback by using Decision tree in the first step and Bayesian classification thereafter.

The first part of the algorithm focuses on finding the best split point for each attribute in order to best optimise the result. In the datasets chosen, it was observed that most of the attributes showed two maxima regions with each of the region being individually normalized. This lead to an idea that Bayesian classifier could be used for the two parts separately with high accuracy. The algorithm mentioned in Listing 2 iterates over the possible values of each attribute in fixed steps. For each instance the respective gain ratio is calculated if split was to occur at this particular instance. At the end of iterations all the calculated values of the gain ratio are compared to find the best possible splitting point. These values are then passed on to the next which then utilises a modified Bayesian approach to find the end results. The detail steps are shown in Algorithm 1.

Algorithm 1. Finds best split point based on Gain Ratio

Input: Frequency set of attribute at each value
Output: Best split point for the attribute

```
splitPoint ← φ
gain ← 0
i ← min(attribute)
while i < max(attribute) do
    Divide attribute set at i
    G ← GainRatio(i)
    if G > gain then
        gain ← G
        splitPoint ← i
    end
end
return splitPoint
```

After successfully deciding the split points for each of the attributes, the second and final part of the model attempts to calculate probability of each test data lying in a particular class. The algorithm mentioned in Listing 3 iterates over the entire test dataset, and for each attributes calculates probability for the respective part depending upon the value of the attribute. The detailed steps are shown in Algorithm 2.

3 Experimental Results and Analysis

Two datasets were considered for the experimental analysis. The first one being from the SONAR [3] imaging, which is a technique that utilizes sound radiations to detect objects underwater. Each record in the dataset comprises of 61 unlabeled attributes each describing a particular output in each SONAR radiation.

Algorithm 2. Predict most Probable label for test case

Input: Frequency and split value for the attribute
Output: label value

$splitPoint \leftarrow \phi$
$Prob_{label0} \leftarrow 1$
$Prob_{label1} \leftarrow 1$
$Result \leftarrow 0$
while *attributes* **do**
  if $attribute_{value} > splitPoint$ **then**
    $Prob_{label0} \leftarrow Prob_{label0} * Prob^{high}_{attributevalue}$
    $Prob_{label1} \leftarrow Prob_{label1} * Prob^{high}_{attributevalue}$
  else
    $Prob_{label0} \leftarrow Prob_{label0} * Prob^{low}_{attributevalue}$
    $Prob_{label1} \leftarrow Prob_{label1} * Prob^{low}_{attributevalue}$
  end
end
if $Prob_{label1} > Prob_{label0}$ **then**
  $Result \leftarrow 1$
end
return *Result*

The last attribute is a binary class describing whether the detected object is a rock or a metal. Figure 1 shows the distribution of frequency for rocks and metals in the dataset. The aim of this classification is to detect presence of wanted metals underwater. 84% of the data was used for training [4, 14] purpose and the rest 16% was used for testing purpose [15].

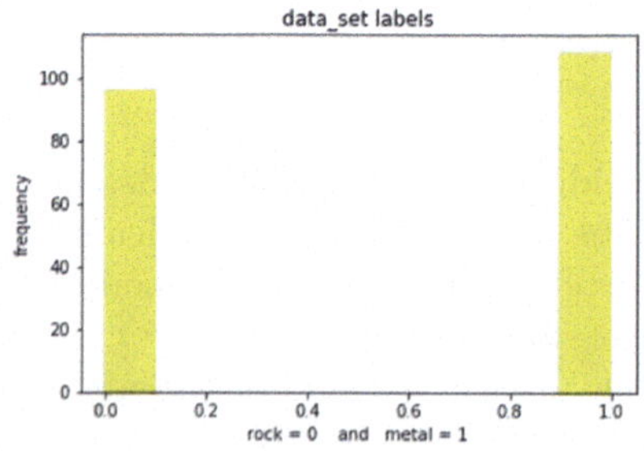

Fig. 1. Distribution of rocks and metals in the SONAR dataset

The second dataset, Credit Fraud Detection dataset contains the transactions done by the credit card users of a bank over a course of two days counting to a rough total of 285k records. Every record in the dataset contains 31 attributes and the last of which describes whether the transaction was done by a genuine source or was fraudulent. The dataset contains records which are already converted into respective float values hence no pre-processing was required. The dataset was highly unbalanced in nature with almost 0.17% records categorized as frauds. The aim is to classify a credit card transaction as fraud or genuine based on a set of different attributes. 65% of the data was used for training purpose and the rest 35% was used for testing purpose.

To analyze the performance of the proposed model, a series of experiments were conducted over the two mentioned datasets and compared the performance of regular Naive Bayes model with the proposed method. Figure 2(a) shows confusion matrix for classification using Bayesian classifier. The bayesian model fails to detect rocks because of poorly normalized data for rocks. Figure 2(b) denotes the confusion matrix in case of classification with the help of decision tree classifier. In this case, it is able to detect rocks but the accuracy of the model turns out to be very low. Finally, Fig. 2(c) shows how the proposed model performs in classifying the test cases into metals and rocks. As visible, the model shows increase in accuracy over Bayesian classification and Decision tree techniques. Table 2 shows the comparison between the three models stated above, the Bayesian approach, the approach utilizing only decision tree [6], and finally the approach used in our proposed model. The Fig. 3 shows the comparison between the accuracy of the three mentioned models i.e. Bayesian, Decision tree, and our proposed model respectively. As shown in the graph, the number of test cases increase the proposed model shown significant boost in the detection of correct outputs.

Table 2. Accuracy comparison for Sonar dataset

Model	Accuracy
Bayesian classifier	32.47
Decision tree	54.05
Proposed model	81.08

In the case of credit fraud detection [2], it has been observed that the data set is highly unbalanced. Only 0.17% data consists of records containing fraudulent transaction. Hence, accuracy has a very little depiction of performance of the system. On the other hand Recall, which is the fraction of detected fraud transactions is used as a performance measurement parameter. Table 3 shows confusion matrix for classification using Bayesian classifier [13]. The bayesian model shows about 74.45% recall and an accuracy of 97.81%. Table 4 shows how the proposed model performs in detecting the fraudulent transactions. The model shows a small increase in accuracy to 98.62%. But, the significant improvement is seen in case of recall with an increase of approx. 13% to 87.27%. Figure 4 provides comparative analysis for recall between the Bayesian model and our proposed model in case of the Credit Fraud database. As the number of test cases increase, it can seen that a significant increase in number of detected fraud transactions.

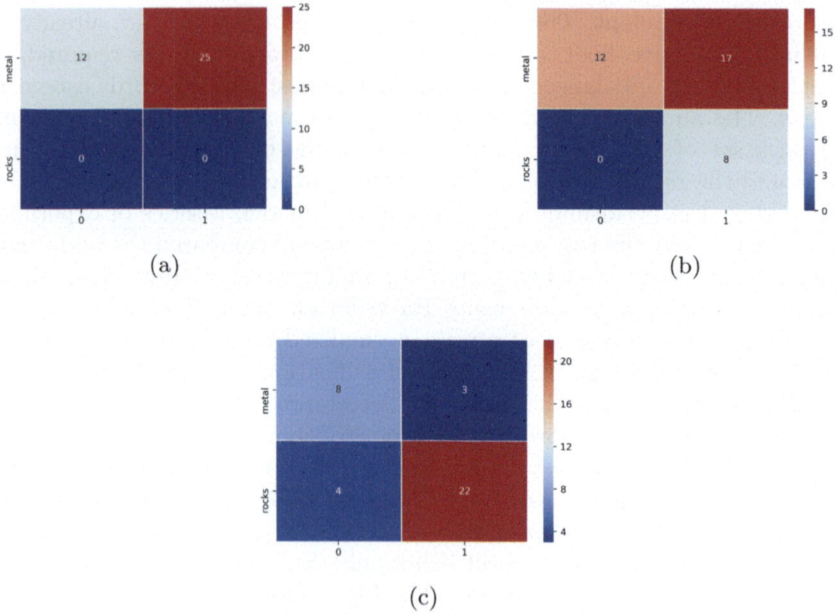

Fig. 2. Confusion matrix: (a) Confusion matrix using Bayes classifier, (b) Confusion matrix using decision tree, (c) Confusion matrix using proposed method.

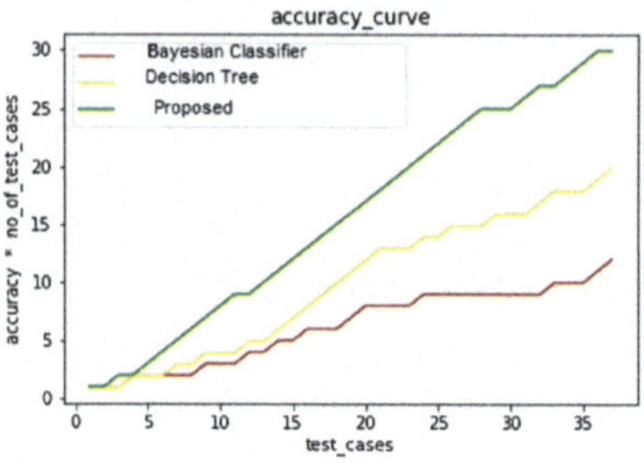

Fig. 3. Accuracy comparison of the Bayesian classifier, decision tree and proposed

Finally, the Table 5 compares accuracy as well as recall for both the models. It is interesting to note here that while for first dataset accuracy is good measure for performance evaluation, for second dataset recall provides for a better measure because of the highly unbalanced nature of the data. In other words catching all the fraud transactions is more important than correctly classifying the genuine transactions.

We present a general proof for why our model performs better than the Bayesian classification in case of locally normalized datasets with multiple maxima. Let us consider a general case with two maxima and one minima. Let us try to find probability at this minima (say v). Let P_1 denote the probability obtained

Table 3. Confusion metric for Bayesian classifier

	Predicted	
Actual	41	967
	14	43785

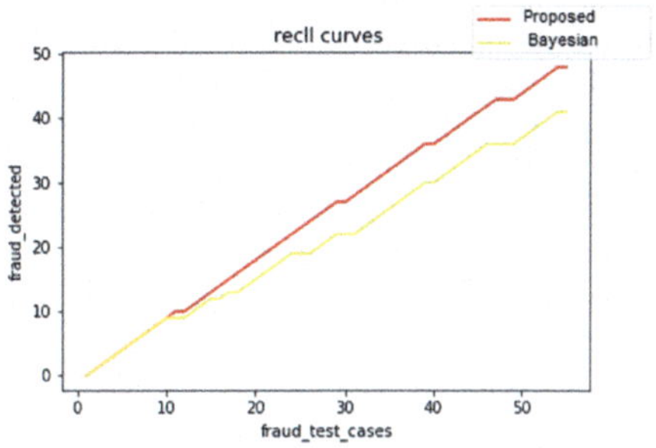

Fig. 4. Accuracy comparison of the Bayesian classifier and the proposed method

Table 4. Confusion metric for proposed model

	Predicted	
Actual	48	610
	7	44142

Table 5. Accuracy and Recall comparison for credit fraud dataset

Model	Accuracy	Recall
Bayesian classifier	97.81	74.45
Proposed model	98.62	87.27

at this point using Bayesian classification. From the formula mentioned in the Gaussian probability, it has been observed that $v \approx \mu$. This results to $P_1 = \frac{1}{\sqrt{2\pi\delta}}$. Now, let us consider the probability at the same point using the proposed approach, in this case $v \neq \mu_1$. Hence, model get a much lower value of probability at v which is correct as the frequency at the point of minima is in-fact negligible.

4 Conclusion

This paper presents an effective two step approach for classifying locally normalized datasets using Bayesian classifier combined with C4.5 decision tree. In step 1, model has used gain ratio to find out the best value to split each attribute in order to optimise the results. The values are then sent over to the second step which uses our proposed modified Bayesian classification approach to classify the test cases into respective classes. Experimental results show that the performance gained by the proposed model against the regular Naive Bayes Classifier.

References

1. Jahromi, A.H., Taheri, M.: A non-parametric mixture of Gaussian Naive Bayes classifiers based on local independent features. In: 2017 Artificial Intelligence and Signal Processing Conference (AISP), pp. 209–212 (2017)
2. Wei, W., Li, J., Cao, L., Ou, Y., Chen, J.: Effective detection of sophisticated online banking fraid on ectremely imbalanced data. World Wide Web **16**, 449–475 (2013). https://doi.org/10.1007/s11280-012-0178-0
3. Au, T., Chin, M.-L., Ma, G.: Mining rare events data by sampling and boosting: a case study. In: Prasad, S., Vin, H., Sahni, S., Jaiswal, M., Thipakorn, B. (eds.) Information Systems, Technology and Management, vol. 54, pp. 373–379. Springer, Heidelberg (2010)
4. Batista, G.E., Prati, R.C., Monard, M.C.: A study of the behavior of several methods for balancing machine learning training data. ACM SIGKDD Explor. Newsletter **6**(1), 20–29 (2004)
5. Freund, Y., Schapire, R.E.: A desicion-theoretic generalization of on-line learning and an application to boosting computational learning theory. In: Vitányi, P. (ed.) Computational Learning Theory, pp. 23–37. Springer, Heidelberg (1995)
6. Cardie, C.: Using decision trees to improve case-based learning. In: Proceedings of the 10th International Conference on Machine Learning, pp. 25–32. Morgan Kaufmann (1993)
7. Ratanamahatana, C.A., Gunopulos, D.: Feature selection for the naive Bayesian classifier using decision trees. Appl. Artif. Intell. **17**(5–6), 475–487 (2003)
8. Zhang, H., Sheng, S.: Learning weighted naive Bayes with accurate ranking. In: Proceedings of the 4th IEEE International Conference on Data Mining, pp. 567–570 (2004)
9. Ching, T., Himmelstein, D.S., Beaulieu-Jones, B.K., Kalinin, A.A., Do, B.T., Way, G.P., Ferrero, E., Agapow, P.M., Zietz, M., Hoffman, M.M., Xie, W., Rosen, G.L., Lengerich, B.J., Israeli, J., Lanchantin, J., Woloszynek, S., Carpenter, A.E., Shrikumar, A., Xu, J., Cofer, E.M., Lavender, C.A., Turaga, S.C., Alexandari, A.M., Lu, Z., Harris, D.J., DeCaprio, D., Qi, Y., Kundaje, A., Peng, Y., Wiley, L.K., Segler, M.H.S., Boca, S.M., Swamidass, S.J., Huang, A., Gitter, A., Greene, C.S.: Opportunities and obstacles for deep learning in biology and medicine. J. R. Soc. Interface **15**(141), 20170387 (2018)
10. Ratanamahatana, C., Gunopulos, D.: Feature Selection for the naive Bayesian classifier using decision trees. Appl. Artif. Intell. **17**, 475–487 (2003)
11. Tsymbal, A., Puuronen, S., Patterson, D.: Feature selection for ensembles of simple Bayesian classifiers. In: Hacid, M.-S., Raś, Z.W., Zighed, D.A., Kodratoff, Y. (eds.) ISMIS 2002. LNCS (LNAI), vol. 2366, pp. 592–600. Springer, Heidelberg (2002)
12. Witten, F.E.: Data Mining: Practical Machine Learning Tools and Techniques with Java Implementations. Morgan Kaufmann, San Mateo (2000)
13. Zheng, Z., Webb, G.I.: Lazy learning of Bayesian rules. Mach. Learn. **41**, 53–84 (2000)
14. Wang, B., Spencer, B., Ling, CX., Zhang, H.: Semi-supervised self-training for sentence subjectivity classification. In: Proceedings of 21st Conference on Advances in Artificial Intelligence, pp. 344–355 (2008)
15. Rosenberg, C., Hebert, M., Schneiderman, H.: Semi-supervised self-training of object detection models. In: WACV/MOTION, pp. 29–36 (2005)

Image Processing and Computer Vision

Implementing Two Recent Techniques for Delivering Nano-robots to Cancer Area

Doaa Ezzat(✉), Safaa Amin, Howida A. Shedeed, and Mohamed F. Tolba

Scientific Computing Department, Faculty of Computer and Information Sciences, Ain Shams University, Cairo, Egypt
doaa.ezzat@gmail.com

Abstract. Recently, it has been proposed to use Nano-robots in cancer treatment. These Nano-robots are injected in the human body. Then they travel through the blood vessels searching for cancer cells. After reaching these cells, the Nano-robots release the drug to destroy only the infected cells while leaving the uninfected ones untouched. In this way we can overcome the drawbacks of the traditional methods for treating cancer (chemo-therapy and radio-therapy). In this paper we apply a powerful algorithm called Teaching Learning Based Optimization (TLBO) to deliver a swarm of Nano-robots to their target. Then we compare this algorithm with the Directed Particle Swarm Optimization (DPSO) algorithm for the same problem.

Keywords: Nano-robots · Cancer treatment · TLBO · DPSO · Swarm intelligence

1 Introduction

In the last decades, the number of deaths caused by cancer increased considerably. According to recent statistics, it is expected that the number of cancer deaths in 2030 will be 13.1 million [1]. Unfortunately, the traditional methods of therapy (radio-therapy and chemo-therapy) for treating this disease have very harmful side effects on the healthy tissues.

To overcome these side effects, we can use a swarm of Nano-robots to deliver drugs into only the cancer area. A Nano-robot is a Nano-scale robot that can be programmed to perform a specific task. The size of a Nano-robot may be from 0.1 to 100 nm (a nanometer = 10^{-9} m). The major components of a Nano-robot are sensors, actuators, power supply, data transmission module and on-board computer. The Nano-robots are injected in the blood vessel. Then they travel through this vessel searching for cancer cells. After reaching the target, the Nano-robots release the drug with a certain concentration [2] to destroy only the tumor cells which can be recognized by their pH value. To control and optimize the movement of the Nano-robots during their journey, they need to use an appropriate swarm intelligence algorithm.

Swarm Intelligence (SI) is a sub-field of Artificial Intelligence that tries to mimic nature to solve search problems in engineering, technology and science. SI uses a group of agents that organize themselves using very basic rules. These rules lead the swarm to

A.-E. Hassanien et al. (Eds.): AICV 2020, AISC 1153, pp. 469–479, 2020.
https://doi.org/10.1007/978-3-030-44289-7_44

behave intelligently. This intelligent behavior can't be achieved by single agents [3–5]. The agents used by SI algorithms make up a population. The general attitude of this population results from the agents interaction among themselves and with the environment. These agents are moving in the search space without any centralized control structure. In other words, there is no leader for the agents and if needed, it will be changed according to its performance in each iteration.

SI has become an interesting area of research due to its simplicity and effectiveness. Many SI algorithms have been proposed in the literature. Most popular examples are Genetic Algorithm (GA), Ant Colony Optimization (ACO), Particle Swarm Optimization (PSO), Bee Colony Optimization (BCO) and the Firefly Algorithm (FA) etc. Some SI algorithms were proposed recently and achieved a good reputation like Cuckoo Search (CS), Jaya Algorithm (JA), Teaching Learning Based Optimization (TLBO) and Bat Algorithm (BA) etc.

Teaching Learning Based Optimization (TLBO) is one of the recently proposed SI algorithms that have been successfully used to solve many optimization problems [6–8]. The common application areas of TLBO are the Travelling Salesman's Problem (TSP), numerical function optimization, vehicle routing and job scheduling etc.

TLBO simulates the natural teaching learning process of a classroom. This process has two fundamental phases: (1) through the teacher (the teacher phase) and (2) through other learners (the learner phase). Beside its power to solve constrained and unconstrained optimization problems, TLBO does not need any problem-specific parameters. It needs only the common controlling parameters like the population size and the maximum number of generations.

Directed Particle Swarm Optimization (DPSO) is another recent swarm intelligence algorithm that introduces a modification in Particle Swarm Optimization (PSO) algorithm [9]. DPSO was essentially proposed for the problem of Nano-robots delivery to cancer area [9]. It was compared with some other swarm intelligence algorithms and it was proved experimentally that DPSO was very successful in delivering a swarm of Nano-robots to the target area in a very short time. DPSO also guarantees the reachability of the tumor region by all Nano-robots [9].

In this paper, we apply TLBO to the problem of Nano-robots delivery to cancer area. Then we compare its results with the DPSO results for the same problem.

2 Related Work

Many researchers proposed to use Nano-robots in detecting and treating diseases [10–16]. Many Nano-robot architectures for medical purposes were proposed by Cavalcanti et al. [11–15, 17]. The authors introduced hardware architecture to diagnose brain aneurysm [11] and diabetes control [12]. In [13], they presented a simulation of Nano-robots with sensors for medical target identification. In this simulation, each Nano-robot has a sensor to measure chemical concentration but there is no communication among Nano-robots.

Lewis and Bekey [18] used Nano-robots with swarm intelligence and chemical signalling techniques for treating tumors. Chandrasekaran and Hougen [19] introduced control strategies for medical Nano-robots that use swarm intelligence, where a

Nano-robot can detect the chemical signal, decide where to move next, and then it reaches its target, which has the maximum or the minimum chemical concentration. Also here, there is no communication among the Nano-robots.

Zhang et al. [20] in 2013 proposed a cooperative control technique for Nano-robots to deliver drugs to the tumor area. In their technique, the Nano-robots cooperate by robot-to-robot communications. They trace the gradient of pH values, and reach the tumor centre which has the minimum pH value.

In 2014, Ahmed et al. [21, 22] proposed a new control strategy for delivering drugs by Nano-robots to the cancer area. They introduced a modification in PSO to deliver the Nano-robots to the tumor environment. They also introduced an obstacle avoidance algorithm for the Nano-robots to avoid collisions with the blood cells. They finally combined these two algorithms to present their control strategy that controls the movement of the Nano-robots.

In 2016, an Improved Bacterial Foraging Optimization Algorithm (IBFOA) was proposed to reach and kill cancer cells using Nano-robots [23]. In 2017, researchers introduced a new computing-inspired bio-detection framework for detecting cancer [24]. In the same year, it was proposed to use Q-learning algorithm for optimizing the path travelled by Nano-robots to reach the target area [25].

In 2018, we presented a modification in PSO [9]. The modified algorithm was called Directed PSO (DPSO). It facilitates Nano-robots delivery to the tumor environment. It can efficiently deliver the whole swarm of Nano-robots to the target area in a very few number of iterations. DPSO was compared with four different SI algorithms and the experiments proved that DPSO outperforms all these algorithms in speeding up the process of Nano-robots delivery.

In 2019, DPSO was applied with a quorum sensing technique to destroy cancer cells [2], and it was proved experimentally that DPSO can speed up the process of killing cancer cells. Also, DPSO was combined with an efficient dynamic obstacle avoidance algorithm and the results were very promising.

3 Implemented Methodology

When applying a Swarm Intelligence (SI) algorithm to the problem of finding the cancer area by a swarm of Nano-robots, each Nano-robot position represents a solution in the search space. The fitness value of a Nano-robot position (a solution) is the pH value measured by this Nano-robot at its position. At normal cells, the pH value equals 7.4 [20]. In the surrounding area of the tumor, this value decreases gradually until the Nano-robot becomes inside the tumor region. Inside this region, the pH value becomes less than or equals to 5.7 [20].

Figure 1 shows a cross-sectional view of a tumor region and the normal tissue surrounding this region. In this figure we can see that the necrotic core of the tumor is the area of $r < r_1$, while the area of $r_1 < r < r_2$ is the layer of proliferating cancer cells. Figure 2 shows how the pH value changes in the tumor environment. The area of $0 < r < r_1$ is the tumor necrotic core with the minimum pH value. From this area to the outside of the tumor ($r_1 < r$) pH value increases continuously.

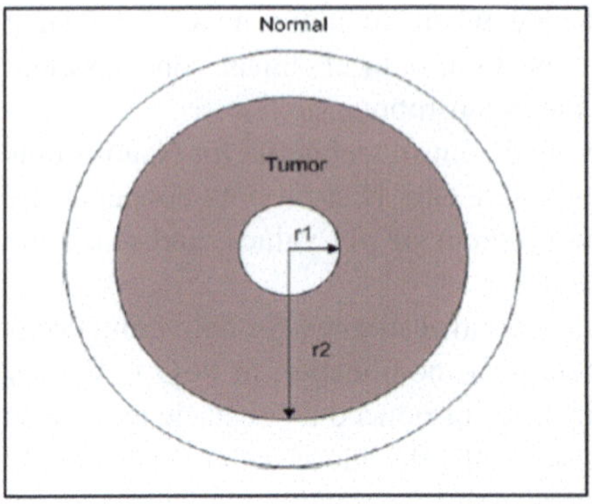

Fig. 1. A cross-sectional view of the tumor environment.

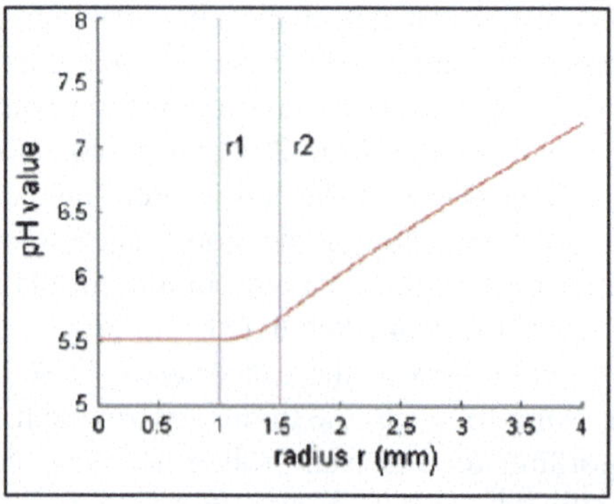

Fig. 2. The profile of the pH value.

In this paper we use two recent and efficient SI algorithms to solve the problem of Nano-robots delivery to cancer area. These algorithms are Teaching Learning Based Optimization (TLBO) and Directed Particle Swarm Optimization (DPSO). The next two sub-sections discuss these algorithms in details.

3.1 Teaching Learning Based Optimization (TLBO)

Teaching Learning Based Optimization (TLBO) is a population based algorithm proposed by Rao et al. [6–8]. It represents the learning process in a classroom where every learner tries to learn something from other learners to improve his knowledge. In TLBO, the teacher represents the fittest solution in the population (the best learner) and the learners are the solutions of the population. The subjects offered to the learners are the design variables of the optimization problem.

The results of the learners represent the fitness values of the solutions. The search process of TLBO contains two phases: the "teacher phase" where the teacher teaches the learners and the "learner phase" where the learners learn from each other.

In the "teacher phase", the teacher transfers knowledge to all learners and tries to increase the mean result of the whole class. At iteration i, $M_{j,i}$ is the mean value of the learners in a certain subject j. And the teacher for iteration i is the most knowledgeable learner in the entire population considering all the subjects. This teacher tries to put maximum effort to increase the knowledge level of the whole class, but the learners will improve their knowledge according to the quality of the teacher and the quality of the learners in the class. So, the difference between the teacher's value and the mean value of the learners in subject j is calculated as follows:

$$Difference\_Mean_{j,i} = r_i\left(X_{j,kbest,i} - T_F M_{j,i}\right) \tag{1}$$

Where $X_{j,kbest,i}$ is the value of the teacher in subject j at iteration i. and r_i is a random number between [0, 1]. TF is the teaching factor and its value can be 1 or 2. T_F is computed randomly with equal probability as follows:

$$T_F = round\,[1 + rand\,(0,\,1)] \quad (2)$$

Using the Difference_Mean$_{j,i}$, each learner in the class (each solution in the population) is updated in the teacher phase according to the following equation:

$$X'_{j,k,i} = X_{j,k,i} + Difference\_Mean_{j,i} \quad (3)$$

Where $X'_{j,k,i}$ is the new value of the solution and $X_{j,k,i}$ is its old value. The algorithm accepts $X'_{j,k,i}$ only if it gives a better fitness value than the old one, otherwise the learner remains unchanged.

In the "learner phase", each learner *P* in the class interacts with other learners by randomly selecting another learner *Q* from the class. If the fitness value of *Q* is higher than the fitness value of *P* then the learner *P* will modify himself by the following equation:

$$X''_{j,P,i} = X'_{j,P,i} + r_i\left(X'_{j,Q,i} - X'_{j,P,i}\right) \quad (4)$$

Where $X''_{j,P,i}$ is the new value of the solution *P* in the learner phase and $X'_{j,P,i}$ is its previous value. If the fitness value of *P* is higher than the fitness value of *Q* then the learner *P* will modify himself by the following equation:

$$X''_{j,P,i} = X'_{j,P,i} + r_i\left(X'_{j,P,i} - X'_{j,Q,i}\right) \quad (5)$$

Again, TLBO accepts $X''_{j,P,i}$ only if it gives a better fitness value than the previous one, otherwise the learner remains unchanged.

3.2 Directed Particle Swarm Optimization (DPSO)

Directed Particle Swarm Optimization (DPSO) represents a modification in the classical Particle Swarm Optimization (PSO) algorithm [9]. This modification makes PSO more efficient to deliver Nano-robots to the cancer area. DPSO can deliver all Nano-robots to the target in a very short time [9].

PSO is a swarm intelligence technique inspired from birds and fish. It was proposed by James Kennedy and Russell Eberhart [26]. It uses a group of particles spread in the search space to find the best solution. Initially, each particle has a random position and a random velocity. In every iteration, the velocity of each particle is updated using the best position found by this particle (*pbest*) and the best position found by all particles (*gbest*). The following equation is used for updating the velocity:

$$V_i^{t+1} = V_i^t + C_1R_1\left(pbest_i^t - P_i^t\right) + C_2R_2\left(gbest^t - P_i^t\right) \quad (6)$$

Where V_i^t represents the velocity of particle *i* at iteration *t*, C_1 and C_2 are acceleration constants related to the importance of pbest and gbest respectively, and R_1 and R_2 are random numbers from 0 to 1. Then, PSO updates the position of each particle by the following equation:

$$P_i^{t+1} = P_i^t + V_i^{t+1} \tag{7}$$

Where P_i^t represents the position of particle i at iteration t.

In DPSO Nano-robots firstly use PSO to move toward the target area. Once only one Nano-robot reaches this target area, the others direct themselves toward the position of this one. Then they take constant steps in this direction until the whole swarm of Nano-robots reaches the target area [9]. It was noticed experimentally that, when using PSO, some Nano-robots may get lost and can't reach the target. But when using DPSO, we can guarantee that all Nano-robots reach the target area. Therefore, the complete drug dose will reach the tumor environment [9].

4 Simulation and Results

We built a simulator using visual C++ to simulate the movement of the Nano-robots after locating them nearby the cancer cells. The simulation is done using 10, 20, 30, 40 and 50 Nano-robots and repeated 5 times for each setup. In each experiment, we applied TLBO and DPSO with the same starting locations of the Nano-robots. Figure 3 shows the initial locations of the Nano-robots.

For each setup, we computed the average number of iterations after which the whole swarm of Nano-robots reaches the tumor region. An iteration of DPSO represents either one PSO iteration or one direct step, but an iteration of TLBO represents either one teacher phase or one learner phase. The results are shown in Table 1.

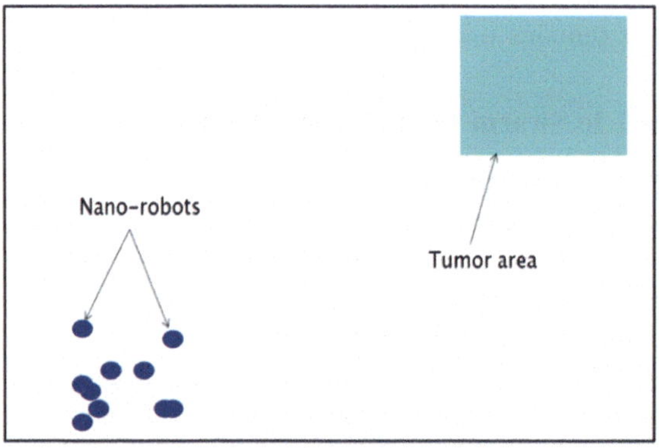

Fig. 3. The Nano-robots in the initial state.

Table 1. The average number of iterations needed to deliver all Nano-robots to the target.

Number of Nano-robots	TLBO	DPSO
10	20	10
20	17	12
30	22	12
40	22	12
50	16	11

From these results, we can conclude that TLBO needs about 20 iterations to deliver the whole swarm of Nano-robots to the target, but DPSO needs only 12 iterations at most to deliver the same number of Nano-robots to the tumor region. This means that DPSO improves the speed of the process of Nano-robots delivery by about 40%.

In each setup, we computed the average of number of Nano-robots that reach the target area after each iteration. Then for both algorithms, we drew the relation between the average number of Nano-robots and the number of iterations elapsed. We found that this relation is almost the same for all setups. Figure 4 shows this relation for 10 Nano-robots. Figure 5 shows this relation for 30 Nano-robots. Figure 6 shows this relation for 50 Nano-robots.

These graphs show that DPSO always outperforms TLBO in the total time needed to deliver a swarm of Nano-robots to the cancer area. But we can easily notice that in the first 5 iterations, DPSO delivers nothing to the target area. On the contrary, TLBO starts to deliver Nano-robots to the target from the third iteration at most. This means that TLBO always outperforms DPSO in the time needed to deliver the first Nano-robot to the target. This may be useful in speeding up the drug release process. Table 2 shows after which iteration the first Nano-robot reaches the target area on average for both algorithms in each setup.

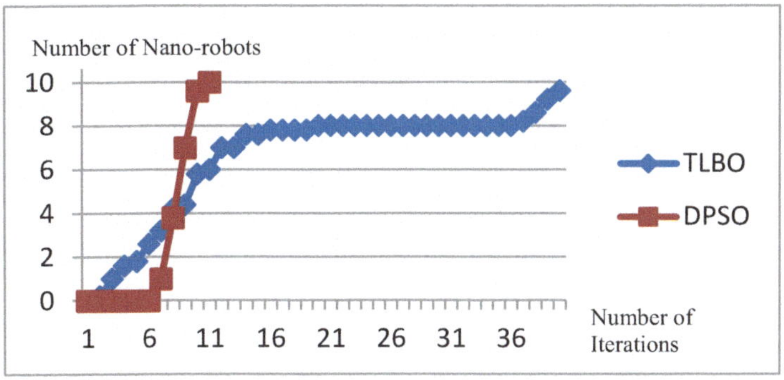

Fig. 4. The average delivery rate for 10 Nano-robots.

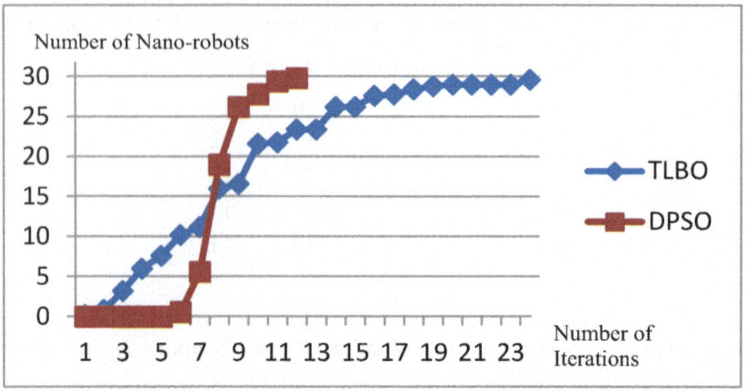

Fig. 5. The average delivery rate for 30 Nano-robots.

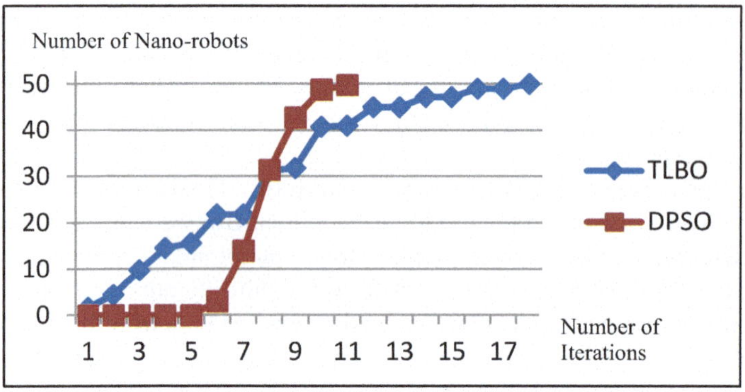

Fig. 6. The average delivery rate for 50 Nano-robots.

Table 2. The average delivery time for the first Nano-robot.

Number of Nano-robots	TLBO	DPSO
10	3	7
20	3	7
30	2	6
40	2	6
50	1	6

From Table 2 we can conclude that for few Nano-robots, both algorithms need more time to deliver the first Nano-robot to the target. But when using more Nano-robots, they need less time for delivering the first Nano-robot. Because for SI algorithms, using more agents usually achieves more cooperation and therefore gives better results.

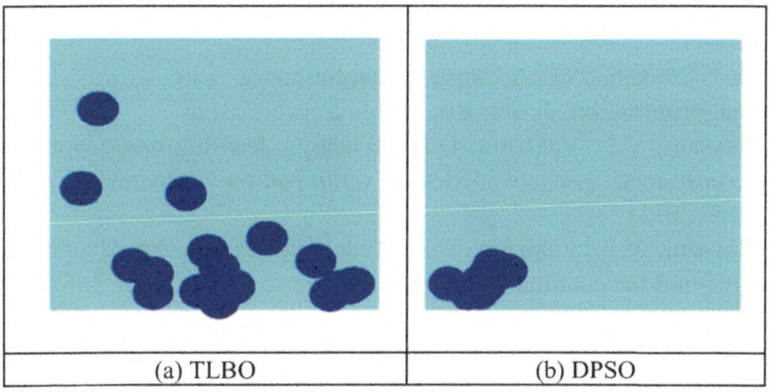

Fig. 7. The distribution of 20 Nano-robots in the cancer area.

After reaching the tumor region, the Nano-robots are grouped together in DPSO. But in TLBO the Nano-robots are scattered in the cancer area. Figure 7 shows the distribution of 20 Nano-robots in the cancer area. Grouping the Nano-robots in this area facilitates the drug release process. Because when using the quorum sensing algorithm that controls the drug release rate, the Nano-robots can't release the drug until their number reaches some threshold [2]. Before this threshold the drug release process can't be initiated and the cancer cells can't be destroyed.

So, we can conclude that for the problem of Nano-robots delivery to cancer area, DPSO always outperforms TLBO in general except for the early delivery rate.

5 Conclusions

Cancer treatment can be done by Nano-robots to deliver drugs into tumor cells without touching the healthy tissues. DPSO is a swarm intelligence algorithm proposed to solve the problem of Nano-robots delivery to the tumor environment. This paper compares its efficiency with another recent swarm intelligence algorithm called TLBO. We found experimentally that DPSO delivers all Nano-robots faster than TLBO. Also, DPSO outperforms TLBO in grouping the Nano-robots together in the cancer area to start the drug release. But TLBO starts to deliver Nano-robots faster than DPSO.

References

1. Devasena Umai, R., Brindha Devi, P., Thiruchelvi, R.: A review on DNA nanobots - a new technique for cancer treatment. Asian J. Pharm. Clin. Res. **11**(6), 61–64 (2018)
2. Ezzat, D., Amin, S., Shedeed, H.A., Tolba, M.F.: A new Nano-robots control strategy for killing cancer cells using quorum sensing technique and directed particle swarm optimization algorithm. In: The International Conference on Advanced Machine Learning Technologies and Applications (AMLTA 2019), pp. 218–226. Springer, Cairo (2019)
3. Binitha, S., Sathya, S.S.: A survey of bio inspired optimization algorithms. Int. J. Soft Comput. Eng. **2**, 137–151 (2012)

4. Kennedy, J., Kennedy, J.F., Eberhart, R.C., Shi, Y.: Swarm Intelligence. Morgan Kaufmann, Burlington (2001)
5. Mahale, R.A., Chavan, S.: A survey: evolutionary and swarm based bio-inspired optimization algorithms. Int. J. Sci. Res. Publ. **2**, 1–6 (2012)
6. Rao, R.V., Savsani, V.J., Vakharia, D.P.: Teaching–learning-based optimization: a novel method for constrained mechanical design optimization problems. Comput. Aided Des. **43**(3), 303–315 (2011)
7. Rao, R.V., Savsani, V.J., Vakharia, D.P.: Teaching–learning-based optimization: a novel optimization method for continuous non-linear large scale problems. Inform. Sci. **183**(1), 1–15 (2011)
8. Rao, R.V., Patel, V.: An elitist teaching–learning-based optimization algorithm for solving complex constrained optimization problems. Int. J. Ind. Eng. Comput. **3**(4), 535–560 (2012)
9. Ezzat, D., Amin, S., Shedeed, H.A., Tolba, M.F.: Directed particle swarm optimization technique for delivering Nano-robots to cancer cells. In: Proceedings of IEEE International Conference on Computer Engineering and Systems (ICCES 2018), Cairo, pp. 80–84 (2018)
10. Ummat, A., Sharma, G., Mavroidis, C., Dubey, A.: Biomedical engineering handbook. In: Bio-Nanorobotics: State of the Art and Future Challenges. CRC Press, London (2005)
11. Cavalcanti, A., Shirinzadeh, B., Fukuda, T., Ikeda, S.: Nanorobot for brain aneurysm. Int. J. Robot. Res. **28**(4), 558–570 (2009)
12. Cavalcanti, A., Shirinzadeh, B., Kretly, L.: Medical nanorobotics for diabetes control. Nanomedicine **4**, 127–138 (2008)
13. Cavalcanti, A., Shirinzadeh, B., Freitas, R., Hogg, T.: Nanorobot architecture for medical target identification. Nanotechnology **19**, 1–15 (2008)
14. Cavalcanti, A., Freitas, R.: Nanorobotics control design: a collective behavior approach for medicine. IEEE Trans. Nanobiosci. **4**, 133–140 (2005)
15. Cavalcanti, A.: Assembly automation with evolutionary nanorobots and sensor-based control applied to nanomedicine. IEEE Trans. Nanotechnol. **2**, 82–87 (2003)
16. Freitas, R.: What is nanomedicine? Nanomedicine **1**, 2–9 (2005)
17. Cavalcanti, A., Freitas, R.: Autonomous multi-robot sensor-based cooperation for nanomedicine. Int. J. Nonlinear Sci. Numer Simul. **3**, 743–746 (2002)
18. Lewis, M., Bekey, G.: The behavioral self-organization of nanorobots using local rules. In: Proceedings of IEEE International Conference on Intelligent Robots and Systems, pp. 1333–1338 (1992)
19. Chandrasekaran, S., Hougen, D.: Swarm intelligence for cooperation of bio-nano robots using quorum sensing. In: Bio Micro and Nanosystems Conference, San Francisco, p. 104 (2006)
20. Zhang, S., Li, S., Guo, Y.: Cooperative control design for Nano-robots in drug delivery. In: Guo, Y. (ed.) Selected Topics in Micro/Nano-Robotics for Biomedical Applications, pp. 101–123. Springer, New York (2013)
21. Ahmed, S., Amin, S., Alarif, T.: Simulation for the motion of nanorobots in human blood stream environment. In: Proceedings of Scientific Cooperation International Workshops on Electrical and Computer Engineering Subfields, pp. 70–75, Koc University, Istanbul (2014)
22. Ahmed, S., Amin, S., Alarif, T.: Efficient cooperative control system for pH sensitive nanorobots in drug delivery. Int. J. Comput. Appl. **103**(1), 0975–8887 (2014)
23. Cao, J., Li, M., Wang, H., Huang, L., Zhao, Y.: An improved bacterial foraging algorithm with cooperative learning for eradicating cancer cells using nanorobots. In: Proceedings of the IEEE International Conference on Robotics and Biomimetics, Qingdao (2016)
24. Chen, Y., Shi, S., Yao, X., Nakano, T.: Touchable computing: computing-inspired bio-detection. IEEE Trans. Nanobiosci. **16**(8), 810–821 (2017)

25. Lambe, A.: Reinforcement learning for optimal path length of nanobots using dynamic programming. In: Proceedings of IEEE International Conference on Industrial and Information Systems, Peradeniya (2017)
26. Eberhart, R., Kennedy, J.: A new optimizer using particle swarm theory. In: Proceedings of the Sixth International Symposium on Micro machine and Human Science, pp. 39–43 (1995)

A Machine Vision Based Automatic Optical Inspection System for Detecting Defects of PCBA

Huan Ngoc Le[1(✉)], Thang Viet Nguyen[1], and Narayan C. Debnath[2]

[1] Mechanical and Mechatronics Department, Eastern International University, Thu Dau Mot, Binh Duong Province, Vietnam
{huan.le, thang.nguyen}@eiu.edu.vn

[2] School of Computing and Information Technology, Eastern International University, Thu Dau Mot, Binh Duong Province, Vietnam
narayan.debnath@eiu.edu.vn

Abstract. The process of manufacturing electronic printed circuit board assembly (PCBA) is a common process in technology companies. In particular, error checking of a PCBA with full components is a difficult job for human inspectors due to the high concentration and persistence. To solve this problem, many solutions have been studied, in which image processing technology has emerged as the best solution with advantages such as fast processing speed and high accuracy. However, the cost of these systems is often quite high for small and medium size companies. This paper presents a solution using image processing technology to check for missing components on complete PCBAs. Accordingly, a suitable algorithm is proposed for low-cost cameras, which allows building a simpler and more economical inspection process for both types of solder joints: through hole and surface mounting device (SMD). Consuming time to test a PCBA with nearly 50 components is less than 1 s. A cheap and easy-to-use Automatic Optical Inspection (AOI) system had been developed and used to test the actual PCBAs at the company.

Keywords: Detecting defects · PCBA · Automatic Optical Inspection (AOI)

1 Introduction

In the past, Image processing technology was commonly applied to test electronic printed circuits, test chips, surface quality of materials, food [1, 2]. However, with the growing technology of camera production along with increasing computer processing power, it is an advantage for the outstanding development of applications using image processing technology. AOI is found in all areas of social life: high quality agriculture, transportation projects, medicine, military and so on.

In PCBA manufacturing process, the quality control methods of PCBAs using a camera are often quite diverse. However, it can be classified into two main groups: the method that references the sample image data and the method that does not refer to the sample image data. The first method requires a standard image (no defects) as a reference. When conducting tests, the images to be tested are included in the analysis

A.-E. Hassanien et al. (Eds.): AICV 2020, AISC 1153, pp. 480–489, 2020.
https://doi.org/10.1007/978-3-030-44289-7_45

with the standard images. Standard image subtraction with test images is a common method used in [3]. Accordingly, the test image is calibrated to be similar to the standard image (due to deviations when taking the picture). Subtraction of the standard image with the test image is then carried out and produces a positive and negative image. These images are again thresholded, binary, and finally filtered and graded. Some other authors used additional transformations such as Fourier transform [4] or Haar Wavelet [4] to analyze different characteristics. Instead of the method of comparison with standard images, the second method uses a set of design rules (obtained when designing the circuit) or standard features of components, solder joints to build the training models for estimating the dimension, position, and orientation of the boards, components or solder joints [5]. In [6], the authors studied 14 features of types of solder joint and conducted training using neural network combined with genetic method. Then, the input image will be included in the algorithm to check whether the solder joints are defective. Besides, features extracted based on the special solder joint image which is acquired by a particular image acquisition system composed of a 3-CCD color digital camera were used to build the models of solder joint types. Then, based on Boolean rules method, the authors can detect eight different types of solder joint types [7]. In [8], the authors also used the neural network to train the border and histogram features of the component for identifying whether a component exists or is rotated. The model needs good light source. If the wrong size of components is collected, the result may not be correct. In [9], the authors developed a new model to train the set of IC solder joint images, and for each IC input image, the model will return the evaluation results. Time requirement for each component or PCBA has not been considered in previous studies.

All of the above studies assumed ideal conditions such as good illumination, less distortion, etc. However, for practical projects, small components mounted on PCBAs are often warped. The relative position between components is not always guaranteed. The components are skewed in 3D space. Therefore, we have further studied the calibration scheme so that we can find the most accurate position of the component. The defective detection of large size through hole ICs is often easier so that we need a simpler algorithm to improve the real-time capabilities of the system. Therefore, it is necessary to develop two independent detection algorithms for large-sized IC and ICs in surface mounting technology (SMT) board.

In keeping with the actual requirements of an efficient system that requires a low cost and can detect a common defect as lack of components, we have proposed a low-cost automatic optical inspection (AOI) system. The main objective of this study is to develop algorithms based on the method required standard images to automatically detect PCBA defects as closely as possible to the result of a human grading expert. A mobile image acquisition and processing system has been developed for on-site applications. This system includes one image acquisition unit and one mobile computer unit attached to the laptop. The developed algorithms have been implemented on mobile processors as an supplementary testing device at one company in HCMC, Vietnam.

2 Proposed Materials and Methods

2.1 Mobile Image Processing Unit

The developed mobile unit consisted of a camera placed in an appropriate position, an illumination system composed of a white LED rectangular-type direct lighting device (see Fig. 1). The system also includes a removable tray to bring the board in and out. The authors experimented with a program running on a computer with a Core I5 CPU configuration, 3 GB RAM, a camera with a resolution of 1920 × 1080 pixels, Fig. 1(b). Experimental selected PCBA is a board with the total number of 48 components. The experiment was conducted on 5 boards with full components and 5 boards without any components. The analysis software was made with C#.

(a) (b)

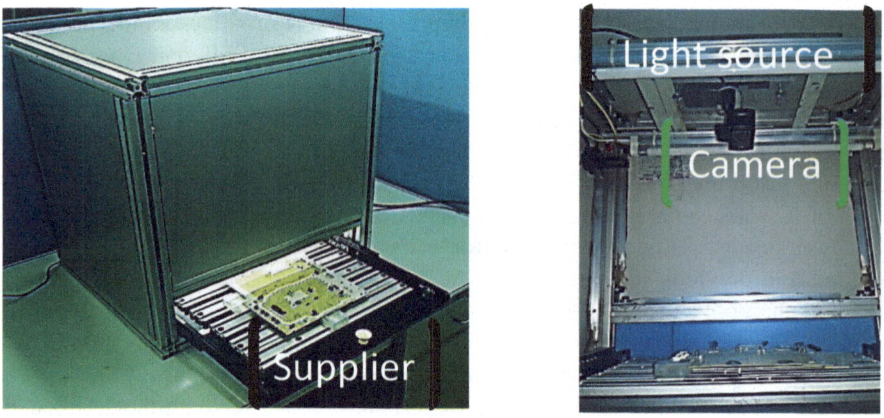

Fig. 1. (a) A full prototype of defective detection system, (b) Lighting system and camera are mounted inside the model.

2.2 Designing Algorithm to Identify PCBAs Missing Components

Algorithm for testing components on PCBAs is proposed based on 2 phases: training phase and test phase. The training phase is the process of creating database sources. In contrast, the test phase is a new data processing phase based on an existing image.

Training Phase: Building a database for distinguishing the standard and non-standard board

The training process is the process of creating a database to help the system clearly distinguish the differences between a defective board with missing components and a standard board. This process is done before starting the testing process. Therefore, it does not require fast processing time, but needs precision in distinguishing whether the board is defective or not. The process of implementation includes the following steps:

- Step 1: Take a image of the empty board (without any components) and the standard board (with full components as designed).
- Step 2: Choose landmarks
- Step 3: Select the area to be tested. The location of the main area is where the components need to be checked. This selection process can be done manually or automatically. However, the automatic selection process is sometime incorrect due to mechanical errors. Therefore, a manual option, with the advantage of high flexibility, is used frequently.
- Step 4: Reduce noise in the standard and non-standard board image by applying a 2-dimensional 5 × 5 Gaussian filter. Two-dimensional Gaussian filter has the effect of reducing noise in the image. In 2-dimensional space, the Gaussian G (x, y) distribution function is defined by the following formula:

$$G(x, y) = \frac{1}{2\pi\sigma^2} e^{-\frac{x^2 + y^2}{2\sigma^2}} \quad (1)$$

  where σ is the distribution variance.
- Step 5: Store data including standard PCBA image, empty PCB image, and parameters related to the area to be checked.

To the boards containing through hole devides, in Step 2, we declare that the number of landmarks is zero.

Test Phase: Test PCBA based on the data available from the training phase

The test phase is the process of identifying a new board to be put into the system based on an existing database. The inspection phase is a process that needs to reduce processing time to meet production requirements while ensuring the inspection quality as required. The proposed process consists of the following steps:

- Step 1: The PCBA image to be checked will be removed noise by using a Gaussian filter.
- Step 2: Redefine the location of the board. Due to mechanical errors (manufacturing errors, errors due to board movement…), the boards that need to be checked may be set not completely in accordance with the standard board setting. Placing the board incorrectly may cause identification errors. To fix this, the template matching method with the correlation coefficient measurement is used to redefine the correct position of the board.

The correlation coefficient calculation for pattern matching method is defined as follows:

$$R(x, y) = \frac{\sum_{x',y'} (T'(x', y').I'(x + x', y + y'))}{\sqrt{\sum_{x',y'} T'(x', y')^2 \sum_{x',y'} I'(x + x', y + y')^2}} \quad (2)$$

with

$$I'(x + x', y + y') = I(x + x', y + y') - \frac{1}{w.h}\sum\nolimits_{x'',y''} I(x + x'', y + y'')$$

$$T'(x', y') = T(x', y') - 1/(w.h) . \sum\nolimits_{x'',y''} T(x'', y'')$$

where T is the standard image, I is the image taken from the camera (see Fig. 2).

The process of detecting the board position on the test image is considered as a correlation map between the image of the standard board image and the test image. In Fig. 3, the two horizontal axes are the two dimensions of the image, while the vertical axis (z axis) is the axis that shows the correlation between the reference board and each region of the test board. The highest value point (corresponding to the largest correlation coefficient) is considered the position of the board in the test image.

- Step 3: Relocate each location of each test area based on the number of landmarks.
- Step 4: Each component (selected in the training phase) will be separated from the large PCBA image and put into the comparator with the original database (created in the training phase). The selected comparison method is to measure the correlation between the histogram of the test image and the standard image, and between the histogram of the test image and empty board image.

The histogram calculation of the image is as follows. H is the histogram of the T image and H[i] is the number of pixels with gray level i in the T image.

The correlation measurement of histogram is performed by the following formula:

$$d(H_1, H_2) = \frac{\sum_I (H_1(I) - \bar{H}_1) . (H_2(I) - \bar{H}_2)}{\sqrt{\sum_I (H_1(I) - \bar{H}_1)^2 \sum_I (H_2(I) - \bar{H}_2)^2}} \tag{3}$$

For $\bar{H}_k = \frac{1}{N}\sum\limits_J H_k(J)$, N is the number of gray levels.

Reposition the Components

For SMT boards, after noise reduction, we have to reposition the components based on relative relationships with the landmarks that have been identified earlier. The position interpolation formula of components is described as follows:

- Considering there are N landmarks with initial coordinates (in the standard image) as sets (x_{Oi}, y_{Oi}), with i from 1 to N. Find the position of these N landmarks in the test image and save them as a set of positions (x_i, y_i) with i from 1 to 10. Then, we calculate the deviation of each marker by the formula:

$$\text{offset}X_i = x_i - x_{Oi} \tag{4}$$

$$\text{and} \quad \text{offset}Y_i = y_i - y_{Oi} \tag{5}$$

with i ranging from 1 to N.

- Setting the distance from each landmark to the component is d_i, with i from 1 to N. We calculate the weight of each landmark for the component as below:

$$w_i = 1/d_i \tag{6}$$

- Calculating the deviation of the component using the following formula:

$$offsetX = \frac{\sum(offsetX_i * w_i)}{\sum(w_i)} \tag{7}$$

and

$$offsetY = \frac{\sum(offsetY_i * w_i)}{\sum(w_i)} \tag{8}$$

with i ranging from 1 to N.

- After having the deviation, recalculate the position of the component according to the deviation: (X_{old}, Y_{old}) is the original position of the component in the test image, and (X_{new}, Y_{new}) is the new position of the component after recalculating.

$$X_{new} = X_{old} + offsetX \tag{9}$$

and

$$Y_{new} = Y_{old} + offsetY \tag{10}$$

- To ensure the exact position of the device, we start scanning around the component location just found according to the rule as shown in the Fig. 4 to find the object most similar to the component.

General Algorithm Using Flowcharts

The solution to detect the missing components on the electronic board consists of two phases: the training phase and the test phase as described above. The algorithm flowchart of the training and test phase is presented as in Fig. 5.

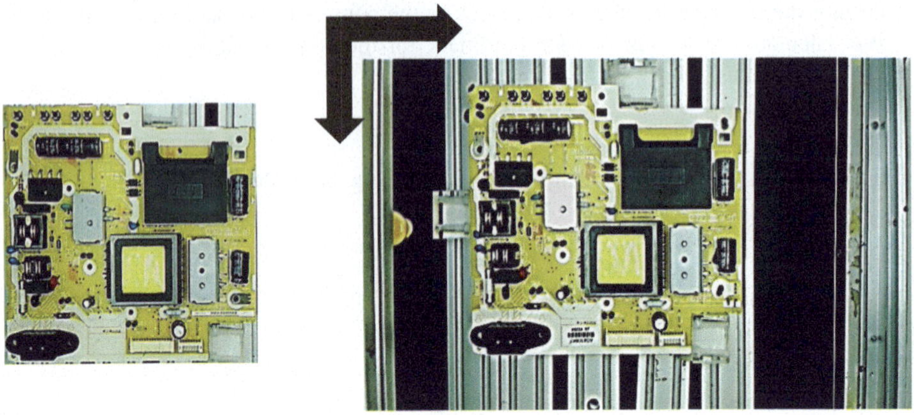

Fig. 2. Locating the board in the input image on left: standard board image and on right: the test image.

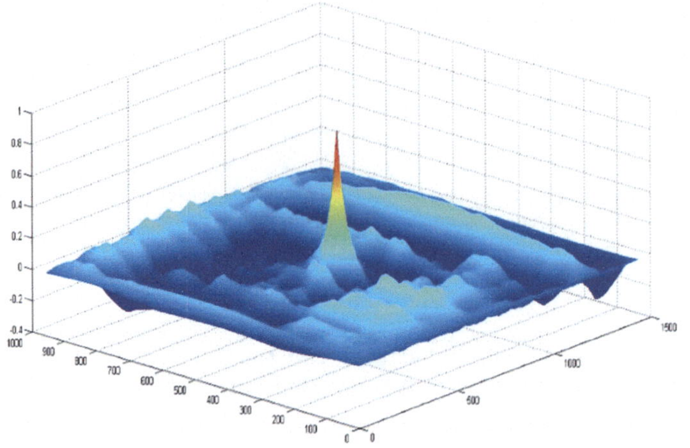

Fig. 3. The correlation map between the image of the standard board and the test image.

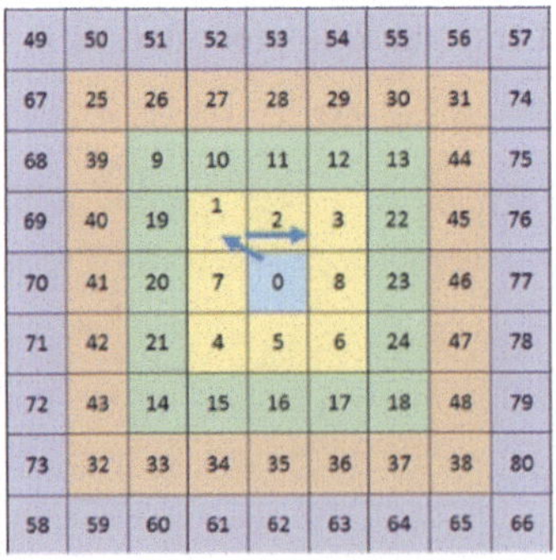

Fig. 4. Scanning scheme for searching component position.

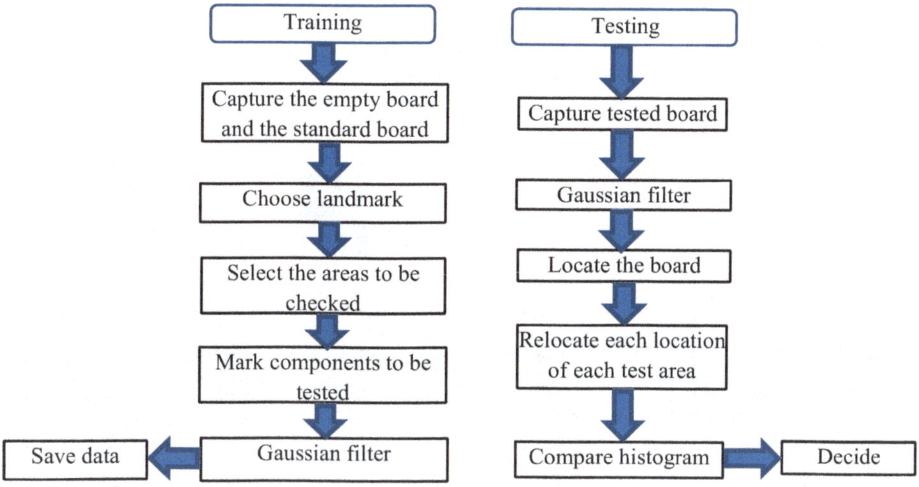

Fig. 5. Algorithms to check for missing components on PCBA: flowchart of the training phase (Left) and flowchart of the test phase (Right).

3 Results and Discussion

Statistics of experimental results:

- With 5 boards with full 48 components: The system reports fake errors twice as shown in Fig. 6. On the first board, the reason is because this component has a much lower gloss than the standard component. On the second fake error board, this is because the component has a wrong position compared to the position in the standard board. Therefore, the detection system has identified that this case lacks components (see Fig. 6).
- With 5 boards lacking components: The system always identifies as a defective board. For each board, the system also shows the correct location as well as the name of each missing component (see Fig. 7).
- Processing time: The time for the computer to receive data from the camera is approximately 0.1 s. The time for processing image data and giving test results is approximately 0.5 s. Thus, the total processing time for each input PCBA is less than 1 s.

The proposed AOI system is an optimal solution from hardware to software to create a device that can be used in practice with the advantages: can run online real-time; can make updating the system much easier compared to the previous researches of using standard images. Compared to the systems that train the features of the component for identifying defective components, the proposed system is extremely simple in terms of algorithm but the result is acceptable for small and medium-sized companies. Besides, at the first stage, we used two fluorescent tubes on both sides of the camera, resulting in poor image quality and inconsistency through the shoots.

For improvement, we replaced them with a 4-row array of LEDs above the board surface and the camera moved to multiple locations to capture the entire board image. The system works in a stable condition until now.

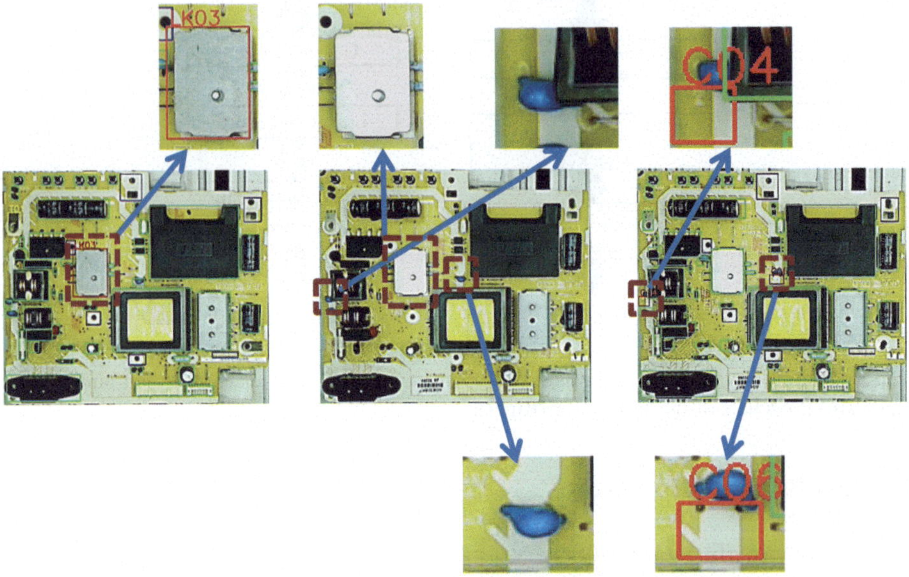

Fig. 6. Some empirical results with board having full components. Middle picture: standard board. The picture on the left: checked board with fake error (at the location of the red rectangle) because the device has been oxidized and dark. Right: checked board with fake alarms due to misplaced components.

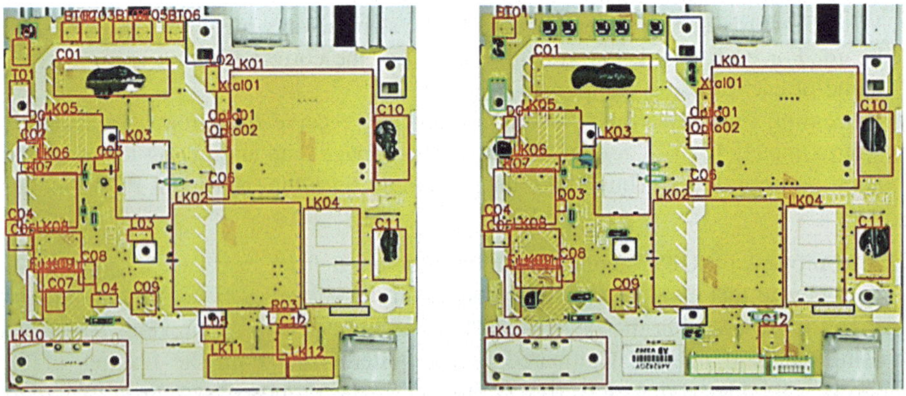

Fig. 7. Images of experimental results with missing components. The blue rectangles indicate that there is a component in that position. The red rectangles indicate the lack of components.

4 Conclusion

The paper describes a solution that helps reduce the cost and processing time for detecting the missing components in PCBAs. The results of the paper can be applied to the development of inspection systems to test for other electronic board errors including errors such as: checking the joint solder on the component pins, deflection of the components, or the circuit lines on the PCB, etc. In addition, the result creates a premise for developing low-price AOI systems to check details, defective products in industrial production lines. However, the improvement of lighting systems and cameras along with the development of accurate and optimal computational algorithms still requires further research to provide more effective inspection process. A mobile computer-based system was developed and it showed significant improvements in the results.

Acknowledgments. PCBAs used in this research were supported by Panasonic Viet Nam in Ho Chi Minh City, Viet Nam. The company is also using our developed AOI system for testing the real products.

References

1. Newman, T.S., Jain, A.K.: A survey of automated visual inspection. Comput. Vis. Image Underst. **61**(2), 231–262 (1995)
2. Chin, R.T.: Automatic visual inspection techniques and application: a bibliography. Pattern Recogn. **15**(4), 343–357 (1982)
3. Ibrahim, I., Rahman, S.A., Bakar, S.A., Mokji, M.M., Mukred, J.A.A., Yusof, Z., Ibrahim, Z., Khalil, K., Mohamad, M.S.: A printed circuit board inspection system with defect classification capability. Int. J. Innov. Manag. Inf. Prod. ISME Int. **3**(1), 82–87 (2012)
4. Chen, C.S., Yeh, C.W., Yin, P.Y.: A novel Fourier descriptor based image alignment algorithm for automatic optical inspection. Elsevier J. Vis. Commun. Image Represent. **20**(3), 178–189 (2009)
5. Cai, Y., Huang, Y., Zhang, S.: Research of defect inspection and processing in PCB automatic optical inspection. In: Proceedings of the 2012 International Conference on Electronics, Communications and Control, ICECC 2012, Zhoushan, China, pp. 803–806 (2012)
6. Hao, W., Xianmin, Z., Yongcong, K., Gaofei, O., Hongwei, X.: Solder joint inspection based on neural network combined with genetic algorithm. Optik **124**, 4110–4116 (2013)
7. Wu, F., Zhang, X.: An inspection and classification method for chip solder joints using color grads and Boolean rules. Robot. Comput. Integr. Manuf. **30**, 517–526 (2014)
8. Alexandre, R.D.M., Marcelo, R.S.: Inspecting surface mounted devices using K nearest neighbor and multilayer perception. In: IEEE International Symposium on Industrial Electronics, Buzios, pp. 950–955 (2015)
9. Cai, N., Lin, J., Ye, Q., Wang, H., Weng, S., Ling, B.W.K: A new IC solder joint inspection method for an automatic optical inspection system based on an improved visual background extraction algorithm. IEEE Trans. Compon. Packag. Manuf. Technol. **6**(1), 161–172 (2016)

Eye Movements Recognition Using Electrooculography Signals

Radwa Reda(✉), Manal Tantawi, Howida Shedeed, and Mohamed F. Tolba

Scientific Computing Department, Faculty of Computer and Information Science, FCIS, Ain Shams University, Cairo, Egypt
radwareda2012@gmail.com

Abstract. Nowadays, the number of people who suffer from severe motor disabilities is increasing remarkably. This disability gradually prevents them from moving all their limbs and they can communicate with their external environment only through the movement of their eyes. Hence, Human Computer Interfaces (HCI) have come out to provide those people with new communication way based on detecting eye movements. Eye movements are recorded by Electrooculogram (EOG). These signals are captured by placing electrodes horizontally and vertically around the eyes. In this work, the five eye movements left, right, up, down and blinking are classified by investigating both horizontal and vertical EOG signals. Statistical and geometrical features are extracted from EOG signals after applying Discrete Wavelet Transform (DWT). Six classifiers are examined in this study using both horizontal and vertical EOG features. The experimental results show the superiority of Naïve Bayes classifier.

Keywords: Electro-oculogram (EOG) signal · Human Computer/Machine Interface (HCI/HMI) · Naïve Bayes · Discrete Wavelet Transform (DWT)

1 Introduction

All over the world, several diseases lead to weakness of the motor neuron muscles which by its role impede some voluntary muscles from performing their functions. Thus, patients are paralyzed forever or for period of time. Unfortunately, the number of patients is increasing, reaching 6,500 new cases each year with 90% occurs without any genetic injury in the family [1–3].

Amyotrophic Lateral Sclerosis (ALS) is a famous disease that causes weakness in the hands and legs muscles and then difficulty in breathing, speaking and swallowing. In the final stage, the limbs and tongue are paralyzed [4, 5], Therefore, ALS patients have only eye movement to communicate with their outer environment. On the other hand, other diseases can also cause death to motor neurons such as Multiple Sclerosis [6], Guillain–Barré Syndrome (GBS) [7], etc. these diseases occur due to some environmental factors such as continuous exposure to electromagnetic waves and some heavy metals including mercury and lead, as well as some wrong habits such as smoking and the use of pesticides [8, 9].

A.-E. Hassanien et al. (Eds.): AICV 2020, AISC 1153, pp. 490–500, 2020.
https://doi.org/10.1007/978-3-030-44289-7_46

Although the limbs of those patients are paralyzed completely, eye movements are usually spared. Consequently, it's the last and only tool for taking action and expressing themselves. Human Computer/Machine Interfaces (HCI/HMI) provide those patients with facility that enables them to do their daily activities through detection of eye movements. For example, patients can select a displayed action on computer screen, select characters to be written (virtual keyboard) or control a wheel chair. The considered eye movements by most existing studies are left, right, up and down movements along with blinking [10]. On the other hand, few studies considered also the four corners. Blinking is used by all interfaces to confirm user selection.

Electrooculogram (EOG) signals record the potential difference between the cornea and the retina [11]. EOG is recorded horizontally and vertically using five electrodes placed around the eyes (two vertical, two horizontal and one for ground). For example, as shown in Fig. 1, when the eyes roll upward or downward, positive or negative pulses will be generated. The positive or negative pulse's amplitude and width will be increased with the increment of rolling angle and duration of the eyeball rolling process [12, 13], respectively.

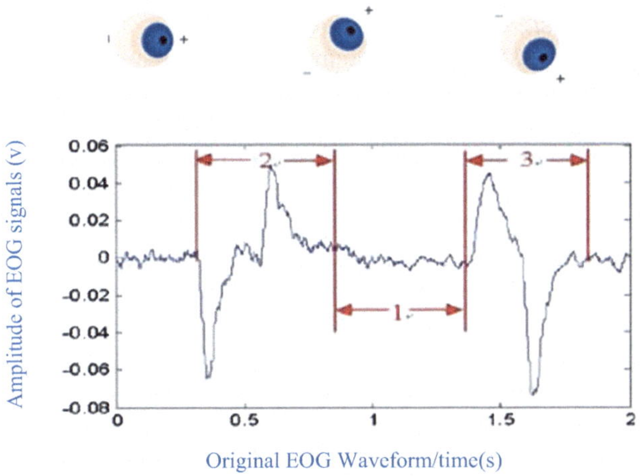

Fig. 1. EOG vertical signal (1) looking straight; (2) rolling eyes upward; (3) rolling eyes downward

EOG signals are utilized to detect the four eye movements (left, right, up and down) along with blinking. After applying Discrete Wavelet Transform (DWT), eight features (mean, variance, energy, peak and valley amplitude value, peak and valley amplitude position and area under the curve) are extracted from horizontal and vertical EOG signals. Linear Discriminant Analysis (LDA), Support Vector Machines (SVM), Multinomial Logistic Regression (MLR), K Nearest Neighbor (KNN), Decision Trees and Naïve Bayes (NB) are examined using both horizontal and vertical features after concatenation. All experiments were carried out using the dataset provided by Usakli and Gurkan [14].

The remaining of the paper is arranged as follows. Section 2 presents the existing studies, the methodology is given in Sects. 3 and 4 discusses the achieved results, and finally the conclusion is given in Sect. 5.

2 Related Work

Several studies have emerged in the last few decades [15–20] for EOG based HCI. EOG based HCI depends on classifying four or eight directional eye movements along with the eye blink. Some of these main studies can be summarized as follows:

Aungsakul et al. [15] proposed fourteen features extracted from characteristics of both vertical and horizontal EOG signals, such as position and amplitude of the peak and the valley. The study considers eight movements along with blinking. The experiments show that only ten features are significant for the task in hand. The classification accuracies reach 100% for some subjects.

Usakli and Gurkan [17] developed a virtual keyboard that allows the user to write texts based on analyzing EOG signals. Four eye movement directions along with blinking are classified using Nearest Neighborhood. By this virtual keyboard, the speed of writing is five letters/25 s and the best result is 95%.

Metin and Hesna [18] suggested a method to increase the writing speed and accuracy of EOG based text entry. Signals are preprocessed with Second-order Butterworth filters to filter the EOG signals, a 0.5–30 Hz band pass filter is used to prevent baseline shift, from two EOG channels. The eye movement is classified by comparing the EOG amplitudes to a specific threshold. For example, regarding vertical EOG, If it is less (greater) than specific threshold then it is identified as up (down) movement, the same way for horizontal EOG signal. The results show that an experienced user can write at average 13.2 words per minute (word of five letters) and the achieved accuracy is 100%.

Kim et al. [20] proposed a new electrooculogram (EOG)-based human-computer interface (HCI) paradigm for binary yes/no communication by patients, For preprocessing, the EOG from each channel are first down-sampled at a rate of 512 Hz and fourth-order Butterworth band-pass filter with cut-off frequencies of 0.2 and 0.4 Hz is applied. The slope of the line connecting the first data point and the first local minimum and maximum are computed as features. Experimental results have reached accuracy of 94%.

It can be observed that the existing studies considered only one classifier in their evaluation. In this study, for more robustness and reliability, extracted features are considered with six different classifiers to distinguish the four eye movement directions along with blinking.

3 The Proposed Approach

The proposed approach includes three stages, which are preprocessing, feature extraction and classification. The following subsections provide details for each stage.

3.1 Preprocessing

The artifacts and noise caused by involuntary EOG movements, power interference and background noise contaminate the acquired EOG signals. Hence, preprocessing for signals is required. The preprocessing includes two steps first applying second order Butterworth band pass filter to keep the EOG band [0.5–20 Hz]. Figure 2 shows an EOG signal before and after filtering. Second step is down sampling signals, such that each one becomes 50 samples to reduce computation.

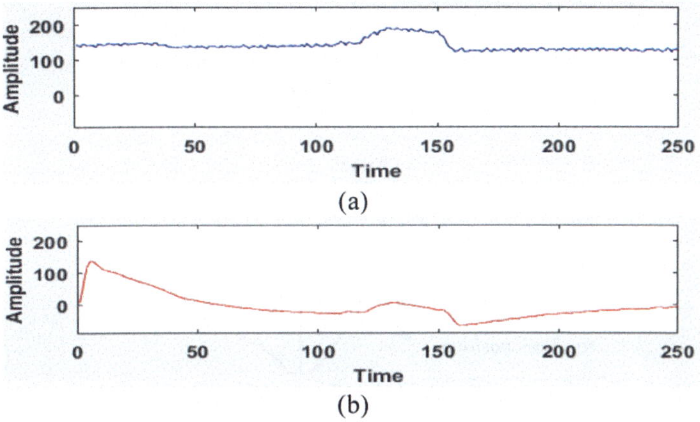

Fig. 2. (a) EOG signal before filtering and (b) EOG signal after filtering

3.2 Feature Extraction

First, both vertical and horizontal EOG signals are decomposed using DWT. Haar wavelet is utilized as the mother wavelet. For each level, the decomposition of each signal results in approximation coefficients and detail coefficients. The projection on the approximation space is kept to calculate the next approximation and detail coefficients of the next level and so on.

Three statistical features are successfully used for representing EOG signals namely mean, standard deviation and energy. The features are calculated from the considered wavelet coefficients as follows:

$$\mu_i = \sum\nolimits_{i=1}^{N} X_i \tag{1}$$

$$\alpha_i = \sum\nolimits_{i=1}^{N} (X_i - \mu_i)^2 \tag{2}$$

$$\mathrm{E} = \sum\nolimits_{i=1}^{N} X_i^2 \tag{3}$$

Where (X) is eye signal, (μ) is the mean of signals, (α) standard deviation of signals and (E) energy of signals.

In addition, the position and amplitude of peak and valley along with area under curve are five geometric features considered by this study as shown in Fig. 3. Finally, a final feature vector of 16 features is constructed by combining both horizontal and vertical features together (eight features for each direction).

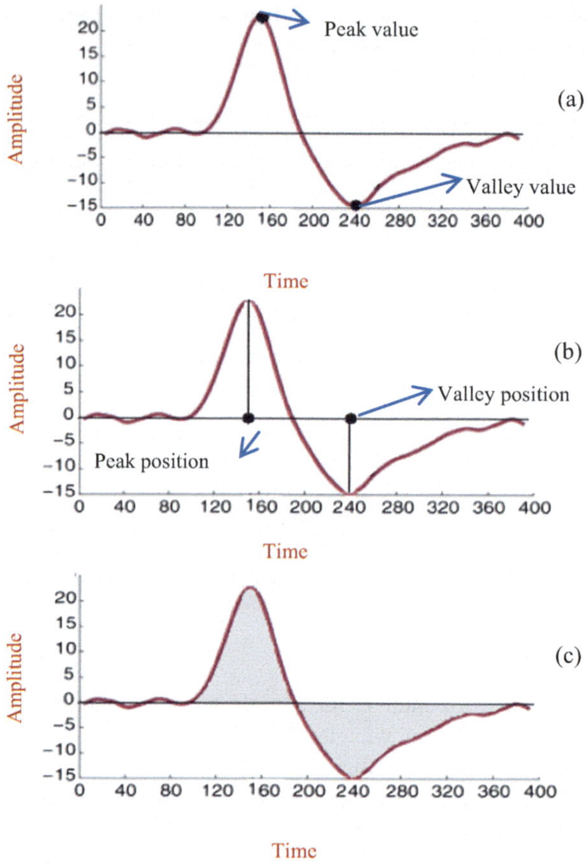

Fig. 3. (a) Peak and valley amplitude values for EOG signal (b) peak and valley amplitude positions for EOG signal (c) area under curve for EOG signal.

3.3 Classification

In the classification stage, the extracted feature vectors from the previous stage are fed into each classifier for training and testing. Linear Discriminant Analysis (LDA), Support Vector Machines (SVM), Multinomial Logistic Regression (MLR), K Nearest Neighbor (KNN), Decision Trees and Naïve Bayes (NB) classifiers are utilized in this study to discriminate between the five eye movements. Each classifier can be summarized as follows:

Linear Discriminant Analysis [21]: The idea of LDA is to find a transformation that discriminates between classes. Thereafter, the discrimination is done in the transformed space using Euclidean distance. LDA is implemented via scatter matrix analysis. In two-class case, transformation from high to low dimensional space is performed under condition that the transformation maximizes the ratio of intra-class scatter to the inter-class scatter. The intra-class matrix and intra-class matrix are computed as follows:

$$S_w = \sum\nolimits_{i=1}^{n} s_i \text{ Where } s_i = \sum\nolimits_{x \in w_i}^{n} \left(\mathrm{x} - x_i^{-}\right)\left(\mathrm{x} - x_i^{-}\right)^T \quad (4)$$

$$s_b = \sum\nolimits_{i=1}^{n} m_i\left(x_i^{-} - \mathrm{x}^{-}\right)\left(x_i^{-} - \mathrm{x}^{-}\right)^T \quad (5)$$

Where (x_i^{-}) is the mean for each class, (x^{-}) is total mean vector and (m_i) is the number of training samples for each class.

Support Vector Machines: It reduces the complexity of the learning mode. Thus, it can efficiently generalize. SVMs are mainly two-class classifiers [22]. SVM builds a hyper plane that maximizes the margin between classes. The adjacent points of these classes are called support vectors. The decision function is built to be used in testing as follows:

$$\mathrm{f(x)} = \sum\nolimits_{i=1}^{n} \alpha_i - \left(\frac{1}{2}\right) \sum\nolimits_{i,j=1}^{n} \alpha_i \alpha_j y_i y_j K\left(x_i, x_j\right) \quad (6)$$

$$K\left(x_i, x_j\right) = \left(x^T y + 1\right)^d \quad (7)$$

Where {(xi, yi), i = 1….N} are the features vector, αi, αj is the Lagrange multiple and k(x, y) is the kernel function.

Multi-class SVM can be constructed from binary SVM classifiers using one-versus-all classification strategy.

Multinomial Logistic Regression [23]: MLR is a one layer neural network [24]; it is the generalization of logistic regression to the multiclass problem. The MLR computes the probabilities for various cases (values) of the dependent variable based on the available independent variables. The "argmax" is the index in the output vector with the highest value is returned as the class name; the main benefit of MLR is that it gives true probability where the sum of all network outputs is 1.

K Nearest Neighbor [25]: It is one of the oldest non-parametric classification algorithms. The new instance is discriminated by computing the distance between it and every other instance in the training set. The k smallest distances are defined; the most represented class by this k nearest neighbors is the winner class.

Decision Trees [26]: According to values of the computed features, decision trees split the training set by a tree to generalize. The algorithm can work on binary or multiclass classification problems. Classification rules are referred by the paths from root to leaf.

Hence, the decision tree can be represented as decision rules where the result is the content of the leaf node and the conditions along the path form a conjunction in the if clause. The tree looks like a flowchart. Leaves are class label sand the inner nodes represent tests for features and the branches represent the result of these tests.

Naïve Bayes [27]: It is a probabilistic classifier based on Bayes' theorem. it can be trained using supervised learning; it is based upon the principle of maximum a posteriori (MAP). This approach can be extended to the case of more than two classes represented as vectors of feature values and the class labels are drawn from some finite set.

$$p(c_k|x_1, \ldots, x_n) \quad (8)$$

$$(posteriori)p(c_k|x) = \frac{p(c_k)p(x|c_k)}{p(x)} \quad (9)$$

Where {x1, ..., xn} are the vectors of features, (n) large number of features and (Ck) is the probability for each class.

4 Experimental Results

Experiments were carried out using the dataset provided by Usakli and Gurkan [14]. This dataset consists of five classes: left, right, up, down and blinking. For each class, 20 pair of horizontal and vertical signals is given. Each signal is 251 samples length (approximately 1.5 s). EOG signals were sampled at 176 Hz. In this work, the data for each class is partitioned into four sets. Three sets for training and one for testing resulting in four trails. Thus, each set exists once as a testing group. Tables 1, 2, 3, 4, 5 and 6 show the results achieved by each considered classifier. In addition, Fig. 4 provides a summary for the average achieved results using the six classifiers. The accuracy of each class is calculated by the following equation:

$$Accuracy = \frac{\text{The number of samples correctly classified}}{\text{Total number of testing cases of specific class}} * 100\% \quad (10)$$

Table 1. The achieved accuracies using LDA classifier

Trial number	Eye movement directions					Average accuracy
	Down	Up	Left	Right	Blinking	
Trial 1	100%	80%	100%	80%	60%	84%
Trial 2	100%	100%	100%	100%	100%	100%
Trial 3	90%	100%	80%	90%	100%	92%
Trial 4	90%	90%	60%	100%	80%	84%
Total the average accuracy: 90%						

Table 2. The achieved accuracies using SVM classifier

Trial number	Eye movement directions					Average accuracy
	Down	Up	Left	Right	Blinking	
Trial 1	90%	90%	70%	90%	80%	84%
Trial 2	90%	92%	100%	98%	100%	96%
Trial 3	80%	84%	80%	88%	88%	84%
Trial 4	88%	100%	82%	82%	88%	88%
Total the average accuracy: 88%						

Table 3. The achieved accuracies using MLR classifier

Trial number	Eye movement directions					Average accuracy
	Down	Up	Left	Right	Blinking	
Trial 1	80%	70%	60%	65%	65%	68%
Trial 2	100%	100%	80%	70%	70%	84%
Trial 3	97%	90%	100%	95%	98%	96%
Trial 4	88%	80%	80%	84%	88%	84%
Total the average accuracy: 83%						

Table 4. The achieved accuracies using KNN classifier

Trial number	Eye movement directions					Average accuracy
	Down	Up	Left	Right	Blinking	
Trial 1	90%	92%	100%	98%	100%	96%
Trial 2	100%	100%	100%	100%	100%	100%
Trial 3	80%	100%	88%	88%	80%	88%
Trial 4	80%	100%	100%	80%	60%	84%
Total the average accuracy: 92%						

Table 5. The achieved accuracies using Decision Trees classifier

Trial number	Eye movement directions					Average accuracy
	Down	Up	Left	Right	Blinking	
Trial 1	94%	100%	96%	98%	72%	92%
Trial 2	95%	90%	95%	90%	70%	88%
Trial 3	98%	90%	80%	98%	70%	84%
Trial 4	80%	70%	90%	100%	60%	80%
Total the average accuracy: 86%						

Table 6. The achieved accuracies using NB classifier

Trial number	Eye movement directions					Average accuracy
	Down	Up	Left	Right	Blinking	
Trial 1	100%	100%	100%	100%	100%	100%
Trial 2	100%	100%	100%	100%	100%	100%
Trial 3	80%	90%	90%	88%	92%	88%
Trial 4	100%	100%	80%	90%	90%	92%
Total the average accuracy: 95%						

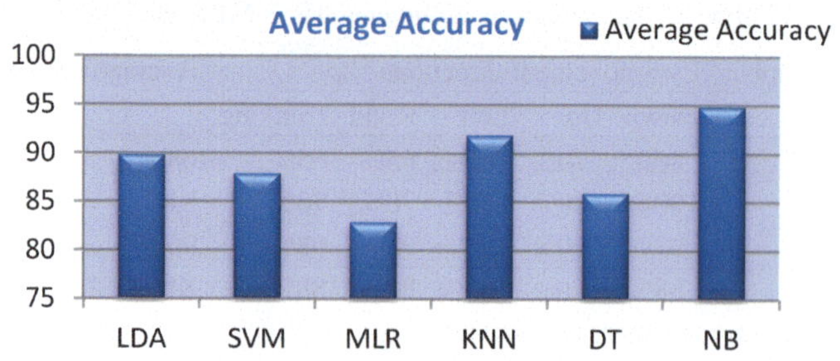

Fig. 4. Summarizes the results achieved by the six classifiers

5 Conclusion

This paper proposes a method for detecting eye movements based on EOG signals. Five classes are considered: up, down, left, right and blinking. The EOG signals are filtered and then eight statistical and geometric features are extracted from each vertical and horizontal EOG signal. Statistical features are mean, standard deviation and energy. On the other hand, position and amplitude of peak and valley along with area under the curve are considered as geometric features. Features of both vertical and horizontal EOG signals are concatenated to form a final Feature vector. Linear Discriminant Analysis (LDA), Support Vector Machines (SVM), Multinomial Logistic Regression (MLR), K Nearest Neighbor (KNN), Decision Trees and Naïve Bayes (NB) are examined for classification. The results have revealed the superiority of Naïve Bayes classifier. The best results achieved for the four trails are 100%, 100%, 88%, and 92%. In the future, we are looking forward to testing different datasets.

References

1. Beleza-Meireles, A., Al-Chalabi, A.: Genetic studies of amyotrophic lateral sclerosis controversies and perspectives. Amyotroph. Lateral Scler. **10**(1), 1–14 (2009)
2. Dion, P.A., Daoud, H., Rouleau, G.A.: Genetics of motor neuron disorders new insights into pathogenic mechanisms. Nat. Rev. Genet. **10**(11), 769–782 (2009)

3. Alonso, A., Logroscino, G., Hernan, M.A.: Smoking and the risk of amyotrophic lateral sclerosis a systematic review and meta-analysis. J. Neurol. Neurosurg. Psychiatry **81**(11), 1249–1252 (2010)
4. Mitchell, J.D., Borasio, G.D.: Amyotrophic lateral sclerosis. Lancet **369**(9578), 2031–2041 (2007)
5. Shaw, P.J.: Molecular and cellular pathways of neurodegeneration in motor neurone disease. J. Neurol. Neurosurg. Psychiatry **76**(8), 1046–1057 (2005)
6. Milo, R., Kahana, E.: Multiple sclerosis geoepidemiology genetics and the environment. Autoimmun. Rev. **9**(5), A387–A394 (2010)
7. Sejvar, J., Andrew, L., Matthew, W., Oliver, W.: Population incidence of Guillain-Barré syndrome: a systematic review and meta-analysis. Neuroepidemiology **36**(2), 123–133 (2011)
8. Wang, H., O'reilly, E.J., Weisskopf, M.G., et al.: Smoking and risk of amyotrophic lateral sclerosis a pooled analysis of 5 prospective cohorts. Arch. Neurol. **68**(2), 207–213 (2011)
9. Miller, R.G., Mitchell, J.D., Moore, D.: Riluzole for amyotrophic lateral sclerosis (ALS)/motor neuron disease (MND). Cochrane Database Syst. Rev. **3**(2), 1–25 (2012)
10. Päivi, M., Kari-Jouko, R.: Twenty years of eye typing systems and design issues. In: Proceedings of the Eye Tracking Research & Application Symposium 2002, ETRA, pp. 15–22. ACM, New Orleans (2002)
11. John, J.W.: Medical Instrumentation: Application and Design, 14th edn. Wiley, Hoboken (2009)
12. Lv, Z., Wu, X.-P., Li, M., Zang, D.-X.: Development of a human computer interface system using EOG. Health **1**(1), 39–46 (2009)
13. Tetsuya, O., Masashi, K.: Development of an input operation for the amyotrophic lateral sclerosis communication tool utilizing EOG. Med. Biol. Eng. **43**(1), 172–178 (2005)
14. Ali, B.U., Serkan, G.: Design of a novel efficient human–computer interface an electrooculagram based virtual keyboard. IEEE Trans. Instr. Measur. **59**, 2099–2108 (2010)
15. Siriwadee, A., Angkoon, P., Pornchai, P., Chusak, L.: Evaluating feature extraction methods of Electroocculography (EOG) signal for human-computer interface. In: 3rd International Science, Social Science, Engineering and Energy, vol. 32, pp. 246–252, Thailand (2012)
16. Aungsakul, S., et al.: Robust eye movement recognition using EOG signal for human-computer interface. In: Proceedings of the 2nd International Conference on Software Engineering and Computer Systems, Kuantan, 180, pp. 714–723, Malaysia (2011)
17. Ali, B.U., Serkan, G., Aloise, F., Vecchiato, G., Babiloni, F.: On the use of electrooculogram for efficient human computer interfaces. Comput. Intell. Neurosci. 5 p. (2010). Article ID 135629
18. Metin, Y., Hesna, Ö.Ü.: A new PC-based text entry system based on EOG coding. Ad. Hum.-Comput. Interact. **2018**(2), 1–8 (2018)
19. Ram, K., Sathesh, K., Emayavaramban, G.: EOG signal classification using neural network for human computer interaction. IJCTA **9**(2), 1–11 (2016)
20. Kim, D.Y., Han, C.-H., Im, C.-H.: Development of an electrooculogram-based human computer interface using involuntary eye movement by spatially rotating sound for communication of locked-in patients. Sci. Rep. **8**(1), 10 p. (2018). 9505
21. Richard, A.J., Dean, W.: Applied Multivariate Statistical Analysis, 6th edn. Prentice Hall, Pearson (2007)

22. Byun, H., Seong-Whan, L.: Applications of support vector machines for pattern recognition: a survey. Pattern recognition with support vector machines, LNCS, vol. 2388, pp. 213–236, Springer, Heidelberg (2002)
23. William, H.G.: Econometric Analysis, 7th edn. Pearson Education, Boston (2012)
24. Christopher, M.: Pattern Recognition and Machine Learning, 1st edn. Springer, USA (2006)
25. Altman, N.S.: An introduction to kernel and nearest-neighbor nonparametric regression. Am. Stat. **46**(3), 175–185 (1992)
26. Quinlan, J.R.: Simplifying decision trees. Int. J. Man Mach. Stud. **27**(3), 221–234 (1987)
27. Martin, M., Janez, D., Michael, W., Zupan, B.: Nomograms for visualization of Naive Bayesian classifier. In: Proceedings of PKDD, pp. 337–348 (2004)

RDB-FSU: Residual Dense Network with Feature Selection Unit for Image Super Resolution

M. Haitham[1(✉)], Alaa Zaghloul[1], and Mostafa AbdEl-Azeem[2]

[1] MUST University, 6th October, Egypt
{mohamed.atif, alaa.zaghloul}@must.edu.eg
[2] AASTMT, Heliopolis, Egypt
melbakary@aast.edu

Abstract. After the great success of deep convolution neural network (DCNN). In practical, residual and dense networks. Most deep methods do not make full use of features from the original low-resolution (LR) images. This paper proposed a novel residual in dense network with feature selection unit (RDB-FSU) that tackle this problem in both image and video SR. RDB-FSU fully use the features from all the convolution layers. Specifically, the residual dense block (RDB) presented to extract features via dense connected convolution layers. Then feature selection unit (FSU) is adaptively learn more effective features from extracted features to be up scaled using pixel-shuffle module. Extensive experiments with various models show that the presented RDB-FSU fulfills appropriate performance versus state-of-the-art techniques on both image and video benchmark dataset.

Keywords: Deep learning · Super resolution · Residual in residual · Dense block

1 Introduction

Surveillance systems have been dramatically increased. A large number of surveillance systems can be observed in real life according to its important in the preservation of public security. To deal with large number of suspicion events captured by cameras whose images are digitized at smaller resolutions than original ones, in order to conserve resources of the system, whereas resolution is a measure of frequency content in an image. Extracted frame from such video is always blurry and noisy, so police authorities are in need of new methods to obtain optimal use of video information in investigational purpose. Because of low vision quality due to the highly compression of surveillance videos.

So the target here is reconstructing an accurate high-resolution (HR) image from low-resolution (LR) one, this process called super resolution (SR). Super resolution image is vastly computer vision task that can be applied in many applications. Since super resolution is ill-posed problem, constructing an effective single image super resolution (SISR) is challenging task, regardless of super resolution difficulties it has

A.-E. Hassanien et al. (Eds.): AICV 2020, AISC 1153, pp. 501–510, 2020.
https://doi.org/10.1007/978-3-030-44289-7_47

been actively studied because it can be applied in variety of services. Recently outstanding performance have been showed by deep learning methods not only by stacking more layers to their models but by designing internal blocks and new network topologies to achieve more results. By generating mapping function between LR and HR image pairs to minimize the loss error between reconstructed image and ground truth image also, maximize the peak-signal-noise-ratio (PSNR) which is common measurement to evaluate super resolution algorithm.

In order to construct a lean architecture, reducing the number of model parameters is not only the main factor. However, number of operations in real-world scenarios is also an essence factor to take in mind. Consider a situation where a SR system applied to a mobile device. Then, the execution speed and capacity of device battery that depend on the number of operations performed is also a crucial importance from a user-experience perspective. In this context, minimizing the number of operations is a challenging and necessary step that most researchers have been ignore it until now.

To handle these requirements and improve recent models, the proposed model based on the ESRGAN [1] architecture, first it needs to be optimized by analyzing and take off not needed modules to simplify the network architecture. Thus, the proposed model trained with suitable loss function and accurate model modification upon training.

Detecting human activity in smart surveillance videos becomes critical problem. It needs real-time detection performance, but generally consuming huge time, due to surveillance clip heavy size and low computation power.

Therefore, reducing the resolution of video clip turn to be remarkable, subsequently many details in the video are lost and it becomes difficult to detect the actual activity. This research will address the issues of generating high-resolution image from low resolution to make it easier in detecting actual activity.

As a conclusion the main contributions are summarized as follows: (1) proposed new feature selection unit (2) propose Residual in Residual Dense network with residual dense blocks (RDB) followed by feature selection unit then pixel-shuffle as upscaling module (3) proposed model achieved good performance in both images and videos dataset benchmarks.

In the following section we will discuss super resolution primarily types given some background on other approaches. In Sect. 3, the presented RDB-FSU explained in details. Then results achieved by the proposed model shown in Sect. 4, which compared with previous state-of-the-art techniques relay on quality performance. At last, we conclude the paper.

2 Related Works and Background

Since the success of deep neural networks in image recognition tasks. Many approaches have been applied to solve computer vision tasks Super resolution is one such a task that can be primarily divided into four types pre-up sampling, post-up sampling, progressive up sampling, and iterative up and down sampling.

2.1 Pre-up Sampling Super Resolution

Commonly uses image interpolation operator such as (Bicubic, nearest neighbor etc…) to match with desired high resolution (HR) image then try to map LR to HR this schema was firstly proposed by SRCNN [2] with simple convolution layers. The advantage of this scheme is up sampling has been done by traditional pre-defined algorithm and deep neural network is try to refine the input image which reduce the learning difficulty, but since the most operations are performed in high-dimension space, the cost of time and space is much higher than other algorithms.

2.2 Post-up Sampling Super Resolution

In order to solve the high cost of time and space, and make full use of deep learning technique to improve image super resolution. Most of operations performed on low-dimension space then attach the network with up sample module at the end of the architectures, which proposed by FRSCNN [3] and ESPCN [4].

2.3 Progressive up Sampling

Although models under post-up sampling reduce time and space computations, it still have short coming, on other hand it may not be suitable for large scale (e.g., 4, 8) and each scale factor requires individual SR model where multi-scale SR cannot be used to address these drawbacks a progressive up sampling proposed by PROSRGAN [5], SCN [6], and LapSRN [7].

2.4 Iterative up and Down Sampling

Models under iterative and down sampling can be better mine the deep relationship between LR-HR images pairs by directly increase the SR features using multiple up and down sample stages as proposed by DBPN [8].

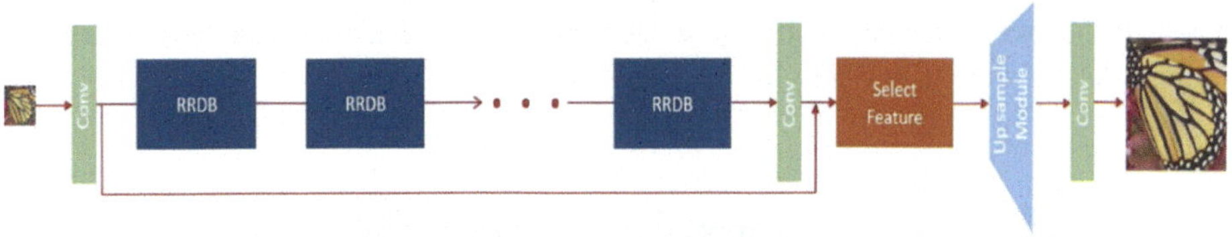

Fig. 1. Proposed model architecture

3 Proposed Model

3.1 Proposed Model Architecture

Proposed model Architecture as shown in Fig. 1 contains five parts: shallow feature extraction, Residual in Residual Dense Block (RRDB) deep feature extractor, Feature

Selection Unit, Upscale and reconstruction part formulated as follows: Let us denote I_{LR}, I_{SR} as input and output of model where

$$F_0 = H_{SFE}(I_{LR}) \tag{1}$$

Where H_{SFE} (.) denote convolution operation that extract shallow features F_0, then feed to the next part Residual in Residual Dense Block (RRDB).

$$F_{DFE} = H_{RRDB}(F_0) \tag{2}$$

Where H_{RRDB} (.) is deep residual in residual dense structure that contain five blocks of convolution and ReLU activation function followed by convolution layer, the output then feed to Feature Selection Unit to extract best feature from the features.

$$F_{SE} = H_{FSE}(F_{DFE}) \tag{3}$$

Where H_{FSE} (.) is Feature Selection Unit that select best features from the image then this features feed to upscale module.

$$F_{UP} = H_{UP}(F_{SE}) \tag{4}$$

Where H_{UP} (.) is upscale module then the up-scaled features reconstructed by one convolution.

$$I_{SR} = H_{REC}(F_{UP}) \tag{5}$$

Where H_{REC} (.) is reconstruction layer.

The model optimized by L1 Loss function, given $\{I_{LR}^i, I_{HR}^i\}_{i=1}^N$ mapped training set between LR and HR images, where N is the number of images.

$$L(\theta) = \frac{1}{N}\sum\nolimits_{i=0}^{N} \left|H_{model}\left(I_{LR}^i\right) - I_{HR}^i\right|_1 \tag{6}$$

Where θ the parameter set of the network that optimized using Adam optimizer [9] training details shown in Sect. 4.2.

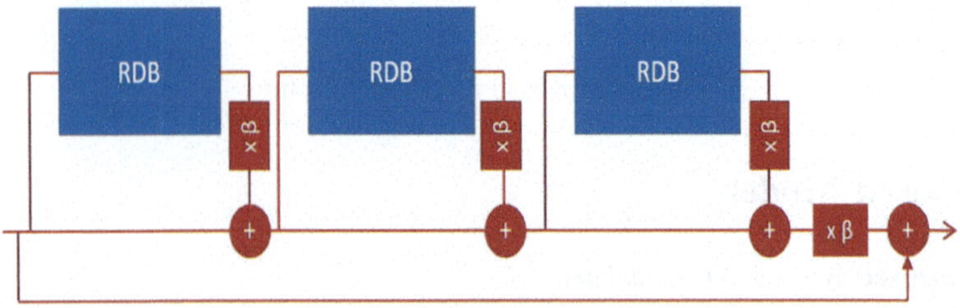

Fig. 2. RRDB architecture

3.2 Residual in Residual Dense Block (RRDB)

As shown in Fig. 2, the proposed RRDB consists of three Residual Dense Block (RDB), which formulated as:

$$F_{RRDB(N)} = H_{RRDB(N)}\left(H_{RRDB(N-1)}\left(\ldots H_{RRDB}\left(1\right)(.)\ldots\right)\right) \tag{7}$$

Where H_{RRDB} (.) contain three RDB as shown in Fig. 3, which formulated as:

$$\begin{aligned} F_{RDB1} &= H_{RDB1}(F_0) \\ F_{RDB2} &= H_{RDB2}(F_0 + F_{RDB1} * \beta) \\ F_{RDB3} &= H_{RDB3}((F_0 + F_{RDB1} * \beta) + F_{RDB2} * \beta) \\ F_{RDB4} &= (F_{RDB3} * \beta) + (F_0 * \beta) \end{aligned} \tag{8}$$

Where β is residual scaling parameter, which is set to 0.2.

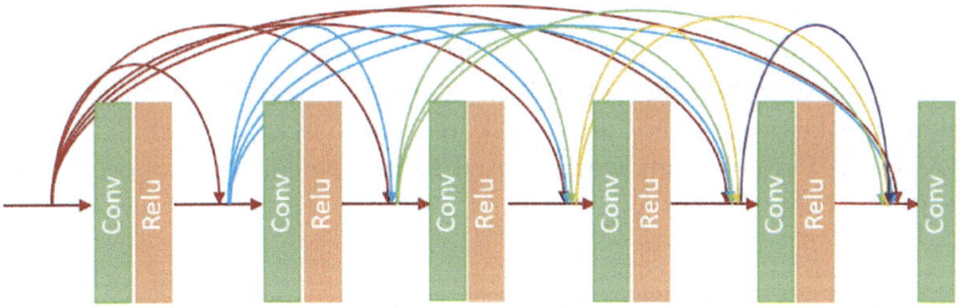

Fig. 3. RDB architecture

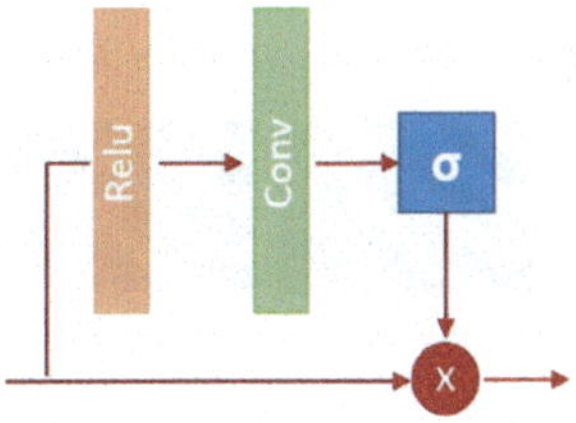

Fig. 4. Feature selection unit architecture

3.3 Feature Selection Unit (FSU)

The proposed Feature selection unit architecture as shown in Fig. 4 used to select best features then feed to upscale module, which formulated as:

$$F_{FSU} = \sigma\left(R\left(WF_{RDDB-N}\right) * W_{Conv}\right) \times F_{RRDB-N} \tag{9}$$

Where σ is sigmoid activation function, R is ReLU activation function, WF_{RRDB-N} and W_{Conv} are the weight set of convolution layers and (*) is convolution operator.

4 Experimental Result

4.1 Datasets

The proposed model trained by DIV2K [10], which is newly proposed high-quality image dataset for many computer vision tasks and flicker2K [11]. Proposed model performance also compared on two categories of benchmark dataset 1- image datasets: Set5 [12], Set14 [13], Urban100 [14] and B100 [15] 2- video dataset: Vid4 [16].

4.2 Training Details

Experiments performed on two scaling factors 4x that applied on both image and video benchmark dataset and 8x that applied only on image benchmark dataset. For training RGB patches are used of size 128×128 from LR images with corresponding HR patches all LR images obtained by Down-sampling corresponding HR images using MATLAB bicubic kernel function. In addition, mini-batch size is set to 16.

The proposed model is trained by Adam optimizer [9] with $\beta 1 = 0.9$ and $\beta 2 = 0.999$. The initial learning rate is set to 10^{-4} and then decreased to half every 2×10^5 iteration. We train the network using L1 loss with 22 Residual in Residual Dense Block (RRDB) and PSNR and SSIM are used to assess SR results on Y channel using MATLAB. Pytorch framework is used to implement the proposed model with Tesla K80 GPU (Fig. 5).

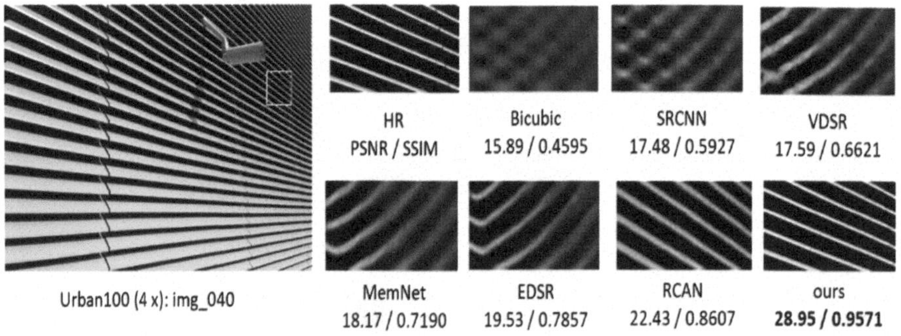

Fig. 5. Super-resolution results on x4 scale

4.3 Results with Bicubic (BI) Degradation Model

The proposed model results compared with 10 state-of-the-art methods: SRCNN [2], FSRCNN [3], VDSR [17], LapSRN [7], MemNet [18], EDSR [19], GMFN [20], D-DBPN [8], SRFBN [21], and RCAN [22].

Table 1. Results with bicubic degradation model on both x4 and x8 scales. PSNR/SSIM

Method	Set5		Set14		B100		Urban 100	
	PSNR	SSIM	PSNR	SSIM	PSNR	SSIM	PSNR	SSIM
x4 scale								
Bicubic	28.42	0.8104	26.00	0.7027	25.96	0.6675	23.14	0.6577
SRCNN [2]	30.48	0.8628	27.50	0.7513	26.90	0.7101	24.52	0.7221
FSRCNN [3]	30.72	0.8660	27.61	0.7550	26.98	0.7150	24.62	0.7280
VDSR [17]	31.35	0.8830	28.02	0.7680	27.29	0.7251	25.18	0.7540
LapSRN [7]	31.54	0.8850	28.19	0.7720	27.32	0.7270	25.21	0.7560
MemNet [18]	31.74	0.8893	28.26	0.7723	27.40	0.7281	25.50	0.7630
EDSR [19]	32.46	0.8968	28.80	0.7876	27.71	0.7420	26.64	0.8033
GMFN [20]	31.69	0.8925	28.35	0.7787	27.49	0.7337	25.68	0.7731
D-DBPN [8]	32.47	0.8980	28.82	0.7860	27.72	0.7400	26.38	0.7946
SRFBN [21]	32.47	0.8983	28.81	0.7868	27.72	0.7409	26.60	0.8015
RCAN [22]	32.63	0.9002	28.87	0.7889	27.77	0.7436	26.82	0.8087
Proposed model	32.12	0.8965	28.53	0.7859	27.59	0.7422	26.56	**0.8089**
x8 scale								
Bicubic	24.40	0.6580	23.10	0.5660	23.67	0.5480	20.74	0.5160
SRCNN [2]	25.33	0.6900	23.76	0.5910	24.13	0.5660	21.29	0.5440
FSRCNN [3]	20.13	0.5520	19.75	0.4820	24.21	0.5680	21.32	0.5380
SCN [6]	25.59	0.7071	24.02	0.6028	24.30	0.5698	21.52	0.5571
VDSR [17]	25.93	0.7240	24.26	0.6140	24.49	0.5830	21.70	0.5710
LapSRN [7]	26.15	0.7380	24.35	0.6200	24.54	0.5860	21.81	0.5810
MemNet [18]	26.16	0.7414	24.38	0.6199	24.58	0.5842	21.89	0.5825
EDSR [19]	26.96	0.7762	24.91	0.6420	24.81	0.5985	22.51	0.6221
D-DBPN [8]	27.21	0.7840	25.13	0.6480	24.88	0.6010	22.73	0.6312
RCAN [22]	27.31	0.7878	25.23	0.6511	24.98	0.6058	23.00	0.6452
Proposed model	27.14	0.7853	25.14	**0.6531**	24.87	0.6044	22.94	**0.6487**

4.4 Results on Video Dataset Benchmark

We compare proposed model results on Vid4 video benchmark dataset with RCAN [22] that achieve better performance on scale x4, D-DBPN [8], B123 + T [23], and VESPCN [24]. As shown in Fig. 6 the proposed model achieve high performance than RCAN and VESPCN in videos benchmark dataset.

Table 2. Results on scale x4 using vid4 video benchmark dataset

Clip Name	Bicubic	RCAN [22]	D-DBPN [8]	B123 + T [23]	VESPCN [24]	Proposed model
Calendar (Y)	20.39/0.5720	22.33/0.7254	22.19/0.714	21.66/0.704	–	**22.46/0.7308**
City (Y)	25.16/0.6028	26.10/0.6960	26.01/0.684	26.45/0.720	–	26.27/**0.7204**
Foliage (Y)	23.47/0.5666	24.74/0.6647	24.67/0.622	24.98/0.698	–	**24.99**/0.6827
Walk (Y)	26.10/0.7974	28.65/0.8719	28.61/0.870	28.26/0.859	–	28.54/0.8697
Average (Y)	23.78/0.6347	25.46/0.7395	25.37/0.737	25.34/0.745	25.35/0.7557	**25.56**/0.7509

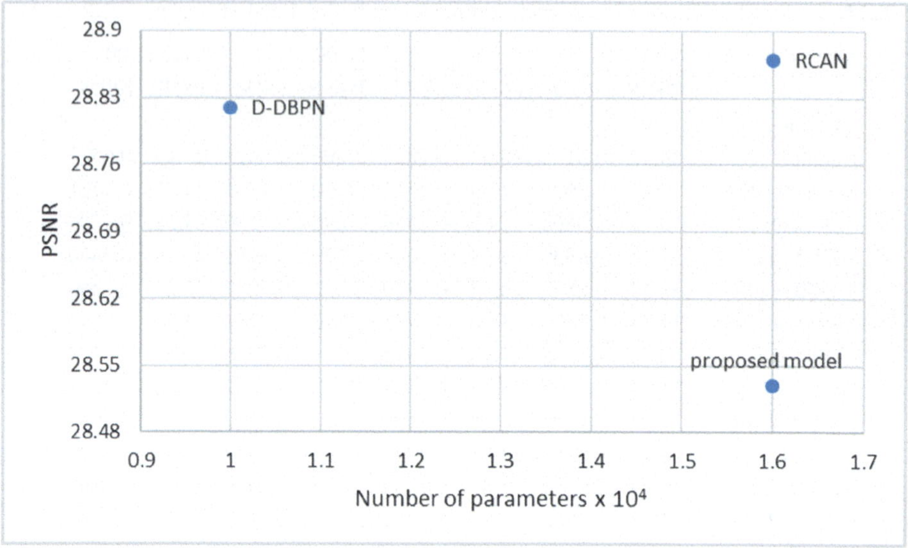

(a)

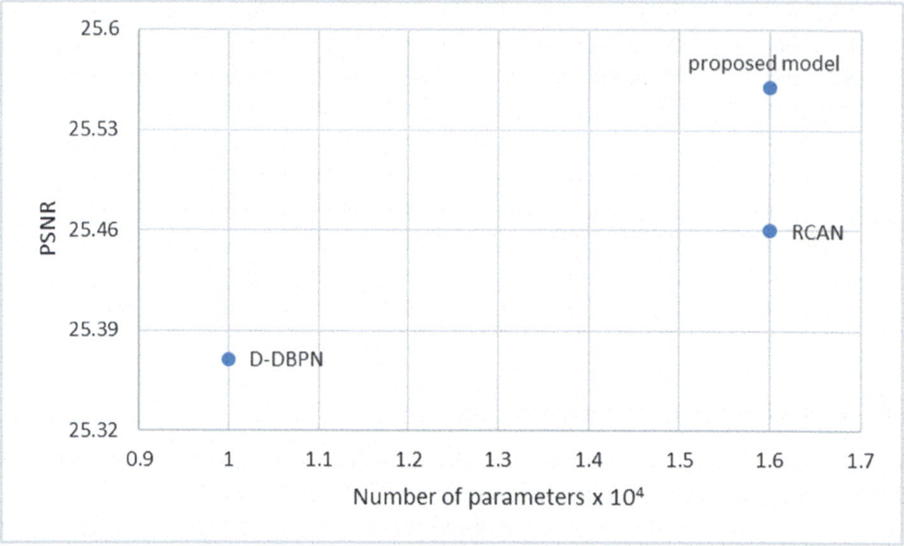

(b)

Fig. 6. Tradeoff between number of parameters and PSNR (a) up, RCAN achieve best performance on Set14 image benchmark dataset (b) down, the proposed RDB-FSU achieve best performance on vid4 video benchmark dataset.

5 Conclusion

We have presented RDB-FSU model that applied on both image and video benchmark dataset as shown in Tables 1 and 2 the proposed model achieves best performance on urban100 benchmark dataset using structural similarity index (SSIM) measurement that quantify image quality on both scales x4 and x8 compared to previous state-of-the-art and also best structural similarity index (SSIM) on Set14 benchmark dataset with x8 scale. Using vid4 benchmark dataset, the proposed model achieve best performance on the average peak signal-to-noise ratio (PSNR) with x4 scale as shown in Fig. 6 We have formulated a novel architecture containing several residual in residual dense blocks with residual dense block followed by feature selection unit which learns to select best features from previous layers to be up scaled.

References

1. Wang, X., Yu, K., Wu, S., Gu, J., Liu, Y., Dong, C., Qiao, Y., Change Loy, C.: ESRGAN: enhanced super-resolution generative adversarial networks. In: Proceedings of the European Conference Computer Vision Workshop (2018)
2. Dong, C., Loy, C.C., He, K., Tang, X.: Learning a deep convolutional network for image super-resolution. In: Proceedings of ECCV (2014)
3. Dong, C., Loy, C.C., Tang, X.: Accelerating the super resolution convolutional neural network. In: Proceedings of ECCV (2016)
4. Caballero, J., Ledig, C., Aitken, A.P., Acosta, A., Totz, J., Wang, Z., Shi, W.: Real-time video super-resolution with spatiotemporal networks and motion compensation. In: Proceedings of CVPR (2017)
5. Wang, Y., Perazzi, F., McWilliams, B., Sorkine-Hornung, A., Sorkine-Hornung, O., Schroers, C.: A fully progressive approach to single-image super-resolution. In: Proceedings of CVPRW (2018)
6. Wang, Z., Liu, D., Yang, J., Han, W., Huang, T.: Deep networks for image super-resolution with sparse prior. In: Proceedings of ICCV (2015)
7. Lai, W.-S., Huang, J.-B., Ahuja, N., Yang, M.-H.: Fast and accurate image super-resolution with deep Laplacian Pyramid networks. In: Proceedings of TPAMI (2018)
8. Haris, M., Shakhnarovich, G., Ukita, N.: Deep back-projection networks for super-resolution. In: Proceedings of CVPR (2018)
9. Kingma, D., Ba, J.: Adam: a method for stochastic optimization. In: Proceedings of ICLR (2015)
10. Agustsson, E., Timofte, R.: NTIRE 2017 challenge on single image super-resolution: dataset and study. In: Proceedings of CVPRW (2017)
11. Timofte, R., Agustsson, E., Van Gool, L., Yang, M.H., Zhang, L., Lim, B., Son, S., Kim, H., Nah, S., Lee, K.M., et al.: NTIRE 2017 challenge on single image super resolution: methods and results. In: CVPRW (2017)
12. Bevilacqua, M., Roumy, A., Guillemot, C., Alberi-Morel, M.L.: Low-complexity single-image super-resolution based on nonnegative neighbor embedding. In: Proceedings of BMVC (2012)
13. Zeyde, R., Elad, M., Protter, M.: On single image scale up using sparse-representations. In: International Conference on Curves and Surfaces (2010)

14. Huang, J.-B., Singh, A., Ahuja, N.: Single image super resolution from transformed self-exemplars. In: Proceedings of CVPR (2015)
15. Martin, D., Fowlkes, C., Tal, D., Malik, J.: A database of human segmented natural images and its application to evaluating segmentation algorithms and measuring ecological statistics. In: Proceedings of ICCV (2001)
16. Liu, C., Sun, D.: On Bayesian adaptive video super resolution. IEEE Trans. Pattern Anal. Mach. Intell. **36**(2), 346–360 (2014)
17. Kim, J., Kwon Lee, J., Mu Lee, K.: Accurate image super resolution using very deep convolutional networks. In: Proceedings of CVPR (2016)
18. Tai, Y., Yang, J., Liu, X., Xu, C.: MemNet: a persistent memory network for image restoration. In: Proceedings of CVPR (2017)
19. Lim, B., Son, S., Kim, H., Nah, S., Lee, K.M.: Enhanced deep residual networks for single image super-resolution. In: Proceedings of CVPRW (2017)
20. Li, Q., Li, Z., Lu, L., Jeon, G., Liu, K., Yang, X.: Gated multiple feedback network for image super-resolution. In: Proceedings of BMVC (2019)
21. Li, Z., Yang, J., Liu, Z., Yang, X., Jeon, G., Wu, W.: Feedback network for image super-resolution. In: Proceedings of CVPR (2019)
22. Zhang, Y., Li, K., Li, K., Wang, L., Zhong, B., Fu, Y.: Image super-resolution using very deep residual channel attention networks. In: Proceedings of ECCV (2018)
23. Liu, D., Wang, Z., Fan, Y., Liu, X., Wang, Z., Chang, S., Huang, T.: Robust video super-resolution with learned temporal dynamics. In: Proceedings of ICCV (2017)
24. Caballero, J., Ledig, C., Aitken, A.P., Acosta, A., Totz, J., Wang, Z., Shi, W.: Real-time video super-resolution with spatio-temporal networks and motion compensation. In: Proceedings of IEEE (2017)

Robust and Real-Time Obstacle Region Detection Based on Depth Feature for Vehicle Detection

Vinh Dinh Nguyen(✉), Anh Quoc Le, Tam Minh Duong, and Narayan C. Debnath

School of Computing and Information Technology, Eastern International University, Thủ Dầu Một, Bình Duong, Vietnam
{vinh.nguyen, anh.le.set15, tam.duong.set15, narayan.debnath}@eiu.edu.vn

Abstract. Obstacle region detection has been one of the most important tasks for detecting and classifying preceding vehicles moving along the road. Traditional U-V disparity-based obstacle detection methods consider only single disparity pixels to extract obstacle regions. They might fail to produce accurate obstacle regions when noise is encountered in the original disparity map. Recently, local patterns have been successfully applied to handle noise in many systems, such as texture classification and vehicle detection. Therefore, this paper introduces a new approach to compute U and V disparity using maximum local density encoding and ternary pattern features. Our method produced better performance than the existing U-V disparity methods in terms of detection rate and accuracy. In addition, we also evaluated the performance of the proposed method with the Faster RCNN-based object detection. Experimental results show that our method improved the detection rate of Faster RCNN by 1.65% on the KITTI dataset, and by 2.28% on the CCD dataset. In addition, the proposed method also improved the running time of Faster RCNN by 47% on the KITTI dataset.

Keywords: Obstacle detection · Region proposal · Deep learning · U-V disparity

1 Introduction

Nowadays, there are a huge number of vehicles travelling on the roads. Therefore, it is necessary to develop a robust driving assistance system (DAS) to help reducing accidents. DAS contains several modules, such as obstacle detection, obstacle classification, obstacle behavior detection, and obstacle tracking. Among these modules, obstacle region detection plays an important role in DAS, which can help to avoid accidents in advance. Many deep learning-based obstacle detection approaches, such as Faster Region Convolutional Neural Network (Faster RCNN) [1], effective region proposals based on 3-channel pattern [2], obstacle detection-based tile convolutional networks [3], and you-look-only-one (YOLO) [4], have been successfully developed to detect vehicles under real driving conditions. However, the processing time of the

A.-E. Hassanien et al. (Eds.): AICV 2020, AISC 1153, pp. 511–519, 2020.
https://doi.org/10.1007/978-3-030-44289-7_48

existing deep learning-based obstacle detection algorithms have not met the real-time requirement of DAS. The running time of the Faster RCNN is slow because it used a region proposal network (RPN) to generate proposal candidates including sky and non-obstacle regions. Therefore, it raises an interesting question whether we can develop an algorithm that can help the existing deep learning-based vehicle detection systems, to reduce not only the processing time but also increase their accuracy. In this paper, we propose a real-time and robust obstacle region detection-based U-V disparity by using the maximum density information and ternary patterns. There are two main contributions of this paper: (1) To introduce a robust U-V disparity algorithm for detecting obstacle regions by using a maximum density encoding and a ternary pattern. (2) To improve processing time and accuracy of the existing vehicle detection systems based on deep learning approaches.

In Sect. 2, we briefly review the existing obstacle detection systems and their limitations. Section 3 proposes a robust obstacle regions detection method using our U-V disparity algorithm. In addition, we also integrate our method to existing deep learning-based vehicle detection systems, to improve both their processing time and accuracy. Section 4 shows experimental results of the proposed system. We conclude this paper, discuss limitations and suggest a research plan in Sect. 5.

2 Related Works

Nowadays, stereo vision has been successfully applied to detect obstacles in DAS. For example. Huang et al. introduced a robust obstacle segmentation algorithm by using the disparity image and bird-eye-view techniques [5]. Motivated by the fact that the obstacles can be represented 3-D model, Franke et al. used stereo vision results to detect obstacles in urban scene [6, 7]. Stereo vision was also investigated to develop obstacle avoidance systems as in [8–10]. More recently, Labayrade et al. proposed V-Disparity technique to detect obstacle regions [11]. The V-disparity map can represent obstacle and road information. In addition, Fu et al. introduced a robust obstacle region detection method by using U-V disparity information [12]. Recently, many deep learning systems, such as Region Convolutional Neural Network (RCNN) [13], Fast RCNN [14], Multi-scale Deep Convolutional Neural Network (MS CNN) [15], Faster RCNN [1], and YOLO [4], have been introduced to detect and classify obstacles appearing along the road. Among these, Faster RCNN obtains the best performance in term of an accuracy. However, Faster RCNN cannot achieve real-time process due to the limitation of the proposal generation step. Recently, Nguyen et al. introduced an adaptive U-V disparity method for obstacle region detection [3]. Experimental results showed that the adaptive U-V disparity method [3] improved the processing time and accuracy of Faster RCNN. However, the adaptive U and V disparity method still has several limitations: (1) Because U and V disparities are computed based on the original disparity map. It cannot provide a robust obstacle information if noise occurs in the disparity map. (2) It does not to consider the relationship between neighboring disparity pixels in the local region. Therefore, this study introduces a robust U and V disparity method to for detecting of obstacle regions. The proposed method improves performance of the existing deep learning-base object detection methods.

3 The Proposed Method

In this section, we briefly review the existing U-V disparity methods and their limitations. To address the limitations of these methods, a robust U-V disparity method is proposed based on maximum density encoding and ternary patterns.

3.1 U-V Disparity

In general, the disparity map is calculated by using 3D information as follows:

$$d = x_l - x_r = f\frac{b}{z} \tag{1}$$

The *U* and *V* disparity are then calculated based on the original disparity map *d*. In general, the *V* disparity is calculated based on row-accumulation techniques from the original disparity map as in [3]:

$$D_v(\mathrm{d,y}) = \sum\nolimits_{x=0}^{W} \Psi_i^j;\ \Psi_i^j = \begin{cases} 1, & D_o(\mathrm{x,\ y}) = d \\ 0, & D_o(\mathrm{x,\ y}) \neq d \end{cases} \tag{2}$$

where $D_v(\mathrm{d,i})$ is the cumulated value of the number of pixels with disparity *d* from the row *y* in the original disparity map D_o. $D_o(\mathrm{x,y})$ is the disparity value at the pixel location (*x*, *y*) in the original disparity map. *H* and *W* are height and width of the disparity image.

The *U* disparity is also calculated based column-accumulation techniques from the original disparity map:

$$D_u(\mathrm{x,\ d}) = \sum\nolimits_{y=0}^{H} \Psi_i^j;\ \Psi_i^j = \begin{cases} 1, & D_o(\mathrm{x,\ y}) = d \\ 0, & D_o(\mathrm{x,\ y}) \neq d \end{cases} \tag{3}$$

3.2 The Proposed Obstacle Detection

Recently, Tan et al. have introduced a local ternary pattern (LTP) to handle noise under uniform regions, such as illumination changes, to improve the performance of texture classification systems [16]. Therefore, motivated by this research [16], this paper introduces a new approach to compute *U* and *V* disparity using a maximum local density encoding and ternary pattern feature from the local region (Fig. 1). Most existing U-V disparity methods did not consider the relationship between neighboring disparity pixels in the local region. Therefore, results of *U* and *V* disparity are sensitive to noise. Given the disparity image *D*, the proposed *U* disparity (U_d) and V disparity (V_d) are calculated as follows:

$$V_d(i,j)_{i\in H, j\in \mathrm{T}} = \sum\nolimits_{k\in W} \Gamma_v(i,j,k) \tag{4}$$

$$\Gamma_v(i, j, k) = \begin{cases} 1 & if|I_d(i, k) - j| \leq \lambda_p \\ 0 & if\lambda_p < I_d(i, k) - j| \leq 2\lambda_p \\ -1 & if\ i|I_d(i, k) - j| > 2\lambda_p \end{cases} \tag{5}$$

$$U_d(i,j)_{i \in T, j \in W} = \sum_{k \in W} \Gamma_u(i,j,k)$$
$$(i, j, k) = \begin{cases} 1 & if|I_d(i,k) - j| \leq \lambda_p \\ 0 & if\lambda_p < I_d(i,k) - j| \leq 2\lambda_p \\ -1 & ifi|I_d(i,k) - j| > 2\lambda_p \end{cases} \tag{6}$$

where $\Gamma_v(i, j, k)$ and $\Gamma_u(i, j, k)$ are the ternary functions that are used to threshold the intensity value of the V and U disparity images, respectively. T is the maximum disparity level. $I_d(i, k)$ is the disparity value of the pixel which has the maximum density in the local region N. $I_d(i, k)$ is calculated as:

$$I_d(i, k) = D(X_\alpha^*); X_\alpha^* = \underset{x_\alpha}{\text{argmax}} \sum\nolimits_{i \in N} f_{density}(|D(Z_i) - D(Z_{x_\alpha}), \lambda_p) \tag{7}$$

The auxiliary function $f_{density}$ is defined as:

$$f_{density}(d, \lambda_p) = \begin{cases} 1, & if\ d \leq \lambda_p \\ 0, & if\ d > \lambda_p \end{cases} \tag{8}$$

where λ_p is a threshold computed as:

$$\lambda_p = \frac{1}{P} \sum\nolimits_{W_p} \frac{1}{W_p - 1} \sum\nolimits_{q \in W_p, q \neq p} |D(Z_p) - D(Z_q)| \tag{9}$$

where Wp is the window size. More details about the calculation of λ_p can be found in [3]. Figure 1 show the result of the proposed U and V disparity method. Existing V disparity algorithms [3] often use Hough transform or RANSAC to find road lines by searching the edges on the entire of the V disparity map. It is time consuming. Therefore, this study introduces a method to reduce the processing time of finding road lines by using the real-time vanish point detection technique [17]. First, the vanishing row is calculated. Then, the obstacle edges, where their locations are above the vanishing row, are then removed from the calculation of road lines. By this way, it can reduce the processing time and increase the accuracy of finding road lines by avoiding obstacle regions. Figure 1 shows the results of calculating the vanishing row by using the method in [17]. In addition, in order to increase features for detecting lines, we also applied a filter to enhance the quality of U and V disparity image as follows:

$$G(x,y) = 9f(x,y) - f(x-1,y-1) - f(x,y-1) - f(x+1,y-1) - f(x-1,y) - f(x+1,y) - f(x-1,y+1) - f(x,y+1) - f(x+1,y+1) \tag{10}$$

Next, we applied a validation stage [3] to remove ground and sky regions. Figure 1 shows the results of removing ground and sky regions of our method.

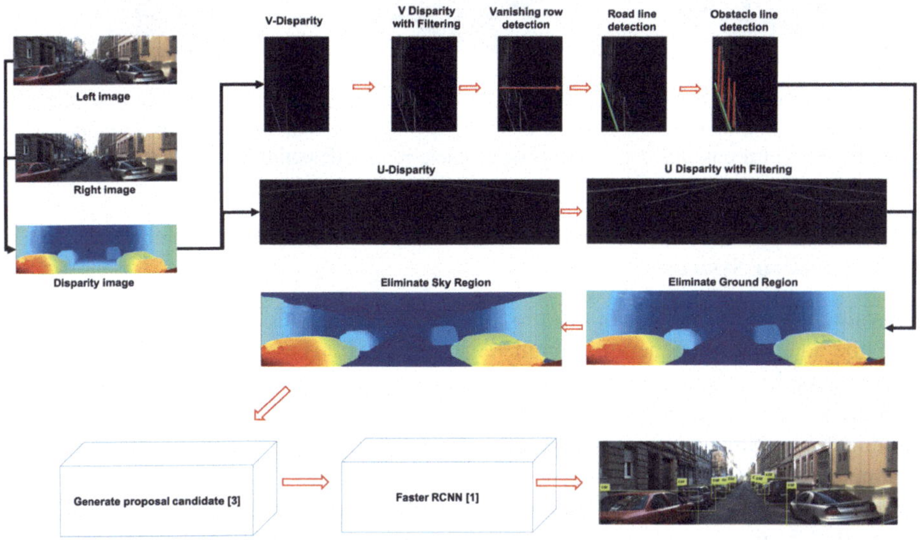

Fig. 1. The proposed vehicle detection model using Faster RCNN [1] and the proposed obstacle region detection.

Fig. 2. Experimental results using the CCD dataset. First column is the original disparity map. Second column is the results of obstacle region detection by using the adaptive U-V disparity [3]. Third column is the results of obstacle region detection of our method.

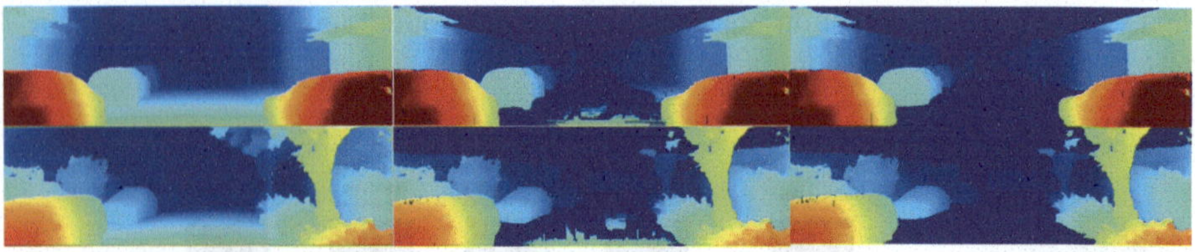

Fig. 3. Experimental results using the KITTI dataset. First column is the original disparity map. Second column is the results of obstacle region detection by using the adaptive U-V disparity [3]. Third column is the results of obstacle region detection of our method.

3.3 The Proposed Region Proposals for Deep Learning-Based Vehicle Detection Systems

To evaluate the performance of our proposed method on vehicle detection systems. We used the idea [3] to generate candidate proposals for Faster RCNN by using our proposed method. Figure 1 shows the workflow of the proposed vehicle detection. We used the same training process for both our method and Faster RCNN, while only the running (or testing) process is different. Please refer to [3] for more detail training parameters. In the testing process, the proposed method (Fig. 1) was first used to generate region candidates by using the disparity information, that is different from Faster RCNN where region candidates were generated from the RPN. The proposed system was implemented on 8 Intel Xeon 3.5 GHz CPUs, 8 GB of RAM, and a Titan X GPU. Faster R-CNN were implemented based on the available source code. Two datasets were investigated in order to evaluate the performance of compared methods: the CCD dataset [3] and the Karlsuhe Institute of Technology and Toyota Technology Institute (KITTI) Car datasets [18]. To evaluate the accuracy of the proposed method, we used two metrics: the detection rate (DR) and false alarm rate (FR) as in [3].

4 Experimental Results

For obstacle region detection, we compared the performance of our method and the adaptive U-V disparity [3]. Figures 2 and 3 show the results of our method and the adaptive U-V [3] using the CCD and KITTI datasets, respectively. The proposed method produced better results than the adaptive U-V disparity due to the use of density information and ternary pattern to extract stable features.

To evaluate how the proposed method can improve the performance of Faster RCNN, we conducted more experiments on the CCD and KITTI datasets. We used the same training process as mentioned in [3]. Figures 4 and 5 show vehicle detection results of the proposed system using the KITTI and CCD datasets. The proposed method accurately detects preceding vehicles in the input images. Table 1 shows a comprehensive experiment of the proposed system and other methods on the KITTI and CCD datasets. Our method achieved the best accuracy. For KITTI dataset, our method, Faster RCNN and adaptive U-V obtained detections rates of 94.15%, 92.50%, and 93.45%, respectively. For CCD dataset, the proposed method, Faster RCNN and adaptive U-V obtained detections rate of 96.35%, 94.70%, and 95.83%, respectively. Our method improved the detection rates of Faster R-CNN by 1.65% on KITTI dataset, and by 2.28% on CCD dataset.

We conducted an experiment on the KITTI dataset to verify the processing time of our method. We used ZFNet, as a base network, to train and evaluate the performance of our proposed method and Faster RCNN. Our proposed method obtained the processing time of 35 ms/frame, while the processing time of Faster RCNN is approximately 75 ms/frame. Our proposed method improved the processing time of Faster R-CNN by 47% on the KITTI dataset.

Table 1. Experimental results using the KITTI and CCD datasets

	KITTI dataset		CCD daylight dataset	
	DR (%)	FA (%)	DR (%)	FA (%)
Faster RCNN [1]	92.50	1.88	94.07	1.75
Adaptive U-V [3]	93.45	1.43	95.90	1.55
The proposed	94.15	1.10	96.35	1.24

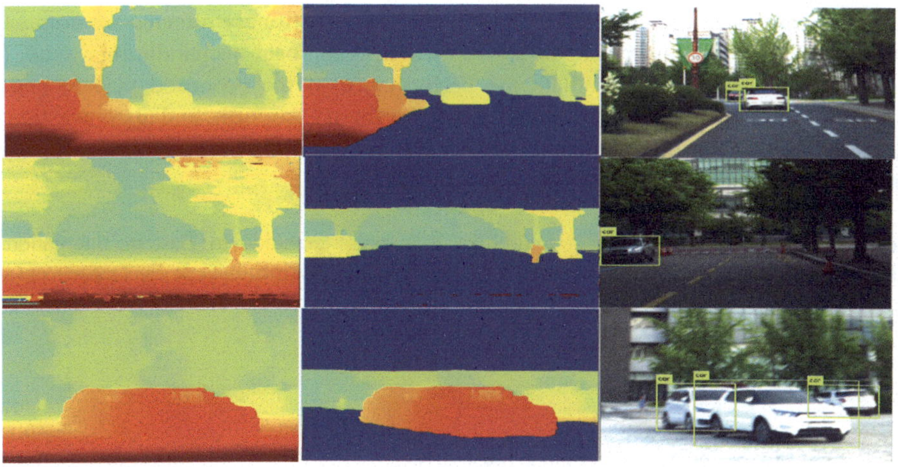

Fig. 4. Experimental results using the CCD dataset. First column is the original disparity map. Second column is the results of obstacle region detection by using our method. Third column is results of vehicle detection using our method.

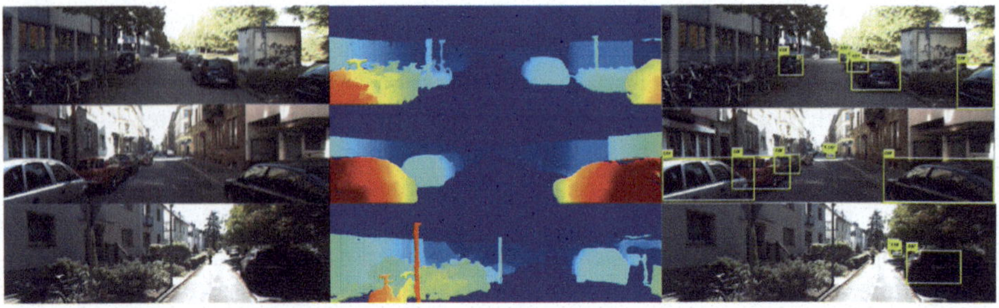

Fig. 5. Experimental results using the KITTI dataset. First column is the left image. Second column is the results of obstacle detection of our method. Third column is the results of vehicle detections using our method.

5 Conclusion

A new approach to compute the U and V disparity map, by taking into account the neighboring pixel information in the original disparity, has been introduced in this paper. The ternary pattern has been successfully applied to help the proposed method to be able to deal with the noise when the quality of the original map is not good enough. Experimental results show that our proposed method improved the performance of the existing deep learning-based vehicle detections. However, currently, the proposed U and V disparities are still computed by using hand-design features. Therefore, it can not handle conditions where the original disparity image is strongly affected by noise. In future works, we aim to investigate a deep learning approach to handle this issue. In addition, we also plan to evaluate the performance of the proposed system under various weather conditions, such as snow, night, rain, and so forth, and also investigate the effect of the proposed method on other deep learning-based object detection methods.

Acknowledgements. The authors would like to express gratitude to Eastern International University (EIU) for their support. A warm thank to all faculty members at the Department of Software Engineering.

References

1. Ren, S., He, K., Girshick, R., Sun, J.: Faster RCNN: towards real time object detection with region proposal networks. IEEE TPAMI **39**(6), 1137–1149 (2017)
2. Nguyen, D.V., Tran, T.D., Byun, J.Y., Jeon, J.W.: Real-time vehicle detection using an effective region proposal-based depth and 3-channel pattern. IEEE Trans. Intell. Transp. Syst. **20**(10), 3634–3646 (2019)
3. Nguyen, D.V., Nguyen, V.H., Tran, T.D., Jeon, J.W.: Learning framework for robust obstacle detection, recognition, and tracking. IEEE Trans. Intell. Transp. Syst. **18**(6), 1633–1646 (2017)

4. Redmon, J., Divvala, S., Girshick, R., Farhadi, A.: You only look once: unified, real-time object detection. In: Proceedings of the IEEE Conference Computer Vision Pattern Recognition, pp. 779–788 (2016)
5. Huang, Y., Fu, S., Thompson, C.: Stereovision-based object segmentation for automotive applications. EURASIP J. Adv. Signal Process. **14**, 1–8 (2015)
6. Franke, U., Heinrich, S.: Fast obstacle detection for urban traffic situation. IEEE Trans. Intell. Transp. Syst. **3**(3), 173–181 (2012)
7. Barth, A., Franke, U.: Estimating the driving state of oncoming vehicles from a moving platform using stereo vision. IEEE Trans. Intell. Transp. Syst. **10**(4), 560–571 (2009)
8. Kramm, S., Bensrhair, A.: Obstacle detection using sparse stereovision and clustering techniques. In: IEEE Intelligent Vehicles Symposium, Alcala de Henares, pp. 760–765 (2012)
9. Khan, M., Hassan, S., Ahmed, S.I., Iqbal, J.: Stereovision-based real-time obstacle detection scheme for unmanned ground vehicle with steering wheel drive mechanism. In: International Conference on Communication, Computing and Digital Systems, Islamabad, pp. 380–385 (2017)
10. Mane, S.B., Vhanale, S.: Real time obstacle detection for mobile robot navigation using stereo vision. In: International Conference on Computing, Analytics and Security Trends, Pune, pp. 637–642 (2016)
11. Labayrade, R., Aubert, D., Tarel, J.P.: Real time obstacle detection in stereovision on non-flat road geometry through "v-disparity" representation. In: IEEE Intelligent Vehicles Symposium, pp. 646–651 (2012)
12. Hu, Z., Lamosa, F., Uchimura, K.: A complete U-V-disparity study for stereovision based 3D driving environment analysis. In: Fifth International Conference on 3-D Digital Imaging and Modeling (3DIM 2005), pp. 204–211 (2015)
13. Girshick, R., Donahue, J., Darrell, T., Malik, J.: Rich feature hierarchies for accurate object detection and semantic segmentation. In: Proceedings CVPR, pp. 580–587 (2014)
14. Girshick, R.: Fast R-CNN. In: Proceedings of ICCV, pp. 1440–1448 (2015)
15. Cai, Z., Fan, Q., Feris, R.S., Vasconcelos, N.: A unified multi-scale deep convolutional neural network for fast object detection. In: Proceedings of the ECCV, pp. 354–370 (2016)
16. Tan, X., Triggs, B.: Enhanced local texture feature sets for face recognition under difficult lighting conditions. In: Analysis and Modeling of Faces and Gestures, LNCS, vol. 4778, pp. 168–182 (2007)
17. Xihaou, L., Jian, Y., Haoang, L., Yahui, L.: 2-line exhaustive searching for real-time vanishing point estimation in Manhattan world. In: IEEE Winter Conference on Applications of Computer Vision (2017)
18. Geiger, A., Lenz, P., Urtasun, R.: Are we ready for autonomous driving? The KITTI vision benchmark suite. In: Proceedings of IEEE International Conference Computer Vision Pattern Recognition, pp. 3354–3361 (2012)

What is the Message About? Automatic Multi-label Classification of Open Source Repository Messages into Content Types

Daniel Campbell, Luis Adrián Cabrera-Diego, and Yannis Korkontzelos(✉)

Department of Computer Science, Edge Hill University, Ormskirk, UK
{daniel.campbell,diegol,Yannis.Korkontzelos}@edgehill.ac.uk

Abstract. Users of Open Source Software (OSS) projects discuss a diverse range of topics online. The content of a post often corresponds to one or more context-sensitive content types, *e.g. a suggestion for a solution*, *a request for further clarification* or *indication that a proposed solution did not work*. The detection of content types can provide several benefits for software developers. For instance, content types can be used as indicators that summarise the content of the messages. These indicators can be exploited as part of a developer-centric knowledge mining platform allowing developers and project managers to create action alerts concerning new bugs found outside of a bug tracker or they can be combined with other metrics to assess the quality of an OSS project. We present a multi-label classifier, able to classify messages exchanged on communication means about OSS, and detailed evaluation results. We experimented with two state-of-the-art multi-label classification approaches HOMER (Hierarchy Of Multilabel classifiER) and RAkEL (RAndom k-labELsets) as these met the technical requirements of the CROSSMINER project. A manually-annotated threaded corpus of posts form newsgroups discussions, bug tracking systems and forums related to Eclipse projects was also used. The results are promising and indicate the potential to attract novel and deeper research for this task.

Keywords: Content classification · Multi-label classification · Open Source Software · Natural language · Text mining · Machine learning · Information retrieval

1 Introduction

In open source software (OSS) engineering, evaluating a project requires considering various perspectives including source code, configuration files, documentation and communications channels. Communication channels are becoming increasingly popular for users to discuss issues, ideas, seek help and share answers to questions posted by other users. These channels are a vital source of rich textual information for developers and project managers to assess, monitor and maintain their code whilst also help understand their community, too.

A.-E. Hassanien et al. (Eds.): AICV 2020, AISC 1153, pp. 520–531, 2020.
https://doi.org/10.1007/978-3-030-44289-7_49

However, manually analysing posts is time consuming. Textual data is often unstructured, thus quickly identifying and classifying context-sensitive information can be complex.

Content classification, also known as content analysis, provides automated context-sensitive organisation of documents, based upon pre-established categories. It has numerous applications in a variety of areas, *e.g.* book recommendations [1], identifying the purpose of forum messages [2] and assigning labels associated with particular types of failures to crash reports [3]. Content classifiers should not be confused with more general classifiers. In the former, the content of an information source, such as a text or image, is split into multiple sub-elements that are analysed and classified individually. In a general classifier, the whole information content is categorised into one or multiple classes. For example, a movie content analyser would predict that the movie *Metropolis* of *Fritz Lang* contains scenes of *love*, *action* and *surrealism*, whereas a general classifier of movie genre, would identify that it is a *sci-fi* movie.

In the context of this work, a content classifier is a tool that determines which elements, such as a *request of clarification* or a *suggestion of a solution*, occur within messages, posts and comments related to OSS projects. The goal content classifier is to be integrated alongside other natural language processing components and metrics within CROSSMINER, a developer-centric knowledge mining platform [4]. It allows software developers and project managers to perform tasks such as the creation of action alerts about new bugs found outside of a bug tracker or assess the quality of an OSS project. From an implementation perspective, CROSSMINER's content classifier has to be implemented in Java and the trained model has to be small to meet the platform's requirements. These requirements define the scope of our experimentation. A tool of this nature also has wider applications and could be adapted to extend the capabilities of existing software mining and analysis platforms such as seCold [5], MARKOS [6], Adinda [7] and OSSMETER [8], CROSSMINER's predecessor.

This paper presents a multi-label content classifier and related evaluation experiments, in context of the technical scope of CROSMINER. Therefore, for evaluation, we utilised a manually annotated dataset of posts made by users participating in discussions within NNTP newsgroups to explore two popular supervised machine learning methods *HOMER* [9] and *RAkEL* [10]. The evaluation results are promising and show the potential to attract novel research.

2 Background

The aim of this work is a content classifier able to identify types of posts, messages and comments in OSS repositories. This section presents state-of-the-art content classifiers in diverse domains, ordered so that those in the beginning perform content classification in broad terms, whereas those towards the end are more related to software engineering.

OSSMETER (ossmeter.org) experimented with single label classification and a naive label combination approach to multi label classification for identifying content-types in OSS forums, issue trackers and NNTP newsgroups [11]. The present work shares the same objectives, but OSSMETER's approach has considerable limitations. A post can often express more than one context-sensitive type, but in a single-label classification approach this cannot be achieved. In addition, the employed multi-label approach considers various label combinations, found within the training data, as a single class and then proceeds as if the task was single-label. The resulting classifier is only aware of known label combinations and unable to assign multiple individual labels to a post.

Bhatia *et al.* [2] studied the problem of classifying individual posts as per their role and/or purpose within online forums in general. Posts are assigned the following labels: *Question*, *Repeat Question*, *Clarification*, *Further Details*, *Solution*, *Positive Feedback*, *Negative Feedback* and *Junk*, with the aim to intelligently improve information extraction, presentation, and the definition of user roles in online forums. Each post was encoded using *content*, *user information*, *structure* and *sentiment/emotion* features. A variety of supervised machined learners were considered: Support Vector Machine, Naive Bayes classifier, Decision tree, multi-layer perceptron and logistic regression. Logistic regression performed best, achieving an accuracy of 72%.

Palau and Moens [12] used a chain of classifiers to identify components, *e.g.* argument, premise and conclusion, in structured text: newspaper articles, parliamentary records and online discussions. *Rhetorical Structure Theory* (RST), a technique that organises natural language based on the relations among its parts [13], was used to identify argumentative posts, to classify argumentative clauses (premise or conclusion) and to determine the limits of an argument and its relations with other arguments nearby. The output of this analysis was transformed into features.

Bacchelli *et al.* classified the contents of emails about software development, by exploiting Naive Bayes and Decision Tree learners [14]. To annotate emails with multiple labels, each line was classified as text, junk, code, patch or stack trace.

Pascarella *et al.* [15] analysed Java source code comments to understand their purpose and intentions. The goal is to improve techniques for measuring the maintainability of a software system. They defined a comment taxonomy of 16 labels, each belonging to 1 of 6 distinct categories. Using the taxonomy, a single label was assigned to each line of a comment, resulting in multiple labels per comment. The experimented with several probabilistic and decision tree based learners, with the latter performing best.

Alfaro *et al.* proposed a combination of supervised and unsupervised methods to detect opinion trends in online blogs comments [16]. Blog comments were identified and then analysed for positive, neutral or negative sentiment. Finally, K-means clustering extracted keywords that represent opinion trends. Xia *et al.* proposed a multi-label genetic algorithm (MLL-GA) for identifying types of software failures in crash reports [3]. The tool can provide developers with contextual knowledge about software failures.

Zhou *et al.* [17] developed a composite method for assigning semantic categories to online discussion posts related to APIs. The categories consisted of terms about particular API components. The output is used to retrieve relevant API information for developers. The present work and the work conducted during the OSSMETER project share the same objective. However, the latter only deals with multi-class classification. Other works have utilised multi-label classification, however, to the best of our knowledge there is no published research that explores multi-label content classification that directly relates to the topic and goal of our work.

3 Dataset and Content Hierarchy

We utilised the "gold standard" corpus and the content hierarchy in [11]. The corpus consists of threaded messages, extracted from NNTP news groups, forums and issue trackers related to Eclipse Projects (projects.eclipse.org). Post length ranges from a few words to several paragraphs. The content hierarchy contains 37 labels arranged into one of eight high level categories that represent content types related to OSS. Each hierarchy node represents a content type, *e.g.* "*Suggestion of a solution*", "*Resolution acceptance*" and "*Request/Report*". Only the leaf nodes may be assigned to individual messages. Super-classes that have descendants provide semantic integrity, grouping related classes together. For example, "*Clarification*" has two sub-classes "*Clarification of request*" and "*Clarification of a solution*". They are leaf nodes and therefore represent 2 labels. The node "*5. Unsuccessful solution notification*" has no descendants and, thus, it represents a label itself. Below are two examples of annotated messages.

1. **Message:** "The compare bug has been fixed. It appears that the object models are being properly garbage collected."
   Annotation: 7.3 - Working solution acknowledgement
2. **Message:** "In this case they're not in a view - it's a wizard, so global action handlers do not apply. Should investigate whether this is an SWT problem."
   Annotations: 1.1 - Clarification of request, 6.1 - Investigation

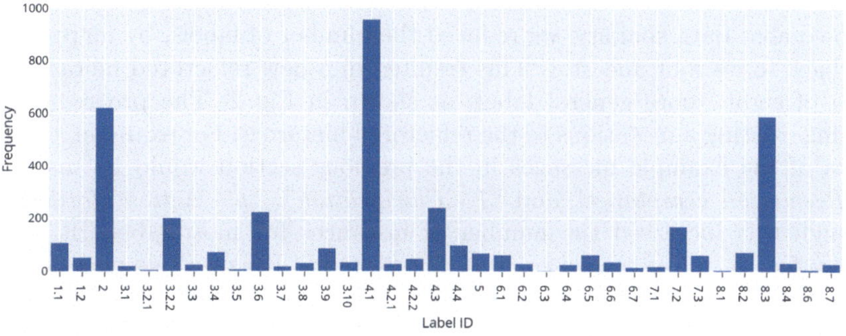

Fig. 1. Content classification label distribution

3.1 Refactoring of the Content Hierarchy

We were concerned about the size of the dataset in comparison to the number of labels within the content hierarchy. Figure 1 depicts the frequency of each leaf node[1]. We observe that the dataset is unbalanced. Labels such as "*4.1 - Initial Request/Report*" are well represented, others such as "*6.3 - Bug Priority Change*" are under-represented and some are not represented at all. Whilst this level of granularity would provide developers with specific knowledge regarding the content of messages, the label distribution within the *gold standard* is extremely unfavourable for under- or un-represented labels, *e.g.* such as "*6.3 - Bug priority Change*" and "*8.5 - Verbal abuse submission*".

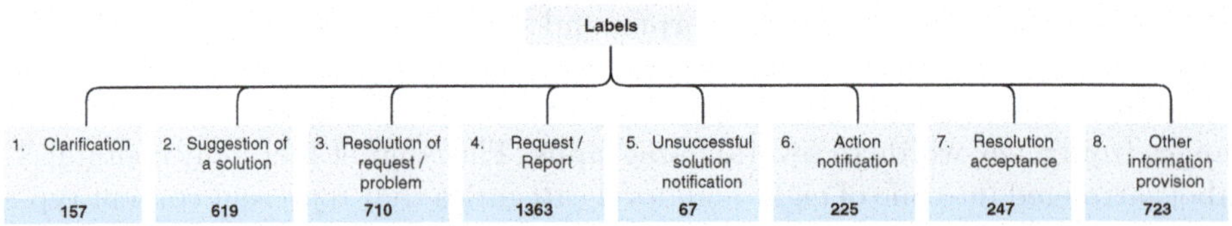

Fig. 2. Refactored hierarchy. Numbers in bold show the frequency of labels in the corpus.

Table 1. Gold standard annotation statistics

Statistic	Total
Threads	1164
Messages	3335
Annotations	4111
Messages annotated with a single label	2659
Messages annotated with multiple labels	676

To combat data sparsity, we reduced the number of labels, by mapping each label back to its root ancestor. This resulted in a new refactored hierarchy consisting of eight, more general labels as shown in Fig. 2. The process required mapping existing annotations to the refactored hierarchy. For example, the annotations of the example messages in the previous section would be mapped to "*7 - Resolution acceptance*" and "*1 - Clarification*", "*6 - Action Notification*". Although this increased the number of instances for most labels, it was not the case for "*5. Unsuccessful solution notification*", because this node had been considered as a leaf node originally.

[1] Labels that have not been assigned to any dataset instances, *e.g.* 8.5, have been omitted.

4 Methodology

Our intention is to design a supervised classifier that would determine the content type(s) of text written by a developer or user of an open source project. Since a message can be assigned to more than one content class labels, the task is multi-label. We have chosen to experiment with two state-of-the-art multi-label classifiers, *HOMER* and *RAkEL*. Each of these classifiers are available in Mulan [18] which is a Java library which supports multi-label classification.

Hierarchy Of Multilabel classifiERs (HOMER) [9] is a machine learner that follows the *divide-and-conquer* principle. It uses multiple multi-label classifiers, arranged hierarchically, each of which deals with a small set of labels only. HOMER uses a Binary Relevance transformation to convert the multi-label task into multiple single-label ones. After the transformation, it applies traditional single-label machine learners.

RAndom k-labELsets (RAkEL) [10] is a multi-label classifier based on the same principle as HOMER, but looks for correlations between labels to improve predictions. RAkEL builds an ensemble of single-label classifiers using a Label Powerset transformation; subsets of labels are selected randomly to improve performance.

We used the default configurations of HOMER and RAkEL in *Mulan*, which use C4.5 Decision Trees as single label classifiers. However, they can be configured to use other single label WEKA classifiers. Prior to feature generation, each message was preprocessed as explained below. Table 1 represents data statistics after pre-processing:

1. **Splitting text into paragraphs:** As a consequence of how the data is parsed, it is important to split text into paragraphs. This enabled greater control of the text retained as a result of subsequent stages of pre-processing.
2. **Text normalisation:** Repetitive spaces, such as new lines or tabulations, are removed. Symbols such as quotes are converted to a common consistent form and other symbols without concrete meaning are deleted.
3. **Removal of HTML Tags:** Newsgroup messages often use HTML tags to format or emphasise text. For example the tags *<b>* ... *</b>* are used to format a sequence as bold. HTML was also used to insert excerpts from other messages to provide context. HTML tags were removed to reduce noise.
4. **Detection and suppression of reply messages (>):** The *greater-than* symbol prefixes a line in newsgroup messages to represent a previous message from a thread. These lines were removed to reduce noise and a flag was triggered.
5. **Detection and suppression of code:** Each paragraph (block) was analysed by a code detector developed in-house. It is a linear classifier trained using FastText [19] on a data set that contained snippets written in Java, Python, C#, Javascript as well as natural language text from Wikipedia. The code detector labels each block as natural language or code. Text marked as code was removed and a related flag was triggered. If suppression would remove all text, the original text was kept.

The feature space was enriched with static and dynamic features. The former represent binary and numerical observations, which remain constant during training and testing. Static features used for experimentation are:

- Contains Code
- Contains Reply
- Position of a message in a thread
- Frequency of question marks
- Percentage of Code
- Frequency of exclamation marks
- Percentage of Natural Language
- Frequency of question words, *e.g. who, what, why.*
- Frequency of thanks words, *e.g. thanks, thank, thanx.*
- Frequency of help words, *e.g. advice*

As *Dynamic Features* we use the maximum n-grams size, the maximum gap between skip-bigrams, the minimum occurrence of n-grams and skip-bigrams and the number of subsets to be used by the classifier. We consider these features as dynamic because the optimal values for each machine learner, were computed using Bayesian Optimisation [20] with a objective function based upon the median or average, whichever is the least, of the *Multi-label Macro F-score* calculated over 10-fold cross validation. The optimised features were (1, 2, 1, 2) for HOMER and (3, 0, 2, 5) for RAkEL.

Experimental Setting: The corpus was split randomly into a training and test set, 80% and 20%, respectively. We ensured that the relative distribution of each label combination in each set follows the distribution in the entire corpus. If a label combination occurs just once, that instance was placed in the training set. If only two instances labelled with a specific combination of labels, one instance was placed in each set.

Each machine learner was trained using the optimised dynamic feature values and for comparison, we considered two baselines. The *Most Frequent Class* baseline assigns the most frequent annotation to all instances, whereas the *Random Class* baseline randomly assigns annotations, from zero to eight labels, to every instance. Combined these will provide an indication of classification performance from two perspectives.

Evaluative Setting: We assessed the performance of each model from different perspectives, employing the following metrics:

- **Multi-label Macro F-Score:** Evaluates how well every label has been predicted by the classifier. It considers all labels to have equal importance.
- **Multi-label Micro F-Score:** Assesses how well each instance and label has been predicted on average.

Table 2. Content classifier experimentation results

	Micro F-Score	Macro F-Score	Instance F-score	Hamming Loss	Subset 0/1 loss
Most Frequent Class	0.360	0.071	0.373	0.177	0.670
Random Class	0.230	0.207	0.218	0.561	0.996
HOMER	0.604	**0.460**	**0.627**	0.121	**0.505**
RAkEL	**0.619**	0.448	0.584	**0.102**	0.521

- **Instance F-Score:** Determines how well the instances were predicted on average.
- **Subset 0/1 Loss:** Indicates the percentage of instances for which a classifier did not correctly predict at least one label.
- **Hamming Loss:** Indication of how different the prediction is from the expected outcome. This metric is based on Hamming Distance.

5 Results and Discussion

Unlike multi-class classifiers, for multi-label classifiers labels are not mutually exclusive, *i.e.*, a classification instance can be assigned none, one or more labels. Therefore, to assess the performance of a multi-label classifier, it is important to consider several evaluation metrics that reflect its classification capabilities from different perspectives. The results per evaluation metric were computed and presented in Table 2.

Baselines: As expected, both the Most Frequent and Random Class baselines yielded poor results across all evaluation metrics. The Subset 0/1 loss indicates the classifiers ability to predict all labels per instance. It is a strict metric, as it only considers an instance as correct if all labels are predicted correctly. Hamming Loss reveals that the Random Class baseline assigned at least one incorrect label to 56% of the instances.

HOMER & RAKEL: Subset 0/1 loss shows that ~50% of the instances were predicted incorrectly. Hamming Loss shows that on average around 10% of the predicted labels are incorrect. The metrics are compatible with each other, because not all instances have more than one label. HOMER and RAkEL out performed the baselines, with HOMER scoring marginally better in three out of five metrics. The classifiers performed almost similarly in terms of Macro F-Score, with HOMER being slightly better. Results are better in terms of Micro than Macro F-score, showing that both methods are better in assigning labels that occur more frequently in the data.

Both methods performed well in terms Hamming Loss, with RAkEL achieving a marginally lower value than HOMER. However, each classifier still predicts a high number of labels incorrectly. Interestingly, the Most Frequent Class baseline performed similarly. However, this value could be considered to be artificially enhanced, since the most frequent label, *i.e.*, *4. Request/Report*, constitutes to ~33% of all annotations.

Since the multi-label metrics showed that HOMER and RAkEL performed similarly, we employ single-label metrics to assess their performance at predicting each label. Table 3 shows for each experiment the Precision, Recall and F-score per label, and the number of null predictions, *i.e.*, instances that were not assigned labels. RAkEL performed better than HOMER in terms of Precision, Recall and F-score, which could have contributed to its better Instance F-score, as it measures how well labels were predicted per instance. Both systems outperformed the baselines and assigned label *4. Request/Report* accurately. Low F-scores indicate that both systems had difficulties predicting labels *5. Unsuccessful*

Table 3. *Precision (P), Recall (R), F-Score (F1)* per label and numbers of null predictions (NP)

		Labels								
		1	2	3	4	5	6	7	8	NP
Most Frequent Class	*P*	0.000	0.000	0.000	0.397	0.000	0.000	0.000	0.000	0
	R	0.000	0.000	0.000	1.000	0.000	0.000	0.000	0.000	
	F1	0.000	0.000	0.000	0.568	0.000	0.000	0.000	0.000	
Random Class	*P*	0.046	0.183	0.202	0.392	0.023	0.068	0.061	0.214	4
	R	0.468	0.505	0.497	0.501	0.494	0.511	0.493	0.502	
	F1	0.084	0.269	0.287	0.440	0.045	0.120	0.109	0.300	
HOMER	*P*	0.167	0.553	0.455	0.788	0.143	0.511	0.600	0.566	55
	R	0.118	0.619	0.539	0.779	0.063	0.500	0.419	0.631	
	F1	0.138	0.584	0.494	0.784	0.087	0.505	0.493	0.597	
RAkEL	*P*	0.200	0.556	0.590	0.802	0.500	0.824	0.630	0.815	143
	R	0.059	0.476	0.440	0.820	0.063	0.304	0.395	0.503	
	F1	0.091	0.513	0.504	0.811	0.111	0.444	0.486	0.622	

solution notification and *1. Clarification.* In both cases, improvements over the baselines were minor, probably due to the rare occurrences of these labels in the corpus. Interestingly, systems perform similarly also for label *7. Resolution acceptance*, perhaps due to the way that this content type is structured.

Being unable to classify instances affects the results detrimentally, due to the small data size. HOMER had significantly less null predictions than RAkEL, ~14%. This could be attributed to Label Powerset, RAkEL's internal mechanism to transform multi-label into single-label instances. It causes partial loss of information about features that belong to each label, individually. Consider the instance " We are investigating this bug. Thank you for telling us.", labelled in the gold-standard as "*Resolution Acceptance*" and "*Action Notification*". RAkEL sees the two labels as a joint one, "*Resolution Acceptance & Action Notification*". This is problematic, as it reduces the number of labels in the training set but increases the number of possible labels to be predicted. RAkEL cannot determine that "*Thank you for telling us.*" should be uniquely related to "*Resolution Acceptance*". This information loss does not happen with HOMER, which transforms the labels using Binary Relevance. The above example would be repeated twice, with each copy annotated with just one label. This does not solve the problem completely, especially in cases where an instance is assigned many labels.

Affected by null predictions, RAkEL achieved a *better* Micro F-Score than HOMER. The metric in a sense favours null predictions because it only considers instances and labels and not their distribution. Hamming Loss is also sensitive to null predictions, as it measures the average difference between predicted labels and the ground truth. For example, in instance 1 in Table 4, HOMER predicts two incorrect labels, resulting in a more negative impact than the null prediction by RAkEL, which affects metrics sensitive to null predictions.

Overall, HOMER and RAkEL performed similarly across all metrics, with minor differences only, and outperformed both baselines. HOMER performed

Table 4. Representative instances from the test corpus and associated predictions.

ID	Post	Actual	HOMER	RAkEL
1	Changing priority better reflect importance defect	6	3 & 8	Null
2	Fixed ages ago. closing pr	3 & 6	3	3 & 6
3	hi adding following line code also solving error	8	8	3

better and generated less null predictions than RAkEL. RAkEL has many null predictions, but is more precise when it predicts labels. For example, in instance 2 in Table 4 HOMER predicted just one label, while RAkEL predicted them both. For label 6, RAkEL is ~30% more precise than HOMER, which, however, achieves ~20% higher recall.

6 Conclusion

Text in messages, posts and comments can be analysed to inform developers about the activity in bug trackers, communication channels and forums associated with Open Source projects. Content, in this work, refers to different functionality types, *e.g.* "*suggestion to a solution*", "*acceptance of a resolution*" and "*clarification*". These types can summarise the characteristics of communication, which in turn can inform developers about the quality of support offered via these means and also suggest ways to improve it. The overall goal of this work was to design and develop a content classifier capable of identifying content types in messages related to Open Source Software. To represent the content in messages, we explored two multi-label methods: *HOMER* and RAkEL, which use Binary Relevance and Label Powerset, respectively. Using a manually annotated corpus and content hierarchy, we reduced the class hierarchy to minimise the impact of under-represented classes. The feature space for each experiment was enriched with heuristics and compared against two baselines. Overall, both methods performed similarly and outperformed the baselines. In terms of the number of correct predictions per label and the number of unpredicted instances, HOMER was clearly better.

Our experimentation indicated that there is indeed potential for a classifier that is capable of automatically assigning multi-labels to messages related to open source software. Although, both methods exceed the baseline, further work is still need to improve the overall performance. Therefore, we aim to expand and further enrich the feature space with additional observations based upon Rhetorical Structure Theory (RST), in order to increase the overall performance. Furthermore, we acknowledge that the size of the corpus is small, regarding the number of labels, and implemented measures to compensate. However, we would like to extend the current set of annotations to increase both the size and the practical application of the dataset.

Acknowledgement. This research work is part of the CROSSMINER Project, which has received funding from the European Union's Horizon 2020 Research and Innovation Programme under grant agreement No. 732223.

References

1. Mooney, R.J., Roy, L.: Content-based book recommending using learning for text categorization. In: Proceedings of DL, pp. 195–204. ACM, New York (2000)
2. Bhatia, S., Mitra, P.: Classifying user messages for managing web forum data. In: WebDB, pp. 13–18 (2012)
3. Xia, X., Feng, Y., Lo, D., Chen, Z., Wang, X.: Towards more accurate multi-label software behavior learning. In: Proceedings of CSMR-WCRE, pp. 134–143 (2014)
4. Bagnato, A., Barmpis, K., Bessis, N., Cabrera-Diego, L.A., Di Rocco, J., Di Ruscio, D., Gergely, T., Hansen, S., Kolovos, D., Krief, P., Korkontzelos, I., Laurière, S., Lopez de la Fuente, J.M., Maló, P., Paige, R.F., Spinellis, D., Thomas, C., Vinju, J.: Developer-centric knowledge mining from large open-source software repositories (CROSSMINER). In: In Proceedings of STAFF, Marburg, Germany, pp. 375–384. Springer (2018)
5. Keivanloo, I., Forbes, C., Hmood, A., Erfani, M., Neal, C., Peristerakis, G., Rilling, J.: A linked data platform for mining software repositories. In: Proceedings of MSR, pp. 32–35 (2012)
6. Bavota, G., Ciemniewska, A., Chulani, I., De Nigro, A., Di Penta, M., Galletti, D., Galoppini, R., Gordon, T.F., Kedziora, P., Lener, I., Torelli, F., Pratola, R., Pukacki, J., Rebahi, Y., Villalonga, S.G.: The market for open source: an intelligent virtual open source marketplace. In: Proceedings of CSMR-WCRE, pp. 399–402 (2014)
7. van Deursen, A., Mesbah, A., Cornelissen, B., Zaidman, A., Pinzger, M., Guzzi, A.: Adinda: a knowledgeable, browser-based IDE. In: Proceedings of ICSE, vol. 2, pp. 203–206 (2010)
8. Di Ruscio, D., Kolovos, D., Matragkas, N., Korkontzelos, I., Vinju, J.: OSSMETER: a software measurement platform for automatically analysing open source software projects. In: Proceedings of ESEC/FSE (2015)
9. Tsoumakas, G., Katakis, I., Vlahavas, I.: Effective and efficient multilabel classification in domains with large number of labels. In: Proceedings of MMD, Antwerp, Belgium, pp. 53–59 (2008)
10. Tsoumakas, G., Katakis, I., Vlahavas, I.: Random k-labelsets for multilabel classification. IEEE Trans. Knowl. Data Eng. **23**(7), 1079–1089 (2011)
11. Korkontzelos, Y., Thompson, P., Ananiadou, S.: Identifying content types of messages related to open source software projects. In: Proceedings of LREC 2016, pp. 1837–1844. European Language Resources Association (ELRA), Portoro (2016)
12. Palau, R.M., Moens, M.F.: Argumentation mining: the detection, classification and structuring of arguments in text. In: Proceedings of BNAIC, pp. 351–352 (2009)
13. Mann, W.C., Taboada, M.: Rhetorical structure theory: looking back and moving ahead. Discourse Stud. **8**(3), 423–460 (2006)
14. Bacchelli, A., Dal Sasso, T., D'Ambros, M., Lanza, M.: Content classification of development emails. In: Proceedings of ICSE, pp. 375–385, June 2012
15. Pascarella, L., Bruntink, M., Bacchelli, A.: Classifying code comments in Java software systems. Empir. Softw. Eng. **24**(3), 1499–1537 (2019)
16. Alfaro, C., Cano-Montero, J., Gómez, J., Moguerza, J.M., Ortega, F.: A multi-stage method for content classification and opinion mining on weblog comments. Ann. Oper. Res. **236**(1), 197–213 (2016)
17. Zhou, B., Xia, X., Lo, D., Tian, C., Wang, X.: Towards more accurate content categorization of API discussions. In: Proceedings of ICPC, pp. 95–105. ACM, New York (2014)

18. Tsoumakas, G., Spyromitros-Xioufis, E., Vilcek, J., Vlahavas, I.: Mulan: a Java library for multi-label learning. J. Mach. Learn. Res. **12**, 2411–2414 (2011)
19. Joulin, A., Grave, E., Bojanowski, P., Mikolov, T.: Bag of tricks for efficient text classification. In: Proceedings of EACL, Valencia, Spain, vol. 2, pp. 427–431 (2017)
20. Mockus, J.B., Mockus, L.J.: Bayesian approach to global optimization and application to multiobjective and constrained problems. J. Optim. Theory Appl. **70**(1), 157–172 (1991)

Path Planning of a Self Driving Vehicle Using Artificial Intelligence Techniques and Machine Vision

Hassan Ashraf Elkholy[3,4], Ahmad Taher Azar[1,2(✉)], Abdalla Saber Shahin[3,4], Omar Ibrahim Elsharkawy[4], and Hossam Hassan Ammar[3,4]

[1] Robotics and Internet of Things Lab (RIOTU), Prince Sultan University, Riyadh, Saudi Arabia
aazar@psu.edu.sa

[2] Faculty of Computers and Artificial Intelligence, Benha University, Banha, Egypt
ahmad.azar@fci.bu.edu.eg

[3] Smart Engineering Systems Research Center (SESC), Nile University, Shaikh Zayed City 12588, Giza, Egypt

[4] School of Engineering and Applied Sciences, Nile University Campus, Sheikh Zayed District, Juhayna Square, 6th of October City 12588, Giza, Egypt
{h.ashraf,a.saber,o.elsharkawy,hhassan}@nu.edu.eg

Abstract. This paper aims to implement an efficient model of the most optimum path to follow an object on a Self Driving Vehicle (SDV). The path of the vehicle is predicted by using Machine Vision (MV) and Neural networks (NN) model. The NN model uses numerous amounts of training data. First the system works by using the MV algorithms to detect objects with predefined colors. Then, the location of the object is fed to the trained NN to get the speeds of the motors needed to reach the object. The training data are obtained from the manual driving of the vehicle in different experiment settings. In this paper, the neural model is compared with two other methods: object detection using MV model and fuzzy logic (FL) model to prove the efficiency of the neural model. All the three models depend on the live record of the camera board and its fast detection of objects using MV algorithms. The three models showed quite similar results; however, the NN model was much more stable and closer to the optimum path.

Keywords: Self-driving vehicle (SDV) · Fuzzy logic (FL) · Neural networks (NN) · Machine Vision (MV) · Machine Learning (ML)

1 Introduction and Literature Survey

Self-driving vehicles use combined sensors and software to control, navigate, and drive the vehicle automatically [10,14,20]. While design details vary, most self-driving systems create and maintain an internal map of their surroundings,

A.-E. Hassanien et al. (Eds.): AICV 2020, AISC 1153, pp. 532–542, 2020.
https://doi.org/10.1007/978-3-030-44289-7_50

based on a wide array of sensors, like radar [19].Various self-driving technologies have been developed by Google, Uber, Tesla, Nissan, and other major automakers, researchers, and technology companies. While design details vary, most self-driving systems create and maintain an internal map of their surroundings, based on a wide array of sensors, like radar [19]. Machine Learning (ML) is an application of artificial intelligence (AI) in machines to provide them the ability of learning and improving through experience or teaching not through direct programming [2,3,6–8,15]. Vehicles with the traditional systems depend on human sensory functions [18] such as vision and sound to watch the road and other cars on the road [16]. These minor tasks of humans can be replaced easily by ML to eliminate human error [11]. Not only these minor tasks, but also driving like human drivers; that means these vehicles need to be provided with the sensory functions, cognitive functions (memory, logical thinking, decision-making and learning) and executive capabilities that humans use to drive vehicles [12]. As the amount of information being fed into IVI (in-vehicle infotainment) units or telematics systems grows, vehicles will be able to capture and share not only internal system status and location data but also the changes in its surroundings, all in real time [15]. Autonomous vehicles are being fitted with cameras, sensors and communication systems to enable the vehicle to generate massive amounts of data which, when applied with artificial intelligence, enables the vehicle to see, hear, think and make decisions just like human drivers do [19].

This paper work is related to three different types of research approaches. In the first category are papers that neglect the camera control or any robot motion and focus on the video imagery and its problem of color-histogram-based person tracking. The second category of papers focuses on person following by using a single fixed camera on a mobile robot. The third category of papers focuses on the problem of manipulating cameras that are trying to stay aimed on the object being followed. In the first category, color distribution represents the head and the hands for the human figure as part of a multi-class statistical blob which is a contribution of the Pfinder [5] for tracking a person in video imagery [12]. There is also the contribution in people tracking in crowded environment using a Kalman filter (KF) combined with nearest neighbor standard filter (NNSF) [13]. In the second category, following a person is done by a mono camera and a laser range imaging sensor which is contributed by the work of Kleinhagenbrock et al. [9]. The third category focuses on the contributions of Alvarez-Santos et al. [1]. They have validated a mobile robot engaging in following people. This work represents a person blob in the images by adopting a color-histogram. In this paper, the whole system depends on what the camera detects. SDV's main problems are high non-linearity and uncertainty, making it very difficult to build a completely descriptive mathematical model which can be employed later in order to control the system. The objective of the paper is to implement algorithms of MV and ML such as object detection and tracking, fuzzy logic model, and the NN model algorithms on a real hardware system and compare the obtained experimental results to determine the most efficient model.

The rest of the paper is organized as follows. In Sect. 2, Hardware Modeling and Implementation is briefly described. In Sect. 3, the proposed methodology is presented. Results and discussions are presented in Sect. 4 Finally, the paper is concluded in Sect. 5.

2 Hardware Modeling and Implementation

The proposed vehicle has four motors connected with a driver to provide communication between the motors and the Arduino Uno board. A raspberry pi is connected serially to the Arduino board which acts as a slave that waits for raspberry pi signals. A camera board, with a resolution of 640×480 and $160°$ angle, is connected to the raspberry pi. Figure 1 (a) shows the connection between all the hardware components of the vehicle. The self driving vehicle body is shown in Fig. 1(b).

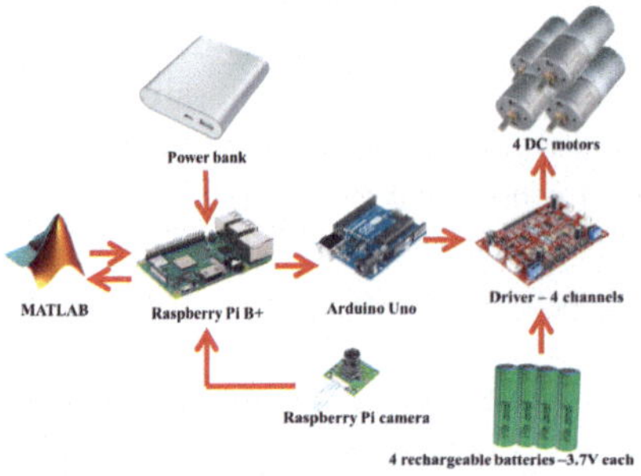

(a) Hardware schematic connection

(b) The self driving vehicle

Fig. 1. The vehicle hardware

3 Proposed Methodology

Three models are implemented on the vehicle system to find the most accurate method that will allow the vehicle to follow the suggested path. A filter on the live camera transmission is added to each of the three models in order to detect only the object to be followed and send the data of the object to the controller that orders the vehicle engines to function in a continuous loop accordingly. The loop stops until the vehicle reaches its destination.

3.1 Object Detection and Tracking Using Machine Vision

The camera has a fish-eye function; in other words, the picture shown is curved and must be measured to work properly. A random camera shot was taken and segmented [21] into four regions: left, right, middle and ignored as shown in Fig. 2.

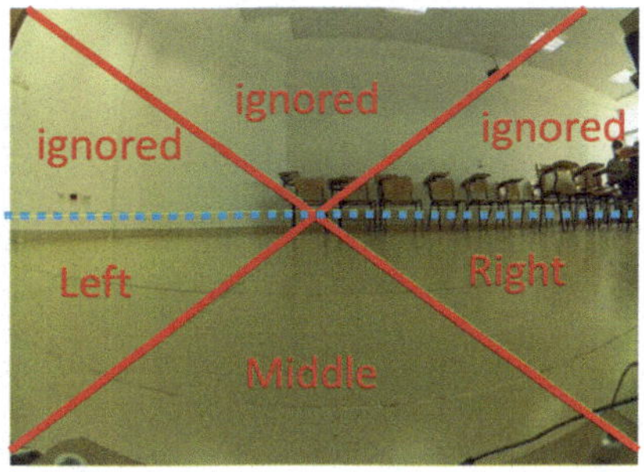

Fig. 2. Image segmentation and calibration

The calibration is accomplished by the equation describing the line between right and medium regions and the line between left and mid-areas. By linear interpolation, the two equations describing the two lines were determined. The equation of the line separating between right and middle regions and the equation separating between left and middle regions are given in Eqs. 1 & 2, respectively.

$$X_p = 13.64 + 1.36 \times Y_p \tag{1}$$

$$X_p = 626.39 - 1.35 \times Y_p \tag{2}$$

where X_p and Y_p are the pixel coordinates. The blind point is the intersection between the two lines, above which is an ignored area where the robot does not act if the camera captures the object.

3.2 Vehicle Modeling and Control Using Fuzzy Logic

The second approach is the fuzzy model that takes the centroid of the object and yields the velocities of the motors on each side of the vehicle. In order to do that, a system of two inputs and two outputs that has 22 firing rules is created according to the image segments shown in Fig. 3 [17]. Table 1 shows the 22 firing rules of the fuzzy model with left motors and right motors, respectively. The membership functions were constructed for the inputs and outputs based on real system data. The inputs data are the x-coordinate and the y-coordinate of the centroid pixel while the outputs data are the Pulse Width Modulation (PWM) given to left motors and right motors. X-coordinates and y-coordinates were divided into five membership functions and three membership functions respectively as shown in Figs. 4 and 5. Each segment of images has certain firing rules that the controller executes and gives output speeds when there is an object. The two fuzzified outputs are the same; they are divided into three triangular membership functions as shown in Fig. 6. The final output of the fuzzy model is two crisp values which represents the PWM values given to the right and left motors [4].

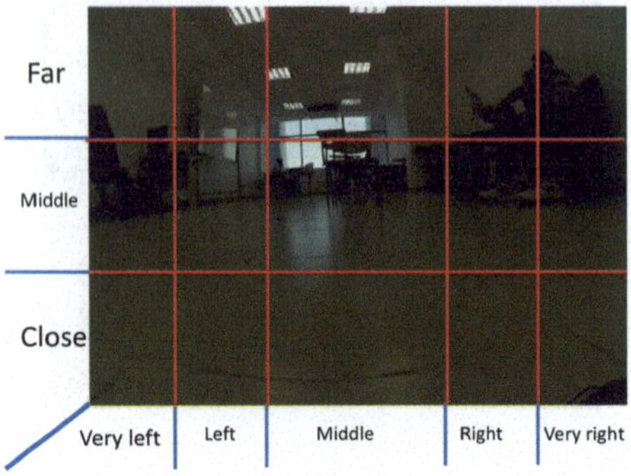

Fig. 3. Main divided sections of the image

Table 1. The firing rules of the two inputs and the two outputs

	Close	Medium	Far
Very left	LM:Slow, RM:Medium	LM:Slow, RM:Fast	LM:Slow, RM:Slow
Left	LM:Slow, RM:Slow	LM:Medium, RM:Fast	LM:Slow, RM:Slow
Middle	LM:Slow, RM:Slow	LM:Medium, RM:Medium	LM:Slow, RM:Slow
Right	LM:Slow, RM:Slow	LM:Fast, RM:Medium	LM:Slow, RM:Slow
Very Right	LM:Medium, RM:Slow	LM:Fast, RM:Slow	LM:Slow, RM:Slow

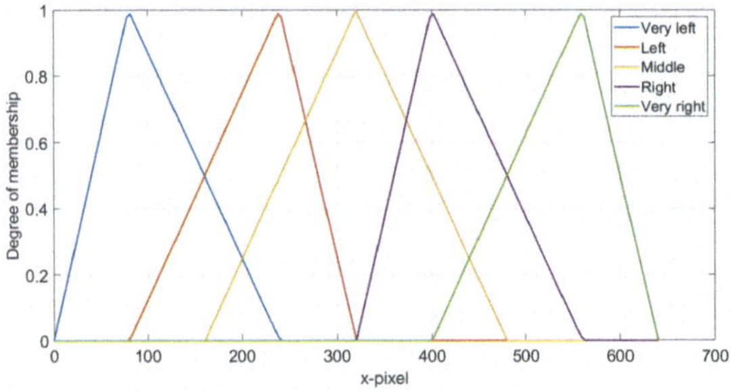

Fig. 4. Fuzzification of x-coordinate of the pixel

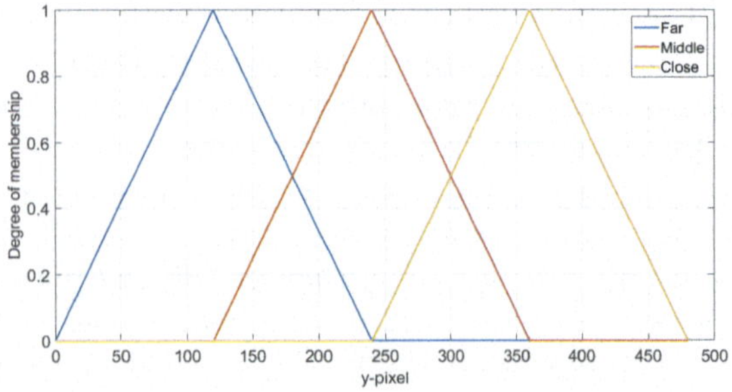

Fig. 5. Fuzzification of y-coordinate of the pixel

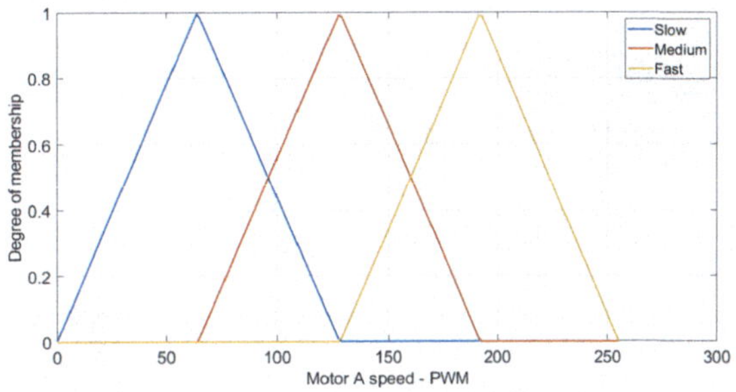

Fig. 6. Fuzzification of left and right motors output-

3.3 Vehicle Modeling and Control Using Neural Networks

The vehicle is trained by an NN model, so that as a normal human it can operate alone. The training data contains the two inputs which are x-pixels and y-pixels of the object's centroid, and the corresponding two outputs which are the motors speeds in each side of the vehicle. The vehicle is first driven manually while the location of the object recorded by the camera and the corresponding motors speeds are saved in a four column matrix. The vehicle's trajectory is compared to the proposed trajectory of error determination by the neural model. Table 2 represents the training goal of the neural model, number of epochs, and maximum fail.

Table 2. Neural model data

Training goal	$10 * e^{20}$
Number of epochs	2000
Max fail	1000

3.4 Experimental Work

After applying each model on the vehicle, it drives from a start point until it reaches the desired object, while the path it takes is drawn by a marker fixed in the vehicle's back and recorded by a fixed vertical camera as shown in Fig. 7. Afterwards, a MATLAB code takes paths images and converts the paths to arrays of numbers that could be plotted and analyzed as shown in Fig. 8.

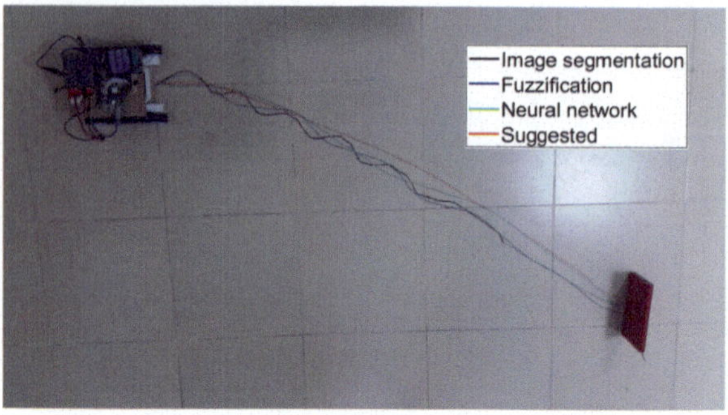

Fig. 7. Actual paths that the vehicle followed

4 Results and Discussion

The results of the neural networks model, object tracking model using machine vision, and fuzzy logic model are presented in this section respectively. The neural model produced a very close system to the actual system. The results of the neural model are shown in Figs. 9, 10, 11 and 12 which represent the

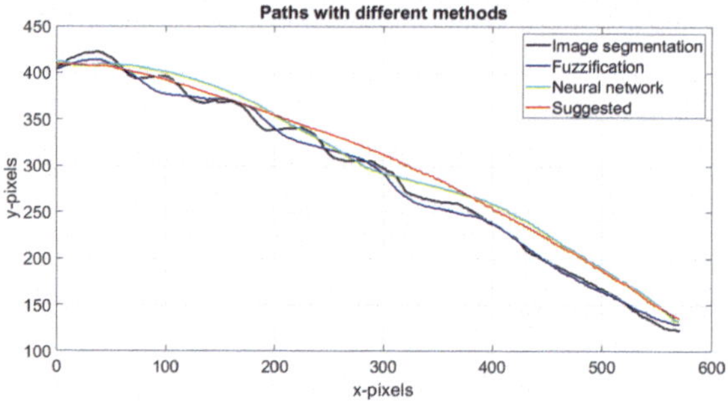

Fig. 8. Paths of each method compared with the suggested path

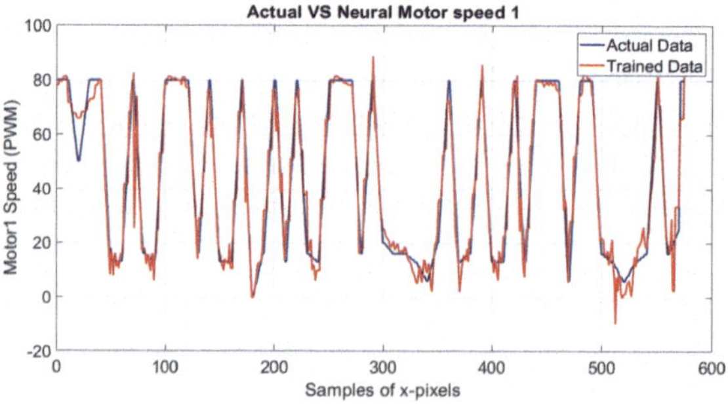

Fig. 9. Actual vs neural of x-pixels and left motors speed

validation of the neural model. Figures 9 and 10 represent the inputs and the corresponding left motors speeds while Figs. 11 and 12 represent the same two inputs and the corresponding right motors speeds.

Object tracking method using MV shows a good path with low steady state error; however, there were so much oscillations around the suggested path. The fuzzy logic model also produced a very well path with lower steady state error and less oscillations than the first method. The comparison between the three methods is based on evaluating the error values between the resulted paths in the three methods and the suggested path using residual sum of squares (RSS) as shown in Eq. 3.

$$RSS = \sum_{i=1}^{n} (y_i - f(x_i))^2 \tag{3}$$

where y_i is the y-pixel data from the suggested path and f(xi) is the y-pixel data from either the paths to be compared.

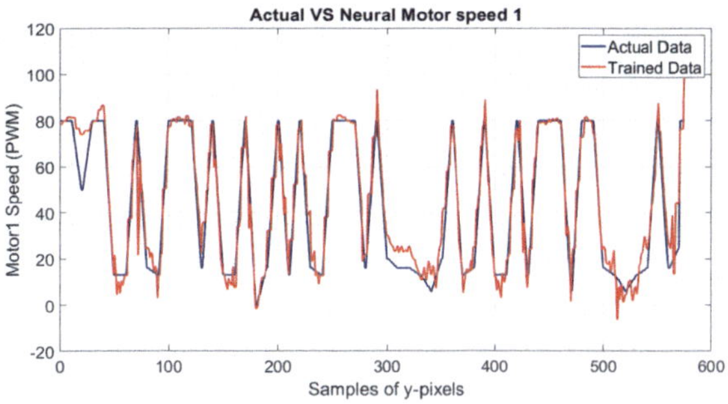

Fig. 10. Actual vs neural of y-pixels and left motors speed

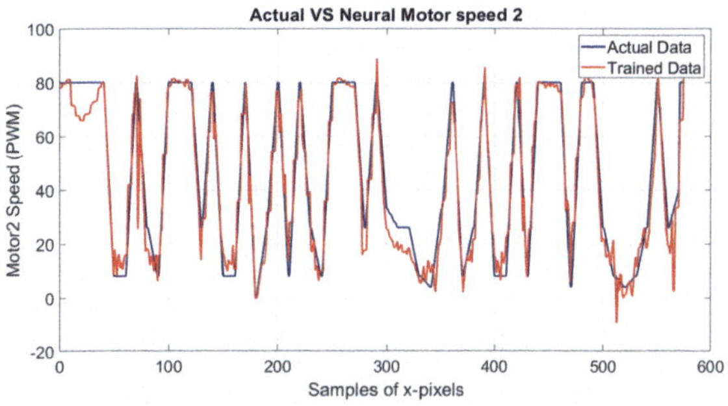

Fig. 11. Actual vs neural of x-pixels and right motors speed

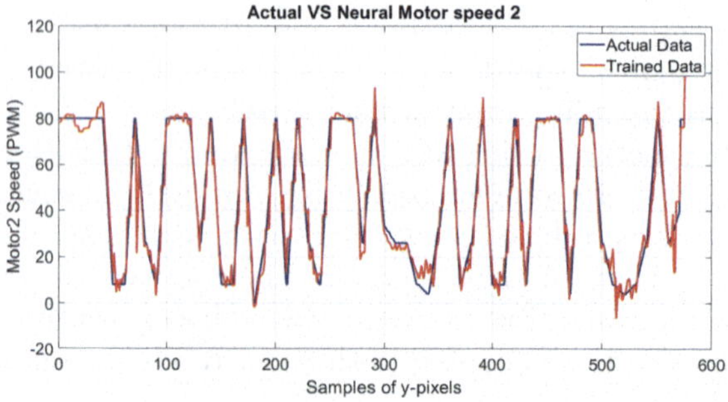

Fig. 12. Actual vs neural of y-pixels and right motors speed

The error values are shown in Table 3. By analyzing the three methods mentioned above, the performance was good. And by comparing the first method and the fuzzy logic method, the results were nearly close but they both have defects of having many overshoots over the suggested path. The neural network model showed better performance than the other two models with minimum overshoots and that is because of the high accuracy of obtaining the training data.

Table 3. The RSS values of each of the three methods

Method	RSS
Object tracking using machine vision	203467
Fuzzy logic model	175332
Neural networks model	42703

5 Conclusion

This paper focuses on making a powerful method to be applied on a SDV to achieve the suggested path between a start point and a followed object. In this paper, three methods were applied on the vehicle to test the response of the vehicle and to compare the results of the three methods. The first method is object detection and tracking using machine vision, in this method, the output image from the camera is segmented so that the vehicle can follow the direction of the object. The second method was the fuzzy logic model which is more accurate because it divides the image to fifteen segments from which the vehicle can accurately determine the direction and speed required to reach the object. The last method and the most powerful one is the NN model by which the vehicle was easily modelled, trained and controlled using lots of data generated from manual driving. The results of the three methods demonstrated the superiority of the neural model, and the trajectory of the vehicle arose from this model with almost no oscillations or errors. For the future work, more recent artificial intelligence techniques will be implemented such as reinforcement learning to enhance the vehicle system and make it more robust and fast so it can be able not only to follow objects but also to avoid obstacles and deliver products.

References

1. Alvarez-Santos, V., Pardo, X., Iglesias, R., Canedo-Rodriguez, A., Regueiro, C.: Feature analysis for human recognition and discrimination: application to a person-following behaviour in a mobile robot. Robot. Auton. Syst. **60**(8), 1021–1036 (2012)
2. Azar, A.T., Aly, A.M., Sayed, A.S., Radwan, M.E., Ammar, H.H.: Neuro-fuzzy system for 3-DOF parallel robot manipulator. In: 2019 Novel Intelligent and Leading Emerging Sciences Conference (NILES), vol. 1, pp. 1–5 (2019)
3. Azar, A.T., Sayed, A.S., Shahin, A.S., Elkholy, H.A., Ammar, H.H.: PID controller for 2-DOFs twin rotor mimo system tuned with particle swarm optimization. In: Hassanien, A.E., Shaalan, K., Tolba, M.F. (eds) Proceedings of the International Conference on Advanced Intelligent Systems and Informatics 2019, pp. 229–242. Springer, Cham (2020)
4. Barakat, M.H., Azar, A.T., Ammar, H.H.: Agricultural service mobile robot modeling and control using artificial fuzzy logic and machine vision. In: Advances in Intelligent Systems and Computing, pp. 453–465 (2019)
5. Davies, R., Wilson, I., Ware, A.: Stereoscopic human detection in a natural environment. Ann. Emerg. Technol. Comput. **2**(2), 15–23 (2018)
6. Giusti, A., Guzzi, J., Ciresan, D.C., He, F.L., Rodriguez, J.P., Fontana, F., Faessler, M., Forster, C., Schmidhuber, J., Caro, G.A.D., Scaramuzza, D., Gambardella, L.M.: A machine learning approach to visual perception of forest trails for mobile robots. IEEE Robot. Autom. Lett. **1**, 661–667 (2016)
7. Gorripotu, T.S., Samalla, H., Jagan Mohana Rao, C., Azar, A.T., Pelusi, D.: Tlbo algorithm optimized fractional-order PID controller for AGC of interconnected power system. In: Nayak, J., Abraham, A., Krishna, B.M., Chandra Sekhar, G.T., Das, A.K. (eds.) Soft Computing in Data Analytics, pp. 847–855. Springer, Singapore (2019)
8. Kamal, N.A., Azar, A.T., Elbasuony, G.S., Almustafa, K.M., Almakhles, D.: PSO-based adaptive perturb and observe MPPT technique for photovoltaic systems. In: Hassanien, A.E., Shaalan, K., Tolba, M.F. (eds.) Proceedings of the International Conference on Advanced Intelligent Systems and Informatics 2019, pp. 125–135. Springer, Cham (2020)
9. Kleinehagenbrock, M., Lang, S., Fritsch, J., Lomker, F., Fink, G., Sagerer, G.: Person tracking with a mobile robot based on multi-modal anchoring. In: Proceedings 11th IEEE International Workshop on Robot and Human Interactive Communication (2002)
10. Liu, P., Ma, Y., Zuo, Y.: Self-driving vehicles: are people willing to trade risks for environmental benefits? Transp. Res. Part A: Policy Pract. **125**, 139–149 (2019)
11. Ohn-Bar, E., Trivedi, M.M.: Looking at humans in the age of self-driving and highly automated vehicles. IEEE Trans. Intell. Veh. **1**(1), 90–104 (2016)
12. Paden, B., Cap, M., Yong, S.Z., Yershov, D., Frazzoli, E.: A survey of motion planning and control techniques for self-driving urban vehicles. IEEE Trans. Intell. Veh. **1**(1), 33–55 (2016)

13. Saito, M., Kitaguchi, K., Kimura, G., Hashimoto, M.: Human detection from fish-eye image based on probabilistic appearance model. Trans. Soc. Instrum. Control Eng. **49**(3), 319–325 (2013)
14. Sallab, A., Abdou, M., Perot, E., Yogamani, S.: Deep reinforcement learning framework for autonomous driving. Electron. Imaging **19**, 70–76 (2017)
15. Sivaraman, S., Trivedi, M.M.: A general active-learning framework for on-road vehicle recognition and tracking. IEEE Trans. Intell. Transp. Syst. **11**, 267–276 (2010)
16. Sivaraman, S., Trivedi, M.M.: Active learning for on-road vehicle detection: a comparative study. Mach. Vis. Appl. **25**(3), 599–611 (2011)
17. Soliman, M., Azar, A.T., Saleh, M.A., Ammar, H.H.: Path planning control for 3-omni fighting robot using PID and fuzzy logic controller. In: Advances in Intelligent Systems and Computing, pp. 442–452 (2019)
18. Stallkamp, J., Schlipsing, M., Salmen, J., Igel, C.: Man vs. computer: benchmarking machine learning algorithms for traffic sign recognition. Neural Netw. **32**, 323–332 (2012)
19. Stilgoe, J.: Machine learning, social learning and the governance of self-driving cars. Soc. Stud. Sci. **48**(1), 25–56 (2018). https://doi.org/10.1177/0306312717741687. pMID: 29160165
20. Wang, G., Makino, K., Harmandayan, A., Wu, X.: Eco-driving behaviors of electric vehicle users: a survey study. Transp. Res. Part D: Transp. Environ. **78**(102), 188 (2020)
21. Zaitoun, N.M., Aqel, M.J.: Survey on image segmentation techniques. Procedia Comput. Sci. **65**, 797–806 (2015)

Improving the Data Quality of the MovieLens Dataset Using Dimensionality Reduction Techniques

Hagar ElFiky(✉), Wedad Hussein, and Rania El Gohary

Information Systems, Faculty of Computer and Information Sciences,
Ain Shams University, Cairo, Egypt
hagar_elfiky@hotmail.com,
wedad.hussein@fcis.asu.edu.eg,
rania.elgohary@cis.asu.edu.eg

Abstract. The presence of recommendation systems within the current era is not controlled or even managed like before, for providing the optimum services to users. For that reason, these services need to be in an ideal and complete state since the real-world data is often not that much perfect. That optimum state could not be achieved without the interaction of preprocessing method. Preprocessing can offer a lot of insights about the data and its structure and quality. Preprocessing techniques ensure that the dataset has complete, consistent and integrant properties for further analysis. Dimensionality reduction is this paper is discussed as one of the preprocessing stages using the most popular matrix factorization methods. We have also made an experiment to show how the dimensionality reduction algorithms can overcome any constrained computational resources and prove better performance and efficiency for such demanding recommendation systems.

Keywords: Preprocessing · MovieLens · Data quality · Dimensionality reduction

1 Introduction

Real word data is often massive, incomplete, inconsistent and suffering from lake of certain trends and most likely contain many errors. Data preprocessing is a proven method for resolving such issues. Preprocessing is any process performed on the raw data to prepare it for another conversion phase. Moreover, it is necessary as it transforms data into a format that will be more easily and effectively processed as per the purpose of the user. There are various methods used for the preprocessing such as sampling, transformation denoising and normalization.

Data quality became the highlight of data scientists in recent technologies as it has a direct impact on the operations, decision making, planning and future maintenance of any system. Therefore, any data that has been collected for research and analysis purposes must be assessed to judge its fitness to serve its purpose in a given context. The quality of data is determined using several factors to ensure its integrity such as availability & accessibility, completeness, comprehensiveness and accuracy. Data

A.-E. Hassanien et al. (Eds.): AICV 2020, AISC 1153, pp. 543–552, 2020.
https://doi.org/10.1007/978-3-030-44289-7_51

quality has succeeded also to prove its important role as being one of the main guides in the market. Moreover, it helped the vendors to observe and know their potential customers as well as for deciding their customers' needs. In other words, data quality has shown it's imprinted in both the market and research areas.

As the data quality contributes in improving the data accuracy, comprehensiveness and completeness, it also concerns with the relevance and the importance of the information these data are holding which is done through the dimensionality reduction. Applying the dimensionality reduction to a dataset before running any machine learning technique could be very beneficial as the space required to store data is reduced leading to less computational and training time. Reducing dimensionalities enables the algorithm to perform well and generate accurate results. Moreover, dimensionality reduction techniques take down the duplicates of features leading to more compact and precise output. It is very difficult to visualize data in high dimensions accordingly reducing the space to two or three dimensions enables us to plot and observe patterns more clearly. To conclude, choosing a preprocessing model and the type of dimensionality reduction technique, a contribution for a better data quality could be achieved to reach final trusted results.

MovieLens web site is a web-based recommender system that recommends personalized movies to its users based on their preferences and interests, in addition to asking the users to give movie ratings in order to receive personalized movie recommendations [3]. MovieLens made a great contribution in multiple domains such as education, industry, research and development. The first release of Movielens datasets was in 1998, describe peoples' movies interests.

These interests were collected by the MovieLens web site [2]. MovieLens datasets are generated as a result of accumulations of users interactions' with MovieLens web-based recommender system for years [1]. Since these datasets are available for public access without any constraints, this encouraged researchers to use them and to get the benefit of their actual and real data. MovieLens enables the researchers to prove their proposed solutions and enhance the quality of either the data or the recommender system. Authors in [3] have found that the MovieLens datasets were aggressively used where more than 140,000 downloads [1] and referenced in the research paper more than 7,500 references to "MovieLens" in Google Scholar happened in 2014.

MovieLens has exported 4 sets of data for researching and analyzing purposes. The first one, is the one that is being used in this research which is known as 1 M dataset movie ratings. It is also known as a Stable benchmark dataset. This dataset of size 6 MB, was released to the public in February 2003 that contains 1 million ratings of approximately 4000 movies using 6000 users. The second dataset was released in January 2009 of size 63 MB. The second dataset contains 10 million ratings of 10,000 movies using 72,000 users. The third dataset was of size 190 MB and released in April, 2015 and updated in October 2016. It contains 20 million movie ratings of 27,000 movies using 138,000 users. The latest and most recent dataset of size 265 MB, is known as 27 M movie ratings as it consists of 27 million records with different ratings of 58,000 movies using 280,000 users, was available to the public in September 2018. We used a 1 M dataset version in this research to verify the basic preprocessing methods used, which can be similarly applied to the rest of the versions.

This paper is organized as follows, the first section is introducing an overview of the data quality and MovieLens website. Section two discusses the preprocessing challenges we faced with the proposed solutions. In section three, the dimensionality reduction and its techniques are discussed. We have applied an experiment and showed the analysis of different dimensionality reduction techniques in section four. Finally, we concluded the importance of applying that the data quality and preprocessing before passing the data to any machine learning algorithm in order to contribute greatly to the performance and the effectiveness of any system especially if it was user-based system.

2 Preprocessing Challenges and Proposed Solutions

In this section we are mainly focusing on the challenges that we have faced in the Movielens dataset. We have listed the challenges we faced in order and their proposed solutions for better data quality and accurate results in the future implementation.

First, we made sure that the dataset files and their attributes are clear and understandable. Therefore, The symbol of % was used as a unique delimiter for the "Movies" file. Moreover, a redundancy check has taken place to avoid any duplication of data or records. Secondly, we moved to the data completeness where a check for null values has taken place to ensure completeness using COUNTBLANK(), a built in function within excel. Data completeness was confirmed using another examination's criteria throughout calculating minimum, maximum, quartile, median and the mean functions on each column of every file. Thirdly, we moved to the data comprehensiveness where we had to split the release year from the movie title column as it would facilitate the insertion of the release year as an important feature in a separate column in the movies table. In the rating table, the timestamp was converted into a readable format. Lastly, we ensured that there were no duplicates in the Users or Movies tables. On the other hand, we ensured also using composite key of UserId and MovieId for each rating record in the ratings table for conforming the data accuracy.

3 Dimensionality Reduction Techniques

The machine learning and data mining classification problems need and require many factors in order to get the final classification output. Most of these factors are called features. The bigger the dataset is, the higher the number of features that could be found and accordingly the harder for the classification algorithms to work. Since these features are sometimes correlated and redundant or even have low variance in the dataset, dimensionality reduction algorithms have been used more in the preprocessing stage. Applying the reduction in the preprocessing phase has showed many advantages in data compression, storage space, computational time, feature redundant and overfitting elimination and finally easier data visualization as reduced to lower dimension [11].

Dimensionality reduction is the process of reducing the number of features taking into consideration keeping the principle and high variance features [9]. In the research domain, it has been used as a preprocessing step preceding the application of data mining algorithms like clustering. When the scientists thought about reducing the dimensions of the dataset, they found that some coordinates do not contribute in the data representation. In addition to subsets of dimensions are highly corelated which may result in bias. Moreover, some of the data mining and clustering algorithms like nearest neighbor is deformed in high dimension space.

Dimensionality reduction can be approached by two different methods which are feature selection and feature extraction/dimensionality reduction. As shown in Table 1, feature extraction methods can be achieved either by component/factor based or by projection based, however projection based is out of scope and irrelevant in the problem ahead of us as it's mainly targeting the strongly non-linear data and visualizations.

Table 1. Dimensionality reduction techniques

Dimensionality reduction techniques		
Feature selection	Feature extraction	
Missing Value Ratio	Components/Factors Based	Projection based
Low Variance Filter	Factor Analysis	ISOMAP
Random Forest	Principal Component Analysis	t-SNE
Backward Feature Elimination	Independent Component Analysis	UMAP
Forward Feature Selection		

In [4] feature selection it is mainly concerned with preserving the most relevant features from the original dataset while feature extraction or dimensionality reduction techniques are concerned with fetching a smaller set of new features. Each feature is composed of mixed values of the other input features, which have the same information as the rest of the input features. Each category will be described briefly in the next paragraphs.

3.1 Feature Selection

Some of the feature selection methods are focusing on dropping features either the features that have large number of missing values or the features that have low variance from the dataset in order to select subset of features from the original dataset like Missing Value Ratio and Low Variance Filter. Other features selection methods like the random forest although it has a good selection of important features capability however, it requires high computational power and resources [5]. On the other hand, the Backward Feature Elimination and the Forward Feature Selection target mainly small datasets as they take a lot of computational time.

3.2 Feature Extraction

The feature extraction usually seeks driving information from original data to build new feature subspace through reducing the number of features in the high dimension space to lower dimension space to reach a new matrix with a smaller number of dimensions. This new matrix will be holding new features that are combination of original features but compressed in a way retaining the most relevant information. The various methods used for dimensionality reduction are Factor Analysis, Principal Component Analysis and Independent Component Analysis.

Factor Analysis
The factor analysis technique groups the variables or features based on the correlation among them. This means that any particular group of features shall have a strong correlation between all it's features and a weak correlation with the features in the other groups. These groups are recognized as factors, which are lower in number as compared to the original data dimensions. However, these factors are holding the related variables only that are very difficult to be observed individually. Accordingly, the major parameter that the factor analysis technique depends on is the "n_components" which represents the number of factors or groups that user is expecting to have in the transformed data.

Principal Component Analysis (PCA)
The principle component analysis or the PCA technique is usually used to map the data matrix into less dimensional through extracting the new subset of variables from the original variables found in the dataset. These newly variables are recognized as principle components. In [9] authors succeeded to reach the fact of the amount of variance extracted by the first component explains higher variance than the variance of the second component and so on. Accordingly, the reference parameter that the PCA technique depends on is the "n_components" which decides the number of principle components or to what extent you need variance in your transformed data.

Independent Component Analysis (ICA)
The independent component analysis or the ICA executes mainly to apply the blind source isolation concept. ICA always aims to find the features that contribute independently in the dataset without depending on the other features. Moreover, the features that have the lowest linkage or connection to the other features in the original dataset in order to isolate those most effective independent sources from the sources in the dataset.

4 Experiment and Analysis

4.1 Dataset Perspective

The dataset that is being used in this research is known as 1 M dataset. The 1 M dataset consists of three files known as Users, Movies and Ratings. The Users file contains 6040 records i.e. distinct users. Each record is associated with unique set of information for this specific user. This file is composed of 5 columns UserID, Gender, Age, Occupation and Zip-code. The UserID is a unique integer number for every user, then the Gender which is char datatype that contains one of two possible values donated by a "F" for female and "M" for male, followed by the Age which is chosen from several ranges. Afterwards the Occupation that is in text datatype and which is chosen from several values such as "1" which is the academic/educator or "2" which is equivalent to artist. Finally, the Zip-code is a numeric integer value which maps to the user country's zipcode. The Movies file contains approximately 3,900 movies. Each movie record is described by three unique columns MovieID, Title and Genres. The MovieID is a unique integer number for every movie, followed by the Title which is text datatype.

Lastly, the Genres are a pipe-separated text such as Action, Adventure, Drama, and comedy and so on.

The last file in this dataset is the ratings which contains 1,000,209 anonymous ratings. Each user has at least 20 ratings. This file is composed of four columns UserID, MovieID, Rating and Timestamp. Rating is an integer value that ranges from 1 to 5. Timestamp is the integer time in which the user X makes rating Y on movie Z represented in seconds. All the above-mentioned columns are automatically captured by the system however, the Ratings are manually inserted by the users according to their interests.

We have applied an experiment on the Movielens dataset that have 1 M records in order to show how the performance differs when changing the number of components from one technique to another from the above, although the three of them are used for feature extraction having the same number of users. As shown in Fig. 1, the factor analysis technique drawn in blue suffers from a performance issue with the positive degradation of the number of components. Factor analysis has showed the highest performance while PCA drawn in yellow has almost performance stability along the increase of components. ICA's performance presented in green in directly proportional with the increase of the number of components. Figure 1 is proving that the nature and the status of the data is main driver of choosing the ultimate dimensionality reduction technique. Factor analysis for example has shown great results with datasets that have inside it groups with strong correlation between their features [15].

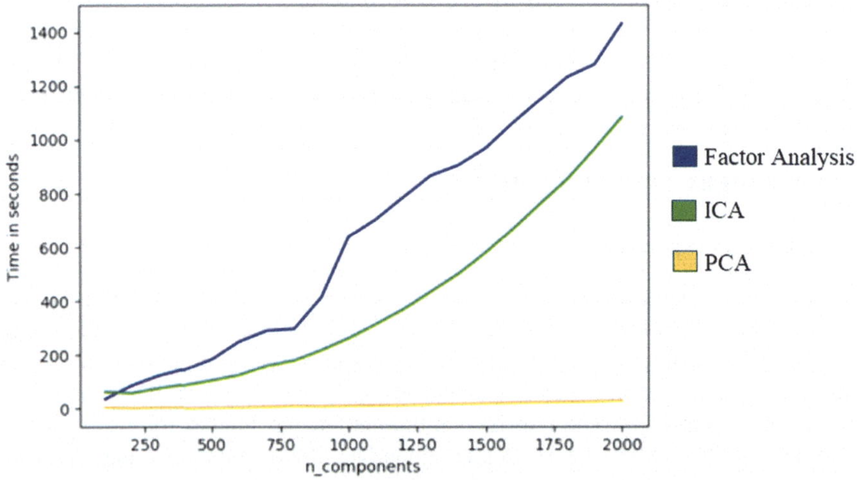

Fig. 1. The Performance of PCA, ICA and Factor Analysis Techniques

4.2 User-Item Interaction Perspective

As Matrix factorization algorithms are mainly used in collaborative filtering recommender systems and especially in the user-item interaction matrices. MF depends on dividing the original user- item matrix into product of two less dimensionality matrices. There are many matrix factorization models, some of them are:

Singular Value Decomposition (SVD)

SVD is ultimately used to get rid of the redundant features from the dataset [6]. As shown in the Table 2 below, using SVD is efficient and numerically robust more than forming the covariance matrix XX^T to begin with, because the creation of XX^T can cause loss of precision.

Table 2. Comparative analysis of PCA and SVD

Point of comparison	PCA	SVD
Description	Eigen-Decomposition Covariance	Singular value decomposition
Eignvalues	Calculated from the covariance matrix	Calculated from XX^T
Pre-requisites	Build the covariance matrix to begin with	Determine optimum number of features
Output	Calculate and rank the importance of features	Reduced data matrix with new values representing the importance of features
Precision	Less	More
Numeric output	Numerical issues	Numerically robust
Data type	Linear	Linear
Recommended size of data	Perform well in case of massive or light data	Perform better with massive data
Sensitivity to outliers	High	High
Loss of information	Low	High
Computational cost	High	High
Purpose	Address generic problems	Address specific problems
Scalable	Yes	Yes
Domain	Anomal detection, bioinformatics and traffic analysis	User-item related recommendation systems
Working environment	Online and offline	Recommended in online
Time consumption	Online: fast Offline: very slow	Fast
Prediction performance	Better	Worse

SVD and PCA are two eigenvalue methods used to reduce a high-dimensional dataset into fewer dimensions while keeping the important information. Eignvalues and eignvectors decomposition are applied to square matrices. For non-square matrices we apply SDV. SVD decomposes the data matrix into multiple matrices that are easier to handle and analyze. SVD's output helps to find the principle and most significant features that we want in PCA.

Principal Component Analysis (PCA)
There are two ways to apply the PCA as shown in Table 2. PCA usually achieved either by using the eigen-decomposition covariance or it can be performed using the singular value decomposition (SVD) of the data matrix. PCA mainly used for data visualization and noise reduction.

After studying the state of the art and reviewing many literature researches, it was found that PCA and SVD work towards the same goal as they are methods that perform the same functionality of dimensionality reduction. However, one can perform better than the other depending on the problem that is being addressed and the choice of the application.

Table 2 is introduced to conclude the variations between two of the most commonly used matrix factorization methods. These methods are the PCA and the SVD. This table has showed the main differences and similarities between PCA and SVD from various comparison points. Although SVD is one of the ways to apply the PCA and both of require liner data to perform adequately, SVD proposes some differences when it comes to the practical implementation than the PCA. Starting from the calculation of the eigenvalues that is followed by the calculation of the eignvectors which in case of the PCA they got calculated from the covariance matrix while in the SVD they are calculated from XX^T. Accordingly, the covariance matrix is a pre- requisite for the PCA to work while selecting the desired number of features to be represented in the final reduced matrix. This pre- requisite is mandatory for the PCA to calculate and rank the importance of features that may have numeric issues and for the SVD to build new precise and numerically robust reduced data matrix holding the new values representing the importance of features. However, in case of SVD it may cause loss of information as the pre- requisite is depending on user input rather than in the other case.

PCA does not have constraints in the size of data dealing with as it performs well in case of massive data or light weight data depending on the application [8] while SVD requires large volume of data to perform well [9]. PCA is better than SVD in finding patterns in dataset, which enabled it to have better prediction performance [9]. Both can scale and handle the missing values as they are very sensitive to outliers which is compensated in the high cost of computation. Although, the most suitable application for the SVD is in the user-item recommendation systems due to its ability to address specific problems [10, 11], however it should be connected to the internet while running in order to gain fast performance [10]. On the other hand, PCA is convenient in anomaly detection [7] and traffic analysis [12, 13] domain and can work in an offline environment but on the account of the time consumption factor. This furtherly got solved with the parallel processing but has led to less precise results.

As per the above comparative analysis, this table has illustrated that the decision making behind choosing the best technique of dimensionality reduction is dependable on multiple factors. These factors involve the choice of problem that is going to be addressed, size of data, domain that will be used and the connection status of the environment. For Movielens dataset and recommendation system that uses a collaborative filtering technique, SVD will be fitting most of our needs that will eventually access further processing applied to this dataset. Dimensionality reduction technique has shown a great intervention in the collaborative filtering recommendation systems. The research in user-item based recommendation systems especially the ones using collaborative filtering is still expanding in terms of movies recommendation domain and its raising issues. Recently the state of the art is towards enhancing the prediction accuracy and trying to overcome the major issues in collaborative flirting like scalability and sparsity [11, 14].

5 Conclusion

The dimensionality reduction techniques have showed a prominent role as a preprocessing step for the machine learning as they eliminate the irrelevant and focus on the weighted data. For that reason, preprocessing became an essential stage before applying any machine learning algorithm. Not only to ensure the integrity, accuracy and completeness, but also to reduce the dimensions of datasets and retaining only the most relevant information only. Feature extractions and feature selections had served that purpose a lot and encouraged the researchers to investigate in this area for better data quality. In this paper we have shed the light on different dimensionality techniques and matrix factorization models that could be used to enhance the quality of data before applying any further work on it. From this research we can conclude that, selecting the proper dimensionality reduction technique depends mainly on the context and the nature of data used in the application.

References

1. https://grouplens.org/datasets/movielens/. Accessed 18 Oct 2019
2. https://movielens.org/. Accessed 18 Oct 2019
3. Harper, F.M., Konstan, J.A.: The MovieLens datasets: history and context. ACM Trans. Interact. Intell. Syst. **5**(4), 19:1–19:19 (2015)
4. https://www.analyticsvidhya.com/blog/2018/08/dimensionality-reduction-techniques-python/. Accessed 20 Oct 2019
5. Al-Garadi, M.A., Mohamed, A., Al-Ali, A., Du, X., Guizani, M.: A survey of machine and deep learning methods for Internet of Things (IoT) security. arXiv preprint (2018)
6. Vozalis, M.G., Konstantinos, G.M.: A recommender system using principal component analysis. In: Published in 11th Panhellenic Conference in Informatics, pp. 271–283. Citeseer (2007)
7. Juvonen, A., Sipola, T., Hämäläinen, T.: Online anomaly detection using dimensionality reduction techniques for HTTP log analysis. Comput. Netw. **91**, 46–56 (2015)
8. Lee, Y.J., Yeh, Y.R., Wang, Y.C.F.: Anomaly detection via online oversampling principal component analysis. IEEE Trans. Knowl. Data Eng. **25**(7), 1460–1470 (2013)
9. Bokde, D., Girase, S., Mukhopadhyay, D.: Matrix factorization model in collaborative filtering algorithms: a survey. In: Proceedings of 4th International Conference on Advances in Computing, Communication and Control, pp. 136–146 (2015)
10. Zhou, X., He, J., Huang, G., Zhang, Y.: SVD-based incremental approaches for recommender systems. J. Comput. Syst. Sci. **81**(4), 717–733 (2014)
11. Zarzour, H., Al-Sharif, Z., Al-Ayyoub, M., Jararweh, Y.: A new collaborative filtering recommendation algorithm based on dimensionality reduction and clustering techniques. In: 9th International Conference on Information and Communication Systems (ICICS), Irbid, pp. 102–106 (2018)

12. Yang, H., Wang, Y.X., Xia, J.J., Zhang, Y.: China GDP prediction on traffic and transportation by PCA. In: Proceeding of the 24th International Conference on Industrial Engineering and Engineering Management, Singapore, pp. 388–394 (2018)
13. Vasan, K., Surendiran, B.: Dimensionality reduction using principal component analysis for network intrusion detection. Perspect. Sci. **8**, 510–512 (2016)
14. Nilashi, M., Ibrahim, O., Bagherifard, K.: A recommender system based on collaborative filtering using ontology and dimensionality reduction techniques. Expert Syst. Appl. **92**, 507–520 (2018)
15. Deldjoo, Y., Constantin, M.G., Eghbal-Zadeh, H., Ionescu, B., Schedl, M., Cremonesi, P.: Audiovisual encoding of multimedia content for enhancing movie recommendations. In: Proceedings of the 12th ACM Conference on Recommender Systems, RecSys 2018, Vancouver, BC, Canada, 2–7 October 2018, pp. 455–459 (2018)

A Novel Spatial Layout Representation for Object Recognition

Noha Elfiky(✉)

Saint Mary's College of California, Moraga, CA, USA
nme5@stmarys-ca.edu

Abstract. Most successful approaches on image classification apply the Bag-of-Words (BoW) approach in the context of category-level image classification. To incorporate spatial image information in the BoW model, Spatial Pyramids (SPs) are used. However, spatial pyramids are rigid in nature and are based on predefined grid configurations. As a consequence, they often fail to coincide with the underlying spatial structure of images from different categories which may negatively affect the classification accuracy.

This study proposes a technique that addresses this problem by presenting a *selective Spatial Pyramid (SP)* representation for automatically learning the most appropriate shape for each category. The proposed approach provides an image representation by inferring the constituent geometrical parts. As a result, the image representation retains the descriptive spatial information to yield a structural description of the image. From large scale experiments on the Pascal VOC2007 and Caltech101, it can be derived that SPs which are obtained by selective search outperforms the standard SPs.

Keywords: Image classification and object recognition · Bag of words · Spatial Pyramids

1 Introduction

For category-level image classification and object recognition, the Bag-of-Words (BoW) approach has been successfully applied [1–4]. The BoW is based on the occurrences of image features. Hence, it treats the image as an order-less collection of local features completely ignoring the spatial image layout. Extending the BoW with spatial information has therefore received considerable attention. Recently, several approaches consider the success of the Spatial Pyramid (SP) approach proposed by Lazebnik et al. [5]. It is shown that the use of SP outperforms the 1×1 image representation on challenging image classification tasks [5], due to the inclusion of image-to-image geometric correspondences. However, in general, SPs are based on rigid image subdivisions (e.g., grids). These rigid spatial configurations are not well suited for freely shaped objects and scenes [31].

The aim of this study is to use the scene geometry to steer the layout of spatial pyramids for category-level image classification. Images within a category usually share similar scene geometries. This study proposes a method to obtain a holistic image representation by inferring the constituent geometrical parts of a scene. In particular,

A.-E. Hassanien et al. (Eds.): AICV 2020, AISC 1153, pp. 553–563, 2020.
https://doi.org/10.1007/978-3-030-44289-7_52

the study proposes a novel approach called the Selective SPs to obtain structural image representations from the scene geometries by obtaining the spatial subdivision representation based on selective search guided by the Agglomerative Information Bottleneck (AIB) theory [8, 9]. The model is used to select the spatial geometry that best suits each category at hand.

The method to generate selective SPs will be compared to existing rigid SP for object recognition tasks and to the Generic SP method recently proposed in [31]. To this end, two benchmark data sets are used in the experiments: Pascal VOC 2007 [6] and Caltech-101 [10]. Furthermore, a large data set is provided (denoted as "stage data set") to learn each stage. This paper is organized as follows. First, in Sect. 2, the related work on spatial pyramids is discussed. In Sect. 3, the method to generate dynamic spatial pyramids is proposed. In Sect. 4, the experimental setup is discussed and the results are given. Finally, the conclusion is provided in Sect. 5.

2 Related Work

In this section, we briefly discuss the method for obtaining standard spatial pyramids [5] and for determining scene geometries (stages) [7] for obtaining Generic pyramids [31].

2.1 Rigid Spatial Pyramids

The spatial pyramid scheme proposed by [5] is a simple and computationally efficient extension of an order-less bag-of-words image representation. This approach represents an image by using weighted multi-resolution histograms which are obtained by repeatedly sub-dividing an image into increasingly finer sub-regions. Histograms are computed over the resulting sub-regions. For each resolution level, the image is subdivided into the cells of a grid. At resolution l, the grid has 2^{2l} cells. The number of points in each grid cell is then recorded.

Marszalek et al. [15] evaluate both regular and irregular grids. Further, they consider a broader set of coarse subdivisions for each dimension, such as a 1×1 grid corresponding to the standard representation of the bag of-words, a 2×2 grid (i.e. four blocks), a horizontal 3×1 grid as well as a vertical 1×3 one. They show that dividing the image plane in three horizontal (i.e. 3×1 grid) regions, provides the highest recognition performance. Further, this approach reduces the dimensionality of the conventional 4×4 (i.e. sixteen blocks) structure; from vocabularysize $\times$ 21 to vocabularysize $\times$ 8.

The use of various image layouts shows the influence of the image configuration and spatial image representation [11–13]. However, unconstrained spatial image representations have not been studied [16, 18]. Moreover, pyramids commonly applied to BoW are not designed for the specific task of categorization, due to the assumption of having fixed rigid grid representation that suits all the dataset categories. The proposed approach resolves the use of rigid spatial pyramids. The aim is to generate more natural spatial pyramids based on the underlying image geometry.

2.2 Image Segmentation to Obtain Depth Layers

For each scene geometry [21–25, 29, 30], the different image segments correspond to a scene part at a certain depth (layer). Each segment represents geometrical entities like walls, ground, and sky. The image divisions provided by the scene geometry models will be used to learn the best geometry that suits each category of concern. Segmentation is based on the occurrence probability in the training set. Ground truth is obtained by manual annotation, thereby dividing the training set according to the scene geometry patterns, and fitting the parameters of each geometry model (horizon, vanishing points) such as to visually best fit the underlying data. For this purpose, the stages data set described in Sect. 4.1 will be used for obtaining the segmentation masks used to represent each scene geometry.

More precisely, suppose that an image belongs to stage S, which is composed of N layers, correspondingly there will be N mask maps. The mask map for the ith partition T_i is obtained by taking the average of the mask maps for each image:

$$T_i(x) = \frac{\sum_{j=1}^{n} M_{j,i}(x)}{n}, \tag{1}$$

where n is the total number of images in the training data set, and $M_{j,i}(x)$ is the mask map of the jth image for i^{th} partition. Note that $M_{j,i}(x)$ is an indicator function: $M_{j,i}(x) = 1$, if $\boldsymbol{x}$ belongs to the ith partition and 0 otherwise.

Segmentation: Mask maps are used to automatically divide the images. Assuming that the images of a stage can be partitioned into N layers, there exist N mask maps corresponding to the partitions in the training data set. Then, the binary mask map is defined as follows:

$$\mathrm{T}'_i(\mathrm{x}) = \begin{cases} 1, & \mathrm{T}_i(\mathrm{x}) = \max_{\mathrm{j=1}}^{\mathrm{N}} \mathrm{T_j}(\mathrm{x}) \\ 0, & otherwise. \end{cases} \tag{2}$$

As a consequence, the values in the mask map are either 0 or 1. In the next section, scene geometry (i.e. scene depth) maps will be used in order to achieve a proper selection of the spatial partitioning that suit each object category of concern.

3 The Proposed Selective Spatial Layout Representation

The SP scheme proposed by [5] represents an image by using weighted multi-resolution histograms which are obtained by repeatedly sub-dividing an image into increasingly finer sub-regions, where the spatial pyramid at level $l \in \{0, \ldots, L\}$ has $R(l) = 2^{2l}$ sub-regions. For image X, all features are assigned to their best visual word v selected from a vocabulary V. The frequency of v inside sub-region i of image X is given by the histogram bin $H_X^i(v)$. The similarity or matching rate between images X and Y at level l, is given by the histogram intersection function [17]:

$$I^l(X,Y) = \sum_{i=1}^{R(l)} \sum_{v=1}^{|V|} \min\left(H_X^i(v), H_Y^i(v)\right). \tag{3}$$

Matches found at finer resolutions are closer to each other in the image space and are therefore more heavily weighted. To accomplish this, each level is weighted to $1/2L - l$ which results in the final SP:

$$\kappa^L(X,Y) = \frac{1}{2^L} I^0(X,Y) + \sum_{l=1}^{L} \frac{1}{2^{L-l+1}} I^l(X,Y). \tag{4}$$

For the geometry-driven pyramid we use the same approach, only for geometric equivalent classes. Formally, for each scene geometry or stage s of n stage types, let $R(s)$ denote the number of sub-regions of s. Instead of using a fixed pyramid, we propose that a spatial stage pyramid is created by computing the similarity between images X and Y for stage s by

$$I^s(X,Y) = \sum_{i_s=1}^{R(s)} \sum_{v=1}^{|V|} \min\left(H_X^{i_s}(v), H_Y^{i_s}(v)\right). \tag{5}$$

Where the different sub-regions i_s for stage s correspond to a scene part at a certain depth (layer). For each stage s there are $R(s)$ sub-regions.

Next, the proposed Selective Spatial Pyramid approach for selecting the appropriate spatial image representation for each category will be explained. This is achieved by learning a proper class-specific spatial model based on information theory. These models encode the proper spatial partitioning for each category; which is used further to obtain class-specific spatial models.

3.1 Selective Spatial Pyramids

In this section, we propose an approach for learning the spatial partitioning based on the Agglomerative Information Bottleneck theory (AIB) [8, 9]. In fact, the main goal of AIB is to reduce the dictionary of visual words V required for representing the categories Z. This means generating a compact set of words $\hat{V}$ from the original dictionary V so that the loss of mutual information to the categories Z is minimal:

$$I(V\,;\,Z) = D_{KL}[p(v,\,z)||p(v)\,p(z)] \tag{8}$$

The functional $D_{KL}[p(v,\,z)||p(v)\,p(z)]$ is the Kullback-Leiber divergence. The joint distribution $p(v,z)$ is estimated from the training set by counting the number of occurrences of each visual word in each category. The information about v captured by z can be measured by the mutual information.

$$I(V,Z) = \sum_{k}\sum_{c} p(v_k, z_c) log \frac{p(v_k, z_c)}{p(v_k)p(z_c)}, \tag{9}$$

Measuring the discriminative power I that one random variable carries about the other. The merging of visual words is achieved by iteratively applying AIB for fusing those two visual words v_k and v_j into $\hat{v}$ that causes the smallest decrease $\mathrm{I}(\hat{V};\mathrm{Z})$ in the original mutual information $I(V;Z)$:

$$\delta I_z(V;\hat{V}) = I(V;Z) - I(\hat{V};Z). \tag{10}$$

At each step, AIB performs the best possible merge $argmin_{\hat{V}} = \delta I_Z(V;\hat{V})$. We extend the original AIB to learn the most appropriate class-specific spatial structure. Concretely, we first over-segment the images as [19]. For each category, we learn on the validation set the most suitable split-up by pruning the least informative image segments based on AIB. Hence, we relate each image segment $((i_s)$ with the data set categories Z, such that the probabilities of each segment $p(i_s)$ is calculated by summing up the probabilities of the vocabularies (v_u) it contains as:

$$p(i_s) = \sum_{i_s=1}^{R} \sum_{u=(i_s-1)|V|+1}^{k|V|} p(v_u), \tag{11}$$

where i_s is the segment index, R is the total number of image segments, u is vocabulary index within segment i_s. To evaluate the Discriminative Power (DP) of the generated segments w.r.t. the categories, we use the information content criteria:

$$I(S,Z) = \sum_{i_s=1}^{R}\sum_{c=1}^{|C|} p(i_s, z_c) log \frac{p(i_s, z_c)}{p(i_s)p(z_c)}, \tag{12}$$

where c is the category index, and C is the total number of data set categories. Finally, the loss in DP is then obtained by:

$$\delta I_z(S;\hat{S}) = I(S;Z) - I(\hat{S};Z). \tag{13}$$

Formally, for each category the best spatial split-up (partitioning) $s \in S$ is learned. Each split-up s has a number of subregions $R(s)$. We propose that the spatial partitioning matching function is created by computing the similarity between images X and Y for partitioning s by:

$$I^s(X,Y) = \sum_{i_s=1}^{R(s)}\sum_{v=1}^{|V|} \min\left(H_X^{i_s}(v), H_Y^{i_s}(v)\right). \tag{14}$$

where the different sub-regions i_s for the partitioning s correspond to a different image segment, such as sky, ground, etc. The whole process is demonstrated by the block diagram in Fig. 1.

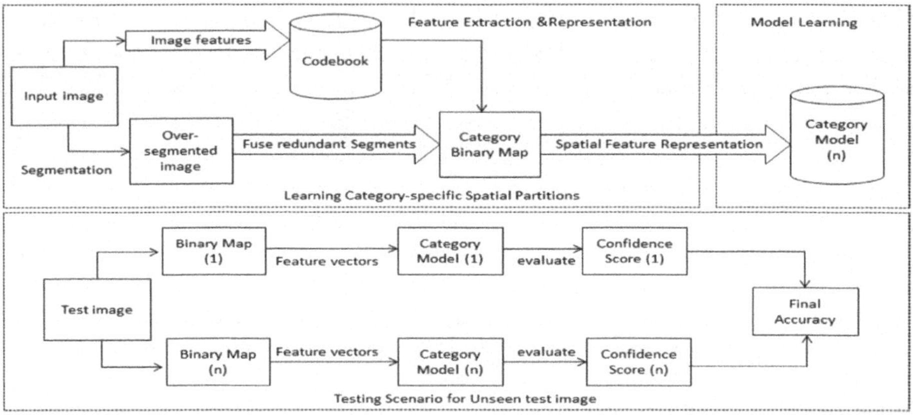

Fig. 1. Outline of class-specific spatial template learning using ground-truth. Note that the codebook models and the class-specific models are obtained off-line. For a test image, we evaluate its performance using the learned class specific models.

4 Experiments

In this section, the proposed method to generate flexible spatial pyramids will be compared to the existing state-of art rigid based pyramids in the context of object recognition. In Sect. 4.1, the data sets used in all experiments are given. The experimental setup used is shown in Sect. 4.2.

4.1 Data Sets

Three independent data sets are used in the experiments. The first data set is a large dataset consisting of 3589 images classified as 15 different categories representing the standard generic scene geometries. 151 "sky + background + ground", 333 "background + ground", 81 "sky + ground", 212 "ground", 139 "ground + Diag-BkgLR", 132 "ground + DiagBkgRL", 75 "diagBkgLR", 71 "diagBkgRL", 84 "box", 57 "1sidewallLR", 69 "1sidewallRL", 266 "corner", 960 "persBkg", 833 "noDepth", and 126 "tabPersBkg". Images are take are under large variety of lighting conditions and imaging conditions (including indoor, outdoor, desert, cityscape, and other settings). We refer to this data set as "stages data set". This data set is used to generate the binary mask maps used by the Generic SPs approach [31]. As shown in Fig. 2, the scene structures of the stage models are shown in different colors. The stage models are used to determine how the image is divided in sub-regions. For instance, images of stage sky +backgnd+gnd are divided into three layers: sky (in blue), background (in yellow) and ground (in brown).

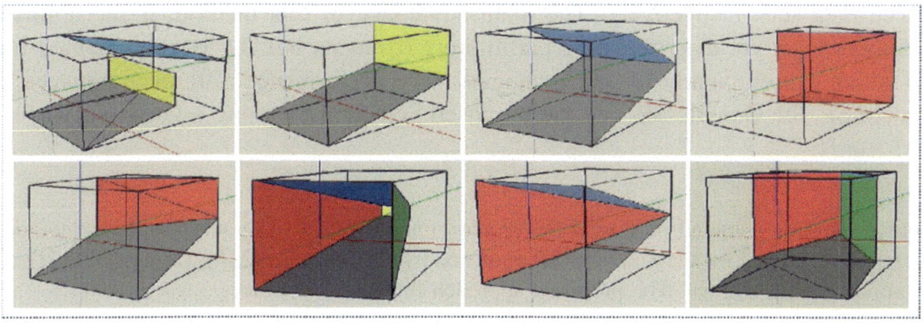

Fig. 2. Stage models and their corresponding instantiations. Top row, from left to right: "sky +backgnd + gnd", "backgnd+gnd", "sky + gnd", "gnd + diagalBackgndLR". Bottom row: "diagalBackgndLR", "box", "1side-wallLR", "corner". This figure is taken from [14].

We also use Caltech 101 [10] and Pascal 2007 [6] datasets as benchmark data sets for evaluating our approach. The Pascal VOC 2007 data set [6] consists of 9963 images of 20 different classes with 5011training images and 4952 testing images. The Caltech-101 data set contains 9144 images of 102 different categories.

4.2 Experimental Setup

To compare the different spatial pyramids, a standard BoW image classification approach is used. SIFT features [20] of 16 × 16 pixel patches are used. For Caltech-101, we use 30 images per category for the training and 50 for testing. The general architecture follows [5]. The SIFT descriptors are extracted on a dense grid rather than interest points, as this procedure has been shown to yield superior performance for scene classification [3]. The codebook used is of size 300. Experiments are conducted over 10 random splits of the data, and the average per-class recognition rates are recorded for each run. The final result is reported as the mean accuracy and its standard deviation from the individual runs. For Pascal 2007, a standard multi-scale grid detector is taken together with a Harris-Laplace point detector [2], and a blob detector. SIFT descriptors are computed for all regions in the feature descriptor step, which are then quantized to a codebook of size 1000 visual-words. We used the standard average precision (AP) criteria to validate the performance on the object recognition task. The average precision is equivalent to the area under a precision-recall curve. Mean average precision (MAP) is used to evaluate the performance of the features over all the data set categories, which is obtained by averaging the AP over all categories. In this study, the proposed Selective SP is compared with the standard three-levels Spatial Pyramid SP proposed by Lazebnik et al. [5] and the Generic SP proposed in [31], as the baseline.

4.3 Selective Spatial Pyramids (Selective SPs)

In this experiment, the Selective SP proposed in Sect. 3 is evaluated, in which the most appropriate "partitioning" is learned by eliminating the least informative partitions. The quantitative results, in Table 1, show an improvement of 6.4% and 8.9% on Pascal and

Caltech data sets, respectively. The main advantage of this method, is the ability to learn a compact yet discriminative partitioning (from its initially oversegmented images) which can efficiently fits each category. The performance difference between the standard SP, the Generic SP, and the proposed Selective SP using a vocabulary of size 1000 words.

Table 1. Results obtained on Pascal (MAP score) and Caltech (Average per class recognition rates) data sets based on the Standard SP proposed by Lazebnik et al. [5], Generic SP [31], and the proposed Selective SP. The proposed approach improves the scores.

Method	Pascal	Caltech
Lazebnik et al. [5]	51.5%	64.6%
Generic SP [31]	55.9% (+4.4%)	70.5% (+5.9%)
Selective SP	**57.9% (+6.4%)**	**73.6% (+8.9%)**

Table 2 shows a relative performance improvement for most of the classes of the PASCAL VOC 2007 data set for having dynamic pyramid representation based on the proposed Selective SP approach compared to those obtained using the standard three-level SP and the Generic SP on object recognition.

Table 2. Average-Precision Results for all classes of the PASCAL VOC 2007 database. Comparison on the average accuracy of the original three level pyramid representation. The third row shows the novel pyramid representation results using the proposed Selective SP approach.

	Plane	Bike	Bird	Boat	Bottle	Bus	Car	Cat	Chair	Cow	Table
Pyramid	68.4	51.8	37.7	64.4	19.8	44.2	71.8	51.9	47.9	31.7	39.7
Generic	74.0	57.1	45.4	66.7	26.2	54.2	75.0	57.2	46.0	38.3	41.6
Selective	75.6	57.9	46.4	67.2	28.0	56.0	77.0	59.0	48.0	40.0	42.6

	Dog	Horse	Mbike	Person	Plant	Sheep	Sofa	Train	Tv	Mean
Pyramid	36.8	71.6	54.5	77.9	17.4	30.5	38.7	66.4	44.9	48.3
Generic	38.4	71.0	57.2	77.7	26.5	40.6	40.1	72.7	46.2	52.5
Selective	39.5	73.0	59.0	79.2	28.3	42.2	42.1	74.2	47.8	54.2

4.4 Comparison with State-of-the-Art

Comparison with previously published results obtained using one type of descriptor are shown in Table 3. We compare our approach with other recent work described in [26–28] which achieves comparable results with our approach using the standard spatial partitioning. The main difference between our work and the work in [26–28] is that we use the classical bag-of-words approach with the standard K-means vector quantization within our novel spatial pooling approach. In contrast, they use an alternative coding scheme called Linear Coordinate Coding (based on sparse code and max pooling), which is shown recently to significantly outperforms the standard vector quantization scheme, leading to state-of-the-art performance on several benchmark datasets using a single type of descriptor. Therefore, we investigate the extension our proposed approach with the aforementioned new coding schemes. We show that the performance of our approach increases by using better coding (i.e., sparse codes) scheme.

Table 3. Results obtained by several recognition schemes using a single type descriptor and intersection kernel on Caltech data set, see text for details. The numbers shown inside brackets in [26] are the codebook sizes used in this work.

	Method	Caltech
Boureau et al. [26]	SP + hard quantization [256]	64.2 ± 1.0
Boureau et al. [26]	SP + hard quantization + max pooling [256]	64.3 ± 0.9
Lazebnik et al. [5]	SP + hard quantization + kernal SVM	64.6 ± 0.8
Boureau et al. [26]	SP + soft quantization [512]	66.1 ± 1.2
Zhang et al. [27]	kNN-SVM	66.2 ± 0.5
Boureau et al. [26]	SP + sparse codes [1024]	70.3 ± 1.3
Boiman et al. [28]	Nearest neighbor + spatial correspondences	**70.4**
This paper	Selective SP + hard quantization [300]	**73.6** ± **1.5**(+4.5%)
This paper	Selective SP + sparse codes [1024]	**79.8** ± **1.2**(+13.4%)

5 Conclusion

Spatial Pyramids have been proposed which are steered by the 3D scene geometry. The geometry of a scene is measured based on image statistics taken from a single image. After the estimation of the scene geometry, the corresponding SP is selected as the geometrical representation. From large scale experiments on the Pascal VOC2007 and Caltech101, it can be derived that SPs which are obtained by selective search outperforms the standard SPs for Pascal VOC 2007 and Caltech101. Selective SP has the highest computational complexity, although it generates compact yet efficient spatial partitioning for each category. On the other hand, Generic SP approach balances between accuracy, dimensionality, and speed on both training and testing phases.

For future work, the proposed system will be extended to automatically learn a hierarchal class-specific adaptive shape model, where the highest levels will incorporate the important localization and/or segmentation knowledge for efficiently capturing the ROI, for restricting the objects spatial location to work with. In addition, the proposed system will be extended to integrate the spatial layout with primitive features like shape, texture and color to reduce the semantic gap. Finally, the proposed system will be extended to improve its computational complexity.

References

1. Dance, L., Fan, J., Willamowski, C., Bray, C.: Visual categorization with bags of keypoints. In: European Conference on Computer Vision Workshop on Statistical Learning in Computer Vision (2004)

2. Mikolajczyk, K., Schmid, C.: A performance evaluation of local descriptors. Trans. Pattern Anal. Mach. Intell. **27**(10), 1615–1630 (2005)
3. Fei-Fei, L., Perona, P.: A bayesian hierarchical model for learning natural scene categories. In: Computer Vision and Pattern Recognition (2005)
4. Zhang, J., Marszalek, M., Lazebnik, S., Schmid, C.: Local features and kernels for classification of texture and object categories: an in-depth study: a comprehensive study. Int. J. Comput. Vis. **73**(2), 213–218 (2007)
5. Lazebnik, S., Schmid, C., Ponce, J.: Beyond bags of features: spatial pyramid matching for recognizing natural scene categories. In: Computer Vision and Pattern Recognition (2006)
6. Everingham, M., Gool, L.V., Williams, C.K.I., Winn, J., Zisserman, A.: The pascal visual object classes challenge 2007 results (2007)
7. Nedovic, V., Smeulders, A.W.M., Redert, A., Geusebroek, J.-M.: Stages as models of scene geometry. IEEE Trans. Pattern Anal. Mach. Intell. **32**(9), 1673–1687 (2010)
8. Slonim, N., Tishby, N.: Agglomerative information bottleneck. In: Neural Information Processing Systems (1999)
9. Fulkerson, B., Vedaldi, A., Soatto, S.: Localizing objects with smart dictionaries. In: European Conference of Computer Vision (2008)
10. Fei-Fei, L., Fergus, R., Perona, P.: Learning generative visual models from few training examples. In: Computer Vision on Pattern Recognition Workshop (2004)
11. Hoiem, D., Efros, A.A., Hebert, M.: Geometric context from a single image. In: International Conference on Computer Vision, pp. 654–661 (2005)
12. Delage, E., Lee, H., Ng, A.Y.: A dynamic bayesian network model for autonomous 3D reconstruction from a single indoor image. In: Computer Vision and Pattern Recognition, pp. 2418–2428 (2006)
13. Sudderth, E., Torralba, A., Freeman, W., Willsky, A.: Depth from familiar objects: a hierarchical model for 3D scenes. In: Computer Vision on Pattern Recognition, pp. 2410–2417 (2006)
14. Nedovic, V., Smeulders, A., Redert, A., Geusebroek, J.-M.: Depth information by stage classification. In: International Conference on Computer Vision (2007)
15. Marszalek, M., Schmid, C., Harzallah, H., van de Weijer, J.: Learning object representation for visual object class recognition. In: International Conference on Computer Vision on Visual recognition Challenge Workshop (2007)
16. van Gemert, J.: Exploiting photographic style for category-level image classification by generalizing the spatial pyramid. In: International Conference on Machine Learning (2011)
17. Grauman, K., Darrell, T.: The pyramid match kernel: discriminative classification with sets of image features. In: International Conference on Computer Vision (2005)
18. Rui, L., Gijsenij, A., Gevers, T., Nedovic, V., De, X., Geusebroek, J.: Color constancy using 3D scene geometry. In: International Conference on Computer Vision (2009)
19. Moore, A.P., Prince, S.J.D., Warrell, J., Mohammed, U., Jones, G., Superpixel lattices. In: Conference on Computer Vision and Pattern Recognition (2008)
20. Lowe, D.: Distinctive image features from scale invariant keypoints. Int. J. Comput. Vis. **60** (2), 91–110 (2004)
21. Gehler, P.V., Nowozin, S.: On feature combination for multiclass object classification. In: International Conference on Computer Vision (2009)
22. Xiao, J., Hays, J., Ehinger, K., Oliva, A., Torralba, A.: Largescale scene recognition from abbey to zoo. In: Computer Vision and Pattern Recognition (2010)
23. Khan, F., van de Weijer, J., Vanrell, M.: Top-down color attention for object recognition. In: International Conference on Computer Vision (2009)
24. Su, Y., Jurie, F.: Visual word disambiguation by semantic contexts. In: International Conference on Computer Vision (2011)

25. Sharma, G., Jurie, F.: Learning discriminative spatial representation for image classification. In: British Machine Vision Conference (2011)
26. Boureau, Y., Bach, F., LeCun, Y., Ponce, J.: Learning midlevel features for recognition. In: Computer Vision and Patten Recognition (2010)
27. Zhang, H., Berg, A.C., Maire, M., Malik, J.: SVM-KNN: discriminative nearest neighbor classification for visual category recognition. In: Computer Vision and Pattern Recognition (2006)
28. Boiman, O., Rehovot, I., Shechtman, E., Irani, M.: In defense of nearest-neighbor based image classification. In: Computer Vision and Pattern Recognition (2008)
29. Yang, J., Yu, K., Gong, Y., Huang, T.: Linear spatial pyramid matching using sparse coding for image classification. In: Computer Vision and Pattern Recognition (2009)
30. van de Sande, K.E.A., Gevers, T., Snoek, C.G.M.: Evaluating color descriptors for object and scene recognition. Trans. Pattern Anal. Mach. Intell. **32**(9), 1582–1596 (2010)
31. Elfiky, N.: Application of analytics in machine vision using Big Data. Asian J. Appl. Sci. **7** (4), 376–385 (2019)

Cell Blood Image Segmentation Based on Genetic Algorithm

A. Y. Ayoub[1], M. A. El-Shorbagy[1,2(✉)], I. M. El-Desoky[1], and A. A. Mousa[1,3]

[1] Department of Basic Engineering Science, Faculty of Engineering, Menoufia University, Shebin El-Kom, Egypt
mohammed_shorbagy@yahoo.com

[2] Department of Mathematics, College of Science and Humanities in Al-Kharj, Prince Sattam bin Abdulaziz University, Al-Kharj 11942, Saudi Arabia

[3] Mathematics and Statistics Department, College of Science, Taif University, Taif, Saudi Arabia

Abstract. There are many methods of clustering that have been used to divide images. *k*-means algorithm is considered the most popular method of cluster analysis. Due to disadvantages of *k*-means algorithm, in this paper, the image is segmented using modified genetic algorithm (GA) by *k*-means algorithm; where *k*-means is used as an initialization of GA. The proposed algorithm was applied to several of cell blood images from microscope and the results showed that the value of *PSNR* for the proposed algorithm is higher than other algorithms, which indicates its efficiency in image segmentation.

Keywords: Image segmentation · Genetic algorithm · Cluster analysis · *k*-means algorithm

1 Introduction

Image Segmentation (IS) is one of the famous ways to split the pixels. The images are divided into different regions so that the pixels are similar in the same area while there is a large difference between regions. However, there is no algorithm that divides images efficiently. IS is very important in many fields such as image processing, medical field, traffic signs, pattern recognition, etc.

In medical field, IS is used in some of the applications [1, 2] such as: border detection in angiograms of coronary, surgical planning, simulation of surgeries, detection and segmentation of tumor, the study of brain development, functional mapping, automated classification of blood cells, mass detection in mammograms, image registration, segmentation of cardiac images analysis, etc. while, in traffic signs, IS is used for automatic traffic sign detection and classification [3, 4].

Many traditional methods are used to perform IS such as threshold based, edge based, cluster based, neural network based [5]. Cluster analysis is one of the traditional methods for IS. *k*-means algorithm is one of the most important methods for cluster analysis. *k*-means algorithm has many disadvantages [6] such as: (1) As a result of using random initial center of clusters may lead to the emergence of some empty

A.-E. Hassanien et al. (Eds.): AICV 2020, AISC 1153, pp. 564–573, 2020.
https://doi.org/10.1007/978-3-030-44289-7_53

clusters, (2) The presence of extreme points in the final clusters leads to non-optimal distribution and (3) There is no optimal method to determine the number of clusters. As a result of these disadvantages, researchers have resorted to evolutionary methods to perform IS such as GA [7, 8], particle swarm optimization (PSO) [9, 10], ant colony optimization (ACO) [11, 12], fuzzy clustering [13, 14], differential evolution (DE) [15, 16], etc.

For instance, Amelio and Pizzuti [8] presented the GA for color image segmentation that adopts the representation of any image as a graph. Dhanalakshmi et al. [10] presented a novel method for image processing using PSO technique. Gupta et al. [11] solved image processing problem based on ACO with a reference to a new automatic enhancement technique based on real-coded particle ant colony. Chen and Ludwig [13] implemented Fuzzy C-Regression Model (FCRM) to color segmentation using Berkeley's segmentation database. Pei et al. [15] proposed a new method based on the maximum between-cluster variance method and DE to perform IS.

Stemming from the above, in this paper an advanced algorithm based on GA to perform IS is proposed. The proposed algorithm is a hybrid between one of the evolutionary methods, GA, and the most important methods for cluster analysis, *k*-means algorithm.

The paper is organized as follows: Sect. 2 illustrates a brief explanation of *k*-means algorithm and GA. In Sect. 3 the modified GA by *k*-means algorithm for cluster analysis is discussed. The proposed algorithm is discussed in Sect. 4. Experimental results are discussed in Sect. 5. Finally, Sect. 6 presents our conclusion.

2 Preliminary

2.1 *k*-means Algorithm

k-means algorithm is one of the most important methods to solve the problem of cluster analysis. The steps of *k*-means algorithm are described as the following: Firstly, the number of clusters (k) is inserted and cluster centers $\left(z_j \forall j = 1, 2, \ldots, k\right)$ are selected randomly [17]. Then, the distances between each element $(x_i \forall i = 1, \ldots, n)$ in the data and the centers of the clusters $\left(z_j \forall j = 1, 2, \ldots, k\right)$ are measured. According to these distances, each element $(x_i \forall i = 1, \ldots, n)$ is assigned to the cluster c_j if the following equation is achieved:

$$\left\|x_i - z_j\right\| \prec \left\|x_i - z_p\right\| \forall p = 1, 2, \ldots, k, \text{ and } \forall j \neq p; \tag{1}$$

Where z_j and z_p are the centers of the clusters c_j and c_p respectively.

Finally, the cluster centers are updated according to the following equation:

$$z_j^* = \frac{\sum_{x_i \in c_j} x_i}{n_i}; \tag{2}$$

Where n_i is the number of elements in the cluster c_j. These steps are repeated until the number of iterations is completed or when the centers are not changed.

2.2 Genetic Algorithm

The genetic algorithm [19] is one of the evolutionary methods to find the optimal solution in the problems of optimization. The initial population of GA is randomly generated and then its GA operators (selection, crossover and mutation) are applied to obtain a new generation of population. This process continues until the maximum number of iterations is reached or there is no improvement in the fitness function for a number of consecutive iterations.

3 Modified Genetic Algorithm by *k*-means Algorithm

This paper presents an advanced algorithm for IS based on GA. The initial population of GA is generated using *k*-means algorithm and the GA operators are applied. The steps of the advanced algorithm are discussed with more details, in the following:

Step 1: Population Initialization
The first generation of the population in the genetic algorithm is created using the *k*-means algorithm so that in successive attempts, the new generation of population is better than all *k*-means solutions [18, 19].

Step 2: Evaluation
The fitness function of clustering is calculated for each individual according to the following equation [17]:

$$\text{Minimization } SSE(C_1, C_2, \ldots, C_k) = \sum_{j=1}^{k} \sum_{x_i \in C_j} \left\| x_i - z_j \right\|; \quad (3)$$

Where $(C_1, C_2, \ldots, C_k)$ are the clusters to which the data is divided, x_i is the data belong to the cluster C_j and Z_j is the center of the cluster C_j.

Step 3: Tournament Selection (TS)
In tournament selection, *m* chromosomes are selected randomly. The chromosome with the appropriate value of fitness function is selected and will be put in the next generation [20]. The number of randomly selected chromosomes is called tournament size. In this algorithm, binary tournament selection is proposed. Two chromosomes are chosen randomly and select the chromosome with the best value of *SSE*. Figure 1 illustrates the mechanism of TS used in this algorithm.

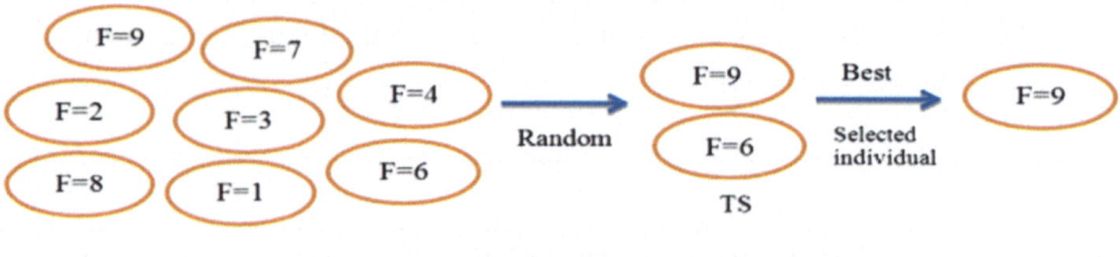

Fig. 1. Binary tournament selection.

Step 4: Crossover Operator
Single point crossover is employed with probability P_c [18, 19].

Step 5: Mutation Operator
This operator starts by choosing random cluster C_R. The nearest cluster C_N to C_R is selected. A point X_R in C_R, which is closer to the center of C_N and far from the center of C_R at the same time, is determined. A point X_N in C_N, which is closer to the center of C_R and far from the center of C_N at the same time, is determined. In the final step, X_R is placed in C_N and X_N in C_R as shown in Fig. 2 [18, 19].

Step 6: Elitist Strategy
A new generation of population is made up from the best population of the old population and the offspring [18, 19].

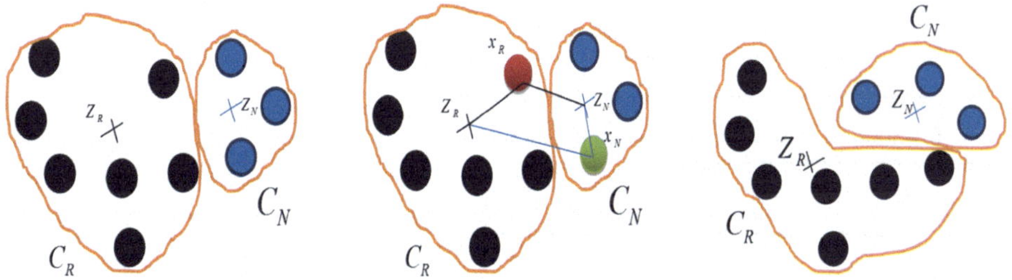

Fig. 2. Mutation operator.

Step 7: Stopping Criterion
The algorithm is terminated when the number of specified iterations expires or when the center of clusters do not update for a number of consecutive iterations [18, 19].

4 Using the Modified Genetic Algorithm by *k*-means to Perform Image Segmentation (IS)

The methodology of using the advanced algorithm to perform IS is shown in Fig. 3. In the first step, the image is loaded and converted it to a set of pixels. These pixels divided into three channels Red (R), Green (G) and Blue (B). After that, the local histogram equalization is used to improve the quality and contrast of each channel separately because IS is often applied to low-contrast and poor quality images according to the following steps:

- Load image.
- Assign window size.
- Padding of zeros.

- Choose first window.
- Find cumulative distributive function value of each pixel value in that window.
- Replace the pixel value with Cumulative distributive function value.
- Continue the process for every window

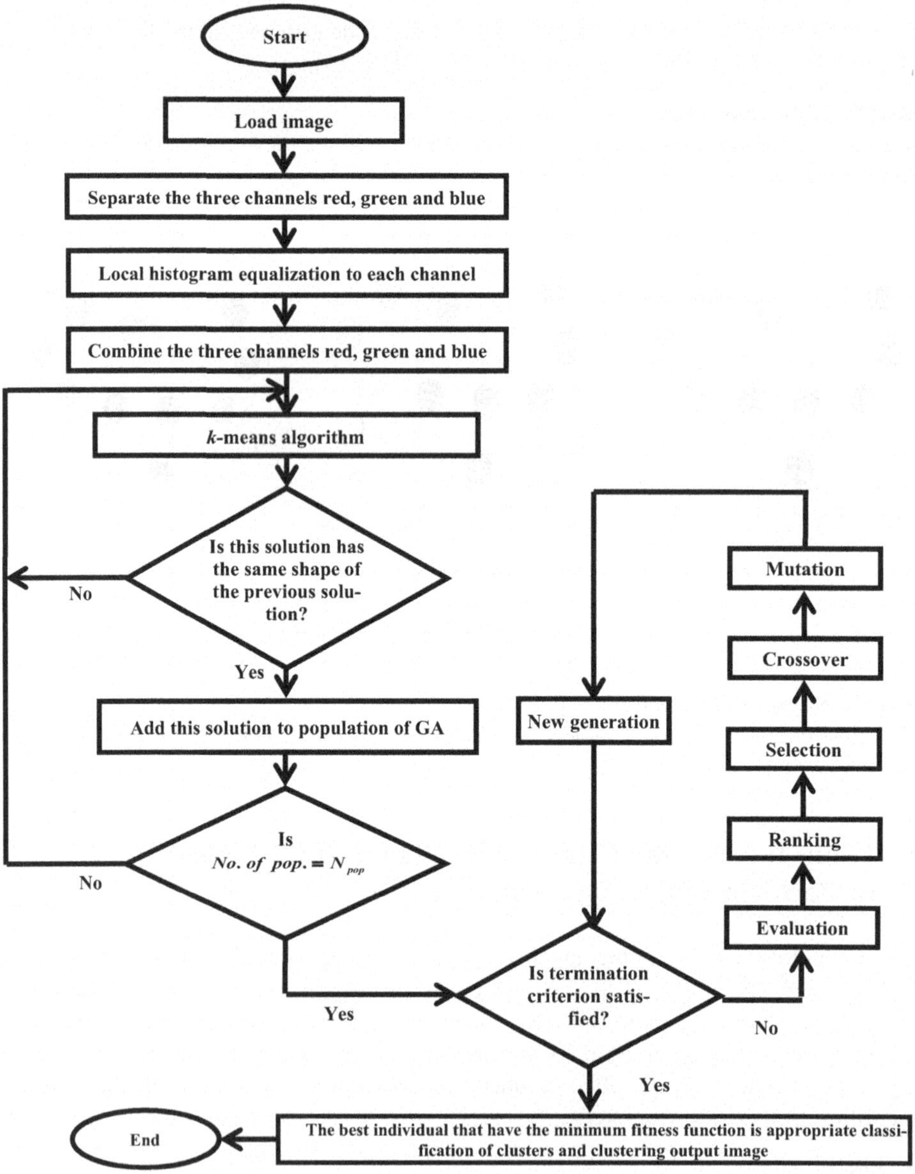

Fig. 3. Flow chart of the proposed algorithm.

The final step in this algorithm, improved image is segmented using the modified GA by *k*-means algorithm that described in Sect. 3. In the final step, reshape the cluster into output image.

5 Experimental Results and Comparisons

The proposed algorithm was applied to a several of cell blood images to evaluate its performance on IS [5]. The measurement values of mean square error (*MSE*) and the peak signal to noise ratio (*PSNR*) are used to evaluate segmented image quality. Equations 4 and 5 illustrate the values of *MSE* and *PSNR*.

$$MSE = \frac{\sum_{m,n} [r(x,y) - t(x,y)]^2}{m.n}; \tag{4}$$

$$PSNR = 10 \log 10\left(\frac{R^2}{MSE}\right); \tag{5}$$

where *r(x, y)* is the original image and *t(x, y)* is the clustering image, *m* and *n* are the number of rows and columns in the original images and *R* is the maximum fluctuation in the original image data type. The higher value of *PSNR* reflects the higher quality of the segmented image.

Figure 4 shows the original images and the segmented images for cell blood images.

The comparison between the values of *PSNR* for different algorithms is shown in Table 1. From Table 1, the value of *PSNR* for the proposed algorithm is higher than other algorithms. This indicates the superiority of the proposed algorithm to other algorithms to perform IS.

Tables 2 and 3 show the rank of the different algorithms used to segmented the images according to the value of *PSNR*. While, Fig. 5 shows a comparison between the different algorithms according to its ranks. According to Table 3 and Fig. 5, the proposed algorithm obtained the first rank to segment the images between the different algorithms. Therefore, we can say that the proposed algorithm gives good results for IS better than others algorithms for all images used.

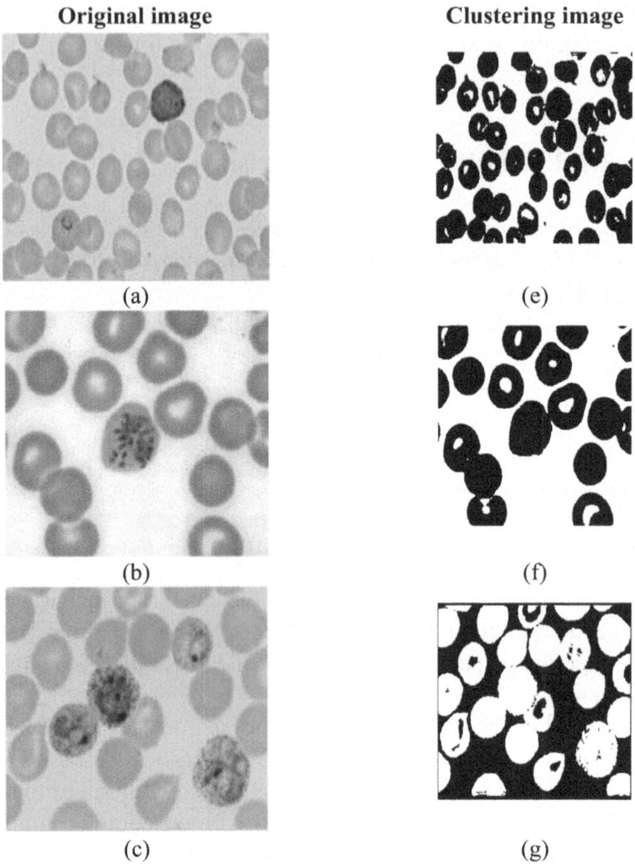

Fig. 4. Original and clustering images for cell blood samples.

Table 1. Comparison between *k*-means algorithm [5], *k*-means clustering algorithm and subtractive clustering algorithm [5] and the proposed algorithm according to *PSNR* value.

Image	*k*-means algorithm [5]	*k*-means clustering algorithm and subtractive clustering [5]	Proposed algorithm
First blood sample	34.63	34.80	34.83
Second blood sample	32.44	34.79	34.81
Third blood sample	33.99	34.43	34.52

Table 2. Ranking of the PSNR values for images for all algorithms

Image	*k*-means algorithm [5]	*k*-means clustering algorithm and subtractive clustering [5]	Proposed algorithm
First blood sample	Rank 3	Rank 2	Rank 1
Second blood sample	Rank 3	Rank 2	Rank 1
Third blood sample	Rank 3	Rank 2	Rank 1

Table 3. Statistics of ranking

Method	Rank 1	Rank 2	Rank 3
k-means algorithm [5]	0	0	3
k-means clustering algorithm and subtractive clustering [5]	0	3	0
Proposed algorithm	3	0	0

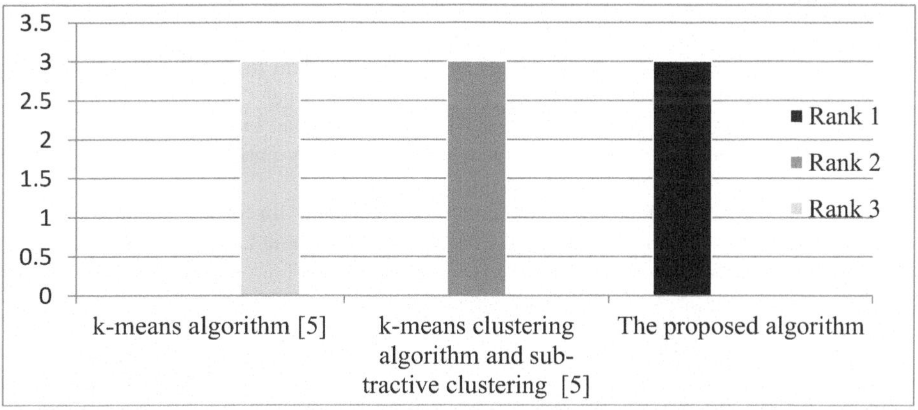

Fig. 5. The comparison between the different algorithms according to its ranks.

6 Conclusion

In this paper, a new method of image segmentation using a genetic algorithm was introduced. The proposed algorithm is a modified genetic algorithm by *k*-means algorithm; where *k*-means is used as an initialization of GA. Also, a new mechanism for mutation in GA is used. Finally, the proposed algorithm was applied on a set of cell blood images. The results proved the superiority of the proposed algorithm to perform image segmentation according to *PSNR* that used to evaluate the segmented image quality.

References

1. Norouzi, A., Rahim, M.S., Altameem, A., Saba, T., Rad, A.E., Rehman, A., Uddin, M.: Medical image segmentation methods, algorithms, and applications. IETE Tech. Rev. **31**, 199–213 (2014)
2. Jose, A., Ravi, S., Sambath, M.: Brain tumor segmentation using k-means clustering and fuzzy C-means algorithms and its area calculation. Int. J. Innov. Res. Comput. Commun. Eng. **2**, 3496–3501 (2014)
3. Khan, J.F., Bhuiyan, S.M.A., Adhami, R.R.: Image segmentation and shape analysis for road-sign detection. IEEE Trans. Intell. Transp. Syst. **12**, 83–96 (2011)
4. Brunoa, L., Parla, G., Celauro, C.: Improved traffic signal detection and classification via image processing. Soc. Behav. Sci. **53**, 811–821 (2012)
5. Dhanachandra, N., Manglem, K., Chanu, Y.J.: Image segmentation using k-means clustering algorithm and subtractive clustering algorithm. Procedia Comput. Sci. **54**, 764–771 (2015)
6. Singh, K., Malik, D., Sharma, N.: Evolving limitations in k-means algorithm in data mining and their removal. IJCEM Int. J. Comput. Eng. Manag. **12**, 105–109 (2011)
7. Bosco, G.L.: A genetic algorithm for image segmentation. In: Proceedings 11th International Conference on Image Analysis and Processing, Palermo, Italy (2001)
8. Amelio, A., Pizzuti, C.: A genetic algorithm for color image segmentation. In: Proceedings of the 16th European conference on Applications of Evolutionary Computation, vol. 1, pp. 314–323 (2013)
9. Gautam, K., Singhai, R.: Color image segmentation using particle swarm optimization in lab color space. Int. J. Eng. Dev. Res. (2018 IJEDR) **6**, 373–377 (2018)
10. Dhanalakshm, L., Ranjitha, S., Suresh, H.N.: A novel method for image processing using particle swarm optimization technique. In: 2016 International Conference on Electrical, Electronics, and Optimization Techniques (ICEEOT), Chennai, India (2016)
11. Gupta, K., Gupta, A.: Image enhancement using ant colony optimization. IOSR J. VLSI Signal Process. (IOSR-JVSP) **1**, 38–45 (2012)
12. Wang, X., Feng, Y., Feng, Z.: Ant colony optimization for image segmentation. In: Proceedings of the Fourth International Conference on Machine Learning and Cybernetics, Guangzhou (2005)
13. Chen, M., Ludwig, S.A.: Color image segmentation using fuzzy C-regression mode. Adv. Fuzzy Syst. **2**, 1–15 (2017)
14. Li, C., Liu, L., Sun, X., Zhao, J., Yin, J.: Image segmentation based on fuzzy clustering with cellular automata and features weighting. EURASIP J. Image Video Process. **2019**, 1–11 (2019)
15. Pei, Z., Zhao, Y., Liu, Z.: Image segmentation based on differential evolution algorithm. In: 2009 International Conference on Image Analysis and Signal Processing, Taizhou, China (2009)
16. Sarkar, S., Patra, G.R., Das, S.: A differential evolution based approach for multilevel image segmentation using minimum cross entropy thresholding. In: International Conference on Swarm, Evolutionary, and Memetic Computing (SEMCCO 2011), Visakhapatnam, India (2011)
17. Maulik, U., Bandyopadhyay, S.: Genetic algorithm-based clustering technique. Pattern Recognit. **33**, 1455–1465 (2000)
18. El-shorbagy, M.A., Ayoub, A.Y., El-desoky, I.M., Mousa, A.A.: A novel genetic algorithm based k-means algorithm for cluster analysis. In: The International Conference on Advanced Machine Learning Technologies and Applications (AMLTA2018), Cairo (2018)

19. El-shorbagy, M.A., Ayoub, A.Y., Mousa, A.A., El-desoky, I.M.: An enhanced genetic algorithm with new mutation for cluster analysis. Comput. Stat. **34**, 1355–1392 (2019)
20. Abdelsalam, A.M., El-Shorbagy, M.A.: Optimization of wind turbines siting in a wind farm using genetic algorithm based. Renew. Energy. **123**, 748–755 (2018)

Discriminative Representation Learning for Face Recognition

Chia-Hao Tang and Gee-Sern Jison Hsu(✉)

National Taiwan University of Science and Technology, Taipei, Taiwan
jison@mail.ntust.edu.tw

Abstract. A significant progress has been made on face recognition in recent years because of the rapid development of deep learning approaches. Deep learning approaches offer a powerful toolbox for tackling many aspects of face recognition, including the search for effective discriminative features. We compare several state-of-the-art face recognition methods and combine different modules from those methods to propose a special approach for discriminative representation learning. Our approach has a special architecture for representation learning combined with a latest design of classification loss function, making it a highly effective solution for uncontrolled face recognition. Experimented on the Labeled Faces in the Wild (LFW), the Celebrities in Frontal-Profile dataset (CFP), and the AgeDB datasets, our approach shows competitive performance to other state-of-the-art methods.

Keywords: Face recognition · Deep learning · Representation learning

1 Introduction

Face recognition can be one of the most widely used biometric technologies, and the applications cover military, security, social network and entertainment. You can use your face to unlock your mobile phone, access an unmanned store, purchase online without entering pass codes. Face recognition has been entering many aspects of our daily life. Face recognition is commonly divided into two protocols: verification and identification, as illustrated in Fig. 1. Face verification, also known as 1:1 comparison, determines whether two face images are from the same person. Face identification, also known as 1:N search, searches for the one that matches to a given face image from a (large) set of face images. A common way to handle verification and identification consists of the extraction of facial representation, and the comparison of the facial representations from the image pair for verification or the computation of a loss function for identification. We focus on the facial representation learning for handling both verification and identification.

The recent progress on face recognition can be generally classified into two categories, one is the architecture designs of the deep learning frameworks

A.-E. Hassanien et al. (Eds.): AICV 2020, AISC 1153, pp. 574–583, 2020.
https://doi.org/10.1007/978-3-030-44289-7_54

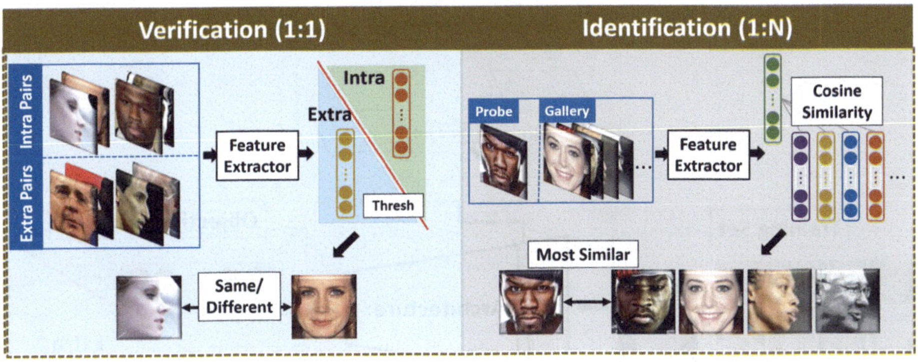

Fig. 1. The test flowchart in face identification and verification repetitively.

[2,14,19,27,35–37] and the other is the special designs for the decision making layer, or the so-called loss functions [5,18,23,29,30]. Many frameworks are modified and improved versions of GANs (Generative Adversarial Networks), and almost all novel loss functions are developed based on the softmax function for computing the cross-entropy loss. We have studied the contemporary work and propose our approach based on some modules of these work for the learning of the discriminative representation good for face identification and verification.

In the following, we first present the recent developments on the network architecture and the objective functions in Sect. 2. The evaluation of the proposed approach on three benchmark databases, the Labeled Faces in the Wild (LFW), the Celebrities in Frontal-Profile dataset (CFP), and the AgeDB datasets is reported in Sect. 3, followed by a conclusion in Sect. 4.

2 Overview

In this section, we give an overview to several novel architectures designed for improving the face recognition performance, and state-of-the-art objective functions which can effectively increase the inter-class discrepancy and decrease the intra-class gap. Our proposed approach is built on the Light CNN [35], combined with residual blocks [10], Max-Feature-Map (MFM and Network-in-Network [17], leading to an effective learning framework for tackling face recognition. To our best knowledge, our approach can be one of the very few that integrates several state-of-the-art modules into a unified network. In the following, we will first introduce the related networks and modules, and then present the contemporary loss functions (Fig. 2).

2.1 Network Backbone

The mainstream architectures for face recognition often follow the deep neural networks proposed for the object classification in the ILSVRC [4], ranging from

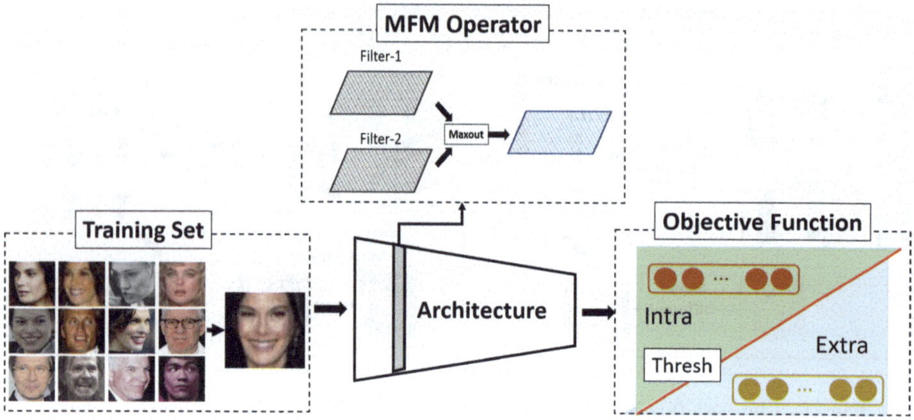

Fig. 2. The proposed framework composed of three important modules, including the large-amount training set, network backbone and latest objective function. We employ MS-Celeb-1M as our training set and exploit A-Softamx [18] as our objective function during the training process. Also, we conduct the IR structure [5] and Max-Feature-Out (MFM) operator [35] to benefit the face recognition.

the classical AlexNet [16] to the contemporary SENet [11]. For example, FaceNet [24] employed the private dataset to trained the GoogleNet [28], and exploited the triplet loss based on the triplets of roughly aligned matched and unmatched face patches generated by a novel online triplet mining method. The VGGFace [21] proposed a large-scale dataset and selected the VGGNet [26] as the network architecture, which achieves the SOTA performance on the LFW. Instead of using the conventional softmax loss function, Deng et al. [5] proposed the additive angular margin loss function, which effectively boosts the discriminative power of the feature embedding learning by using the ResNet [10].

Several architectures proposed recently have been verified effective for tackling face recognition under uncontrolled settings. The ResNet [10] proposed shortcut connections for blending the layer input with shortcut input rather than layers of convolutions to avoid degradation problem and ease the training of deep networks. It does not only perform well for the ImageNet classification but can handle the training of a very deep network with compelling performance. On top of that, Deng et al. [5] conducted a survey on more advanced residual unit settings [8], and provided an effective way to distill facial information and dispel facial variances for face recognition. The proposed Improved Residual (IR) unit is composed of a Batch Normalization (BN) [13] followed by convolution, followed by another BN, then the Parametric ReLU (PReLU [9]) activation, and then one more BN, denoted as BN-Conv-BN-PReLU-BN. Such a unit outperforms the previous unit explored in [8].

Apart from the conventional activate functions, as the variants of the maxout activation function [6], the Light-CNN [33–35] introduces the concept of competitive relationship to activate a small number of neurons, and builds the max-feature-map (MFM) into each convolutional layer. Because the noisy labels in large-scale datasets severely deteriorate the training, the MFM plays an important role to not only separate noisy and informative signals but also select the characteristic features between two feature maps. Additionally, the network includes the residual blocks [10], the Network in Network (NIN) [17] and the structure with small convolution kernel with the MFM, which makes the model light and robust. We follow the IR structure [5] and employ the MFM operator [35] to further improve the residual units and extract the 512-D embedding features.

2.2 Evolution of Objective Functions

Existing objective function for face recognition usually based on Softmax function with cross-entropy loss, however, it has some drawbacks. For instance, the size of the fully connected layer grow with the identities number which significant increase computation costs. And the learned features are not discriminative enough for open-set face recognition since the network-updated according to the closed-set classification. Several variants have been proposed to tackle these issue, the evolution of the objective function is discussed in the following contents.

Since most of the previous studies take Softmax function with cross-entropy loss (Softmax Loss) as the core ideas. We firstly revisit the Softmax by looking into the decision criteria and the Softmax Loss can be written as follows:

$$L_s = -\frac{1}{N_b}\sum_{i=1}^{N_b} \log \frac{e^{W_{y_i}^T x_i + b_{y_i}}}{\sum_{j=1}^{n} e^{W_j^T x_i + b_j}} \tag{1}$$

where $x_i \in \mathbb{R}^d$ denotes the d–dim deep feature of the i-th element, belonging to the y_i-th class. $W_j \in \mathbb{R}^d$ denotes the j-th column of the weight $W \in \mathbb{R}^{d \times n}$ and $b_j \in \mathbb{R}^n$ is the bias term. N_b and n are the batch size and the class number, respectively. The Softmax loss is widely used in deep face recognition [3]. However, the softmax loss function does not optimize the feature embedding to enhance higher similarity for intra-class samples and diversity for inter-class samples. This motivates the developments of other loss functions.

***L_2* Softmax Loss.** The L_2-softmax loss [23] was proposed to add a simple but effective restrict to the regular softmax loss for training face recognition system. It provides a symptomatic treatment that constraints the L_2-norm of the feature descriptor to a constant value α, and the L_2-softmax loss is defined as follows:

$$\begin{gathered} L_{l2} = -\frac{1}{N}\sum_{i=1}^{N} \log \frac{e^{W_{y_i}^T x_i + b_{y_i}}}{\sum_{j=1}^{n} e^{W_j^T x_i + b_j}} \\ \text{where } ||x_i||_2 = \alpha, \forall \text{i} = 1, 2, ..., \text{N} \end{gathered} \tag{2}$$

where x_i is the sample features in a mini-batch of size N, where the features extract from the second to last layer of DNN which is the most representative vector to various subjects. The rest of the parameter is as same as in (1). The authors have argued that quality of images leads to different level of L_2-norm.

The additional L_2-constraint aims to bounds the signal of the features on the scaling parameter α, which boost the sample consistency on both good quality and blurry images.

Angular Softmax Loss. The Angular Softmax Loss (A-Softmax) [18] was proposed to improve the issues with the bias $b_j = 0$ and $\|W_j\| = 1$ [18]. The issue of the bias $b_j = 0$ is handled by transforming the logit [22] as $W_j^T x_i = \|W_j\| \|x_i\| \cos\theta_j$, where θ_j is the angle between the weight W_j and the feature x_i. The issue with the individual weight $\|W_j\| = 1$ is handled by taking the l_2 normalization to make the prediction only depend on the angle between the feature vector and the weight vector. To make it discriminative, the authors generalize it to the following Angular Softmax (called A-Softmax in short) Loss L_{AS}, and name their solution "SphereFace".

$$L_{as} = -\frac{1}{N}\sum_{i=1}^{N} \log \frac{e^{\|x_i\| \cos(m\theta_{y_i})}}{e^{\|x_i\| \cos(m\theta_{y_i})} + \sum_{j=1, j\neq y_i}^{n} e^{\|x_i\| \cos\theta_j}}. \tag{3}$$

where $\theta_{y_i} \in [0, \frac{\pi}{m}]$. A-Softmax loss has stronger requirements for a correct classification when $m \geq 2$, which generates an angular classification margin between the learned features of different classes. A-Softmax loss imposes a discriminative power to the learned features via angular margin, equivalent to learning features that are discriminative on a hypersphere manifold, while Euclidean margin losses learn features in Euclidean space. To put it briefly, we employ SphereFace directly on our network to replace the common Softmax with cross-entropy loss. Empirically, using SphereFace as a softmax replacement on the network can further improve performance.

3 Experimental Evaluation

The experiments aim to make the fair comparison on different techniques and highlight the advantages of assembling the concepts from SOTA architectures modules with latest objective functions. In additions, the recent works mainly performed experiments on LFW [12] and YTF [32], which are nearly saturated. More unconstrained qualitative and quantitative experiments with comprehensive and complicated environment variations are needed to verify the efficacy of the proposed methods, e.g., cross-age database AgeDB [20]. In the following, we will present the datasets with the test protocol considered in our experiments in Sect. 3.1, followed by the ablation study and quantitative evaluation on benchmarks in Sect. 3.2 to analyse respective roles, which aim to demonstrate the superiority of our proposed method.

3.1 Experimental Settings

Datasets. To examine our survey on our proposed aggregation methods, we employ MS-Celeb-1M [7] as training sets and evaluated on multiple test sets, including, AgeDB [20], CFP [25] and LFW [12]. We would make a brief introduction within the protocols in the following:

1. **MS-Celeb-1M.** The MS-Celeb-1M offers about 10 millions images of 100k subjects taken *in the wild*, and present the subject's distribution in different aspects including profession, ethnicity, age, and gender. Owing to the label noise exists in most large-scale databases, we have cleaned it and extracted 5.8 million face images for 85k subjects by applying semi-automatic procedure.
2. **LFW.** The Labeled Faces in the Wild (LFW) dataset offers 13,323 images of 5,749 celebrities. It divided into 10 splits with 6,000 video pairs. Performance is compared the mean recognition accuracy and evaluated by several independent protocols, including (A) the labeled outside data, denoted as *restricted* protocol and (B) the additional training pairs are in training, denoted as *unrestricted* protocol. We follow the latter setting which report the classification accuracy.
3. **CFP.** The Celebrities in Frontal-Profile dataset (CFP) dataset contains 7,000 images of 500 subjects. The subjects were imaged under 10 frontal images and 4 profile images per subject. The evaluation protocol consists of two different phases including frontal to frontal (FF) and frontal to profile (FP) face verification.
4. **AgeDB.** The AgeDB is one of the most popular cross-age database which contains 12,240 images of 440 subjects, and the average age range in each individual is 49 yeas. The evaluation protocol composes of 10 splits and each split contains 300 positive pairs and 300 negative pairs respectively.

Implement Details. Throughout the experiments, the input set are collected from MS-Celeb-1M and we flip the non-near frontal faces ($e.g., yaw > 15°$) and remain the number of near frontal faces to balanced the pose distribution. Following the setting in [5], the training and test images are processed by the off-the-shelf Face Alignment Network [1] for face and landmark detection. All the face images are applied affine transformation to normalize faces to 128×128 according to the 5 points landmark under eyes, nose and mouse corners, and randomly cropped to 112×112 as inputs; the network initialized with zero-centered normal distribution with a standard deviation of 0.02. The network is trained via Adam optimizer [15] with initial learning rate of 0.1 and divided by 10 at 35th and 65th epochs. The batch size sets to be 128 during the training process and the proposed network is implemented based on the publicly available Pytorch platform under CUDA 9.0 with cuDNN 7.6, and using single NVIDIA TITAN RTX GPU with 24G memory.

Since the setting differences in equipment, training/testing sets, the feature dimensions, the pre-processing and other parameters, making the comparison hard to confirm. In this section, we first investigate different architectures and objective functions under the unified data pre-processing and hyper-parameter settings to see their respective roles in face recognition. And we compare three

Table 1. Verification performance (%) of variants of our proposed method.

Methods	LFW	AgeDB-30	CFP-FF	CFP-FP
Light-CNN	99.07	88.93	96.14	90.42
IR-50	99.22	90.70	96.38	91.33
IR-50$_{\mathrm{M}}$	99.43	90.93	97.44	94.26
IR-50$_{\mathrm{As}}$	99.57	91.72	97.86	94.54

Table 2. Performance evaluation (%) on LFW, AgeDB-30 and CFP datasets.

Methods	LFW	AgeDB-30	CFP-FF	CFP-FP
FaceNet [24]	99.63	–	–	–
Center Loss [31]	99.28	–	–	–
SphereFace [18]	99.42	91.70	–	94.38
Proposed	99.71	92.23	98.42	95.21

variants of our proposed method, i.e., different architecture under same objective function, the effect of architecture aggregation, and the results under the same recognition framework, in each case. Then, we will show the final proposed framework performance compare with other works. The verification performance in the AgeDB-30, YTF, CFP and IJB-B are shown in Table 1.

3.2 Performance Evaluation on Benchmarks

Our aggregation network is redesigned to distill the identity information, while providing more discriminative representation for synthesizing the normalized faces. To facilitate the model to extract better and more delicate features, we make a comparison on the improved residual unit (IR) [5] and Light-CNN architecture under the unified hyper-parameter settings and employ Softmax functions with cross-entropy loss to explore the benefits on two of them. From the first and second column in Table 1, we can obviously observe that the IR structure could extract more discriminative features. Despite the Light-CNN employ the shortcut mechanism, the deeper network in IR leads to a better learnt representation. To facilitate the model to extract better and more delicate features, we exploit MFM operator [35] to separate the noisy and informative signals, denoted as IR_{M} from large-scale training set and performs the feature selection between two feature maps. The residual unit in IR becomes BN-MFM-PReLU-MFM-BN structure. As shown in third column in Table 1, although we have already applied the semi-automatic procedure to eliminate the noisy signals in training set, the MFM operator seems still apparent to improve the recognition performance. By comparing the results from various decision boundaries strategies, the common way to classify the face to specific identity is by using cross entropy loss with softmax function, but softmax loss is not discriminative enough. From Table 1, we employ

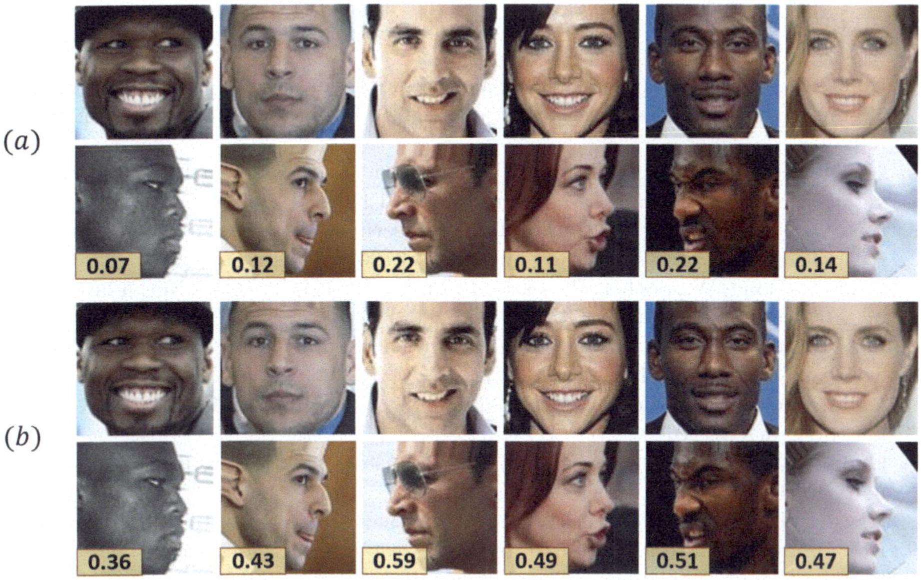

Fig. 3. The similarity score given by (a) IR structure and (b) proposed method. The number on the left-bottom shows the corresponding cosine similarity compare with the frontal faces. All use the A-Softmax [18] as the objective function for training.

Angular Softmax Loss (A-Softmax) [18] and denoted **IR-50**$_{\mathrm{AL}}$. Empirically, using A-Softmax can achieves improved performance. Then, we evaluate our proposed method to compare with SOTA methods in Table 2, and Proposed refers to the composite of different approaches mentioned above.

To better understand the benefit obtained by our proposed methods, as shown in Fig. 3, the similarity computed by using the cosine similarity between the genuine pairs feature vectors extracted by using IR framework and our proposed framework. We can directly observe that our proposed method could better cluster the same identity features distribution, especially the cross-pose face images, and boost the verification performance.

4 Conclusion

Our proposed approach, built on the Light CNN and combined with the improved residual blocks, the MFM operators, the Network-in-Network design and an highly effective loss function, leads to a competitive learning framework for tackling face recognition. Because the modules embedded in the framework collaborate to improve each other, the framework demonstrates a competitive performance to other SOTA approaches, as shown on the three benchmark databases, the LFW, the CFP and the AgeDB.

References

1. Bulat, A., Tzimiropoulos, G.: How far are we from solving the 2D & 3D face alignment problem? (and a dataset of 230,000 3D facial landmarks). In: ICCV (2017)
2. Cao, K., Rong, Y., Li, C., Tang, X., Change Loy, C.: Pose-robust face recognition via deep residual equivariant mapping. In: CVPR (2018)
3. Cao, Q., Shen, L., Xie, W., Parkhi, O.M., Zisserman, A.: VGGFace2: a dataset for recognising faces across pose and age. In: FG (2018)
4. Deng, J., Dong, W., Socher, R., Li, L.-J., Li, K., Fei-Fei, L.: ImageNet: a large-scale hierarchical image database. In: CVPR (2009)
5. Deng, J., Guo, J., Xue, N., Zafeiriou, S.: ArcFace: additive angular margin loss for deep face recognition. In: CVPR (2019)
6. Goodfellow, I.J., Warde-Farley, D., Mirza, M., Courville, A., Bengio, Y.: Maxout networks. arXiv (2013)
7. Guo, Y., Zhang, L., Hu, Y., He, X., Gao, J.: MS-Celeb-1M: a dataset and benchmark for large-scale face recognition (2016)
8. Han, D., Kim, J., Kim, J.: Deep pyramidal residual networks. In: CVPR (2017)
9. He, K., Zhang, X., Ren, S., Sun, J.: Delving deep into rectifiers: surpassing human-level performance on ImageNet classification. In: ICCV (2015)
10. He, K., Zhang, X., Ren, S., Sun, J.: Deep residual learning for image recognition. In: CVPR (2016)
11. Hu, J., Shen, L., Sun, G.: Squeeze-and-excitation networks. In: CVPR (2018)
12. Huang, G.B., Mattar, M., Berg, T., Learned-Miller, E.: Labeled faces in the wild: a database for studying face recognition in unconstrained environments (2008)
13. Ioffe, S., Szegedy, C.: Batch normalization: accelerating deep network training by reducing internal covariate shift. arXiv (2015)
14. Kang, B.-N., Kim, Y., Kim, D.: Pairwise relational networks for face recognition. In: ECCV (2018)
15. Kingma, D.P., Ba, J.: Adam: a method for stochastic optimization. arXiv (2014)
16. Krizhevsky, A., Sutskever, I., Hinton, G.E.: ImageNet classification with deep convolutional neural networks. In: NIPS (2012)
17. Lin, M., Chen, Q., Yan, S.: Network in network. arXiv (2013)
18. Liu, W., Wen, Y., Yu, Z., Li, M., Raj, B., Song, L.: SphereFace: deep hypersphere embedding for face recognition. In: CVPR (2017)
19. Masi, I., Chang, F.-J., Choi, J., Harel, S., Kim, J., Kim, K., Leksut, J., Rawls, S., Wu, Y., Hassner, T., et al.: Learning pose-aware models for pose-invariant face recognition in the wild. TPAMI **41**, 379–393 (2018)
20. Moschoglou, S., Papaioannou, A., Sagonas, C., Deng, J., Kotsia, I., Zafeiriou, S.: AgeDB: the first manually collected, in-the-wild age database. In: CVPRW (2017)
21. Parkhi, O.M., Vedaldi, A., Zisserman, A., et al.: Deep face recognition. In: BMVC (2015)
22. Pereyra, G., Tucker, G., Chorowski, J., Kaiser, Ł., Hinton, G.: Regularizing neural networks by penalizing confident output distributions. arXiv (2017)
23. Ranjan, R., Castillo, C.D., Chellappa, R.: L2-constrained softmax loss for discriminative face verification. arXiv (2017)
24. Schroff, F., Kalenichenko, D., Philbin, J.: FaceNet: a unified embedding for face recognition and clustering. In: CVPR (2015)
25. Sengupta, S., Chen, J.-C., Castillo, C., Patel, V.M., Chellappa, R., Jacobs, D.W.: Frontal to profile face verification in the wild. In: WACV (2016)

26. Simonyan, K., Zisserman, A.: Very deep convolutional networks for large-scale image recognition. arXiv (2014)
27. Song, L., Gong, D., Li, Z., Liu, C., Liu, W.: Occlusion robust face recognition based on mask learning with pairwise differential Siamese network. In: ICCV (2019)
28. Szegedy, C., Liu, W., Jia, Y., Sermanet, P., Reed, S., Anguelov, D., Erhan, D., Vanhoucke, V., Rabinovich, A.: Going deeper with convolutions. In: CVPR (2015)
29. Wang, F., Cheng, J., Liu, W., Liu, H.: Additive margin softmax for face verification. IEEE Sig. Process. Lett. **25**, 926–930 (2018)
30. Wang, H., Wang, Y., Zhou, Z., Ji, X., Gong, D., Zhou, J., Li, Z., Liu, W.: CosFace: large margin cosine loss for deep face recognition. In: CVPR (2018)
31. Wen, Y., Zhang, K., Li, Z., Qiao, Y.: A discriminative feature learning approach for deep face recognition. In: ECCV (2016)
32. Wolf, L., Hassner, T., Maoz, I.: Face recognition in unconstrained videos with matched background similarity (2011)
33. Wu, X.: Learning robust deep face representation. arXiv (2015)
34. Wu, X., He, R., Sun, Z.: A lightened CNN for deep face representation. arXiv (2015)
35. Wu, X., He, R., Sun, Z., Tan, T.: A light CNN for deep face representation with noisy labels. IEEE Trans. Inf. Forensics Secur. **13**, 2884–2896 (2018)
36. Yang, J., Ren, P., Zhang, D., Chen, D., Wen, F., Li, H., Hua, G.: Neural aggregation network for video face recognition. In: CVPR (2017)
37. Yin, B., Tran, L., Li, H., Shen, X., Liu, X.: Towards interpretable face recognition. In: ICCV (2019)

Towards Real-Time Edge Detection for Event Cameras Based on Lifetime and Dynamic Slicing

Sherif A. S. Mohamed[1(✉)], Mohammad-Hashem Haghbayan[1], Jukka Heikkonen[1], Hannu Tenhunen[1,2], and Juha Plosila[1]

[1] University of Turku (UTU), 20500 Turku, Finland
samoha@utu.fi
[2] Royal Institute of Technology (KTH), 11419 Stockholm, Sweden

Abstract. Retinal cameras, such as dynamic vision sensors (DVF), transmit asynchronous events with ultra-low latency (~10 µs) only at significant luminance changes, unlike traditional CMOS cameras which transmit the absolute brightness of all pixels including redundant backgrounds. Due to these significant characteristics, they offer great potential to obtain efficient localization of high-speed and agile platforms. Moreover, event cameras have a high dynamic range (~140 dB), which makes them suitable for platforms that operate indoors in low-lighting scenarios and in outdoor environments, where the camera might be pointing at a strong light source, e.g. the sun. In this paper, we propose an algorithm to detect edges in event streams coming from retinal cameras. To do that, an algorithm is developed to extract edges from events by augmenting a batch of events with their lifetimes. The lifetime of each event is computed using a local plane fitting technique. We use a batching technique to increase the frame rate of generated images since events with a high sample rate cause the processing of a single event to be computationally expensive. The size of the batch will be adjusted based on the mean optical flow of the previously generated batch. The obtained experimental results show that our proposed technique can significantly reduce the response time with the same sharpness in generating the edges.

Keywords: Edge detector · Event camera · Dynamic Vision Sensor (DVS) · SLAM · Visual odometry

1 Introduction

Event cameras, such as the Dynamic Vision Sensor (DVS) [1] and the Dynamic and Active Pixel Vision Sensor (DAVIS) [2], are bio-inspired vision sensors that use a temporal contrast pixel technology to capture the dynamic changes in the environment. They publish event streams in an asynchronous fashion based on brightness changes in every pixel, unlike traditional cameras that use CMOS

A.-E. Hassanien et al. (Eds.): AICV 2020, AISC 1153, pp. 584–593, 2020.
https://doi.org/10.1007/978-3-030-44289-7_55

sensors to generate full images, i.e., the absolute brightness at a constant rate. In these cameras, when the log-intensity of a pixel changes above or below a certain threshold, an event is triggered consisting of three values: the 2D position of the pixel (x, y), time-stamp (t), and polarity of the change (p). The polarity is indicated using a single bit, i.e., brightness decreases ("OFF" or '0') or brightness increases ("ON" or '1'). The main advantages of event cameras are: low-latency (in order of microseconds), very high dynamic range (up to 140 dB compared to 60 dB of traditional cameras) [3], and ultra-low power consumption of less than 25 mW. These significant characteristics make event cameras have great potential for self-localization, i.e., visual odometry (VO) and visual localization and mapping (VSLAM) [4] in challenging scenarios, e.g., high-speed movement, low lighting, and high-dynamic ranges. In addition, other applications can benefit from event cameras such as augmented reality (AR) and video game applications. However, most vision-based algorithms cannot be applied to the data acquired by event cameras, and therefore new methods are required to process the output and unlock their remarkable potential.

One of the most common ways of obtaining the position and orientation of a vehicle is to use feature-based visual odometry methods. Feature-based or *indirect* approaches extract points of interest in each captured image using feature detectors, such as corner or edge detectors. The simplest technique to detect edges from asynchronous events generated by event cameras is to accumulate events in a fixed time window [5] or a specified number of events [6]. However, these techniques are inefficient when the velocity of the camera or the complexity of the environment varies at runtime. This comes from the fact that the number of events generated in an event based camera is proportional to changes of surroundings information over time, i.e., a combination of the amount of surroundings information and the velocity of the camera [7]. For example, to detect an edge by an event-based camera when the camera velocity is too low/high, if each frame is constructed by accumulating events in a fixed time epoch, the edge in the constructed frame becomes very noisy, i.e., sparse/bleeding. Figure 1c shows the situation when the time epoch is too long resulting in a bleeding image that is an accumulation of events in one frame.

In this paper, we propose a novel methodology to select the events produced by 1D features (e.g. edges) in a scene to make a proper frame. The events will be selected based on their lifetime, i.e., the maximum amount of time before the event shifts by 1 pixel, and event colonization. The lifetime is obtained by fitting a local plane to the surface of an active event using the linear least square minimization, and then computing the velocity of the event (V_x, V_y). Event colonization is used to reduce the computing penalty imposed on the system by the heavy processing of events. It significantly improves the performance of the system. The proposed approach provides a fast edge detection method for event camera output, which is robust against environmental variations. Moreover, we will show how the technique can be tailored for resource-constrained platforms due to its performance benefits compared with other similar techniques.

The rest of this work is structured as follows. In Sect. 2, we illustrate the related work in the state-of-the-art. In Sect. 3, we explain the methodology of the proposed edge detector for event cameras. In Sect. 4, we present the experimental results. Finally, in Sect. 5, we draw a conclusion of the article.

2 Related Work

Edge detection is one of the most important concepts in visual odometry techniques, especially for textureless scenes. Edges are natural and robust features prominent in most environments. There is a variety of algorithms to extract edges from images captured by frame-based cameras. The majority of these approaches are gradient or Laplacian of Gaussian (LoG) based methods [8]. A Laplacian edge detector, e.g. the Marr–Hildreth algorithm [9], uses one kernel to search for the zero crossings in the second derivative of the image. Unlike a Laplacian detector, a gradient edge detector, such as the Canny edge detector [10], uses two kernels to detect the edges by looking for the maximum and minimum in the first derivative of the image.

There are a few methods developed for detecting edges from data acquired from event cameras [3]. The approaches can be classified into static and dynamic methods. In static methods, edges are detected either by accumulating events in a fixed time-window or by defining a specific number of events per batch, i.e., a fixed number of events. Both techniques have some major drawbacks for detecting high-quality edges. The quality of edges in the time-window technique depends on the speed of the camera and objects in the scene. For instance, if the camera starts to move faster, the number of generated events increases, which leads to low-quality edges, i.e., blurry edges. On the other hand, the quality of edges in the event batch method depends on the amount of information in the environment. For example, if the camera moves from a simple textureless environment with a small gradient to a more complex environment, the number of events increases dramatically, which leads to low-quality edges, i.e., bleeding edges. Therefore, it is difficult to set a fixed time-window or number of events that would be suitable for all scenarios.

Dynamic methods, on the other hand, seek to detect robust edges against environmental variations and different camera speeds. Some approaches focus on extracting line segments from events. One of the most common methods to detect lines and curves from images is based on the Hough transform [11] due to its simplicity. However, applying such a technique to event streams and to preserve low latency of events is highly infeasible for applications that run on embedded processors, because calculating the Hough projection for each new event is computationally expensive. For example, in [12] the authors have proposed a line detection algorithm based on the Hough transform and implemented it using the third generation of the Artificial Neural Networks (ANN) model, i.e., a Spiking Neural Network (SNN). The main limitations of this technique are its performance degradation in areas where multiple lines meet and its high demand for computational power. Another way is to extract lines from events based on the line segment detector (LSD) [13] algorithm to reduce the computational cost. A LSD detects line segments by computing the orientation of the spatial derivative of each pixel and clustering pixels with a similar orientation into support regions, which are later used to fit a line. For instance, in [14] the authors have proposed an algorithm to detect lines from asynchronous events based on the original LSD algorithm. A Sobel filter is used to compute the orientation of the Surface of Active Events (SAE). Then, a clustering process is performed within neighbor events. A drawback of this method is that accumulation of events leads to incorrect line fitting in some scenarios.

A few approaches in the event-based vision sensors literature have focused on detecting edges. Barranco et al. [15] present a method to detect the boundary of objects in the scene and perform border ownership, i.e., segmentation. The learned Structured Random Forest (SRF) method is used to estimate boundaries and ownership using the extracted features such as orientation, time-texture, motion, and time-stamps from the generated frame, i.e., accumulated events. However, the overall edge extraction performance is not great due to the focus on segmentation. In [16] the authors have proposed an algorithm to estimate the lifetime of each incoming event using a local plane-fitting algorithm, after which events with their lifetimes are accumulated to detect edges. However, this algorithm is based on an event-by-event scheme, where events are processed one at a time, which dramatically increases the computational complexity of the system. We address the drawbacks of the state-of-the-art methods by proposing a fast and robust algorithm to detect edges from event streams using the lifetime technique and dynamic slicing based on the optical flow of events.

3 Methodology

In this section, the operation of event-based cameras and the surface of active events (SAE) are discussed first. Subsequently, we devise our algorithm to continuously obtain the edges from the output of an event camera.

3.1 Event Camera Mechanism

Traditional CMOS cameras capture the absolute intensity of each pixel and process all the pixels in the frame at a rate specified by an external clock, e.g., 60 fps. The frames in event cameras, such as DVS, are composed of independent pixels that indicate only the brightness changes in a scene. Thus, the output is a stream of asynchronous events at a variable data rate, and each event represents a change of brightness, i.e., *log intensity*. If the (logarithmic) brightness of a pixel exceeds a predefined threshold (typically 15%), an event $e = (x, y, t, p)$ is triggered consisting of the location of the pixel (x, y), the time (t) of the event, and the polarity p of the change, where "1" or positive means an increase in brightness and "0" or negative means a decrease in brightness.

The Surface of Active Events (SAE) is presented in three-dimensional coordinates, i.e., the spatio-temporal domain which is often used to represent data

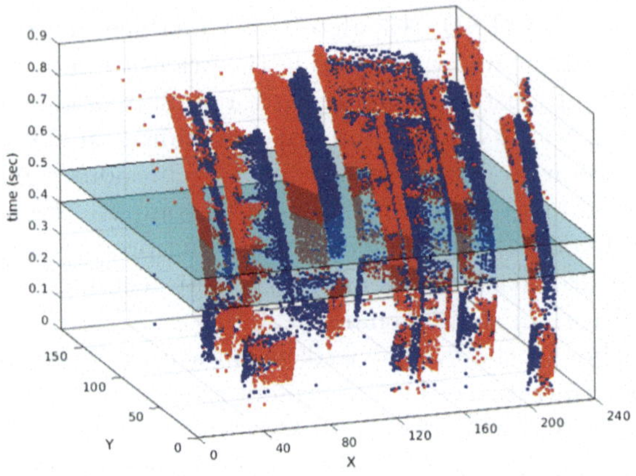

(a) Surface of Active Events, i.e., 3D event stream with a slice shown in (b)

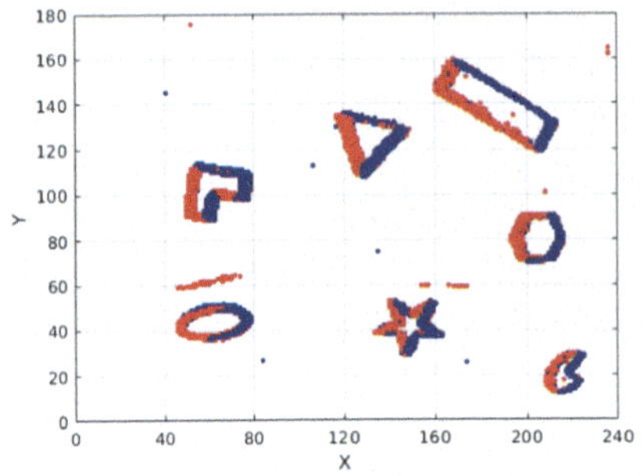

(b) Accumulated events between t= 0.4s and 0.5s.

(c) Traditional camera frame at the same moment in time.

Fig. 1. Surface of Active Events (SAE) of shapes on a wall captured by a DVS camera. The camera is moving to the right which causes the events to shift to the left. Events caused by sensor noise are visible as isolated dots. A 100 ms slice starting at t = 0.4 s is shown in (b), which corresponds to the naive method that accumulates events over a fixed time interval (here 100 ms). As the camera speed increases the edges becomes several pixels wide, which corresponds to motion blur.

that is captured across space and time (see Fig. 1a). The SAE is composed of the two-dimensional sensor frame and one dimension to represent the time. When an event is triggered, a point is generated or updated on the surface with time t assigned to it. For each pixel on a certain location, the time value of the surface is equal to the timestamp of the most recent event:

$$SAE : (p_i) \rightarrow t_i \tag{1}$$

Here p is the 2D location of a pixel in location (x, y) and, therefore, each point of the SAE can be represented by a 3D vector $S(p_i) = (x_i, y_i, t_i)$.

Figure 1a illustrates the SAE of events captured by a DVS camera in the spatio-temporal domain. The recorded sequence consists of different static objects, i.e., geometric shapes mounted on the wall, while the camera is moving freely in 6-DoF with various speeds. Isolated points represent the sensor noise, which will be explained in detail in the next section. Figure 1b shows the generated event frame using the simple fixed time-window technique. In this example, a 100ms time epoch is used for each frame.

3.2 Algorithm

The overall system architecture of the proposed edge detection algorithm is illustrated in Fig. 2. In the first phase, events are sliced to batches based on the mean optical flow of the previous batch. Then, inliers are detected for each event inside the batch using the random sample consensus (RANSAC) algorithm. Afterwards, we use the inliers obtained from the previous step to fit a local plane to the SAE of each event using linear least square minimization. Finally, we estimate the normal vector of the plane and calculate the lifetime of events to detect live edges.

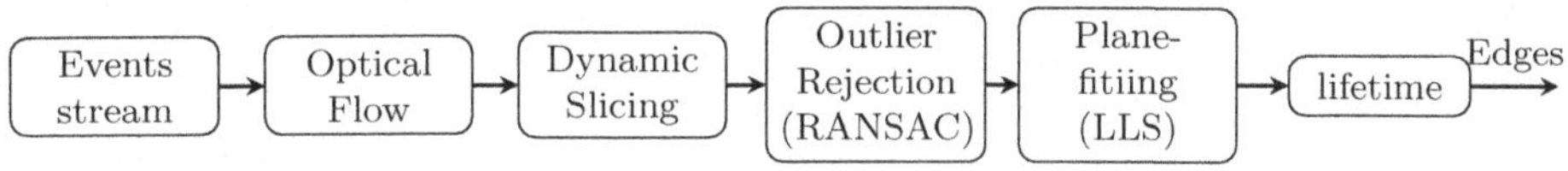

Fig. 2. Overview of the proposed edge detection algorithm for event cameras.

3.2.1 Dynamic Slicing

Generating a suitable batch of events highly depends on the richness of information captured by the camera. To estimate this richness, we use the optical flow, i.e., the velocity of the events in the previous batch. During the initial phase, since the velocity of the events is not obtained yet, a fixed time window is used (in our case 1 ms). The following equation is used to estimate the number of events considered in the next batch:

$$\Delta t_i = \frac{1}{\bar{v}_{i-1}} \tag{2}$$

$$\bar{v}_{i-1} = \frac{\sum_{j=1}^{M} v_j}{M} \tag{3}$$

where Δt_i is the time window that will be used to generate the current batch and v_{i-1} denotes the mean optical flow of M number of events in the previous batch.

3.2.2 Outlier Rejection

RANSAC is an algorithm developed by Fischler et al. [17] that enables plane fitting without trying all possibilities by removing outliers from a set of data. It is essentially based on a probabilistic model of finding the best set required to estimate the model. To detect inliers we select an NxN local window around each new event, where N has to be an odd number, thus the current event is in the center of the window. Then we choose 3 points (the minimal number of points needed to estimate a plane), two random events and the current event. We obtain the plane parameters with these 3 points. Then we give each model a score which is basically the number of points belonging to the plane. A point is considered to belong to the plane when the Euclidean distance is less than a parameter δ (here, we use $\delta = 0.001$). At the end, we keep the plane with the best score, i.e., the number of inliers belonging to the current event. Using Eq. 4 we then compute the number of iterations N_{iter} to get a desired probability:

$$N_{iter} = \frac{\log(1-p)}{log(1-(1-p_{out})^s} \tag{4}$$

where p denotes the desired probability ≈ 0.99, p_{out} is the fraction of outliers in the neighborhood, and s is the number of points in the model.

3.2.3 Plane-Fitting (LLS)

We use the linear least square technique to fit a plane to the 3D points, i.e., events in the local neighbourhood of an event. Given a set of inlier events $e_i = \{x_i, y_i, t_i\}_{i=1}^{m}$ in the local window, we compute the parameters A, B, and C so that the plane $t = Ax + By + C$ best fits the data set. In other words, the sum of squared error between the t_i and the plane value $Ax_i + By_i + C$ is minimized.

3.2.4 Lifetime

We estimate the lifetime of the events using the algorithm proposed in [18]. The lifetime of an event defines the time until the corresponding 3D point in the world that has generated the current event generates a new event. In other words, it is the maximum amount of time from the current event to the moment when a new event is triggered in the neighbourhood:

$$lifetime[e_i(x_i, y_i)] = max\Delta t \qquad subject\ to\ ||(x_i, y_i)|| = 1\ pixel \tag{5}$$

where Δt is the estimated time it will take for an event to move to the neighbour pixel (i.e. a 1-pixel shift) and can be obtained as follows: $SAE(p+\Delta p) - SAE(p)$, in which p is the 2D pixel coordinates of the event in (x,y).

The direction of the steepest slope is used to the find the normal vector $v = [n_1 \quad n_2 \quad n_3]^T$. The optical flow, i.e., velocity and lifetime of the event are computed as follows:

$$v_x = \frac{-n_3}{n_1}, \quad v_y = \frac{-n_3}{n_2} \tag{6}$$

$$lifetime[e_i(x_i, y_i)] = \frac{1}{v_i} = \frac{1}{\sqrt{v_{xi}^2 + v_{yi}^2}} = 1/n_{3i}\sqrt{n_{1i}^2 + n_{2i}^2} \quad (7)$$

4 Experimental Results

We evaluate the proposed algorithm using a sequence provided by a DVS camera (128 × 128) and the Event-Camera Dataset [19] which contains many sequences captured with a DAVIS-240C camera. DAVIS cameras can capture events and intensity images at the resolution of 240 × 180. In a normal motion scenario, the DAVIS camera generates up to 10^5 events per second. On the other hand, in rapid camera movements, it can generate up to 1.5 million events per second. The proposed algorithm is tested on an Intel i5 (1.6 Ghz) laptop with 8 GB of RAM.

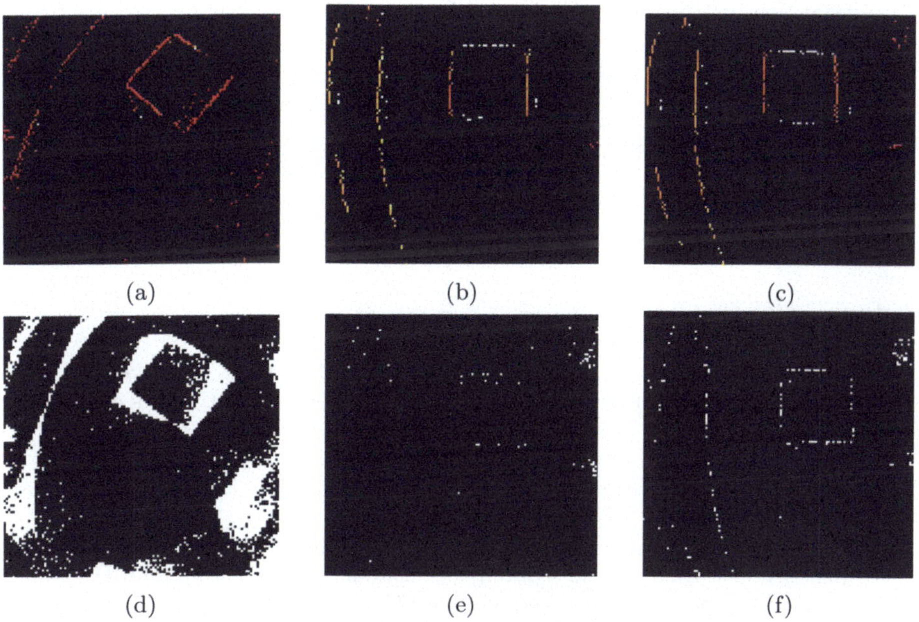

Fig. 3. Results of proposed system a, b, and c compared with a time fixed window algorithm (t = 30 ms) d, e, and f in different camera speed.

Figure 3 shows that our algorithm generates sharp and accurate edges in different camera speeds. In a high speed, the fixed time interval causes blur due to the huge amount of events generated by the camera movement; our algorithm, however, shows consistency and robustness against the high camera speed (see Fig. 3a). In another scenario, where the camera is moving with a slow speed (see Fig. 3b and e), the time-fixed algorithm shows many isolated events, i.e., sparse edges due to the low amount of events generated from the scene; on the other hand, our algorithm manages to generate an informative frame. Finally, in an average speed, both algorithms show good results.

5 Conclusion

In this paper, a fast and robust edge detection algorithm is proposed to generate smooth and sharp edges based on data captured by an event-based camera. The main contribution of the approach is its ability to flexibly change the number of events that are processed as a frame. To do this, we employed the concept of the lifetime of an event to generate frames with sharp edges. The algorithm tackles the issues of sparse edges due to a low number of events accumulated within a fixed time-window and edge bleeding, i.e. blur, due to a high velocity of the camera or an increase in environmental complexity. To reduce the computational complexity, we implemented a novel dynamic batch-to-batch algorithm based on the optical flow of the previous batch. The optical flow of the previous frame is used to dynamically determine the number of events that are considered for the current batch. The experimental results show that the proposed scheme outperforms the traditional approaches that are based on considering either a fixed number of events per frame or a fixed time interval to accumulate the events. It is also faster than existing algorithms that use the lifetime and event-to-event based techniques. Our algorithm can operate efficiently and accurately with different camera velocities and in different environmental conditions.

References

1. Lichtsteiner, P., Posch, C., Delbruck, T.: A 128 × 128 120 dB 15 µs latency asynchronous temporal contrast vision sensor. IEEE J. Solid-State Circuits **43**(2), 566–576 (2008)
2. Brandli, C., Berner, R., Yang, M., Liu, S., Delbruck, T.: A 240 × 180 130 dB 3 µs latency global shutter spatiotemporal vision sensor. IEEE J. Solid-State Circuits **49**(10), 2333–2341 (2014)
3. Gallego, G., Delbrück, T., Orchard, G., Bartolozzi, C., Taba, B., Censi, A., Leutenegger, S., Davison, A., Conradt, J., Daniilidis, K., Scaramuzza, D.: Event-based vision: a survey. *CoRR*, vol. abs/1904.08405 (2019)
4. Mohamed, S.A.S., Haghbayan, M., Westerlund, T., Heikkonen, J., Tenhunen, H., Plosila, J.: A survey on odometry for autonomous navigation systems. IEEE Access **7**, 97466–97486 (2019)
5. Liu, M., Delbruck, T.: Adaptive time-slice block-matching optical flow algorithm for dynamic vision sensors, September 2018
6. Rebecq, H., Horstschaefer, T., Gallego, G., Scaramuzza, D.: EVO: a geometric approach to event-based 6-DOF parallel tracking and mapping in real time. IEEE Robot. Autom. Lett. **2**(2), 593–600 (2017)
7. Mohamed, S.A.S., Haghbayan, M., Heikkonen, J., Tenhunen, H., Plosila, J.: Towards dynamic monocular visual odometry based on an event camera and IMU sensor. In: Intelligent Transport Systems, From Research and Development to the Market Uptake (INTSYS 2019). Springer (2020)

8. Mlsna, P., Rodriguez, J.: Gradient and Laplacian edge detection, pp. 495–524. Elsevier Inc. (2009)
9. Marr, D., Hildreth, E.: Theory of edge detection. Proc. R. Soc. Lond. Ser. B **207**, 187–217 (1980)
10. Canny, J.: A computational approach to edge detection. IEEE Trans. Pattern Anal. Mach. Intell. **PAMI–8**(6), 679–698 (1986)
11. Leavers, V.: Which hough transform? CVGIP Image Underst. **58**(2), 250–264 (1993)
12. Seifozzakerini, S., Yau, W.-Y., Zhao, B., Mao, K.: Event-based hough transform in a spiking neural network for multiple line detection and tracking using a dynamic vision sensor. In: Proceedings of the British Machine Vision Conference (BMVC), pp. 94.1–94.12. BMVA Press, January 2016
13. von Gioi, R.G., Jakubowicz, J., Morel, J., Randall, G.: LSD: a line segment detector. IPOL J. **2**, 35–55 (2012)
14. Brändli, C., Strubel, J., Keller, S., Scaramuzza, D., Delbruck, T.: ELiSeD – an event-based line segment detector. In: 2016 Second International Conference on Event-Based Control, Communication, and Signal Processing (EBCCSP), pp. 1–7, June 2016
15. Barranco, F., Teo, C.L., Fermüller, C., Aloimonos, Y.: Contour detection and characterization for asynchronous event sensors. In: 2015 IEEE International Conference on Computer Vision (ICCV), pp. 486–494, December 2015
16. Mueggler, E., Forster, C., Baumli, N., Gallego, G., Scaramuzza, D.: Lifetime estimation of events from dynamic vision sensors. In: IEEE International Conference on Robotics and Automation (ICRA 2015), Seattle, WA, USA, 26–30 May 2015, pp. 4874–4881 (2015)
17. Fischler, M.A., Bolles, R.C.: Random sample consensus: a paradigm for model fitting with applications to image analysis and automated cartography. Commun. ACM **24**(6), 381–395 (1981). http://doi.acm.org/10.1145/358669.358692
18. Benosman, R., Clercq, C., Lagorce, X., Ieng, S., Bartolozzi, C.: Event-based visual flow. IEEE Trans. Neural Netw. Learn. Syst. **25**(2), 407–417 (2014)
19. Mueggler, E., Rebecq, H., Gallego, G., Delbrück, T., Scaramuzza, D.: The event-camera dataset and simulator: event-based data for pose estimation, visual odometry, and SLAM. I. J. Robot. Res. **36**(2), 142–149 (2017)

An Automatic Detection of Military Objects and Terrorism Classification System Based on Deep Transfer Learning

Doaa Mohey El-Din[1(✉)], Aboul Ella Hassanein[2], and Ehab E. Hassanien[1]

[1] Information Systems Department, Faculty of Computers and Artificial Intelligence, Cairo University, Cairo, Egypt
d.mohey@alumni.fci-cu.edu.eg, E.Ezat@fci-cu.edu.eg
[2] Information Technology Department, Faculty of Computers and Artificial Intelligence, Cairo University, Cairo, Egypt
aboitcairo@gmail.com

Abstract. This paper presents an automated detection of military objects and terrorism classification system based on deep transfer learning. It constructs a new structure of neural networks based on AlexNet pre-trained transfer learning model. This network is designed by six neural networks for six spectrums (Intensified visual images, Near-infrared spectroscopy (NIR) images, thermal images, LWIR (long wave infrared images), DHV, and RGB). It uses for detecting objects and actions in various spectrums in night mode depends on two sensory data types of images and videos. The system proposes an automated description layer for improving the classification domain whether military or terrorism domain. The detection and classification result reaches 92% for objects and actions detection and classifying the compatible domains whether terrorism or military. The experiments are applied to three datasets due to the lack of critical data (images or videos) in these domains. This dataset reaches 7992 images in multiple spectrums.

Keywords: Object detection · Classification · Spectrums · Military objects · Terrorism actions · Sensors

1 Introduction

Terrorism is defined by the illegal use of violence and threat, especially against civilians, in the tracking of political objectives [1]. Terrorism becomes spreading in many countries, however it poses a threat to any country, especially in remote areas in the country such as Sinai in Egypt [2]. According to real statistics, this average number of victims from terrorism's attacks are increasing annually [3]. In 2017, terrorism is caused by 26,445 death people annually in the world. Victims have been rising year on year for the past three years, from 11 deaths in 2017, 26 deaths in 2018, to 77 deaths by the end of September 2019 [4]. The military is responsible to save lives and protect the country. Till now, the military finds an obstacle to discover all terrorists and their actions because terrorists use some methods to disappear and hide the weapons. So, the

A.-E. Hassanien et al. (Eds.): AICV 2020, AISC 1153, pp. 594–603, 2020.
https://doi.org/10.1007/978-3-030-44289-7_56

military goes forward to use weapons holding various spectrum sensors, such as infrared, that provide them in detecting any strange objects or actions.

Object detection, segmentation, classification are the main uses tasks of transfer learning [5]. Deep transfer learning uses for improving the performance of tasks' challenges or applications by reusing the pre-trained models [6]. The power of transfer learning is represented with hierarchical architecture and a large-scale training set that usually contains millions of images. However, in military and terrorism domains, object detection is faced a big challenge due to the hardness of finding enough training or real data whether in actions or objects such as types of snippers, weapons. So, the output will not reach enough accuracy and performance [7]. There is also a blurred classification between terrorism and military domains.

This paper proposes an automatic detection and classification system of military and terrorism in objects and actions. This system consists of three layers (1) The automated detection uses the Alexnet transfer neural network model. The neural network is designed based on six neural networks for six spectrums, (Intensified visual images, Near-infrared spectroscopy (NIR) images, thermal images (long-wave infrared images), DHV, and RGB) in night mode. (2) This system creates a description layer to improve the quality of datasets and classification results. (3) The automated classification uses to remove the blurred vision between two domains terrorism and military based on creating the description layer. The proposed system improves the classification between objects and actions and between military and terrorism domains.

The rest of this paper is organized as follows. Section 2 examines the background and previous related works about classification, object detection, and transfer learning. Section 3 presents automatic terrorism and military System. Section 4 introduces the experiments and discusses the results. Finally, conclusion and future works of this work in Sect. 5.

2 Background and Related Works

This section discusses previous motivations in object detection invariant data types as shown in Table 1. It also demonstrates the examination study of classification importance for improving the accuracy results in the military domain as shown in Table 2. Some motivations aim to solve data augmentation and overfitting.

2.1 Detection and Classification in Multi-specrtal

The object detection challenge is considered one of computer vision challenges that have several motivations to detect the objects related to the specific domain [6, 7]. Multi-Spectral is a hot area of research, especially in night mode [8]. There are several spectrum types of vision such as RGB, DHV, Thermal, or infrared. Image and video classifications are complex interpretation and processing to reach the object's classes. So, there are several techniques to reach a suitable accuracy for the classification. Previous researches are discussed in Table 1.

Table 1. A comparison between previous motivations of classification in the military domain

Paper no.	Classification	Used technique	Dataset type	Results
[7]	Radio access technologies (RAT)	Traditional Machine learning and CNN	Signals	CNN achieved 100%
[8]	Automatic object detection in military and space images	Convolutional Neural Network (CNN)	Images	95.6%
[9]	Classify military objects from images	Deep transfer learning	Images	Approximately more than 95%
[10]	Convolutional Long Short-term Deep Neural Network (CLDNN), and a deep Residual Network (ResNet)	Automatic modulation classification	Images	Around 90%
[11]	Faster RNN algorithm	Classify objects based on a combination of human visual salience and visual psychology	Images	90.7% improve accuracy by 27 to 33%

There are several motivations in image or video classifications as shown in Table 1. Researchers in [7–9], reach high accuracy results through using neural network approaches in various data types (signals, and images) and domains. Other researches as [10, 11] achieve less accuracy results for automatic classification that require improving. It has several methods for training based on previous training to improve future training on various datasets or limited datasets [10–14]. In Table 2, previous motivations are shown in a comparison between classification problems using transfer learning functions with different targets and domains. Transfer learning includes several algorithms [9, 15–21].

Table 2. A comparison between motivations using deep transfer learning.

Paper	Use	Target	Dataset	Results
[9]	Alexnet	Image classification	Alcoholism Identification	97%
[15]	VGG 16	Yelp Restaurants	Image classification Kaggle	60%
[16]	VGG 19	Image classification	Gender classification in age estimation	98%
[17]	Inception V3	Image classification	CIFAR-10	70%
[18]	Xception	Image classification	Microsoft Malware Dataset	99%
[19]	ResNet-34	Brain Abnormally Image	MRI Images	100%
[20]	AlexNet	Ear Recognition	Ear Image dataset	100%
[21]	ResNet	Classify military objects	Small dataset	90%

The essential objective of transfer learning is to learn a target predictive function for an objective task with support to the target domain and other domains for various tasks. Previous motivations refer to finding a challenge of object detection and classification in small datasets, overfitting, and non-expert training domains such as military and terrorism. These challenges lead the accuracy results to small value and inefficiency. This research targets solving these open challenges to reach good accuracy in terrorism and military domains.

3 The Military Detection and Terrorism Classification System

The proposed solution is automated detection of military objects and terrorism classification system based on deep transfer learning. İt uses for improving accuracy results however, its disadvantage appears in adapting the model on the new tasks.

3.1 The Proposed Terrorism Classification System Neural Network

The input includes two format types of images and videos for several objects and actions, such as fire, or armed attack, in six spectrums in night mode [22]. This network is designed by six neural networks for six spectrums (Intensified visual images, Near-infrared spectroscopy (NIR) images, thermal images, LWIR (long wave infrared images), DHV, and RGB). It uses for detecting objects and actions in various spectrums in night mode depends on two sensory data types of images and videos [23]. Each neural network includes two neural networks as shown in Fig. 1. The proposed neural network for detecting objects and another one for detecting actions extracted from images and videos. The experiment's results can be achieved to high performance and high accuracy by utilizing this type of network. The system consists of the following steps, as shown in Fig. 2, Constructing the six neural networks based on AlexNet, AlexNet pre-trained

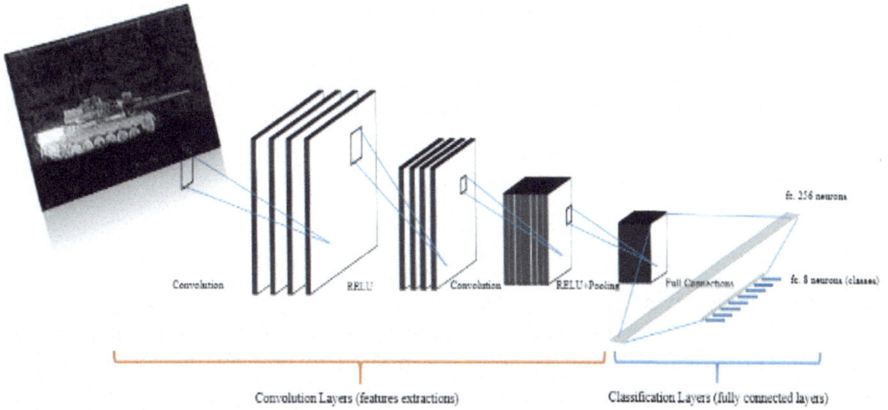

Fig. 1. Alexnet transfer learning model for terrorism and military detection and classification system

model acts as a feature extractor, Freezing the weights of all the other layers, Deleting the last layer of the network and replacing it with our classifier, Training the network normally, The Accuracy assessment, and deploying results.

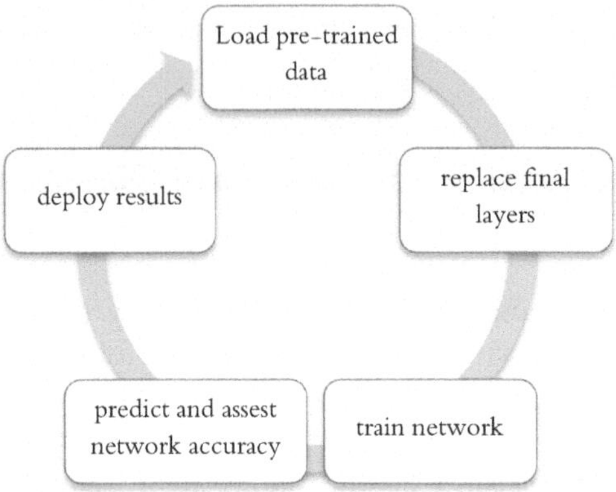

Fig. 2. A detailed transfer learning architecture

The Alexnet pre-trained model is based on Convolutional neural networks (CNN). The following study, the constructed topology of the CNN architecture to discover a convenient model, declaration image classification through Transfer Learning that consists of four layers like the following.

1. Convolutional layer
   Convolution has a great performance in object detection, semantic segmentation, and feature extraction. In this layer, all weights are trainable. The process is interpreted as,

$$x_J^L = \left(\sum\nolimits_{i \in M_J} y_I^{L-1} \otimes K_{ij}^l + b_j^l\right) \tag{1}$$

   In which x_J^L is the input of the jth feature map in the l layer, y_I^{L-1} is the output of the i the feature map in the *l−1* layer. K_{ij}^l is the convolutional kernel between the two feature maps and b_j^l is the corresponding bias. F(*) is the active function and using RELU in the network. It extracts from the following form,

$$f(x) = \max(x, 0) \tag{2}$$

2. Down-sampling layer
   After convolution, the number of feature maps and their dimensions increases rapidly. The down-sampling layer is constructed to prevent this problem. It can minimize the dimension of the features and enhance the strong to rotation or other geometrical translation while maintaining the information contained in the authentic feature at the same time. The process of down-sampling can be described as,

$$x_j^l = \text{f} \ (\beta_j^l g_{down}\left(x_j^{l-1}\right) + b_j^l) \tag{3}$$

$g_{down}(*)$ Represents the down-sampling function, which describes an area of n × n in the former feature maps. Therefore, the output will be 1/n of the original input.

3. Fully connected layer
   Every unit in the fully connected layer is connected to all the units in the feature maps of former layer, and the output layer. The transfer learning process includes two steps. The first step, the network requires to train on big dataset to enable learning the ability to extract common characteristics. Then, the connections have several weights that should be tuned to learn the specific characteristics of military objects with a small training set. The phase of Supervised pre-training, the neural network is trained with a large dataset ImageNet which includes more than 1,000,000 images about 1000 classes with image-level annotations for common object classification. When the input layer receives a training sample. The AlexNet can enhance the terrorism classification model with high accuracy results.
4. Create two neural networks
   This system presents two neural networks in each AlexNet pertained neural network for classifying objects and actions in each image or video's frame. That also improves the classifying domains whether military or terrorism. The domain Classification is an important and complex process for classifying the domain based on extracting features, especially in multi-model system.

   The proposed system works on a supervised classification to enhance the accuracy of classifiers and utilize training samples which known as the ground truth. The spectral signatures of the training areas are used to inspect for similar signatures in the remaining pixels of the image. This system classifies by AlexNet transfer neural network and construct and description layer for is used in this study for improving the results.

3.2 A Description Layer

This system extracts the terrorism domain with actions and objects from a set of military datasets due to the lack of information in this domain. The description layer is created to fill the gap between military and terrorism domains with interpreting a description for images and videos. For example, the different description of armed people whether a terrorist or the army. It is designed based on targeting the semantic features and processing images and videos more adaptive to many objects features. It requires constructing a new annotated lexicon for improving the descriptions. This lexicon enhances the accuracy percentage of domains' classification.

3.3 Data Augmentation

The lack of training data of military and terrorism domains is a still challenge that is faced in applying the neural network. This system introduces a solution for this challenge through increasing the number of images used for training to reduce the overfitting of the small datasets. It applies the label-preserving transformations. It uses four data augmentation techniques used in this research are: (a) Reflection around the X-axis, (b) Reflection around the Y axis, (c) Reflection around the X−Y axis, and (d) gaussian. The used techniques of augmentation can be adaptive to increase the number of images by a factor of 6 times compared with the original dataset. The dataset increased a training dataset and testing phases. The proposed system was developed using a software package (Matlab). The implementation was graphical processing unit (GPU) specific.

4 Experiment and Results

4.1 Dataset

The experiments are applied to three datasets; however, it suffers from a lack of critical data (images or videos) in military or terrorism domains. The used dataset makes depends on 1332 images that is enlarged by six times for reaching 7992 images and videos for six spectrums. It provides the classification between objects and actions (Tables 3 and 4).

Table 3. The used datasets includes six spectrums for objects and actions

Dataset	Ref	Description	Dataset Size
TNO Image Fusion	[24]	Visual (0.4–0.7 μm), near-infrared (NIR, 0.7–1.0 μm) and long-wave infrared (LWIR, 8–14 μm)	579 images
Gun objects dataset	[25]	Images dataset in real images	333 images
Multi-spectral images	[26]	Dataset of seven objects into three spectrums: visible, near-infrared and thermal spectrum	420 images

Table 4. The accuracy measurements

	Relevant	Non-relevant
Retrieved	TP	FP
Not retrieved	FN	TN

$$Recall = \frac{TP}{TP + FN} \quad (4)$$

$$Precision = \frac{TP}{TP + FP} \quad (5)$$

$$F\text{-}measure = \frac{2Recall \cdot Precsision}{Recall + Precision} \tag{6}$$

4.2 Results

The system improves the accuracy of detection and classification by applying the Eqs. 4, 5, and 6 [27]. The classification accuracy results rough 75% for two spectrums (NIR and VIS) and thermal spectrum reaches 89% approximately for thermal spectrum, 91% for LWIR and DHV spectrums, RGB reaches around 92%. Table 5 shows the precisions, recall, and F1-score for the accuracy computation in object detection and classification.

Table 5. Performance metrics of classified spectrums

Metrics/Model	NIR	VIS	Thermal	LIWR	DHV	RGB
Precision	74.5%	75%	89.3%	91.4%	90.3%	92.8%
Recall	75.2%	74.6	91.2%	92.2%	90.4%	91.7%
F1-score	75.1%	75.6%	89%	91.9%	91.1%	92.1%

5 Conclusion and Future Work

The proposed military detection and terrorism classification system is designed based on three steps, (1) constructing six AlexNet transfer neural network learning. They use for classifying six spectrums in night mode. Each neural has two neural for detecting objects and classifying actions. They also classify the domain whether terrorism or military. (2) A description layer is generated for combining extracting features and interpretation description for information in the image to strengthen the ability of classification and differing between variant domains. (3) the Evaluation accuracy of classification depends on precision, recall, and f-measure assessments. The experiments are applied to three datasets for six spectrums in night mode. These datasets include objects and actions in videos and images. Using data augmentation, for improving the classification with enlarging the dataset by six times. The classification accuracy result reaches 92%. The future work refers to build a fusion system that works on multiple sensory data with variant data types to improve the accuracy of detection and classification results. It can make up the lack of datasets to draw the full vision of data for improving the classification accuracy.

References

1. Gohar, F., Haider, W., Qamar, U.: Terrorist group prediction using data classification. In: The International Conference on Artificial Intelligence and Pattern Recognition (AIPR 2014). Asia Pacific University of Technology and Innovation (APU) (2014)

2. Tinguria, D., Kumar, B.: Detecting terror- related activities on the web using neural network. Orient. J. Comput. Sci. Technol. **3**(2), 331–336 (2010)
3. Ritchie, H., Hasell, J., Appel, C., Roser, M.: Terrorism. OurWorldInData.org (2019). https://ourworldindata.org/terrorism
4. Institute for Economics and Peace: Global Terrorism Index 2019, Measuring the Impact of Terrorism, Sydney (2019). http://visionofhumanity.org/reports
5. Felzenszwalb, P.F., Girshick, R.B., McAllester, D., Ramanan, D.: Object detection with discriminatively trained part-based models. IEEE Trans. Pattern Anal. Mach. Intell. **32**(9), 1627–1645 (2010)
6. Nascimento, I., et al.: Deep learning in RAT and modulation classification with a new radio signals dataset. In: Simpósio brasileiro de telecomunicações e processamento de sinais - sbrt, 16–19 de setembro de 2018
7. Yang, J., et al.: Hyperspectral image classification using two-channel deep convolutional neural network. In: Proceedings of the 2016 IEEE International Geoscience and Remote Sensing Symposium (IGARSS), Beijing, China, pp. 5079–5082 (2016)
8. Mormont, R., Geurts, P., Mar, R.: Comparison of deep transfer learning strategies for digital pathology. In: IEEE/CVF Conference on Computer Vision and Pattern Recognition Workshops (CVPRW) (2018)
9. Patil, C.J., Shinde, S.V.: Object detection in military and space image by deep learning with convolutional neural network. Int. J. Comput. Sci. Eng. **6**(11), 363–368 (2018)
10. Yang, Z., et al.: Deep transfer learning for military object recognition under small training set condition. Neural Comput. Appl. **31**(10), 6469–6478 (2018)
11. Remjee, S., Yang, D., Elgamal, A.: Fast deep learning for automatic modulation classification. arXiv:1901.05850v1 (2019)
12. Wu, Y., Ji, Q.: Constrained deep transfer feature learning and its applications. In: The IEEE Conference on Computer Vision and Pattern Recognition (CVPR), pp. 5101–5109 (2016)
13. Liu, X., Liu, Z.-T., Wang, G.: Ensemble transfer learning algorithm. IEEE Access **6**, 2389–2396 (2017)
14. Wang, S., et al.: Alcoholism identification based on an AlexNet transfer learning model. Front. Psychiatry **10**, 205 (2019)
15. Singh, D., Garzon, P.: Using Convolutional Neural Networks and Transfer Learning to Perform Yelp Restaurant Photo Classification (2018)
16. Smith, P., Chen, C.: Transfer learning with deep CNNs for gender recognition and age estimation. arXiv:1811.07344v1 (2018)
17. Hussain, M., Bird, J.J., Faria, D.R.: A study on CNN transfer learning for image classification. In: 18th Annual UK Workshop on Computational Intelligence (UKCI), Nottingham (2018)
18. Kwan, H.K., Lee, C.K.: A neural network approach to pulse radar detection. IEEE Trans. Aerosp. Electron. Syst. **29**(1), 9–21 (1993)
19. Talo, M., et al.: Application of deep transfer learning for automated brain abnormality classification using MR images. Cogn. Syst. Res. **54**, 176–188 (2018)
20. Almisreb, A.A., Jamil, N., Din, N.Md.: Utilizing AlexNet deep transfer learning for ear recognition. In: Fourth International Conference on Information Retrieval and Knowledge Management (CAMP) (2018)
21. Signoroni, A., et al.: Deep learning meets hyperspectral image analysis: a multidisciplinary review. J. Imaging **5**, 52 (2019)
22. Zhan, Y., et al.: Hyperspectral band selection based on deep convolutional neural network and distance density. IEEE Geosci. Remote Sens. Lett. **14**, 2365–2369 (2017)
23. Zhao, W., et al.: On combining multiscale deep learning features for the classification of hyperspectral remote sensing imagery. Int. J. Remote Sens. **36**, 3368–3379 (2015)

24. Toet, A., Pinkus, A.R., Hogervost, M.A.: The TRICLOBS dynamic multi-band image data set for the development and evaluation of image fusion. PLoS ONE **11**, e0165016 (2016)
25. Chandan, J.G., et al.: Real time object detection and tracking using Deep Learning and OpenCV. In: International Conference on Inventive Research in Computing Applications (ICIRCA) (2018)
26. Zukal, M., et al.: Interest points as a focus measure in multi-spectral imaging. Radio Eng. **22** (1), 68–81 (2013)
27. Khalifa, N.E.M., et al.: Deep iris: deep learning for gender classification through iris patterns. Acta Informatica Medica **27**(2), 96–102 (2019)

An Efficient Multi Secret Image Sharing Scheme Using Hill Cipher

Mervat Ahmed[1], Hatem S. Abdul-kader[1], Arabi kishk[1], and Azza A. Abdo[2(✉)]

[1] Faculty of Computers and Information, Menoufia University, Shebeen El-Kom, Egypt
[2] Computer Department, College of Science and Humanities, Imam Abdulrahman Bin Faisal University, Jubail, Saudi Arabia
aaaali@iau.edu.sa

Abstract. This paper presents a multi secret sharing scheme using Hill cipher and Blakley secret sharing. The proposed scheme can reconstruct secret images with only k or more of shadow images. It is compared with the scheme proposed by another work, which produced a visual secret sharing scheme for gray scale images with a (n, n)-threshold. The proposed scheme decreases the threshold to (n, n) and it's compared with other work. Besides, the proposed scheme shares more than one secret key using the Hill Cipher algorithm without depending on a random grid as a shared image. Using Hill Cipher and Blakley to share more than secret image increases the security of secret images with lossless image recovery and no pixel expansion. Experimental results demonstrate the efficiency of the scheme.

Keywords: Visual Secret Sharing · Hill Cipher · Blakley secret sharing

1 Introduction

Security is playing an important role for protecting information when it flows over in a network. A visual secret sharing (VSS) was by Naor and Shamir for secure communication of secret images. Secret sharing is a technique by which the dealer shares the secret key into t secret keys, in the way that only a threshold number n, such that $1 < n \leq t$ of shared keys must be collected to retrieve the original secret key. Secret sharing schemes are important in cryptography and the key for many secure protocols such as threshold cryptography, access control, and attribute-based encryption.

Building of secret sharing images schemes is depending on minimizes the communication overhead. While using of multiple shares for encoding one secret image increases, so there are multi-secret sharing schemes have been introduced to be a better solve to the problem [2–4].

In 2013 Chen et al. proposed a scheme for multi secret image sharing [5]. Chen scheme used Hill cipher scheme to divide a secret gray image into sub shared images, and then each shared image is XORed with a random grid matrix to encrypt the these shared images, so, n secret images are encoded into $n + 1$ meaningless shares. The scheme claimed to provide secure communication with easy decryption. However, in

A.-E. Hassanien et al. (Eds.): AICV 2020, AISC 1153, pp. 604–613, 2020.
https://doi.org/10.1007/978-3-030-44289-7_57

2014, Bunker [6] the weakness of Chen and et al. scheme by guessing the hill matrix encryption key even if the attacker does not have Hill cipher keys, secret image is partly revealed if random grid is simply XOR'ed with the encrypted sub images. Bunker used a system of linear equations with secret keys (coefficients) to divide a secret image into sub images of smaller size.

In 2015, Jothi [7] presented a secure VSS scheme based on linear equations and random grids. Jothi' scheme was an extension of the Chen scheme [9]. But unlike Chen's scheme, Jothi's scheme achieved a multi-secret sharing. Jothi's scheme claimed to provide a perfect security than Chen's scheme, so Jothi implemented the cipher grids which were almost random. But using a linear equations and random grids, n secret images are encoded into n + 1 meaningless shares in Jothi scheme; i.e. it use $(n, n+1)$ threshold. In 2018, Hodeish et al. [10] proposed two-Key Hill Cipher (TKHC)-based image sharing scheme is proposed. The idea of the new Hill Cipher-based algorithm (TKHC) is based on the use of two key matrices. The first key matrix is randomly generated, and the second key matrix is generated by computing the cofactors of the first matrix.

In this paper, we propose a multi secret sharing scheme using Blackley secret sharing and Hill cipher as encoded algorithm. In comparison with Jothi [7] scheme we find that we do not use random grid as share image. Also, a (k, n)-threshold secret sharing method using Blakely concept which only k or more of shadow images can reconstruct all secret images so it will decrease the network overflow. Also using of Hill cipher will increase the security to the scheme.

2 Related Work

2.1 Hill Cipher

Matrix multiplication is the core of Hill cipher. To encrypt the secret image I, a secret key matrix K and is used, and the encryption process will be as follows. The image I is divided into equal size blocks and Hill cipher is applied on each of these blocks separately. Consider a block $\mathrm{P} = \begin{bmatrix} \mathrm{p}_1 \\ \mathrm{p}_2 \end{bmatrix}$ of two consecutive pixels from the image I. Block P is encoded into its equivalent cipher block $C = \begin{bmatrix} c_1 \\ c_2 \end{bmatrix}$ using the following linear equations:

$$\begin{aligned} c_1 &= (k_{11}p_1 + k_{12}p_2) \ mod\ 256, \\ c_2 &= (k_{21}p_1 + k_{22}p_2) \ mod\ 256 \end{aligned} \tag{1}$$

Where k_{11}, k_{12}, k_{21}, *and* k_{22} denote the encryption key. The above equations can also be written as $\mathrm{C} = \mathrm{KP}\ mod\ 256$

Where $C = \begin{bmatrix} c_1 \\ c_2 \end{bmatrix}$, $k = \begin{bmatrix} k_{11} & k_{12} \\ k_{21} & k_{22} \end{bmatrix}$ and $P = \begin{bmatrix} p_1 \\ p_2 \end{bmatrix}$

The secret can be retrieved as $C = K^{-1}P\ mod\ 256$. It is sensible to note that K must be invertible and the entries in inverse matrix must be positive integers.

2.2 Blakley's Secret-Sharing Scheme

Blakley's secret-sharing scheme invented by George Blakley in 1979 [8]. Blakley scheme is built on a point on k-dimensional hyper planes. Hyper planes are intersecting at one point, and each share defines a hyper plane. Secret is the point of intersection of any k hyper planes. First, the secret is divided into n hyper planes. A hyper plane is the set of solutions $x = (x_1, \ldots.., x_n)$ to an equation of the form $a_1x_1 + \ldots a_nx_n = b$. By finding the intersection of any n of these hyper planes, the secret can be obtained.

2.3 Jothi Scheme

Jothi scheme [7] was built on the following two phases:

2.3.1 Sharing Phase

Given two secret images S_1, S_2 the pixel blocks are constructed by fetching pixel from secret images in the same position as $p_1 = S_1(i,j)$ and $p_2 = S_2(i,j)$. The sharing phases are lists as the following:

- **Step-1:** Given two secret images $\boldsymbol{S}_1, \boldsymbol{S}_2$, each of size $M \times N$. Construct *i'th* pixel block $C(i) = \begin{bmatrix} c_1 \\ c_2 \end{bmatrix}$ using $(i) = \begin{bmatrix} p_1 \\ p_2 \end{bmatrix}$, where p_1 and p_2 are the i'th pixels from S_1 and S_2 respectively.
- **Step-2:** The ciphered pixels $C(i,1)$ and $C(i,2)$ are generated by encrypting pixels $P(i,1)$ and $P(i,2)$ using Hill-cipher as follows:

$$\begin{aligned} C(i,1) &= (k_{11}p_1 + k_{12}p_2)\ mod\ 256, \\ C(i,2) &= (k_{21}p_1 + k_{22}p_2)\ mod\ 256 \end{aligned} \tag{2}$$

- **Step-3:** Repeat step-2 for all blocks and construct the intermediate images $I_1\ and\ I_2$ respectively where $I_1(i) = C(i,1)\ and\ I_2(i) = C(i,1)$.
- **Step-4:** Construct a random grid $\boldsymbol{R}$, where $\boldsymbol{R}$ has the same size with the secret image is built through a random number generator function.
- **Step-5:** A second layer of encryption is performed through the random grid to obtain the two final ciphered grid images $E_1\ and\ E_2$ as follows:

$$\begin{aligned} E_1(i) &= I_1(i) \otimes R(i) \\ E_2(i) &= I_2(i) \otimes R(i) \end{aligned} \tag{3}$$

  Where $\otimes$ is Bit-wise XOR operation and R is the random grid.
- **Step-6:** Distribute, $E_1\ and\ E_2\ and\ R$ as the share images.

2.3.2 Reconstruction Phase

The secret is retrieved by following the reverse process. The XOR operation is applied between random grid R and the ciphered grid images $E_1\ and\ E_2$, the result will give intermediate images $I_1\ and\ I_2$ respectively. Then these images are deciphered using inverse of the key matrix of hill cipher to bring forth the secret image.

3 The Proposed Scheme

The proposed scheme contains two stages, sharing stage and recovering stage. The proposed scheme is shown in Fig. 1. In this section, a multi-secret sharing (k, n) algorithm based on Blakley VSS and Hill cipher is proposed.

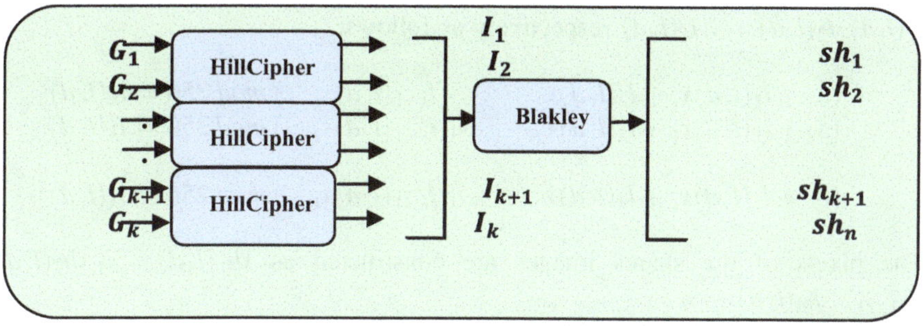

Fig. 1. Diagram of sharing stage.

In the proposed scheme, even k secret images $G_1, G_2, \ldots, G_k$ of size $L \times D$ are encoded in to n share images $Sh_1, Sh_2, \ldots, Sh_n$. The proposed encoding scheme is described in detail in two stages sharing stage and recovery stage.

3.1 Sharing Stage

- Step-1: For each two secret images $\mathrm{G_i}$ and $\mathrm{G_{i+1}}$:
  For $l = 1 : L$
  For $d = 1 : D$
  Construct p_1 and p_2 pixels block $P = \begin{bmatrix} p_1 \\ p_2 \end{bmatrix}$, where p_1 and p_2 are chosen from the same position as $G_i(l, d)$ and $G_{i+1}(l, d)$ respectively.
- Step-2: Hill cipher will encode the block P into the cipher block $C = \begin{bmatrix} c_1 \\ c_2 \end{bmatrix}$ as follows:

$$C = \begin{bmatrix} c_1 \\ c_2 \end{bmatrix} = \begin{bmatrix} k_{11} & k_{12} \\ k_{21} & k_{22} \end{bmatrix} \begin{bmatrix} p_1 \\ p_2 \end{bmatrix} \quad (4)$$

  Where $I_i(l, d) = c_1, I_{i+1}(l, d) = c_2$
- Step-3: Until step-2 has been finished we have the intermediate secret shares $I_1, I_2, \ldots.., I_k$ for all secrets $G_1, G_2, \ldots, G_k$.
- Step-4: Construct disjoints sections from intermediate shares images $I_1, I_2, \ldots.., I_k$. Each section has k of pixels. Each pixel has the same position of the intermediate images as $I_1(l, d), I_2(l, d), \ldots, I_k(l, d)$.

- Step- 5: Construct a polynomial equation:

$$\left(S_j + a_1x_1 + a_2x_2 + \cdots + a_{k-1}x_{k-1}\right) mod\, 256 = a_k \tag{5}$$

where $(a_1, a_2, \ldots, a_k)$ are the pixel values of each section, $x_1, x_2, \ldots, x_{k-1}$ are randomly selected to obtain S_j shared values, and $j = 1{:}n$.
- Step-6: Substitute the coefficients $a_1, a_2, \ldots, a_k$ in Eq. 5 by $I_1(l,d), I_2(l,d), \ldots, I_k(l,d)$ respectively as follows:

$$\begin{array}{l} \left(S_1 + I_1(l,d)x_1 + I_2(l,d)x_2 + \ldots + I_{k-1}(l,d)x_{k-1}\right) mod\, 256 = I_k(l,d) \\ \left(S_2 + I_1(l,d)x_1 + I_2(l,d)x_2 + \ldots + I_{k-1}(l,d)x_{k-1}\right) mod\, 256 = I_k(l,d) \\ \vdots \\ \left(S_n + I_1(l,d)x_1 + I_2(l,d)x_2 + \ldots + I_{k-1}(l,d)x_{k-1}\right) mod\, 256 = I_k(l,d) \end{array} \tag{6}$$

The pixels of the shares images are constructed as $sh_1(l,d) = s_1, sh_2(l,d) = s_2, \ldots\ldots sh_n(l,d) = s_n$.
- Step-7: Successively take the pixels of the unprocessed sections in Eq. (6) and obtain n shared images $sh_1, sh_2, \ldots\ldots, sh_n$.

3.2 Recovery Stage

- Step-1. Collect any k of shared images $sh_1, sh_2, \ldots\ldots, sh_n$.
- Step-2. Collect the values of parameters $a_1, a_2, \ldots, a_k$ from shared images $\mathrm{sh}_1, \mathrm{sh}_2, \ldots\ldots, \mathrm{sh}_n$.
- Step-3. Form the equation (Eq. 5) by using those parameters and solve equations to get pixel values of different images.
- Step-4. Collect all the pixels of these images to get the intermediate shared images $I_1, I_2, \ldots I_k$.
- Step-5. For $i = 1 : 2 : k$ do the following steps:
  For each two images I_i and I_{i+1}:
  Construct c_1 and c_2 pixels block $C = \begin{bmatrix} c_1 \\ c_2 \end{bmatrix}$, where c_1 and c_2 are chosen as $I_i(1,d)$ *and* $I_{i+1}(1,d)$ respectively.
  Take first pixel of sub-images $I_i\, and\, I_{i+1}$ into Eq. (7) and obtain two values $P = \begin{bmatrix} p_1 \\ p_2 \end{bmatrix}$ as follows:

$$C = \begin{bmatrix} p_1 \\ p_2 \end{bmatrix} \begin{bmatrix} k_{11} & k_{12} \\ k_{21} & k_{22} \end{bmatrix}^{-1} \begin{bmatrix} c_1 \\ c_2 \end{bmatrix} \tag{7}$$

  Successively take the pixels $p_1\, and\, p_2$ into Eq. (7) and recover the original secret images $G_i(1,d)$ and $G_{i+1}(1,d)$ respectively without distortion.
- Step-6. Until step-4 has been finished we successively take the original secret images $G_1, G_2, \ldots G_k$ without distortion.

4 Results and Discussion

In this section, results of the proposed sachem are presented. The results is implemented on the gray scale images of size 256 × 256. The gray-scale images Babon and Baby are used to be the secret images as shown in Fig. 2(a–b) respectively using encryption key with size [3 2;4 3] and two shared images are shown in Fig. 2(c–d). The recovered images in Fig. 2(e–d). The gray-scale images Cameraman, lena, girl and eye are used to be the secret images as shown in Fig. 3(a–d) by using the encryption key with size 4*4 is [2 1 1 5, 253 253 252 6, 3 3 3 2,4 4 4 3] and four shared images are shown in Fig. 3(e–h). The recovered images in Fig. 2(i–l) are constructed using six shares.

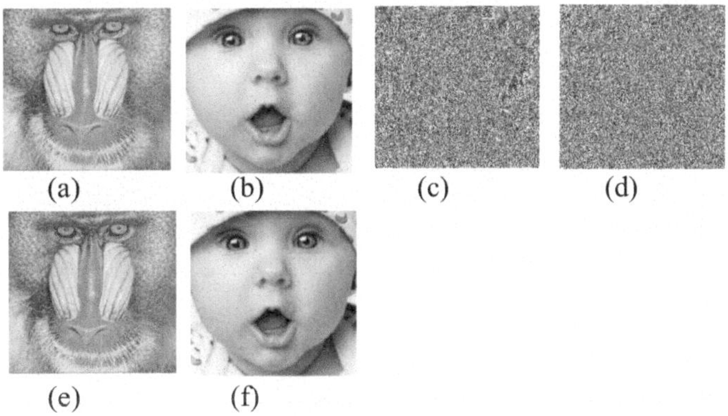

Fig. 2. Secret image sharing: (a–b) secret images, (c–d) 2 shares, (e–f) constructed images.

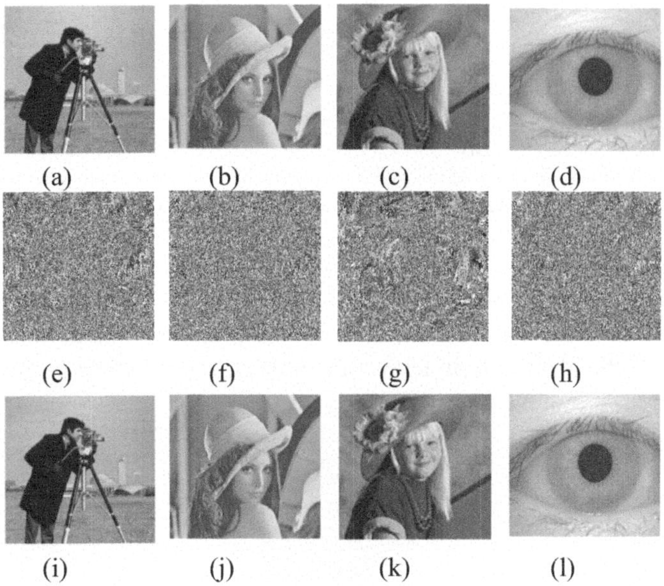

Fig. 3. Secret image sharing: (a–d) secret images, (c–d) 4 shares, (i–l) constructed images.

The gray-scale images Cameraman, lena, girl and eye are used to be the secret images as shown in Fig. 3(a–d) by using the encryption key with size 4*4 is [2 1 1 5, 253 253 252 6, 3 3 3 2, 4 4 4 3] and six shared images are shown in Fig. 3(e–j). The recovered images in Fig. 2(i–n) are constructed using only four shares from Six shares. The results of the experiments are visualized evidences to the merit of pixel expansion, as the size of every share is equal to the size of the secret image. The proposed scheme is ideal and indeed satisfies the pixel expansion precision (Fig. 4).

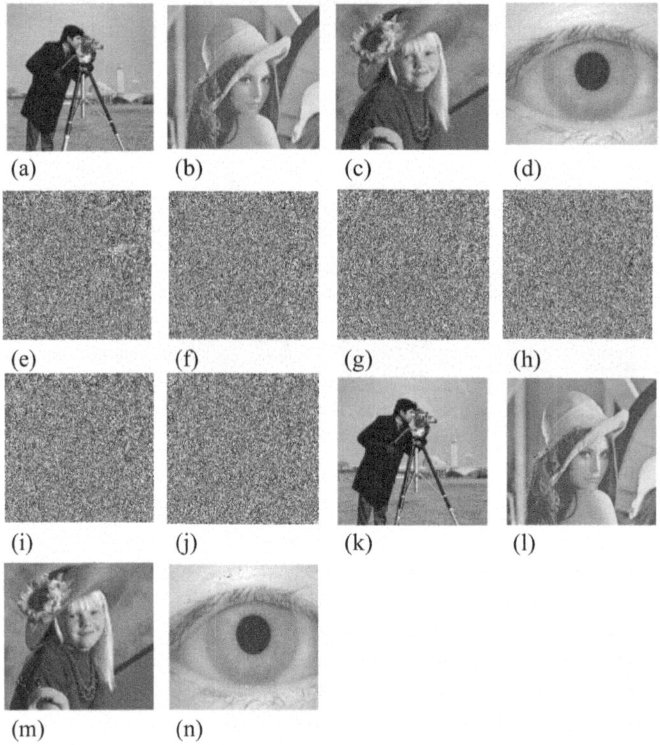

Fig. 4. Secret image sharing: (a–d) secret images, (e–j) shares, (k–n) constructed image.

5 Analysis

This section demonstrates several objectives of the proposed scheme such as security, and it discusses the identically in terms of confusion and diffusion properties.

Fig. 5. Recovered image when Sh_3 is unknown

(A) Security: The proposed scheme provides the security criteria based on the secret key used, where the secret images can't be recovered without having a full information about all shared images and hill secret key matrix. That is, preventing any single shared image from leaking any information about the original secret image. Figure 5 shows the experiments of the proposed scheme, with (4, 4)-threshold. By assuming that one of the shared images Sh_3 is unknown, the recovered secret image used only three shared images. The recovered secret image is shown in Fig. 5. In this case no information about the secret image is obtained. The mean square error between the secret image and recovered image is equal to 0, which is acceptable value.

(B) Confusion and Diffusion Properties: To improve the confusion and diffusion properties of the proposed scheme, statistical analysis such as correlation between adjacent pixels of the shared images and histogram analysis is applied. Histogram of the secret image compared with histograms of its shared images is shown in Fig. 6. It is observed that the histograms of the shares are fairly uniform compared with the significantly different histogram of the original image. Also, the correlation analysis is by randomly selected 1000 pairs of two horizontally, vertically, and diagonally adjacent pixels, for the original image as well as for its shares. In each case, the results for the correlation coefficient of each pair are shown in Table 1. In Table 1, it is shown that the correlation coefficient for two horizontally adjacent pixels for the secret image is 0.8590, which is very near to 1. Nevertheless, coefficients are 0.0154, and 0.0066, respectively for two shared images which are very close 0. Also the similar results for vertically and diagonally adjacent pixels are exploited. These properties allow the scheme to strongly resist statistical attacks. Finally, correlation between adjacent pixels in the shared images compared with the correlation between the adjacent pixels' in the secret images is presented in Fig. 7. The correlation between two diagonal vectors of pixels and tow horizontal vector of pixels are computed and visualized as in Fig. 7. It is seen that there is a random distribution of the shared images (refer to Fig. 7(c, d)), are compared with regular correlation between pixels for the secret images (Fig. 7(a, b)). This means that the proposed scheme is efficient in terms of confusion and diffusion properties.

Table 1. Correlation coefficients of two adjacent pixels in the secret images and its shares

	Original image1	Original image2	Shr1	Shr2
Horizontal	0.8590	0.6605	0.0154	0.0066
Vertical	0.7371	0.8305	0.0232	0.0050
Diagonal	0.7404	0.6659	0.0738	0.0121

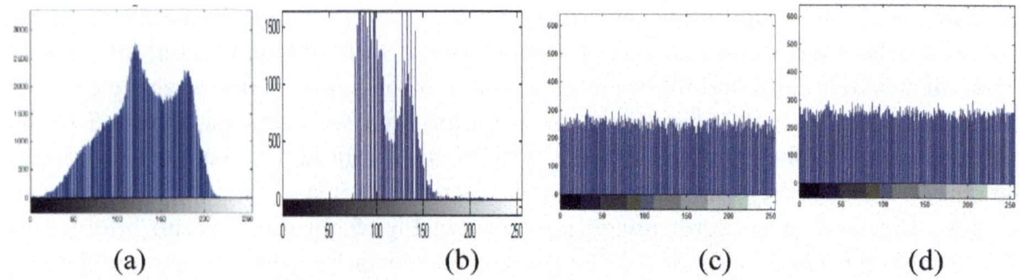

Fig. 6. Histogram of the two secret images (a–b) and its shares (c–d).

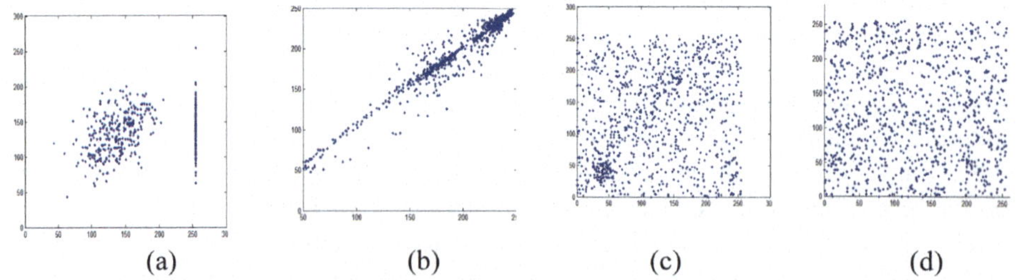

Fig. 7. Correlation of the two secret images (a–b) and its shares (c–d).

6 Conclusion

In this paper, when using Hill Cipher with Blakley sharing scheme we observed that it is a method for multi secret sharing which (k, n)-threshold secret sharing method i.e. only k or more of shadow images can reconstruct secret images and $k - 1$ cannot obtain information about secret images and we did not need to use random grid step that is used in scheme [7] and this increase security to sharing scheme in [8]. Given secret images to be encrypted using hill cipher technique into ciphered images and then construct polynomials of ciphered images to get shares. It is observed that security is improved when multiple images are encrypted using Hill cipher so shares of Blakley do not reveal any information about the original secrets without need to random grids.

References

1. Shamir, A.: How to share a secret. Commun. ACM **22**(11), 612–613 (1979)
2. Wu, H.C., Chang, C.C.: Sharing visual multi-secrets using circle shares. Comput. Stand. Interfaces **28**, 123–135 (2005)
3. Shyu, S.J.: Sharing multiple secrets in visual cryptography. Pattern Recogn. **40**, 3633–3651 (2007)
4. Chen, T.-H., Tsao, K.-H., Lee, Y.-S.: Yet another multiple-image encryption by rotating random grids. J. Sig. Process. **92**, 2229–2237 (2012)

5. Chen, T.-H., Wu, C.-S.: Efficient multi-secret image sharing based on boolean operations. J. Sig. Process. **91**, 90–97 (2011)
6. Bunker, S.C., Barasa, M., Ojha, A.: Linear equation based visual secret sharing scheme. In: Advance Computing Conference (IACC), pp. 406–410. IEEE International (2014)
7. Jothi, R., Ojha, A.: On Multi-secret sharing using Hill cipher and random grids. In: International Conference on Advances in Computer Engineering and Applications (2015)
8. Tso, H.-K.: Sharing secret images using Blakley's concept. Opt. Eng. **47**, 077001 (2008)
9. Chen, W.-K.: Image sharing method for gray-level images. J. Syst. Softw. **86**, 581–585 (2013)
10. Hodeish, M.E., Bukauskas, L., Humbe, V.T.: A new efficient TKHC-based image sharing scheme over unsecured channel. https://doi.org/10.1016/j.jksuci.2019.08.004

Intelligent Systems and Applications

Cyber-Medicine Service for Medical Diagnosis Based on IoT and Cloud Infrastructure

Safia Abbas(✉)

Department of Computer Science, Faculty of Computer and Information Sciences, Ain Shams University, Cairo, Egypt
Safia_abbas@cis.au.edu.eg, safia_abbas@yahoo.com

Abstract. Nowadays, medical misdiagnose propagates widely, due to among other reasons, a serious lack of cloud support software. Many patients have difficulty finding either their condition or the appropriate treatment. This paper presents a cyber-medicine service for medical diagnosis based on IoT and cloud infrastructure. It is elastic, on-demand, and reliable medical service offered for both patients and specialists. For patients, it can diagnose a disease based on symptoms and provide a convenient treatment based on a patient's conditions. The analysis of patients' conditions is started by optionally taking medical measurements of heart rate using sensors. After then, the service starts to interact with the patients to get the symptoms and search for a diagnosis. For prentice doctors, many features are provided as an on-demand disease description, scientific paper search, and symptoms displaying. Finally, to get an unrecognized combination of symptoms, web crawling and multiple classifiers are used.

Keywords: Cyber-medicine · IoT · e-Health · Cloud infrastructure

1 Introduction

For a long time, lots of people suffer, in a forthright way, from a misdiagnose of their medical conditions. Death and lifelong disabilities can come from a misdiagnosis error by not properly diagnosing results and not being able to give the correct treatment in a timely manner. There are many reasons that lead to such misdiagnosis process, such as advanced diseases that doctors are not completely aware of, lack of knowledge about specific diseases, confusing symptoms of multiple diseases that occur more often, and disordering the proper laboratory tests. Accordingly, many researchers have been devoted to harnessing data science, internet and communication technologies to find a drastic solution for such a crisis that increases as time pass. Consequently, many scientific terms have appeared to indicate different strategies in the health domain that deploys such technology; eHealth [3, 14, 25] and e-Health [16, 20, 21, 24] are examples of those terms. However, recently, terms such as cyber medicine, IoT and cloud-based infrastructure propagate and are deployed widely [1, 2, 4, 15]. Cyber medicine, as a metaphor for eHealth [1, 3, 12, 17], considered the deployment of internet communication to provide a health care system based on a disparate computer databases and patient-management applications together via the internet; patients can access and store their data and medical conditions to consult their specialist and ask for

A.-E. Hassanien et al. (Eds.): AICV 2020, AISC 1153, pp. 617–627, 2020.
https://doi.org/10.1007/978-3-030-44289-7_58

drug prescriptions online, without face to face clinical visits [3, 18, 19]. Many applications appeared to deploy such concepts in the health domain and have invested in the business domain successfully; WebMD is an example for these applications [24]. Later on, IoT appeared as an ecosystem for the objects implanted with sensors, software, and network connectivity to communicate with each other for data collection and exchange [7, 8, 11, 23]. Recently, cloud computing played an important role in the eHealth domain [5, 9, 10, 19]; IoT eHealth applications prosper the remote monitoring of patient health and fitness information, trigger alarms when critical conditions are detected, and, in some cases, allow the remote control of certain medical treatments or parameters [8–10]. Such systems usually consist of multiple layers including cloud infrastructure as a service (IaaS) and software as a service (SaaS) layers and provides a scalable, on-demand, flexible, and secure environment for the health care domain that assures low cost and secure services [3, 6, 8, 12]. Such characteristics are considered the main shortcomings and major weak points in the cyber medicine field. Therefore, It is expected that applications deploying IoT based Cloud IaaS dedicates a reliable decision-making system that enjoys certain advantages, such as (1) data collection and adaptation based on sensors, data repositories, and others; (2) data availability that both specialists and patients can access freely; (3) an easy data sharing and exchanging; and (4) securing data transmission process between different parties. This paper introduces a novel diagnostic service that deploys the IoT based cloud IaaS in order to benefit from the previously mentioned features. It entails different layers and modes for both patients and doctors. Moreover, it deploys multiple classifier agents to guarantee accurate decision making in the diagnosis process.

The paper is organized as follows: Sect. 2 reviews the relevant work performed in the cross fields IoT and eHealth based cloud. In Sect. 3, the Proposed Service Oriented Architecture of the medical environment is presented associated with the entailed main Models. Section 4 provides an illustrative example including screenshots of the prototype. Finally, concluding remarks and future work are presented in Sect. 5.

2 Related Work

Recently, numerous studies and applications have been conducted on and provided as Home-diagnosis service over WWW-based cloud infrastructure using big data concepts and IoT. Some of these studies and applications are concerned with how to deploy the cloud infrastructure (IaaS) and big data in their applications to achieve an on-demand, pervasive service with secure data transmission channels. On the other hand, others are concerned with how to utilize the IoT to analyze and monitor the patients' conditions and to send alerts for critical cases.

In [26], authors target people in Suboptimal Health Status (SHS), also known as "the third state" (between being healthy and falling sick). They developed home self-caring services to help people get disease precaution knowledge easily. They designed a distributed Lucene-based search cluster that deploys the cloud IaaS to achieve scalability and privacy protection. In [5], the authors propose a Mobile Health care application, as a decision support system, to monitor and diagnose the diabetes severity, based on experimental data from the UCI repository. They propose a new hybrid technique called Fuzzy Rule-based Neural Classifier for diagnosing and use the Hadoop files to store data. In [19], the authors consider the IoT main aspect that is to connect heterogeneous entities and gather a huge amount of data. They shed the light on the inevitable potential of healthcare personnel, as a mutual knowledge sharing for the ubiquitous accessibility for patients' data. Finally, the authors reveal a baseline vision for a smart, remote diagnose decision support model that deploys the provided insights of the IoT concepts to provide a ubiquitous e-health environment. In [13], the authors introduced a survey on the existing remote monitoring technologies, and how it could be deployed in the IoT and cloud environment. They discussed different challenges that could be faced during the implementation, such as heterogeneous and transfer latency of data. Finally, the authors discuss the benefits and the impact of health monitoring and management using IoT sensing with cloud processing. In [22], the authors provide an industrial IoT-driven healthcare ecosystem in which two different types of stakeholders are connected together to form a complex Health IIoT ecosystem, where parties of stakeholders varied between patients, specialists, and drug and health industry organizations. Authors expected that such an ecosystem will allow fast and secured data transmissions between parties, offer real-time patient monitoring, and avoid hospital errors. They focused on the ECG analysis using wavelet transformations and watermarking processes to secure the analog and digital analyzed data.

In the proposed service, the IoT is an environment that encompasses multi-users' types, such as patients and doctors. It is deployed through different types of sensors, remotely or M2M, for patient health monitoring, such as the ECG for heartbeat disorder detection. Moreover, the diagnostic process is done based on the bagging bootstrapping concept, where multiple classifiers reveal the interoperability between different symptoms and crawl for the most relevant and convenient diagnose. The search is usually done over the immense and varied repositories of patients' data to assure the high confidence and accuracy for the decision making the process. The environment will be able to manipulate different types of data, analogue, scans (images), and digital, in addition to offering various services, such as disease prediction through patient data monitoring, diagnosis through symptoms analysis, searching for specific information about specific symptoms or disease, this aid not only the patients but also the specialists. Table 1 shows a comparison between different systems representing the advantages and disadvantages of each study.

Table 1. Comparison between IoT medical environments

App	Ref	Diagnoses	Advantages	Disadvantages
Health Status (SHS)	[26]	Disease precaution knowledge	- Deploys Lucene-based cluster to achieve scalability, privacy and protection -Provides an on-demand elastic scalable data model -Uses bloom filter signature, for symptoms overlapping	- Needs an offline Lucene data file preparation (since automatic preparation for such files seems difficult) - Looses self adaptation in the online mode
Mobile Health	[5]	Diabetes severity	- Uses Hadoop data file system to verify scalability - Considers confidentiality of patients' medical records	- Lacks prediction of new cases that do not entailed in the Hadoop DB file system - Depends only on fuzzy logic as decision making rather than bagging bootstrapping concept for more accuracy
Smart eHealth system	[19]	Monitoring system of heart pluses	- Reveals insights of the IoT ubiquitous e-health environment	- Lacks specific technique or methodology
Survey on remote monitoring in IoT based cloud environment	[13]	Health management in general	- Introduces several architectures for the IoT health care based cloud - Discuss challenges that could be faced during the implementation Offers a proactive framework for prognosis of diseases at an incipient stage	- No specific disease is diagnosed, it considers general health management - No clear predictive strategy or techniques Focus on sensors and wearable devices rather than the general view of the diagnoses processes
Health IIoT ecosystem	[22]	ECG	- Connects stakeholders such as patients, specialists, and health industrial organizations - Deploys watermarking for data transmission of ECG signals through cloud	- Focus on the ECG analysis was proposed previously by many authors - It was better to verify the expectation on much more complex data such as brain or kidney disorder detection

3 Proposed Service Oriented Architecture

This section introduces a cyber-medicine service in the e-health domain that co-operates the IoT and cloud computing concepts, aiming to help both patients and specialists by (i) providing a medical service in the cloud layers; (ii) integrate multiple techniques to manipulate various types of input data; and (iii) cultivate the decision making process by using multiple classifiers. The service has two different modes, either to sign in as a doctor or as a patient, and contains three main layers as shown in Fig. 1.

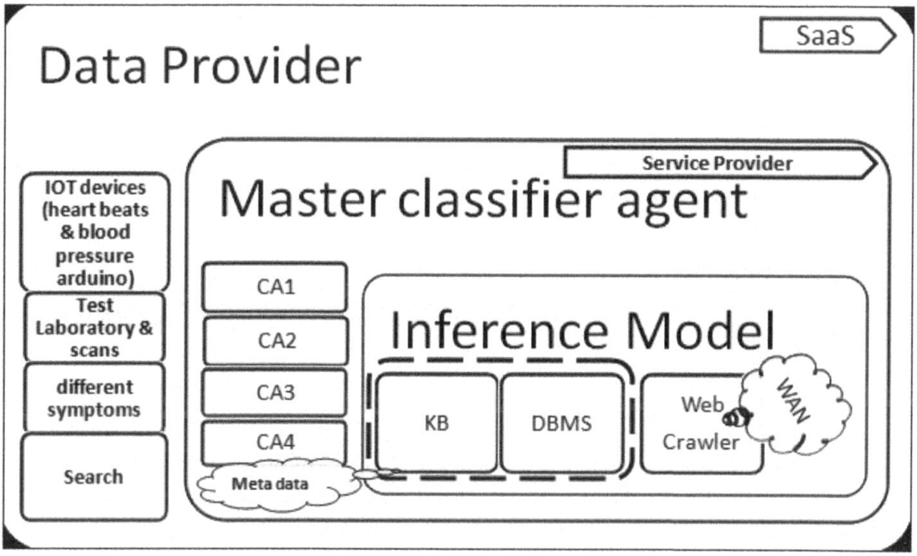

Fig. 1. Service Oriented Architecture

3.1 Data Provider (DP)

DP manipulates the heterogeneous data types' interactions through the whole architecture, such data is generated from multiple sources as described below:

- **Arduino Measurements (optional phase):** Usually, heartbeats and blood pressure considered as an important pre-analysis tests for patient severity conditions. It aids in predicting and discovering the critical cases such as myocardial infarction, if the patient suffers from chest pain. The Arduino kit converts the analog data into digital one and draws the corresponding BPM diagram similar to the ECG plot, an example is provided in Fig. 3.
- **Test laboratory and scans:** This service allows the user to choose a specific part of the body, and the type of data to be entered: symptoms, laboratory tests or CT scans. After then, it detects patient's abnormal scans and marks the disorder, seen Fig. 2 for kidney's stone detection example.

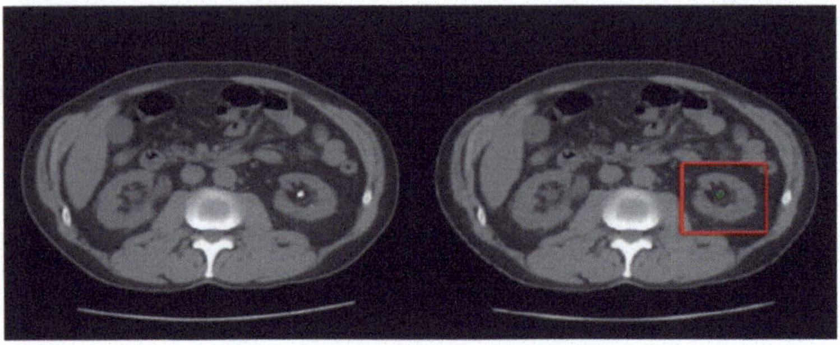

Fig. 2. CT before and after stone detection

- **Search:** the specialist/doctor is allowed to search in previous cases by diagnosis, symptoms, or both, and follow up specific patient conditions. Moreover, doctor can search for scientific papers on a specific topic, in a specific duration, or more information for specific diseases.

3.2 Master Classifier Agent

The agent is responsible of mastering multiple classifiers based on the type of data provided from DP as follows:

- **Analogue and scans data:** the analogue data and the scanned data, which is gathered from sensors and entered by the users, respectively, are both sent by the master classifier agent to the classifier CA1. CA1 deploys and implements image processing techniques in order to reveal the disorder information.
- **Digital data:** any other data types, other than analogue and scans, are considered digital and manipulated using different classification algorithms. The master classifier agent gathers different digital data from the DP and distributes the required tasks between classifiers from CA2 to CA4. Each classifier would then perform its task by implementing classification techniques based on an inference model, and return back with the results to the master agent. In this case, the master agent would have three choices: 1. if the returned decision has high accuracy, the agent will admit and show out the diagnose result. 2. If the result accuracy is low, the master agent will ask other classifiers to perform the same task, and will make a voting between the results to choose the most accurate one. 3. If the results do not match any previous case, the master agent will send to the inference model asking for crawling the internet to get a decision.

3.3 Inference Model

The inference model consists of two main phases, meta-data phase and crawling phase. A description of these models is as follows:

- **Meta-data phase:** in order to facilitate and fasten the search, triggers are sent to the inference model by the master agent, meta-data files that describe both the entailed knowledge base (KB) and database management system (DBMS) are used for that purpose. The importance of the KB is to provide the required explanations for the diagnosis process. The importance of the DBMS is to provide the details and the required analytical analysis in the form of extracted reports, plot diagrams, or charts.

- **Web Crawler phase:** in order to cultivate the master agent results and reveal an on-demand new information, a web crawler is provided to the inference stage. To illustrate, if the master agent could not find a satisfying results as an answer for the trigger under the search process, it sends to the web crawler. The web crawler crawls the websites searching for relevant information, copies the return information from the internet, preprocesses, transforms, and stores them in both DBMS and KB. Finally, it provides the meta-data with an abbreviated representation that describes the added information.

4 Illustrative Example for System-User Interaction

This section represents an incipient run for the proposed prototype development. It contains screenshots for the graphical user interface that interacts with different types of users, either patients or doctors.

User ≫ run the program

Service ≫ start reading the heart pulses and do the analysis, see Fig. 3.

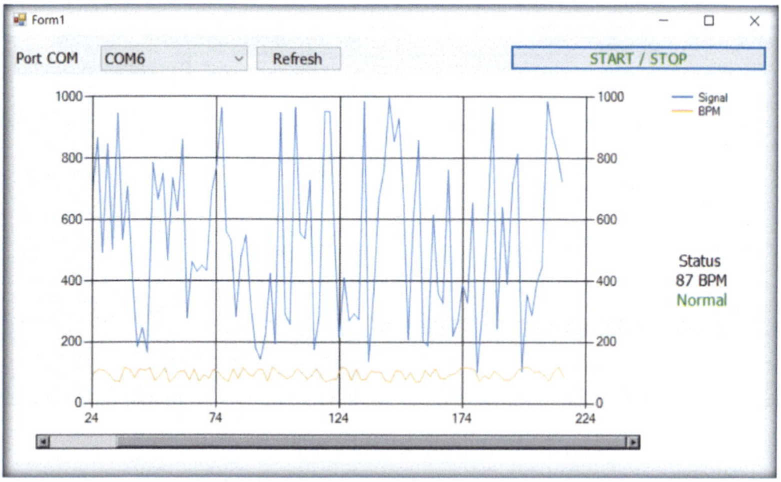

Fig. 3. ECG Plot chart

User ≫ press Diagnose button
Service ≫ show the screen in Fig. 4.

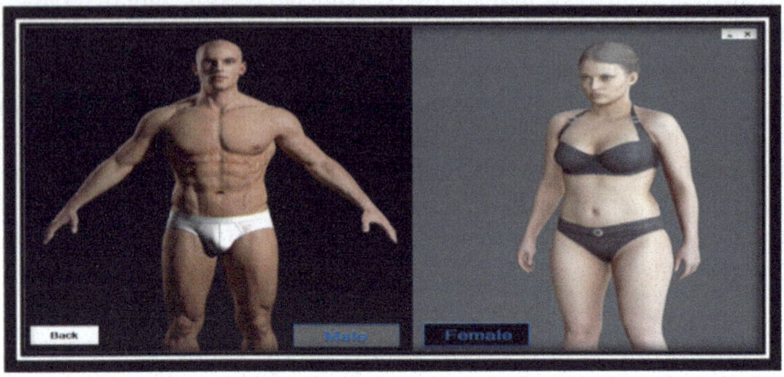

Fig. 4. Select gender

User ≫ select female, then select specific parts of the body, where he/she feels the pain.

Service ≫ common symptoms will be showed to the user.

User ≫ select the symptoms, see Fig. 5.

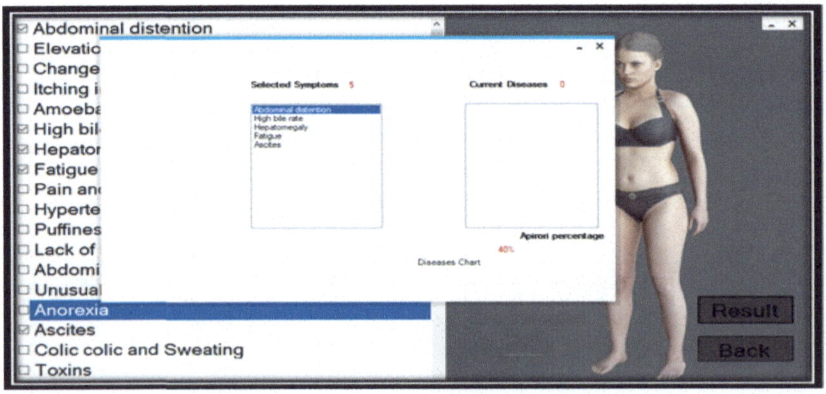

Fig. 5. Body part selection and symptoms

Service ≫ shows the results using crawler, see Fig. 6.

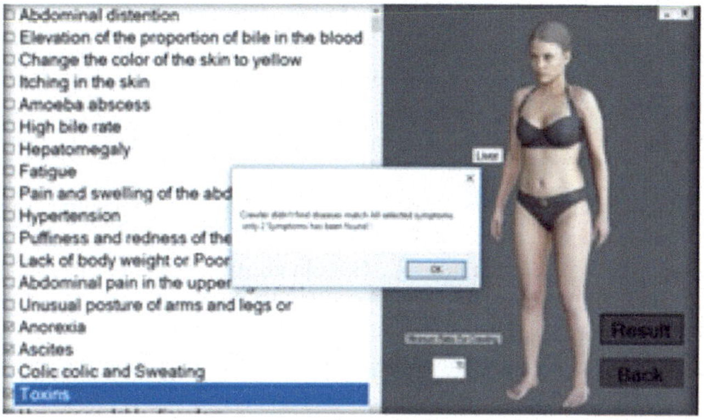

Fig. 6. Results based on crawler

5 Conclusion and Future Work

Recently, misdiagnose propagates widely due to the tremendous entwine and ambiguity of symptoms that overwhelm the diagnosis process and hinder the specialists from providing a reliable diagnosis. This leads to a fallacious treatment that might cause crucial harm. Accordingly, the e-health is an inevitable domain that deploys multiple disciplines such as IoT and cloud to assure accuracy, ubiquity, and availing of the diagnostic process. In this paper, a novel diagnostic service based on IoT and Cloud IaaS is proposed. The core of the service is a cyber-medicine software application that consists of three main layers and deploys different classifiers for the diagnostic process. Moreover, it is elastic, on-demand, and reliable medical diagnostic service offered for both patients and prentice doctors. Despite that our proposed service is in the early implementation phases, it is expected to be broadly availing for multifarious users since it enjoys certain advantages over others, such as (i) it is an adaptive extendable online diagnostic service that deploys crawler to search for none pre-existing data. (ii) it is flexible enough to manipulate heterogeneous data (analog, scans, or context), (iii) it cultivates the diagnosis process by deploying various classifiers and guarantee high accuracy, and (iv) it offers a new search features for both specialists and patients, to search for specific scientific paper, topic or specific disease and symptoms. In the future, the details of the preprocessing, classification, and decision-making processes are going to be discussed, showing the implemented techniques, data processing methodologies, and decision-making approach.

References

1. Riyadh, M., Tandjaoui, D., Romdhani, I.: Architecting the Internet of Things: state of the art. In: Robots and Sensor Clouds, vol. 36, pp. 55–75. Springer (2015)
2. Eysenbach, G., Ryoung, E., Diepgen, T.L.: Shopping around the internet today and tomorrow: towards the millennium of cybermedicine. Br. Med. J. (BMJ) **319**, 1294 (1999)
3. Arvanitis, T.N., Eysenbach, G., Woodall, J.: Towards the millennium of cybermedicine. In: The 4th World Congress on the Internet in Medicine (MEDNET 1999), Heidelberg, Germany (1999)
4. Kreps, G.L., Neuhauser, L.: New directions in eHealth communication: opportunities and challenges. Patient Educ. Couns. **78**(3), 329–336 (2010)
5. Kumar, P., Lokesh, S., Varatharajan, R., Gokulnath, C., Parthasarathy, P.: Cloud and IoT based disease prediction and diagnosis system for health care using fuzzy neural classifier. Future Gener. Comput. Syst. **86**, 527–534 (2018)
6. Hu, Y., Bai, G.: A systematic literature review of cloud computing in eHealth. Health Inform. Int. J. (HIIJ) **3**, 11–20 (2014)
7. Indrajit, I., Nangpal, S.: "Cybermedicine and cyberhealthcare" review of medicine on the internet. Med. J. Armed Forces India **57**, 215–220 (2001)
8. Ishaq, I., Hoebeke, J., Moerman, I.: Internet of Things virtual networks: bringing network virtualization to resource-constrained devices. In: IEEE International Conference on Green Computing and Communications (GreenCom), pp. 293–300. IEEE Computer Society, Besançon, Paris (2012)
9. Islam, S.R., Kwak, D., Kabir, M.H., Hossain, M., Kwak, K.S.: The Internet of Things for health care: a comprehensive survey. IEEE Access **3**, 678–708 (2015)
10. Kumar, P.J., Chaithra, M.A.: A survey on cloud computing based health care for diabetes: analysis and diagnosis. IOSR J. Comput. Eng. (IOSR-JCE) **17**(4), 109–117 (2015)
11. Hu, L., Qiu, M., Song, J., Hossain, M.S., Ghoneim, A.: Software defined healthcare networks. IEEE Wirel. Commun. **22**(6), 67–75 (2015)
12. Li, Y., Guo, L., Guo, Y.: Enabling health monitoring as a service in the cloud. In: IEEE/ACM 7th International Conference on Utility and Cloud Computing, pp. 127–137. IEEE, London (2014)
13. Page, A., Soyata, T., Sharma, G., Aktas, M., Hassan, M., Mateos, G.: Health monitoring and management using Internet-of-Things (IoT) sensing with cloud-based processing: opportunities and challenges. In: IEEE International Conference on Services Computing (SCC 2015), pp. 285–292. IEEE Computer Society, New York (2015)
14. Oh, H., Rizo, C., Enkin, M., Jadad, A.: What is eHealth (3): a systematic review of published definitions. J. Med. Internet Res. (JMIR) **7**(1), 1–11 (2005)
15. Lake, D., Milito, R., Morrow, M., Vargheese, R.: Internet of Things: architectural framework for eHealth security. J. ICT **3&4**, 301–328 (2014)
16. Tran, N.: WebMD: transforming the U.S. health care industry. Technical report (2000). Accessed January 2019
17. Vermesan, O., Friess, P.: Digitizing the Industry Internet of Things Connecting the Physical, Digital and Virtual Worlds, vol. 49. River, Denmark (2016)
18. Kumar, S., Nilsen, W., Pavel, M., Srivastava, M.: Mobile health: revolutionizing healthcare through transdisciplinary research. Computer **46**(1), 28–35 (2013)
19. Chatterjee, P., Armentano, R.L.: Internet of Things for a smart and ubiquitous ehealth system. In: 7th International Conference on Computational Intelligence and Communication Networks (CICN), Jabalpur, India, pp. 178–183 (2015)

20. Verma, P., Sood, S.K.: Cloud-centric IoT based disease diagnosis healthcare framework. J. Ambient Intell. Hum. Comput **9**(5), 1293–1309 (2017)
21. Hailey, D., Dales, R., Mensinkai, S.: A review of guidelines for referral of patients to sleep laboratories [technology report no 55]. Technical report, Canadian Coordinating Office for Health Technology Assessment, Ottawa (2005)
22. Shamim, M., Muhammad, G.: Cloud assisted Industrial Internet of Things (IIoT) enabled framework for health monitoring. Comput. Netw. J. **101**(C), 192–202 (2016)
23. Kirtava, Z., Gegenava, T., Gegenava, M.: mHealth for cardiac patients telemonitoring and integrated care. In: 15th International Conference on e-Health Networking, Applications and Services (HealthCom 2013), Lisbon, Portugal, pp. 21–25. IEEE Explore (2013)
24. Mohammed, J., Thakral, A., Ocneanu, A.F.: Internet of Things: remote patient monitoring using web services and cloud computing. In: IEEE International Conference on Internet of Things (iThings 2014), Taipei, Taiwan, pp. 256–263. IEEE (2014)
25. Uribe, S., Marino, R.J.: Internet and information technology use by dental students in Chile. Eur. J. Dent. Educ. **10**(3), 162–168 (2006)
26. Lin, W., Dou, W., Zhou, Z., Liuc, C.: A cloud-based framework for home-diagnosis service over big medical data. J. Syst. Softw. **102**, 192–206 (2014)

Predicting Stock Market Trends for Japanese Candlestick Using Cloud Model

Magda M. Madbouly[1], Mohamed Elkholy[2], Yasser M. Gharib[1](✉), and Saad M. Darwish[1]

[1] Institute of Graduate Studies and Researches, Alexandria University, Alexandria, Egypt
{mmadbouly, saad.darwish}@alexu.edu.eg, yasrmmg@yahoo.com
[2] Faculty of Engineering, Pharos University in Alexandria, Alexandria, Egypt
mohamed.elkhouly@pua.edu.eg

Abstract. Cloud model covers the randomness gap in fuzzy logic model and represents the uncertainty transformation between two different concepts. First concept is the linguistic term that represent the qualitative mean. While the second is the crisp term which represent the quantitative mean. The proposed work presents promising model which combines cloud model, fuzzy time series, and Heikin-Ashi candlestick to predict and confirm accurate stock trend. The model solves several challenging such as: nonlinearity, uncertainty and noises in stock market trend. Heikin-Ashi Candlesticks are an extended branch of Japanese candlesticks, such candlestick filters out stock noise and effort to highlight the trend. Heikin-Ashi Candlestick is constructed by calculating averages of the previous and current period prices. Cloud model handle the ambiguous and uncertainty in the Japanese candlestick definitions (qualitative information) and actual stock prices (quantitative data). It is applied to build membership functions by handling the uncertainty and vagueness of the stock historical data. Then the suggested model constructs dynamic weighted fuzzy logical relationships based on the membership functions to predict the next open and close prices of the stock as well as the high and low values. Finally it constructs the next Heikin-Ashi Japanese candlestick pattern that clarify the trend direction based on the patterns sequence. The imperial evaluation proves that the proposed model has high forecasting accuracy and is feasible to be implemented.

Keywords: Cloud model · Fuzzy time series · Stock trend · Heikin-Ashi candlestick · Japanese candlestick

1 Introduction

The two main targets in the stock market are "Get high financial gain", and "when?" [1]. That mean in nonlinear, noisy and uncertainty situation of stock price environment, predicting the price of the stock and determining the best time sell or buy stocks is challenging [1]. Forecasting stock prices has two main approaches: fundamental and technical analysis. Fundamental analysis is concerned by finding the economic variables affecting the earnings future of asset [2, 3]. Technical analysis (TA) forecasts the

A.-E. Hassanien et al. (Eds.): AICV 2020, AISC 1153, pp. 628–645, 2020.
https://doi.org/10.1007/978-3-030-44289-7_59

future price movements by studying the action market itself (historical stock price–volume–other related data) through derivative chart data. TA looks for specific patterns, or rules on the charts. TA has three main assumptions: the market discounts everything, prices move in trends and, history repeats itself [2]. Mainly TA use statistical methods and/or Artificial Intelligent (AI) techniques for stock forecasting. Statistical methods have drawbacks such as the need to more normal distribution data, not deal with linguistic historical data. AI techniques such as data mining, machine learning, or pattern recognition have problems. One of these problems is that the training process is nontrivial. Also training results can-not be further used for another target. Traders also have problems like prediction results are ambiguous, the model is a black box, the ambiguous definition of the patterns itself. Therefore, in several situations, understanding the forecasting models is important to the traders to get better prediction results. Observing the candlestick pattern helps in predicting the stock price pattern dynamically rather than predict the stock price every day [4].

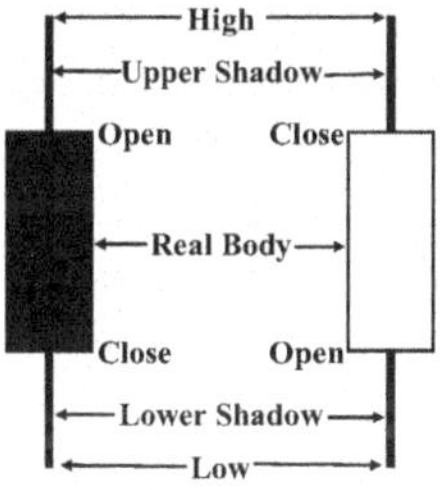

Fig. 1. The dark and white candlestick.

Several forecasting models do not contain the qualitative information that would help about predicting the future. Traditional Japanese candlestick technique is believed to show both quantitative information, such as market trend movements, and qualitative information, such as the market psychology [5]. Therefore, traditional Japanese candlestick technique is a unique charting tool for developing financial forecasting models [6], several traders and investors use the candlestick (Fig. 1) chart system and analyze the pattern visually. Each candlestick is constructed from four values which are Open price, High price, Low prices and Close price (OHLC) in that time period. Candlestick pattern has crisp values: body length and two shadows length (upper and lower). Body length is the absolute different between open and close prices. The upper shadow length is represented by difference between high and maximum value of open and close prices. While the lower shadow length is represented by the difference between low and minimum value of open and close prices [4]. Traditional Japanese candlestick was effected with noisy stock price [7]. Heikin-Ashi (HA) Candlesticks are an extending branch of Japanese candlesticks, and is widely used in candlestick charts [8]. In Japanese, "Heikin" means "average" and "Ashi" means "pace". HA produces simpler and smoother candlestick that filter out stock noise and effort to highlight the trend. The HA candlestick (HaCndl) technique uses modified OHLC values and represent them as a candlesticks form [9]. HA transforms the default and usual bars into other bars in a sense of avoiding noise, false signals or spikes. From the visual graph, the trading in HA is based on the effect of the color and size of the candlestick bodies [10]. HA like traditional candlestick patterns have no crisp or standard definitions. These patterns are defined in uncertain fuzzy linguistic values like "equal", "short", "long" and other. This definition is ambiguous, only expert traders can analysis it accurate.

Cloud Model (CM) is a promising method with the ability to converse between qualitative and quantitative values. Cloud is formed by several cloud drops. The cloud shape presents several features of the quantity concept. CM is successfully applied and shows better performances result in several fields as data mining [11], time series forecasting [12, 13], decision making [14]. CM is also used in risk assessment [15–18], image segmentation [19, 20], Air Combat Situation Assessment [21], robots [22], clustered wireless sensor networks [23] and Power quality [24, 25], supply chain management [26].

1.1 Contribution

The main contribution of the proposed model is to build an accurate stock trend prediction model through utilizing a combination of CM, HA, and Fuzzy Time Series (FTS) in a unified model. The purpose of CM is to add the randomness and uncertainty to the fuzziness linguistic definition of HA and FTS. FTS is utilized to abstract linguistic values from historical data instead of numerical ones, to find internal relationship rules. HA is employed to give easier readability of the candle' features by reduction of noise, eliminates the gaps between candles, and smooth the movement of the market.

The reminder of this paper is organized as follows. Section 2 presents a background for CM, FTS, and HAs. Section 3 looks at problem definition. Section 4 provides a detailed description of the proposed model. The experimental results are discussed in Sect. 5. The conclusion and future work is given in Sect. 6.

2 Background

2.1 Cloud Model

CM as illustrated in (see Fig. 2) depends on probability statistics and traditional theories of fuzzy models [30]. CM is an uncertainty model that converse between the linguistic terms that represents qualitative concepts and the quantitative numerical representation. It covers randomness gap in the fuzzy logic model [11]. The cloud includes three digital features (CM(*Ex*, *En*, *He*)): Expected value (*Ex*), Entropy (*En*) and Hyper entropy (*He*). *Ex* represents the best qualitative concept, which is subordinated to the qualitative concept. It is usually coordinated to the cloud center. While En presents the uncertainty measure associated with the qualitative concept. It is decided by the concept randomness as well as the concept fuzziness. *En* is considered as Gaussian function that represents the quantitative range ambiguity. *It* represents the size of cloud's thickness which reflects the dispersion degree of cloud [11, 27, 30].

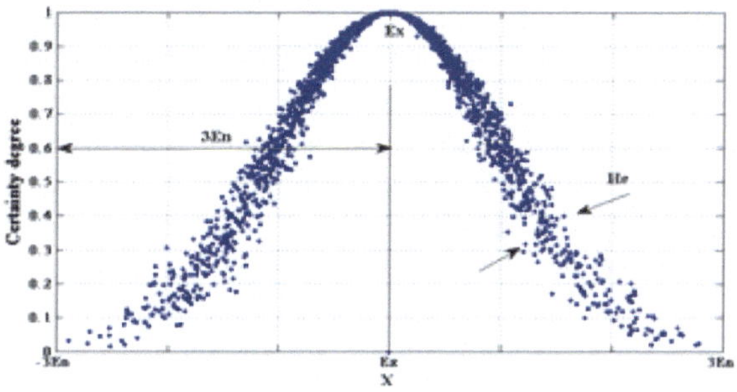

Fig. 2. Cloud model.

2.2 The Fuzzy Time Series Model

FTS which is based on fuzzy theory was firstly introduced in Song et al. [27, 28]. FTS concepts is divided into two types related to the time. The two types are time variant and time invariant [28]. Song et al. construct fuzzy rule matrix that is able to predict the enrollments of Alabama University. They used complicated max–min composition operations which need high computational power to deal with large the fuzzy rule matrix. Chen [31] used the arithmetic operations to reduce the computational complexity of max-min operations. Hwang [32] utilized the similarity of current time variation with previous time variation to construct a window basis in enrollment prediction. Yu [33] introduced a model to tackle recurrence in TALEX forecasting using weighted fuzzy time series (WFTS). Vovan [34] used the variations of data to improved FTS forecasting model. All these methods improve the accuracy of FTS but didn't take the uncertainty and randomization in his mind which give more accurate result.

2.3 Review of Heikin-Ashi Candlestick Pattern

Several forecasting models do not contain the qualitative information that would help about predicting the future. Nison [5] introduced Japanese candlestick concepts as a promising forecasting model. Japanese candlestick patterns show both quantitative information like price and trend as well as qualitative information like the psychology of the market. It considers the close values and the candlestick body information to provide a trading sessions information [35].

Japanese candlesticks is widely used in creative methods in the last decades. Lee et al. [36] suggest a system with IF-THEN rules. The system detects candlestick patterns as well as flag sell and buy orders in the Korean market. This work fuzzify both the candle line and the candle lines relationship, and implement intelligent tools for investment knowledge learning and sharing [37]. Then Lee et al. [4] proposed a decision system that is based on candlestick models for the prediction. Another prediction approach creates personal candlestick pattern ontologies by Lee et al. [38]. The ontology allow users to have their own explanation of a candlestick model. Naranjo et al. [39] apply fuzzy methods to the gap between candles and add it as extended elements in candlesticks patterns. A handle specification of several known candlestick models is introduced in [40].

Since the revolution of Japanese candlestick in western markets [5], traders used it as the most popular and easier charting tools [41]. However it has contradictory information due to market's noise [40], and challenging of interpreting candlesticks [9]. The Heikin-Ashi (HA) technique is a promising modification to candlestick traditional chart. HA allows easier reduction of noise, eliminates the gaps between candles, and smooth the movement of the market. The Heikin-Ashi graph provides easier readability features than traditional candles, and can be considered as a real trading system [10].

3 Problem Definition

The purpose of this study is to predict accurate stock future trend from Heikin-Ashi patterns. The existing forecasting models lacks of accuracy and certainty. Several problems appear in the existing models such as problems in data, prediction model, traders, and Candlestick Patterns. The main problems in data are associated with uncertainty, noise, and nonlinearity. On the other hand, prediction model shows low efficiency due to complex calculations. Traders also have problems associated with ambiguous results. Thus a gap exists between forecasted result and the decision of investment. Moreover traditional candlestick patterns have problems that can be listed as follows: (1) Interpreting candlesticks can be challenging. (2) Large number of the patterns. (3) Definition of patterns itself is ambiguous. (4) Effected with noisy stock price.

4 Methodology

The main steps of the suggested model for stock trend prediction of HaCndl based on CM is shown in Fig. 3 that contains: (1) Preparing historical data (HD). (2) Prepressing of HaCndl data. (3) HaCndl representation using CM. (4) Forecasting the next day price (open, high, low, close) using cloud-based fuzzy time series (5) Formalize the next day HA candle stick features. (6) Forecast the trend.

4.1 Preparing HD

Four different values (open price, high price, low price, close price) time series for several companies were downloaded from (http://finance.yahoo.com) and used as HD.

4.2 Prepressing of HaCndl Data

The extended candlestick (Heikin-Ashi) patterns has one candlestick line. Herein, the suggested model constructs several HA' features that are derived from traditional candlestick patterns. The HaCndl' features are computed by: [9]

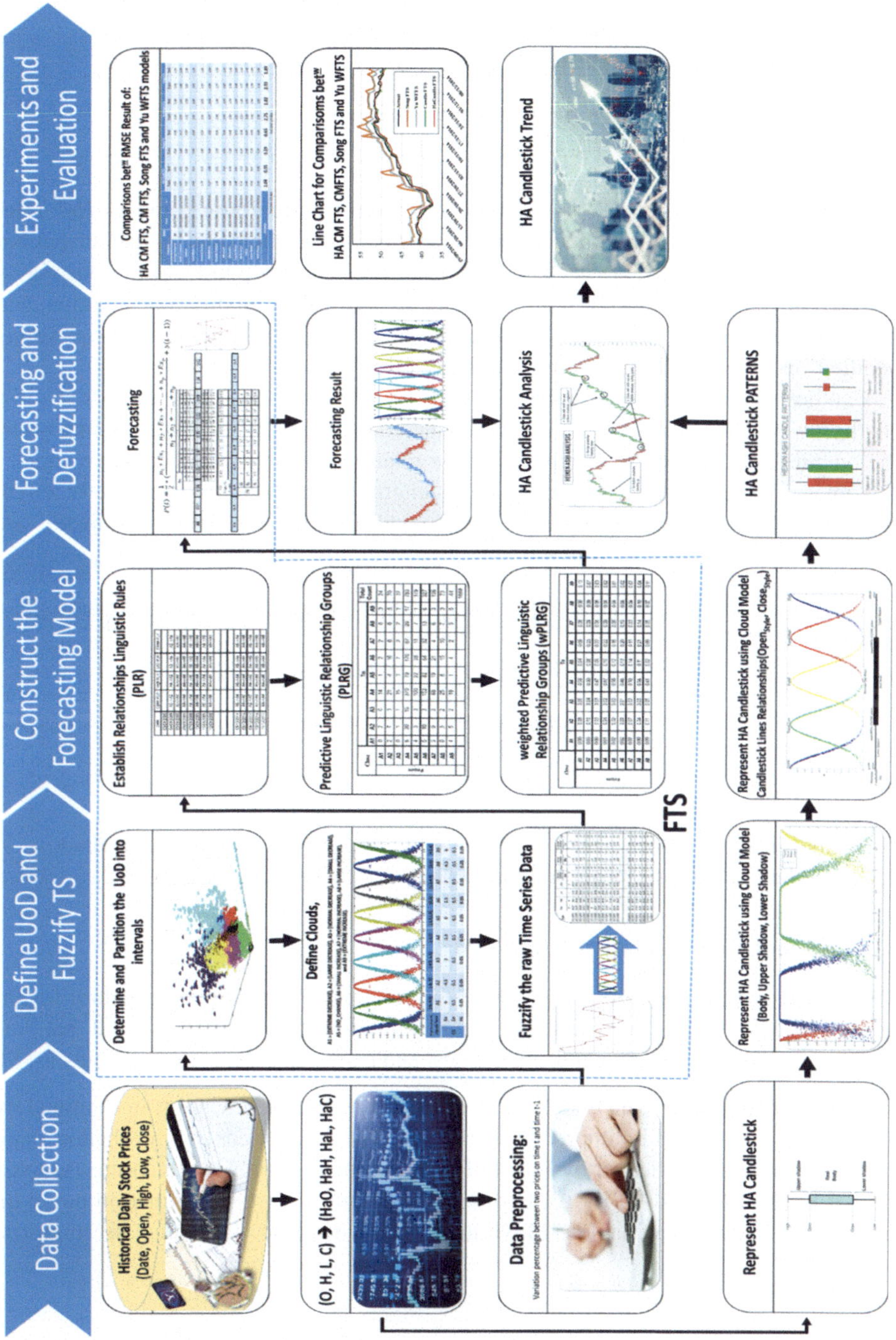

Fig. 3. The procedure of proposed model.

$$\left.\begin{array}{r} HaClose = \frac{(Open + High + low + Close)}{4} \\ HaOpen = \frac{HaOpen(PreviousBar) + HaClose(PreviousBar)}{2} \\ HaHigh = Max(High, HaOpen, HaClose) \\ HaLow = Min(Low, HaOpen, HaClose) \end{array}\right\} \tag{1}$$

where "*Open*", "*High*", "*Low*" and "*Close*" refer to the current price, prefix "*Ha*" refer to the cross ponding HA modified values. The *HaClose* is calculated as the current bar average value of the current bar. The *HaOpen* represents midpoint of previous bar.

HaHigh is the maximum of set that includes {high value, modified open value, and modified close value}. *HaLow* is minimum value of set that includes {Low value, modified open value (*HaOpen*), and modified close value (*HaClose*)}.

Candlestick lines can be described by a set of parameters which are: upper shadow length, lower shadow length, body length, line color, opening style and closing style. The opening and closing style represents the relationship between successive candlestick lines. The crisp value of upper shadow length, lower shadow length, body length and body color identifies a candlestick pattern and its efficiency. These parameters are directly calculated with Eq. (2).

$$\left.\begin{array}{l} \mathrm{HaL_{body}} = \frac{\max(\mathrm{HaOpen, HaClose}) - \min(\mathrm{HaOpen, HaClose})}{\mathrm{HaOpen}} * 100 \\ \mathrm{HaL_{upper\ shadow}} = \frac{\mathrm{HaHigh} - \max(\mathrm{HaOpen, HaClose})}{\mathrm{OPEN}} * 100 \\ \mathrm{HaL_{lower\ shadow}} = \frac{\min(\mathrm{HaOpen, HaClose}) - \mathrm{HaLow}}{\mathrm{HaOpen}} * 100 \\ \mathrm{HaColor} = \mathrm{HaClose} - \mathrm{HaOpen} \end{array}\right\} \tag{2}$$

where the character "*HaL*" indicates the body length, upper and lower shadow of HaCndl. The *HaColor* is mean the body color of the HaCndl.

4.3 HaCndl Representation Using CM

CM is used in transform the crisp candlestick parameters (quantitative value) to linguistic variable define the candlestick (qualitative value). Fuzzy HaCndl patterns ontology is built and contain candlestick lines, and candlestick lines relationships [4, 8].

(a) Candlestick Lines: The body and shadows length of the HaCndl is not defined by crisp value; it is defined by imprecise and vague. The cloud model of the shadows and the body length are indicated by four different linguistic variables which are (LONG, MIDDLE, SHORT, and EQUAL). Figure 4 shows the cloud model function based on $\mu(x)$ of the linguistic variables. Maximum $\mu(x)$ is used to determine its linguistic variable.

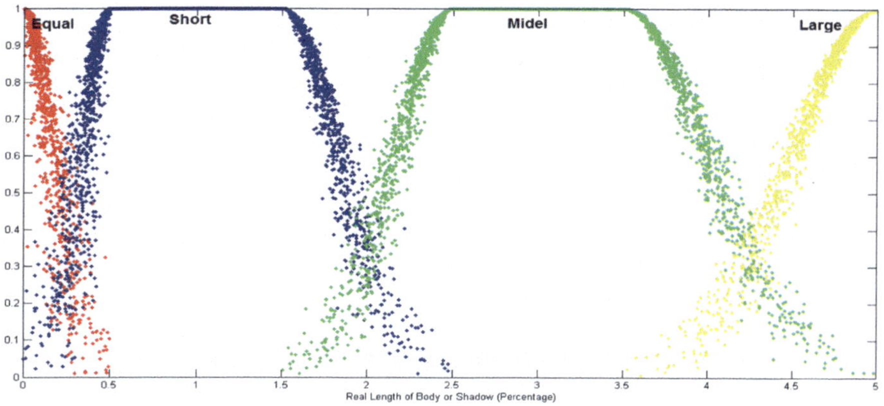

Fig. 4. The membership function of the body and shadows length based on cloud model.

The domain of HaL_{body}, $HaL_{lower\text{-}shadow}$, and $HaL_{upper\text{-}shadow}$ are set to (0, p) to represent the stock price see-sawing percentage. In Fig. 4, the x-axis represent the real value HaL_{body}, $HaL_{lower\text{-}shadow}$, and $HaL_{upper\text{-}shadow}$, and percentage is y-axis unit.

$$EQUAL(x:\ a,b) = \begin{cases} 0, & x<a \\ e^{\left(\frac{-(x-Ex)^2}{2En^2}\right)}, & a\le x\le b \\ 0, & x>b \end{cases} \tag{3}$$

$$SHORT/MIDEL(x:\ a,d) = \begin{cases} 0, & x<a \\ e^{\left(\frac{-(x-Ex_1)^2}{2En_1^2}\right)}, & a\le x\le b \\ 1, & b<x<c \\ e^{\left(\frac{-(x-Ex_2)^2}{2En_2^2}\right)}, & c\le x\le d \\ 0, & x>d \end{cases} \tag{4}$$

$$LARGE(x:\ a,b) = \begin{cases} 0, & x<a \\ e^{\left(\frac{-(x-Ex)^2}{2En^2}\right)}, & a\le x\le b \\ 1, & x>b \end{cases} \tag{5}$$

Another important candlestick line feature is body color (*BODY_COLOR*). It is defined by three cases BLACK, WHITE, and DOJI. A DOJI case is described when candlestick body is horizontal bar. i.e., open price equals close price. The body color is defined in Eq. (6):

$$\left.\begin{array}{c} If(OPEN - CLOSE) > 0\ then\ BODY\_COLOR = BLACK \\ If(OPEN - CLOSE) < 0\ then\ BODY\_COLOR = WHITE \\ If(OPEN - CLOSE) = 0\ then\ BODY\_COLOR = DOJI \end{array}\right\} \tag{6}$$

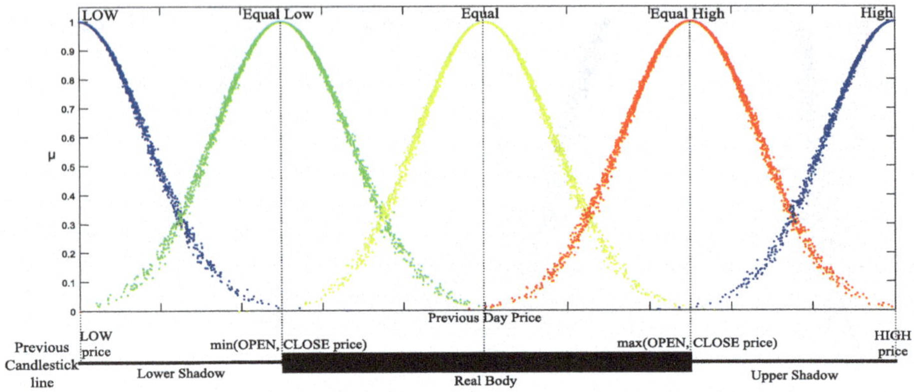

Fig. 5. The HA open and HA close styles membership function.

(b) Candlestick Lines Relationships: It defines the place of the HaCndl with the previous one to form in open style and close style linguistic variable and showed as Fig. 5. At the bottom of Fig. 5, the previous trading day candlestick line is represent. The previous day trading prices is the x axis and the Cloud member (Cm) function $\mu(x)$ is the y axis for each linguistic values. Five linguistic values (LOW_Style, Equal-Low_Style, Equal_Style, EqualHigh_Style, and High_Style) are constructed to represent the relationships style (*X_style*). For any *X* Style, cloud is defined by Eq. (7).

$$X\_Style(x:\ a, b) = \begin{cases} 0, & x < a \\ e^{\left(\frac{-(x-Ex)^2}{2En^2}\right)}, & a \le x \le b \\ 0, & x > b \end{cases} \tag{7}$$

Combining the HaCndl line description and the HaCndl lines relationship, a HaCndl pattern is completely defined.

4.4 Forecast the Next Day Price (Open, High, Low, Close) Using FTS Prediction Method Based on CM

FTS method based on CM consists of six steps [14]:

(a) Determine and Partition the Universal of Discourse (UoD) into intervals,
(b) Define clouds, classify the HD to its cloud,
(c) Establish the Predictive Logical Relationships (PLR).
(d) Establish the Predictive Linguistic Relationship Groups (PLRG),
(e) Forecasting method (Defuzzifying and calculating the predicted value)
(f) Transform the forecasting results (open, high, low, close) to the next HaCndl.

(a) **First step, Determine and Partition the UoD into intervals (define clouds):**

i. HD should be prepared and defining the domain of the UoD. The open, high, low, and close prices should be prepared as data price set for each one.
ii. For each data price set, the variation percentage should be calculated between two prices on time t and time $t + n$ and collected into variation percentage (VP) set by using Eq. (8).

$$\frac{(\mathrm{Pr}_{(t+n)} - \mathrm{Pr}_t)}{\mathrm{Pr}_t} \times 100 \tag{8}$$

where Pr_t mean the price on day t and $Pr_{(t+n)}$ mean the price after n trading days from day t.
iii. Partition the UoD set into intervals, from VP set find the minimum VP (D_{min}) and the maximum VP (D_{max}). Define U based on D_{min} and D_{max} as $U = [D_{min} - D_1, D_{max} + D_2]$ where D_1 and D_2 are appropriate positive numbers. Several UoD ($[D_{min}, D_{max}]$) values from several paper like [−6, 8] in [4] was checked and found the best result with [−6, 6] in [42].

(b) **Define the clouds and classify the historical data to its cloud:**

Based on CM, defined and determines linguistic variables as a set of linguistic terms, and represented as:

$$A\{Cm_{A1}(Ex_1, En_1, He_1), \ldots Cm_{Am}(Ex_m, En_m, He_m)\}.$$

On the data based, the predictive Linguistic rule sets can be discovered as:

$$\begin{aligned} A_1 &= 1.0/Cm_1 + 0.5/Cm_2 + 0.0/Cm_3 + \ldots + 0.0/Cm_{m-1} + 0.0/Cm_m \\ A_2 &= 0.5/Cm_1 + 1.0/Cm_2 + 0.5/Cm_3 + \ldots + 0.0/Cm_{m-1} + 0.0/Cm_m \\ &\vdots \\ A_m &= 0.0/Cm_1 + 0.0/Cm_2 + 0.0/Cm_3 + \ldots + 0.5/Cm_{m-1} + 1.0/Cm_m \end{aligned}$$

where Cm is cloud member function. By set the number of clouds (m) to 9 as in (Fig. 6), the nine clouds are:

$$\begin{aligned} A_1 &= (\text{EXTREME DECREASE}), & A_2 &= (\text{LARGE DECREASE}), \\ A_3 &= (\text{NORMAL DECREASE}), & A_4 &= (\text{SMALL DECREASE}), \\ A_5 &= (\text{NO\_CHANGE}), & A_6 &= (\text{SMALLI NCREASE}), \\ A_7 &= (\text{NORMAL INCREASE}), & A_8 &= (\text{LARGE INCREASE}), \\ A_9 &= (\text{EXTREME INCREASE}). \end{aligned}$$

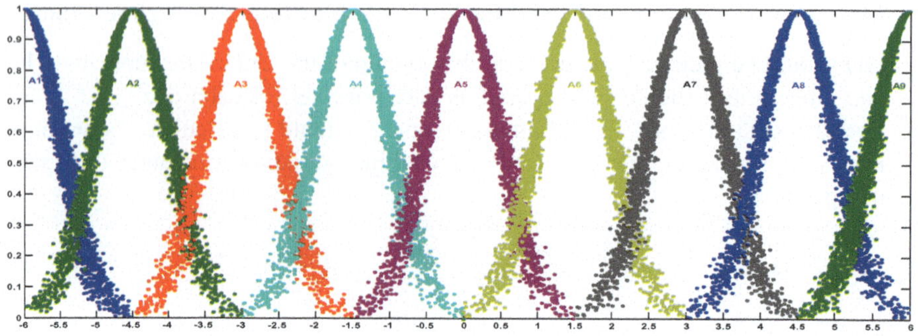

Fig. 6. The clouds of the linguistic terms.

Table 1. The cloud member function digital features for each linguistic term

Linguistic terms		A_1	A_2	A_3	A_4	A_5	A_6	A_7	A_8	A_9
Cm	*Ex*	−6	−4.5	−3	−1.5	0	1.5	3	4.5	6
	En	0.5	0.5	0.5	0.5	0.5	0.5	0.5	0.5	0.5
	He	0.05	0.05	0.05	0.05	0.05	0.05	0.05	0.05	0.05

The cloud member function digital features *Cm(Ex, En, He)* for each linguistic term show in Table 1.

(c) **Establish Predictive Linguistic Relationships (PLR):**

Based on historical data cloud models, create soft inference linguistic rule ($A_{t-1} \rightarrow A_t$) by searching with the pattern ($A_{t-1} \rightarrow A_t$) in all clouds in time series, where A_{t-1} and A_t are linguistic concept clouds.

(d) **Establishment of Predictive Linguistic Relationship Groups (PLRG):**

Create Predictive Logical Relationships Groups (PLRG) by grouping the same "current state" of training dataset PLRs into the same PLRG. For example, if "A_1, A_2, ···, A_m" is the "current state" of one PLR in the training dataset and there are *r* PLRs in the training dataset as follows:

$$A_1 \rightarrow A_1; \quad A_1 \rightarrow A_2; \quad ; \quad ; A_1 \rightarrow A_m$$

Then the *r* PLRs are grouped into the same PLRG as follows:

$$A_1 \rightarrow A_1; A_2; \cdots; A_m$$

Assign the weight values (w) for each PLRG as follows:

$$\begin{aligned} R &= \{\text{number of repetition of the particular PLRs}\} \\ &= \{n_1; .. n_m\} \end{aligned}$$

Total number of PLRs $n = n_1 + n_2 + \ldots + n_\mathrm{m}$, then,
w_m = (number of repetition of A_i)/(total number of PLRs),
thus, $w_1 = n_1/n$, $w_2 = n_2/n$, $w_3 = n_3/n, \ldots.$, $w_m = n_m/n$, so that,

$$\sum_{i=1}^{m} w_i = \frac{n_1}{n} + \frac{n_2}{n} + \ldots + \frac{n_m}{n} = \frac{n}{n} = 1$$

Presented the weight values (w) in the weight matrix W as follows:

$$W(A_i) = W(t) = [w_1 w_2 \ldots w_m].$$

(e) **Forecasting method:**

Use the following rules to Forecast value $P(t)$ at time t or to next day (open, high, low, close) prices:

Rule 1: If there is one PLR in the PLRG, then,

$$P(t) = \frac{Ex_i + S(t-1)}{2} \tag{9}$$

Rule 2: If there is r PLR in the PLRG, ($A_1 \rightarrow A_1$, A_2, ..., A_p) then,

$$P(t) = \frac{1}{2}\left(\frac{n_1 Ex_1 + n_2 Ex_2 + \ldots + n_p Ex_p}{n_1 + n_2 + \ldots + n_p} + S(t-1)\right) \tag{10}$$

Rule 3: If there is no PLR in the PLRG, then,

$$P(t) = \frac{Ex_i + S(t-1)}{2} \tag{11}$$

Where in Eqs. (9), (10), and (11), Ex_i is A_i cloud expected value, n_i is the number of A_i repetition in the PLRG, $1 \leq I \leq r$, $S(t-1)$ is the observed value at time t − 1.

(f) **Transform the forecasting results (open, high, low, and close) to the next HaCndl.**

4.5 Formalize the Next Day HaCndl Features, Forecast the Trend [9]

Rule 1: **If** *BODY_Color* is WHITE and HaL_{body} is LONG **then, UP Trend.**
Rule 2: **If** *BODY_Color* is BLACK and HaL_{body} is LONG **then, Down Trend.**
Rule 3: If *BODY_Color*WHITE and HaL_{body} is LONG and $HaL_{lower\ shadow}$ is EQUAL **then, STRONG UP Trend.**

Rule 4: **If** *BODY_Color* is BLACK and HaL_{body} is LONG and $HaL_{upper\ shadow}$ is EQUAL **then, STRONG Down Trend.**
Rule 5: **If** (HaL_{body} is EQUAL) and ($HaL_{upper\ shadow}$ and $HaL_{lower\ shadow}$) is LONG **then, Change of Trend.**
Rule 6: **If** (HaL_{body} is SHORT) and ($HaL_{upper\ shadow}$ and $HaL_{lower\ shadow}$) is not EQUAL **then, Consolidation Trend.**
Rule 7: **If** (HaL_{body} is (SHORT or EQUAL)) and (*HAOpen_Style* and *HAClose_Style*) is (LOW_Style or EqualLow_Style) and $HaL_{lower\ shadow}$ is EQUAL **then, Weaker Trend.**

5 Experimental Results

MATLAB language to implement and test the efficiency and validity of proposed prototype model. The efficient prototype model was built in a Dell™ Inspiron™ N5110 Laptop machine with the following features: Intel(R) Core(TM) i5-2410 M CPU@ 2.30 GHz, and 4.00 GB of RAM, 64-bit Windows 7. Dataset was collected from 13 real time stocks series of New York Stock Exchange (NYSE) companies, each one with the four prices (open, high, low, and close). Time series dataset were downloaded from yahoo finance web site (http://finance.yahoo.com), Table 2 show the companies' name, its symbol and starting date, ending date for the selected data set. Dataset was divided to 2/3 of dataset for training and other 1/3 for testing.

Table 2. Selected time series datasets

Company	Symbol	From	To
Boeing Company	BA	02/01/1962	27/06/2018
Bank of America	BAC	03/01/2000	12/12/2014
DuPont	DD	03/01/2000	12/12/2014
Ford Motor Co.	F	03/01/2000	12/12/2014
General Electric	GE	03/01/2000	12/12/2014
Hewlett-Packard	HPQ	03/01/2000	12/12/2014
Microsoft	MSFT	03/01/2000	12/12/2014
Monsanto	MON	18/10/2000	12/12/2014
Toyota Motor	TM	03/01/2000	12/12/2014
Wells Fargo	WFC	01/06/1972	27/06/2018
Yahoo	YHOO	03/01/2005	12/12/2014
Exxon Mobil	XOM	02/01/1970	21/05/2018
Walt Disney	DIS	02/01/1962	27/06/2018

The Mean Squared Error (*MSE*) that is used by academicians and practitioners [4], was used to evaluate the accuracy of the proposed model. Equation (12) is defined *MSE* as follows:

$$MSE = \frac{\sum_{i=1}^{n} (forcasted\ value - actual\ value)^2}{n} \tag{12}$$

where n is the number of trading days needed to be predicted.

Table 3 show the Comparisons result between our two model, first one Open, high, low and close price was used as initial price in cloud FTS model (Cloud FTS), the second model, HaOpen, HaHigh, HaLow and HaClose prices was used as initial price in cloud FTS model (HA Cloud FTS) and other two model Song FTS [12] and Yu WFTS [35] model in the same period time as Table 2. Figure 7 represent an example of the line chart between prediction models for training data of Yahoo (YHOO) from 29/09/2014 to 12/12/2014.

Table 3. Comparison MSE for CLOSE price prediction between HA Cloud FTS, Cloud FTS, Yu WFTS and Song FTS models.

Company		MSE							
		HA Cloud FTS		Cloud FTS		Yu WFTS		Song FTS	
		Train	Test	Train	Test	Train	Test	Train	Test
Boeing Company	BA	0.048	0.672	0.078	0.960	5.290	3.460	5.954	3.725
Bank of America	BAC	0.941	0.023	1.124	0.029	6.503	2.592	2.756	0.960
DuPont	DD	0.270	0.116	0.397	0.152	5.336	2.496	14.516	7.076
Ford Motor Co.	F	0.168	0.020	0.203	0.026	5.905	2.690	4.080	1.588
General Electric	GE	3.204	0.023	3.423	0.036	8.526	2.403	9.425	2.074
Hewlett-Packard	HPQ	1.392	0.096	1.769	0.130	7.182	2.756	6.605	2.372
Microsoft	MSFT	0.740	0.048	0.922	0.068	5.905	2.403	7.129	2.372
Monsanto	MON	1.904	0.314	2.528	0.476	8.009	3.028	6.052	1.588
Toyota Motor	TM	1.166	0.449	1.369	0.504	6.300	2.856	19.272	9.303
Wells Fargo	WFC	0.023	0.102	0.040	0.144	4.928	2.624	3.133	1.638
Yahoo	YHOO	0.203	0.073	0.250	0.090	5.664	2.624	6.052	2.496
Exxon Mobil	XOM	0.040	0.221	0.068	0.314	4.580	2.560	6.656	3.572
Walt Disney	DIS	0.023	0.130	0.036	0.194	5.198	2.723	4.580	2.250
Average		0.779	0.176	0.939	0.240	6.102	2.709	7.400	3.155

The experimental result (Table 3) shows that using HA Cloud FTS and Cloud FTS have lower average than the other methods used in Literatures. This low average proves the high accuracy of the proposed model. HA Cloud FTS provide MSE = 0.779 for train data and 0.176 for test data and Cloud FTS gives MSE 0.939 for train data and 0.240 for test data such result mean that HA Cloud FTS method which uses HaOpen, HaHigh, HaLow, HaClose prices as initial price has significant improvement in stock market trend prediction.

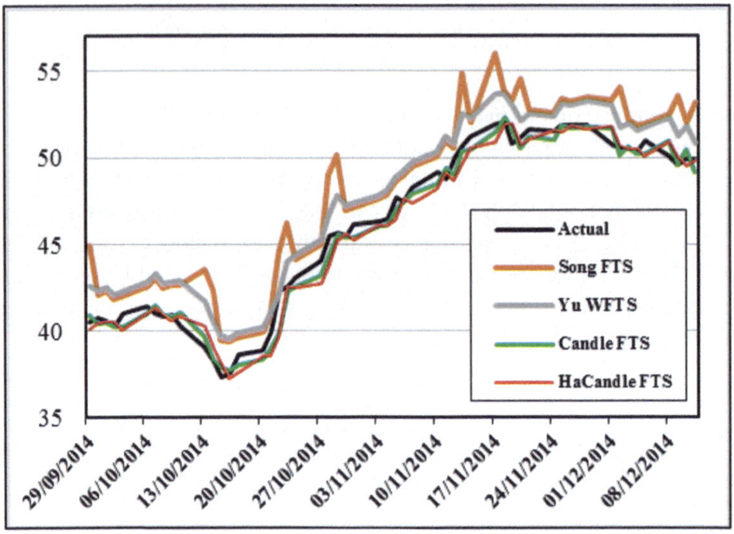

Fig. 7. Forecasting result in line chart

6 Conclusion and Future Work

The proposed work shows an improved fuzzy time series prediction model based on cloud model and Heikin-Ashi Japanese candlestick to predict and confirm accurate stock trend. The MSE of the prediction results of the proposed model is smaller than that of the other model. That is, the proposed model can get a higher prediction accuracy rate for predicting stock prices than Song FTS and Yu WFTS model.

There are several ways to continue this research:

1. Connecting online stock data prices for forecasting the stock trend.
2. Applying high order fuzzy time series to get more accurate model.
3. Use Heikin-Ashi model in deferent AI model.
4. Apply Cloud model in define Heikin-Ashi and Japanese candlestick in different model.

References

1. Gunduz, H., Yaslan, Y., Cataltepe, Z.: Intraday prediction of Borsa Istanbul using convolutional neural networks and feature correlations. Knowl.-Based Syst. **137**, 138–148 (2017). https://doi.org/10.1016/j.knosys.2017.09.023
2. Abbad, J., Fardousi, B., Abbad, M.: Advantages of using technical analysis to predict future prices on the amman stock exchange. Int. J. Bus. Manag. **9**(2), 1–16 (2014). https://doi.org/10.5539/ijbm.v9n2p1

3. Goswami, M.M., Bhensdadia, C.K., Ganatra, A.P.: Candlestick analysis based short term prediction of stock price fluctuation using SOM-CBR. In: 2009 IEEE International Advance Computing Conference, Patiala, India, pp. 1448–1452. IEEE (2009). https://doi.org/10.1109/IADCC.2009.4809230
4. Lee, C.-H.L., Chen, W.S., Liu, A.: Pattern discovery of fuzzy time series for financial prediction. IEEE Trans. Knowl. Data Eng. **18**(5), 613–625 (2006)
5. Nison, S.: Japanese Candlestick Charting Techniques. New York Institute of Finance, USA (1991). ISBN 0-13-931650-7. http://pdfs.semanticscholar.org/e1b0/56855d8725748a2a2500 48418f7846b2f9c3.pdf. Accessed 15 Nov 2019
6. Kamo, T.: Integrated computational intelligence and Japanese candlestick method for short-term financial forecasting (2011). Doctoral dissertations (1908). http://scholarsmine.mst.edu/doctoral_dissertations/1908. Accessed 15 Nov 2019
7. Chandrinos, S.K., Lagaros, N.D.: Construction of currency portfolios by means of an optimized investment strategy. Oper. Res. Perspect. **5**, 32–44 (2018). https://doi.org/10.1016/j.orp.2018.01.001
8. Di Lorenzo, R.: Basic technical analysis of financial markets. Perspect. Bus. Cult. **38**, 86–101 (2013). https://doi.org/10.1007/978-88-470-5421-9
9. Valcu, D.: Using the Heikin-Ashi technique. Tech. Anal. Stocks Commodities Mag. **22**(2), 16–29 (2004). http://iticsoftware.com/media/upload/forex-e-books/Using_The_Heikin_Ashi_Technique_D_Valcu.pdf
10. Di Lorenzo, R.: Heikin Ashi. In: How to Make Money by Fast Trading. Perspectives in Business Culture, pp. 165–169. Springer, Milano (2012). https://doi.org/10.1007/978-88-470-2534-9_34
11. Wang, S., Li, D., Shi, W., Li, D., Wang, X.: Cloud model-based spatial data mining. Ann. GIS **9**(1–2), 60–70 (2003). https://doi.org/10.1080/10824000309480589
12. Bao, Y.P., Li, X., Wang, M.: A novel method for endpoint temperature prediction in RH. Ironmaking Steelmaking **46**, 1–4 (2017). https://doi.org/10.1080/03019233.2017.1392104
13. Yan, G., Jia, S., Ding, J., Xu, X., Pang, Y.: A time series forecasting based on cloud model similarity measurement. Soft. Comput. **23**, 6443–6454 (2018). https://doi.org/10.1007/s00500-018-3190-1
14. Wang, P., Xu, X., Cai, C., Huang, S.: A linguistic large group decision making method based on the cloud model. IEEE Trans. Fuzzy Syst. **26**(6), 3314–3326 (2018). https://doi.org/10.1109/tfuzz.2018.2822242
15. Yang, X., Xu, Z., He, R., Xue, F.: Credibility assessment of complex simulation models using cloud models to represent and aggregate diverse evaluation results. In: Intelligent Computing Methodologies. ICIC 2019. Lecture Notes in Computer Science, vol. 11645, pp. 306–317. Springer, Cham (2019). https://doi.org/10.1007/978-3-030-26766-7_28
16. Xu, X., Huang, Q., Ren, Y., Sun, H.-B.: Condition assessment of suspension bridges using local variable weight and normal cloud model. KSCE J. Civil Eng. **22**(10), 4064–4072 (2018). https://doi.org/10.1007/s12205-018-1819-3
17. Li, T., Yang, X.: Risk assessment model for water and mud inrush in deep and long tunnels based on normal grey cloud clustering method. KSCE J. Civil Eng. **22**(5), 1991–2001 (2017). https://doi.org/10.1007/s12205-017-0553-6
18. Wu, X., Duan, J., Zhang, L., AbouRizk, S.M.: A hybrid information fusion approach to safety risk perception using sensor data under uncertainty. Stoch. Env. Res. Risk Assess. **32**(1), 105–122 (2017). https://doi.org/10.1007/s00477-017-1389-9
19. Li, W., Li, F., Du, J.: A level set image segmentation method based on a cloud model as the priori contour. SIViP **13**(1), 103–110 (2018). https://doi.org/10.1007/s11760-018-1334-5

20. Li, W.S., Du, J., Zhao, Z., Long, J.: Fusion of medical sensors using adaptive cloud model in local laplacian pyramid domain. IEEE Trans. Biomed. Eng. **66**(4), 1172–1183 (2018). https://doi.org/10.1109/tbme.2018.2869432
21. Ji, H., Han, Q., Li, X., You, H., Ye, Z.: Air combat situation assessment based on improved cloud model theory. In: 2019 IEEE 8th Joint International Information Technology and Artificial Intelligence Conference (ITAIC), Chongqing, China, pp. 754–758. IEEE (2019). https://doi.org/10.1109/itaic.2019.8785869
22. Lü, X., Chen, C., Wang, P., Meng, L.: Status evaluation of mobile welding robot driven by fuel cell hybrid power system based on cloud model. Energy Convers. Manag. **198**(111904), 1–18 (2019). https://doi.org/10.1016/j.enconman.2019.111904
23. Zhang, T., Yan, L., Yang, Y.: Trust evaluation method for clustered wireless sensor networks based on cloud model. Wireless Netw. **24**(3), 777–797 (2016). https://doi.org/10.1007/s11276-016-1368-y
24. Wang, T., Shang, L., Ma, X.: Application of cloud model and matter element theory in transformer fault diagnosis. In: 2018 IEEE 3rd Advanced Information Technology, Electronic and Automation Control Conference (IAEAC), Chongqing, China, pp. 2089–2092. IEEE (2018) https://doi.org/10.1109/iaeac.2018.8577253
25. Lv, C., Tian, L., Wang, Z.: A fuzzy comprehensive evaluation model of power quality based on normal cloud model. In: EEET '18 Proceedings of the 2018 International Conference on Electronics and Electrical Engineering Technology, Tianjin, China, pp. 34–38. ACM (2018). https://doi.org/10.1145/3277453.3277459
26. Li, J., Fang, H., Song, W.: Sustainable supplier selection based on SSCM practices: a rough cloud TOPSIS approach. J. Clean. Prod. **222**, 606–621 (2019). https://doi.org/10.1016/j.jclepro.2019.03.070
27. Song, Q., Chissom, B.S.: Forecasting enrollments with fuzzy time series - Part I. Fuzzy Sets Syst. **54**(1), 1–9 (1993). https://doi.org/10.1016/0165-0114(93)90355-l
28. Song, Q., Chissom, B.S.: Forecasting enrollments with fuzzy time series - part II. Fuzzy Sets Syst. **62**(1), 1–8 (1994). https://doi.org/10.1016/0165-0114(94)90067-1
29. Bose, M., Mali, K.: Designing fuzzy time series forecasting models: a survey. Int. J. Approx. Reason. **111**, 78–99 (2019). https://doi.org/10.1016/j.ijar.2019.05.002
30. Li, D.Y.: Knowledge representation in KDD based on linguistic atoms. J. Comput. Sci. Technol. **12**(6), 481–496 (1997). https://doi.org/10.1007/BF02947201
31. Chen, S.-M.: Forecasting enrollments based on fuzzy time series. Fuzzy Sets Syst. **81**(3), 311–319 (1996). https://doi.org/10.1016/0165-0114(95)00220-0
32. Hwang, J.R., Chen, S.-M., Lee, C.H.: Handling forecasting problems using fuzzy time series. Fuzzy Sets Syst. **100**(2), 217–228 (1998). https://doi.org/10.1016/s0165-0114(97)00121-8
33. Yu, H.-K.: Weighted fuzzy time series models for TAIEX forecasting. Physica A: Stat. Mech. Appl. **349**(3:4), 609–624 (2005). https://doi.org/10.1016/j.physa.2004.11
34. Vovan, T.: An improved fuzzy time series forecasting model using variations of data. Fuzzy Optim. Decis. Making **18**(2), 151–173 (2018). https://doi.org/10.1007/s10700-018-9290-7
35. Fiess, N.M., MacDonald, R.: Towards the fundamentals of technical analysis: analysing the information content of high, low and close prices. Econ. Model. **19**(3), 353–374 (2002). https://doi.org/10.1016/S0264-9993(01)00067-0
36. Lee, K.H., Jo, G.S.: Expert system for predicting stock market timing using a candlestick chart. Expert Syst. Appl. **16**(4), 357–364 (1999). https://doi.org/10.1016/s0957-4174(99)00011-1

37. Leon, C.-H., WenSung, L., Liu, C.A.: Candlestick tutor: an intelligent tool for investment knowledge learning and sharing. In: Fifth IEEE International Conference on Advanced Learning Technologies (ICALT 2005), Kaohsiung, Taiwan, pp. 238–240. IEEE (2005). https://doi.org/10.1109/icalt.2005.82
38. Lee, C.L.: Modeling personalized fuzzy candlestick patterns for investment decision making. In: 2009 Asia-Pacific Conference on Information Processing, Shenzhen, China, pp. 286–289. IEEE (2009) https://doi.org/10.1109/apcip.2009.207
39. Naranjo, R., Arroyo, J., Santos, M.: Fuzzy modeling of stock trading with fuzzy candlesticks. Expert Syst. Appl. **93**, 15–27 (2018). https://doi.org/10.1016/j.eswa.2017.10.002
40. Hu, W., Si, Y.W., Fong, S., Lau, R.Y.K.: A formal approach to candlestick pattern classification in financial time series. Appl. Soft Comput. **84**, 1–28 (2019). https://doi.org/10.1016/j.asoc.2019.105700
41. Li, D., Di, K., Li, D.: Knowledge representation and uncertainty reasoning in GIS based on cloud model. In: 2000 Proceedings 9th International Symposium on Spatial Data Handing, Beijing, pp. 3a.3—14. http://www.researchgate.net/publication/237515742_Knowledge_representation_and_uncertainty_reasoning_in_GIS_based_on_cloud_models. Accessed 15 Nov 2019
42. Thi, N.: Forecasting and Trading Stock Using Technical Analysis and Neural Fuzzy Network, Master of Science in Information Technology Thesis, King Mongkut's University of Technology North Bangkok, Bangkok, Thailand (2006). http://www.scribd.com/document/233931961/4870180041. Accessed 15 May 2015

Prediction of the Hitec Molten Salt Convective Heat Transfer Performance Using Artificial Neural Networks

Ahmed Ibrahim ElShafei[1(✉)], Omar Khaled Sallam[2], Mohammed A. Boraey[1,3], and Amr Guaily[1,4]

[1] Smart Engineering Systems Research Center (SESC), Nile University, Giza 12588, Egypt
{AElshafei,mboraey,aguaily}@nu.edu.eg
[2] School of Engineering and Applied Sciences, Nile University, Giza 12588, Egypt
okhaled@nu.edu.eg
[3] Mechanical Power Engineering Department, Zagazig University, Zagazig 44519, Egypt
[4] Department of Engineering Math and Physics, Faculty of Engineering, Cairo University, Giza 12613, Egypt

Abstract. Hitec molten salt is a ternary eutectic mixture salt that is used as an energy storage medium in concentrated solar power plants to improve the system performance and reduce the operational cost. Thus, the heat transfer performance represented in Nusselt number has been investigated numerically under different inlet temperature and velocity conditions with constant uniform side heat flux. Also, friction factor and mass flow rate are studied numerically. CFD input/output data with 40 studied cases are used as a training dataset of a 2-layer Neural Network for thermo-hydro fields' accelerated results predictions. Bayesian regularized Neural Network showed a satisfactory agreement for thermo-hydro fields' predictions compared to the CFD results for the testing dataset not included in the training of the network.

Keywords: Artificial Neural Network (ANN) · Hitec molten salt · Heat transfer · Forced convection heat transfer

1 Introduction

Renewable energy has been recently a hot topic in research especially solar power as it is a clean abundant, sustainable, easily accessible source of energy and the shortage of other resources such as fossil fuel and natural gas. The problem with renewable energy is the lack of continuous supply despite the fluctuations in solar input; as solar radiation is uneven and varies because of daily and seasonal variations and weather conditions. Thus, to lessen this irregularity (mismatch between energy supply and demand), Concentrated solar power (CSP) is introduced as a competitive option beside the Photovoltaic systems as they can provide a continuous supply of power in the absence of sun via Thermal energy storage (TES) technology which has become one of the most recent

A.-E. Hassanien et al. (Eds.): AICV 2020, AISC 1153, pp. 646–654, 2020.
https://doi.org/10.1007/978-3-030-44289-7_60

and promising technologies because the energy is stored in tanks often filled with phase change materials, which regulates the supply power according to the demand. Molten salts are often considered as a medium for energy storage and heat transfer due to their wide range of operational temperature, non-toxicity, low vapor pressure, high chemical stability at high temperatures, high specific heat capacity, and low price. Types of mostly used Molten salts are FLiNaK, Hitec and Solar Salt (60%-$NaNO_3$/40%-KNO_3). However molten salt comes with several drawbacks which are the required high pumping power for the feeding of the fluid in addition to clogging problems due to local solidification of the salt at sites with a temperature lower than the melting point of the salt. These drawbacks negatively affect the efficiency of TES and retrieval.

In CSP plants, a large array of reflecting and sometimes revolving mirrors are used to concentrate the solar power towards a solar receiver which transfers the concentrated solar rays in the form of sensible heat to the heat transfer fluid which is stored and sent to the power block to convert the thermal energy into electricity. There are two CSP storage configurations, direct storage or indirect storage. The former has the same media for heat transfer and storage while the latter uses a different fluid for the heat transfer from the storage media. Gemasolar in Southern Spain is the first solar power station that produces electricity 24 h a day. Thousands of Heliostats (revolving mirrors) reflecting and concentrating the sunlight energy towards the top of the tower at the middle of the Heliostats field. CSP plants should be best constructed in sites with great potential where the integral direct normal irradiance (DNI) is at least 1800 kW h/m^2.

The cost projections in 2050 show an average Levelized cost of electricity (LCOE) of 71 USD/MW h and an average installation cost of 4000 USD/kW [1], which have made the CSP a cost-competitive technology.

Few studies had been recently carried out to simulate the flow and predict the heat transfer characteristics of molten salt using Computational Fluid Dynamics (CFD) and validating them with previous experimental studies that essentially measured the physical and thermal characteristics of molten salt, Bin et al. [2] numerically assessed the convective heat transfer namely, the thermal-hydraulic characteristics, of turbulent flow of molten salt in circular pipe using different Reynolds, averaged Navier-Stokes (RANS) models and showed that k-ε model with Non-Equilibrium Wall Treatment (NEWT) has good agreement with the well-known correlations and the experimental data. Li et al. [3], analyzed the factors affecting the performance of the Molten salt receivers through numerical modeling of the heterogeneous heat flux boundary on the heat receiver. Lu et al. [4], studied the laminar natural convection heat transfer properties of molten salt around multiple cylinders with different diameters. Feng et al. [5], investigated the properties of FLiNaK salt as a Heat exchanger fluid through a CFD model and validated them by comparison with correlations and experimental data of friction factor, Nusselt number, and thermal and hydraulic lengths with varying Reynolds number in a circular tube. Qiu et al. [6], Investigated the heat transfer performance and flow friction of several molten salts in turbulent conditions using CFD with a comparison of experimental data and correlations. In addition to a comparison between the results of applying a non-uniform wall heat flux and uniform wall heat flux and the results were nearly the same.

The investigation of the thermal and hydrodynamical performance of molten salt numerically has been mainly studied in many studied. However, no study has considered predicting the molten salt performance by an artificial neural network.

This study focuses on investigating numerically the performance of flowing molten salt under various thermal and hydrodynamical inlet conditions with constant uniform side heating then feeding the output data of the numerical simulations to ANN where LM & BR algorithms are used and tested the system parameters.

2 Physical Model and Numerical Analysis

In this section, the physical, numerical models, assumptions and operational conditions used for the numerical analysis are reported.

2.1 Physical Model

The hydrodynamic and thermal performance of Hitec molten salt, a ternary nitrate molten salt with the eutectic composition ($NaNO3$–$KNO3$–$NaNO2$, 7–53–40 wt%). The salt thermophysical properties [6] are defined by the following equations:

$$\rho = 2280.22 - 0.636T, \quad 450 < T < 800 \tag{1}$$

$$C_p = 1560, \quad 450 < T < 800 \tag{2}$$

$$\lambda = 2.2627 - 0.01176T + 2.551e - 5T^2 - 1.863e - 8T^3, \quad 450 < T < 536 \tag{3}$$

$$\lambda = 0.7663 - 6.47e - 4T, \quad 536 < T < 800 \tag{4}$$

$$\mu = 0.93845 - 5.4754e - 3T, \quad 450 < T < 500 \tag{5}$$

$$\begin{gathered}\mu = 0.23816 - 1.2768e - 3T + 2.6275e - 6T^2 - 2.4331e- \\ 9T^3 + 8.507e - 13T^4, \\ 500 < T < 800\end{gathered} \tag{6}$$

Where T is the salt temperature in K, ρ is the salt density in kg/m^3, C_p is the specific heat capacity in J/kg · K, λ is the thermal conductivity in W/m · K and μ is the dynamic viscosity in Pa s.

The temperature-dependent properties molten salt flows in the pipe of length 120 times the diameter, large enough to reach a fully developed state with inlet Reynolds number varying from 10,000 to 100,000 and varying inlet temperature in the range of 500 to 700. The pipe is subjected to a constant uniform heat flux of 10,000 W/m^2.

2.2 Mathematical Model

The governing equations for non-polar, incompressible, steady and temperature-dependent fluids are:

Continuity equation

$$\frac{\partial u_i}{\partial x_i} = 0 \tag{7}$$

Momentum equation

$$\frac{\partial}{\partial x_i}\left(\rho u_i u_j\right) = -\frac{\partial P}{\partial x_i} + \frac{\partial}{\partial x_j}\left[\left(\mu_t + \mu\right)\left(\frac{\partial u_i}{\partial x_j} + \frac{\partial u_j}{\partial x_i}\right)\right] \tag{8}$$

Energy equation

$$\frac{\partial}{\partial x_i}\left(\rho u_i h\right) = \frac{\partial}{\partial x_i}\left[C_P\left(\frac{\mu}{Pr} + \frac{\mu_t}{Pr_t}\right)\frac{\partial T}{\partial x_i}\right] \tag{9}$$

Where u_i are the Cartesian velocity components $(m\,s^{-1})$, ρ as the fluid density $(kg\ m^{-3})$, P as the static pressure Pa, $\mu\,\&\,\mu_t$ are the dynamic and turbulent dynamic viscosity respectively (Pas) and h as the convective heat transfer coefficient $(Wm^{-2}K^{-1})$.

2.3 Meshing

A structured H-grid was used, which gives better accuracy over the triangular mesh, with quadratic elements order to have a higher number of nodes per element and thereby higher accuracy. Y+ of values ranging between 20 to 25 where used as recommended for the $K - \epsilon$ turbulence model with non-equilibrium wall function turbulence model as per [7], the y+ value was modified for each Reynolds number which required regenerating the mesh for every case. To obtain a proper mesh quality to obtain accurate results it was important to achieve an element aspect ratio less than 2.

2.4 Boundary Conditions and Solution Setup

A 2D axisymmetric steady pressure-based solver was used. The $K - \epsilon$ turbulence model with Non-Equilibrium wall function. The SIMPLE scheme algorithm was used to solve pressure-velocity coupling, the momentum, turbulent kinetic energy, and turbulent dissipation rate were discretized using a second-order scheme and pressure gradient was discretized using least-squares cell-based method. The residual convergence criteria for the continuity, momentum, and energy equations were set to 10-6. Boundary conditions values are given in Table 1.

Table 1. Operational conditions

Inlet Freestream velocity V_{in}	$\frac{Re\mu}{\rho D}$, $10{,}000 < Re < 100{,}000$
Inlet temperature T_{in}	$500K \rightarrow 700K$
Outlet gauge pressure	$0\ Pa$
Surface heat flux Q	$10{,}000\ W/m^{-2}$
Turbulence model	Realizable $K\epsilon$ NEWT
Turbulent kinetic energy (k_i)	$1.5\left(0.16V_{in}Re^{-0.125}\right)^2$
Turbulent dissipation rate (ϵ_i)	$\frac{c_\mu^{0.75}k_i^{1.5}}{0.07D}$

All solution setup is shown in Table 2.

Table 2. Solution setup

Pressure-velocity coupling	SIMPLE scheme	
Spatial discretization	• Pressure gradient	Least squares cell based
	• Momentum	Second order
	• Turbulent kinetic energy	Second-order upwind
	• Turbulent dissipation rate	Second-order upwind

2.5 Mesh Independence Study

First, to make sure that our results are independent of the mesh grid, mesh independence study was carried out. The refinement is repeated multiple times. The mesh sensitivity analysis was carried out, with different element sizes with maintaining the same y+ values. Comparing the important parameters for the study which are the Nusselt number and the friction factor as seen below, which shows that finer meshes with more than 85k elements have no significant impact on the parameters. Thus, to save the computational time, 85k element grid is used in the present study (Fig. 1).

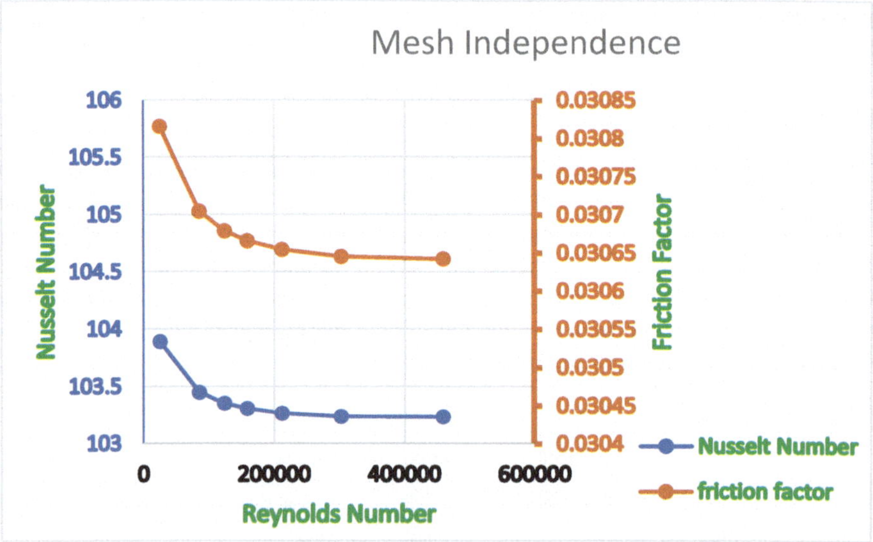

Fig. 1. Grid independence

2.6 Performance Indicators

The Nusselt Number (Nu) and friction factor (f) is non-dimensional performance parameters of interest to give insight on the flow characteristics heat transfer performance defined as:

$$Nu = \frac{hD}{\lambda} \tag{10}$$

$$h = \frac{Q}{\Delta T} \tag{11}$$

where ΔT is the mean temperature difference between the wall and the fluid bulk temperature.

$$f = \Delta p \frac{D}{L} \frac{1}{\frac{1}{2}\rho v^2} \tag{12}$$

2.7 Numerical Results

The following figures depict the simulation results and salt thermal performance due to variation in its thermophysical properties of one of the cases studied. Figure 2 illustrates the pressure distribution contours across the pipe. Figure 3 depicts the molten

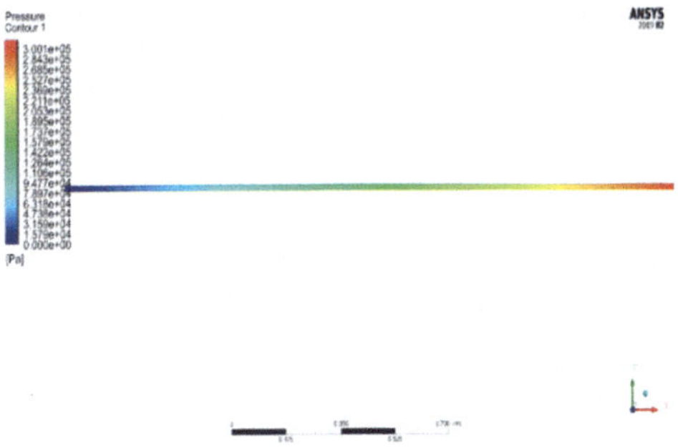

Fig. 2. Pressure contours of the molten salt along the pipe

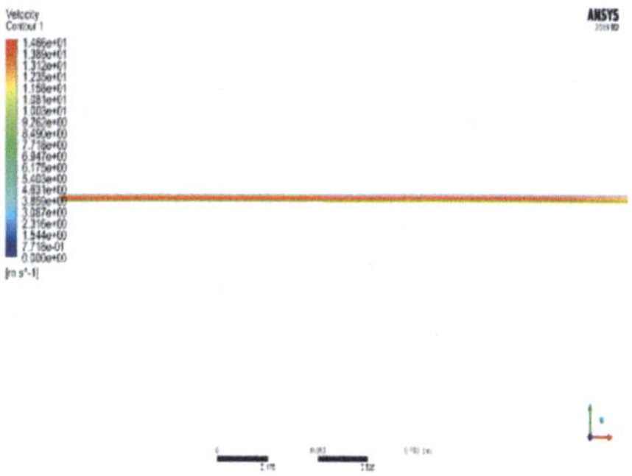

Fig. 3. Velocity contours of the molten salt along the pipe

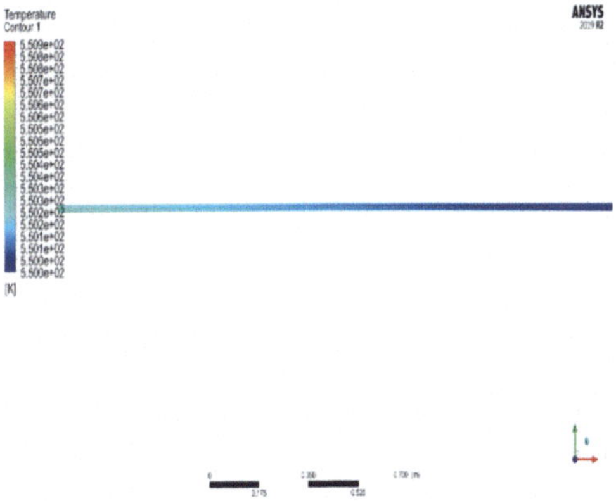

Fig. 4. Temperature contours of the molten salt at the pipe mid-section

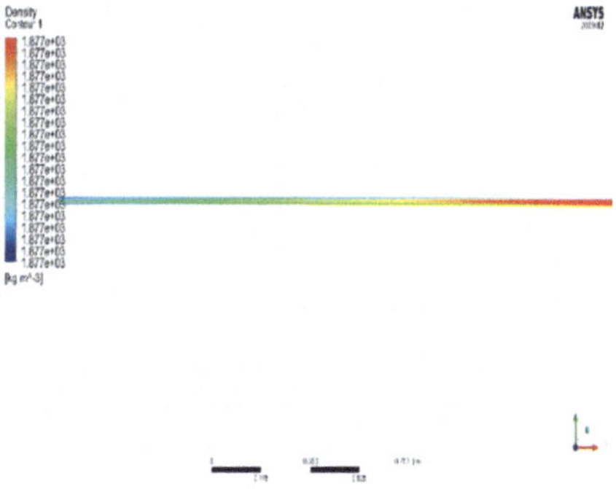

Fig. 5. Density contours of the molten salt at the pipe mid-section

salt's velocity variation. While Fig. 4 shows the variation of salt's temperature along the pipe. While Fig. 5 illustrates the molten salt's density variation along the pipe as an effect of the change in temperature due to the salt's temperature-dependent properties.

3 Neural Network Results and Discussion

A two hidden layer Bayesian regularized technique [8, 9] Neural Network with 5 hidden neurons is designed to predict the three output features of interest (Friction factor, Nusselt number, and mass flow rate) for the 2 input features represented by (inlet temperature and inlet velocity). By the aid of CFD simulations, 40 Input/output cases are obtained, 33 cases as a dataset are used to train the Neural Network while the remaining 7 datasets are used to test the network performance and reliability without being trained on.

Figure 6 shows the Neural Network predictions compared to the CFD simulations for the 33 training datasets, while Fig. 7 shows the Neural Network predictions for the remaining 7 datasets to test the network. Table 3 shows the RMSE for the 3 output features where the RMSE is increased for the testing cases slightly as predicted but still very negligible and acceptable.

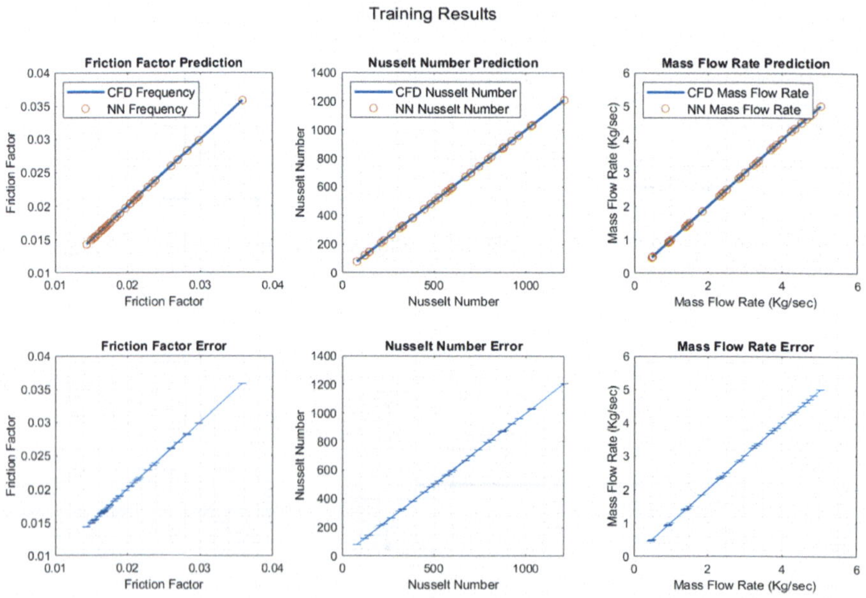

Fig. 6. Neural network predictions for training data

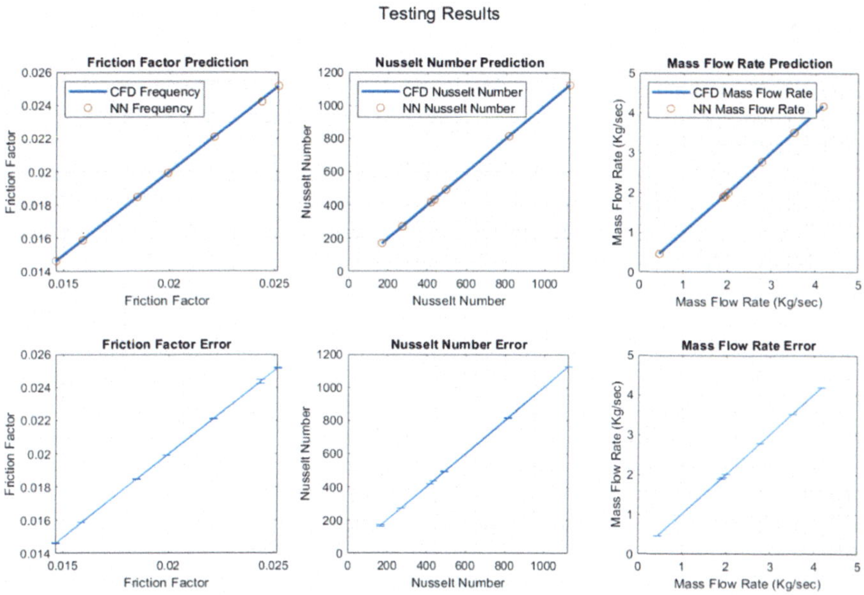

Fig. 7. Neural network predictions for testing data

Table 3. RMSE for testing and training neural network predictions

	Training RMSE	Testing RMSE
Friction factor	1.68×10^{-5}	4.93×10^{-5}
Nusselt number	0.21	1.16
Mass flow rate	1.34×10^{-5}	2.35×10^{-5}

4 Conclusion

The turbulent forced heat transfer and hydraulic characteristics of molten salt in a smooth tube under various temperature and Reynolds number is studied numerically. Nusselt number, friction factor and mass flow rate of the molten salt are calculated in the numerical simulations. Bayesian regularized Neural Network is trained on the input/output CFD data to accelerate the thermo-hydro field predictions. The testing results' root means square error for the network is found to be with in the same order of magnitude for the training data that shows a satisfactory implementation of the use of Neural Network.

References

1. Capocelli, M., Caputo, G., De Falco, M., Balog, I., Piemonte, V.: Numerical modeling of a novel thermocline thermal storage for concentrated solar power. J. Sol. Energy Eng. **141**(5), 051001 (2019)
2. Bin, L., Yu-ting, W., Chong-fang, M., Meng, Y., Hang, G.: Turbulent convective heat transfer with molten salt in a circular pipe. Int. Commun. Heat Mass Transf. **36**(9), 912–916 (2009)
3. Li, P.J., Cheng, Y.T., Liu, S.: Factors affecting the performance of molten salt heat receiver. In: Advanced Materials Research, vol. 365, pp. 8–13 (2012)
4. Lu, Y.W., Li, X.L., Du, W.B., Wu, Y.T., Ma, C.F.: Laminar natural convection heat transfer characteristics of molten salt around the horizontal cylinder. Energy Procedia **69**, 681–688 (2015)
5. Ferng, Y.M., Lin, K.Y., Chi, C.W.: CFD investigating thermal-hydraulic characteristics of FLiNaK salt as a heat exchange fluid. Appl. Therm. Eng. **37**, 235–240 (2012)
6. Qiu, Y., Li, M.J., Li, M.J., Zhang, H.H., Ning, B.: Numerical and experimental study on heat transfer and flow features of representative molten salts for energy applications in turbulent tube flow. Int. J. Heat Mass Transf. **135**, 732–745 (2019)
7. ANSYS FLUENT User's Guide, "Ansys Fluent Theory Guide," ANSYS Inc., November 2013
8. MacKay, D.J.: A practical Bayesian framework for backpropagation networks. Neural Comput. **4**(3), 448–472 (1992)
9. Okut, H.: Bayesian regularized neural networks for small n big p data. In: Artificial Neural Networks-Models and Applications (2016). https://doi.org/10.5772/63256

Multi-switching Combination Synchronization of Fractional Order Chaotic Systems

Shikha Singh[1] and Ahmad Taher Azar[2,3](✉)

[1] Department of Mathematics, Jesus and Mary College, University of Delhi, New Delhi, India
sshikha7014@gmail.com

[2] Robotics and Internet-of-Things Lab (RIOTU), Prince Sultan University, Riyadh, Saudi Arabia
aazar@psu.edu.sa

[3] Faculty of Computers and Artificial Intelligence, Benha University, Benha, Egypt
ahmad.azar@fci.bu.edu.eg

Abstract. This paper describes multi-switching synchronization schemes between three similar chaotic fractional-order systems. The Rössler fractional system is considered to apply the technique. With multi-switch combination synchronization, two drive system state variables are synchronized with different response system state variables at the same time. The multi-switching combination synchronization of three fractional-order equivalent systems was investigated based on the stability of fractional-order chaotic systems. Appropriate controllers have been configured to synchronize three similar fractional-order chaotic systems. Numerical calculations were carried out in Matlab to support the theoretical argument.

Keywords: Fractional-order chaotic system · Multi-switching synchronization · Stability theory · Combination synchronization

1 Introduction

A large number of applications of chaos synchronization in many research areas make this subject an important topic in nonlinear systems [7,8,16,18]. Such technologies have evolved through the implementation of a fractional version of chaotic systems. The key benefit of the fractional calculus, distinct from the integer calculus, lies in the fact that the fractional order system represents complex systems more elegantly in interdisciplinary fields compared to the integer order system as it has memory and has proven to be a very suitable method for defining the memory and hereditary properties of various materials and processes [12,13]. Fractional dynamics is a scientific idea that has been developed. Fractional dynamics analysis is a 300-year-old topic that can be traced back

A.-E. Hassanien et al. (Eds.): AICV 2020, AISC 1153, pp. 655–664, 2020.
https://doi.org/10.1007/978-3-030-44289-7_61

to Leibniz, Riemann, Liouville, Grunwald and Letnikov [17]. For quite a while, however, the fractional calculus did not draw much attentionb due to its unique benefits, it has attracted more interest from scientists and has broader potential until implementation.

There has been substantial progress in fractional computing over the last three decades such that the application of fractional calculus in increasing numbers of scientific areas covering physics, materials science, engineering, computer science and materials [3,4,14]. These days, given that fractional calculus applications have been identified in an ever-increasing number of scientific fields covering physics, material science, engineering, computer science and materials, tremendous progress has been made on fractional calculus over the past three decades. Fractional differential equations have effectively demonstrated diffusion-identified wonders, visco-elastic structures, signal processing, dielectric polarization, and so on. Further examples for complex fractional structures can be found in [2,15]. In addition, in many areas, uses of fractional calculus have been accounted for, for example, image processing, signal processing and automatic control [17]. These and many other comparative examples splendidly illustrate the essence of complex fractional-order analysis and review. The combination synchronization scheme between three chaotic systems was recently introduced by Runzi et al. [20]. The design consists of two drive systems and one response system in this form of synchronization. This method of synchronization is an extension of traditional synchronization schemes and possesses potential applications in electrical engineering, physics, theory of communication and many other fields [5,26]. The signal to be transmitted by two separate chaotic systems using a combination synchronization method, for example, is more stable in secure communication than the signal transmitted through another synchronization scheme. The signal transmitted by this system may therefore have a greater anti-attack capability than the one transmitted by the normal model. A new synchronization system has been further discussed in [1,6,9,10,19]. The principle of multi-switching synchronization suggested in [21–25] has made some progress in recent years. For multi-switching synchronization schemes, the various drive system states are synchronized with the desired response system status. In the wide range of possible synchronization directions that exist due to multi-switching synchronization, the relevance of such types of synchronization studies to information security is evident. Although these schemes clearly provide enhanced resistance and anti-attack ability to secure contact, only a few such studies have been recorded in the literature. Motivated by the discussion above, this paper presents the synchronization of the multi-switching combination of non-identical fractional-order chaotic systems.

The following is the structure of this report. The definition of the fractional order Rössler chaotic model is given in Sect. 2. The problem formulation is defined in 3. The key results are provided in Section 4 where controllers are measured. Numerical tests are given in Sect. 5. Finally, the short conclusion is presented in Sect. 6.

2 System Description

The fractional-order Rössler [11] chaotic system is given by:

$$\begin{cases} \dfrac{d^q u_1}{dt^q} = -(u_2 - u_1) \\ \dfrac{d^q u_2}{dt^q} = u_1 + au_2 \\ \dfrac{d^q u_3}{dt^q} = u_3(u_1 - m) + b \end{cases} \tag{1}$$

where u_1, u_2, u_3 are the state variables, q is the fractional order satisfying $0 < q < 1$ and a, m, b are the parameters.

For the parameters values $a = 0.5$, $m = 10$, $b = 0.2$ and $q = 0.9$, the phase portraits of system showing chaotic behavior is shown in Figure 1.

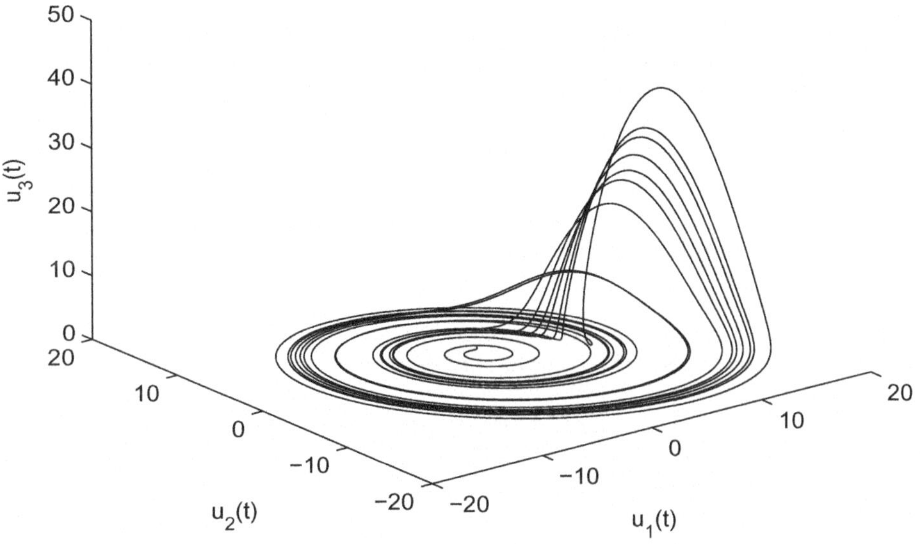

Fig. 1. 3D phase portrait of the fractional-order Rössler chaotic system

3 Methodology

In this section, methodology is proposed to obtain the multi-switching combination synchronization of fractional-order chaotic systems. For this purpose, we have considered two drive systems and one response system. The two drive systems are represented by:

$$\frac{d^q u(t)}{dt^q} = Q_1 u + J_1(u) \tag{2}$$

$$\frac{d^q v(t)}{dt^q} = Q_2 v + J_2(v) \tag{3}$$

where the state vectors for the drive systems (2) and (3) are given by $u = (u_1, u_2, ..., u_n)^T$ and $v = (v_1, v_2, ..., v_n)^T \in \mathbf{R}^n$, respectively.

$Q_1, Q_2 \in \mathbf{R}^n$ *and* $J_1, J_2 : \mathbf{R}^n \rightarrow \mathbf{R}^{nxm}$ are non-linear vector functions. Let us consider the response system represented by:

$$\frac{d^q w(t)}{dt^q} = Q_3 w + J_3(w) + \sigma \tag{4}$$

where the state vectors for the response system (4) is given by $w = (w_1, w_2, ..., w_n)^T$.

$Q_3 \in \mathbf{R}^n$ *and* $J_3 : \mathbf{R}^n \rightarrow \mathbf{R}^{nxm}$ are non-linear vector functions. The control vector $\sigma = (\sigma_1, \sigma_2, ..., \sigma_n)^T$ is designed to accomplish the desired result.

Definition 1: The multi-switching combination synchronization is achieved between the drive systems (2) and (3) and the response system (4) if the following condition is satisfied.

$$\lim_{t\rightarrow\infty} \|e\| = \lim_{t\rightarrow\infty} \|Rw - Pu - Qv\|$$

where $P, Q, R \in \mathbf{R}^{nxn}$ are constant diagonal matrices and $R \neq 0$.

$\| * \|$ symbolize the matrix norm and $e = Rw - Pu - Qv$ is the error synchronization vector.

Let's assume $P = diag(\alpha_1, \alpha_2, ..., \alpha_n)$, $Q = diag(\beta_1, \beta_2, ..., \beta_n)$ and $R = diag(\gamma_1, \gamma_2, ..., \gamma_n)$ then the components of error vector e are expressed as:

$$e_{klm} = \gamma_k w_k - \alpha_l u_l - \beta_m v_m$$

The possible choices for the indices of the error states to obtain the error states are:

$$k = i = m,\ k = l \neq m,\ k \neq m = l \ and\ k \neq l \neq m\ (k,\ l,\ m = 1,\ 2,\ 3)$$

So, the error dynamical system is transformed with the help of Eqs. (2), (3) and (4), which is expressed as:

$$\frac{d^q e(t)}{dt^q} = R(Q_3 w + J_3(w) + \eta) - P(Q_1 u + J_1(u)) - Q(Q_2 v + J_2(v)) \tag{5}$$

The objective is to model the efficient $\sigma(u, v, w)$ controller to achieve multi-switch combination synchronization in compliance with Definition 1, and the dynamics of errors are reduced to:

$$\frac{d^q e(t)}{dt^q} = -Ke(t) \tag{6}$$

in such a manner that the eigenvalues of $-K$ satisfy $\|arg(\lambda_i)\| > \frac{\alpha\pi}{2}, (i = 1, 2, 3, ..., n)$, which in turn ensures the achievement of the multi-switching combination synchronization of (2) and (3) systems with the (4) system.

Remark 1: In case where $P = 0$ or $Q = 0$, the multi-switching modified projective synchronization is observed.

Remark 2: In case where $P = Q = 0$, the chaos synchronization is observed.

4 Illustration

The first drive system is represented by:

$$\begin{cases} \frac{d^q u_1}{dt^q} = -(u_2 - u_1) \\ \frac{d^q u_2}{dt^q} = u_1 + au_2 \\ \frac{d^q u_3}{dt^q} = u_3(u_1 - m) + b \end{cases} \tag{7}$$

The second drive system is represented by:

$$\begin{cases} \frac{d^q v_1}{dt^q} = -(v_2 - v_1) \\ \frac{d^q v_2}{dt^q} = v_1 + av_2 \\ \frac{d^q v_3}{dt^q} = v_3(v_1 - m) + b \end{cases} \tag{8}$$

Let's consider the response system represented by:

$$\begin{cases} \frac{d^q w_1}{dt^q} = -(w_2 - w_1) + \eta_1 \\ \frac{d^q w_2}{dt^q} = w_1 + aw_2 + \eta_2 \\ \frac{d^q w_3}{dt^q} = w_3(w_1 - m) + b + \eta_3 \end{cases} \tag{9}$$

where $\sigma_1, \sigma_2, \sigma_3$ are the controllers to be designed.

The control vector $\sigma = (\sigma_1, \sigma_2, ..., \sigma_n)^T$ is designed to accomplish the desired result. Under the appropriate conditions for the $k, l, m = 1, 2, 3$ indices as defined in Definition 1, there are various possible switching combinations to define the error status for the (7)–(9) drive response systems as mentioned below.

For $k \neq l \neq m$, we have $e_{123}, e_{132}, e_{213}, e_{231}, e_{312}$ *and* e_{321}.
For $k = m \neq l$, we have $e_{121}, e_{131}, e_{212}, e_{232}, e_{313}$ *and* e_{323}.
For $k \neq l = m$, we have $e_{122}, e_{133}, e_{211}, e_{233}, e_{311}$ *and* e_{322}.
For $k = m \neq l$, we have $e_{112}, e_{113}, e_{221}, e_{223}, e_{331}$ *and* e_{332}.

In this paper, we define results from different switching possibilities for three randomly selected error state vector combinations.
By considering the error system as follows:

$$\begin{cases} e_{123} = \gamma_1 w_1 - \alpha_2 u_2 - \beta_3 v_3 \\ e_{231} = \gamma_2 w_2 - \alpha_3 u_3 - \beta_1 v_1 \\ e_{312} = \gamma_3 w_3 - \alpha_1 u_1 - \beta_2 v_2 \end{cases} \tag{10}$$

The error dynamical system is obtained as follows:

$$\begin{cases} \dfrac{d^q e_{123}}{dt^q} = \gamma_1 \dfrac{d^q w_1}{dt^q} - \alpha_2 \dfrac{d^q u_2}{dt^q} - \beta_3 \dfrac{d^q v_3}{dt^q} \\ \dfrac{d^q e_{231}}{dt^q} = \gamma_2 \dfrac{d^q w_2}{dt^q} - \alpha_3 \dfrac{d^q u_3}{dt^q} - \beta_1 \dfrac{d^q v_1}{dt^q} \\ \dfrac{d^q e_{312}}{dt^q} = \gamma_3 \dfrac{d^q w_3}{dt^q} - \alpha_1 \dfrac{d^q u_1}{dt^q} - \beta_2 \dfrac{d^q v_2}{dt^q} \end{cases} \tag{11}$$

The error system (11) is translated into the following form by using Eqs. (7)–(9):

$$\begin{cases} \dfrac{d^q e_{123}}{dt^q} = \gamma_1(-(w_2 - w_1) + \eta_1) - \alpha_2(u_1 + a u_2) - \beta_3(v_3(v_1 - m) + b) \\ \dfrac{d^q e_{231}}{dt^q} = \gamma_2(w_1 + a w_2 + \eta_2) - \alpha_3 u_3 - \beta_1 v_1 \\ \dfrac{d^q e_{312}}{dt^q} = \gamma_3 w_3 - \alpha_1(u_3(u_1 - m) + b) - \beta_2(v_1 + a v_2) \end{cases} \tag{12}$$

We propose the following theorem in order to realize the control gain:

Theorem 1: If the control function $\sigma_1, \sigma_2, \sigma_3$ is selected as mentioned below, the drive systems (7), (8) will achieve multi-switching combination synchronization with response (9).

$$\begin{cases} \sigma_1 = w_2 - w_1 + \dfrac{\alpha_2}{\gamma_1}(u_1 + a u_2) + \dfrac{\beta_3}{\gamma_1}(v_3(v_1 - m) + b_3) - k_1 e_{123} \\ \sigma_2 = -(w_1 - a w_2) + \dfrac{\alpha_3}{\gamma_2}(u_3(u_1 - m) + b) + \dfrac{\beta_1}{\gamma_2}(-v_1 - v_2) - k_2 e_{231} \\ \sigma_3 = -(w_2(w_1 - m) + b) + \dfrac{\alpha_1}{\gamma_3}(-(u_2 - u_1)) + \dfrac{\beta_2}{\gamma_3}(v_1 + a v_2) - k_3 e_{312} \end{cases} \tag{13}$$

where $-K = (k_1, k_2, k_3)$ is a gain matrix. $-K$ should have eigenvalues all satisfying the above stability condition of Sect. 3.

The following corollary are obtained on the basis of the Theorem 1:

Corollary 1: Assume $P \neq 0$ and $Q = 0$, then we will have the following theorem-based Corollary 1. If the σ_1, σ_2 and σ_3 control functions are designed as follows:

$$\begin{cases} \sigma_1 = w_2 - w_1 + \dfrac{\alpha_2}{\gamma_1}(u_1 + au_2) - k_1 e_{123} \\ \sigma_2 = -(w_1 - aw_2) + \dfrac{\alpha_3}{\gamma_2}(u_3(u_1 - m) + b) - k_2 e_{231} \\ \sigma_3 = -(w_2(w_1 - m) + b) + \dfrac{\alpha_1}{\gamma_3}(-(u_2 - u_1)) - k_3 e_{312} \end{cases} \tag{14}$$

In this corollary multi-switching modified projective synchronization is observed.

Corollary 2: Suppose $P = 0$ and $Q \neq 0$, then we have the following corollary which is based on Theorem 1. If the control functions σ_1, σ_2 and σ_3 are designed as follows:

$$\begin{cases} \sigma_1 = w_2 - w_1 + \dfrac{\beta_3}{\gamma_1}(v_3(v_1 - m) + b_3) - k_1 e_{123} \\ \sigma_2 = -(w_1 - aw_2) + \dfrac{\beta_1}{\gamma_2}(-v_1 - v_2) - k_2 e_{231} \\ \sigma_3 = -(w_2(w_1 - m) + b) + \dfrac{\beta_2}{\gamma_3}(v_1 + av_2) - k_3 e_{312} \end{cases} \tag{15}$$

In this corollary multi-switching modified projective synchronization is observed.

Corollary 3: Assume $P = 0$ and $Q = 0$, then we have got the following theorem-based Corollary 1. If the σ_1, σ_2 and σ_3 control functions are designed as follows:

$$\begin{cases} \sigma_1 = w_2 - w_1 - k_1 e_{123} \\ \sigma_2 = -(w_1 - aw_2) - k_2 e_{231} \\ \sigma_3 = -(w_2(w_1 - m) + b) - k_3 e_{312} \end{cases} \tag{16}$$

In this corollary the multi-switching combination synchronization problem turns into chaos control problem of (9).

5 Numerical Simulations

Matlab is used to simulate the performance of the controllers designed.

Let us consider the values as follows: $(\alpha_1,\ \alpha_2,\ \alpha_3 = (1,\ 1,\ 1))$, $(\beta_1,\ \beta_2,\ \beta_3 = (1,\ 1,\ 1))$, $(\gamma_1,\ \gamma_2,\ \gamma_3 = (1,\ 1,\ 1))$ and the fractional-order $q = 0.9$.

The parameter value for which (7), (8) and (9) shows chaotic behavior is taken as $(a,\ m,\ b) = (0.5,\ 10,\ 0.2)$, respectively. Drive and response system initial states are randomly selected as $u_1(0),\ u_2(0),\ u_3(0) = (0.5,\ 1.5,\ 0.1)$ $v_1(0),\ v_2(0),\ v_3(0) = (0.8,\ 1,\ 0.5)$ and $w_1(0),\ w_2(0),\ w_3(0) = (1,\ 2,\ 0.8)$. When we set all $-K$'s own values to be -2, which satisfy $\|arg(\lambda_i)\| > \frac{\alpha\pi}{2}$, $(i = 1, 2, 3, ..., n)$. Thus, using pole placement technique, we can obtain the gain matrix as $-K = diag(-2,\ -2,\ -2)$. Figure 2 displays the time response of e_{231}, e_{312}, e_{123} synchronization errors.

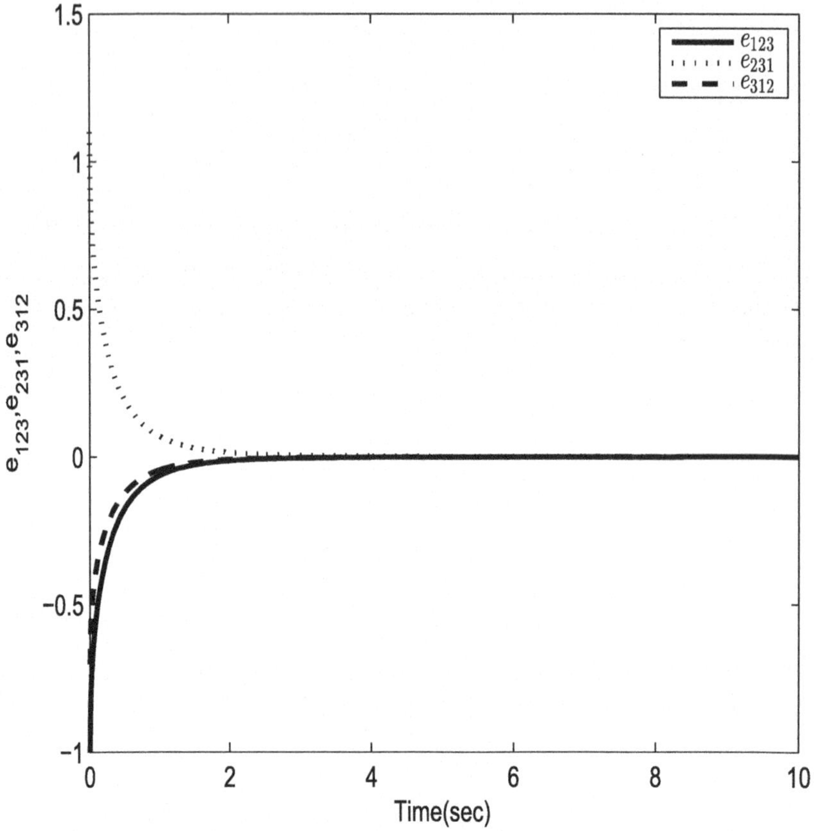

Fig. 2. Synchronization error between states of drive and response systems

6 Conclusion

In this paper, we have proposed multi-switching combination synchronization scheme among fractional-order chaotic systems. Stability criteria of fractional-order chaotic systems is taken into account to derive results for the multi-switching combination synchronization between two drive fractional-order chaotic systems and one response fractional-order chaotic system. Eventually, numerical results are given to validate the efficacy of the proposed synchronization scheme. There is excellent consensus between theoretical and numerical results.

References

1. Ayub, K., et al.: Dynamical behavior and reduced-order combination synchronization of a novel chaotic system. Int. J. Dyn. Control **6**(3), 1160–1174 (2018)

2. Azar, A.T., Serrano, F.E.: Fractional order sliding mode PID controller/observer for continuous nonlinear switched systems with PSO parameter tuning. In: International Conference on Advanced Machine Learning Technologies and Applications, pp. 13–22. Springer, Cham (2018)
3. Caponetto, R.: Fractional Order Systems: Modeling and Control Applications, vol. 72. World Scientific, Hackensack (2010)
4. Ghoudelbourk, S., Dib, D., Omeiri, A., Azar, A.T.: MPPT control in wind energy conversion systems and the application of fractional control (PI^{α}) in pitch wind turbine. Int. J. Model. Identif. Control. (IJMIC) **26**(2), 140–151 (2016)
5. Kassim, S., Hamiche, H., Djennoune, S., Bettayeb, M.: A novel secure image transmission scheme based on synchronization of fractional-order discrete-time hyperchaotic systems. Nonlinear Dyn. **88**(4), 2473–2489 (2017)
6. Khan, A., Shikha: Robust adaptive sliding mode control technique for combination synchronisation of non-identical time delay chaotic systems. Int. J. Model. Ident. Control **31**(3), 268–277 (2019)
7. Khan, A., Singh, S., Azar, A.T.: Combination-combination anti-synchronization of four fractional order identical hyperchaotic systems. In: Hassanien, A.E., Azar, A.T., Gaber, T., Bhatnagar, R., F. Tolba, M. (eds.) The International Conference on Advanced Machine Learning Technologies and Applications (AMLTA2019), pp. 406–414. Springer, Cham (2020)
8. Khan, A., Singh, S., Azar, A.T.: Synchronization between a novel integer-order hyperchaotic system and a fractional-order hyperchaotic system using tracking control. In: Hassanien, A.E., Azar, A.T., Gaber, T., Bhatnagar, R., F. Tolba, M. (eds.) The International Conference on Advanced Machine Learning Technologies and Applications (AMLTA2019), pp. 382–391. Springer, Cham (2020)
9. Khan, A., et al.: Combination synchronization of time-delay chaotic system via robust adaptive sliding mode control. Pramana **88**(6), 91 (2017)
10. Khan, A., et al.: Combination synchronization of Genesio time delay chaotic system via robust adaptive sliding mode control. Int. J. Dyn. Control **6**(2), 758–767 (2018)
11. Li, C., Chen, G.: Chaos and hyperchaos in the fractional-order Rössler equations. Physica A **341**, 55–61 (2004)
12. Lyapunov, A.M.: The general problem of the stability of motion. Int. J. Control **55**(3), 531–534 (1992)
13. Machado, J.T., Kiryakova, V., Mainardi, F.: Recent history of fractional calculus. Commun. Nonlinear Sci. Numer. Simul. **16**(3), 1140–1153 (2011)
14. Magin, R.L.: Fractional Calculus in Bioengineering. Begell House Publishers, Redding (2006)
15. Ouannas, A., Grassi, G., Azar, A.T., Singh, S.: New control schemes for fractional chaos synchronization. In: International Conference on Advanced Intelligent Systems and Informatics, pp 52–63. Springer, Cham (2018)
16. Pano-Azucena, A.D., de Jesus Rangel-Magdaleno, J., Tlelo-Cuautle, E., de Jesus Quintas-Valles, A.: Arduino-based chaotic secure communication system using multi-directional multi-scroll chaotic oscillators. Nonlinear Dyn. **87**(4), 2203–2217 (2017)
17. Podlubny, I.: Fractional Differential Equations: An Introduction to Fractional Derivatives, Fractional Differential Equations, to Methods of Their Solution and Some of Their Applications, vol. 198. Elsevier, Amsterdam (1998)
18. Prajapati, N., Khan, A., Khattar, D.: On multi switching compound synchronization of non identical chaotic systems. Chin. J. Phys. **56**(4), 1656–1666 (2018)

19. Runzi, L., Yinglan, W.: Finite-time stochastic combination synchronization of three different chaotic systems and its application in secure communication. Chaos Interdisc. J. Nonlinear Sci. **22**(2), 023109 (2012)
20. Runzi, L., Yinglan, W., Shucheng, D.: Combination synchronization of three classic chaotic systems using active backstepping design. Chaos Interdisc. J. Nonlinear Sci. **21**(4), 043114 (2011)
21. Singh, S., Azar, A.T., Ouannas, A., Zhu, Q., Zhang, W., Na, J.: Sliding mode control technique for multi-switching synchronization of chaotic systems. In: 9th International Conference on Modelling, Identification and Control (ICMIC), pp. 880–885. IEEE (2017)
22. Singh, S., Azar, A.T., Bhat, M.A., Vaidyanathan, S., Ouannas, A.: Active control for multi-switching combination synchronization of non-identical chaotic systems. In: Advances in System Dynamics and Control, pp. 129–162. IGI Global (2018)
23. Singh, S., Azar, A.T., Vaidyanathan, S., Ouannas, A., Bhat, M.A.: Multiswitching synchronization of commensurate fractional order hyperchaotic systems via active control. In: Mathematical Techniques of Fractional Order Systems, pp. 319–345. Elsevier (2018)
24. Singh, S., Azar, A.T., Zhu, Q.: Multi-switching master-slave synchronization of non-identical chaotic systems. In: Innovative Techniques and Applications of Modelling, Identification and Control, pp. 321–330. Springer (2018)
25. Ucar, A., Lonngren, K.E., Bai, E.W.: Multi-switching synchronization of chaotic systems with active controllers. Chaos, Solitons Fractals **38**(1), 254–262 (2008)
26. Vaidyanathan, S., Azar, A.T., Rajagopal, K., Sambas, A., Kacar, S., Cavusoglu, U.: A new hyperchaotic temperature fluctuations model, its circuit simulation, FPGA implementation and an application to image encryption. Int. J. Simul. Process Model. **13**(3), 281–296 (2018)

Economic Diversification in a Digital Economy

Amer Al-Roubaier, Allam Hamdan(✉), and Adel M. Sarea

Ahlia University, Manama, Bahrain
allamh3@hotmail.com

Abstract. In many developing countries export earnings account for a large percentage of the country Gross Domestic Product (GDP), government revenues and expenditures and public investment. High dependency on a limited number of commodities for export increases the economy vulnerability to price fluctuations in the global markets. Not only instability in export earnings reduces the ability of the economy to finance development, but also increases uncertainty about future growth. Economic diversification underscores the importance of productivity growth to balance development and reduce dependency on international markets.

The new economy, driven by digital technologies could help developing countries diversify output and foster economic growth. Building capacity for digital development enhances the country capabilities to promote innovation, create knowledge and disseminate information. Digital networks increase communication as well as allow people, regions and nations to collaborate and share information aiming at fostering growth and sustaining development. For developing countries, external knowledge enhances the economy readiness to diversify output, create employment opportunities and improve global competitiveness.

Keywords: Digital economy · Economic diversification · Developing countries

1 Introduction

Digital technologies are changing our world by shrinking geography and bringing people closer to each other. From development perspective, communication plays an important role in decision making and policy construction. Digital technologies empower individuals, institutions and nations to foster economic growth and diversify output. In particular, the developing countries, are in a favorable position to take advantage of the opportunities offered by the digital economy and participate in Fourth Industrial Revolution. Making use of digital technologies will boost innovation, create new knowledge, enhance competitiveness and generate linkages needed to stimulate growth and improve the quality of life. Responding to the challenges of the digital economy will require countries to increase investment in Information and Communication technologies (ICTs), human capital and infrastructure.

Technologies of the Fourth Industrial Revolution are expected to disrupt market activities by making the economy unstable, unequal and uncertain. Rapid technological change influences market demand for labor causing in the process structural

A.-E. Hassanien et al. (Eds.): AICV 2020, AISC 1153, pp. 665–671, 2020.
https://doi.org/10.1007/978-3-030-44289-7_62

unemployment, inequalities and redistribution of income among and within nations. Countries with inadequate capabilities to participate in the new economy will be left behind. Similarly, countries with narrow economic base, including resource-based economies will not be able to support development due to financial vulnerability and low economic diversification. It is important that these countries take serious initiatives to close the digital gap and participate in the global economy to strengthen linkages creation, knowledge acquisition and information dissemination.

Building sound physical and ICT infrastructure is crucial in the process of structural transformation into a digital economy. Infrastructure reduces the cost of development which encourages both local and foreign enterprises to participate in market activities. For example, ICTs infrastructure produces positive externalities driven by digital networks, which include the internet, mobile phone, and satellite technologies. The main purpose of this paper is to shed light on the role that digital technologies plays in development. The digital economy offers new opportunities for countries with limited resources to enhance global linkages and gain access to skills, knowledge and technology for development. Connectivity facilitates knowledge sharing and increases collaboration which supports local entrepreneurship and fosters growth. The developing countries can use digital services to bridge the knowledge gap and speed up the process of social and economic transformation. Acquiring existing knowledge elsewhere reduces the cost of development and encourages investments in the domestic economy.

2 Digital Economy

The digital economy can contribute to development by transforming the socio-economic structure into a diversified productive system, driven by greater access to knowledge and information. Digital technologies increase the ability of the economy to acquire knowledge and disseminate information which is necessary for decision making and policy construction. Building infrastructure for supporting digital services will empower the economy to strengthen knowledge creation and innovation diffusion. Developing countries must undertake structured efforts to cultivate the benefit of the digital economy to create new job opportunities, increase economic diversification and improve global competitiveness (Al-Roubaie 2019b).

The inclusiveness of digitization provides new opportunities for all members of the community to benefit from the new economy and diversify output. In many of the developing countries, young people, women, indigenous population and people with disabilities are not given equal access to resources, including funding, to participate in the economy. Inclusiveness of digital technologies broadens market access to both local and foreign enterprises to participate in the new economy. In countries where people are isolated because of geography, digital technologies allow governments to provide services across wide range of geography including.

The concept of the digital economy is derived from the power and services provided by digital technologies including international networks, driven by the internet, mobile telephone and satellite technologies. Connectivity to global networks has increased the country ability to communicate, absorb and acquire knowledge and information for development. Developing countries should make use the existing knowledge elsewhere instead of producing knowledge locally. Such tendency allows the country to make use of the existing knowledge to build capacity for development and foster economic growth. Production of new knowledge could be expensive due to weak infrastructure, lack of research facilities and shortages of human capital. There are several names are used to describe the digital economy including the Internet economy, the web economy and the new economy. Transactions in the digital economic are conducted through networks linking individuals, enterprises and institutions. These networks enable the country to conduct financial and trade transactions with minimum costs. The Economist Intelligence Unit defines the digital economy as one that "can provide a high quality of ICT infrastructure and harness the power of ICTs to benefit consumers, businesses and governments" (World Bank Group 2016, PP. 2–3). The United Nations defines the digital economy to represent "That part of economic output derived solely or primarily from digital technologies with a business model based on digital goods or services" (UNCTAD 2019, P. 5) Digital Economy Report 2019).

Economic activities in the digital economy are electronically conducted so that businesses, governments and individuals are able to make deals without being physically available in one locations. In developing countries, such electronic transactions encourage productivity growth by allowing local enterprises to adapt to market conditions and improve competiveness. The inclusiveness nature of the digital economy brings together entrepreneurs, decision makers and global players to set up strategies and build capacity for development. However, realization of development objectives will require countries to create appropriate business environment and invest in education, ICTs infrastructure, and innovation to strengthen linkages creation and support economic diversification. According to the World Bank "Countries that are able to swiftly adjust to this evolving digital economy will reap the greatest digital dividends, while the rest are likely to fall behind" (P. 5 World Bank Group World Development Report 2016, Digital Dividends). In the new economy, innovation is chital for development. Developing countries can use the internet to increase the economy readiness for technological learning and innovation dissemination. Connectivity to external markets enhances knowledge sharing and facilitates technology transfer to build platform innovation and economic diversification. By making the economy more productive, the internet increases the prospect for job creation and global integration. In addition, government services are to become more efficient due to reduction in costs and time required for delivery of services. "Digitalization of economic activities and transactions can help to overcome certain barriers to more inclusive development" (UNCTAD 2017, P. 3).

The digital economy could speed up the process of economic transformation by enhancing productivity, improving competiveness, encouraging entrepreneurship and creating new job opportunities. Connection to global markets stimulates production linkages by allowing local firms to participate in international markets and gain access to knowledge and information for development. Most developing countries are lacking to technical, financial and managerial resources to accelerate productivity and support development. At the national level, the digital economy connects regions and communities so that to increase communication and coordination to bridge the information gaps and diversifies output.

The new technologies are driving the new industrial age The Fourth Industrial Revolution representing “a transition to a new set of systems that bring together digital, biological, and physical technologies in new and powerful combinations” (World Economic Forum 2016, P. 3). For developing countries, the Fourth Industrial Revolution could represent a new model for development, driven by social networks which facilitate knowledge sharing, technological learning and innovation diffusion. Digital technologies increase development capabilities by empowering individuals and enterprises to communicate, absorb and acquire knowledge for development. Many of the developing countries are ill-equipped with physical, scientific, technological and financial resources to stimulate growth and speed up the process of transformation. Building capacity for development requires institutional reforms and investment in higher education to increase the economy readiness and promote innovation. A significant contribution of digital technologies to development is their impact on innovation. It is important for developing countries to increase productivity and foster growth. Innovation allows the country to generate linkages, encourage investment, diversify output and improve competitiveness. Domestic firms will be under pressure to continue to innovate in order to survive in highly competitive markets. The new technologies, including artificial intelligence, robotics, and automation, are to cause disruption, particularly in labor markets, which require countries to pursue effective strategy for training workers and respond to changing labor markets.

Digital technologies are expected to be disruptive which require public policy and decision makers to strengthen market stability and participate in the Fourth Industrial Revolution. There several ways that digital technologies affect innovation including R&D and basic research, business model innovation, product and process innovation, increasing market size, reducing barriers to entry, and acquiring and leveraging knowledge of consumer preferences. (World Economic Forum, Technology report 2016). Deepening integration into the digital economy could help developing countries to surpass several stages in their development. The digital economy is offering new opportunities by which countries can increase productivity and strengthen knowledge production and innovation dissemination. This will unleash new economic dynamics that encourage entrepreneurship, increase labor market stability, create job opportunities and enhance productivity. The role of the government is instrumental in digital transformation by supporting a business environment that encourages investment and promotes innovation.

Today, it is easy to access to knowledge produced somewhere else and make use of as a starter to support economic diversification and reduce dependency on production and exports of a limited numbers of commodities. Constructing an effective strategy to benefit from digitization requires re-structuring the productive system to address labor market demand and enhance integration into the global economy. Promoting local development must focus on harnessing the new digital technologies to increase the share of the domestic economy in global trade. Access to trade enhances the country capabilities to attract foreign direct invest and encourage technology transfer. Manufacturing production in most developing countries is inadequate to generate linkages and stimulate diversification. Digital technologies facilitate access not only to knowledge and skills, but also to global value chains and meet international standards. Global cooperation also help developing countries building their digital capacities so that to support economic diversification and participate in the digital economy.

3 Economic Diversification

Most developing countries, including the Arab world, rely on a limited number of commodities, mainly natural resources, to finance development and support public expenditures. Heavy dependence on revenues from exports makes the economy vulnerable due to the instability in export earnings and uncertainty about future demand in international markets. Economic diversification is to broaden the productivity base by allowing the economy to produce multiple products which reduce the risk of price fluctuations and stabilize earnings. Recent trends in energy prices have had substantial impact on development forcing oil producing countries to restructure their public expenditures and cut back on investment allocations. It has been realized that new development model is needed not only to reduce economic dependency and diversify output, but also to support integration into the digital economy. Knowledge and information are becoming the key drivers of the global economy which makes it urgent that countries rethink development and respond to the challenges of the new economy. Developing countries can initiate policies to take advantage of the digital technologies and cultivate the benefit of the Fourth Industrial Revolution. Economic diversification is to transform the economy from producing a limited number of products to multiple sources aiming at reducing economic vulnerability and supporting sustainable growth. Diversification enhances linkages creation which stimulates investment in predictive activities and improves global competitiveness. Structural factors such as, human capital, population, and quality of institutions, is among the important forces that support rapid diversification. According to the United Nations "Diversification increases with increasing population as local firms have access to a large market and thus benefit from economies of scales. Human capital allows economies to change their specialization patterns from primary commodities towards more knowledge-intensive manufactured goods. Political and economic institutions foster business confidence and cultivate the development of new business activities by creating a friendly investment environment" (United Nations (2016), P. 17).

The new economy, also referred to as a knowledge-based economy, depends more on the power of the brain than the natural capital. Dynamism is among the important features of the knowledge economy, which requires continuous investment in human capital, ICTs, R&D and innovation. Lifelong learning and training are necessary for reducing the risk of labor market disruption and strengthening the capacity for knowledge creation and innovation. Governments must build effective strategy to diversify the economy by investing in the new economy (Al-Roubaie 2019a).

In recent years, a substantial amount of literature has been published with specific reference to economic diversification as an alternative approach to development. Countries with limited resources can increase productivity by investing in human capital and innovation to enhance the economy capabilities developing new methods, create new knowledge and produce new products. Economic conversion entails restructuring the productive system to produce new products and stabilize export earnings. Economic Complexity Index (ECI) measures the number of goods and services produced by a country that are not produced by many other countries. A low economic diversification implies that resources are mobilized for production of limited products, mainly for exports, and not allocated for production of high value-added activities such as manufacturing and services. The index also measures the knowledge intensity of an economy or production of a product. In 2017, Japan ranked on top by scoring 2,309 compared to Germany 2,075, Qatar 0.396, UAE 0.128, Jordan −0.149, Tunisia −0.287, Egypt −0.319, Algeria −0.811 and Sudan −1.457 (Economic Complexity Index) The majority of developing counties, including the Arab world, are ranked low in the complexity index reflecting their weak capabilities to support rapid diversification. Diversification fosters economic growth through allocation of productive resources across a wide range of sectors allowing the economy to produce and exports products that require a contain complexity.

4 Conclusion

This paper briefly discusses the impact of economic diversification on development. Many of the developing countries are lacking to some of the basic managerial, technical, scientific and financial resources needed to restructure the economic system and reduce high dependency on international trade. High economic dependency subjects the economy to high degree of instability, and therefore, there is urgent need for building resilient economy to stand against external shocks. Economic diversification broadens the productive structure by enabling the economy to produce variety of goods and services instead of a single commodity. Recent advancement in digital technologies provides new opportunities for developing countries to participate in the global economy and sustain development. Digital technologies allow knowledge sharing and exchange ideas which encourage entrepreneurship and attract foreign direct investment. Developing countries can utilize digital technologies not only to bring people and regions together, but also to promote collaboration and gain access to global value chains. Digital technologies empower individuals and enterprises to benefit from the new economy through participation in the global markets and acquire knowledge, information and technology. Integration into the global economy allows developing countries to accelerate the process of economic transformation and speed up the process of participation in the digital economy.

Digital technologies could increase the country readiness and support economic diversification. Knowledge sharing and information dissemination provide new opportunities for decision makers to build capacity for development. Connectivity enables the economy to make appropriate choices to ensure that global integration facilitates technology transfer, innovation diffusion and knowledge creation. Weak productive structure has been responsible for economic volatility in most developing countries. In particular, countries in the Middle East and North Africa, lack of economic diversification has contributed to high unemployment and low level of industrialization. Digital technologies allow governments to formulate effective strategies and construct policies that ensure inclusiveness and improve equity in allocation of resources among regions for the benefit of all people.

References

Al-Roubaie, A.: Building capacity for knowledge economies in the Arab world: the role of human capital. Int. J. Eng. Technol. **7**(4.9), 55–62 (2018). https://www.sciencepubco.com/index.php/IJET

Al-Roubaie, A.: Knowledge absorption and economic diversification: the case of bahrain. Int. J. Recent Technol. Eng. (IJRTE) **8**(IC2), 450–455 (2019a). https://www.ijrte.org/

Al-Roubaie, A.: Building capacity for digital development in the Arab world. Role Educ. Int. J. Eng. Adv. Technol. (IJEAT) **8**(2249–8958, 5C), 1530–1537 (2019b). https://www.ijeat.org/

United Nations Conference on Trade and Development (UNCTAD): Digital Economy Report 2019 (2019). https://unctad.org/en/PublicationsLibrary/der2019_overview_en.pdf

United Nations: The Concept of Economic Diversification in the Context of Response Measures (2016). https://unfccc.int/sites/default/files/resource/Technical%20paper_Economic%20diversification.pdf

International Monetary Fund: Economic Diversification in oil Exporting-Arab Countries (2016) https://www.imf.org/external/np/pp/eng/2016/042916.pdf

Mckinsey: Digital Middle East: Transforming the Region into a Leading Digital Economy (2016) https://www.mckinsey.com/featured-insights/middle-east-and-africa/digital-middle-east-transforming-the-region-into-a-leading-digital-economy

World Economic Forum: The Global Information Technology Report 2016 (2016). https://www.weforum.org/reports/the-global-information-technology-report-2016

World Bank Group: Digital Economy concepts, Trends and Visions: Towards a Future Proof Strategy (2016). http://pubdocs.worldbank.org/en/513361482271099284/Digital-Economy-Russia-Discussion-paper-2016-12-20-eng.pdf

United Nations Conference on Trade and Development (UNCTAD): Information Economy Report 2017 (2017). https://unctad.org/en/PublicationsLibrary/ier2017_en.pdf

Economic Complexity Ranking. https://oec.world/en/rankings/country/eci/

The Fourth Industrial Revolution: Challenges and Opportunities for Mena Region

Amer Al-Roubaie(✉) and Mansoor Alaali(✉)

Ahlia University, Manama, Bahrain
{aalroubaie,malaali}@ahlia.edu.bh

Abstract. The new technologies driving the Fourth Industrial Revolution are expected to cause substantial impact on human societies. Recent advancement in digital technologies is changing market demand for labor by introducing new tasks that require new skills. The new technologies including automation, robotics and artificial intelligence are disruptive causing structural unemployment and inequalities among and within nations. In particular, countries in MENA region need to respond to the challenges of the new technologies by formulating strategies aiming at increasing participation in the Fourth Industrial Revolution and sustaining economic growth. This paper sheds light on the challenges and opportunities facing developing countries in general and MENA region in particular by discussing the prospects that these countries can benefit from the new digital economy.

Keywords: Fourth Industrial Revolution · Opportunities · Mena region

1 Introduction

Recent literature on development economics highlights the impact of digital technologies on economic growth of nations. The Fourth Industrial Revolution (4IR) is identified with disruptive technologies causing structural unemployment and labor market instability. Digital technologies, including automation, robotics and artificial intelligence, are expected to have far reaching consequences not only on production and distribution of goods and services, but also on the way people work and live. Responding to these challenges will require countries to adapt programs and formulate strategies to lessen the impact of technological disruption on society. Educational institutions are to play an important role in responding to the challenges of the 4IR by undertaking scientific research and training graduates with adequate skills for satisfying market demand.

The main objective of this paper is to examine some of the opportunities and challenges that the Middle East and North Africa (MENA) face for participating in the 4IR. Digital technologies are improving networking by improving communication and providing greater access to global markets. Developing countries in general and MENA in particular, can benefit from the new economy to diversify productivity and improve global competiveness. Digital technologies empower individuals, enterprises and institutions to improve communication and make productive decisions that respond to technology disruption and sustain growth. MENA region needs to encourage public-private

A.-E. Hassanien et al. (Eds.): AICV 2020, AISC 1153, pp. 672–682, 2020.
https://doi.org/10.1007/978-3-030-44289-7_63

partnership to support mobilization of resources and increase competitiveness. Networks facilitate collaboration and cooperation with foreign forms which help local enterprises joining the Global Value Chains (GVCs) and benefiting from technology transfer and skills acquisition. Digital readiness in most MENA countries remains inadequate to meet the challenges of the digital economy, and therefore, there is urgent need for structural reforms to enhance capabilities of the economy producing knowledge, disseminating innovation and participating in the Fourth Industrial Revolution. This paper sheds light on some of the causes and consequences of the new economy on MENA region.

2 The Fourth Industrial Revolution

The last two centuries have witnessed substantial socio-economic transformation marred by technological development and scientific discoveries. Output per person increased sharply reflecting the rise in productivity and the contribution of knowledge and innovation to development. Western countries were able to cultivate the fruits of scientific and technological advancement to foster economic growth and build modern civilization. Science, technology and innovation were among the most distinguished features of Western civilization that transformed societies into welfare states enjoying high consumption status. The First Industrial Revolution, a period in the second half of the 18th century, was able to transform the European societies into industrialized countries replacing mainly the agrarian society with machines and factories producing goods and services. Fueled by the steam power, the First Industrial Revolution began in Britain and later spread into the rest of Europe and the United States. The Second Industrial Revolution, a period occupied most of the second half of the 19 century and early 20st century. This era of rapid industrialization has witnessed the production of automobiles, steel, electricity, chemicals and manufactured goods. Unlike the First Revolution, the Second Revolution spread over wider geographical space to include the rise of Japan and other Asian countries. The second half of the 20th century gave birth to a Third Industrial Revolution with the emergence of new technologies including nuclear energy, electronics and biotechnology. This era also witnessed the rise of automations and digital technologies which enhanced human communication and networking. Semiconductors, the internet and personal computers have had considerable impact on our ability to work, communicate and live. The technologies of the Fourth Industrial Revolution, also known industry 4.0, smart industry, intelligent industry and smart factory, are to make substantial impact on human societies by changing the dynamism of the labor market with emphasis on such technologies as artificial intelligence, the internet of things, 3-D printing, robotics, material science, nanotechnologies and automation. These technologies are to cause structural unemployment as well as to reduce the demand for people with low skills. It requires highly flexible labor market to meet demand requirements and quickly adjust to changes in technology. As Schwab points out: The Fourth Industrial Revolution is characterized by a range of new technologies that are "fusing the physical, digital, and biological worlds, impacting all disciplines, economies and industries, and even challenging ideas about what it means to be human" (Schwab, p. 9). Our world is becoming highly connected due to rise of digital technologies and the internet of things. People across geographical boundaries are able to communicate and exchange ideas that concern human affairs and impact future trends. In particular, for developing countries the 4IR provides new opportunities to close the income, knowledge, and technological gaps. On their parts, developing countries must respond to the challenges of the 4IR by investing in education, building digital infrastructure and promoting innovation. Digitization increases the economy capabilities to diversify productivity and improve labor market flexibility. In this age of information, access to international markets facilitate the flow of FDI and encourage entrepreneurship to participate in the digital economy. The 4IR has the tendency to improve the quality of live and enhance economic productivity. So far, the benefit from the 4IR has been limited to countries with advanced digital technologies which enabled them to access to global markets and acquire skills and knowledge required to close the digital gap and participate in the new economy. Technological advancement has made it possible for countries to produce new products and services that increased the productivity of the economy and improved the society welfare. E-services which made possible through the internet of things and other digital devices have encouraged individuals, institutions and corporations to engage in productive activities and expand the frontier of knowledge. The 4IR has the potential to promote development and speed up the process of socio-economic transformation. Investment in education, science, technology and innovation increases the economy readiness to participate in the digital economy and support rapid transformation. Not only technology has become accessible through FDI and Global Value Chains (GVCs), but also can be produced domestically through entrepreneurship and collaboration with foreign firms. Technological innovation has also the potential to improve efficiency and diversify productivity by allowing countries to produce goods and services and compete in both local and global markets. The cost of production is expected to decline as new technologies increase the ability of the economy to create knowledge, acquire skills and diffuse innovation. The 4IR is driven by the power of the brain which entails investment in research and development, higher education, science and technology, employability skills and lifelong learning to respond to the challenges posed by the new technologies. The knowledge economy, for example, relies on intangible assets to produce goods and services and create economic value. Countries that are still use natural capital as an input in production may not be able to continue to generate growth without knowledge creation and innovation. Almost all countries in the Middle East are still dependent on natural resources to support development. In the digital economy, driven by the 4IR, digital technologies broaden business activities by giving access to marketing channels, credit, logistic and other basic services (World Economic Forum 2019, p. 8) In the digital economy, networks play an important role in supporting business activities by allowing newcomers to gain access to resources necessary for business. “The fundamental and global nature of this revolution means it will affect and be influenced by all countries, economies, sectors and people” (Schwab, The Fourth Industrial Revolution, p. 9). Rapid growth in technological development could become disruptive due to shift in market needs, especially for skilled labor. In the 4IR, machines are expected to substitute for labor by preforming tasks much faster and cheaper. Such tendency will increase the risk of job losses creating structural unemployment, especially for those working in traditional sectors of the economy. On average, it is estimated that 42% of the core skills in the job market will be new by

2022. (World Economic Forum 2019, p. 11) In the new digital economy, the job market requires workers with analytical thinking, critical thinking, emotional thinking, complex problem solving, system design and evaluation and leadership. This implies that by 2022, the market need for skills required to perform most tasks will shift significantly towards people with specific skills. By 2025, the share of machines in in total market task is to reach 58% leaving 42% for labor. However, digital technologies are offering on line job opportunities which will benefit people with less mobility, including women in conservative societies. Not only on line jobs provide new opportunities, but also improve efficiency through access to information and improving ways for workers to communicate. Unemployment is perhaps the greatest challenge will be facing countries in the era of the 4IR. In MENA region not only the rate of unemployment is high in countries across the region, but also most graduates from universities and educational institutions are not adequately trained to participate in the 4IR. Worldwide, there are about 2 billion youth, the future of home will be influenced by the new technologies that require certain skills for joining the labor market. More than half of this nearly 2 billion youth will not be able to have the right skills or qualifications to participate in the digital economy, according to the Education Commission (See Deloitte, Preparing Tomorrow's Workforce for the Fourth Industrial Revolution 2018) The majority of these youth will be in countries that are lacking to some of the basic fundamentals for preparing young people to take part in the 4IR. (Al-Roubaie 2018). Developing countries need to rethink the model of education currently employed and revise new ones so that to diversify educational opportunities and graduate students with relevant skills for the labor market. The complex nature of the new digital technologies requires private-public partnership to improve decision-making and support the participation of the youth in the Fourth Industrial Revolution. The replacement of labor with robotics and automation technologies could have a devastating impact on labor market forcing millions of young people out of the job market. In countries with high unemployment, the 4IR will widen income inequalities by providing opportunities for a few skilled workers. Lack of jobs could result in brain drain forcing human capital to flee outside the country in search for better working conditions. Globalization has led to the growth of an international labor market coupled with free movement of factor inputs, mainly labor and capital. Migration of human capital could have severe consequences on development in sending countries. The role of education and training for responding to the challenges of technological change becomes instrumental to reduce the risk of market disruption and improve the flexibility of labor market. Universities need offer curricula with specific programs to increase the supply of skills and meet market demand. Most universities in MENA are still graduating students with inadequate technical knowledge to qualify them participating in the digital economy. Continuous training and lifelong learning underscore the need for a dynamic educational system capable of supplying relevant skills for the new economy. Involvement of the state is necessary for providing funding for research and development, incentives and training programs to support labor market flexibility.

3 Human Capital for the Fourth Industrial Revolution

The new economy, also described as a knowledge-based economy, underscores the importance of knowledge as input in production and its contribution to productivity growth and economic transformation. A knowledge-based economy is an economy directly linked to production, distribution and use of knowledge and information. Modern economic growth theories are using knowledge and technology to examine the process of economic growth and its impact on development. Knowledge empowers society by increasing the economy's capabilities to diversify output and improve global competitiveness. Investment in knowledge stimulates linkage creations and encourages entrepreneurship which strengthens the ability of the economy to foster growth and sustain development. Production of knowledge is critical for participation in the 4IR which fuels the dynamism of the productive system through creativity and innovation. Knowledge usually embodied in human minds and technology which can be mobilized to foster economic growth and diversify output. Investment in the 4IR is being directed to robotics, artificial intelligence, nanotechnology, automation and computer related services. "The ability of economies to meet 4IR requirements is largely determined by the competency of their human capital, which is the product of the educational process; and the ability to utilize the economic returns of RDI in various areas of production. Without these, human capital cannot contribute effectively to the production process and bring about a transition to a competitive knowledge economy" (UNDP (2018) p. 73). In addition, investment in research and development is equally important for generating new ideas and promoting innovation. This requires production of knowledge workers equipped with skills and technical know-how required to work in a wide range of computer related activities. In the new model, knowledge is no longer treated as an external factor in production function and therefore, investment in knowledge is to have direct impact on the productivity of labor and capital. Production of knowledge enhances innovation capabilities which represent a major force in the economic dynamism driving the 4IR. The economy benefits from knowledge production to increase the stock of knowledge which stimulates business activities and create new job opportunities. From macroeconomic perspective, knowledge can be easily shared among productive agents to ease the constraints on economic growth and diversification of output. Knowledge linkages facilitate innovation and rapid technological development which support market dynamism and improve capabilities to participate in the 4IR. In the case of the MENA, investment in knowledge is vital for speeding up the process of global integration. In this age of information, knowledge production and acquisition are much easier to achieve through a wide range of means including FDI, GVCs, partnership, and collaboration and cooperation with foreign firms. In recent years, brain gain has been widely recognized as an important source of skills and knowledge that developing countries can make use of to strengthen readiness and participate in the global economy. According to the World Bank there are 20 million diaspora members of MENA origin working outside their countries of origin. Diaspora represents important human capital assets which can be mobilized to encourage entrepreneurship and contribute to the digital economy. Information communication technologies are providing access to global markets allowing countries to acquire a range of information, skills and knowledge. In the knowledge-based economy, knowledge networks increase the economy ability to diffuse and use information, promote innovation, develop new products and create new knowledge. Similarly, networks can efficiently be used to share, communicate and distribute knowledge among users both locally and globally. In this regard, government policies need to focus on strengthening the fundamentals for building a knowledge-based economy by investing in higher education, research and development and science and technology. Policies must focus on the capacity to learn and lifelong learning to improve the matching of skills supply and increase labor market flexibility. Building human capital capacity requires effective educational system not only capable of producing relevant skills, but also stimulates new research and contributes to the digital economy, driven by the 4IR. In the knowledge economy, university/industry collaboration is necessary to meet skill requirements and support market demand for employability. Universities are at the center for development for contributing to technological advancement and scientific research. Universities could become a powerful vehicle for change by increasing the society capabilities to move forward and catch up with the industrialized countries. Thus, to ensure the development of new ideas and support innovation, universities need be integrated into the productive sector. Across MENA region, the educational system must be transformed into one that is flexible and graduate students to match the skills that they need to participate in digital economy. Curriculums currently in use need to be upgraded and reformed to include courses in critical thinking, problem solving, science and technology-based development and leadership in order to respond to market change and foster innovation. The faculty must be trained to become aware of the rapid changes in the labor market and introduce new programs to boost technology-based skills. The faculty also needs to frequently change the teaching methods by adapting new models that address the challenges of the 4IR. Unfortunately, large number of universities in MENA region are not adequately funded and technologically equipped to respond to market needs and produce quality students for taking part in the digital economy. Countries in MENA region must invest in science and technology to enable them building local capacity for knowledge creation and technological advancement. So far, it does not seem that universities in MENA region are responding to educational need and produce quality students for building knowledge society. There is a need for educational system not only to promote creativity and innovation, but also to permit learning that adapts to the changing demand of the digital economy. The new paradigm in education is to be driven by new opportunity, creativity, problem solving and critical thinking. There is a need for changing the role of government in education to encourage independent thinking and ease pressure on educational institutions by public management. The public sector must introduce measures that provide incentive for universities to promote quality, conduct research and support development. This makes investment in human capital the most valuable investment that a nation can achieve to produce positive externalities and sustain growth. Human capital contributes to society development by creating new knowledge and promoting innovation. According to the World Bank, human capital contribution to global wealth reached to 64% in 2014 reflecting the importance of investment in people as a source of wealth and support for development. This is compared to 35% in the Middle East and North Africa, 60% in East Asia and 61% in Latin America. (World

Bank Group 2018) "The challenge facing the higher education sector in the Arab world is that, even as it has expanded in recent years to try to accommodate a historically large youth contingent, it has not kept pace with the shifting demand for human capital. Governments have educated a growing cohort of students without ensuring that they have the skills needed to succeed and without laying the economic groundwork to employ them. As a result, many countries in the region are suffering from a chronic oversupply of university graduates, while experiencing a shortage of workers with the skills sought by employers in the marketplace" (The Brooking Project (2011), p. 1).

4 Digital Technologies and the Digital Economy

The digital economy, also called the new economy, the web economy, the internet economy, is an economy driven by the power of the digital technologies. It enables individuals; business and institutions conduct communication and acquire information through electronic devices on the internet. OECD defines the digital economy as "the single most important driver of innovation, competitiveness, and growth in the world." (World Bank Group 2016, p. 2) Digital technologies empower society to foster economic growth and enhance capabilities to participate in the global economy. An important feature that distinguishes the 4IR is the advancement in digital technologies. The new economy depends on social networks and the internet services which are shaping the 21st century development. Digital technologies are connecting individuals, business and institutions so that to increase communication and gain access to international markets. For many of the developing countries, connectivity facilitates knowledge transfer and information dissemination which are needed to support the economy capacity to produce goods and services. The potential impact of digitization is expected to contribute to employment, especially among the youth and women. For countries with high unemployment rates as in the case of MENA region, investment in digital technologies will provide new opportunities for entrepreneurship to participate in the digital economy and benefit from the 4IR. "ICT will continue to play an essential role in building and maintaining resilient infrastructure, in promoting inclusive and sustainable industrialization, and in fostering innovation in the emerging information and knowledge societies, which depend on open access to academic research, transparency to make informed decisions and the power of online collaboration to support cross -sector and in-house co-reaction learning and work." (United Nations 2019, p. 23)

Digital economies, including the internet, are transforming societies in profound ways. It increased the economy capabilities to communicate, acquire knowledge and information, and participate in international trade. However, rapid advancement in technology is changing the demand for people with specific skills to perform complex tasks and develop new products. On line services could reduce the gender digital gap allowing women in conservative societies to join the labor force and contribute to the well-being of their communities. Making use of the internet will encourage women to make use of their tacit knowledge and participate in market activities. Harnessing tacit knowledge enhances the economy readiness to diversify productivity as well as to strengthen knowledge creation and innovation diffusion. Digital technologies facilitate business start-up and provide incentives for enterprises to compete by developing new products and inventing new methods. To strengthen links to global market and embrace the 4IR, countries in MENA region should cultivate the knowledge and adopt innovation deployed in other countries. Not only is knowledge costly to produce, but also requires physical and financial resources many of the countries in the region are still lacking. Use of online services brought by digitization could facilitate knowledge transfer and innovation diffusion at lower costs than those produced at home. For MENA countries, speeding up the process of development require putting the existing knowledge, including foreign knowledge, for use. Use of external knowledge is valuable for building local knowledge system and supporting innovation. In other words, MENA countries "do not need to devote the bulk of their time and attention to building institutions that produce new knowledge. What they need instead are institutions that have the capacity to find and help governments and private enterprises utilize the existing knowledge embedded in the thousands of patents that expire every month around the world." (The World Bank 2008, p. 52). Table 1 illustrates the gaps in Networked Readiness Index (NRI) and the Global Innovation Index (GII) for several, mainly Arab countries in MENA region. Both indices represent important indicators of the current state of development in Arab countries and their readiness to participate in the 4IR. Not only digital technologies and innovation represent the main components of these indices, but also represent key drivers of the digital economy. Most Arab countries, as data in Table 1 show, performed lower in 2017 compared to 2011 reflecting the inability of these countries to mobilize resources in productive manner and deepen integration into the digital economy. Arab countries are facing acute development challenges that require capacity building programs to exploit opportunities offered by the digital economy and participate in the 4IR. There is urgent need for reducing the digital divide and support long term development. Digital technologies provide numerous external benefits for being inclusive as well as offering opportunities that empower all society members.

Table 1. Networked Readiness Index (NRI) and Global Innovation Index (GII) ranking evolution, Arab Countries.

Country	NRI 2010–2011	NRI 2016	GII 2011	GII 2017
UAE	24	26	34	36
Qatar	25	27	26	49
Bahrain	30	28	46	66
Saudi Arabia	33	33	54	55
Tunisia	35	81	66	74
Oman	41	52	57	77
Jordan	50	60	41	83
Egypt	74	96	87	105
Kuwait	75	61	52	56
Morocco	83	78	94	72
Lebanon	95	88	49	81

(continued)

Table 1. (*continued*)

Country	NRI 2010–2011	NRI 2016	GII 2011	GII 2017
Algeria	117	117	125	108
Mauritania	130	138	NA	NA
Yemen	NA	NA	123	127

Source: United Nations (2019) United Nations, Arab Horizon United Nations: Arab Horizon 2030, Digital technologies for Development, Beirut (2019).

5 Role of Institutions

Institutions are critical for fostering economic growth and supporting socio-economic transformation. In the new economy, the contribution of the institutions is essential to ensure labor market flexibility and enhance the economy capabilities to produce goods and services. Economic development is influenced by the effectiveness of government institutions and their ability to promote growth and economic stability. Not only institutions provide incentives for productive agents and market support, but also strengthen market confidence and encourage investments. The 4IR will increase uncertainty about future market trends due to rapid changes in technology and increase competitiveness among nations. Risk management and market stability become necessary to strengthen confidence and minimize the impact of disruption of the new technologies. Government institutions can support the market by promoting research and development, providing financial incentives, enhancing connectivity and introducing regulatory measures that regulate new technologies. Governments need to rethink the current institutions and create new types that can contribute and adapt quickly to market changes, especially with regard to the new technologies. The risk of market disruption requires institutions that enjoy high degree of flexibility to accommodate changes in technology and testing new methods. "Disruptive technologies call for completely radical institutional structures and approaches." (World Bank 2016, p. 6) governments must contribute to the performance of the digital economy by setting up new institutional structure. Governments should formulate strategies and encourage research to predict future trends and reduce the risk of market disruption. Gathering data information about market trends can provide important guidance for policy construction and strategic planning. In the digital age, gathering data and information about current local and global trends increase government capabilities to reduce market disruption and sustain market stability. The impact of digital technologies, including automation, robotics and artificial intelligence, on countries in MENA region are likely to cause substantial social and economic consequences. Low technical, managerial, institutional, and educational capabilities will weaken the ability of governments to respond to the challenges of technological change and minimize the risk of market disruption. Youth unemployment rate in MENA region is among the highest in the world which requires government support programs to lessen the impact of the new digital technologies on job losses and labor market stability. Educational programs and training become essential for preparing students to participate in the new economy and contribute to development. Government institutions can support the creation of "liquid workforce" which is quickly adaptable to technological change and market need. Building such liquid workforce requires good market assessment of skills need in both local and global markets. The impact of digital technologies is expected to have broad dimensions affecting all sectors of the economy. Various government institutions can contribute to market demand by monitoring market requirements and establishing mechanisms to respond to such requirements. In the new economy, different kind of leadership and management will be required to make decisions and sustain market stability. As Professor Schwab notes, "We need leaders who are emotionally intelligent and able to model and champion co-operative working. They'll coach, rather than command; they'll be driven by empathy, not ego. The digital revolution needs a different, more human kind of leadership". (Schwab, p. 21) The disruptive nature of the Fourth Industrial Revolution requires leaders who understand not only how to manage people working in different locations, but also how to virtually connect divers skills to enhance the firm's performance. In MENA region, leadership must decide on what policies and strategies are needed to manage the disruption in the labor market. Also, leadership should be in a position to design and implement models and approaches that technology will not replace people creativity and their contribution innovation. Technology should be human-centered to strengthen the economy capabilities and foster economic growth. Leadership can create conditions that integrate technology into the productive system without impacting the society's values.

6 Conclusion

The technologies driving the Fourth Industrial Revolution are expected to have a substantial impact on human societies. Technologies, including artificial intelligence, robotics and automation, are to disrupt labor markets causing structural unemployment, inequalities and uncertainty about future trends. Many of the developing countries, including the Arab world, are still lacking to some of the basic scientific and technological requirements to meet the challenges of the Fourth Industrial Revolution. In the new economy, technological innovation and knowledge creation are important for supporting market competitiveness and deepening integration into the digital economy. Digital technologies empower individuals and enterprises to participate in the global economy and gain access to skills and information for development. Responding to the challenges of the Fourth Industrial Revolution underscores the importance of investment in human capital, R&D, ICTs technologies and institutional reforms. Governments should formulate strategies that integrate digital services into the national policy to make the economy resilient against technological disruption. Governments should take advantage of the new technologies not only to increase basic services to the public, but also to eradicate poverty and reduce inequality among and within regions. Building capacity for digital economy must be inclusive so that equal opportunity is provided to all individuals to participate in the Fourth Industrial Revolution. Involvement of the Private sector in market activities could enhance the economy capabilities to promote innovation and produce goods and service for the global markets. In the Middle East and North Africa (MENA) region, the public sector still exercises sizable influence over market activities which hinder private sector participation in the new economy. Digital technologies are expected to have positive impact on the quality of life and people welfare. However, digital technologies are also creating inequalities among and within nations by generating unequal development opportunities. Benefit from digitization is linked to access to the internet and the availability of skills to operate digital technologies. Most developing countries are still lacking to internet services to gain access to knowledge and information.

References

Al-Roubaie, A.: Building capacity for knowledge economies in the Arab world: the role of human capital. Int. J. Eng. Technol. **7**(4.9), 55–62 (2018). https://www.sciencepubco.com/index.php/IJET

Al-Roubaie, A.: Knowledge absorption and economic diversification: the case of bahrain. Int. J. Recent Technol. Engineering (IJRTE) **8**(IC2), 450–455 (2019a). https://www.ijrte.org/

Al-Roubaie, A.: Building capacity for digital development in the Arab world: the role of education. Int. J. Eng. Adv. Technol. (IJEAT) **8**(2249–8958, 5C), 1530–1537 (2019b). https://www.ijeat.org/

Schwab, K.: The Fourth Industrial Revolution, p. 9. https://www.weforum.org/about/the-fourth-industrial-revolution-by-klaus-schwab

Deloitte: Preparing Tomorrow's Workforce for the Fourth Industrial Revolution (2018)

The Brooking Project: Higher Education Reform in the Arab World, p. 1 (2011). https://www.brookings.edu/wp-content/uploads/2016/06/08_education_reform_wilkins.pdf

United Nations: Arab Horizon 2030, Digital technologies for Development, p. 23, Beirut (2019)

UNDP: Knowledge and the Fourth Industrial Revolution (2018). http://knowledge4all.com/en/125/Pages/Knowledge-and-the-Fourth-Industrial-Revolution-2018

World Economic Forum: Dialogue Series on New Economic and Social Frontier, Shaping the New Economy in the Fourth Industrial Revolution (2019). https://www.weforum.org/whitepapers/dialogue-series-on-new-economic-and-social-frontiers-shaping-the-new-economy-in-the-fourth-industrial-revolution

The World Bank: Science, Technology and Innovation. p. 52, The World Bank, Washington (2008)

World Bank Group: Digital Economy, Concept, Trends and Vision: towards a Future-Proof Strategy (2016). http://pubdocs.worldbank.org/en/513361482271099284/Digital-Economy-Russia-Discussion-paper-2016-12-20-eng.pdf

World Bank: The Changing Wealth of Nations (2018). https://www.worldbank.org/en/news/feature/2018/01/30/the-changing-wealth-of-nations-2018

Rethinking Economic Development in Muslim Societies in the Context of the Fourth Industrial Revolution

Amer Al-Roubaie and Adel M. Sarea(✉)

College of Business and Finance, Ahlia University, Manama, Kingdom of Bahrain
Adelsarea@yahoo.com

Abstract. Economic development is a process of structural transformation not only involves productivity growth, but also improvement in the quality of life. In most developing countries, including Muslims, economic productivity remains low to foster growth and meet the challenges of the 21st century. The new economy, driven by the Fourth Industrial Revolution, underscores the importance of investment in human capital, knowledge creation and innovation so that to increase competitiveness and deepen integration in the global markets. Digitization facilitates connectivity and enhances the country capabilities to gain access to technology and skills needed for promoting innovation and diversifying output. Rethinking development requires adapting new strategies for development to support labor market flexibility and strengthen digital readiness.

With the exception of a few countries, economic productivity in the Muslim world remains low to foster rapid growth and catch up with the rest of the world. In Muslim countries, economic development is driven by traditional sectors, mainly agriculture, with little contribution of manufacturing production. The Fourth Industrial Revolution requires investment in digital technologies, artificial intelligence, nanotechnology, green energy and robotics to increase the country innovative capabilities and diversify output. The Muslim world should rethink development to develop strategies that strengthen connectivity, collaboration, technological learning and digitization. The aim of this paper is to highlight the importance of knowledge and technology for economic development in Muslim countries. The paper focuses on the challenges facing Muslim countries in the 21st century and benefit from the Fourth Industrial Revolution.

Keywords: 4th IR · Development · Knowledge · Digital · Value chains

1 Introduction

Economic development is a process of structural transformation aimed at fostering productivity growth and improving the quality of human endeavors. Building capacity for development increases the society's capabilities to diversify output and leap forward. The failure of nations to promote development is hindered by both internal and external forces representing economic, financial, cultural, political, and environmental factors. In general, development usually refers to the transformation of a society from

A.-E. Hassanien et al. (Eds.): AICV 2020, AISC 1153, pp. 683–695, 2020.
https://doi.org/10.1007/978-3-030-44289-7_64

agricultural-based to manufacturing-based economy, and general increase in per capita income. Currently, the income gap between developed and developing countries is so wide that requires substantial efforts to be made by developing countries to meet the challenges and close the development gap. The 21st century, driven by the Fourth Industrial Revolution will require these countries to initiate policies aimed at increasing the stock of human capital, building ICT capacity and deepening integration in the global economy. The digital economy provides new opportunities for rapid socio-economic transformation by allowing countries to gain access to knowledge, skills information and technology.

The state of economic development in most Muslim countries suffers from several impediments reflecting the inability of these countries to manage resources productively and meet people's basic needs. Economic development in these countries is driven by traditional sectors, mainly agriculture, providing goods and services primarily for domestic markets. Investments in manufacturing production and technology have not been adequate to stimulate linkages and encourage industrialization. Historically, these countries were not able to take advantage of the First, Second and Third Industrial Revolutions and build technological capacity for development. However, the Fourth Industrial Revolution, driven by digital technologies, is offering new opportunities for developing countries to participate into the digital economy. Integration into the 21st century economy underscores the importance of investments in human capital, research and development, knowledge creation, information and communication technologies and innovation to strengthen the economic fundamentals and participate in the Fourth Industrial Revolution.

The aim of this paper is to shed light on the capabilities of Muslim countries to participate in the Fourth Industrial Revolution. Adaptation of the new digital technologies will require radical reforms that involve restructuring the productive system and diversify output. Muslim countries must initiate policies and formulate strategies to enhance human capabilities and deepen integration into the digital economy. The Fourth Industrial Revolution is offering challenges and opportunities for countries with limited resources to adapt the new technologies and participate into the 21st century economy.

2 The Fourth Industrial Revolution

Substantial amount of literature has been written in recent years about the role that digital technologies can play in the development of nations. The Fourth Industrial Revolution, driven by digital technologies, is expected to transform economies from traditional to knowledge-based economies causing profound changes in the way people work, communicate, and live. Responding to the challenges of the new economy entails countries to enhance global competitiveness by investing in the educational systems – especially higher education and technical training to reduce the risk of structural unemployment. This is extremely important for countries with high rates of population growth, including Muslims, and facing complex sets of internal and external challenges in fostering productivity to participate in the Fourth Industrial Revolution. These countries need to stimulate innovation and increase the stock of human capital to enhance sectoral productivity and contribute to the Fourth Industrial Revolution. In addition, connectivity and the internet are crucial for integration of the economy into the global economic system so that to enable countries meeting the rapidly changing needs of labor market. ICT increases the economy capabilities to gain access to global skills and knowledge for development. Linking various sectors of the economy with global markets through ICT is important for restructuring productivity and diversifying output. This will strengthen the capacity of the economy to integrate into the new economy and participate in the Fourth Industrial revolution.

Developing countries, including Muslims, must construct strategy to adapt to changes in digitization by increasing the country technological, scientific, managerial and technical capabilities. Investment in human capital is critical for enhancing human capabilities to upgrade foreign technologies and build indigenous capacity for development. Muslim countries should take advantage of the opportunities offered by the Fourth Industrial Revolution to restructure the productive system and diversify output. Building capacity for digital economy entails investment in Information and Communication Technologies (ICTs) and research and development to increase the country readiness and meet the challenges of the new economy. Not only digital technologies facilitate access to international markets and enhance competiveness, but also strengthen the economy linkages creation via knowledge acquisition and technology transfer. It is expected that improvement in efficiency will reduce costs and support entrepreneurship.

Digitization has the prospect to increase productivity and diversify output which reduces the country dependency on production and export of limited primary commodities. Bridging the technological and knowledge gaps allows countries restructuring the productive system and supporting innovation. Economies with low productivity can diversify output and build technological capacity for development. Technologically, Muslim countries benefited little from the previous industrial revolutions to restructure the productive system and facilitate integration into the global economic system. Participation in the digital economy underscores the importance of institutional reforms to provide the necessary incentive for knowledge creation, technology transfer and innovation dissemination. Muslim countries should also integrate digital solutions into export promotion to deepen global integration and enhance competitiveness. "The digital technologies, such as IOT, and the data obtained from their use, can provide new sources of knowledge, innovation and profits if utilized correctly and effectively" (UNCTAD 2017, p. 5).

The new technologies driving the Fourth Industrial Revolution are expected to distrust the labor market by demanding greater labor market flexibility and highly skilled workforce. These technologies, including artificial intelligence, automation, robotics and nanotechnologies are expected to substitute for labor across the economy which may result in structural unemployment, especially in countries with low industrial productivity. The Fourth Industrial Revolution requires high degree of labor market flexibility to adapt to the changes in technology and gain access to global markets. In the new economy, the demand for labor will require talent, creative thinking and problem-solving skills that institutions, corporatizing and universities must integrate into their strategies and increase supply of labor skills. Not able to provide the required skills could increase social tensions by segregating the labor market into "low-skill/low pay" and "high skills/high pay". The existence of such tendency could increase inequality, which widens the income gap between highly-skilled labor and low-skilled workers. As Klaus Schwab points out that "The Fourth Industrial Revolution is evolving and emerging in ways that are creating new challenges and concern for the world at time when concerns about inequality, social tension and political fragmentation are rising, and where vulnerable populations are increasingly exposed to economic uncertainty and the threat of natural disasters" (Schwab 2018, pp. 12–13).

In the Fourth Industrial Revolution, those benefits most from the new economy are providers of intellectual capital and creative ideas, which are necessary for knowledge creation and innovation. It is no longer labor is the core input in production; rather it is knowledge and creativity. The new technologies of the Fourth Industrial Revolution could contribute to inequalities by requiring specific skills many workers may not be able to acquire. Technological changes offer new opportunities which nations must harness to build capacity for manufacturing production and speed up the process of transformation. Digital technologies are expected to have impact across the economy affecting employment, income, market structures, gender, and inequality. In the Fourth Industrial Revolution, the labor market is to experience rapid changes due to the rise of jobs performed by machines including robots. To reduce the risk of market failure, there is a need for good management of the labor market to protect jobs and reduce the risk of inequalities. Workers lacking to adequate skills will be excluded from the labor market causing structural unemployment. "The outcomes of this revolution—which is occurring in an environment of unparalleled global connectivity and demographic change—are likely to perpetuate and worsen existing chasms between those who are prepared for 4IR and those who are not" (Deloitte 2019, p. 5).

Making initiatives to increase integration into the 21st century digital economy, will require countries to focus on human capital development and life-long learning to enhance workers creativity and gain comparative advantage in the global economy. Thus, increasing the country readiness to meet the challenges of the 21st century and catch up with the industrialized countries should be integrated into the national strategy for digital development. By 2030 the number of young people worldwide is expected to reach 2 billion half of them are not adequately equipped with skills required by the Fourth Industrial Revolution. Population growth in Muslim countries are among the highest in the world and therefore, the number of young people looking for job opportunities will be rising in the coming decades. Integrating the youth into the Fourth Industrial Revolution becomes necessary to reduce the risk of social unrest and maintain the social equilibrium. Young people need to be trained in technical, skills, soft skills, managerial skills and entrepreneurial skills.

3 New Model for Development

The failure of Muslim countries to speed up the process of development and catch up with the rest of the industrialized world is largely due to the economic model currently in use in most Muslim countries. Over the last several decades, development in most Muslim countries has been shaped by production and export of primary commodities which subjected the domestic economy to changes in global trade. This model makes the economy vulnerable to changes in global trade and commodity prices in the global markets. Economic and financial vulnerabilities reduce confidence in the economy by discouraging both local and foreign investors from investing in the economy.

Muslim countries did not benefit from globalization of finance and trade to participate in international trade and gain access to skills, knowledge and technology. Export-led growth model limits the ability of these countries to support development by imposing constraints on government commitment to finance development projects. Instability in export earnings undermine the ability of the government to plan for future financing. In the case of nonrenewable resources, export-led growth models are unsustainable because of depletion of resources overtime. This could be true with regard to oil and gas resources the reserves of which are expected to decline as more and more of these resources will be consumed (Al-Roubaie 2019a).

The new economy, also known as 'knowledge-based economy' is offering challenges and opportunities for developing countries including Muslims. Knowledge facilitates structural transformation and diversifies output allowing countries to surpass several stages in development. Unlike the traditional economy in which capital and labor are the main inputs, the knowledge economy depends on knowledge and innovation as the main drivers of growth and wealth creation. Taking advantage of the new opportunities offered by the knowledge economy require nations to strengthen the fundamentals for building capacity for knowledge creation and innovation diffusion. Muslim countries need new model for development to enhance technological and knowledge capabilities so that to increase global linkages and encourage both local and foreign investment in the economy. Rapid growth in population and rising unemployment, especially among young people and women, could hinder growth and increase the risk of social and political instability. However, collaboration, cooperation and partnership among key players such as universities, entrepreneurs, business groups and institutions could accelerate growth and improve productivity.

Rapid technological change requires quick response and high degree of policy flexibility to ensure that the economy is not left behind. "The Fourth Industrial Revolution is sometimes described as an incoming thunderstorm, a sweeping pattern of change visible in the distance, arriving at a pace that affords little time to prepare. While some people are ready to face the challenge, equipped with the tools to have the change and take advantage of its effects, others do not even know a storm is brewing" (Deloitte 2019, p. 8).

In the new model, economic growth depends more on knowledge, creativity and innovation as drivers of market activities and entrepreneurship. The internet and digital services are instrumental in knowledge creation and innovation diffusion. Connectivity facilitates access to global markets providing the country with knowledge, skills and information that support technological learning and knowledge localization. Similarly, investment in digital technologies helps the economy to adapt to the new technologies and facilitate integration into the global economy. By adapting the new model, the economy will be in better position to create linkages and foster economic growth. Unlike the traditional model, driven by export earnings, the new model increases the country capabilities to start new businesses, create new knowledge, build new industries, and create new opportunities for people to participate and contribute to the economy (Al-Roubaie 2018).

As an input in production, knowledge increases productivity and fosters economic growth. Not only creation of new knowledge offers new employment opportunities, but also encourages entrepreneurship and promotes innovation. Countries with limited resources can benefit from investment in knowledge creation to build capacity for development and strengthen innovation dissemination. Muslim countries should initiate strategy to strengthen the pillars of the knowledge economy by investing in human capital, introducing institutional reforms, building ICTs infrastructure, increasing expenditure on research and development and promoting innovation. Governments could contribute to the knowledge economy through better management and greater incentives for the future direction of the country.

Finland is a good example of government contribution to knowledge development and innovation dissemination. In the case of Finland, the "roles of various public organizations and other actors have been important. The Finish system of innovation is characterized by public-private partnerships, good public governance, consensus building and dense networking among companies, universities, and research organizations" (Dahlman 2006, p. X). Muslim countries should draw on the best practices from other countries and build foundations for knowledge-based economy to diversify output and strengthen linkages to the digital economy. Governments could speed up the process of knowledge acquisition, absorption and creation by providing institutional regime capable of making efficient use of global knowledge and encouraging entrepreneurship. Rethinking development must involve knowledge creation and innovation dissemination to stimulate growth and create new opportunities for entrepreneurship (Table 1).

Table 1. Model of knowledge-based economy

	Economic (KBE)	Social	Natural
Education (development of human capital)	Education for a skilled workforce	Education for total human development	Education for sustainable development
Innovation (development of structural capital)	Systems, processes, and technological innovation	New institutions and protocols for peace, equity and human rights	Environmental technologies, e.g. renewable energy technologies
Building networks (development of stakeholder capital)	Financial and physical networks e.g. ICT infrastructure	Social networks, social trust, cultural integrity	Agreements to protect and sustain planetary life support system

Source: ADB, Moving Toward Knowledge-based Economies: Asian Experience

The objective of development is to make changes in the economic structure so that to increase output and create new opportunities for people to work and earn income. Historically, economic development in Muslim countries is shaped by the colonial trading system in which developing countries were supplier of primary products earning foreign exchange to pay for imported capital goods and raw material for development. Samir Amin, an Egyptian economist, argued that dependency on the world system has led to the deterioration of the terms of trade by allowing resources to flow from poor countries to rich countries.

4 Knowledge Deficit in Muslim Countries

Building readiness for the knowledge economy requires the construction of an effective knowledge strategy aimed at increasing investments in human capabilities, creative thinking and innovation. Investment in information and communication technologies (ICTs) is critical for building knowledge capacity. ICT facilitates connectivity so that local enterprises gain access to the global knowledge system for strengthening linkages and creating knowledge domestically. Institutional reform represents another important pillar of the knowledge economy which requires restructuring the country's institutions to contribute to knowledge development and innovation dissemination. Institutional reforms facilitate investments in education, research and development, strengthen entrepreneurship and improve labor market flexibility. In most Muslim countries, the public sector remains heavily involved in market activities which may violate market freedom that the knowledge economy rests upon.

Data in Table 2 represent values of several indicators for selected countries. The table illustrates that scores of indices, including Economic Complexity Index (ECI), Global Competitiveness Index (GCI), Innovation index and Human Capital Index, are lower for Muslim countries reflecting the weaknesses of the productive system and its ability to support rapid structural transformation. These indicators can also be interpreted as measures of the economy readiness to participate in the digital economy and benefit from the Fourth Industrial Revolution. In the case of ECI, which measures the knowledge intensity of an economy or a product, the majority of Muslim countries are far behind in knowledge contents in their productive activities. Economic development is a reflection of the country productive capabilities including the human, physical, scientific, financial, economic and technological inputs. These capabilities influence the country's ability to produce quality products for the global markets.

According to the ECI, the products of a country either diversified and export competitively, or ubiquitous produced by large number of countries. Production in most Muslim countries falls into the second category by producing ubiquitous products with little knowledge required to produce them, whereas Japan, Germany and Singapore rank high in ECI due to their exports of highly sophisticated products very few countries can produce. Similar trends can be seen in GCI, HCI and innovation index reflecting the structural weakness of the productive systems in Muslim countries. These countries need new model for development driven by new ideas, creative thinking, knowledge creation, and innovation. The new knowledge economy underscores the importance of investments in research and development, human capital and ICT infrastructure. This will enable the economy to diversify production, encourage entrepreneurship, attract FDI and enhance global competitiveness. Cooperation and collaboration among Muslim countries will accelerate the process of transformation through knowledge sharing, ideas exchange, policy coordination and project financing.

Table 2. Economic Complexity Index, Global Competitiveness Index, Innovation Index and Human Capital Index for selected countries

Country	ECI (2017)	GCI (2018)	Innovation index 2017	Human capital index (HCI) 2018
Japan	2.309	5.49	5.43	72.05
Germany	2.075	5.65	5.58	74.30
Singapore	1.865	5.71	5.33	73.28
U.S.A	1.775	5.85	5.64	74.84
Malaysia	0.971	5.17	4.72	68.29
Saudi Arabia	0.747	4.83	3.69	58.52
Qatar	0.396	5.11	4.87	63.97
Turkey	0.175	4.42	3.31	60.33
Oman	−0.060	4.31	3.26	——
Jordan	−0.149	4.30	3.73	58.15
Iran	−0.158	4.27	3.19	54.97
Tunisia	−0.287	3.93	3.03	50.76
Indonesia	−0.305	4.68	3.99	62.19
Egypt	−0.319	3.90	2.75	55.99
Pakistan	−0.859	3.67	3.28	46.34
Morocco	−0.892	4.24	3.11	49.47
Sudan	−1.457	—	——	——
Bangladesh	−1.714	3.90	2.80	51.75

Sources: Economic Complexity Index (ECI) (2017) https://atlas.media.mit.edu/en/rankings/country/eci/, Global Competitiveness Index (GCI), World Economic Forum (2018) Global Competitiveness Report 2017–2018 http://www3.weforum.org/docs/GCR2017-2018/05FullReport/TheGlobalCompetitivenessReport2017%E2%80%932018.pdf

Data for innovation obtained from World Economic Forum (2017) Global Competitiveness Report http://www3.weforum.org/docs/GCR2016-2017/05FullReport/TheGlobalCompetitivenessReport2016-2017_FINAL.pdf.

Export diversification helps countries to reduce trade shocks and stabilize export earnings (Al-Roubaie 2019a). In other words, diversification helps the economy to make progress by producing sophisticated products which contributes to socioeconomic transformation. In Muslim countries both concentration index and diversification index of exports are high reflecting the high share of trade in total GDP. For example, in 2017 the concentration index for the Organization of the Islamic Conference (OIC) accounted for 0.257, 0.380 for Arab states, 0.376 for GCC countries and 0.113 for ASEAN. This is compared with 0.066 for OECD countries and 0.087 for NAFTA. As with regard to diversification index, Muslim countries are ranked among low-diversified economies compared to other regions in the world. The diversification index accounted 0.440 for OIC countries, 0.593 for Arab states, 0.619 for GCC countries and 0.303 for ASEAN. This is compared to 0.150 for OECD countries and 0.193 for NAFTA.

The relationship between trade and growth has been widely debated in the literature, however high degree of trade concentration could increase cyclical fluctuations and increase the risk instability. Manufacturing production in most Muslim countries was discouraged due to lack of competition, low FDI inflows and weak linkages. In 2017, the share of the Organization of Islamic Conference (OIC) in total world FDI accounted for 5.7%, League of Arab States 1.5%, the Cooperation Council for the Arab States of the Gulf (GCC) 0.8%, and Association of Southeast Asian Nations (ASEAN) 7.2%. This is compared to 41% for OECD countries, and 17.6% for North American Free Trade Agreement (NAFTA) (UNCTAD STAT).

The quality of education and investments in research and development is among the important pillars for building knowledge capacity. In Muslim countries, the quality of education needs to be improved to increase skills requirement and meet labor market demand for highly-skilled workers. High unemployment in some Muslim countries reflects the inability of the educational system to provide young people with adequate opportunities to find employment. In the new economy, not only education is concerned with rising people basic literacy, but also with the improvement of people technical, scientific, linguistic, and managerial capabilities. In this regard, Muslim countries require new curricula to address these qualifications needed by the labor markets both locally and globally. Education also increases the country readiness to innovate which is essential for speeding up the process of knowledge creation and enhancing global competitiveness. Investment in research and development, which is important for creativity and innovation, remains low across the Muslim world weakening the ability of many countries to build appropriate environment for innovation. Muslim countries spend less than 1 percent of their total GDP on R&D which is inadequate to stimulate knowledge creation and support innovation. Low FDI inflows also affect the ability of these countries to integrate into global value chains and benefit from tapping global knowledge.

Building capacity for knowledge development will require rethinking development by introducing new vision and formulating new strategy to implement policies and endorse reforms that are rapidly facilitate integration in the digital economy (Al-Roubaie 2019b). These strategies must be linked to the four pillars of the knowledge economy to provide support for leapfrogging and diversifying output. There is a need for greater cooperation and collaboration among Muslim countries in all activities related to knowledge creation for enhancing the domestic capacity for development. The current development model employed in most Muslim countries is not flexible to facilitate knowledge absorption and technological learning. Heavy dependence on production of limited primary products for exports has hindered the economy capabilities to generate linkages and stimulate growth. Muslim countries participation in the global trading system has been limited to low productivity products with little manufacturing production. Most countries in the Muslim world suffer from lack of diversification to absorb external shocks and maintain economic stability.

Technologically, the economies of these countries were dependent on the global markets for imported capital goods used mainly for demotic needs. Thus, the new model must incorporate the technologies of the Fourth Industrial Revolution to increase productivity and strengthen the fundamentals a knowledge economy. In the knowledge economy, productivity is linked to human capital and learning allowing linkages creation and innovation to take place and stimulate growth. In this age of connectivity, knowledge is treated as a global good available for users with little or no costs. Building absorption capacity for access to global knowledge becomes essential to enhance localization of knowledge and support endogenous development. In other words, knowledge policies must be adapted to the local needs in order to increase utilization of resources and foster growth of productive sectors of the economy. On their parts, governments should encourage private enterprises to participate in the knowledge economy by introducing structural reforms and regulatory measures to make it attractive for both local and foreign firms to set up business operations. Without finance, new entrepreneurs may not succeed because of high cost of capital and high risk in investment in new businesses. In this regard, risk sharing investment, as in case of Islamic finance, could provide alternative financing for private firms in Muslim countries.

These new technologies could speed up the process of development by stimulating economic growth and creating new opportunities for employment. It is important however, to learn how to unlock the benefit of these technologies to include the whole society. Government should introduce regulatory measures to ensure that the distribution of income and new opportunities will be distributed in a manner that lead to structural transformation. Among the important challenges of the new technologies are job losses which must be replaced to meet labor market demand. "Policymakers will need to renew their focus on antitrust regulations – and increase their understanding of the impact of the new technologies and business models on costs, production functions and industry structure – to preserve dynamic markets, broad based- participation and the power of creative destruction" (World Economic Forum 2017/2018, p. 4). In Muslim countries, the public sector remains in control of investment activities, which require governments to formulate flexible strategy capable of integrating the new technologies into the local economy. The new development model must increase private sector participation to take advantage of the new economy.

5 Global Value Chains

Muslim countries should take advantage of the new opportunities offered by Global Value Chains (GVCs) to strengthen manufacturing productivity and gain skills and know- how from foreign firms. A value chains represent the "full range of activities that firms undertake to bring a product or a service from its conception to its end use by final consumers. At each step in the chain, value is added in some form or another" (African Development Bank 2014, p. 126). Globalization has provided greater access to global trade allowing countries to produce intermediate or finished goods for multinational corporations. In this age of digitization, global value chains or production networks provide the developing countries an opportunity to integrate into global trading system and participate in production of manufactured products for foreign firms. Countries with low manufacturing production can participate in the GVCs to diversify output and strengthen linkage creations in the domestic market. In other words, developing countries can be linked to the GVCs to enhance market competitiveness by providing multinationals with skills and services. Such access to global value chains transforms productivity in the local economy into higher quality products for exports.

Involvement in these chains facilitates technology transfer and knowledge acquisition that helps domestic firm upgrade technology for local production. Countries with low manufacturing output can increase the share of industrialization in national production and compete in the global markets. Each stage in the chains could provide not only learning opportunities for local enterprises, but also create new jobs, introduce new skills, transfer new technologies and increase government income. "Today global value chains, or international networks of production, span many countries, often the entire globe and are closely intertwined with shifting wealth and the rise of the South" (African Development Bank Group 2014, p. 127). From economic development perspective, global value chains could facilitate structural transformation by increasing the country capabilities to produce diversified products for both local and global markets. Increasing production for global export enables the country to diversify output and encourage foreign firms to invest in the economy. However, the ability of the country to acquire knowledge and stimulate innovation could strengthen the economy capabilities to build capacity for industrialization. For example, the integration of the Chines economy into the global chains over the past few decades has helped China not only to attract foreign direct investment, but also to 'build a competitive supply base of intermediate goods' through linkages creation and technological learning. In Muslim countries, upgrading of foreign technology to the local conditions could move up the value chain into higher value activities that enables local firms to 'capture a higher share of value in the global value chain and enhances competitiveness' (African Development Bank Group 2014, p. 130). In doing so, economic upgrading not only boost structural transformation and promote innovation, but also create new job opportunities for local workers. To this end, integration into the chain and benefiting from knowledge transfer will depend on the country readiness capabilities. The small share of Muslim trade in global trade reflects the weakness of participation of Muslim economies in global value chains. This could hinder the attempt to join the digital economy and take advantage of the Fourth Industrial Revolution. Muslim countries have attractive human, financial and natural resources endowments for participation in global value chains, however, barriers in the form of infrastructure, corruption, and sound business environment are holding it back. To this end, "GVCs can facilitate access to global markets and integration in the global economy for developing countries, which no longer have to develop an entire industry to generate exports, but can focus on fewer tasks within industry value chain" (UNCTAD 2013, p. 24).

6 Conclusion

Recent advancement in digital technologies has provided new opportunities for nations to participate in the digital economy and contribute to development. The developing countries, including Muslims, must make efforts to build digital capacity and participate in the Fourth Industrial Revolution. Economic development in many Muslim countries is still in a state of low productivity reflecting the inability of these economies to adapt the scientific and technological advancement of the pervious industrial revolutions. The development model employed generated very limited linkages to foster economic growth and diversify the productive system. High dependency of these countries on international trade has increase economic and financial vulnerability by keeping the economies of these countries swinging following changes in commodity prices in the global markets. Low level of commodity diversification caused the terms of trade to deteriorate overtime reducing the ability of these countries to meet their international obligations and support development. Export-led growth model discouraged capital inflows and technology transfer to build local capacity for industrialization. Non-active private sector has discouraged entrepreneurship from participating in the economy and contributes to economic diversification.

Technologies of the 21st century are expected to have substantial impact on the way people work, communicate and live. The new technologies such as automation, robotics, artificial intelligence, and nanotechnology will require continuous learning and taring to maintain labor market flexibility and adapt to new technologies. In most Muslim countries, the educational system produces skills and specialties not suitable for market needs. This hindered the economy's capabilities to participate in the First, Second and Third Revolutions. Not being able to promote industrialization, the productive capacity of the economy continued to be lacking to economic stimulus to encourage local entrepreneurship and promote competitiveness. Access to the global market was hindered by weak productive capacity and production of quality goods as well as by low levels of FDI. The duality of the economic system in most Muslim countries created weak linkages to support economic diversification and provide incentives for local enterprises to compete in both domestic and global markets.

Muslim countries should rethink development and restructure productivity to stimulate market activities and participate in the digital economy. Investments in human capital and building digital infrastructure are essentials to benefit from the Fourth Industrial Revolution. Rethinking the economic structure requires radical reforms to facilitate technology transfer, knowledge creation and innovation dissemination. The digital economy increases interconnectivity within and among nations allowing greater cooperation and collaboration among key players in both local and global markets. Such interconnection provides new opportunities for individuals and organizations to participate in market activities and accelerate the process of structural transformation.

References

African Development Bank Group: African Economic Outlook (2014). https://www.un.org/en/africa/osaa/pdf/pubs/2014afrecooutlook-afdb.pdf

Al-Roubaie, A.: Building capacity for knowledge economies in the Arab world: the role of human capital. Int. J. Eng. Technol. **7**(4.9), 55–62 (2018). www.sciencepubco.com/index.php/IJET

Al-Roubaie, A.: Knowledge absorption and economic diversification: the case of Bahrain. Int. J. Recent. Technol. Eng. (IJRTE) **8**(IC2), 450–455 (2019a). https://www.ijrte.org/

Al-Roubaie, A.: Building capacity for digital development in the Arab world: the role of education. Int. J. Eng. Adv. Technol. (IJEAT) **8**(5C), 1530–1537 (2019b). ISSN: 2249-8958. https://www.ijeat.org/

Asian Development Bank (ADB): Moving Toward Knowledge-Based Economies: Asian Experience (2007). https://www.adb.org/sites/default/files/publication/29699/knowledge-based-economies.pdf

Brookings Institution: A New Generation of Economic Development, 24 February 2015. https://www.brookings.edu/blog/the-avenue/2015/02/24/a-new-generation-of-economic-development/

Dahlman, C.J. (ed.): Finland as a Knowledge Economy. World Bank, Washington (2006)

Deloitte: Preparing Tomorrow's Workforce for the Fourth Industrial Revolution (2019). https://www2.deloitte.com/global/en/pages/about-deloitte/articles/gx-preparing-tomorrow-workforce-for-the-fourth-industrial-revolution.html

Schwab, K.: Shaping the Future of the Fourth Industrial Revolution. Penguin Random House, New York (2018)

United Nations Conference on Trade and Development (UNCTAD): Information Economy Report (2017). https://unctad.org/en/PublicationsLibrary/ier2017_en.pdf

United Nations Conference on Trade and Development (UNCTAD): World Investment Report (2013). https://unctad.org/en/pages/PublicationWebflyer.aspx?publicationid=588

United Nations Conference on Trade and Development UNCTAD STAT. https://unctadstat.unctad.org/wds/TableViewer/tableView.aspx?ReportId=96740

United Nations Conference on Trade and Development (UNCTAD): Global Value Chains and Development (2013). https://unctad.org/en/PublicationsLibrary/diae2013d1_en.pdf

World Economic Forum: The Global Competitiveness Report 2017–2018. https://www.weforum.org/reports/the-global-competitiveness-report-2017-2018

Innovative Technology: The Aviation Industry and Customers Preference

Ahmed Fuad[1], Mukhtar Al-Hashimi[2], and Allam Hamdan[2(✉)]

[1] Faculty of Engineering Management and Systems Engineering, The George Washington University, Washington, D.C., USA
[2] College of Business and Finance, Ahlia University, Manama, Bahrain
Allamh3@Hotmail.Com

Abstract. This paper has investigated the impact of innovative technology on customer preference. Innovative technologies used by other carriers in the world were explored and populations from the Kingdom of Bahrain were asked if these technologies have any impact on their selection. The majority of the respondent considered innovative technology to have an influence on their selection of an airline. The technologies that had the most significance are radio frequency identification and the Internet availability onboard the flights. Where these two impacted the decision of selecting an airline on the consumer. The literature focuses on the research journals and articles along with figures and theories that are linked to the main variables of the research. It also includes the literature of innovation, technologies and customer satisfaction, and the history of aviation along with the history of some of the technologies.

Keywords: Innovative technology · Airline industry · Airline passengers · Bahrain

1 Introduction

Comparing this industry to other business sectors the aviation industry is unlike the lengthy history and rather has got a short one looking back at it. It is almost impossible to provide a specific date which is the starting point of the aviation business it is compulsory to mention that the Wright brothers who conducted air flights which are powered back in 1903 and only by that it sets the basis for personal aviation (Sheehan and Oclott 2003).

Several years later the previously mentioned flight, a German manufacturer generalized in aircrafts Zeppelin was the founder of the first aviation company. November the sixteenth 1909 the Deutsche Luftschiffahrts Aktiengesell schaft (DELAG) has been established to operate flights for passengers from frankfurt which the base of the company which is now called the HUB (Sheehan and Oclott 2003) Agosti and Bieger (2005) have developed a four stage model which is consisting of the how airline business evolves and also this has been divided into four parts. In the first initial phase which is called the technical stage and it end at the of the Second World War. Traveling by air in those years was rather considered as an adventure. The business was a supply side business which aviation industry were hardly making any profit. The following

A.-E. Hassanien et al. (Eds.): AICV 2020, AISC 1153, pp. 696–707, 2020.
https://doi.org/10.1007/978-3-030-44289-7_65

second stage is called the political stage which consists of establishing the international agreements, standards and regulations which were set for the industries of transport in 1950 an important achievement in technology was made, it was the jet planes which allowed aviation to start a significant growth.

The aviation industry started to put more effort and concentration on the cost and quality back in the 1970s. This has shaped the third stage by introducing policies of the open sky pricing schemes, services and the new competitors entering the market with concepts considered to be new for the business. The final stage which consists of alliance and network started in the 1990s which has introduced the lost cost carriers and the alliance formation process among carriers that were already established, this stage has changed the market conditions completely. In the European continent, new low-cost airlines that were established took advantage of the open European skies and has successfully managed positioning themselves in the aviation market. Accordingly, and at the same period several old firms and carriers faced severe problems, which were due to finance (Wilfing 2012).

2 Literature Review

2.1 The Innovation

Innovation today is almost used for business models by almost every company. It is noticeable how the worlds leading firms and companies such as Starbucks and amazon benefit from such model to the unique method they follow by marketing ordinary products with a special added value (Trends Magazine 2013). Accordingly, innovation has changed its aim from products new development or a new process to focusing on increasing the customers experience by accumulating a value to a product that already exists. Innovation is the development process of introducing products with new specifications or a new process to the world in a specific field, market, or a specific industry (Reilly and Tushman 2004). The aviation industries are well known for being innovative, as they must develop strategies and implement innovative technologies to gain competitive advantage. In the past decade, the competition in the airline industry has been increasing rapidly when it comes to offering innovative products and innovative services along with technologies introduced in both airports and aircrafts (Southwest 2012). Although this specific industry is not always stable, its profit is considered as the highest among other industries. Therefore, companies must employ new strategies, which contain lots of flexibilities in order to keep along with the market that is constantly changing market and to be different from the competitors. This what makes understanding trends in innovation very important among other airlines (Wensveen 2008). Industries are now in need for flexible business strategies in order to survive. In the past decade, aviation industries entered are focusing on ways to maximize their revenue. Since 2005 airline industries continuously thinking of ways to be innovative. Airlines focus in being different so it enables them in becoming market leaders. Most of the innovations raised during this period came from the several low-cost strategies (Wensveen 2008). In order for companies to survive, they need an essential tool, which is Innovation. Nowadays firms have an easy access to new technologies used, and information that can be categorized as a very important factors for the economic growth.

2.2 Definition of Innovation

The researcher Agrela (2013) stated in his research that the first definition of innovation was by Joseph Schumpeter back in 1932 was the first to define innovation "a new combination of things that come to market", which is categorized also as products and the methodologies of producing them and offering them to the market in new ways. (Agrela 2013; Sundbo 1998). The Researcher Agrela (2013) commented further as innovation should be constraint under the term new things and it claims that is has to be considered as more then implementation of new concepts. If a firm practices an knowledge that already exists or an existing invention for its detailed needs, it should also considered innovation since the product or service appears in a new ways or a better shape than it was originally introduced. The firm can also implement innovative idea and structure to its business in order to tackle a specific need. Let's take an example of how innovation can be introduced in business activities within the aviation industry itself. Since most airline travel to close by destinations which do not require a layover and that the aircraft used for the sector itself returns back to its base directly after the flight, this process actually decreases the expenses on the airline on the airline company and provides the operating cabin crew a normal lifestyle. Hence, is it not essentially important for innovation to come into a market as long as it covers the need for adding value either for customers or the company (Agrela 2013). The ultimate tool to make innovation possible is invention. It is the initial existence of an idea for a something new which could either be a product or service, product development or even a new process. That's when we know that an innovation is possible. However, a different combination of capabilities, resources, skills, and knowledge are what determines the possibility of innovation. An innovative process or a product had to be introduced and offered to the public or to be used within a production process to covers the need of the market or the for the benefit of use within the firm. Thus, an invention can only be considered as innovation, when the market or other corporates benefit from the service or product (Sundbo 1998). Another can consider another example in the same field, when the famous inventor Leonardo da Vinci had idea about airplanes which at the age were completely ahead of its time, which back seemed reckless. The lack of materials, power source and other sources, Da Vinci was not able move forward with ideas he had and apply them to reality. Furthermore, his ideas and philosophies were very complicated if not impossible to apply, until the invention that made it all closer to reality and more possible to achieve, its the combustion engine. It is now more obvious that innovation is concerned within the manner of abstracting the value from ideas or in other words gaining beneficiary outcome regardless to the customers or developers. Accordingly, some innovations can only be done due to previous ideas and inventions (Agrela 2013). Returning to the meaning of innovation it is now becomes clearer that it's difficult to set a definition and that such may not exist. In his research, Agrela (2013) states that each author has his own explanation to the term innovation, keeping into mind specific factors and attributes. He also elaborated further by combining several concepts from litterateurs and different author, which lead to conclude the following definition: Innovation is adding value to improve the products or services within a certain company or their market approach, by the development of new ideas or improvement of an existing understanding. A new invention is only considered as an innovation when it has success and an impact on the market or as explained earlier that it is customers or developers benefits from. A firm is considered innovative when it is the leading firm in taking advantage of that knowledge in their specific market. In general, an innovative company is normally spotted as the frontrunner in a specific market as it is the first who converts a successful idea and takes advantage of its success. However, an industry where its competition considered to be very high, such as the aviation industry, as explained earlier, innovation is a very important factor for survival. We also need to keep in mind that implementing an innovative idea does not suggest incomes at the exact moment or instantly. Implementing an innovative idea takes time to start and subsequently, generates revenues (Boston Consulting Group 1983).

2.3 The Innovation Cycle

The Innovation Cycle is a structure that is set by the researcher Schoen et al. (2005) and was recognized for the Model for the Innovation Process of invention. Targeting the thoughts directly at the technological incubators manager and the decisions makers, (Opt 2005) explained that the previously done models of the project management were not complete and rated them as in the innovation cycle as incomplete representations. Schoen et al. (2005) further explained that these processes are not always necessarily defined strictly. Project management models, which have been done previously, the authors, argued that there should be outcomes, which are well defined, and therefore it helps into proceeding towards the other outcomes. The researcher continued and later then introduced the new model, which is named as the Spiral Model. This was again however borrowed from the systems development spiral model, which was explained further by another researcher. Schoen et al. (2005) continued and discussed that best used model and more sufficient is the Spiral Model is used best and more sufficient when used in management models done previously in the development cycles as the development process the outcomes are not well defined and unclear. Schoen et al. (2005) noted that the decision makers when it comes to technology often deal with technologies that do not have an outcome which is certain and may not meet the project goals, expectation or even the timelines; and that these technologies which have been adopted are some recommended from university researcher's dissertation. This means that the starting points of these projects are unclear. Schoen et al. (2005: 8) have described further that innovation "is not a step-by-step, set the pins up and knock them down type of operation and requires mating a good idea with an even better concept". This concept was initially from Albert Einstein which believed that innovation was isn't a product thought of the logic, even though the result was linked to structural logic. Schoen et al. (2005) had thought that in this regard the other researchers Hamel's (2000) assertion that innovation represents the accomplishment of breaking free from mental constraints.

2.4 Customers Satisfaction

Customer satisfaction in the term of business is identification of how much the customer is satisfied towards the product or service the company offers. In an era where the competition is intense, all businesses try to achieve customer satisfaction tries where in such market it is considered as the key elements of success for all businesses. (Tahir Naveed 2013). In order to gain maximum, the satisfaction of the customers, the firm has assured the customers safety as each customer requests safety. Customer satisfaction has a positively influenced positively (Anderson and Sallivam 1993). Customers tend to become loyal and satisfied if the company provides a service or a product according to the customer's requirements. One of the key components to customer satisfaction is innovation, the main aim of adopting innovative technologies by companies is to satisfy their customers and meet their expectation. Most of the time customers are aware of the company's behavior related towards their customers complaints and whether these complaints are taken seriously and followed up by corrective actions or not (Tahir Naveed 2013). Anderson and Sullivan (1993), argues that in order to keep the customer coming back to your service or product you need to keep him satisfied. Handling the complaints quickly and efficiently will lead subsequently to loyalty and minimize the customer's negative impression. Although many establishments provide products and services to the customer which are considered as high value, they somehow undergo some difficulties and this arises due to the lack of customer interaction which subsequently leads to failure in understanding the exact information about customer needs due to lack of understanding in the change occurring in technology and trends. According to the researchers that elaborated further that sometimes even though companies provide a very high quality of features in their products or services, these companies fail to achieve customer satisfaction. And the reason behind that is basically that these feature are not important or do not have any value to the customer (Lacobucci et al. 1995). This problem rises due to the lack of customer communication. As companies lack direct interaction with customers about the services and products they offer to the market and how much the customer is satisfied when it comes to the product. Instead, they transform the product or modify it according to initial studies which are incomplete (Tahir Naveed 2013). In many situations if the company interacts very little or has no communication with its customers then the retailer or in this case the travel agent or local office authorized plays a vital role in enhancing the customer satisfaction. The local office authorized or the travel agent should always treat the customer as a representative of the company itself (Tahir Naveed 2013). Customer satisfaction has a strong influence when the firm's employees interact with its customers (Boshoff and Tait 1996), the examination of the behavior of the employees is considered very critical. On the same hand the operating organizational culture has a high influence on the behavior (Schein 1996), a focused assessment on the quality of service reflects the perception of the consumer in five specific service dimensions. On that same hand, a comprehensive view of the satisfaction of the customer is determined by the perception of service and product quality along with the prices, situation factors and personal factors (Zeithaml and Bitner 2001) (Fig. 1).

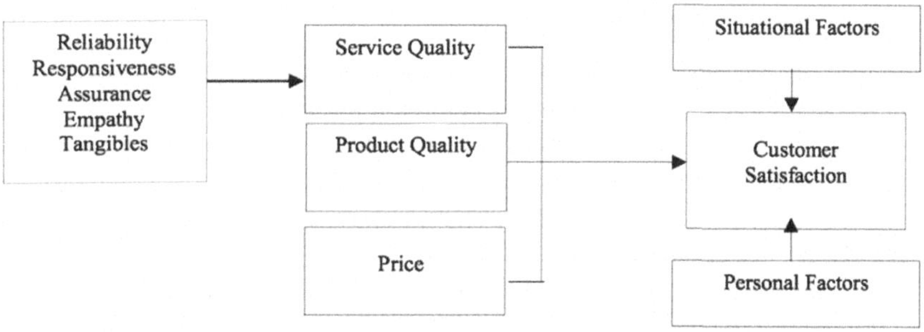

Fig. 1. Customer satisfaction model, Source Zeithaml and Biner (2001)

Customer satisfaction has become an essential component in the operations of service because of the huge value it produces to organizations (Ranaweera and Prabhu 2003). Researchers Shin and Elliott (2001) elaborated further and stated in their research that the importance of customer satisfaction is consequential result of a philosophy that has been frequently approved on that Satisfying the customers is the key factor in order for a business to be successful and profitable. Previous research has verified that satisfaction is strongly associated with re-purchase intentions Customer satisfaction also serves as an exit barrier, helping a firm to retain its customers (Fornell 1992). Another Research conducted by Blodgett et al. (1995) have concluded that it costs more to gain a new customer than it does to retain an existing one. In addition, gaining customer satisfaction will leads to favorable word-of-mouth publicity that provides valuable indirect advertising for an organization (Fornell 1992). Many companies define having satisfied customers as that the organization receives fewer complaints. in most fields it means that subsequently the organization is tend to receive less complaints. Hence, reducing costs in handling service failures. Researchers also have observed that satisfied customers are willing to pay more for the benefits they receive and are more likely to be tolerant of an increase in price concluded that, through satisfying customers, organizations could improve profitability by expanding their business and gaining a higher market share as well as repeat and referral business (Fornell 1992). In a very aggressive environment the key to gaining a competitive advantage is being innovative. The ability to innovate or adopt innovative technology has a direct impact on the organizations ability to compete against its rivals (Neely 1998). As researcher Mostaghel (2006) described that in order to have a thorough customer happiness and loyalty, it is important to have working condition which is described as above average which is needed to have employees which are satisfied which will subsequently lead to loyalty of employees. Only by preparing such environment the good production will subsequently follow and accordingly have an affection on customer the satisfaction of the firms customers and earn their loyalty and only by retaining the customer the company can retain its high revenue and build a profit on.

2.5 Technological Innovation

Technology is converting innovation at its core; industries now can implement new ideas and prices that were impossible to believe a decade ago. Features can now be updated on a firm's website and receive a respond from results of how customers respond to it within hours. Results are also available to preview from promotions offered, productivity of a certain service and the efforts to boost it. There will be even better payoffs for customers: Their likes and dislikes will have much more impact on companies' decisions. In globally competitive markets, they will ultimately end up getting products and services better tailored to their needs (Brynjolfsson 2009). Already, this powerful new capability is changing the way some of the biggest companies in the world do business, inspiring new strategies and revolutionizing the research-and-development process. Growth of companies by the use of innovation is essential. Globalization has caused the increase of international competition and subsequently it changed the innovation needs. Companies nowadays have access to information easier than before and also have easier access to market and technologies; this knowledge is categorized as the economics growth and innovation great factor (Oslo Manual 2005). New technology obvious factors of adoption and its benefits received by the user and the efficient cost of adion. Multiple cases indicate that these benefits are simply considered as the main factor of the difference in profits and revenue generated when a company transfer from an old technology to a more advanced one, of course when it comes to consumers, the increased utility from the new product is the main factor of benefit, (Hall and Khan 2012) as long as customers distinguish the product or service as new one then the product can be considered as innovative produce (Schiffman and Kaunk 2001). Company with innovation capabilities will replace the market if the business doesn't conduct product innovation (Millett 1990). Innovation in products and services are becoming very important in order for an organization to continue it growth and its sustainable existence. From all perspectives of different scholars, both products with innovative technology and existing products which are introduced, copied and developed are considered as product innovation as long as the new product could meet the new demand of a market and be differentiated from existing products such as different style, material and procedure. Product innovation does not necessarily mean a new invention. Agrela (2013) has identified innovation and agreed to the definition of Tseng (2010) which both stated that it is not important for innovation to be a new invention. However, as long as it adding a value for either for customers or the firm by adding a new service that was not previously introduced.

2.5.1 Improvements in the Old Technology

The researcher Rosenberg (2006) had argued and some concerns regarding the act of substituting older technologies for an updated one or a new module. When a new technology at times is an innovation that's close for an existing technology substitution, then the innovation for itself may encourage providers of technologies that are old to make improvements or engagement in other types of competitive behavior in an effort to retain their market position. This in turn will slow the diffusion of the new technology.

2.6 Technologies

2.6.1 Radio Frequency Identification RFID Historical Overview

Radio frequency Identification RFID was originally used as military identification systems and was used and implemented during the tragic Second World War, it main purpose was to identify planes. The system used is known as the identification system of Friend or Foe IFF (Bilginer and Ljunggren 2011). The Radio frequency identification technology has been developed further to unable the use low cost commercial use. It initial developments tags that has been used for surveillance. It consisted of two main statuses, which was on and off, if the state has not been switched off an alarm would go off if the tag passes an electronic reader. The existence of the first tags were Passive but with time, and introduced also the active tags. RFID is nowadays a well-known and spread to almost every mega or small business. This technology is being used in all over the world (Bilginer and Ljunggren 2011).

The Use of the Technology in the Aviation Industry

Radio Frequency Identification (RFID) is a wireless information exchange over short distances technology. This technology has been invented for almost five decades ago, recent development in organizations has made the RFID devices to finally start showing its potential (Bite 2010). The researcher Bite (2010) elaborated further on RFID, explaining that it's a technology merged into a silicon chip that produces radio signals which matches a user-defined serial number with an item. In this case the item is a piece of check-in baggage. This number can be read at a distance by an antenna. The RFID also has many following characteristics as it enables the baggage to be sorted automatically and loaded faster than with barcode systems, while reducing the number of mishandled baggage and its associated costs at the same time (Bilginer and Ljunggren 2011). Wyld (2005) focused on a specific airline in the united stated and how RFID technology could improve the service provided to the customer through better operational efficiency in baggage handling, which has been demonstrated to be an essential component of an airline's customer service rating. Even though the value of this technology in handling passenger bags is accepted in general at the industry, the adoption of these it is stalled by concerns relating to their inadequate return on investment (Bite 2010). The researcher Bit has compared the barcode and Radio Frequency Identification which has been listed in the table below in detail comparing several main attributes such as the optic view, reading possibility, read rates accuracy and where each is placed in term of baggage. While the barcode is easy to remove the RFID is almost impossible to remove without destroying it. The below table also compares the speed of response (Fig. 2).

The researcher Bite (2010) has observed the difference between the traditional Barcode, which is mostly used by major airlines and the RFID in the terms of being implemented in chain of handling baggage at the airports.

Attributes	Bar-code	RFID
Optic view	Necessary	Antenna is reading from the distance
Reading possibility	Scanner points	Active tag: always, Passive tag: access points
Read rates accuracy	80-90 %	95-99%
Read - Write	Read only	Read-Write
Up-dating	1 time	Anytime
Real time bag matching	No	With the people
Data	Definite	Indefinite
Location	Top of bags	Everywhere
Removable, Vulnerability	Easily	Impossible
Reading after Vuln.	Mishandling	It can be identified correctly
Configuration	Long Paper	Can be embedded in everything
Technical equipment	Paper, Printer Scanner	Tag Read-Writer Antenna for the reading International Database possibility
Environments	Disposable	Re-usable
Speed	Slow	Fast
Automated (Manpowered)	Manually	Automated
Price	6-8 cent	20-42 cent
Cost	Cheap	The tags are expensive, + implementation costs
Maintenance	Has to be wiped daily	Little

Fig. 2. A comparison between the traditional Bar-code and Radio Frequency Identification. Source: (Bite 2010).

2.7 Near Field Communications NFC

Two major companies Philips along with Sony initiated the initial development of near field communication. The technology consisted of a protocol and an interface which were developed on top of Radio frequency identification that makes the device which is connected NFC a part of this standard and compatible with a radio frequency technology that already exists (Bilginer and Ljunggren 2011). The main key feature that makes a difference between Near Field Communication NFC from Radio Frequency Identification RFID is the possibility of transferring information which allows communication between two NFC devices. In order to connect between one device to another the two devices must be close to each other or physically touch each other devices physically. The Near Field communication NFC then automatically configures itself and communicates in a peer-to-peer network (Bilginer and Ljunggren 2011). Near Field Communications (NFC) is a short-range wireless technology that allows mobile devices to actively interact with passive physical objects and other active mobile devices, connecting the physical world to mobile services in ways that empower and benefit users. We will also be using the term "Tap 'n Go" because it clearly conveys a visual image in which this technology is intended to be used for. The aviation industry is one, which has a lot of levels in its processes between passengers and the airport and passengers or the carrier itself. IATA conducted a Corporate Air Travel Survey (2009) concluded that 50% of passengers around the world prefer more self-service options to fast track the travel processes. IATA identifies below (Fig. 3) many potential points for Near Field Communications NFC Technology:

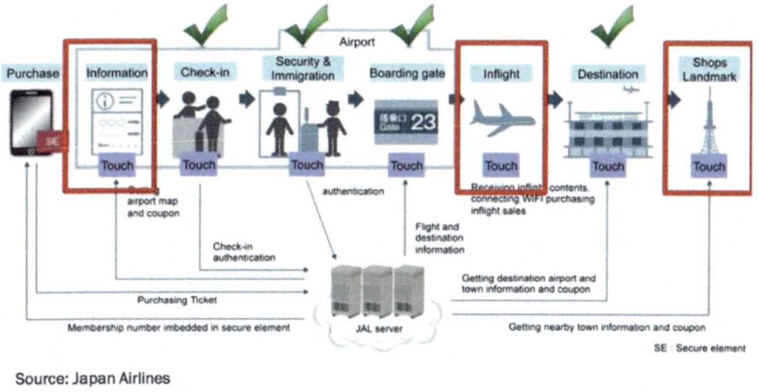

Fig. 3. The use of NFC at international airports, Source (Japan Airlines 2013)

2.8 Always Staying Connected

Mobile technology has impacted all phases of the travel life cycle. With advanced wireless networks, growing worldwide adoption of web-enabled full feature phones, smartphones and tablets; passengers are no longer tethered to a desktop PC, but now have instant access to information anywhere, anytime from multiple devices. Mobile technology provides airlines an opportunity for continuous engagement across the travel life cycle (Fig. 4).

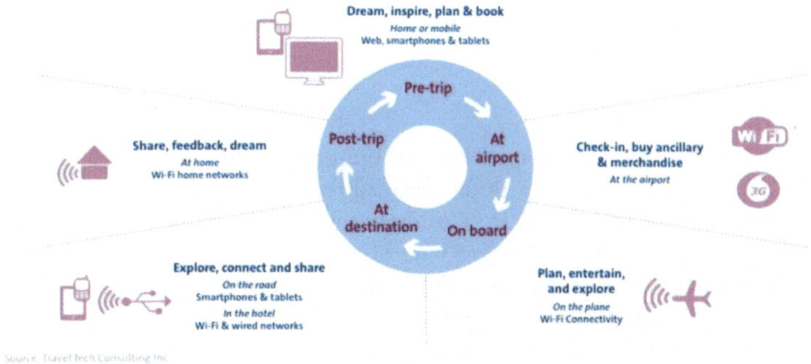

Fig. 4. The use of internet and stays constantly connected Source: (Amadeus 2014).

The always-connected passenger has an evolving set of requirements and expectations for information and services. This paper reviews how mobile technology is being used today by major carriers worldwide and what is coming in the near, mid and longer-term future to improve passenger efficiency, increase loyalty and drive incremental revenue. (Amadeus 2014). The literature shows how innovation and innovative technology has changed the way organizations compete, the research will be conducted in order to understand the relationship between customer satisfaction and the innovative technology provided, subsequently knowing if the innovative technology is the reason most people select a carrier for their travel.

3 Conclusion

The review of the literature related to the Impact of Innovative Technology on the Aviation Industry and on Customers Preference, which is the topic of the study, the literature review has established simple knowledge in the innovative technology. It significantly helped in understanding the technologies used by other carriers and how to adopt them by the carrier which the study is based on. The literature also helped in developing the questionnaire which is the main tool of this research. Moreover, literature explained the beginning of the aviation industry and how innovation impacted its growth. Followed that is the definition of innovation which then focused more on the innovative technology used by other carriers and the history of these technologies. Each researcher had his own set of factors some had a similar point of veiw. The below table summarizes the main innovative technology listed in the literature and are classified by the name of some of the key researchers and the measurement tool used in their research. The following matrix is a demonstration of the sources listed and used in the literature review it focuses on the researchers and highlights the added value to this research the matrix helped the researcher in the construction of a methodology along with a research instrument which are relevant. It is clear that the above mentioned studies that have been reviewed in the literature have had a significant contribution in the study. These studies at first provided suggestions and recommendations on the knowledge and research gap from their studies. The articles and researches previewed have also contributed to the researcher and provided a good insight on how to construct the appropriate and valid methodology that can be suited to the research objectives.

References

Agrela, M.B.: Innovation in the airline market – a competitive study on delta and southwest Poul Wolffsen. Business Administration Department (2013)

Anderson, E., Fornell, C., Lehmann, D.R.: Customer satisfaction, market share, and profitability: findings from Sweden. J. Mark. **58**(1), 53–66 (1994)

Anderson, E.W., Sullivan, M.W.: The antecedents and consequences of customer satisfaction for firms. Manag. Sci. J. **12**(2), 125–143 (1993)

Ann Cavoukian Mobile Near Field Communications (NFC): "Tap 'n Go" Keep it Secure & Private Information and Privacy Commissioner, Ontario, Canada

Alfardan, H.: The foreigner are 51% of the whole population in Bahrain. Alwasat (2013). http://www.alwasatnews.com/4315/news/read/900150/1.html. Accessed 9 May 2015

Bilginer, B., Ljunggren, P.-L.: Near Field Communication Sony Ericsson Mobile Communications AB (2011)

Brynjolfsson: The New, Faster Face of Innovation (2009). http://www.wsj.com/articles/SB10001424052970204830304574130820184260340

Boyd, P.: By The Numbers: A sample size table. Quirks Marketing Research Media (2006). http://www.quirks.com/articles/2006/20061209.aspx. Accessed 10 May 2015

Boshoff, C., Tait, M.: Quality perceptions in the financial services sector: the potential impact of internal marketing. Int. J. Serv. Ind. Manag. **7**(5), 5–31 (1996)

Boston Consulting Group: The Growth Share Matrix or the Product Portfolio (1983). Reprint no. 135

Blodgett, J.G., Wakefield, K.L., Barnes, J.H.: The effects of customer service on consumer complaining behavior. J. Serv. Mark. **9**(4), 31–42 (1995)
Bite, K.E.: Improving on Passenger and Baggage Processes at Airports with RFID. Faculty of Transportation Engineering, Department of Transport Economics Hungary (2010)
Cooper Donald, R., Schindler Pamela, S.: Business Research Methods, 8th edn. McGraw-Hill higher Education, New York (2003). ISBN: 0- 07- 249870-6
Fornell, C.: A national customer satisfaction barometer: the Swedish experience. J. Mark. **56**, 6–21 (1992)
Halstead, D., Page Jr., T.J.: The effects of satisfaction and complaining behavior on consumers repurchase behavior. J. Satisf. Dissatisfaction Complain. Behav. **5**, 1–11 (1992)
Hall, Khan: Adoption of New Technology/The Economy Book. University of California at Berkeley, Berkeley (2012)
Kothari, C.: Research Methodology. New Age International (P) Ltd, New Delhi (2004)
Kossmann, M.: Delivering Excellent Service Quality in Aviation: A Practial Guide for Internal and External Service Providers. Ashgate, Hampshire (2006)
Iacobucci, D., Ostrom, A., Grayson, K.: Distinguishing service quality and customer satisfaction: the voice of the consumer. J. Consum. Psychol. **4**(3), 277–303 (1995)
IATA: Airline passengers call for more selfservice (2009). www.iata.org/pressroom/pr/Pages/2007-20-11-01.aspx. Accessed 12 Feb 2012
Millett, S.M.: The strategic management of technological R&D: an ideal process for the 1990s. Int. J. Technol. Manag. **5**(2), 153–163 (1990)
Neely, A., Hii, J.: Innovation and business performance. The Judge Institute of Management Studies, University of Cambridge (1998)
Rose, N.: The always-connected traveller: how mobile will transform the future of air travel (2014)
Mostaghel, R.: Customer Satisfaction: service quality in online purchasing in Iran. Luleå University of Technology (2006)
Ranaweera, C., Prabhu, J.: On the relative importance of customer satisfaction and trust as determinants of customer retention and positive word of mouth. J. Target. Meas. Anal. Mark. **12**(1), 82–98 (2003)
Reilly, T.O., Tushman, M.L.: The ambidextrous organizations. Harv. Bus. Rev. (2004). http://iic.wiki.fgv.br/file/view/the+ambidextrous+organization.pdf. Accessed Mar 27 2015
Wilfing, R.: Passenger Decision Making Behavior and Implications for Airline Marketing: Case Emirates. Model Vienna University (2012)
Rosenberg, N.: Factors affecting the diffusion of technology. Explor. Econ. Hist. **10**(1), 3 (2006)

The Role of IT Governance in Enhancing the Performance of Smart Universities

Fatema Abdulrasool(✉) and Stephen Turnbull(✉)

University of Tsukuba, Tsukuba, Ibaraki 305-8577, Japan
fabdulrasool.85@gmail.com,
turnbull.stephen.fw@u.tsukuba.ac.jp

Abstract. IT is the backbone for all organizations, and it has challenged the traditional way of doing business. Universities now are competing globally, and they are virtually crossing their borders. University governance in general plays a very important role in the running of any university. It directs the university towards achievement of its objectives, and it is affected by the surrounding environment. IT Governance is a subset of the university corporate governance and it is it implemented to govern and manage IT effectively. People, information, technology service, infrastructure, applications, culture ethics, processes, principles, policies, frameworks and organizational structure are the tools for IT Governance. The environmental challenges are threatening the existence of many universities and requiring a prompt attention. IT Governance may be the key for the universities to reposition itself and overcome the problems that threaten their existence. AI is a very important strategic tool that can help smart universities in fulfilling their strategic goals. AI presents challenges and opportunities to the IT Governance function. Literature shows that universities are not fully utilizing their IT capabilities and this a clear evidence that they are having serious issues with their IT Governance practices. This exploratory research aims to investigate the role that IT Governance in enhancing universities performance in areas such as: research, teaching and learning, competitiveness, and resources utilization. We develop a conceptual framework that identifies the environmental factors that impact the university IT Governance structure; and strategical and political enablers and benefits of implementing IT Governance system in universities.

Keywords: IT Governance · University IT Governance · Business/IT alignment · Smart universities · Artificial Intelligence · Strategic use of IT

1 Introduction

Funding and policies issues reshaped the higher education systems in the whole world. University agility and ability to react and adapt to the changing and competing environment is under examination (Howells et al. 2014). Financial constraints, the spread of information, communication technology, growing professionalization of academic administration as well as competition and marketization require a full attention (Sporn 2007; Yamada 2018; Goldfinch 2006; Howells et al. 2014; Carnegie and Tuck 2010; Abugabah et al. 2015; Sporn 2001). Universities has overcome major challenges over

A.-E. Hassanien et al. (Eds.): AICV 2020, AISC 1153, pp. 708–720, 2020.
https://doi.org/10.1007/978-3-030-44289-7_66

time due to their unique form of administrative and governance structure which enabled them to successfully adapt to changing environments (Sporn 2007; Yonezawa and Shimmi 2015).

Universities are considered smart organizations since they are leveraging technology in learning, management, governance, social, health, and green to enhance their performance and utilize resources. AI in general and Knowledge Management specifically is the foundation of a smart university because it is the corner stone to fulfill business goals.

The operations of colleges and universities were affected by new technologies. Functions of academic institutions such as: student learning, research outcomes and institutional functioning can be improved through the using of information technology. Academic programs and courses have been supported by the Internet with tools such as groupware, communication software, and end-user tools. Research has been improved through providing electronic access to the scientific community internationally and to databases and online publications as well. Technology has also altered the way universities are managed. Major universities' functions like enrollment processes, accounting, reporting, international relations, and decision support can be automated (Sporn 2007).

Nowadays universities are under a huge pressure. Despite the financial constraints they are undergo, they have been asked "to do more with less," enhance market position, and graduate quality alumni as well as research results (Sporn, Governance and Administration: Organizational and Structural Trends, 2007; Yonezawa and Shimmi 2015). Universities are looking for innovative ways to address these issues.

Computing capacities and the level of technology utilization defined the future of HEIs. The advances in Artificial Intelligence brings new possibilities and challenges for teaching and learning, research, and governance of HEIs. The innovative use of AI in universities presents the capability to utilize resources. It can help in reducing the number of administrative staff and teaching assistants. AI techniques can be used in many aspects of universities administrative functions such as: helpdesk, automated timetable, and decision making. It also helps in the university core missions; teaching and learning, and research; by means like: AI data analysis tools and personalized online learning (Popenici and Kerr 2017).

Despite the huge amount of money that the universities invest in IT, it never seeming to achieve its full potential. Nowadays, universities recognized the importance of IT governance in supporting their mission in attaining the full potential of their IT spending (Hicks et al. 2012). The concept of IT Governance focuses on the sustainability in controlling, managing and monitoring IT activities through five driven mechanisms which are strategic IT/business alignment, value delivery, IT resource management, IT risk management, and performance measurement (Subsermsri et al. 2015) IT governance is often the weakest link in a corporation's overall governance structure. It represents one of the fundamental functional governance models receiving a significant increase in attention by business management (Brown and Grant 2005). The difference in corporation governance private and national universities required a unique IT Governance schemes for each sector. Research shows that the private sector is more efficient in the development, implementation and governance of IT while the public sector is lagging behind. Usually the IT is viewed as a service not as a value creator in public institutions (Campbell et al. 2009)

In this study, due to the importance of IT Governance as part of the overall university corporate governance, we develop a conceptual framework that identifies the environmental factors that impact the university IT Governance structure. The implementation IT Governance is empowered by different strategical and political enables which will in return create strategical and political benefits. Entrepreneurial universities who are able to identify the benefits that the IT Governance may provide for the university would invest in developing the enables that will allow them to improve their IT Governance practices.

2 University Governance

"Higher Education Institutions (HEIs) are driven by a complex set of cultural and motivational factors, arising from their status as non-profit organizations, which directly affect their management and governance" (Fernández and Llorens 2009). Despite the huge research interest explaining how universities perform their everyday practices, the relationship between leadership, governance, and management in universities is still vague (Blaschke et al. 2014).

Rosca et al. (2010) define university governance as the set of policies, laws, processes, customs, and departments which affect the way a university is directed, administered or controlled. University governance includes the relationships among the stakeholders and the business goals through which the university is governed. University stakeholders include: shareholders, management, the board of directors, employees, customers, creditors, suppliers, students, professors, regulators, and members of the community (Rosca et al. 2010).

The traditional structure of universities is a bottom-heavy form of structure that is in essence a loosely coupled organization. Loosely coupled organizations are marked by fuzzy goals, unclear understandings of the root and influencing factors to an arising problem (Maassen 2017). The loosely coupled structure affect the decision-making process in a way that made the process fairly chaotic and it does not necessarily follow formal regulations and hierarchical structures. The interaction between different departments and units is infrequent, random, slow, or otherwise weak (Blaschke et al. 2014; Wilmore 2014).

"HEIs are facing a growing governance crisis which might cause a decrease in staff and students retention rate. 'Good' university governance does not simply happen. It is the product of continuous efforts to find the most appropriate governance structures, protocols and processes. It is also about timing and judgement: it requires boards of governors to recognize when a governance model is not working, why and how to repair it. Ultimately, governance models are created by people to govern people. They are only as good as they who devise and apply them, as well as those who live by them" (Trakman 2008).

3 IT Governance

IT Governance concept is defined as "a set of relationships and processes designed to ensure that the organization's IT sustains and extends the organization's strategies and objectives, delivering benefits and maintaining risks at an acceptable level" (Iliescu 2010; Gunawan et al. 2018).

IT Governance is a broad concept that inherited much from the corporate governance discipline and then developed into a discipline of its own rights that has little in common with the overarching discipline and lacks a shared view on important issues and how they should be handled (Rubino and Vitolla 2014). IT governance is often the weakest link in a corporation's overall governance structure. It represents one of the fundamental functional governance models receiving a significant increase in attention by business management (Brown and Grant 2005).

IT governance can be deployed by using a mix of structures, processes and relational mechanisms. Structures involve the existence of responsible functions within the organizational structure with clear roles and responsibilities. Processes refer to strategic decision making and monitoring activities. The relational mechanisms rely on active participation and collaboration between principle stakeholders that include business/IT participation and correlations, strategic dialogue and cross-functional business/IT training and rotation (De Haes and Van Grembergen 2005).

For successful implementation of IT Governance, organization should consider all the following factors Strategic alignment between business and IT; Value delivery from an investment in IT which is driven by the needs of the investing entity; Resource management that ensures that IT has sufficient, competent, and efficient resources to meet the organization's demands; Risk management which ensures that the strategic objectives of the business are not jeopardized by IT failures; and performance measurement that helps the board and senior management in evaluating IT performance towards achieving the organization objectives (Ramlaoui et al. 2015).

4 University IT Governance

The advent of the personal computer in the early 1980s reshaped the HEIs. The higher management of universities (presidents, provosts, and trustees) are leveraging information technology to achieve the institutions' strategic visions (Grajek 2014; Hicks et al. 2012). Universities are suffering from multiple issues resulting from the decentralized nature of HEIs organization structure. The identified issues are: lack of control and accountability in IT, executive pressure to reduce or justify expenditure on IT, insufficient IT risk management, duplication of IT resources, and inconsistent IT architecture (Hicks et al. 2012).

4.1 IT Governance Structure

There are three different styles of IT Governance structures which are: centralized, decentralized and federal. With a view to analyze the best-fit IT Governance structure for universities; Bianchia et al. (2017) conducted an exploratory study in six universities located in three different countries (Brazil, the Netherlands and Portugal). The research shows a consensus that the a totally decentralized IT Governance model which is characterized by higher risks, duplication and waste of resources, as well as control and communication difficulties; has been perceived as inappropriate for HEIs. Moreover, the size of HEIs will drive the selection between totally centralized and federal IT Governance structures. Large and extra-large universities tend to adopt the federal IT Governance structure where the strategy, infrastructure, roles and procedures are centralized to avoid resource wasting and the execution and operations are decentralized. In order to increase the efficiency, the decentralized IT units should have a strong interaction with central IT, and they must communicate and work as partners. As for medium size university with only one campus, centralized governance structure is the best since it allows economizing on skills and applications which lead to cost reduction and standardization (Bianchia et al. 2017).

Traditionally, the skills of IT managers to the greatest extent were technology oriented while now the C-level positions are responsible for the entire IT strategy creation, alignment, and communication (Hotzel et al. 2015). In universities, a respected and engaged CIO can make all the difference in helping functional areas use technology effectively and foster collaboration (Grajek, Top 10 IT Issues, 2019: The Student Genome Project 2019).

Committees are widely used IT Governance mechanism that provide a formal and effective way for organizations to manage its IT. Different committees contains different memberships and has a specific authority and are subjected to organizational culture. Weill and Ross (2004) suggested to limit the number of governance committees to increase the overall efficiency. One of the most commonly used committee is the IT strategy committee which is accountable for reviewing and approving IT strategy. The aforementioned strategy provides high-level direction and control over IT to deliver value and mitigate risks. The IT council committee considers different levels of policies and investments (Ko and Fink 2010). IT Governance committee is the high-level body that sets overall IT strategy, approves, and prioritizes IT projects within the available resources. Since the IT is an enabler for the overall university objectives; governance decisions for IT should not be exclusive to IT professionals only. The committee should involve members from other business units.

4.2 IT Governance Processes

Several processes were identified to improve the outcomes of IT Governance structure which were: a formal strategic planning process, an IT Steering Committee, appointment of a CIO or equivalent position, a comprehensive system for IT risk identification and management, and clear roles and responsibilities, specifying accountability for IT functions and decisions, transparency of IT decision making, stakeholders involvement in IT decision making, and central control or coordination of IT (Hicks et al. 2012).

ISO/IEC 38500, ISO/IEC 27001, COBIT, and ITIL are internationally accepted best practices guides that are implemented and followed to Govern IT including the governance of Information Security (Jairak and Praneetpolgrang 2013; Ribeiro and Gomes 2009).

AI brought new challenges and opportunities to IT Governance. AI intensifies cyberattacks and it also can be utilized to improve cyber defenses. It can be also used for enhancing decision making process by analyzing historical data which can be beneficial to predict future trends (Coccoli et al. 2014).

The Joint Information Systems Committee (JISC) for universities in the United Kingdom took the lead and designed the first IT Governance model which provides a reference for the whole university system. Along with the IT Governance reference model, the committee developed a toolkit for the self-evaluation of IT Governance maturity, which aims to help universities in the process of identifying and defining the IT role within the planning and governance of the university corporate governance. This framework was designed to be highly flexible and it considered different contingency factors like the size of the university, governing style (public/private) and the university culture (Fernández and Llorens 2009).

Performance measure is important to justify the use of IT, however, it did not receive the required attention because of the lack of consensus about different perspectives within the organizations. The success of the assessing or measurement IT Governance function in universities relies on firstly the gathering of performance measures and secondly the assessment of those measures by someone with the authority to act on the results or to forward results to an authorized body that can action them. These measures include intangible and qualitative factors such as feedback from faculties and other constituents, exception reporting when governance issues arise, internal and external audit reports, and IT strategic progression compared to plan (Hicks et al. 2012; Ko and Fink 2010).

4.3 IT Governance Relational Mechanisms

Research shows that successful implementation of IT Governance relies not only on the design of the IT Governance framework, but also on how well-communicated the IT strategy and IT policy from the board is (Jairak and Praneetpolgrang 2013).

Participatory leadership and shared governance that involves employees in decision-making might be the key to foster compliance in HEIs. Information Security Policy enforcement which is achieved using coercive force does not blend well with higher education culture (Kam et al. 2016).

The top management involvement headed by the president is important to enhance communication between different university units and foster collaboration. Their role is extended to make decisions on major matters related to IT construction and orienting the IT development (Zhen and Xin-yu 2007). Stakeholders involvement, employee's integration, business process improvement, and flexibility of the IT infrastructure are critical success factors for the implementation of IT Governance framework. The effect of active commitment from the top management in driving the capabilities to manage IT resources and improving internal process-level performance cannot be neglected (Jairak and Praneetpolgrang 2013).

In order to implement IT Governance framework in any organization, an IT Governance training must be conducted for the IT managers (Van Grembergen and De Haes 2005).

5 Benefits

Universities' missions are centered to cover three main objectives: teaching and learning, research, and knowledge transfer. While information processing and communication are crucial to achieve the missions of a university; codifying, processing, storing, and communicating the information using Information Systems can enhance the ability of the university to achieve its goals (Wilmore 2014). The strategic use of IT has been increased in the core academic activities. Online course delivery, high-performance computing tools for research, and large data sets analysis tools are examples to name a few. This increase is a result of mounting globalization, and the rising competition to attract and retain quality students and researchers (Wilmore 2014; Bianchi et al. 2017).

Universities with successful entrepreneurial investment in technology and knowledge transfer, and businesses were able to earn money through patents and joint ventures (Sporn 2007). Inadequate IT capabilities may hinder the universities ability to partner with industry and hence affect research (Woolgar 2007).

An effective IT Governance model covers business/IT alignment, risk management, information security, Regulatory and institutional policy compliance, service quality, and IT resources utilization. Information Security Governance (ISG) has been identified as a part of the IT governance umbrella (Jairak and Praneetpolgrang 2013; Grama 2015). The alignment between IT and business must bidirectional. IT leaders must understand and consider institutional priorities and timelines while setting IT priorities. Accordingly, institutional leaders (Grajek 2019).

IT governance helps higher education in attaining regulatory compliance specially when it is facing higher scrutiny from the public. Even though the scholars advocated top-down management as an effective technique to enforce information security policy (ISP); the unique structure of HEIs necessitates the need for another approach to inspire compliance (Kam et al. 2016).

6 Changing Environment of Higher Education

The higher education sector has unique characteristics which makes it incomparable with any other types of organization (Ferlie et al. 2008). The value function for universities is qualitative and cannot be specified clearly (Massy 2004).

Historically, higher education sector was not prone to the competitiveness in the market place due to the power of prestige. But as costs have climbed, online programs have become an increasingly attractive choice for whom a college education is too expensive. The new entrant to the Higher Education who offered these kinds of programs has change the DNA of the Universities (Christensen and Eyring 2011).

Internationalization of HEIs is growing rapidly. The number of students studying cross national borders are increasing. Universities are opening branch campuses in other countries which are mostly operated by local partners while some institutions have established their own premises. E-learning distance education has not grown as expected, partly because of the limited ICT capacities especially in the third world countries (Marginson 2006).

Universities are exposed to different external (strategic issues) and internal power issues (political issues). Strategic issues further classified into core, peripheral, direct support, and indirect support issues. The two core issues for any university are the teaching and research whereas, peripheral issues include the extra additional services that the universities offer to the public like knowledge and technology transfer and lifelong learning. The mission of universities is supported directly through measures like tenure appointment, academic promotion, personnel development, and information management. The other elements that support the achievement of the mission but in an indirect way are: statutory issues, finance and controlling, organization and decision support, planning and strategy, facility management, marketing and public relations, gender and diversity, quality management and evaluation, and internationalization (Blaschke et al. 2014).

The online technology that threatens the traditional university may also expands their capacities. Fortunately, traditional universities have natural advantages in delivering online learning and won the competition against the new entrants by simply utilizing the available resources. They already have the required assets, experience, and human resources, whereas new online degree providers must start from scratch (Christensen and Eyring 2011). Information management directly aid the accomplishment of university mission (Blaschke et al. 2014).

The real strength of the traditional universities is their ability to blend online and on class learning experiences which has proven more effective than adopting one method. Hybrid instruction has the potential to take traditional universities to new levels by allowing them not only to respond to competition but also to serve more students with their existing resources. Universities that fail to employ online learning technology will loss the opportunity to grow and may also lose students as the cost disparity between the traditional model and the technology-enabled model increases (Christensen and Eyring 2011).

World class universities strength doesn't rest in their reputation, endowment, or its extraordinarily faculty. It is certainly a product of clinging to the culture of change and innovation. The key is to honor their strengths and build upon their past achievements while being forward-looking (Christensen and Eyring 2011).

7 Conceptual Framework for IT Governance Considering Changing Environmental Factors

Universities are under lots of pressure to enhance their corporate governance and efficiency since they are the remedy to revitalizing economy. They are the main source for future innovators and leaders; and they are also the source for knowledge.

Innovative universities who can lead the way, are those who can sense and adapt to any external or internal environmental changes. World class universities are the advocates for this concept which put them on the top of the pyramid. Sporn (2001) identified five environmental factor (restructuring of economies, changing role of the state, shifting demographics, new technologies, and globalization) that may necessitate a change in the university corporate governance, management, and leadership structure. IT Governance is part of overall university governance and hence it these factors will have the same impact on the IT Governance. IT Governance implemented in universities to ensure that the IT is properly managed; universities' strategic objectives and IT objectives are aligned; and the IT investment creates business value for the university stakeholders (Yaokumah et al. 2015; Hotzel et al. 2015).

Universities are now searching for other way to recover the shortage that may threaten their existence. The competition to attract students is fierce. Internationalization, life-long learning, and e-learning are potential solution to the demographic crises.

Information technology and modern telecommunication revolutionized higher education, it introduced new threats and opportunities for universities. Universities with strong IT infrastructure and governance may utilize this capability to offer Internet-based degree programs, enhance their international research collaborations, university-industry collaborations, networking, marketization, and improve teaching and learning processes (Sporn 2001).

The competition in higher education is increasing drastically. Universities are competing to attract students, faculties, and resources. Collaborations is a very powerful strategic tool that could help universities in gaining more resources, however, bureaucratic and collegial nature of university structures often hinder collaboration, adaptation and entrepreneurial behavior. "A culture of mutual understanding, respect and trust, together with more entrepreneurial and self-regulating norms, can be the basis for successful adaptation" (Sporn 2001).

Universities are operating in a very complex environment, they are exposed to different IT external strategic issues which arise from external environment and many other IT political issues resulted from the internal environment. IT Governance is implemented using three main mechanisms as stated by De Haes and Van Grembergen (2005). These mechanisms are: structures, processes and relational management.

Chief Information Officer (CIO) spearheads the IT Governance structure who supposed to report directly to the university president and should have a seat in the university council committee. Top management involvement is very essential element in guaranteeing effective implementation of IT Governance as they will direct, evaluate, and monitor the performance of the adopted IT Governance scheme, further, their involvement will shape the culture of the organization. Accountability framework must be in place which can be achieved through a clear roles and responsibilities for each individual in the universities. Reporting line and authority must be explicitly stated as well as the internal and external communication requirements with other stakeholders. IT Governance Committee, IT Strategic Committee, and IT Steering Committees are examples of committees that organizations form to govern IT.

Policies, laws, and procedures guides the planning, implementation and monitoring of IT projects. Best practices are used to enlighten the management about the best way to maximize their profit from their IT investment while the risk is controlled, and resources are optimized.

IT Governance cannot accomplish its goals by implementing the ultimate IT Governance structure with solid policies, laws, and procedures only. Relational mechanisms are the most challenging element of the IT Governance implementation. Building relations with academic units is not an easy endeavor. Academics perceive any suggestions from units other than their academic department as a threat that may jeopardize their autonomy. Communication channels must be utilized till the maximum to ensure that all stakeholders are aware of the university direction and governing rules. Training and awareness programs are useful tools to change the culture to more collaborative atmosphere (Fig. 1).

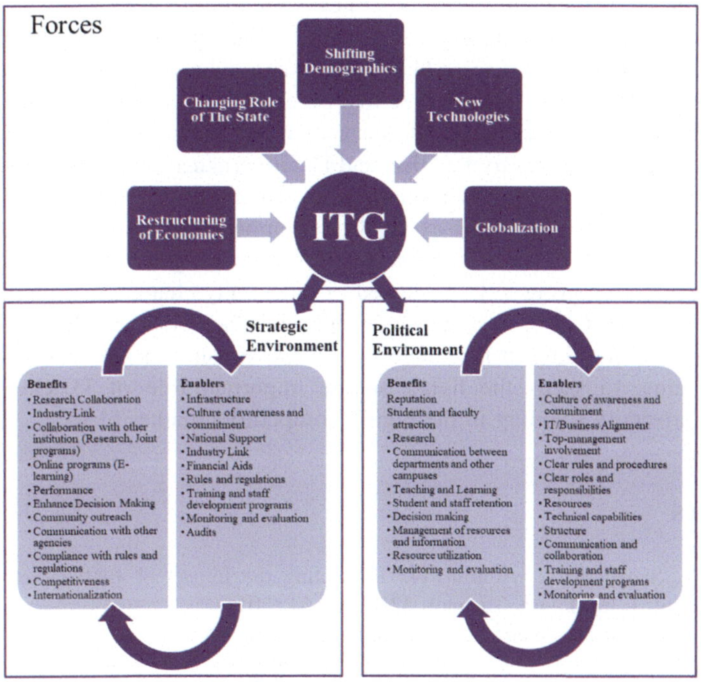

Fig. 1. Conceptual framework for IT governance considering changing environmental factors

Governments understood the strategic role of IT in enhancing its standing in the global market. They enacted variety of laws to govern and mitigate the risk of IT. In the context of universities, many corporations arise to enhance the university IT utilization and governance performance such as EDUCAUSE organization in America. Some governments developed tools for universities based on IT Governance best practices frameworks such as ISO. These tools help universities to assess their IT Governance performance.

IT is a strategic tool that can help universities to be better equipped to face the turbulent environment. IT Governance can help universities in elevating its status. It enhances university management and performance by providing high quality information to different stakeholders. Although universities has invested heavily in the IT infrastructure with the aim to improve management, research, and teaching and learning in universities, universities utilization for these resources is deemed to be low.

8 Conclusion

IT is the backbone for all business and universities are no exception to this fact. The operations of universities are affected by technologies. Despite the huge amount of money that the universities invest in IT, it never seeming to achieve its full potential. Universities recognized the importance of IT governance in supporting their mission in attaining the full potential of their IT investment to achieve their strategic goals. IT Governance is implemented using three main mechanisms which are: structures, processes and relational management. AI is a very important strategic tool which brings opportunities and challenges to the HEIs and IT Governance function. Universities are operating in a very complex environment, they are exposed to different IT external strategic issues which arise from external environment and many other IT political issues resulted from the internal environment. Effective IT governance is dynamic and flexible in nature, requiring to be reviewed to reflect any change in technology or the business environment. In this study a conceptual framework that identifies the environmental factors that impact the university IT Governance structure was developed. This study aims to sheds the lights on the important role of IT Governance in improving university standing in the highly competitive market place.

References

Abugabah, A., Sanzogni, L., Alfarraj, O.: Evaluating the impact of ERP systems in higher education. Int. J. Inf. Learn. Technol. **32**(1), 45–64 (2015)

Bianchi, I.S., Sousa, R.D., Pereira, R.: IT governance mechanisms at universities: an exploratory study. In: Twenty-Third Americas Conference on Information Systems, Boston (2017)

Bianchia, I.S., Sousab, R.D., Pereirac, R., Luciano, E.: IT governance structures in Brazilian, Dutch and Portuguese universities. Procedia Comput. Sci. **121**, 927–933 (2017)

Blaschke, S., Frost, J., Hattke, F.: Towards a micro foundation of leadership, governance, and management in universities. High. Educ. **68**, 711–732 (2014)

Brown, A.E., Grant, G.G.: Framing the frameworks: a review of IT governance research. Commun. Assoc. Inf. Syst. **15**, 696–712 (2005)

Campbell, J., McDonald, C., Sethibe, T.: Public and private sector IT governance: identifying contextual differences. Australas. J. Inf. Syst. **16**, 5–18 (2009)

Carnegie, G.D., Tuck, J.: Understanding the ABC of university governance. Aust. J. Public Adm. **69**(4), 431–441 (2010)

Christensen, C., Eyring, H.J.: The innovative university: changing the DNA of higher education. Forum for the Future of Higher Education, pp. 47–53 (2011)

De Haes, S., Van Grembergen, W.: IT governance structures, processes and relational mechanisms: achieving IT/Business alignment in a major belgian financial group. In: Proceedings of the 38th Hawaii International Conference on System Sciences, p. 237b. IEEE, Hawaii (2005)

Ferlie, E., Musselin, C., Andresani, G.: The steering of higher education systems: a public management perspective. High. Educ. **56**, 325–348 (2008)

Fernández Martínez, A., Llorens-Largo, F., Hontoria Hernández, E.: UNIVERSITIC: IT survey in Spanish and Latin American universities. EUNIS Research and Analysis Initiative-ERAI, European University Information Systems (EUNIS), Thessaloniki (2015)

Fernández, A., Llorens, F.: An IT governance framework for universities in Spain. European University Information Systems Organization (EUNIS), Santiago de Compostela (2009)

Goldfinch, S.: Rituals of reform, policy transfer, and the national university corporation reforms of Japan. Gov. Int. J. Policy Adm. Inst. **19**(4), 585–604 (2006)

Grajek, S.: Top-ten IT issues, 2014: be the change you see. EDUCAUSE Rev. **49**(2), 10 (2014)

Grajek, S.: Top 10 IT issues, 2019: the student genome project. EDUCAUSE Rev., 4–41 (2019)

Grama, J.: Understanding IT GRC in higher education: IT governance. EDUCAUSE Rev. (2015). https://er.educause.edu/articles/2015/2/understanding-it-grc-in-higher-education-it-governance

Gunawan, W., Kalensun, E.P., Fajar, A.N., Sfenrianto: Applying COBIT 5 in higher education. IOP Conf. Ser. Mater. Sci. Eng. **420**, 12108 (2018)

Hicks, M., Pervan, G., Perrin, B.: A study of the review and improvement of IT governance in Australian universities. In: CONF-IRM 2012 Proceedings. AIS Electronic Library (AISeL) (2012)

Hotzel, H., Wimmer, M., Heyde, M., Lang, U.: IT governance – role of a CIO in German universities – a survey by ZKI. PIK **38**, 121–126 (2015)

Howells, J.R., Karataş-Özkan, M., Yavuz, Ç., Atiq, M.: University management and organisational change: a dynamic institutional perspective. Camb. J. Reg. Econ. Soc., 251–270 (2014). https://doi.org/10.1093/cjres/rsu005

Jairak, K., Praneetpolgrang, P.: Applying IT governance balanced scorecard and importance-performance analysis for providing IT governance strategy in university. Inf. Manag. Comput. Secur. **21**, 228–249 (2013)

Kam, H.-J., Katerattanakul, P., Hong, S.: IT governance framework: one size fits all? In: AMCIS 2016: Surfing the IT Innovation Wave - 22nd Americas Conference on Information, San Diego (2016)

Ko, D., Fink, D.: Information technology governance: an evaluation of the theory-practice gap. Corp. Governance Int. J. Bus. Soc. **10**, 662–674 (2010)

Iliescu, F.-M.: Auditing IT governance. Informatica Economică **14**(1), 93–102 (2010)

Maassen, P.: The university's governance paradox. High. Educ. Q. **71**(3), 290–298 (2017). https://doi.org/10.1111/hequ.12125

Marginson, S.: Dynamics of national and global competition in higher education. High. Educ. **52**, 1–39 (2006)

Massy, W.F.: Collegium economicum: why institutions do what they do. Mag. High. Learn. **36**, 26–35 (2004)

Popenici, S.A., Kerr, S.: Exploring the impact of artificial intelligence on teaching and learning in higher education. Res. Pract. Technol. Enhanc. Learn. (2017). https://doi.org/10.1186/s41039-017-0062-8

Ramlaoui, S., Semma, A., Dachry, W.: Achieving a balance between IT governance and agility. IJCSI Int. J. Comput. Sci. Issues **12**, 89–96 (2015)

Ribeiro, J., Gomes, R.: IT governance using COBIT implemented in a high public educational institution - a case study. In: Proceedings of the 3rd International Conference on European Computing Conference, pp. 41–52. World Scientific and Engineering Academy and Society (WSEAS), Tbilisi (2009)

Rosca, I.G., Nastase, P., Mihai, F.: Information systems audit for university governance in Bucharest Academy of Economic Studies. Inf. Econ. **14**(1), 21–31 (2010)

Rubino, M., Vitolla, F.: IT governance, risk management and internal control system: the role of the COBIT framework. In: International OFEL Conference on Governance, Management and Entrepreneurship, pp. 174–188. ProQuest, Dubrovnik (2014)

Sporn, B.: Building adaptive universities: emerging organisational forms based on experiences of European and US universities. Tert. Educ. Manag. **7**, 121–134 (2001)

Sporn, B.: Governance and administration: organizational and structural trends. In: Forest, J.J.F., Altbach, P.G. (eds.) International Handbook of Higher Education, vol. 18, pp. 141–157. Springer, Dordrecht (2007). https://doi.org/10.1007/978-1-4020-4012-2_9

Subsermsri, P., Jairak, K., Praneetpolgrang, P.: Information technology governance practices based on sufficiency economy philosophy in The Thai University Sector. Inf. Technol. People **28**, 195–223 (2015)

Trakman, L.: Modelling university governance. High. Educ. Q. **62**, 63–83 (2008)

Van Grembergen, W., De Haes, S.: Measuring and improving IT governance through the balanced scorecard. Inf. Syst. Control J. **2**, 35–42 (2005)

Wilmore, A.: IT strategy and decision-making: a comparison of four universities. J. High. Educ. Policy Manag. **36**, 279–292 (2014)

Woolgar, L.: New institutional policies for university–industry links in Japan. Res. Policy **36**, 1261–1274 (2007)

Yamada, R.: Impact of higher education policy on private universities in Japan: analysis of governance and educational reform through survey responses. High. Educ. Forum **15**, 19–37 (2018)

Yaokumah, W., Brown, S., Adjei, P.O.-M.: Information technology governance barriers, drivers, IT/Business alignment, and maturity in Ghanaian universities. Int. J. Inf. Syst. Serv. Sect. **7**, 66–83 (2015)

Yonezawa, A., Shimmi, Y.: Transformation of university governance through internationalization: challenges for top universities and government policies in Japan. High. Educ. **70**, 173–186 (2015). https://doi.org/10.1007/s10734-015-9863-0

Zhen, W., Xin-yu, Z.: An ITIL-based IT service management model for Chinese universities. In: International Conference on Software Engineering Research, pp. 493–497 (2007)

Outlook on Security and Privacy in IoHT: Key Challenges and Future Vision

Moustafa Mamdouh[1], Ali Ismail Awad[2,3(✉)], Hesham F. A. Hamed[1,4], and Ashraf A. M. Khalaf[1]

[1] Department of Communications and Electronics, Faculty of Engineering, Minia University, Minia, Egypt
moustafa.mmdouh@outlook.com, hfah66@yahoo.com, ashkhalaf@yahoo.com
[2] Department of Computer Science, Electrical and Space Engineering, Luleå University of Technology, Luleå, Sweden
ali.awad@ltu.se
[3] Faculty of Engineering, Al-Azhar University, P.O. Box 83513, Qena, Egypt
[4] Department of Telecommunications Engineering, Egyptian Russian University, Cairo, Egypt

Abstract. The Internet of Things (IoT) security and privacy have received considerable research attention due to the IoT applicability in various domains. IoT systems have several applications, such as smart homes, smart cities, e-Health, industry, agriculture, and environmental monitoring. One of the most important applications is the Internet of Healthcare Things (IoHT) because it helps humans and patients obtain rapid diagnoses, remote monitoring, and home rehabilitation. IoHT security can be classified into four categories: applications, architecture, communication, and data security. This paper presents a short, but focused, review on IoHT security and privacy. It also explores recent security algorithms and protocols that are used to secure personal patient data, clinicians, and healthcare information. The future vision of IoHT challenges and countermeasures is given at the end of this study. Blockchain healthcare technology provides secure digital payment and privileged data access.

Keywords: Security · Privacy · IoHT · Threats · Blockchain

1 Introduction

The Internet has evolved from peer-to-peer (P2P) networking to mobile Internet then Internet of Things (IoT) in the last few decades. Through the IoT paradigm, people can connect things or "objects" from anywhere anytime with anyone with IoT technology using local or global networks for different services [1]. The IoT paradigm allows things to connect, communicate, share and exchange useful information providing a smart system controlled based on the published (transferred) data in nonreal time and real-time services [2].

A.-E. Hassanien et al. (Eds.): AICV 2020, AISC 1153, pp. 721–730, 2020.
https://doi.org/10.1007/978-3-030-44289-7_67

Increasing objects that are connected to the Internet realize the idea of the IoT. The Internet of Healthcare Things (IoHT) is one of the IoT applications and most directly affects human life. The IoHT system collects patient health information (PHI), which can be added to electronic healthcare records (EHR) for patients to communicate with the system anywhere, everywhere, and anytime. This remote connection helps clinicians observe patients remotely. Therefore, EHR and PHI need to be secured for patients' safety and privacy protection purposes [3]. Security and privacy are the core of any strength or weakness of IoT applications, services, and systems. Patients have sensitive personal or health-related information; thus, they may be unable to use IoHT services unless they are aware of the system's security. Blockchain technology for healthcare data management can reduce the computational and communication overhead due to its decentralization [4].

Driven by the fact that IoHT is one of the most important systems that improve the quality of life and human health, this paper explores the IoHT system and presents a short but focused outlook on the current challenges and proposes a future vision regarding security and privacy challenges. Additionally, the review reports the IoHT security algorithms and protocols that can be used to fulfill the security requirements for protecting patients' personal data and healthcare information in both transmission and steady states.

The rest of this paper is organized as follows. IoT enabling technologies are introduced in Sect. 2. In Sect. 3, an overview of communication protocols used on the IoT network and their weaknesses are presented. Section 4 introduces the whole IoHT architecture and highlights security challenges and defenses in the IoHT system. Discussions and future vision on several important security- and privacy-related issues in the IoHT system are presented in Sect. 5. Finally, conclusions and future work are written in Sect. 6.

2 IoT Enabling Technologies

IoT infrastructure should be well designed to enable connectivity for smart devices to communicate with end users through the internet. However, there are some limitations within the current research and development attempts to produce an entire IoT network. Applications and services of IoT are sporadically developed from various vendors but not the usage of the standard technology. Also, there is no standard networking protocol for IoT applications, although there are different networking protocols, such as Wireless Fidelity (Wi-Fi), Bluetooth Low Energy (BLE), ZigBee, Z-wave, low-power wide-area (LPWA) and different mobile generations communication. In other words, no device can communicate with all networking protocols unless you build your own device using an embedded system software package; however, it has a value issue. To solve this drawback, a gateway or an IoT device that can support heterogeneous networks is incredibly important for creating one complete IoT system [5–7].

LPWA technologies can be splitted into two types by whether the frequency spectrum is authorized or unauthorized. The technologies that work in an unauthorized frequency band such as long-range (LoRa), SIGFOX are nonstandard and released as user-defined. The authorized frequency bands such as different mobile generations [8]. The basic security algorithm for the LoRaWAN network architecture is an encryption algorithm with 128 bits key length, where an application session key (AppSKey) and a network session key (NwkSKey) are used. These keys provide data confidentiality and data integrity between the end device and the application and between the end device and the network server, respectively. Navarro-Ortiz et al. in [9] proposed a modified gateway for LoRaWAN to integrate it with the LTE mobile network, providing end-to-end security requirements using AppSKey and KwSKey, which use the advanced encryption standard (AES)-128 bits algorithm. The AES algorithm can be implemented with high speed of encryption, low power consumption, and area optimization [10].

3 IoT Communication Protocols

IoT has more than one architecture design, and these architectures mainly use three standard layers. These layers are called the perception layer (physical layer), network layer, and application layer, as shown in Fig. 1 with security issues and security countermeasures [11]. The application layer provides the services that users need [12]. There are several protocols used in the IoT networks; these protocols responsible for share and update information through the internet between different devices in the IoT network with the required data and carry out commands from application layer to the IoT devices [13].

The recently used protocols are the constrained application protocol (CoAP), message queue telemetry transport (MQTT), extensible messaging and presence protocol (XMPP), advanced message queuing protocol (AMQP), data distribution service (DDS), and hypertext transfer protocol (HTTP). These protocols use different transport layer protocols and security algorithms, such as transport control protocol (TCP) and user datagram protocol (UDP) [14].

Machine-to-machine (M2M) communication uses MQTT designed to provide a publish/subscribe messaging protocol with a minimum bandwidth requirement, code footprint size, power consumption, and message data overhead [15]. Many researchers have reported that the MQTT protocol is suitable for IoT devices. MQTT contain three main elements: a subscriber, a publisher, and a broker. The publisher responsible for send and receive data through the broker and subscribers. The broker acts as a security mechanism for both publisher and subscriber [13].

To achieve messaging quality of service (QoS), there are three options: at most (no acknowledgment), at least (acknowledgments required), and exactly (high level of QoS). MQTT provides higher throughput and lower latency compared to CoAP. MQTT is a lightweight protocol, and it is the most recommended TCP/IP communication protocol, which has a small frame size format, so, it does not need high CPU or memory usage [16]. MQTT has major advantages in achieving better security because of its simplicity and small message format size. MQTT supports user authentication and identification [17]. The MQTT broker achieves security authorization to identify publishers and subscribers [18]. MQTT is better for large data size real-time-based applications

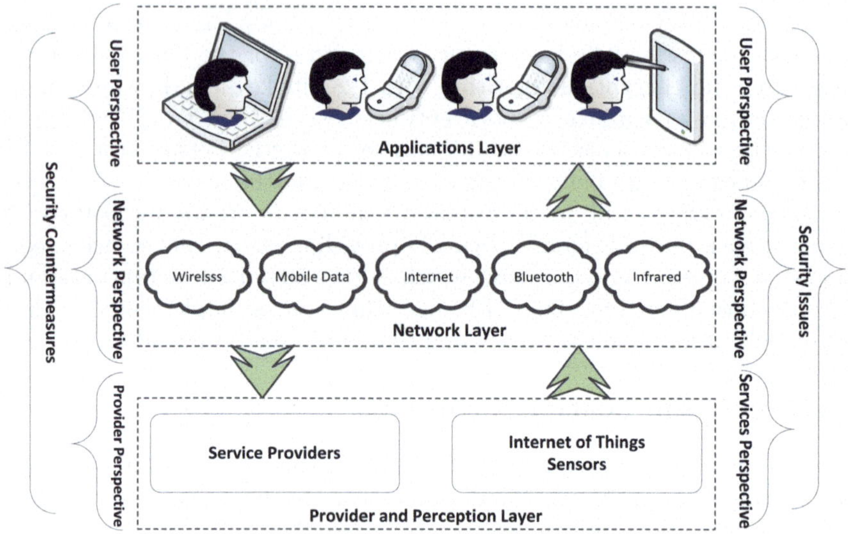

Fig. 1. IoT architecture showing the three common layers [11].

with low latency, overhead, memory usage, and energy consumption [19]. Additionally, MQTT achieves resistance to attacks, and user privacy [20].

4 Internet of Healthcare Things (IoHT)

The IoHT system architecture consists of four layers: healthcare sensing devices, networking, cloud computing, and healthcare applications. The IoHT collects temperature, blood glucose levels, heart rate, pressure, or any historical data called EHR or PHI that help doctors define human status. Applications on MQTT brokers, Web, iOS, and Android devices enable the doctors to monitor and make decisions based on calculations without being near the patients [21]. The heterogeneous IoHT system architecture has four layers: an application layer for medical applications, a cloud computing layer, a networking layer, and a sensing layer, which includes patient sensors or medical devices [22], as shown in Fig. 2. It is noteworthy that each component and layer in the system may be attacked from the interceptor or hacker to collect PHI.

There are many security algorithms and protocols to prevent and protect patient privacy and information. Ali and Awad [11], identified several critical threats, assets, security risks and vulnerability for IoT-based smart home systems. Threat models could be one or more of the following types: ransomware, distributed denial-of-service (DDoS) attacks, insider attacks, email compromise, fraud scams, collusion attacks, eavesdropping, impersonation, or leakage and destruction [23]. This paper presents a short survey on recent and selected research papers using different security algorithms and protocols to secure EHR

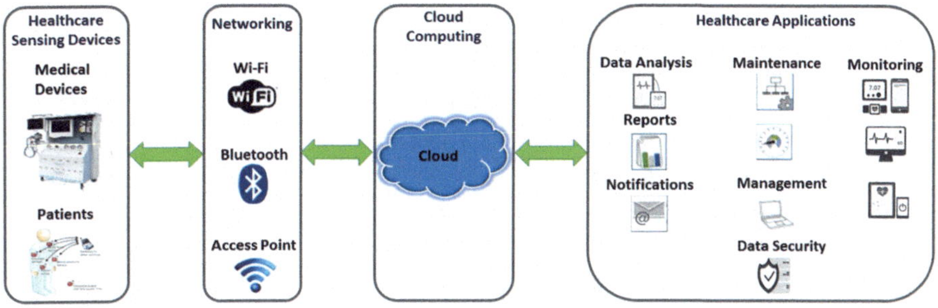

Fig. 2. Heterogeneous IoHT system architecture.

and PHI. Islam et al. in [24], introduced IoHT system technologies and a state-of-the-art network architecture/platform, applications, and industrial solutions. Additionally, they proposed IoT security and privacy requirements, threat models, and attacks to minimize security risks. Several issues, such as network type, application model, QoS, and data protection, have been discussed to provide good healthcare services and technologies.

4.1 IoHT Application

Kougianos et al. [25] proposed a complete design for automatic tracking applications for image and video exchange over IoT to enhance real-time compression. Additionally, secure digital camera architecture platform integrated with secure BPG compression algorithm was illustrated. Yeh in [26] proposed an implementation of a secure IoT-based healthcare system with a body sensor network to achieve system efficiency and robustness of transmission with the Raspberry PI platform using the SHA-2 algorithm technique to perform two-way authentication. Elmisery, Rho, and Botvich [27] proposed privacy preserving of the end-users' cloud-based healthcare services to generate accurate health information. Fog node solutions were used between IoHT devices and cloud-based healthcare services to prevent middleware attacks. Monrat, Schelén, and Andersson [28] proposed a large survey for blockchain technology applications, such as healthcare applications, which allows the tracking of products and immune EHR and EPI against threats.

4.2 Network Architecture

Kwabena et al. [29] introduced a novel MSCryptoNet framework that enables the scalable execution and conversion of state-of-the-art learned neural networks for the privacy-preservation setting. This model proved data security for the MSCryptoNet framework as a privacy-preserving deep learning scheme over the aggregated encrypted data. MSCryptoNet architecture consists of a data provider, computation party, data service provider, and data request (physicians). MSCryptoNet, a novel collaborative privacy-preserving, computationally efficient, multischeme fully homomorphic encryption-based architecture, can

Table 1. A comparison of the selected healthcare research papers. The ● symbol represents deployed. The ○ symbol represents not deployed.

Method	2016				2017	2018						2019		
	[25]	[26]	[27]	[32]	[30]	[38]	[36]	[35]	[39]	[33]	[34]	[29]	[40]	[19]
Application	●	●	●	●	○	●	○	○	●	●	●	●	○	●
Architecture	○	○	○	○	●	○	●	○	○	○	●	●	○	○
LoRaWAN	○	○	○	○	○	○	○	●	○	○	○	●	●	○
Data	○	○	○	●	○	○	●	○	●	●	●	●	●	●

resolve the major challenges in [30]. Daraghmi, Daraghmi, and Yuan [31] proposed a MedChain software component architecture for a blockchain network. Additionally, they showed how to generate, validate and append new blocks in each node provider.

4.3 Data Security

Upadhyay, Borole, and Dileepan in [32] proposed an encryption method for transferring and receiving sensor data and monitoring it on network devices using the MQTT protocol for home automation applications. The named data networking (NDN) secure system was proposed by Boussada et al. [33] for security and a privacy-preserving IoHT system solution. The IoHT system meets the security and data privacy requirements when using the AES-128 bits algorithm. Elhoseny et al. [34] proposed a hybrid encryption algorithm using AES 128-bit key length and Rivest, Shamir, and Adleman (RSA) algorithms by encrypting the secret data and then used the results as a cover image. This model was compared to state-of-the-art methods and proved that confidential patient data could be hidden in a transmitted cover image with high imperceptibility, capacity, and minimal deterioration in the received image.

Tsai et al. [35] introduced simplified 128-bit AES with 5 rounds only for low power consumption for the LoRaWAN communication network. The analysis claimed to save 26.2% of power consumption and resist known-key attacks, replay attacks, and eavesdropping attacks. Hassan and Awad in [36] proposed a framework for ensuring humans' personal right to privacy in the IoT urban paradigm. Additionally, they reviewed the prospects of and barriers to IoT implementation in the urban context from the perspectives of security and privacy. Jin et al. [37] proposed a survey for how to provide security and privacy preserving for data transactions and access control for the EHR and EPI.

EHR and PHI should be secured over the whole network to provide immunity against different types of inside and outside threats and attacks. Table 1 shows a comparison of the selected IoHT research papers that apply security and privacy algorithms and use different security protocols on various components, such as application, architecture, network, and data.

5 Discussions and Future Vision

Artificial intelligence (AI) can be used for attack prediction and changes the encryption algorithm and the key size for more security and privacy preservation. AI can predict and prevent zero-day inside and outside known and unknown threats and attacks to prevent impersonating a patient's personality. IoHT networks, services, applications, and databases should be scalable for individuals and healthcare organizations. Employing convolutional neural networks (CNNs) can predict critical vulnerabilities and automatically send medical notification messages or warnings to the patient and physicians who observe the patient's health to ensure a safe state. The more wide-scale the IoT application is, the more affected it is by attackers and new threats and vulnerabilities that need to be predicted by up-to-date machine learning techniques to prevent unauthorized persons from obtaining valuable patient information.

A considerable quantity of data is shared between the network and the cloud that need to use cloud services for data storage and data analysis. IoT cloud security should be performed and increased in the future by all security vendors. IoT cloud security provides data loss prevention and protection from inside and outside threats and attacks for full visibility and continuous availability. Cloud computing services with high processing capabilities can be used to apply AI techniques for predicting critical cases for some patients before any decline or anticlimax. Although cloud computing offers high scalability and wide deployments of IoHT systems, it imposes new security and privacy challenges linked to data storage, servers, backup, data transmission, and data processing in the cloud. Cloud service providers ensure the security of sensitive data stored online using different methods, such as penetration testing, virtual private networks (VPNs), tokenization, and firewalls. Web application firewalls (WAFs) are used to secure services application in the cloud to prevent different types of attacks such as DDoS attacks. This led us to protected cloud connections (PCCs), such as HTTP(s) protocols and TCP/UDP services. Protected cloud access (PCA) uses a fingerprint or eye to authenticate the end user.

Blockchain technology provides patients with decentralization, tracking, data transparency, security, and privacy for their EHR. Healthcare blockchain applications such as medical recording, financial transactions, medical products and services, and patients' digital identities should be secured. Blockchain provides high security and privacy preservation of transactions and high integrity. Additionally, blockchain provides high management for healthcare systems by using privileged digital identities and access management [41,42]. To date, there has been no research in the field of the learning process for computational statistics on blockchain technology [43].

6 Conclusions and Future Work

The IoT paradigm spans many applications, such as smart homes, smart cities, smart industry, and healthcare. IoHT services have become crucial for patients and human life due to their capability to support patients in many critical cases, speed up the diagnosis process, and improve the given medical care. Despite the need for IoHT services, data shared among the IoHT networks can transmit sensitive information that should be accessible by authorized persons. Therefore, security and privacy issues should be considered. This paper focused on IoHT security and presented a short but focused review on different layers, including applications, architectures, communication protocols, and data security in transition and steady states. Furthermore, it explained different security algorithms and protocols and performed a comparative analysis between them. Through this review, it was found that MQTT is the most commonly used protocol for improving power consumption, memory usage, latency, and delay in terms of security and privacy issues; thus, it is recommended for IoHT systems and should be subject to security-related investigations.

In the future, IoHT services will increase due to their effects on human life. AES and RSA encryption algorithms and MQTT protocol should be optimized for further deployments in IoHT systems for data confidentiality. Intensive application of machine and deep learning techniques should be used for creating adaptive security mechanisms to prevent well-known attacks and predict unknown attack patterns. Blockchain technology can be used to enhance privileged access management and secure payment transactions.

References

1. Porambage, P., Okwuibe, J., Liyanage, M., Ylianttila, M., Taleb, T.: Survey on multi-access edge computing for Internet of Things realization. IEEE Commun. Surv. Tutor. **20**(4), 2961–2991 (2018)
2. Al-Fuqaha, A., Guizani, M., Mohammadi, M., Aledhari, M., Ayyash, M.: Internet of Things: a survey on enabling technologies, protocols, and applications. IEEE Commun. Surv. Tutor. **17**(4), 2347–2376 (2015)
3. Okoh, E., Awad, A.I.: Biometrics applications in e-health security: a preliminary survey. In: International Conference on Health Information Science, pp. 92–103. Springer (2015)
4. Ismail, L., Materwala, H., Zeadally, S.: Lightweight blockchain for healthcare. IEEE Access **7**, 149935–149951 (2019)
5. Miao, Y., Li, W., Tian, D., Hossain, M.S., Alhamid, M.F.: Narrowband Internet of Things: simulation and modeling. IEEE Internet Things J. **5**(4), 2304–2314 (2017)
6. Mroue, H., Nasser, A., Hamrioui, S., Parrein, B., Motta-Cruz, E., Rouyer, G.: MAC layer-based evaluation of IoT technologies: LoRa, SigFox and NB-IoT. In: 2018 IEEE Middle East and North Africa Communications Conference (MENACOMM), pp. 1–5. IEEE (2018)
7. Yang, K., Blaauw, D., Sylvester, D.: Hardware designs for security in ultra-low-power IoT systems: an overview and survey. IEEE Micro **37**(6), 72–89 (2017)
8. Lauridsen, M., Vejlgaard, B., Kovacs, I.Z., Nguyen, H., Mogensen, P.: Interference measurements in the European 868 MHz ISM band with focus on LoRa and SigFox. In: 2017 IEEE Wireless Communications and Networking Conference (WCNC), pp. 1–6. IEEE (2017)

9. Navarro-Ortiz, J., Sendra, S., Ameigeiras, P., Lopez-Soler, J.M.: Integration of LoRaWAN and 4G/5G for the industrial Internet of Things. IEEE Commun. Mag. **56**(2), 60–67 (2018)
10. Fathy, A., Tarrad, I.F., Hamed, H.F.A., Awad, A.I.: Advanced encryption standard algorithm: issues and implementation aspects. In: International Conference on Advanced Machine Learning Technologies and Applications, pp. 516–523. Springer (2012)
11. Ali, B., Awad, A.I.: Cyber and physical security vulnerability assessment for IoT-based smart homes. Sensors **18**(3), 817 (2018)
12. Polianytsia, A., Starkova, O., Herasymenko, K.: Survey of the IoT data transmission protocols. In: 2017 4th International Scientific-Practical Conference Problems of Infocommunications, Science and Technology (PIC S&T), pp. 369–371. IEEE (2017)
13. Florea, I., Rughinis, R., Ruse, L., Dragomir, D.: Survey of standardized protocols for the Internet of Things. In: 2017 21st International Conference on Control Systems and Computer Science (CSCS), pp. 190–196. IEEE (2017)
14. Solapure, S.S., Kenchannavar, H.: Internet of Things: a survey related to various recent architectures and platforms available. In: 2016 International Conference on Advances in Computing, Communications and Informatics (ICACCI), pp. 2296–2301. IEEE (2016)
15. Babovic, Z.B., Protic, J., Milutinovic, V.: Web performance evaluation for Internet of Things applications. IEEE Access **4**, 6974–6992 (2016)
16. Yassein, M.B., Shatnawi, M.Q., Al-zoubi, D.: Application layer protocols for the Internet of Things: a survey. In: 2016 International Conference on Engineering & MIS (ICEMIS), pp. 1–4. IEEE (2016)
17. Benson, K., Fracchia, C., Wang, G., Zhu, Q., Almomen, S., Cohn, J., D'arcy, L., Hoffman, D., Makai, M., Stamatakis, J., Venkatasubramanian, N.: SCALE: safe community awareness and alerting leveraging the Internet of Things. IEEE Commun. Mag. **53**(12), 27–34 (2015)
18. King, J., Awad, A.I.: A distributed security mechanism for resource-constrained IoT devices. Informatica **40**(1), 133–143 (2016)
19. Sultana, T., Wahid, K.A.: Choice of application layer protocols for next generation video surveillance using Internet of video things. IEEE Access **7**, 41607–41624 (2019)
20. Kaedi, S., Doostari, M.A., Ghaznavi-Ghoushchi, M.B.: Low-complexity and differential power analysis (DPA)-resistant two-folded power-aware Rivest-Shamir-Adleman (RSA) security schema implementation for IoT-connected devices. IET Comput. Digital Tech. **12**(6), 279–288 (2018)
21. Alamelu, J., Mythili, A.: Design of IoT based generic health care system. In: 2017 International conference on Microelectronic Devices, Circuits and Systems (ICMDCS), pp. 1–4. IEEE (2017)
22. Musaddiq, A., Zikria, Y.B., Hahm, O., Yu, H., Bashir, A.K., Kim, S.W.: A survey on resource management in IoT operating systems. IEEE Access **6**, 8459–8482 (2018)
23. Elrawy, M.F., Awad, A.I., Hamed, H.F.: Intrusion detection systems for IoT-based smart environments: a survey. J. Cloud Comput. **7**(1), 21 (2018)
24. Islam, S.R., Kwak, D., Kabir, M.H., Hossain, M., Kwak, K.S.: The Internet of Things for health care: a comprehensive survey. IEEE Access **3**, 678–708 (2015)
25. Kougianos, E., Mohanty, S.P., Coelho, G., Albalawi, U., Sundaravadivel, P.: Design of a high-performance system for secure image communication in the Internet of Things. IEEE Access **4**, 1222–1242 (2016)

26. Yeh, K.H.: A secure IoT-based healthcare system with body sensor networks. IEEE Access **4**, 10288–10299 (2016)
27. Elmisery, A.M., Rho, S., Botvich, D.: A fog based middleware for automated compliance with OECD privacy principles in Internet of healthcare things. IEEE Access **4**, 8418–8441 (2016)
28. Monrat, A.A., Schelén, O., Andersson, K.: A survey of blockchain from the perspectives of applications, challenges, and opportunities. IEEE Access **7**, 117134–117151 (2019)
29. Kwabena, O.A., Qin, Z., Zhuang, T., Qin, Z.: MSCryptoNet: multi-scheme privacy-preserving deep learning in cloud computing. IEEE Access **7**, 29344–29354 (2019)
30. Li, P., Li, J., Huang, Z., Li, T., Gao, C.Z., Yiu, S.M., Chen, K.: Multi-key privacy-preserving deep learning in cloud computing. Future Gener. Comput. Syst. **74**, 76–85 (2017)
31. Daraghmi, E.Y., Daraghmi, Y.A., Yuan, S.M.: MedChain: a design of blockchain-based system for medical records access and permissions management. IEEE Access **7**, 164595–164613 (2019)
32. Upadhyay, Y., Borole, A., Dileepan, D.: MQTT based secured home automation system. In: 2016 Symposium on Colossal Data Analysis and Networking (CDAN), pp. 1–4. IEEE (2016)
33. Boussada, R., Hamdaney, B., Elhdhili, M.E., Argoubi, S., Saidane, L.A.: A secure and privacy-preserving solution for IoT over NDN applied to e-health. In: 2018 14th International Wireless Communications & Mobile Computing Conference (IWCMC), pp. 817–822. IEEE (2018)
34. Elhoseny, M., Ramírez-González, G., Abu-Elnasr, O.M., Shawkat, S.A., Arunkumar, N., Farouk, A.: Secure medical data transmission model for IoT-based healthcare systems. IEEE Access **6**, 20596–20608 (2018)
35. Tsai, K.L., Huang, Y.L., Leu, F.Y., You, I., Huang, Y.L., Tsai, C.H.: AES-128 based secure low power communication for LoRaWAN IoT environments. IEEE Access **6**, 45325–45334 (2018)
36. Hassan, A.M., Awad, A.I.: Urban transition in the era of the Internet of Things: social implications and privacy challenges. IEEE Access **6**, 36428–36440 (2018)
37. Jin, H., Luo, Y., Li, P., Mathew, J.: A review of secure and privacy-preserving medical data sharing. IEEE Access **7**, 61656–61669 (2019)
38. Yang, Y., He, D., Kumar, N., Zeadally, S.: Compact hardware implementation of a SHA-3 core for wireless body sensor networks. IEEE Access **6**, 40128–40136 (2018)
39. Tao, H., Bhuiyan, M.Z.A., Abdalla, A.N., Hassan, M.M., Zain, J.M., Hayajneh, T.: Secured data collection with hardware-based ciphers for IoT-based healthcare. IEEE Internet Things J. **6**(1), 410–420 (2018)
40. Xu, Z., Xu, C., Liang, W., Xu, J., Chen, H.: A lightweight mutual authentication and key agreement scheme for medical Internet of Things. IEEE Access **7**, 53922–53931 (2019)
41. Nguyen, D.C., Pathirana, P.N., Ding, M., Seneviratne, A.: Blockchain for secure EHRs sharing of mobile cloud based e-health systems. IEEE Access **7**, 66792–66806 (2019)
42. Shahnaz, A., Qamar, U., Khalid, A.: Using blockchain for electronic health records. IEEE Access **7**, 147782–147795 (2019)
43. Kim, H., Kim, S.H., Hwang, J.Y., Seo, C.: Efficient privacy-preserving machine learning for blockchain network. IEEE Access **7**, 136481–136495 (2019)

Dynamic Analysis of a Fractional Map with Hidden Attractor

Amina Aicha Khennaoui[1], Adel Ouannas[2], Giuseppe Grassi[3], and Ahmad Taher Azar[4,5](✉)

[1] Department of Mathematics and Computer Sciences, University of Larbi Ben M'hidi, Oum El Bouaghi, Algeria
kaminaaicha@yahoo.fr

[2] LAMIS Laboratory, Department of Mathematics, University of Larbi Tebessi, 12002 Tebessa, Algeria
ouannas.a@yahoo.com

[3] Dipartimento Ingegneria Innovazione, Università del Salento, 73100 Lecce, Italy
giuseppe.grassi@unisalento.it

[4] Robotics and Internet-of-Things Lab (RIOTU), Prince Sultan University, Riyadh, Saudi Arabia
aazar@psu.edu.sa

[5] Faculty of Computers and Artificial Intelligence, Benha University, Benha, Egypt
ahmad.azar@fci.bu.edu.eg

Abstract. In this paper, a novel three dimensional fractional-order map is proposed. This fractional map has no fixed point, but it can also exhibit rich and complex dynamical behavior. The dynamical properties of the new model are investigated by applying numerical tools such as bifurcation diagram, maximum Lyapunov exponent, phase portraits, and evolution of states. It shows that the fractional order map is more complex when the fractional order is small.

Keywords: Chaos · Bifurcation diagram · Discrete fractional calculus · Hidden attractors · Lyapunov exponent

1 Introduction

Fractional calculations have recently become a great tool to model many physical phenomena and engineering problems [2,4,8,10,11,15,16,19,22,26,27]. Discrete fractional calculus is a new attractive field that has received considerable attention in nearly all kinds of fields [5,6]. Some researchers have started proposing fractional maps based on discrete fractional calculations [14,17,18,20,23–25,28,30] and consider the application in encryption as soon as their synchronization becomes feasible [7,21,31].

In recent years, the study of the complex dynamics of chaotic systems with no equilibrium point has become a very interesting topic due to its importance in engineering application. The attractors associated with these systems have no fixed point and are called hidden attractors. Several nonlinear continuous models

A.-E. Hassanien et al. (Eds.): AICV 2020, AISC 1153, pp. 731–739, 2020.
https://doi.org/10.1007/978-3-030-44289-7_68

with hidden attractors have been proposed in the literature [9,12,29]. On the other hand, only little effort has been made in discrete models [13]. In this paper, a new generalized order map without fixed point is proposed by applying the Caputo-like difference operator. By the help of numerical simulation, we try to find out whether it is possible to have new hidden chaotic attractors in the generalized order map.

The rest of the paper is organized as follows. In Sect. 2, a novel three dimensional fractional map without fixed point is proposed. In Sect. 3, chaos in the fractional map is verified by numerical simulation and its dynamical behavior are investigated with varying fractional order. Finally, the paper is concluded in Sect. 4.

2 Description of the Proposed System

In this section, we propose a new three dimensional discrete-time system with fractional order by:

$$\begin{cases} {}^C\Delta_a^\nu x(t) = -\alpha_1 x(t-1+\nu)^2 + \alpha_2 z(t-1+\nu)y(t-1+\nu) \\ \qquad\qquad + \alpha_3 z(t-1+\nu)x(t-1+\nu) + \alpha_4 - x(t-1+\nu), \\ {}^C\Delta_a^\nu y(t) = x(t-1+\nu) - y(t-1+\nu), \\ {}^C\Delta_a^\nu z(t) = y(t-1+\nu) - z(t-1+\nu), \end{cases} \tag{1}$$

where x, y, z are the state variables, $\alpha_1, \alpha_2, \alpha_3, \alpha_4$ are system parameters, and $0 < \nu < 1$ is the fractional order. Operator ${}^C\Delta_a^\nu$ is the Caputo like difference operator with order ν [1], which is defined as Eq. (2):

$${}^C\Delta_a^\nu x\,(t) = \frac{1}{\Gamma\,(1-\nu)} \sum_{s=a}^{t-(1-\nu)} \frac{\Gamma(t-s)}{\Gamma(t-s-\nu)} \Delta_s x\,(s)\,, \tag{2}$$

or Γ denotes the gamma function, $t \in \mathbb{N}_{a+1-v}$, and a is the starting point. If we set a to 0, the fractional Eq. (1) with an initial condition (x_0, y_0, z_0) is equivalent to the discrete volterra integral equation [3]:

$$\begin{cases} x(t) = x(0) + \frac{1}{\Gamma(\nu)} \sum_{s=1-\nu}^{t-\nu} (t-s-1)^{(\nu-1)}(-a_1 x(s-1+\nu)^2 + a_2 z(s-1+\nu) \\ \qquad y(s-1+\nu) + a_3 z(s-1+\nu)x(s-1+\nu) + a_4 - x(s-1+\nu)), \\ y(t) = y(0) + \frac{1}{\Gamma(\nu)} \sum_{s=1-\nu}^{t-\nu} (t-s-1)^{(\nu-1)}(x(s-1+\nu) - y(s-1+\nu)), \\ z(t) = z(0) + \frac{1}{\Gamma(\nu)} \sum_{s=1-\nu}^{t-\nu} (t-s-1)^{(\nu-1)}(y(s-1+\nu) - z(s-1+\nu)). \end{cases} \tag{3}$$

Now, replace the discrete kernel function $\frac{(t-s-1)^{(\nu-1)}}{\Gamma(\nu)}$ by $\frac{\Gamma(t-s)}{\Gamma(t-s-\nu+1)}$, t by n, and set $s+\nu = j$. The numerical formula of Eq. (3) can be represented as follows:

$$\begin{cases} x(n) = x_0 + \sum_{j=1}^{n} \frac{\Gamma(n-j+\nu)}{\Gamma(n-j+1)} (-a_1 x(j-1)^2 + a_2 z(j-1)y(j-1) + a_3 z(j-1) \\ \qquad x(j-1) + a_4 - x(j-1)), \\ y(n) = y_0 + \sum_{j=1}^{n} \frac{\Gamma(n-j+\nu)}{\Gamma(n-j+1)} \left(x(j-1) - y(j-1)\right), \\ z(n) = z_0 + \sum_{j=1}^{n} \frac{\Gamma(n-j+\nu)}{\Gamma(n-j+1)} \left(y(j-1) - z(j-1)\right). \end{cases} \tag{4}$$

In the following section, we will investigate the dynamics of the proposed map (1) by using theory and numerical simulation.

3 Dynamics Analysis

3.1 Fixed Point

The fixed point of the new model (1) can be obtained by setting ${}^{C}\Delta_a^{\nu}x = {}^{C}\Delta_a^{\nu}y = {}^{C}\Delta_a^{\nu}z = 0$. Thus, we have:

$$\begin{cases} -\alpha_1 x^2 + \alpha_2 zy + \alpha_3 zx + \alpha_4 - x = 0, \\ x - y = 0, \\ y - z = 0. \end{cases} \tag{5}$$

From Eq. (5) we have,

$$-\alpha_1 x^2 + \alpha_2 x^2 + \alpha_3 x^2 + \alpha_4 - x = 0. \tag{6}$$

Consequently, Eq. (5) has no solution when $-\alpha_1 + \alpha_2 + \alpha_3 \neq 0$ and $\Delta < 0$. Although there are no fixed point for the fractional order map (1). By choosing suitable parameters, the proposed fractional map (1) can generate hidden chaotic attractor. For instance, when $\alpha_1 = -1, \alpha_2 = -0.7, \alpha_3 = -1, \alpha_4 = -0.7$ and $\nu = 0.9962$ with initial condition $(x_0, y_0, z_0) = (-1.5, -2.84, 1.2)$, the hidden chaotic attractor is illustrated in Fig. 1.

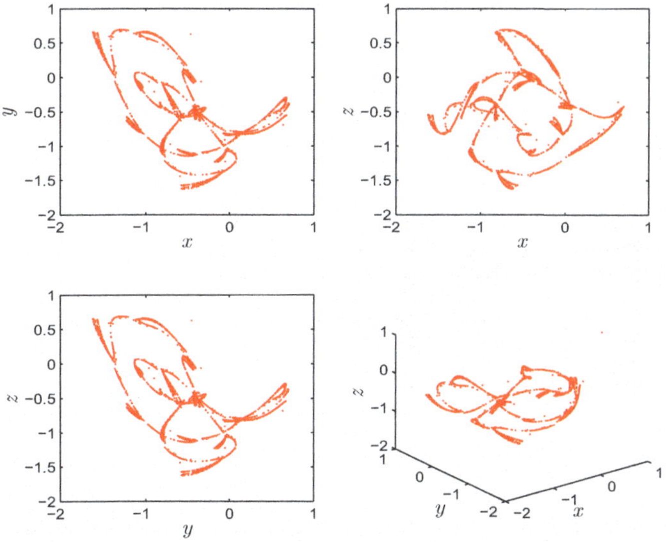

Fig. 1. Hidden chaotic attractor of the fractional-order map (1) for $\nu = 0.9962$ with initial state $(x_0, y_0, z_0) = (-1.5, -2.84, 1.2)$, and system parameters $(\alpha_1, \alpha_2, \alpha_3, \alpha_4) = (-1, -0.7, -1, -0.7)$.

3.2 Numerical Simulation

In order to find the dynamic property of the three dimensional fractional model (1), we analyze its bifurcation diagram, maximum Lyapunov exponent, and phase portraits for different values of ν. Four cases are considered:

- **Case 1: Effect of varying fractional order ν**
Here, we choose to fix system parameters as $\alpha_1 = -1, \alpha_2 = -0.7\alpha_3 = -1, \alpha_4 = -0.7$, and vary the value of ν from 0.81 to 1. The corresponding bifurcation diagram of state variable $y(t)$ is shown in Fig. 2(a), and the maximum Lyapunov exponent diagram with respect to ν is shown in Fig. 2(b). The results show that the fractional map (1) changes from chaos state to periodic state and then to a chaotic state as the order ν decreases. Obviously, when $\nu \in [0.9463, 1]$, the system (1) displays chaotic behavior with periodic windows. It's also noticed that when $\nu \in [0.8465, 0.9463]$, the maximum Lyapunov exponent is negative and sometimes equal to zero, which implies that the system (1) displays periodic windows. Also, when $\nu \in [0.81, 0.8465]$, the new system (1) displays chaotic behavior which confirmed by the largest Lyapunov exponent. It's observed that the smaller the fractional order ν, the more complex the fractional order map (1) is. To further observations of the dynamical behavior and phase portrait of the fractional order map (1) in $y - z$ plane are considered. Figure 3(a)–(f) showed the phase diagrams with $\nu = 0.9962$, $\nu = 0.9707$, $\nu = 0.9395$, $\nu = 0.9279$, $\nu = 0.9$, $\nu = 0.8313$. As shown, six different hidden attractors are observed.
  The time evolution of the states is depicted in Fig. 4 for $\nu = 0.9962$. The time series in Fig. 4 confirms the irregular behavior.

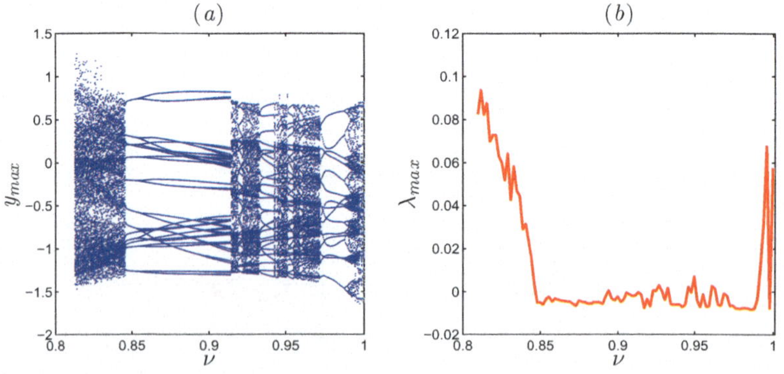

Fig. 2. Bifurcation diagram and maximum Lyapunov exponent of the fractional order map (1) versus $\nu \in [0.81, 1]$ with $(\alpha_1, \alpha_2, \alpha_3, \alpha_4) = (-1, -0.7, -1, -0.7)$. (a) Bifurcation diagram, (b) maximum Lyapunov exponent.

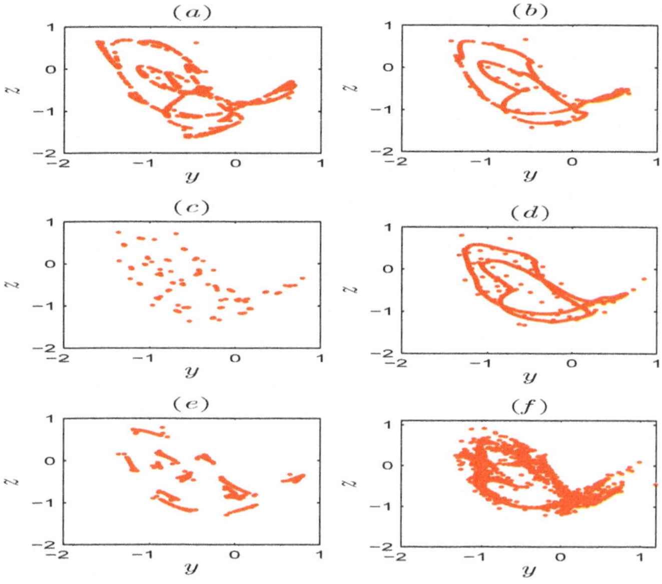

Fig. 3. Hidden attractors of the fractional-order map (1) for: (a) $\nu = 0.9962$, (b) $\nu = 0.9707$, (c) $\nu = 0.9395$, (d) $\nu = 0.9279$, (e) $\nu = 0.9$, (f) $\nu = 0.8313$.

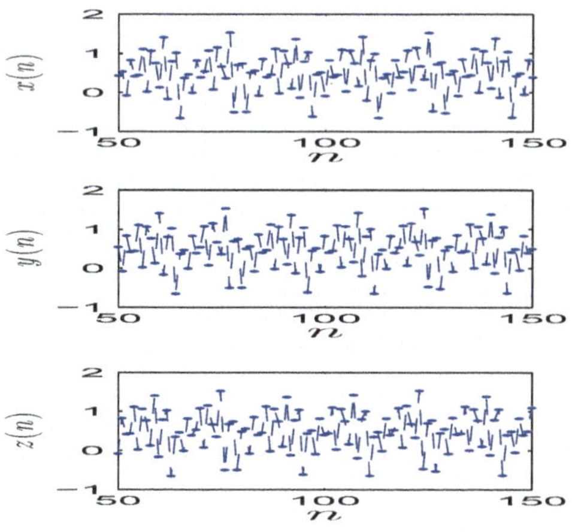

Fig. 4. Time evolution of the fractional order (1) map for $\nu = 0.9962$

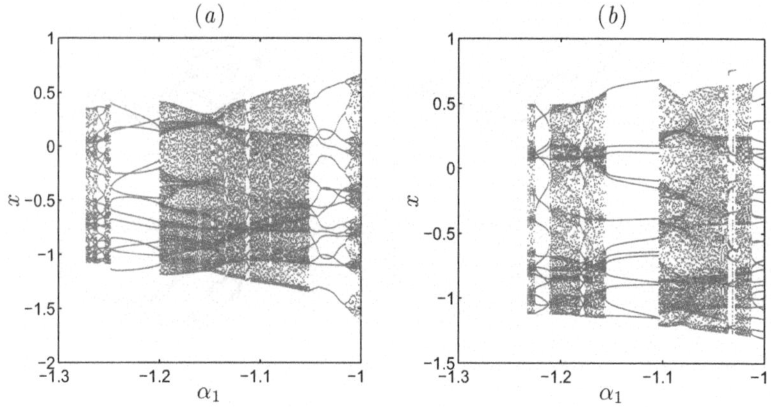

Fig. 5. Bifurcation diagrams of the fractional order map with hidden chaotic attractors versus α_1 for: (a) $\nu = 0.9962$, (b) $\nu = 0.9395$.

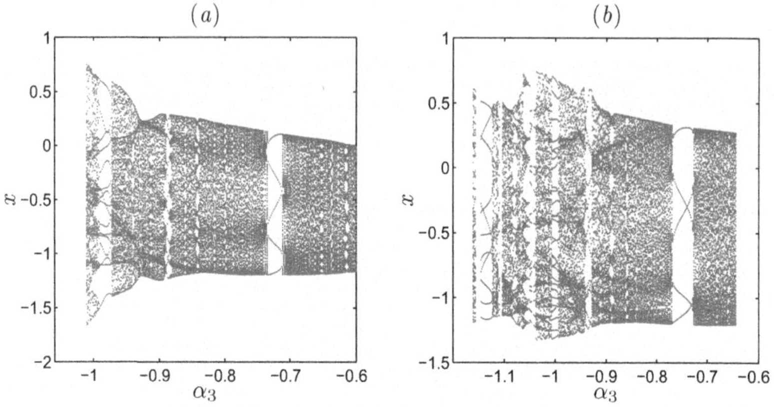

Fig. 6. Bifurcation diagrams of the fractional order map with hidden chaotic attractors versus α_3 for: (a) $\nu = 0.9962$, (b) $\nu = 0.9395$.

- **Case 2: Effect of varying system parameter α_1**
  Here, we fix the parameters $\alpha_2 = -0.7$; $\alpha_3 = -1$; $\alpha_4 = -0.7$ and vary the system parameter α_1 from -1.3 to -1 with the step size 0.01. Bifurcation diagram with two different ν are presented in Fig. 5(a) with $\nu = 0.9962$ and Fig. 5(b) with $\nu = 0.9395$. It's illustrated that the states of the discrete fractional model (1) changed according to the variation of parameter α_1. From Fig. 5, it was found that when $\nu = 0.9962$, chaotic motion exists in the interval $[-1.009, -1[\cup] - 1.077 - 1.053]$. As the value of ν decreases to 0.9395, the chaotic motion decreases to $[-1.026, -1.015]$.

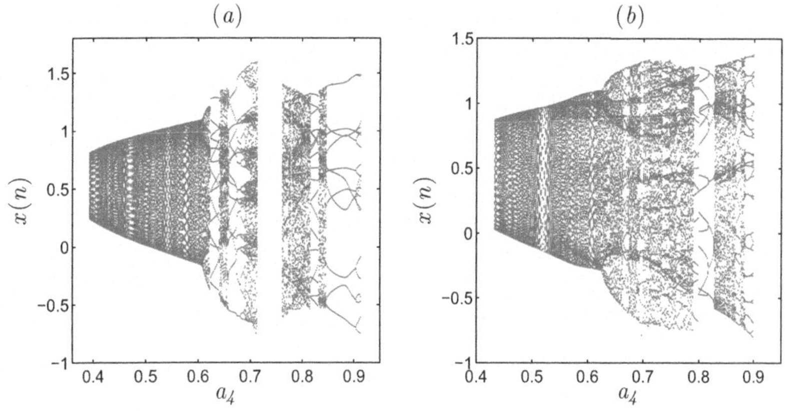

Fig. 7. Bifurcation diagrams of the fractional order map with hidden chaotic attractors versus α_4 for: (a) $\nu = 0.9962$, (b) $\nu = 0.9395$.

- **Case 3: Effect of varying system parameter** α_3
  Here, bifurcation diagrams are calculated against $\alpha_3 \in [-1.2, -0.6]$. The bifurcation analysis in Fig. 6 shows that the two diagrams are similar. Moreover, it's observed that decreasing the value of the fractional order ν from 0.9962 to 0.9395 leads to shifting the bifurcation diagram to the left.
- **Case 4: Effect of varying system parameter** α_4
  Here, bifurcation diagrams are calculated against $\alpha_4 \in [0.4, 0.9]$. Bifurcation diagrams in Fig. 7 show that chaos exist in the fractional order map (1), and periodic areas increases with decreases of order ν. In particular, when $\nu = 0.9962$ the trajectories converge into an unbounded attractor at $[0.7139, 0.7702]$, whereas when $\nu = 0.9395$, the fractional map (1) exhibits periodic behavior.

4 Conclusion

In this paper, a chaotic discrete fractional model with hidden attractors was proposed. Chaotic dynamics are observed to vary according to the fractional order ν. When $\nu \in [0.81, 0.833]$, the proposed model exhibit hidden chaotic attractors with larger values of Lyapunov exponent. Generally speaking the new map exhibits different types of attractors of different fractional orders.

References

1. Abdeljawad, T.: On Riemann and Caputo fractional differences. Comput. Math. Appl. **62**(3), 1602–1611 (2011)

2. Ammar, H.H., Azar, A.T.: Robust path tracking of mobile robot using fractional order PID controller. In: Hassanien, A.E., Azar, A.T., Gaber, T., Bhatnagar, R., Tolba, M.F. (eds.) The International Conference on Advanced Machine Learning Technologies and Applications (AMLTA 2019), pp. 370–381. Springer, Cham (2020)
3. Anastassiou, G.A.: Principles of delta fractional calculus on time scales and inequalities. Math. Comput. Model. **52**(3), 556–566 (2010)
4. Azar, A.T., Serrano, F.E.: Fractional order two degree of freedom PID controller for a robotic manipulator with a fuzzy type-2 compensator. In: Hassanien, A.E., Tolba, M.F., Shaalan, K., Azar, A.T. (eds.) Proceedings of the International Conference on Advanced Intelligent Systems and Informatics 2018, pp. 77–88. Springer, Cham (2019)
5. Azar, A.T., Vaidyanathan, S.: Advances in Chaos Theory and Intelligent Control, vol. 337. Springer, Berlin (2016)
6. Azar, A.T., Vaidyanathan, S., Ouannas, A.: Fractional Order Control and Synchronization of Chaotic Systems. Studies in Computational Intelligence, vol. 688. Springer, Berlin (2017)
7. Bendoukha, S., Ouannas, A., Wang, X., Khennaoui, A.A., Pham, V.T., Grassi, G., Huynh, V.V.: The co-existence of different synchronization types in fractional-order discrete-time chaotic systems with non-identical dimensions and orders. Entropy **20**, 1–13 (2018)
8. Djeddi, A., Dib, D., Azar, A., Abdelmalek, S.: Fractional order unknown inputs fuzzy observer for Takagi-Sugeno systems with unmeasurable premise variables. Mathematics **7**(10), 984 (2019)
9. Dudkowski, D., Jafari, S., Kapitaniak, T., Kuznetsov, N.V., Leonov, G.A., Prasad, A.: Hidden attractors in dynamical systems. Phys. Rep. **637**, 1–50 (2016)
10. Ghoudelbourk, S., Dib, D., Omeiri, A., Azar, A.T.: MPPT control in wind energy conversion systems and the application of fractional control (PIα) in pitch wind turbine. Int. J. Model. Ident. Control **26**(2), 140–151 (2016)
11. Gorripotu, T.S., Samalla, H., Jagan Mohana Rao, C., Azar, A.T., Pelusi, D.: TLBO algorithm optimized fractional-order PID controller for AGC of interconnected power system. In: Nayak, J., Abraham, A., Krishna, B.M., Chandra Sekhar, G.T., Das, A.K. (eds.) Soft Computing in Data Analytics, pp. 847–855. Springer, Singapore (2019)
12. Jafari, S., Sprott, J.C., Nazarimehr, F.: Recent new examples of hidden attractors. Eur. Phys. J. Spec. Top. **224**(8), 1469–1476 (2015)
13. Jiang, H., Liu, Y., Wei, Z., Zhang, L.: Hidden chaotic attractors in a class of two-dimensional maps. Nonlinear Dyn. **85**(4), 2719–2727 (2016)
14. Jouini, L., Ouannas, A., Khennaoui, A.A., Wang, X., Grassi, G., Pham, V.T.: The fractional form of a new three-dimensional generalized Hénon map. Adv. Differ. Equ. **122**, 1–12 (2019)
15. Khan, A., Singh, S., Azar, A.T.: Combination-combination anti-synchronization of four fractional order identical hyperchaotic systems. In: Hassanien, A.E., Azar, A.T., Gaber, T., Bhatnagar, R., Tolba, M.F. (eds.) The International Conference on Advanced Machine Learning Technologies and Applications (AMLTA 2019), pp. 406–414. Springer, Cham (2020)
16. Khan, A., Singh, S., Azar, A.T.: Synchronization between a novel integer-order hyperchaotic system and a fractional-order hyperchaotic system using tracking control. In: Hassanien, A.E., Azar, A.T., Gaber, T., Bhatnagar, R., Tolba, M.F. (eds.) The International Conference on Advanced Machine Learning Technologies and Applications (AMLTA 2019), pp. 382–391. Springer, Cham (2020)

17. Khennaoui, A.A., Ouannas, A., Bendoukha, S., Wang, X., Pham, V.T.: On chaos in the fractional-order discrete-time unified system and its control synchronization. Entropy **20**(7), 530 (2018)
18. Khennaoui, A.A., Ouannas, A., Bendoukha, S., Grassi, G., Lozi, R.P., Pham, V.T.: On fractional-order discrete-time systems: chaos, stabilization and synchronization. Chaos, Solitons Fractals **119**, 150–162 (2019)
19. Meghni, B., Dib, D., Azar, A.T., Ghoudelbourk, S., Saadoun, A.: Robust adaptive supervisory fractional order controller for optimal energy management in wind turbine with battery storage, pp. 165–202. Springer, Cham (2017)
20. Ouannas, A., Khennaoui, A.A., Bendoukha, S., Vo, T., Pham, V.T., Huynh, V.: The fractional form of the Tinkerbell map is chaotic. Appl. Sci. **8**(12), 2640 (2018)
21. Ouannas, A., Khennaoui, A.A., Grassi, G., Bendoukha, S.: On the Q–S chaos synchronization of fractional-order discrete-time systems: general method and examples. Discrete Dyn. Nat. Soc. **2018**, 1–8 (2018). Article ID 2950357
22. Ouannas, A., Grassi, G., Azar, A.T., Singh, S.: New control schemes for fractional chaos synchronization. In: Hassanien, A.E., Tolba, M.F., Shaalan, K., Azar, A.T. (eds.) Proceedings of the International Conference on Advanced Intelligent Systems and Informatics 2018, pp. 52–63. Springer, Cham (2019)
23. Ouannas, A., Khennaoui, A.A., Bendoukha, S., Grassi, G.: On the dynamics and control of a fractional form of the discrete double scroll. Int. J. Bifurcat. Chaos **29**(06), 1950078 (2019)
24. Ouannas, A., Khennaoui, A.A., Grassi, G., Bendoukha, S.: On chaos in the fractional-order Grassi-Miller map and its control. J. Comput. Appl. Math. **358**, 293–305 (2019)
25. Ouannas, A., Khennaoui, A.A., Odibat, Z., Pham, V.T., Grassi, G.: On the dynamics, control and synchronization of fractional-order Ikeda map. Chaos, Solitons Fractals **123**, 108–115 (2019)
26. Ouannas, A., Grassi, G., Azar, A.T.: Fractional-order control scheme for Q-S chaos synchronization. In: Hassanien, A.E., Azar, A.T., Gaber, T., Bhatnagar, R., Tolba, M.F. (eds.) The International Conference on Advanced Machine Learning Technologies and Applications (AMLTA 2019), pp. 434–441. Springer, Cham (2020)
27. Ouannas, A., Grassi, G., Azar, A.T.: A new generalized synchronization scheme to control fractional chaotic systems with non-identical dimensions and different orders. In: Hassanien, A.E., Azar, A.T., Gaber, T., Bhatnagar, R., Tolba, M.F.(eds.) The International Conference on Advanced Machine Learning Technologies and Applications (AMLTA 2019), pp. 415–424. Springer, Cham (2020)
28. Ouannas, A., Grassi, G., Azar, A.T., Khennaouia, A.A., Pham, V.T.: Chaotic control in fractional-order discrete-time systems. In: Hassanien, A.E., Shaalan, K., Tolba, M.F. (eds.) Proceedings of the International Conference on Advanced Intelligent Systems and Informatics 2019, pp. 207–217. Springer, Cham (2020)
29. Pham, V.T., Jafari, S., Volos, C., Kapitaniak, T.: Different families of hidden attractors in a new chaotic system with variable equilibrium. Int. J. Bifurcat. Chaos **27**(09), 1750138 (2017)
30. Wu, G.C., Baleanu, D.: Discrete chaos in fractional delayed logistic maps. Nonlinear Dyn. **80**(4), 1697–1703 (2015)
31. Wu, G.C., Baleanu, D., Xie, H.P., Chen, F.L.: Chaos synchronization of fractional chaotic maps based on the stability condition. Physica A Stat. Mech. Appl. **460**, 374–383 (2016)

A Comparative Analysis of Different Feature Extraction Techniques for Motor Imagery Based BCI System

Sahar Selim[1](✉), Manal Tantawi[2], Howida Shedeed[2], and Amr Badr[3]

[1] Faculty of Media Engineering and Technology, German University in Cairo, Cairo, Egypt
s.selim@grad.fci-cu.edu.eg
[2] Faculty of Computer and Information Science, Ain Shams University, Cairo, Egypt
[3] Faculty of Computers and Artificial Intelligence, Cairo University, Giza, Egypt

Abstract. Brain Computer Interface (BCI) system based on Motor Imagery (MI) signals is one among the most prospering systems in the field of BCI. Motor imagination does not involve motor output from the human. This can be used to help motor disabled people to accomplish elementary tasks by themselves. One of the primary difficulties facing MI-based BCIs is extracting discriminative features from the EEG signal. This study seeks to minimize the number of features input to classifier by finding out the best feature extraction techniques that can achieve high system accuracy with minimum computation cost. To achieve this purpose, this study compares five different feature extraction techniques; Root Mean Square (RMS), Renyi entropy, Shannon entropy, Katz fractal dimension and Common Spatial Patterns (CSP), regarding accuracy and execution time. To ensure that these techniques are reliable; they are applied on three benchmark datasets; BCI Competition III datasets IVa and IIIa, and BCI Competition IV dataset IIa. The extracted features are examined with two classifiers; Support Vector Machine (SVM) and Linear Discriminant Analysis (LDA). Experimental results showed that CSP is the best feature extraction technique compared to other examined techniques, yet Renyi entropy has the least computation time which is a critical issue for an online system.

Keywords: Brain Computer Interface · Root Mean Square · Renyi entropy · Shannon entropy · Katz FD · Common Spatial Pattern · Feature extraction

1 Introduction

Brain Computer Interface (BCI) acts as an output channel to help disabled patients to do primitive tasks by themselves. It employs control signals emitted from electroencephalography (EEG) brain activity [1]. Motor Imagery based BCIs (MI-BCIs) relies on motor imagery (MI) EEG signals. These signals are emitted from the brain when a person imagines the motor action, whether he/she actually performs the action or not [2]. This can be used by motor disabled people to control devices like a wheelchair or

A.-E. Hassanien et al. (Eds.): AICV 2020, AISC 1153, pp. 740–749, 2020.
https://doi.org/10.1007/978-3-030-44289-7_69

robotic arm [3]. For an MI-BCI system to be used in everyday life outside of the laboratory environment, it must be highly reliable and robust.

Feature extraction and classification are essential components in the BCI system. Too many features can cause the classifier to be unable to classify the input signal correctly. Therefore, a feature extraction method is needed to minimize the number of features input to the classifier. There are various commonly used feature extraction techniques for BCI systems. Wavelet was used widely in many studies to extract features from the EEG signal [2]. Several studies used Root Mean Square (RMS) feature for dimension reduction [4, 5]. Common Spatial Patterns (CSP) is considered a robust feature extractor for a binary classification problem in MI systems as illustrated in [6–9]. Renyi entropy was used recently by researchers as an alternative to CSP [10, 11]. In [12], they examined Higuchi, Katz and Renyi entropy with the fuzzy k-nearest neighbor, SVM and LDA classifiers for the binary classification problem, where their results showed that Katz outperformed other fractal dimension methods. The selection of classifier plays a key role in the efficiency of a BCI system. Among the commonly used classifiers in BCI systems are Support Vector Machine [9, 11, 13], Linear Discriminant Analysis [7, 13], Naïve Bayes [6, 13], Hidden Markov Model [14], k-Nearest Neighbors [15], Artificial Neural Network [16]. This study focuses mainly on the feature extraction process. To the best of our knowledge, Shannon entropy is employed for the first time as a feature extraction technique for MI-BCI system in this study. Also, the comparison of the examined feature extraction techniques is held for the first time in one study.

2 Materials and Methods

2.1 Methodology

The MI-BCI system encloses preprocessing (noise removal, band-pass filtering, and epoch extraction), feature extraction, dimension reduction, and classification. The workflow of the proposed methodology is shown in Fig. 1. This approach is applied to the three motor imagery datasets. Only the left and right hand signals of BCI Competition IV dataset IIa and BCI Competition III dataset IVa are used, whereas, for BCI Competition III dataset IVa, the right hand and foot are utilized. Channel selection is applied for BCIC III dataset IVa which is recorded from 118 channels. For the three datasets, the EEG signals are filtered with a fifth-order Butterworth filter between 10 Hz and 30 Hz for the acquisition of mu (μ) and beta (β) frequency bands. Epochs are extracted by segmenting the filtered signal (as used by most studies for these three datasets) [2.5–5.5] s for dataset BCIC IV dataset IIa, [3–6] s for BCIC III dataset IIIa and [0.5–3.5] s for BCIC III dataset IVa [17]. Five different feature extraction techniques are applied to extracted epochs; Root Mean Square (RMS), Renyi entropy, Common Spatial Patterns (CSP), Shannon entropy, and Katz fractal dimension. Two linear classifiers (SVM and LDA) are employed to assist the reliability of outcomes acquired from the five techniques. For evaluation purposes, classification accuracy and computation cost are considered to be the key metrics.

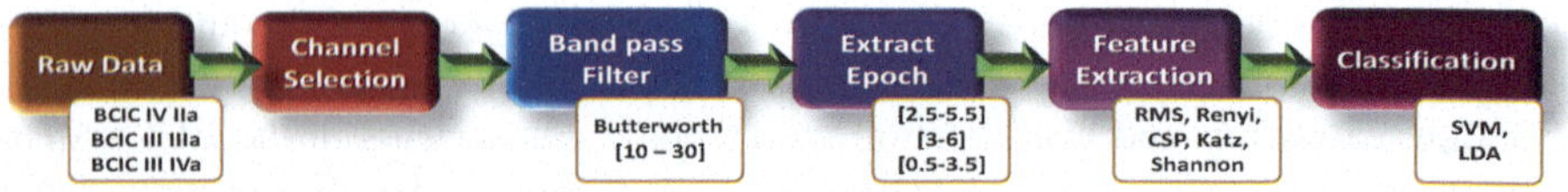

Fig. 1. Proposed methodology.

2.2 Datasets

In this study, BCIC III dataset IVa, BCIC III dataset IIIa, and BCIC IV dataset IIa are used. Only the left and right hand signals of BCIC IV dataset IIa and BCIC III dataset IIIa are included.

BCI Competition III Dataset IVa. The dataset is recorded with a sampling rate of 100 Hz from 118 electrodes. The signals of the left hand and foot were captured from five subjects (aa, al, av, aw, ay). Every subject has 280 trials, which are unevenly distributed among training and testing [18]. This study employs only the 18 sensorimotor cortex channels proposed by Wang et al. [19] which are; "C5, C3, C1, C2, C4, C6, CP5, CP3, CP1, CP2, CP4, CP6, P5, P3, P1, P2, P4, and P6".

BCI Competition III Dataset IIIa. The data is captured from three subjects, sampled at 250 Hz, through 60 electrodes [20]. The subjects performed four motions; left, right hand, foot, and tongue. For each class of subject k3b, 90 trials are recorded. Meanwhile, 60 trials are recorded for each class of subjects k6b and l1b [20].

BCI Competition IV Dataset IIa. The signals are captured from 22 electrodes for nine subjects, performing motor imagery signal of the left, right hand, foot, and tongue, and sampled at 250 Hz. Each session of training and testing includes 288 trials divided equally on the four class [21].

2.3 Feature Extraction

One of the weaknesses facing classification techniques is not being able to function efficiently with high dimension data. The generalization capacity of any learning process decreases with the increase of the input data dimension. Moreover, the increase in the number of input data channels causes a boost in the complexity and execution time of the used algorithms. Reaching informative features can lead to better classification accuracy. Accordingly, the BCI system should acquire these informative features without excess in computational costs, which is a key problem for real-time BCI systems [4]. Thus, dimension reduction techniques are required to decrease the dimension of the input vector to a classifier. The following subsections present the feature extraction techniques examined in this study.

Root Mean Square (RMS). It is represented as a random Gaussian amplitude-modulated method Eq. (1). The efficiency of this feature is obtained by computing the average power of the motor imagery signal.

$$RMS_t = \sqrt{\frac{1}{N}\sum\nolimits_{i=1}^{N} x_i^2} \tag{1}$$

where t is the present epoch, N represents the number of epochs, x_i identifies the present signal point and finally i represents the present point index.

Renyi Entropy. Renyi utilizes brain activity to distinguish the complexity of time series. Thus, measuring the chaoticity in EEG signal [22] as defined in Eq. (2).

$$R_q = \frac{1}{1-q}\log_2\left(\sum\nolimits_{i=1}^{n} p_i^q\right) \tag{2}$$

q represents the order of Renyi, $q > 0$ where $q \neq 1$. p_i is the resulting probability of distribution of $i = 1, \ldots, n$.

Shannon Entropy. Shannon entropy is a unique case of Renyi, where $q = 1$. It measures the degree of irregularity in the system. It can detect changes in the signal spectrum and amplitude. Given X is a finite discrete variable where $X = (x_1, x_2, \ldots, x_n), x_i \in R_d$ and Shannon entropy is given in Eq. (3) [23]:

$$\mathrm{H(x)} = -\mathrm{c}\sum\nolimits_{i=1}^{n} p(x_i)\ln p(x_i) \tag{3}$$

where c is a positive constant serving as a measuring unit and $p(x_i)$ is the probability of $x_i \in X$.

Katz Fractal Dimension. Katz proposed the Fractal Dimension (FD) as the ratio of the curve length L, which is measured as the sum of Euclidean distances between two consecutive points, divided by maximum distance d far from the first point [24]. FD is given by the following equation:

$$FD = \frac{\log_{10}(L)}{\log_{10}(d)} \tag{4}$$

Where L is the Euclidean distance between two consecutive points as follows:

$$L = \sum\nolimits_{i=1}^{N} dist(s_i, s_{i+1}), i = 1, \ldots, N-1 \tag{5}$$

The distance between the first and furthest points in the sequence is defined as the diameter of the curve d defined by:

$$d = \max(dist(s_i, s_{i+1})), i = 1, \ldots, N-1) \tag{6}$$

Common Spatial Patterns (CSP). It can be briefly outlined as follows: Spatial Filters are implemented on a two-class problem, to satisfy the maximization of the variance of one class whereas the variance of the other is minimized. c represents the two classes. Let $X^1, \ldots, X^m \in \mathbb{R}^{n \times l}$ represents the trials of class x, and $Y^1, \ldots, Y^m \in \mathbb{R}^{n \times l}$ represents the trials of class y, n denotes the number of channels while l denotes the total number of samples in a single trial. The sample covariance matrix S is the normalized form of the total variance of a given trial E, as revealed in Eq. (7) where the transpose of matrix is denoted by "T", and the matrix trace is given by $tr(\cdot)$ [8].

$$S = \frac{EE^T}{tr(EE^T)} \tag{7}$$

2.4 Classification

Linear Discriminant Analysis (LDA). LDA accomplished remarkable success in the classification of various mental tasks [25]. The distribution of data is supposed to be Gaussian with equivalent covariance matrices for the two classes [26]. The scattering of the middle and the inner classes are acquired by the following equations:

$$S_w = \sum\nolimits_j p_j \times \left(cov_j\right) \tag{8}$$

$$S_b = \sum\nolimits_j \left(\mu_j \times \mu_3\right) \times \left(\mu_j \times \mu_3\right)^T \tag{9}$$

where S_w is the scatter within the class and S_b is the scatter between classes, p_j is the parameter of prior probability, cov_j represents the covariance matrix of class j. μ is considered the total average of classes, μ_j is the average of class μ_j.

Support Vector Machine (SVM). SVM identifies the optimal hyperplane that can differentiate between two classes. The decision function of the hyperplane is shown in Eq. (10) [27].

$$f(x) = sign\left(\sum\nolimits_{i=1}^{N} \alpha_i y_i K(x_i, x) + b\right) \tag{10}$$

where α denotes the Lagrange multipliers which varies between $0 \le \alpha^i \le C, i = 1, 2, \ldots, N$, C represents the penalty parameter while $K(x_i, x)$ is the kernel function of the support vectors.

2.5 Evaluation

The efficiency of the implemented approach is assessed by classification accuracy which is calculated as shown in Eq. (11).

$$ACC = \frac{Correctly\ Classified\ Samples}{Total\ Number\ of\ Samples} \tag{11}$$

3 Experimental Results

This study aims to extract a minimum number of the most discriminant features by comparing the effectiveness of five different feature extraction techniques; RMS, Renyi entropy, Shannon entropy, Katz FD and CSP. Using RMS, Renyi, Shannon or Katz FD have the advantage of reducing feature vector to only one dimension vector, whereas, the CSP has a higher dimension vector (the top eight features were selected) compared to others. These techniques are examined with three datasets; BCIC III dataset IVa, BCIC III dataset IIIa, and BCIC IV dataset IIa. Furthermore, the loading of data, feature extraction and classification time are presented in milliseconds.

BCIC III Dataset IVa. Based on the study of Selim et al. [5], which compared different combinations of channels, only 18 channels were used instead of 118 channels for this dataset. The five feature extraction techniques are applied to the aforementioned selected channels. RMS, Renyi, Shannon and Katz FD can achieve dimension reduction by depicting the 18 channels of the input signal by only one feature. Results in Table 1 show that Renyi has nearly the same accuracy as RMS, where RMS has minimum computation time. The Shannon entropy gave less accuracy compared to Renyi and RMS, especially with LDA. As illustrated in Fig. 2, RMS, Renyi, and CSP gave comparable results, yet, CSP gave minimum execution time among all techniques for this dataset. On the contrary, Katz FD has the disadvantage of having the highest computation time with less accuracy compared to others.

Table 1. Results of feature extraction techniques for BCIC III dataset IVa

Technique	RMS		Renyi		Shannon		Katz FD		CSP	
Subject	SVM	LDA	SVM	LDA	SVM	LDA	SVM	LDA	SVM	LDA
aa	66.96	69.64	68.75	66.96	66.96	69.64	69.64	67.86	79.46	79.46
al	92.86	89.29	89.29	92.86	89.29	91.07	94.64	92.86	94.64	100
av	61.22	59.18	59.69	60.20	61.22	56.63	58.67	55.61	54.08	57.14
aw	80.80	88.84	83.04	87.95	80.36	79.02	77.68	80.8	91.52	92.41
ay	86.11	86.90	87.30	82.54	84.52	75.79	80.95	78.57	70.41	69.84
Accuracy	77.59	78.77	77.61	78.10	76.47	74.43	76.32	75.14	78.02	79.77
Time	14.16	14.00	33.60	33.02	35.84	35.65	58.32	58.11	10.01	09.93

BCIC III Dataset IIIa. In this experiment, all the set of 60 channels of this dataset were used. As presented in Table 2, SVM gave the best average classification accuracy with RMS, achieving approximately 15% higher than LDA with nearly the same computation time. The SVM classifier also outperformed LDA when using Renyi entropy. Using Renyi with SVM enhanced accuracy by $\approx$5% when compared to RMS. The Shannon entropy gave less accuracy compared to RMS and Renyi with higher computation time. As shown in Fig. 3, the SVM classifier outperformed LDA for almost all feature extraction techniques. CSP gave the best results in comparison with all other examined techniques. Meanwhile, Renyi gave the least execution time compared to other techniques. Though Katz has comparable results to Renyi with the SVM classifier, yet it has a remarkable increase in computation time.

Table 2. Results of feature extraction techniques for BCIC III dataset IIIa

Technique	RMS		Renyi		Shannon		Katz FD		CSP	
Subject	SVM	LDA	SVM	LDA	SVM	LDA	SVM	LDA	SVM	LDA
K3b	91.11	80.00	92.22	87.78	88.89	73.33	92.22	87.78	98.89	97.78
K6b	66.67	48.33	75.00	36.67	60.00	63.33	76.67	46.67	68.33	71.67
L1b	76.67	61.67	81.67	63.33	81.67	45.00	80.00	51.67	90.00	90.00
Accuracy	78.15	63.33	82.96	62.59	76.85	60.55	82.96	62.04	85.74	86.48
Time	10.56	10.38	09.81	09.57	12.19	12.01	157.01	156.88	10.74	10.76

BCIC IV Dataset IIa. Table 3 presents the results of the five techniques where Renyi with the SVM classifier gave higher results compared to RMS. Katz FD achieved results similar to that of Renyi and RMS with SVM, yet, it suffers from high computation time. As concluded from Fig. 4, the SVM is the best classifier for almost all techniques. CSP and Renyi gave the best classification accuracies, while Shannon did not achieve any significant results. By comparing the feature extraction time of all examined techniques, it is observed that Renyi has the least computation time, while Katz FD has the worst computation time compared to others.

Table 3. Results of feature extraction techniques for BCIC IV dataset IIa

Technique	RMS		Renyi		Shannon		Katz FD		CSP	
Subject	SVM	LDA	SVM	LDA	SVM	LDA	SVM	LDA	SVM	LDA
1	91.61	90.91	93.01	93.01	91.61	83.92	84.63	74.86	90.62	93.01
2	58.45	63.38	59.15	61.27	61.27	64.08	61.27	59.15	52.71	51.41
3	93.71	92.31	92.31	93.01	94.41	90.91	88.11	86.71	93.31	92.41
4	70.83	76.39	73.53	75.08	67.36	70.83	77.08	76.39	68.06	66.67
5	53.52	53.52	60.56	53.52	58.45	59.15	64.79	61.27	59.15	59.86
6	63.64	63.64	62.24	65.73	61.54	60.84	67.13	65.73	74.73	75.52
7	61.81	66.67	69.19	68.06	58.33	60.42	65.97	70.83	78.47	77.78
8	95.14	95.14	96.53	93.06	92.36	93.06	92.36	92.36	97.22	95.31
9	86.11	84.72	88.19	84.03	86.81	84.72	86.11	83.33	84.03	80.11
Accuracy	74.98	76.30	77.19	76.31	74.68	74.21	76.38	74.51	77.59	76.90
Time	13.38	12.14	8.31	7.26	9.96	9.00	141.32	140.33	12.93	12.63

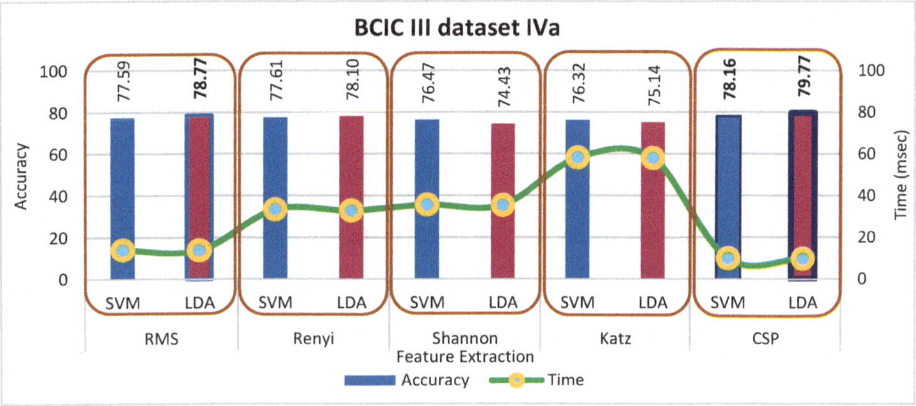

Fig. 2. Accuracy & execution time of the five techniques for BCIC III dataset IVa

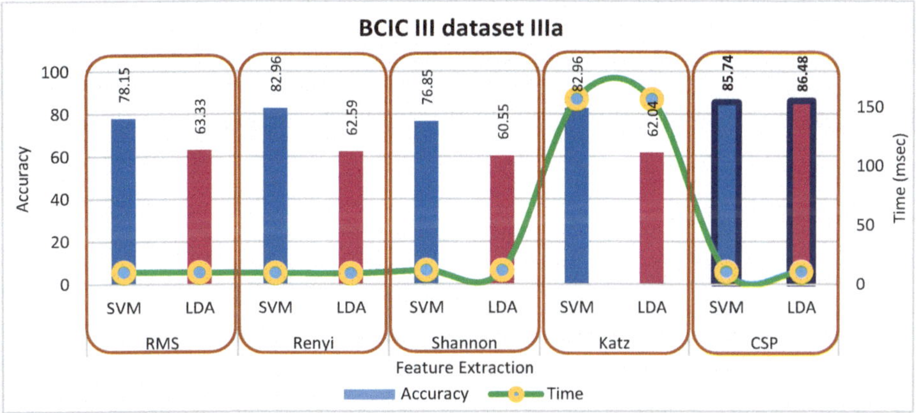

Fig. 3. Accuracy & execution time of the five techniques for BCIC III dataset III

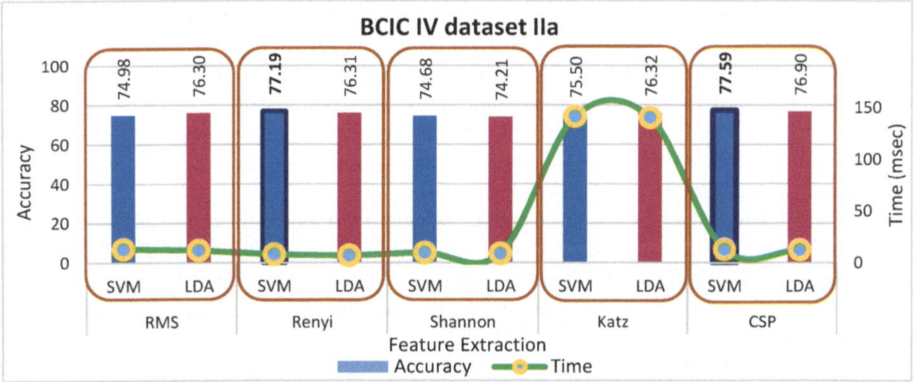

Fig. 4. Accuracy & execution time of the five techniques for BCIC IV dataset IIa

4 Conclusion

The high dimension of the input signal is one of the major problems in the motor imagery based BCI systems. This study aims to reduce the dimension of the processed signal by finding out the most effective features and thereby, reducing computational costs, that are essential for real-time BCIs. Hence, this study compares the results of applying the five feature extraction techniques (RMS, Renyi, Shannon, Katz FD, and CSP) on three datasets (BCIC IV-IIa, BCIC III-IIIa, and BCIC III-IVa), using two linear classifiers (SVM and LDA). This study examines Shannon entropy for the first time as a feature extraction technique in MI-BCI. The efficiency of these methods is assessed with regard to classification accuracy and execution time.

The RMS and the three entropies have an advantage over CSP. They reduce any input dimension to only one feature. The results of the three datasets illustrated that CSP gave the best results followed by Renyi compared to other techniques. Renyi proved its superiority in having minimum computation cost with comparable results to other techniques. Though CSP has slightly higher computational time compared to Renyi, yet it is still the best feature extraction technique for a binary classification problem. For real-time BCIs it is advisable to maintain a balance between accuracy and speed. Therefore, CSP and Renyi will be compared with the multi-class problem in future work.

References

1. Kim, Y.K., Park, E., Lee, A., Im, C.H., Kim, Y.-H.: Changes in network connectivity during motor imagery and execution. PLoS One **13**, 1–18 (2018)
2. Kumar, S.U., Inbarani, H.H.: PSO-based feature selection and neighborhood rough set-based classification for BCI multiclass motor imagery task. Neural Comput. Appl. **28**, 3239–3258 (2016)
3. Ofner, P., Schwarz, A., Pereira, J., Wyss, D., Wildburger, R., Müller-Putz, G.R.: Attempted arm and hand movements can be decoded from low-frequency EEG from persons with spinal cord injury. Sci. Rep. **9**, 7134 (2019)
4. Hamedi, M., Salleh, S.-H., Noor, A.M., Mohammad-Rezazadeh, I.: Neural network-based three-class motor imagery classification using time-domain features for BCI applications. In: Presented at the 2014 IEEE Region 10 Symposium (2014)
5. Selim, S., Tantawi, M., Shedeed, H., Badr, A.: Reducing execution time for real-time motor imagery based BCI systems. In: Hassanien, A., Shaalan, K., Gaber, T., Azar, A., Tolba, M. (eds.) Proceedings of the International Conference on Advanced Intelligent Systems and Informatics 2016, AISI 2016. Advances in Intelligent Systems and Computing, vol. 533. Springer, Cham (2017)
6. Ang, K.K., Chin, Z.Y., Wang, C., Guan, C., Zhang, H.: Filter bank common spatial pattern algorithm on BCI competition IV Datasets 2a and 2b. Front. Neurosci. **6**, 39 (2012)
7. Lindig-León, C., Bougrain, L.: A multi-label classification method for detection of combined motor imageries. In: IEEE International Conference on Systems, Man, and Cybernetics, Hong Kong, China, pp. 3128–3133 (2015)
8. Ramoser, H., Muller-Gerking, J., Pfurtscheller, G.: Optimal spatial filtering of single trial EEG during imagined hand movement. IEEE Trans. Rehabil. Eng. **8**, 441–446 (2000)

9. Selim, S., Tantawi, M., Shedeed, H., Badr, A.: A CSP\AM-BA-SVM approach for motor imagery BCI system. IEEE Access **6**, 49192–49208 (2018)
10. Kee, C.-Y., Ponnambalam, S.G., Loo, C.-K.: Binary and multi-class motor imagery using Renyi entropy for feature extraction. Neural Comput. Appl. **28**, 2051–2062 (2017)
11. Selim, S., Tantawi, M., Shedeed, H., Badr, A.: Comparing multi-class approaches for motor imagery using Renyi entropy. In: Hassanien, A., Tolba, M., Shaalan, K., Azar, A. (eds.) Proceedings of the International Conference on Advanced Intelligent Systems and Informatics 2018, AISI 2018. Advances in Intelligent Systems and Computing, vol. 845. Springer, Cham (2019)
12. Loo, C.K., Samraj, A., Lee, G.C.: Evaluation of methods for estimating fractal dimension in motor imagery-based brain computer interface. Discrete Dyn. Nat. Soc. **2011**, 8 (2011)
13. Krishna, D.H., Pasha, I.A., Savithri, T.S.: Classification of EEG motor imagery multi class signals based on cross correlation. Procedia Comput. Sci. **85**, 490–495 (2016)
14. Saa, J.F.D., Çetin, M.: Discriminative methods for classification of asynchronous imaginary motor tasks from EEG data. IEEE Trans. Neural Syst. Rehabil. Eng. **21**, 1–8 (2012)
15. Mirnaziri, M., Rahimi, M., Alavikakhaki, S., Ebrahimpour, R.: Using Combination of μ, β and γ bands in classification of EEG signals. Neuroscience **4**, 76–87 (2013)
16. Singla, R., Sharma, N.: Function classification of EEG signals based on ANN. Int. J. Soft Comput. Eng. **3**, 158–163 (2014)
17. Bai, X., Wang, X., Zheng, S., Yu, M.: The offline feature extraction of four-class motor imagery EEG based on ICA and Wavelet-CSP. In: The 33rd Chinese Control Conference, Nanjing, China (2014)
18. Müller, K.-R., Blankertz, B.: BCI Competition III Dataset IVa. http://www.bbci.de/competition/iii/
19. Wang, S., James, C.J.: Extracting rhythmic brain activity for braincomputer interfacing through constrained independent component analysis. Comput. Intell. Neurosci. **2007**, 1–9 (2007)
20. Pfurtscheller, G., Schlögl, A.: BCI Competition III Dataset IIIa. http://www.bbci.de/competition/iii/
21. Tangermann, M., Müller, K.R., Aertsen, A., Birbaumer, N., Braun, C., Brunner, C., et al.: Review of the BCI competition IV. Front. Neurosci. **6**, 55 (2012)
22. Rényi, A.: On a new axiomatic theory of probability. Acta Mathematica Academiae Scientiarum Hungaricae **6**, 285–335 (1955)
23. Kannathal, N., Choo, M.L., Acharya, U.R., Sadasivan, P.K.: Entropies for detection of epilepsy in EEG. Comput. Methods Programs Biomed. **80**, 187–194 (2005)
24. Katz, M.J.: Fractals and the analysis of waveforms. Comput. Biol. Med. **18**, 145–156 (1988)
25. Bashashati, A., Fatourechi, M., Ward, R.K., Birch, G.E.: A survey of signal processing algorithms in brain–computer interfaces based on electrical brain signals. J. Neural Eng. **4**, R32–R57 (2007)
26. Balakrishnama, S., Ganapathiraju, A.: Linear Discriminant Analysis - A brief tutorial. Institute for Signal and Information Processing (1998)
27. Vapnik, V.N.: The Nature of Statistical Learning Theory. Springer, Wiley, New York (1995)

Reversible Watermarking for Protecting Patient's Data Privacy Using an EPR-Generated QR Code

Alaa H. ElSaadawy[1(✉)], Ahmed S. ELSayed[1], M. N. Al-Berry[1], and Mohamed Roushdy[2]

[1] Faculty of Computer and Information Sciences, Ain Shams University, Cairo, Egypt
alaa_elsaadawy@cis.asu.edu.eg
[2] Faculty of Computers and Information Technology, Future University in Egypt, New Cairo, Egypt

Abstract. As a result of the modern communication technology and the transmission of medical images from one place to another, watermarking has become urgent to protect copyright, authentication and integrity of medical images and patient's information. This paper proposes a solution to the authentication and integrity problems by proposing a fragile reversible watermarking scheme. The proposed scheme generates a Quick Response (QR) code from the patient's data and uses it as a watermark hidden in the medical image. The proposed scheme resulted in an average Peak Signal to Noise Ratio (PSNR) of 51.25 dB for a payload of one bit per pixel (1 bpp), Structural Similarity Index Measure (SSIM) of 0.99 and Mean Squared Error (MSE) of 0.488 indicating that the watermarked images obtained are of high visual quality.

Keywords: Watermarking · QR code · Fragile · Medical imaging · Tamper detection

1 Introduction

In recent years, modern communication technologies availability increased with low price, which leads businesses to change their way of communication. The medical sciences filed has been also affected by this change and now doctors exchange patients' data via the internet [1, 2].

Securing the patient's documents against any act of tampering by unauthorized persons is the main priority in handling medical images [3]. Confidentiality and authenticity are the two main issues when transferring a patient's data from one place to another. Confidentiality means to be sure that only authorized users access the patient's data while authenticity is to be sure that the data belongs to the correct patient [1, 4]. Thus, developing standard solutions to preserve the authenticity and confidentiality of medical images and patient's information is the main goal of the electronic medical systems [3, 5, 6].

A.-E. Hassanien et al. (Eds.): AICV 2020, AISC 1153, pp. 750–761, 2020.
https://doi.org/10.1007/978-3-030-44289-7_70

Hiding data or adding a watermark is one of the methods used in medical image security. Watermarking information is usually inserted in the medical image in binary format using the sender. On the other side, this information is extracted by the receiver to check whether the transmitted image is hacked or not [3, 7]. Watermarking algorithms are used to detect if there is any tampering done on the transmitted image by unauthorized users [1].

For confidentiality and authenticity, an Electronic Patient Record (EPR) is embedded in the medical image [1, 8–10]. The embedded EPR should be imperceptible in the medical image. Doctors can retrieve the EPR data from the transmitted embedded medical image and use it in diagnosis.

Watermarking algorithms can be classified based on different perspectives. For example, based on the domain in which the medical image is processed, they can be classified into spatial and transform domain techniques. In the spatial domain, embedding the watermark is fast and simple, as the watermark is embedded directly in the medical image pixels [11, 12]. In the transform domain, the medical image is transformed to another domain using a suitable transform such as Integer Wavelet Transform (IWT) [13], Discrete Wavelet Transform (DWT) [14] and Discrete Cosine Transform (DCT) [15], before embedding the watermark [1, 16]. Watermarking techniques can also be classified based on the visibility of the watermark. The watermark can be visible or invisible. Visible watermarks are commonly used as logos for copyright protection, while invisible watermarks are used for authentication and integrity [1, 17]. Invisible watermarking can be classified according to its robustness into four categories: fragile, semi-fragile, robust and hybrid [18, 19]. In case the watermark is easily destroyed by any modification, it is considered a fragile watermark. semi-fragile watermarks cannot protect medical images from malicious attacks like noise addition and geometrical distortion, but it can protect them from unintentional attacks like cropping and resizing. Robust watermarking can be verified against different attacks. Finally, hybrid watermarking is used in authentication, integrity verification and copyright protection. It is a mixture of robust and fragile techniques [3, 18, 20].

Reversibility is another criterion of classification. Reversible watermarking techniques can restore hidden data and medical images without any distortion. They provide tamper detection and authentication [1, 21]. In irreversible techniques, the extracted watermark is different from the original one.

In this paper, a watermarking technique is used to protect medical images. The technique is fragile to help in detecting any manipulation in the original image. It is also reversible, invisible and works in the spatial domain. The rest of the paper is organized as follows: Sect. 2 presents the literature review, Sect. 3 explains the proposed technique, Sect. 4 shows the experimental results and finally Sect. 5 contains the conclusions and future work.

2 Literature Review

Spatial domain and transform domain are the two main categories of watermarking algorithms [3]. There are several types of research works that have been done in both categories. Most of the works are frequency domain-based watermarking and some of them are spatial domain-based that depend on directly changing the value of the image pixels [22, 23]. Copyright protection, data hiding, copy protection and integrity check are the common objectives of watermarking techniques [24]. A brief literature review of watermarking techniques for medical images is presented in this section.

Shehab et al. [25] presented a study of a fragile Singular Value Decomposition (SVD)-based watermarking technique. Their proposed scheme was used to protect sensitive medical images by tamper localization and self-recovery. Two approaches were implemented where they used block authentication in the first approach and block information in the second approach. They showed that neighbouring pixel information hides at a distant location and provides better self-recovery. They used Arnold transform [26] to find out the embedding position, which made the scheme more reliable and secure for hiding image information.

Abbasi et al. [1] developed a lossless data-hiding algorithm based on dividing the medical image into the Region Of Interest (ROI) and the Region Of Non-Interest (RONI). Binary Pattern (BP) watermark was embedded in the spatial domain of the ROI. To embed the BP in ROI the first bit of the ROI was replaced with the BP using a simple Least Significant Bit (LSB). BP watermark was used to check the integrity of the ROI area as it helped in the identification of the tampered area in the ROI. A Composed Watermark (CW), which is a concatenation of the patient record, hospital logo, doctor's ID and the Least Significant Bit (LSB) information, was used in the authentication of the medical image. The CW watermark was embedded in the frequency domain of the RONI. To embed the CW in the RONI, first the wavelet horizontal (HL), vertical (LH) and diagonal (HH) coefficients were obtained by applying IWT to the RONI, then the HL, LH and HH coefficients were compressed by using Arithmetic Encoding (AE) technique to get the region for embedding payload. Finally, the CW was embedded in the RONI in addition to the compressed HL, LH and HH coefficients. The method has been tested on 65 slices captured of patients and a PSNR of 31.6209 was recorded.

A region-based multiple watermarking algorithm using DWT and DCT was proposed by Sharma et al. [27]. The medical image was divided into ROI and RONI. Rivest–Shamir–Adleman (RSA) cryptography algorithm was used to embed the EPR in the RONI. In order to make the text watermark robust against signal processing attacks, error correction codes were applied to the encoded EPR data. Message Digest 5 (MD5) was used to hash the watermark image before embedding it in the ROI. To protect the generated watermarked image from unauthorized access, an encryption step was applied to it. The reported scheme was tested on Magnetic Resonance Imaging (MRI), Computed Tomography Scan (CT-Scan) and ultrasonic images. PSNR was used to evaluate the proposed method using with an image size of 512×512 and a watermark size of 256×256. The average value of PSNR was 51.833 for MRI images and 50.39 for CT-Scan images.

Gull et al. [24] proposed a fragile watermarking technique, which was capable of tamper detection and localization. The input of the reported technique was the medical image and the watermark image both in greyscale. First, the medical image was divided into 4×4 non-overlapping blocks, then the watermark pixels were encrypted. The LSB algorithm was then used to embed the encrypted watermark in the upper half of the block and the mean of the block was embedded in the lower half of the block.

Their algorithm has been tested using 14 medical images of size 265 × 265 and a grey image of size 64 × 64 was used as a watermark. They obtained an average PSNR of 51.26 dB and an SSIM of 0.9950.

QR code was used as a watermark image by Kavitha and Shan [28] to reduce the data payload. Firstly, the medical image was decomposed into four frequency components Approximation Coefficients (CA), Horizontal Coefficients (CH), Vertical Coefficients (CV) and Diagonal Coefficients (CD) using IWT. Next, a compression technique was used to compress all the planes selected from each of the frequency components. The QR code was generated by reading the EPR of the patient and transforming it into the QR code. Then the QR code bits were embedded in the original medical image. Finally, the Inverse Integer Wavelet Transform (IIWT) was applied to get the watermarked medical image. Kavitha and Shan [28] used different bit planes of the United State (US) medical images and obtained PSNR is 52.007, SSIM is 0.64 and Mean Square Error (MSE) is 0.474. Results of their method indicated that the difference between the watermarked image and the original medical image is huge.

Kaya and Elbasi [22] used two methodologies to embed the watermark in the spatial domain using LSB and in the transform domain using Digital Wavelet Transform (DWT), Digital Fourier Transform (DFT) and Digital Cosine Transform (DCT). The proposed approaches have embedded two different watermarks into three different medical images, the approaches have been evaluated against 11 different attacks and evaluated using PSNR and Similarity Ratio (SR), after analyzing the results they found that DFT has the best PSNR value of 48 dB and the worst SR value, while LSB, DWT and DCT recorded almost the same results.

Discrete Wavelet Transform (DWT) was used by Singh and Dwivedi [29] to decompose the medical image into low and high-frequency sub-bands. As a result, four non-overlapping coefficient sets LL, HL, LH and HH were obtained. Secondly, the LL band was divided into 3 × 3 non-overlapping blocks and the centre pixel was considered as a threshold. The difference between the centre pixel and the neighbours (d_i) was calculated and if the d_i is less than zero, the value of the result (b_i) was set to zero, otherwise, the value is set to one. Then, XOR between difference results (bi), $0 \leq i \leq 8$, was used to generate a logistic map. A binary image was taken for the patient information to be used as a watermark. The watermark was encrypted by XORing it with the generated logistic map. The encrypted watermark was embedded into the LL block. Finally, Inverse Wavelet Transform (IWT) was applied to the generated image to generate the watermarked image. Different sizes of different medical images and different binary images have been used to evaluate the proposed method against 4 attacks (Blur, Average, Gaussian and Crop), the performance values of Normalized Cross-Correlation (NCC) for Blur was 1.005, Average was 1.007, Gaussian was 1.0009 and for Crop was 1.005.

After analyzing the literature work we have noticed that most of the discussed techniques do not consider the patient's information and used any grey image as a watermark. This paper proposes using the QR Code generated from patient's information as a watermark, embedding patient's data in the medical images will reduce the size of transferred data and protect patients' data from being changed.

3 Proposed Watermarking Technique

This paper proposes a reversible watermarking technique to embed a QR Code that contains the patient's data into the medical image. The proposed method consists of two main blocks: the embedding block and the extraction block as shown in Fig. 1.

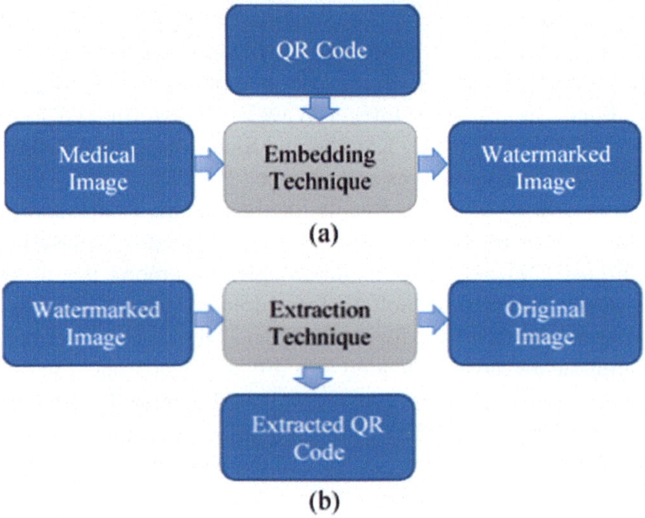

Fig. 1. Main architecture (a) Embedding (b) Extraction watermarking technique

3.1 Embedding Technique

The used embedding technique consists of mainly seven steps as shown in Fig. 2. We have divided the medical images into main blocks of size 4 × 4 pixels; each block is divided into two halves: upper half and lower half. The mean of the main block after setting the last 2 bits to zero (M_w) is embedded into the LSB of the lower half. We used QR Code as a watermark, before embedding the QR Code pixel into the upper half we encrypted it using mean of the block (Mw)

Here we go through the steps:

Step 1: Divide medical image pixels into 4 × 4 blocks (B).
Step 2: Split each block (B) into upper half (UH) and lower half (LH).
Step 3: Set the least two significant bits of the block (B) to zero to obtain the modified block (Bn).
Step 4: Compute the mean (m) of the modified block (Bn).
Step 5: Encrypt the QR code pixel by XORing it with the mean (m).
Step 6: Embed the encrypted QR code pixel in the upper half (UH).
Step 7: Embed the block mean (m) in the lower half (LH).

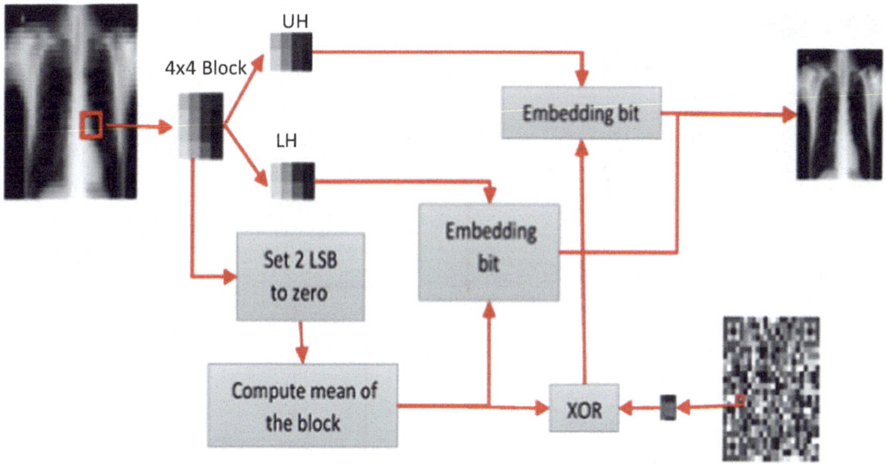

Fig. 2. Embedding technique steps

3.2 Extraction Technique

The used extraction technique consists of mainly eight steps as shown in Fig. 3. The purpose of the extraction technique is to check and localize the tamper of the watermarked image and to extract the watermark image. Watermarked image is divided into 4 × 4 blocks (Bx) and the mean of each block is computed after setting the last 2 bits to zero (m). To extract the watermark pixel, the embedded bit in the upper half of the block (B) is extracted and XORed with the mean (m). To detect and localize the tamper we XOR the mean (m) with the extracted bit.

Here we go through the steps:

Step 1: Divide watermarked image pixels into 4 × 4 blocks (Bx)
Step 2: Split the block (Bx) into upper half (UH) and lower half (LH).
Step 3: Set the least two significant bits of the block (Bx) to zero to obtain the modified block (Bn).
Step 4: Compute the mean (m) of the modified block (Bn).
Step 5: Extract the encrypted QR Code pixel from the upper half (UH).
Step 6: Extract block mean (Mb) from the lower half (LH).
Step 7: Decrypt the encrypted QR Code pixel by XORing it with the mean of the block (m).
Step 8: Check if the block (Bx) is tampered or not by comparing the computed mean (m) by the one extracted from lower half (Mb), in case they match, the block has not tampered otherwise it has tampered.

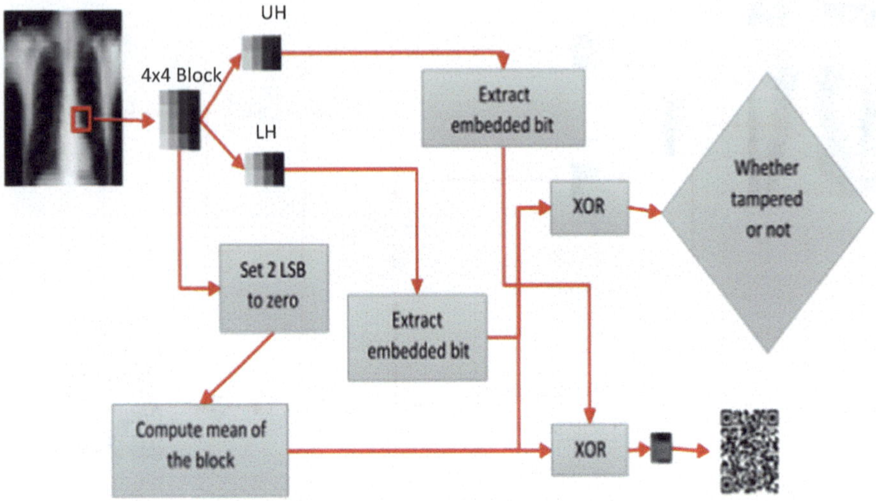

Fig. 3. Extraction technique steps

A simple example of the Encryption steps of the proposed method is shown in Fig. 4.

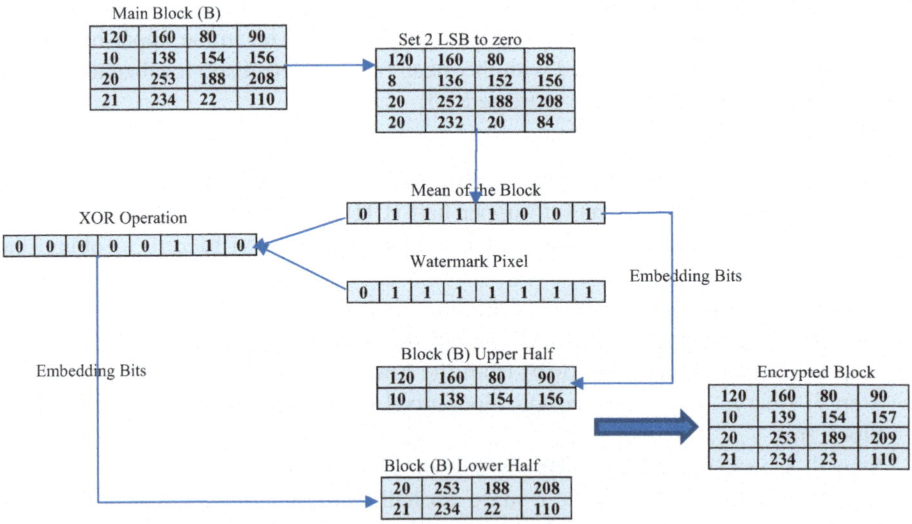

Fig. 4. Example of Encryption steps

4 Results and Discussion

The proposed method is evaluated by utilizing watermark invisibility. The watermark invisibility is utilized to measure the similarity between the original medical image and the watermarked one. It is used to measure the quality of the watermarked image. Three measures are used in this paper to measure the watermark invisibility: (1) Mean Square Error (MSE); (2) Peak Signal to Noise Ratio (PSNR) and (3) Structural Similarity Index Matrix (SSIM) [31].

MSE measures the average of squared errors between the watermarked M_W image and the original medical image (M). MSE is defined as follows [30]:

$$MSE = \frac{1}{(Kl)^2}\left(\sum_{i=1}^{k}\sum_{j=1}^{l}[M_w(i,j) - M(i,j)]^2)\right) \quad (1)$$

where k, l is the width and height of watermarked M_W image and original medical image (M).

PSNR is the most commonly used measure of the quality of the watermarked image [29]. The value of PSNR is directly proportional to the similarity between the watermarked image and the original medical image (M) which means that the watermark is more imperceptible [31]. The measure of PSNR is defined as follows [32]:

$$PSNR(M, Mw) = 10\log_{10}\frac{255^2}{MSE} \quad (2)$$

SSIM is used for comparing the similarity between the original medical image (M) and the watermarked image. It gives an average of the whole image and may be considered as PSNR but more suitable for measuring quality [33].

The measure of SSIM is defined as follows:

$$SSIM(M, M_w) = l(M, M_w)c(M, M_w)s(M, M_w) \quad (3)$$

where, M_w is the watermarked image, $l(M, M_w)$ is the luminance comparison function, $c(M, M_w)$ is the contrast comparison function and $s(M, M_w)$ is the structure comparison function. The values of SSIM is in the range [0–1]. SSIM value of 1 means that $M = M_w$ and a value of 0 means $M \neq M_w$ [31].

$$l(M, M_w) = \frac{2\mu_M\mu_{M_w} + C1}{\mu_M^2 + \mu_{M_w}^2 + C1} \quad (4)$$

$$c(M, M_w) = \frac{2\sigma_M\sigma_{M_w} + C2}{\sigma_M^2 + \sigma_{M_w}^2 + C2} \quad (5)$$

$$s(M, M_w) = \frac{\sigma_{MM_w} + C3}{\sigma_M\sigma_{M_w} + C3} \quad (6)$$

Where μ_M is the average of M, μ_{M_w} is the average of M_w, σ_M^2 is the variance of M, $\sigma_{M_w}^2$ is the variance of M_w and σ_{MM_w} is the covariance of M and M_w.

Also, $C1 = (K_1L)^2$, $C2 = (K_2L)^2$ and C3 = C2/2. Where $K_1 = 0.01$, $K_2 = 0.03$ and L is the dynamic range of the pixel-values.

To validate and evaluate our work, the proposed method is tested on 138 greyscale medical images of sizes (64 × 64, 128 × 128, 256 × 256, 512 × 512, 1024 × 1024, 2048 × 2048 and 4096 × 4096). The images are taken from the OPENi medical image database [34]. QR code is generated using the medical data in the OPENi database to be used as a watermark.

The performance of the proposed method is evaluated on different sizes of the medical images and the QR code and the average of the results of the 138 medical images is recorded in Table 1.

Table 1. Average performance under medical image of different size

Medical image size	QR code size	PSNR	SSIM	MSE
64 × 64	**16 × 16**	51.259	0.996	0.48698
128 × 128	**32 × 32**	51.2494	0.994	0.4879
256 × 256	**64 × 64**	51.243	0.9928	0.4886
512 × 512	**128 × 128**	51.2485	0.991	0.488
1024 × 1024	**256 × 256**	51.2484	0.9893	0.488
2048 × 2048	**512 × 512**	51.2493	0.9886	0.4879
4096 × 4096	**1024 × 1024**	51.2489	0.9881	0.48799

In order to validate the proposed method, we have compared it with the other two state-of-the-art methods described in [21] and [24]. The comparison is recorded in Table 2. The proposed method and [24] are common in the source of the dataset and the size of the medical image. On the other hand, both the proposed method and [28] use a QR code that contains the patient's data as a watermark.

Table 2. Comparison between the proposed method and other methods

	Ref [24]	Ref [28]	Proposed
Dataset	14 medical images from OPENi database	US medical images	138 medical images from OPENi database
Size of medical image	256 × 256	Not mentioned	256 × 256
Watermark image	Grey-scale image	QR code	QR code generated from the patient's data
Performance measures	–PSNR: 51.26 dB –SSIM: 0.9950 –MSE: Not mentioned	–PSNR: 52.007 dB –SSIM: 0.6400 –MSE: 0.4704	–PSNR: 51.243 dB –SSIM: 0.9928 –MSE: 0.4886

In addition, the performance of the proposed method has been analyzed from the time perspective for each size of the medical image as shown in Table 3.

Table 3. Time analysis of the proposed method

Medical image size	QR code size	Encryption time (s)	Decryption time (s)
64 × 64	**16 × 16**	0.13	0.076
128 × 128	**32 × 32**	0.106	0.145
256 × 256	**64 × 64**	0.414	0.412
512 × 512	**128 × 128**	2.002	1.47
1024 × 1024	**256 × 256**	7.024	5.74
2048 × 2048	**512 × 512**	27.95	22.27
4096 × 4096	**1024 × 1024**	110.017	149.053

After analyzing performance measures of the three methods we have found that our proposed method has improved the value of SSIM, which implies that the watermarked image is similar to the original image which indicates that the change in the values of the pixels of the medical image is very small.

5 Conclusion

Nowadays, securing medical image by embedding a watermark has become necessary to protect a patient's data while transforming the medical image from one place to another through modern communication technology. Large numbers of approaches have been implemented in this research area some used spatial domain and other frequency domain.

In this paper, a reversible fragile watermarking technique is proposed. QR code is generated from the patient's data and embedded in the medical image. Medical image is divided into blocks of size 4 × 4. Encrypted watermark, which is QR code pixel XORed with the mean of the block after setting least significant 2 bits to zero, is embedded in the upper half of the block. The mean of the block is embedded in the lower half of the block. Bits are embedded by changing LSB of the pixel with the new value. This scheme is found to be efficient with average PSNR of 51.25 dB, SSIM of 0.99 and MSE of 0.488. The proposed scheme helps in protecting the patient's data and make sure of both the authenticity and the integrity of transmitted data as it ensures that the QR code extracted contains the patient's data.

However, the proposed method has limitations on the size of the medical image, when the size of the medical image is larger than 4096 × 4096 it will take much time in encryption and decryption.

In the future, the proposed technique will be tested against geometric attacks, like scaling and rotation, signal processing attacks, like Gaussian and median filters, and for tamper detection and localization attacks like a copy, paste and text addition. Coloured medical images can be also used instead of greyscale ones.

References

1. Abbasi, F., Memon, N.A.: Reversible watermarking for the security of medical image databases. In: IEEE (2018)
2. Memon, N.A., Gilani, S.A.M.: NROI watermarking of medical images for content authentication. In: Proceedings of 12th International IEEE Multi-topic Conference (INMIC 2008), pp. 106–110 (2008)
3. Mousavi, S.M., Naghsh, A., Abu-Bakar, S.A.R.: Watermarking techniques used in medical images: a survey. J. Digit. Imaging **27**, 714–729 (2014)
4. Coatrieux, G., Montanger, J., Huang, H., Roux, Ch.: Mixed reversible and RONI watermarking for medical image reliability protection. In: 29th IEEE International Conference of EMBS, Lyon (2007)
5. Kuang, L.Q., Zhang, Y., Han, X.: A medical image authentication system based on reversible digital watermarking. In: 2009 1st International Conference, pp. 1047–1050 (2009)
6. Bhatnagar, G., Jonathan, W.U.Q.M.: Biometrics inspired watermarking based on a fractional dual tree complex wavelet transform. Future Gen. Comput. Syst. **29**, 182–195 (2013)
7. Pan, W., Coatrieux, G., Cuppens-Boulahia, N., Cuppens, F., Roux, C.: Medical image integrity control combining digital signature and lossless watermarking. In: Garcia-Alfaro, J., Navarro-Arribas, G., Cuppens-Boulahia, N., Roudier, Y. (eds.) Data Privacy Management and Autonomous Spontaneous Security, pp. 153–162. Springer, Heidelberg (2010)
8. Agrawal, R., Sharma, M.: Patient record data hiding in medical images using M-ary modulation. Res. J. Eng. Sci. **5**(9), 5–10 (2016)
9. Chang, Y.J., Wang, R.Z., Lin, J.C.: Sharing based fragile watermarking method for authentication and self recovery of image tampering. EURASIP J. Adv. Signal Process. (2008)
10. Memon, N.A., Keerio, Z.A., Abbasi, F.: Dual watermarking of CT scan medical images for content authentication and copyright protection. In: Shaikh, F.K., et. al. (eds.) IMTIC 2013, Communications in Computer and Information Science, vol. 14, pp. 173–183. Springer (2014)
11. Zain, J.M., Clarke, M.: Reversible region of non-interest (RONI) watermarking for authentication of DICOM images. Int. J. Comput. Sci. Netw. Secur. **7**(9), 19–28 (2007)
12. Wu, N.-I., Hwang, M.-S.: Data hiding current status and key issues. Int. J. Netw. Secur. **4**(1), 1–9 (2007)
13. Punidha, R., Sivaram, M.: Integer wavelet transform based approach for high robustness of audio signal transmission. Int. J. Pure Appl. Math. **116**(23), 295–303 (2017)
14. Ramos, R., Valdez-Salas, B., Zlatev, R., Wiener, M.S., Rull, J.M.B.: The discrete wavelet transform and its application for noise removal in localized corrosion measurements. Hindawi Int. J. Corros. **2017**, 1–7 (2017)
15. Ahmed, N., Natarajan, T., Rao, K.R.: Discrete cosine transform. IEEE Trans. Comput. **C-23** (1), 90–93 (1974)
16. Heylena, K., Dams, T.: An image watermarking tutorial tool using matlab. In: Mathematics of Data/Image Pattern Recognition, Compression, and Encryption with Applications XI, Proceedings of SPIE, pp. 1–12 (2008)
17. Mohanty, S.P., Ramakrishnan, K.R.: A dual watermarking technique for images. In: Proceedings of the 7th ACM International Multimedia. ACM Press (1999)
18. Memon, N.A., Chaudhry, A., Ahmad, M., Keerio, Z.A.: Hybrid watermarking of medical images for ROI authentication and recovery. Int. J. Comput. Math. **88**(10), 2057–2071 (2011)

19. Jabade, V.S., Gengaje, S.R.: Literature review of wavelet-based digital image watermarking techniques. Int. J. Comput. Appl. **31**(1), 28–35 (2011)
20. Memon, N.A., Gilani, S.A.M.: Watermarking of chest CT scan medical images for content authentication. Int. J. Comput. Math. **88**(2), 265–280 (2010)
21. Kaur, M., Kaur, R.: Reversible watermarking of medical images authentication and recovery-a survey. J. Inf. Oper. Manag. **39**(1), 241–244 (2012)
22. Kaya, V., Elbasi, E.: Robust medical image watermarking using frequency domain and least significant bits algorithms. In: IEEE (2018)
23. Jane, O., Elbasi, E.: A new approach in non-blind watermarking method based on DWT and SVD via LU decomposition. Turkish J. Electr. Eng. Comput. Sci. **22**(5), 1354–1366 (2013)
24. Gull, S., Loan, N.A., Parah, S.A., Sheikh, J.A., Bhat, G.M.: An efficient watermarking technique for tamper detection and localization of medical images. J. Ambient Intell. Humaniz. Comput. (2018)
25. Shehab, A., Elhoseny, M., Muhammad, K., Sangaiah, A.K., Yang, P., Huang, H., Hou, G.: Secure and robust fragile watermarking scheme for medical images. Digit. Object Identifier **6**, 10269–10278 (2018)
26. Li, M., Liang, T., He, Y.: Arnold transform based image scrambling method. In: 2013, 3rd International Conference on Multimedia Technology (ICMT 2013) (2013)
27. Sharma, A., Singh, A.K., Ghrera, S.P.: Robust and secure multiple watermarking for medical images. Wireless Pers. Commun. **92**, 1611–1624 (2016)
28. Kavitha, K.J., Shan, P.B.: An efficient medical image watermarking technique using integer wavelet transform and quick/fast response codes. Int. J. Intell. Syst. Technol. Appl. **18**(3), 271–280 (2019)
29. Singh, R.P., Dwivedi, S.S.: Advanced medical image watermarking technique of hiding patient information for medical image authentication. Int. J. Eng. Tech. Res. (IJETR) **8**(5), 22–29 (2018)
30. Singh, C., Ranade, S.K.: Geometrically invariant and high capacity image watermarking scheme using accurate radial transform. Opt. Laser Technol. **54**, 176–184 (2013)
31. Hosny, K.M., Darwish, M.M., Li, K., Salah, A.: Parallel multi-core CPU and GPU for fast and robust medical image watermarking. IEEE Access **6**, 77212–77225 (2018)
32. Zhangetal, H.: Affine Legendre moment invariants for image watermarking robust to geometric distortions. IEEE Trans. Image Process. **20**(8), 2189–2199 (2011)
33. Tsougenis, E.D., Papakostas, G.A., Koulouriotis, D.E., Tourassis, V.D.: Performance evaluation of moment-based watermarking methods: a review. J. Syst. Softw. **85**(8), 1864–1884 (2013)
34. https://openi.nlm.nih.gov/. Accessed 31 Aug 2019

Dimensions of Agility Capabilities Organizational Competitiveness in Sustaining

Asma Y. Zainal[1], Hana Yousuf[2], and Said A. Salloum[2,3](✉)

[1] Faculty of Business Management, The British University in Dubai, Dubai, UAE
[2] Faculty of Engineering & IT, The British University in Dubai, Dubai, UAE
[3] Machine Learning and NLP Research Group, Department of Computer Science, University of Sharjah, Sharjah, UAE
ssalloum@sharjah.ac.ae

Abstract. The uncertain global changes in the market industry that can unfold potential opportunities and risk it became vital for organizations to be agile to sustain its market position and be able to face challenges and maintain the success journey ultimately. Accordingly, it's essential to explore organizational agility capabilities as enablers to maintain organizational competitiveness. Flexibility, learning, speed, innovation, and strategy were addressed as the primary organizational capabilities to be considered for successful sustainable organizational competitiveness. The paper also identified relevant theoretical framework linked to the address capability including Organizational Change Theory by Lewin, Organizational Change Model by Kotter, Organizational Learning Theory by Oslen & March and Theory of Dynamic Capabilities by Teece was it discussed how those capabilities are linked. Moreover, the paper addressed a case study on the impact of Deep Learning on Organizational Agility Lastly; the paper is concluded by the illustration of findings, gaps, and suggestions for future research.

Keywords: Organizational agility · Organizational capabilities · Competitiveness · Organizational sustainability · Agility capabilities · Deep learning

1 Introduction

In today's changing world of business and growing market demand along with an ambiguous future that is filled with opportunities and risks, it becomes essential for organizations to be able to face those changes through coping and being able to adapt to new situations and market needs continuously. Therefore, many organizations tend to use organizational tools and management processes to empower them to be more flexible and agile. The paper aspires to explore the main drivers of agility capabilities that have significant impacts on organizational competitiveness, in addition to attempting to portray recommendations on the possible future directions of agility capabilities based on the attained literature review as the next focus for organizations to sustain its organizational competitiveness. The paper will attempt to answer the

A.-E. Hassanien et al. (Eds.): AICV 2020, AISC 1153, pp. 762–772, 2020.
https://doi.org/10.1007/978-3-030-44289-7_71

following main questions on agility capabilities and its impact on organizational competitiveness which are: what are the dimensions of agility capabilities of organizations? And how do applied agility capabilities and dynamic organizational capabilities have a significant relationship with each other?

To conduct this research, some variables need to be considered to be able to answer research questions effectively, which revolves around agility capabilities found in several literature reviews where they are referred to in the upcoming sections, including flexibility learning, speed, innovation, and strategy.

2 Literature Review

Although the term agility was not typical in the previous decades, the essence of it was discussed within the same context in different terms. Some literature argues that agility has started in the manufacturing industry where organizations utilized agility as an empowering force to sustain them with changing and uncertain growing market demands and needs [1]. Within the same context, agility as a concept was first argued and published as literature by the Institute of Iacocca in the United States (USA) as it was agreed also that the first industry to implement the concept of agility is the manufacturing industry [2]. Along the way with market developments, agility as a concept was well-identified in the services industries along with the growing international market fierce competitions [3].

There are several definitions and explanations of agility across different industries as obvious efforts were made by the scholar to come up with an organic explanation that facilitates a better understanding of agility for future researches. The ability of organizations to reconfigure towards the new change in the shadows of uncertainty [4] were one of the most generic definition themes as throughout the research it combined the fundamental essence of agility as a concept. Agility is also defined as the continuity of sustaining organization flexibility, performance and speed towards changes in a dynamic manner [5]. Another definition related agility to the quickness of organizational response towards changes [6] whereas Prater defines agility as the utilization of organizational resources, in attaining outcomes with speed and flexibility [3]. Another similar definition related agility to the speed and organizational flexibility in changing in alignments with market needs [7]. Consequently, agility is not only limited to a certain sector or a specific business industry as it represents a holistic approach that includes different aspects and dimensions related to capabilities including people learning, organizational flexibility, Artificial Intelligence and innovative culture in order to enable the organization to be prepared to perform efficiently towards future faced uncertainties [2]. Therefore, in the purse of establishing a clear understanding of the organizational dimensions of applied agility capabilities, we need to explore further though the published literature this field in order to learn from efforts conduct by other scholars to be able to serve the ultimate goal of this paper questions on the definitions of organizational agility capabilities which are flexibility learning, speed, innovation and strategy.

2.1 Agility Capability: Flexibility

In 1930, economist Stigler defined flexibility as the organizational efforts in adaptation to manage demand change ambiguities [8], a group of scholars [9] also referred to agility has as a very close meaning to flexibility where they also stated that other scholars identified strategic agility as the ability of the organization to adapt and change their strategic directions and relevant value according to changes in order to create the aspired value to the organization [9]. According to [10], several researches have clearly identified agility in terms of definitions and implications to flexibility where most of their literature linking it to agility in its core essence [10]. On the other hand, it's was also evident from the available knowledge in the aspect of defining agility that not all scholars agreed on the same identification of agility as other scholars came to conclusion that agility cannot be a mirror definition to flexibility or the term of responsiveness as they differ in the essence of their core meaning and related concepts although they run in the same vein of impressions [11]. To further elaborate on the same context, this paradigm argues that the difference between agility and flexibility is in the core of managing situations that require change, which means flexibility is simply about a retort of an organizations towards known situations where there are set of established process to be undertaken in order to enable the organization to manage the change, whereas agility is similar concept to flexibility with an additional main element of ambiguity [11]. To emphasize further, backhouse and burns published a work on the international journal of agile management systems where they clearly reasoned that agility is different than flexibility because agility mainly relevant to the future uncertainty while flexibility is usually applicable to the known and identified changes [12]. In the same vein, while scholars disregard that agility is an alternative word for flexibility, they suggested that flexibility can be considered as a capability for agility [13] which throughout literature it makes a clear logic in terms of considering flexibility as an enabler to agility where organizations that are flexible tend to have aspects of agility as it was argued that agility consists of flexibility and structure where organizations are aided to face changes and seize further opportunities [14].

2.2 Agility Capability: Learning

Several studies linked agility to learning where Katayama and Bennett argue that there are three main principles, which include, delivering customer values, readiness towards change, valuing knowledge and skills, and establishing collaborations [15]. Learning agility is also a concept introduced by organizations where they link it to the ability of individuals within organizations to learn and apply the learning outcomes to develop the progress of organizational learning [16]. Therefore, it is obvious that agility learning unfolds the entire cycle of people learning in their organizations. Learning is concerned with the capability for organizations to increase their agility through learning [17]. Learning culture and agility are linked in terms of providing a successful framework for organizations where it will lead to developing a vision of fast and flexible execution towards occurred changes [1]. Moreover, Sharp, Irani, and Desai argue that agility capabilities for organizations consist of learning creativity and innovation, along with other competences [18]. Furthermore, researchers concluded that organizational learning is an essential element in organizations to be developed to have agile-ready employees where they can perform towards sustaining organizational agility [19]. Other models were also developed by scholars in terms of linking organizational learning with agility but had no practical implementations to support their research efforts [20]. Hence, the above literature in this section has clearly emphasized that many scholars refereed learning as an important agility capability for organizations to be able to enhance its readiness for future changes and adaptably towards eliminating risks and seizing potential opportunities.

2.3 Agility Capability: Speed

Considering the fierce market competition and ambiguous changes along with risks that the future can unfold, agility as a concept works at bridging the existed gap between learning and speed necessary for organizational change [5]. One of organizational agility inputs is speed as its part of the business environment swiftness in addition to the results related to the study conducted by three scholars [21] that all high dynamic organizational culture tends to have high speed and acceleration in their environments [21]. Additionally, as mentioned in the above sections that organizational agility is linked with speed and flexibility as capabilities for attaining successful outcomes.

2.4 Agility Capability: Innovation

Several contributions founded in literature by scholars claimed that organizational agility consists of several skills, including learning and innovation [18]. Innovation processes and implementations sustain organizational agility through newness and breakthrough outcomes in the form of product, services or innovative organizational processes [22]. Innovation can lead organizations to be agile through its ability to develop opportunities and unique outcomes, which can come in the form of business model reconfiguration as this consequently plays a vital role in assisting organizations in facing changes and market needs [23]. Moreover, other scholars argued through their case study and survey that the existence of a strategic agility framework has the impact to accelerate organizational business model through employing innovation for model enhancement redevelopment which can lead to radical innovative outcomes [24]. The concept of open innovation has also an impact on organizational agility since it opens new market opportunities [9]. From this section, it is evident that there is indeed a clear link between innovation and organizational agility as a capability that can open new horizons of opportunities that can sustain the organization in the market competition in addition to providing an added-value niche to put the organization into a leading position entirely or in a certain aspect especially when it comes to producing innovative outcomes be it products, services, redeveloping business models or process.

2.5 Agility Capability: Strategy

The strategy is considered the heart of an organization where all different efforts made by organizations in terms of agility, innovation, and the latest technologies will have no clear outcomes of meaningfulness if it was not aligned with clear directions and heading towards achieving strategic objectives [9]. In the same vein, the strategy is considered successful if it was built upon an agile strategic framework when it facilitates achieving organizational objectives along with cope with market changes [25]. Today's organizations have to sustain their performance through certain dynamic capabilities including flexibility, strategic directions and adaptability [5]. Strategic agility is defined as the capability of organizations to perform their strategy in a progressive manner along with its ability to face the changing business environment in order to cope with market needs. Moreover, strategic agility differs from the traditional strategy in terms of its ability of responsiveness in addition to the capability to reconfigure and adapt itself [26]. Therefore, from understanding relevant available literature in this part of the paper, it's obvious that strategy is a crucial component of any organization to be able to set clear direction, however, in today's changing business market organizations have to ensure that strategy is not only implemented traditionally but also an element of agility is injected into the framework of strategic master plan as this will ensure organizational sustainability and capability to adapt and reconfigure itself according to different changes of market demand along with coping with evolving trends in various fields of organizational outcomes towards customers of other patterns that can reshape and influence the approach of business within the industry.

2.6 Competitive Advantage

Competitive advantage has become a significant research area after it was published and understood by scholars as it can be defined by the approach in which collaborations between firms and other industries and other parties as facilitators in enhancing competitive advantage [27]. Competitive advantage also has an impact on the national and international levels [28]. In the context of organizations, it can be referred to the efforts that organizations do in order to gain & develop unique attributes that allows it to perform better than competitors [29]. Therefore, it is clear that for organization to sustain its success and industry presence consistency leadership, it has to develop specific capabilities to be addressed, developed and readjust for organizations to transform it from a traditional organizational to a competitive organization where it will be enabled to be more agile to face future challenges along with corresponding with market needs [30]. This will not only effect organizations outcomes but also on a national and international level where it will enhance the countries international indicators and its participation in global indexes and reports.

2.7 Relevant Theoretical Frameworks

This part will attempt to link relevant theories on the themes discussed in this paper. One of the appropriate approaches to agility is Organizational Change Theory, where it was introduced by Kurt Lewin, who was considered as a pioneer in the fields relevant to organizational development [31]. Lewin explained change theory as change occurs when there is an obvious sense of the need for change occurs as this phase called (unfreezing), followed by stepping towards new sets of behavior and values (moving) and finally stabilizing into a new set of behavior [32]. This theory portrays the vitality of performing organizational changes through sensing and creating the feel of the need to change organizational-wide which reflects upon individuals working within the organization, then emphasize the new learning experiences and values for the upcoming phase of organizational work, and finally solidify and set the new organizational values and experiences. This will enable organizations to adapt to changes swiftly when linking learning as an organizational learning capability to this theory. Organizational Change Model by Kotter in 1996 is another significate model in the literature review to be considered for this research where the model offers eight stages for a chance to successfully occur in an organization which are established sense of urgency, create guiding alliances, develop vision and strategy for change, communication of the new changed vision and strategy, empower and enable action taking, execute quick wins, make further changes and set the new approach within the culture [33]. This showcases that change goes through phases of ensuring the vitality of flexibility for organizations to cope with new changes through creating a sense of need followed up by setting a new direction and communicate to emphasize the organizational new learning of values while setting the experience through creating success stories.

Another theory that is linked to the theme of this paper is Organizational Learning Theory by Oslen and March in 1975 through four sets on learning stages which cascades from individual-level into effecting the organization and environment, where it starts with individual beliefs leads to different actions leads to organizational activities which directly producing environment response [34]. Where this theory conceptualizes the process of learning progresses of individuals within the organization through learning by observing which can emphasize their actions and decision making and hence influence change [35]. This theory showcases that the faster organizational learning through role models and showcasing best practices the more learning occurred whereas mentioned in the above sections the relations between learning and organizational agility capability for organizations to attain the desired outcomes.

Theory of Dynamic Capabilities by Teece in 1997 is another important theory to be showcased in this paper where it was defined as the organizational ability to create, reconfigure and assimilate towards a highly changing business environment [36]. Organizational dynamic capabilities revolves around three main aspects which are sensing to identify opportunities, seizing where redesigning and reconfiguring of business models and related activities occur as it's also linked to strategy in terms of readiness for competition and saving innovative intellectual properties, and lastly transformation where realignment of structure and culture towards existed capabilities [37]. The dynamic theory was initiated to assist the organization in managing competition and information to build a competitive advantage [38]. Based on the illustrated literature review on this theory, it can be understood that there is a clear linkage between maintaining organizational competitiveness through capabilities that can reflect upon agility in the ultimate outcomes where organizations readiness towards competition and sustaining its position and performance in the constant changing market demands.

3 Sustainability and Research Impact

Sustainability has been one of the critical concerns of organizations to ensure business endurances in the past decade, where it plays a role in organizational progress and success as a sustainable agile organization tends to attain better organizational outcomes [5]. It has been also depicted that flexibility with proper collaborations with partners can lead to innovative sustainable results [39]. This research is linked directly to main sustainability global themes that reflect efforts leading international institutions in them pursue to make a difference where agility contributes in some essence to the sustainability of organizations which ultimately reflects on the national performance that is measured internationally. Global Competitiveness Report is a yearly published report from World Economic Forum establish in 1971 in Switzerland as a global organization that engages political, cultural, industrial agendas such as, the world economic forum agenda meeting in addition to publishing an annual global competitiveness report [40]. that present countries' status based on certain themes, some of the main presented index in the global competitiveness report are sustainability and agility [41]. This showcases the importance of organizations working to sustaining agility as it not only reflects upon its internal performance but the global performance of the country enabling it to compete economically with others. United Nation's Sustainable Development Goals is another consideration not only for organizations but also on an international scale as agility and sustainability play major roles for organizations to consider its high scale strategic linkage for 2030 agenda for sustainable development that cascades from international goals, national goals and organizational goals [42]. This showcased that institutions including government take major considerations to attain those goals. Therefore, organizations have to adapt and reconfigure to certain capabilities to be able to reach towards sought international standards, agility capabilities such as innovation, flexibility, and strategy can assist in sustaining organizational performance in the long-term to achieve the aspired standards and results.

4 Organization Agility and Intelligence Systems

In today's organizations, Artificial Intelligence considers a main variable for the organizations to measure their flexibility, adoptability and resiliency to change working environment [43–48]. [49] Studied the impact of deep learning on organizational agility. They highlighted the changes that may occurred on the organization if they adapted the functions of deep learning such as data collection, data analysis and algorithms. Deep learning can complete each tasks automatically where in the absence of deep learning set of sequential action, which may take time and space in performing it. The Study done in a Chinese company that already implemented a deep learning using multiple qualitative methods. The authors addressed the impact of implementing the deep learning in the organizations and the results of the three main variables will be completed in the first quarter of 2020.

5 Methodology

The research methodology mainly used available literature reviews from existed knowledge in the field of agility topic in addition to taking and attempt to creating meaningful linkage to assist in answering the questions of this paper which were to investigate the dimensions of organizational agility capabilities as a form of organizational enablers towards managing its competitive advantage. Moreover, the article also explored and reviewed different relevant well-established theories about the topic such as change theory by Kurt Lewin, organizational change model by Kotter, learning theory by Oslen, and March in addition to dynamic theory by Teece. All obtained knowledge from this paper was from assed articles in established journals as sources of the approved source of information and proper scholars' contributions of knowledge and wisdom.

6 Conclusion and Future Work

In conclusion, agility is the ability of organizations that enable them to move dynamically in terms of changes where from undertaking this research, the literature review depicted clearly that organizations with undoubtedly noticed growing demands, competitions, and future uncertainty requires organizations to have specific capabilities to survive in the fierce competition, therefore, agility provide the capability needs to sustain organizational readiness toward those changes where it allow organization to act swiftly towards new horizons. In addition, organizational agility comes with certain skills to be considered by other organizations to attain successful and fruitful outcomes. This paper identified the selection of capabilities based on an existed literature review which are flexibility, learning, speed, innovation, and strategy.

Many scholars referred to flexibility as one of the main enablers for successful agility where other thoughts there is a similarity between both terms. This paper found through literature that there is indeed a difference between flexibility and agility in terms of the level of uncertainty involved in both where flexibility is more systematic as there as clear tools to enable organization to utilize it to be flexible while agility is related to facing organizational uncertainty with the presence of various capabilities that runs across all various organizational levels. Moreover, learning is an essential part of agility capability as it assists in organizational ability to learn and be able to implement the learning into agile aspects which can result in innovative solutions in forms of products, services or re-configuration of processes. Speed is another capability where it allows the organization to act fast to seize opportunities and open new horizons through market leadership. In addition to that innovation is another capability for the agile organization where the transformation of knowledge into innovative outcomes is vital to come up with solutions to challenges and risks in addition to provide unique results. Finally, a strategy where agile strategic plan is essential for the organization to be able to adjust according to different changes, risk and competition. All these capabilities are considered as enablers to provide organizations with the required tool to have a competitive advantage in the industry competition. This is reflected in different aspects including national and international levels in the form of global indexes and reports along with other standards that enhance countries' reputations and consequently result in stronger economies.

The paper also provides a brief illustration of theoretical frameworks of relevant theories that were also important to briefly present in order to explore the relation between the main two components of this paper which are agility capabilities and organizational competitiveness. The paper presented Organizational Change Theory by Lewin, Organizational Change Model by Kotter, Organizational Learning Theory by Oslen & March and Theory of Dynamic Capabilities by Teece. In addition, the paper also showcased how sustainability is connected to the paper's topic and its relation to the global influence of its results be it in the form of the global report, indexes and international indicators such as Global Competitiveness Report, World Economic Forum, and United Nation's Sustainable Development Goals. Finally, the conducted review of literature in this paper based on the addressed above research questions have directed to recognize two main future research gaps which were noticeable variances in existed knowledge on the research topic, one is related to clear identifications of agility as definitions and it's capability, some scholars took capabilities as synonym of agility. Moreover, organizational agility was not also similar by scholars in aspects of components and enablers. This paper contributes to the current existed knowledge in terms of making an attempt to identify the main elements of organizational agility capabilities in relation to competitive advantage, which reflects upon organizational sustainability and consequently reflects upon the national and international economy. This paper also suggests further research be explored in the topic in addition to creating a model for organizational agility showcasing different links between capabilities and elements on internal and external forces to sustain a competitive advantage based on practical solid and evident research ground.

References

1. Pantouvakis, A., Bouranta, N.: Agility, organisational learning culture and relationship quality in the port sector. Total Qual. Manag. Bus. Excell. **28**(3–4), 366–378 (2017)
2. Eltawy, N., Gallear, D.: Leanness and agility: a comparative theoretical view. Ind. Manag. Data Syst. **117**(1), 149–165 (2017)
3. Pantouvakis, A., Dimas, A.: The role of corporate agility and perceived price on the service quality–customer satisfaction link: some preliminary evidence from the port industry. Int. J. Shipp. Transp. Logist. **5**(4–5), 412–431 (2013)
4. Prince, J., Kay, J.M.: Combining lean and agile characteristics: creation of virtual groups by enhanced production flow analysis. Int. J. Prod. Econ. **85**(3), 305–318 (2003)
5. Appelbaum, S.H., Calla, R., Desautels, D., Hasan, L.: The challenges of organizational agility (Part 1). Ind. Commer. Train. **49**(1), 6–14 (2017)
6. Kidd, P.T.: Agile manufacturing: a strategy for the 21st century (1995)
7. Christopher, M.: The agile supply chain: competing in volatile markets. Ind. Mark. Manag. **29**(1), 37–44 (2000)
8. Stigler, G.: Production and distribution in the short run. J. Polit. Econ. **47**(3), 305–327 (1939)

9. Teece, D., Peteraf, M., Leih, S.: Dynamic capabilities and organizational agility: risk, uncertainty, and strategy in the innovation economy. Calif. Manage. Rev. **58**(4), 13–35 (2016)
10. Fayezi, S., Zutshi, A., O'Loughlin, A.: Understanding and development of supply chain agility and flexibility: a structured literature review. Int. J. Manag. Rev. **19**(4), 379–407 (2017)
11. Bernardes, E.S., Hanna, M.D.: A theoretical review of flexibility, agility and responsiveness in the operations management literature: toward a conceptual definition of customer responsiveness. Int. J. Oper. Prod. Manag. **29**(1), 30–53 (2009)
12. Backhouse, C.J., Burns, N.D.: Agile value chains for manufacturing–implications for performance measures. Int. J. Agil. Manag. Syst. **1**(2), 76–82 (1999)
13. Flynn, B.B., Sakakibara, S., Schroeder, R.G., Bates, K.A., Flynn, E.J.: Empirical research methods in operations management. J. Oper. Manag. **9**(2), 250–284 (1990)
14. Marlow, P.B., Casaca, A.C.P.: Measuring lean ports performance. Int. J. Transp. Manag. **1** (4), 189–202 (2003)
15. Katayama, H., Bennett, D.: Agility, adaptability and leanness: a comparison of concepts and a study of practice. Int. J. Prod. Econ. **60**, 43–51 (1999)
16. DeRue, D.S., Ashford, S.J., Myers, C.G.: Learning agility: in search of conceptual clarity and theoretical grounding. Ind. Organ. Psychol. **5**(3), 258–279 (2012)
17. Arun, N., Coyle, P.T., Hauenstein, N.: Learning agility: still searching for clarity on a confounded construct. Ind. Organ. Psychol. **5**(3), 290–293 (2012)
18. Sharp, J.M., Irani, Z., Desai, S.: Working towards agile manufacturing in the UK industry. Int. J. Prod. Econ. **62**(1–2), 155–169 (1999)
19. Alavi, S., Abd. Wahab, D., Muhamad, N., Arbab Shirani, B.: Organic structure and organisational learning as the main antecedents of workforce agility. Int. J. Prod. Res. **52** (21), 6273–6295 (2014)
20. Pereira, V., Mellahi, K., Temouri, Y., Patnaik, S., Roohanifar, M.: Investigating dynamic capabilities, agility and knowledge management within EMNEs-longitudinal evidence from Europe. J. Knowl. Manag. **23**, 1708–1728 (2018)
21. Park, Y., El Sawy, O.A., Fiss, P.: The role of business intelligence and communication technologies in organizational agility: a configurational approach. J. Assoc. Inf. Syst. **18**(9), 1 (2017)
22. Carmeli, A., Dothan, A.: Generative work relationships as a source of direct and indirect learning from experiences of failure: implications for innovation agility and product innovation. Technol. Forecast. Soc. Change **119**, 27–38 (2017)
23. Battistella, C., De Toni, A.F., De Zan, G., Pessot, E.: Cultivating business model agility through focused capabilities: a multiple case study. J. Bus. Res. **73**, 65–82 (2017)
24. Kosonen, M., Doz, Y.: Fast Strategy: How Strategic Agility Will Help You Stay Ahead of the Game. Pearson Education, London (2007)
25. Alon, I., Madanoglu, M., Shoham, A.: Strategic agility explanations for managing franchising expansion during economic cycles. Compet. Rev. Int. Bus. J. **27**(2), 113–131 (2017)
26. e Cunha, M.P., Gomes, E., Mellahi, K., Miner, A.S., Rego, A.: Strategic agility through improvisational capabilities: implications for a paradox-sensitive HRM. Hum. Resour. Manag. Rev. **30**(1), 100695 (2020)
27. Porter, M.E.: Competitive advantage, agglomeration economies, and regional policy. Int. Reg. Sci. Rev. **19**(1–2), 85–90 (1996)
28. Grant, R.M.: Porter's 'competitive advantage of nations': an assessment. Strateg. Manag. J. **12**(7), 535–548 (1991)

29. Dobbin, F., Baum, J.A.C.: Introduction: economics meets sociology in strategic management. In: Baum, J., Dobbin, F. (eds.) Economics Meets Sociology in Strategic Management, pp. 1–26. Emerald Group Publishing Limited, Bingley (2000)
30. Oliver, C.: Sustainable competitive advantage: combining institutional and resource-based views. Strateg. Manag. J. **18**(9), 697–713 (1997)
31. Shirey, M.R.: Lewin's theory of planned change as a strategic resource. JONA J. Nurs. Adm. **43**(2), 69–72 (2013)
32. Coghlan, D., Brannick, T.: Kurt Lewin: the "practical theorist" for the 21st century. Irish J. Manag. **24**(2), 31 (2003)
33. Appelbaum, S.H., Habashy, S., Malo, J.-L., Shafiq, H.: Back to the future: revisiting Kotter's 1996 change model. J. Manag. Dev. **31**(8), 764–782 (2012)
34. March, J.G., Olsen, J.P.: The uncertainty of the past: organizational learning under ambiguity. Eur. J. Polit. Res. **3**(2), 147–171 (1975)
35. Mouritsen, J.: Rationality, institutions and decision making: reflections on March and Olsen's rediscovering institutions. Account. Organ. Soc. **19**(2), 193–211 (1994)
36. Teece, D.J., Pisano, G., Shuen, A.: Dynamic capabilities and strategic management. Strateg. Manag. J. **18**(7), 509–533 (1997)
37. Teece, D.J.: A dynamic capabilities-based entrepreneurial theory of the multinational enterprise. J. Int. Bus. Stud. **45**(1), 8–37 (2014)
38. Teece, D.J.: Dynamic capabilities as (workable) management systems theory. J. Manag. Organ. **24**(3), 359–368 (2018)
39. van de Wetering, R., Mikalef, P., Helms, R.: Driving organizational sustainability-oriented innovation capabilities: a complex adaptive systems perspective. Curr. Opin. Environ. Sustain. **28**, 71–79 (2017)
40. Pérez-Moreno, S., Rodríguez, B., Luque, M.: Assessing global competitiveness under multi-criteria perspective. Econ. Model. **53**, 398–408 (2016)
41. Our Mission, World Economic Forum. https://www.weforum.org/about/world-economic-forum
42. Biermann, F., Kanie, N., Kim, R.E.: Global governance by goal-setting: the novel approach of the UN sustainable development goals. Curr. Opin. Environ. Sustain. **26**, 26–31 (2017)
43. Al-Emran, M., Arpaci, I., Salloum, S.A.: An empirical examination of continuous intention to use m-learning: an integrated model. Educ. Inf. Technol. 1–20 (2020)
44. Alhashmi, S.F.S., Salloum, S.A., Mhamdi, C.: Implementing artificial intelligence in the United Arab Emirates healthcare sector: an extended technology acceptance model. Int. J. Inf. Technol. Lang. Stud. **3**(3), 27–42 (2019)
45. Alghizzawi, M., Habes, M., Salloum, S.A., Ghani, M.A., Mhamdi, C., Shaalan, K.: The effect of social media usage on students' e-learning acceptance in higher education: a case study from the United Arab Emirates. Int. J. Inf. Technol. Lang. Stud. **3**(3), 13–26 (2019)
46. Alomari, K.M., AlHamad, A.Q., Salloum, S.: Prediction of the digital game rating systems based on the ESRB. Opción **35**(19), 1368–1393 (2019)
47. Salloum, S.A., Alhamad, A.Q.M., Al-Emran, M., Monem, A.A., Shaalan, K.: Exploring students' acceptance of e-learning through the development of a comprehensive technology acceptance model. IEEE Access **7**, 128445–128462 (2019)
48. Alhashmi, S.F.S., Salloum, S.A., Abdallah, S.: Critical success factors for implementing artificial intelligence (AI) projects in Dubai Government United Arab Emirates (UAE) health sector: applying the extended technology acceptance model (TAM). In: International Conference on Advanced Intelligent Systems and Informatics, pp. 393–405 (2019)
49. Wang, H., Huang, J., Zhang, Z.: The Impact of Deep Learning on Organizational Agility (2019)

Skill Gaps in Management Information Systems Alumni

Mukhtar Al-Hashimi, Allam Hamdan(✉), Anjum Razzaque, Abdalmuttaleb Al-Sartawi, and Sameh Reyad

College of Business and Finance, Ahlia University, Manama, Bahrain
allamh3@hotmail.com

Abstract. The aim of this paper is to epitomize on the outcomes of the skill gap questionnaire of Management Information Systems (MIS) alumni from Ahlia University, Bahrain. This study evaluated the importance of the job skills and the skill gaps which are associated with 58 technical and non-technical (soft) skills. These skills were assembled into five classifications in relevance with prior literature. The outcomes of this study can help instructors in the MIS programs of the higher education institutions to improve the MIS curricula, hence the programs, as per the current needs of MIS graduates.

Keywords: Skill gap · Management Information Systems · Alumni · Bahrain

1 Introduction

Currently, Information system and information technology careers are considered as the highest bands for knowledge employees within organizations. In the United States, knowledge employees form over half of the country's workforce. IT professionals consist of systems analysts, programmers, web designers, database administrators and network specialists (Laudon 2007). The required skills to be successful within the Management Information Systems (MIS) job market are developing constantly due to the rapid growth of technology trends. Therefore, periodic assessments of skills requirements are crucial in order to make sure that MIS programs are offering graduates required skills in order to be successful within the workplace (Janicki et al. 2004). The purpose of this paper is to evaluate the crucial skills that contribute in creating successful career levels within MIS related jobs, and to investigate required adjustments within the curricula. This study critiques a review of past literature related with management information system skill gaps and skill importance. The objectives of this study are to identify the crucial skills which contribute to the success of the early MIS career positions, identify existed gaps among actual and required skill levels, and prioritize skills so to be included within the MIS curricula.

2 Literature Review

There are ample studies related to MIS skills and gaps, conducted through surveys distributed to MIS alumni, educators and students. Further, analysis of ads for online jobs and focus groups was conducted with the sole purpose to analyze and identify the

A.-E. Hassanien et al. (Eds.): AICV 2020, AISC 1153, pp. 773–782, 2020.
https://doi.org/10.1007/978-3-030-44289-7_72

required skills for the IT professionals (Wilkerson 2012). Golding et al. (2008) stated that surveys of current students assist in evaluating students' skill levels within MIS program, however, are limited by such studies collecting information that is related to required skills for successful workplace because of scant experience of the respondents. Similarly, conducting surveys of graduates are crucial in order to assess the curricula of courses because they target individuals with work experience which include students who will be influenced by changes within the curricula. Ample studies have been conducted on MIS alumni. E.g. (Davis et al. 2008; Koopi 2009; and Reinig 2007).

Plice and Reinig (2007) distributed a survey to alumni to identify whether the balance between technical content and business should be modified within MIS programs. They stated that alumni are moving into jobs that require additional management responsibilities, thus, knowledge and management skills should acquire higher attentions than technical skills. Their findings were consistent with other studies (McMurtrey et al. 2008a, b, c; Davis and Woodward 2006a, b; Merhout et al. 2009; Golding 2008; Noll and Wilkins 2002a, b). Sumner and Yager (2008) indicated that soft skills are more crucial than technical skills for the MIS alumni, however a balanced curriculum contributes to prepare the alumni to attain important technical skills which are also required. Further, Sumner and Yager specified that skills and knowledge within emerging application development are specifically crucial technical skills for MIS alumni. Fang et al. (2005) argue that previous studies related to IS job skills were using number of classifications IS skills of job, thus, obtaining comparison of job skill studies is hard. However, studies were built on the previous work of (Todd 1995; Lee et al. 1995) to offer a classification scheme composed of interpersonal skills, organizational knowledge, core technical IS knowledge and IS knowledge, and personal skills. Core IS knowledge consists of knowledge that differentiates an IS personnel from other company colleagues. Organizational knowledge consists of knowledge of functional areas of business including marketing, accounting, etc. Interpersonal skills are communication and team skills. Personal skills are related to personal capabilities and traits including critical thinking and creative skills as well as personal motivation. Subsequent studies organized their questionnaires based to Fang et al. classifications of job skill scheme (McMurtery et al. 2008a, b, c; Golding et al. 2008; Tesch et al. 2008). Previous studies used the analysis of skill gap to determine the differences between actual and expected skill levels of alumni and recommend curricula changes (McMurtrey et al. 2008a, b, c; Richards et al. 2011; Trauth et al. 1993; Cappel 2001; Fang et al. 2005; Nelson 1991; Tang et al. 2001). Tesch et al. (2008) indicated that the three skills that have high gap between actual and expected skill levels include written communication, ability to listen and self-motivation. These three skills were ranked top five most importance skills in the employer job skills survey. Moreover, they are considered as soft skills. Cappel (2001) conducted survey on employers on soft and technical skills and revealed skill gaps in sixteen to nineteen technical skills. While, Lee and Han (2008) focused on analyzing hundreds of online job ads posted on Fortune 500 organizations' websites and reported that technical skills of programmers and analysts are more crucial than soft skills. From past literature it is noticeable that employers and MIS graduates believe that soft skills are more crucial than technical skills. However, both skills are significant. Further, significant skill gaps exist, and should be addressed in both skill aspects. Soft skills, including communication ability, ability of effective teamwork, and personal motivation are important skills of all professionals, and contribute to create a successful workforce. However, these skills do not differentiate MIS professionals from other business professionals including marketing and accounting professionals who should attain soft skills for career success (Downey et al. 2008a, b).

3 Research Methodology

A survey was searched out from MIS literature which was reviewed for related job skills and skill gaps. The most relevant survey that meets the aims and objectives of this study was adopted from Wilkerson (2012), An alumni assessment of MIS related job skill importance and skill gaps by Wilkerson. The adopted survey items from Wilkerson (2012) were further modified to ensure consistent wording of the remaining items. Additional items were added, and the resulting survey was reviewed by two MIS faculty members who further advised few modifications and refinements. The resulting survey consists of 78 items. The first 16 items consist of demographic and employment questions. The second part of the survey consists of 62 MIS related skill and knowledge questions organized into categories according to the Fang et al. (2005) classification scheme. An additional category—Technical Competencies—was added. The method of Meier et al. (2000) was used for all the 62 skills of the survey. Respondents were asked to rate their level of competence and importance from a scale of 1 to 5. 5 = 'Very competent' or 'Very important' and 1 = 'Not at all competent' or 'Not at all important'. To develop the results in terms of comprehension and to conserve the accuracy with prior results (Meier et al. 2012; Wilkerson 2012), the scores of both the competency and importance were transfigured into a 100-point scale, which was achieved by subtracting the score by 1 and then multiplying it by 25. To calculate the skill gap of the competency, the scored mean of the current competency is subtracted from the scored mean of the current importance. The positive scores designate the skills which contain skill gaps as the importance exceeds the competency. The higher the number, the bigger the skill gap. Meier et al. (2000) stated that results alone are not adequate enough to prioritize the skills with accuracy to make curriculum changes. Making inaccurate changes in the curriculum coulee result in new skill gaps. Thus, Meier et al. came up with a formula to calculate the priority taking into consideration the skill gap and importance:

$$\text{Priority} = (\mathrm{i} + (\mathrm{i} - \mathrm{c}))/2$$

where 'i' is skill importance and 'c' is skill competency. The priority score highlights the items with the highest importance and the highest skill gaps, which are the items that should receive the greatest attention when making curricular adjustment decisions.

Limitation

The limitation that was faced is that access was not permitted for international calls. Therefore, the study was conducted on Bahraini students only. Another limitation is that sample size is small as the research targets only one private university in Bahrain and it only graduated 123 MIS students and the researchers were only able to reach 90.

Demographic Results
The total respondents were 90, 51% of them were females, and the remaining 39% were male. Taking into consideration that overall number of females in the department is 72 out of 123. This could mean that females are more attracted to MIS than men. In other words, the department is female dominant.

4 Results

The results are presented category wise in order of importance (Table 1).

Table 1. Descriptive analysis

Category	Scaled mean	Rating question mean
Interpersonal	83.39	
Personal	81.5	
Organizational	72.42	
Technical competency	67.5	
Core technical IS knowledge	66.75	

Table 2. Functional analysis

Functional topic	Scaled mean	p-value
Marketing	77.75	.068
Operations	76.5	.567
Accounting	67.5	.161
Finance	67.5	.508
Economics	66.75	.741

Tables 3, 4, 5, 6 and 7 do show the scaled mean importance of all the five main categories. That in turn had the results sorted by descending order, in other words they're organized from most important to least important as highlighted by their necessities within the Tables above. The categories are depicted in order of importance as shown in Table 2. It is also notable that an item with a scaled mean of 50 is equivalent to earning a score of 3 (on a score of 1 to 5 that is), and an item with a scaled mean of 75 is equivalent to earning a score of 4 on the survey. However, it is notable that items earning a score of less than 50 would be tabled as unimportant to career success. As evident in Table 3, the nine items that fall under the interpersonal category ranged in importance from a scaled mean of 89 to 79.5, attesting that all interpersonal skill items measured in the survey are highly imperative to any career success. The survey also showed that the average of all interpersonal items is 81.5.

Table 3. Skills analysis

Functional topic	Scaled mean
Oral communication skills	89
Working effectively in teams	86
Persuading others	84.75
Teaching/training others	83.25
Giving effective presentations	83.25
Writing clearly and effectively	82.25
Resolving conflicts	81.75
Effectively leading a team or group	80.75
Listening to others	79.5

In Table 4 it shows the scaled mean importance of the nine items from the personal category. These items ranged in importance from 86.5 to 77 with an average of 81.5. The survey determines that 'Generating new ideas' was the most important skill in this category, followed closely by the skill 'Working under pressure'.

Table 4. Personal skills

Functional topic	Scaled mean
Generating new ideas (creative thinking skills)	86.5
Working under pressure	83
Accomplishing assignments	83
Analyzing problems and developing solutions (critical thinking skills)	82.5
Perform multiple tasks at the same time	81.75
Making decisions	80.5
Working independently to accomplish a goal/objective	79.75
Learning new skills and concepts	79.5
Managing time effectively	77

Table 5 shows the scaled mean importance of items within the Organizational category. These items ranged in importance from 85.5 to 64.75. While the survey showed the two most important items in this category are none other than 'Business ethics and privacy issues' and 'Marketing'.

Table 5. Organizational category

Functional topic	Scaled mean
Business ethics and privacy issues	85.5
Marketing	77.75
Operations	76.5
General business functions and principles	75.75
Knowledge of a specific business industry	69.75
Accounting	67.5
Finance	67.5
Economics	66.75
Knowledge of a specific company or organization	64.75

Table 6 depicts the scaled mean importance of the items in the Core Technical IS Knowledge category. These items ranged in importance from 67 to 49.75; meaning that this category contained a combination of skills associated with systems analysis, system design and implementation. All the aforementioned skills scored above 50: except 'Designing system architecture'.

Table 6. Knowledge category

Functional topic	Scaled mean
Creating web pages and web sites	67
Anticipating implementation problems	60
Creating or evaluating computer security and privacy policies	59.25
Designing user interfaces	59.25
Computer programming	59
Developing applications in multiple environment/platforms	57.75
Performing feasibility analysis	57.25
Using a specific computer operating system	55.25
Performance tuning of databases	54.25
Using a specific database management system (Oracle, Sybase, SQL Server, MySQL, etc.)	53.75
Applying software design patterns	53.5
Using software testing tools and strategies	51.75
Programming in object-oriented language	50.25
Data modelling	50
Designing system architecture	49.75

Table 7 depicts the scale's mean importance of the technical competencies ranging from 68.5 to 54. The most important technical skill is '*Creating Flow Charts*' with a scaled importance of 68.5. The next most important skill was '*Using MS Access*' with a scaled importance of 67.75. The two lowest scoring items are '*Programming in Java*' and '*Programming in C++*' (Tables 8 and 9).

Table 7. Technical competencies

Functional topic	Scaled mean
Creating Flow Charts	68.5
Using Ms Access	67.75
Creating HTML pages	66.5
Creating Data Flow Diagram (DFDs)	65.75
Programming in PHP	63.5
Using project management tools (MS Project, etc.)	62.75
Using static tools (SAS, SPSS, Minitab, etc.)	62.75
Programming in Visual Basic	62.25
Using Oracle	62.25
Using Spreadsheet tools (Excel, Lotus, Quattro Pro, etc.)	62
Processing XML documents	61.5
Using MySQL	60.75
Using data analysis and data mining tools	60.75
Using integrated development environments (Visual Studio, Eclipse, etc.)	60
Creating or reading UML diagrams	59.75
Using Enterprise Resource Planning (ERP) tools (SAP, etc.)	59.75
Using SQL Server	58.5
Developing applications using ASP	57.75
Programming in Java	55.75
Programming in C++	54

Table 8. Professional skills

Functional topic	Scaled mean	p-value
Ms Access	67.75	.012
Oracle	62.25	.058
MySql	60.75	.113
SQL Server	58.5	.561

Table 9. Advance professional skills

Functional topic	Scaled mean	p-value
PHP	63.5	.016
Visual Basic	62.5	.200
Java	55.75	1.000
C ++	54	.426

As shown in the table above, PHP is the most important programming language included in the survey, which is then followed by Visual Basic. The least important programming language is C++.

5 Conclusion

Consistent with prior studies, these results indicate that 'soft' skills are more important to MIS professionals' career success than technical skills. The below table shows skills with gaps. The results of this paper will be compared with the results of Wilkerson (2012). The Interpersonal category has skill gap in 5 skills out of 9. 'Teaching/training others' has the highest skill gap in the category which is 5.49 and the lowest skill gap was 'Resolving conflicts' with a score of 0.25. Wilkerson measured the same skills but got different skills except for two which are 'Persuading others' and 'Writing clearly and effectively'. The Personal category had 9 skills and gaps were found in 8 of them, meaning, only one skill did not have a gap. This shows that participants think that all these soft skills are important but don't have the required competency in them. Wilkerson evaluated 15 skills and found skill gaps in the same skills except for one skill 'Working independently to accomplish a goal/objective'. The study found skill gaps in the organizational category. It has 5 skill gaps with the highest score of 5 in 'Knowledge of a specific business industry' and the lowest score of 1 in 'Finance'. Wilkerson did not get any skill gaps in the organizational category.

The Technical Competency category bares 20 skills; 13 of which have skill gaps. The item with the highest skill gap is '*Creating Flow Charts*' with a score of 5.25. Whereas, the minimum skill gap was '*Using Enterprise Resources Planning (ERP) tools (SAP, etc.)*' with a score of 0.5. The last category, Core Technical IS Knowledge had 11 items with skill gaps. More than one item is regarding databases; 'Performance tuning of databases' and '*Using a specific database management system (Oracle, Sybase, SQL Server, MySQL, etc.).*'

Majority of the skills that has skill gaps highest priority items are from the Interpersonal and Personal categories. The highest item based on curriculum priority are: Generating new ideas (creative thinking skills), accomplishing assignment, oral communication skills, working effectively in teams, persuading others, and writing clearly and effectively. The skills in the Technical competency category with the highest priority: using project management (MS project, etc.), programming in PHP, and creating flowchart. Only these three skills were in the top 1/3 of the curriculum priority items. Only one item from the Core Technical IS Knowledge category; creating web pages and websites.

This study also concluded that soft skills are more important than technical skills. Emotions could also be an important factor that needs to be taken into consideration in the future (Alalawi and Razzaque 2016).

References

Wilkerson, J.W.: An alumni assessment of MIS related job skill importance and skill gaps. J. Inf. Syst. Educ. **23**(1), 85–98 (2012)

Golding, P., Tennant, V., Donaldson, O.: An empirical classification of employability skills for entry level IS/IT professionals. Paper presented at the AIS SIGED IAIM Conference, Paris, France (2008)

Davis, D.C., Woodward, B.: An analysis of skills required of graduates of an information systems program. Inf. Technol. Learn. Perform. J. **24**(2), 11–21 (2006a)

Noll, C.L., Wilkins, M.: Critical skills of IS professionals: a model for curriculum development. J. Inf. Technol. Educ. **1**(3), 143–154 (2002a)

Lee, S.M., Lee, C.K.: IT managers' requisite skills: matching job seekers' qualifications with employers' skill requirements. Commun. ACM **49**(4), 111–114 (2006)

McMurtrey, M., Downey, J., Zeltmann, S., Friedman, W.: Critical skill sets of entry-level IT professionals: an empirical examination of perceptions from field personnel. J. Inf. Technol. Educ. **7**, 101–120 (2008a)

Fang, X., Lee, S., Koh, S.: Transition of knowledge/skills requirement for entry-level IS professionals: an exploratory study based on recruiters' perception. J. Comput. Inf. Syst. **46** (1), 58–70 (2005)

McMurtrey, M.E., Downey, J.P., Zeltmann, S.M., Friedman, W.H.: Critical skill sets of entry-level IT professionals: an empirical examination of perceptions from field personnel. J. Inf. Technol. Educ. **7**, 101–120 (2008b)

Janicki, T.N., Kline, D., Gowan, J.A., Konopaske, R.: Matching employer needs with IS curriculum: an exploratory study. Inf. Syst. Educ. J. **2**(21), 3 (2004)

Hecker, D.E.: High-technology employment: a NAICS-based update (2005)

Downey, J.P., McMurtrey, M.E., Zeltmann, S.M.: Mapping the MIS curriculum based on critical skills of new graduates: an empirical examination of IT professionals. J. Inf. Syst. Educ. **19** (3), 351–363 (2008a)

Gill, T.G., Hu, Q.: Information systems education in the USA. Educ. Inf. Technol. **3**, 119–136 (1998)

Todd, P.A., McKeen, J.D., Gallupe, R.B.: The evolution of IS job skills: a content analysis of IS job advertisements from 1970–1990. MIS Q. **19**(2), 1–27 (1995)

Wu, J., Chen, Y., Chang, J.: IS knowledge gaps: an industrial perspective. In: Proceedings of the 39th Hawaii International Conference on System Science (2006)

Davis, D.C., Woodward, B.: An analysis of skills required of graduates of an information systems program. Inf. Technol. Learn. Perform. J. **24**(2), 11–21 (2006b)

Noll, C.L., Wilkins, M.: Critical skills of IS professionals: a model for curriculum development. J. Inf. Technol. Educ. **1**(3), 143–154 (2002b)

Merhout, J.W., Havelka, D., Hick, S.N.: Soft skills versus technical skills: finding the right balance for an IS curriculum. Paper presented at the Americas Conference on Information Systems, San Francisco, California (2009)

Sumner, M., Yager, S.E.: An investigation of preparedness and importance of MIS competencies: research in progress. Paper presented at the ACM SIGMIS CPR Conference on Computer Personnel Doctoral Consortium and Research, Charlottesville, Virginia, pp. 97–100 (2008)

Plice, R.K., Reinig, B.A.: Aligning the information systems curriculum with the needs of industry and graduates. J. Comput. Inf. Syst. **48**(1), 22–30 (2007)

Downey, J.P., McMurtrey, M.E., Zeltmann, S.M.: Mapping the MIS curriculum based on critical skills of new graduates: an empirical examination of IT professionals. J. Inf. Syst. Educ. **19** (3), 351–363 (2008b)

McMurtrey, M.E., Downey, J.P., Zeltmann, S.M., Friedman, W.H.: Critical skill sets of entry-level IT professionals: an empirical examination of perceptions from field personnel. J. Inf. Technol. Educ. **7**, 101–120 (2008c)

Cappel, J.J.: Entry-level IS job skills: a survey of employers. J. Comput. Inf. Syst. **42**(2), 76–82 (2001)

Tesch, D.B., Braun, G.F., Crable, E.A.: An examination of employers' perceptions and expectations of IS entry-level personal and interpersonal skills. Inf. Syst. Educ. J. **6**(1) (2008)

Trauth, E.M., Farwell, D.W., Lee, D.: The IS expectation gap: industry expectations versus academic preparation. MIS Q. **17**(3), 293–307 (1993)

Alalawi, M., Razzaque, A.: Peak-end rule to promote social capital in a virtual community environment. In: International Conference on Emerging Research for Sustainable Economic Development (ERSED), Dubai, UAE (2016)

Internet of Things for Learning Styles and Learning Outcomes Improve e-Learning: A Review of Literature

Anjum Razzaque and Allam Hamdan(✉)

College of Business and Finance, Ahlia University, Manama, Kingdom of Bahrain
allamh3@hotmail.com

Abstract. Past scholars highlighted the dissatisfactions of learning in traditional teachings, which is a vital concern for the Higher Education Institutions (HEIs). This is especially due to the numerous concerns of scholars pertaining to the declining teaching learning quality. This study reviewed literature to explore what role Internet of Things (IoT) and students' Learning Style (LS) has on learner expectations, through the Learning Outcomes (LOs) of HE students. Though this stage is a research in progress, instructors can improve curriculums using complimentary LS and LOs through the help of IoT, after reading the reviews of literature portrayed in this article. And HE instructors can extend their reach to students via the e-learning environments as the model was this study was examined within the e-learning context to confirm that IoT and LS do facilitate the achieving of LOs; thru an interdisciplinary model proposed in this paper. Implications to theory and practice are also expressed.

Keywords: Higher education · Internet of Things · Learning outcomes · Learning styles · e-Learning

1 Introduction

Higher Education Institutions (HEIs) suffer from declining quality in Teaching and Learning (T&L), particularly due to poor curriculum design as indicated in student surveys (Akalu 2016; Harrison and Risler 2015; Subbiondo 2013). Though some researchers identify some remedial measures, they point out that more should be done to address sustainability and delivery of HEIs quality (Lu et al. 2017; Fiselier et al. 2018); an essential literature gap for addressing this issue is to improve HEI T&L quality. Scholars have identified various factors like Learning Outcomes (LOs), and Learning Style (LS), in multinational HEIs context can enhance T&L quality (Caspersen et al. 2017a, b; Hill 2012). However, the nature of the relation between these factors and T&L quality is poorly understood; implying that investigating such factors can help arrest the declining HEI T&L quality. With the prevalence of the use of technology in the T&L activities, several technology-related factors can also have been heavily researched in this context. Internet of Things (IoT), have emerged as strong contender factor to integrate in T&L activities (Davalos et al. 2016). One area that

A.-E. Hassanien et al. (Eds.): AICV 2020, AISC 1153, pp. 783–791, 2020.
https://doi.org/10.1007/978-3-030-44289-7_73

attracted researchers and practitioners are LOs for their high potential to reverse the decline in HEI T&L quality, and the other is understanding the role of IoT within all this phenomenon (Farhan et al. 2018). Moreover, researchers argue that the relation between IoT, LSs and LOs, for improving T&L quality, is under researched with contradicting current knowledge underpinned by lacking consensus (Farhan et al. 2018; Caspersen et al. 2017a, b; Hill 2012). Such a holistic integration of IoT, LSs and LOs can deepen the under researched understanding of LOs for solving the declining HEI T&L quality. Moreover, a lack of consensus and contradictions amongst researchers pertains to factors affecting LOs and HEI T&L quality, an uncertain and inconclusive knowledge (Huang et al. 2018). For instance, literature reveals that some contributing factors in enhancing T&L quality, other than LOs, are teaching methodology, faculty development, learning strategies, self-efficacy, learner and instructor interaction, in addition to IoTs. From the perspective of IoT, this is not surprising since education sector tends to be slow at adopting information and communication technologies (Sultana and Haque 2018; Huang et al. 2018; Reich et al. 2016; Canfield et al. 2015; Tong et al. 2015; Caspersen et al. 2017a, b; Hill 2012). However, limited literature was found on how and to what extent those factors affect LOs and hence T&L quality, or if these factors can reliably predict how HEIs deliver better T&L quality. This research studies the relation between students' perceived use of IoT, and its effect on LS, and LOs. Such a model is motivated by the concerns of the past scholarly who report that HEIs' LOs generally do not account for LS, or via IoT, since IoT is considered by scholars as in its blossoming stage since everyone is trying to define this field as per their own interpreted vision; making this field terms also as web of everything, though such a technology (WoT), Cloud of Things (CoT), Internet of Everything (IoE), or Machine to Machine (M2M); already faces several challenges pertaining to its interoperability, scalability, efficiency, storage, availability, and security (Čolaković and Hadžialićb 2018; Jesus 2018; Mircea et al. 2017; Miraz et al. 2015). This is not surprising since from a theoretical perspective, LOs are weakly grounded on established theories, thus applying current knowledge into practice is questionable (Rajaram and Bordia 2013). Furthermore, the literature expresses no comprehensive explanations to link curriculums LO, on the one hand, with students perceived use of IoT and LS, on the other; using currently available theories; hence any new knowledge linking LOs and LS can shed insight on HEI T&L quality.

2 Literature Review

2.1 Effect of IoT on LSs, and LOs via e-Learning

By 2022, it is forecasted that 29 billion devices will be connected, and 18 billion of these devices will be IoT. Costing around 263 billion USD, will get connected on the internet to enable the analysis of data for artificial intelligent decision-making (Ericsson 2019; Gartner 2014) to connect smart technologies (devices) anytime and anywhere (Ashton 2009a, b). In an IoT environment various processes will be integrated to add value to services by grouping various IoT applications to support domains like healthcare, traffic, retail, etc. Still, IoT is considered by scholars in its blossoming stage since everyone is trying to define this field as per their own interpreted vision; making this field terms also as web of everything, though such Web of technology (WoT), Cloud of Things (CoT), Internet of Everything (IoE), or Machine to Machine (M2M); already faces several challenges pertaining to its interoperability, scalability, efficiency, storage, availability, and security (Čolaković and Hadžialićb 2018; Jesus 2018; Mircea et al. 2017; Miraz et al. 2015).

Within our evolving society people feel the need for knowledge and learning through their cognitive and social skills by integrating personally and professionally: through their familiar social and interaction ties (Razzaque et al. 2013). This is possible using technologies like knowledge management tools (e.g., virtual communities) (Razzaque and Eldabi 2018), which is one success story of a technology since the technological revolution of the 1990s; when the World Wide Web went live (Budd 2019). It was not long before the adoption of such technologies got manifested within the education sector though there are scholars who consider the education sector to be slow at adopting information and communication technologies; for instance (Sultana and Haque 2018). The most important development within the education sector is the ability for individuals to choose to accept the adoption of any technological application. Changes only occur when most of the individuals accept to adopt a technology like e-learning solutions which are successful when facilitated by intelligent system. These are based on the standards of quality teaching blended with ICT for achieving a ubiquitous learning experience when IoT and cloud computing, also known as cloud learning in education research, assimilate mutually benefiting intelligent systems for form smart learning furnished as a Smart Education as a Service (SEaaS), still a concept which is under scholarly debate (Razzaque 2019; Stoica et al. 2018; Sánchez et al. 2016). Technological developments are bringing upon the 21st century economic, political, cultural, and social changes (Bodura et al. 2019). In the 1990s, Kevin Ashton who was one of the developers of the Radio Frequency Identification (RFID) community introduced the concept of IoT, and practicalized it with increasing mobile phones extended with sensors (i.e., things that reside virtually to collect and process information, through cloud computing, and data analytics for decision making. Internet of Things connects various smart objects since it enables objects (devices/things) that were not meant to talk to each other and interact smartly (Ashton 2009a, b). Since the 1990s several scholars have expressed interest in the application of IoT within the education sector, such as (Magrabi et al. 2018; Mershad and Wakim 2018; Shyr et al. 2018; Tan et al. 2018; Burd et al. 2017; Bagheri and Movahed 2016). Such a research is essential considering that the deployment IoT is foreseen to generate 300 billion USD through 26 billion IoT devices on the Internet by 2020, which will surpassing the total population on Earth (Ager 2019; Elyamany and AlKhairi 2015). In the education sector, the use of IoT has shown prosperous practical implications from the prism of its influential connectivity for educating learners, and for the instructors to improve this teaching methods blended with the T&L infrastructure (Elyamany and AlKhairi 2015). This is possible through the use of sensors, e.g., Super Mechanical's Twine7 like ubiquitous objects in HEIs (Cisco Consulting Services and Cisco EMEAR Education Team 2013). The adoption of IoT is also quicker for instructors and learners since IoT by both HEI stakeholders is an attractive concept (Chin and Callaghan 2013). For instance, in the English language curricula, IoT is interfaced with learners by their LSs achieved through visual and audio sensors which correct the teered (promouncations) (Chin and Callaghan 2013). Another example is the aligning of IoT with learning strategies within the program language curriculas, through wireless sensor networks that are interfaced with cloud computing and data mining, so learners can analytically learn while they tap into the educational resources of their intelligent/digital campuses (Davalos et al. 2016; Cheng and Liao 2012; Wang 2010). Cheng and Liao provide an excellent analysis when differentiating intelligent from a digital HEI campus through the application of cloud computing and IoT such as smartboards, tablets, smart phones, e-books, attention trackers which is integrated with scored face detection via the learner monitoring through video camera system for tracking learnings connection/s or interaction/s with their instructor, learning management systems like ILIAS, Sakkai, or Moodle (Abdel-Basset et al. 2018; Farhan et al. 2018; Malik et al. 2015). From our experience and observation, instructors are involved in various decisions when they refine their curriculums to suit students' needs for achieving learning outcomes by assuring appropriate learning styles. Decisions are based on choices that lead to cascading actions. Here, a Decision Support System (DSS), i.e., computerized algorithm driven by analyzed data, from various sources, supports decision-making (Sauter 2014). The term data here is a composition of its two time-periods. Initially it was referred as internal data, residing in database/s and understood in structure format, i.e., in the form of specialized reports, and during the second period the same data has been terms as "big data" which is when new approaches, like Instructive Information Mining (IIM) and e-learning assessment tools for understanding learners LS and formulating an environment for instructors to work out how to accordingly improve learning materials to reach the desired LOs, to be furnished external practical implications of this data through external sources for data analysis for business processes for decision-making (Sallis et al. 2016; Garrick and Pende 2014; Hess and Saxberg 2014; Sauter 2014; Arenas-Gaitán et al. 2011). Within the T&L context, IoT aid through its human-centric and smart environment by smarten up the surrounding environments by personalizing learning via personalized hybrid learning method (blending of formal and information learning activities outside classrooms: e.g., learning through gaming); through an arrange of IoT devices is also referred as "ubiquitous computing," and "smart learning" (Patel and Patel 2016; Abualkishik and Omar 2013). Previous research has indicated the importance of IoT is learning, such that scholars identified how motivation and learner experience are important factor for students to participant in e-learning environments (Hess and Saxberg 2014; Krause et al. 2014). Accordingly, the three propositions, P2 and P2: Fig. 1 are:

P1: IoT plays a positively significant role on LSs.
P2: IoT plays a positively significant role in LOs.

2.2 Effect of LS on LOs

LOs is recognized in HEIs for efficient and effective coursework quality, through student- centralized teaching statements outlining learner-expected knowledge/skills (Havnes and Prøitz 2016). Accreditation agencies assure such evidenced-assessment tools, like grades, for curriculum development (Michelsen et al. 2017; Prøitz et al.

2017; Davis 2016; Reich et al. 2016; Canfield et al. 2015). LOs is demanded in research, and by employees expecting social, critical thinking, and soft skills from graduates (Caspersen et al. 2017a, b; Reich et al. 2016; Fatmi et al. 2013; Hill 2012; Ogundokun 2011; Osters and Tiu 2012). Havnes and Prøitz (2016) are one of few reporting LOs shortcomings, testifying that the two-fold LOs implementation (i.e. managerial quality assurance and course designing for T&L) harvests multiple LOs meanings, at different HEI levels. Therefore, curriculum designers are skeptical of LOs since there meaning vary at a cognitive, socio-cultural, and behavioral perspective of T&L. Limited research articulated a mutual understanding of LOs from T&L perspectives and therefore calls for a thorough investigation in LOs from the lens of perspectives since the meaning of LOs varies by geographical regions; making such investigation vital since HEI are currently realizing a high intake of international students (Jacob and Gokbel 2018; James-MacEachern and Yun 2017). Furthermore, limited research quantitatively understands how LSs predict LOs for better curriculum design through many scholars contributed new theories and applicable models (Moylan and Razzaque 2015; Strang 2010). From 1970s, LSs research first expressed the Kolb's Model of Experiential Learning (KMEL) includes concrete experience, reflective experience, abstract conceptualization and active experimentation (Knoll et al. 2016; DiBartola et al. 2001; McLeod 2013). This model investigates LSs, though criticized for theoretical problems as Kolb rejected the notion that LS depend on objective measures and LOs. To optimize LOs, a learner should know their preferred LS (Heffler 2010). A weak underpinning KMEL instrument's research methodologies highlight statistical problems, making it invalid (Koob and Funk 2002; Henson and Hwang 2002). Later, Honey's and Lobley's (1986) LS theory was adapted by Charlesworth's (2008) who quantitatively assess LSs on Chinese, Indonesian and French HE students, to recognize the importance of their cultural backgrounds, since classrooms function differently in different cultures. Charlesworth's LS model, adapted from Honey's and Lobley, was four-dimensional: theorist, activist, reflector and pragmatist. Charlesworth's findings were aligned with Honey and Lobley: i.e., activist learners flourish when challenged with new experiences, but uninterested in implementing such experiences. While reflector learners stand back and listen to consider all possibilities before concluding, pragmatist learners deliberately analyze and search new ideas. They dissect and reassemble problems while applying new learnt ideas. Theorist learners keenly learn theories, thus detached analytically and cannot appreciate ambiguous experiences (Kappe et al. 2009; Charlesworth 2008; Honey and Lobley 1986). A strong link resonates between LS and LOs since learning occurs in four stages. In each stage, every LS plays a vital role. The first stage is gaining experience, appreciated by activist learners. The second is reviewing experience, appreciated by reflector learners. The third stage is concluding an experience, appreciated by theorist learners. The fourth is the planning stage; appreciated by pragmatist learners (Goldfinch and Hughes 2007). LOs are cognitive traits of learning performance by absorbing knowledge. To optimize HE quality, instructors should design curriculum with clear LSs customized to what is being studied (Ogundokun 2011) since teaching styles interact with students' preferred LS. Curriculum designing can incorporate student preferred learning approaches or use contradicting LSs to improve learners' weak LS/s or apply a wide range of LSs catering all students (Yu et al. 2010). The literature reported four LSs: active/reflective, sensing/intuitive, visual/verbal and sequential/global. Active learners solve problems while reflective learners loiter thinking and learning. Sensemaking learners memorize to comprehend. Intuitive learners mathematically solve problems. Visual learners learn through diagrams while verbal learners learn through interpretations. Sequential learners' micro-understand before comprehending the big picture. Global learners learn from a holistic sense for problem solving (Yu et al. 2010; Ogundokun 2011; Yousef 2016; Knoll et al. 2016). Holt et al. (1974) is one of first few scholars expressing the importance of LS for LOs. Various scholars express contradicting results towards this role. McCann (2006)'s quantitative assessed LS on students' performance (LOs) in face-to-face interaction, e-learning with minimal instructor interaction and e-learning with high instructor interaction; evidenced that students score higher in face-to-face-classrooms and e-learning environments with high instructor interactions; indirectly indicating a strong affiliation between LS and LOs. Lu et al. (2007) assess the role of Kolb's LS Inventory (KLSI) on (1) endured time-on-e-learning behavior (ELB), (2) KLSI on LO and (3) LO and ELB by experimenting forty undergrad students, revealing: (1) no significance between KLSI and ELB, (2) no significant between LS and LO and (3) significance between LO and ELB. Ogundokun (2011) also assess the role of LS, school environment, and test anxiety, on LOs by surveying 300 senior secondary school students, revealing all significant relations. Further, Liew et al. (2015) assessed LS on LO by (1) identifying LS of preclinical undergrads and (2) assessed the role of learning preference (LSs and learning approaches) on performance (LOs) via 419 international preclinical undergrad survey participants from a Kuala Lumpur university; revealing that majority of learners preferred unimodal LS, and minority-preferred multimodal LS. Also indicated no relation between LS and LOs. The best definition of LS, driven from Kappe et al. (2009) and Charlesworth (2008), categorizes LS in four categories: theorist, activist, reflector and pragmatist LS. Students with activist LS enjoy challenges while enjoying new experiences. Such socially inclined students enjoy experiences, though bored when implementing them. Students with reflector LS act slowly, since they observe processes and predict outcomes while learning. Pragmatists are deliberate and analytic since they are practical and down to earth; enjoy dissecting/reassembling problems though disliking deep learning. Although Charlesworth reported that Chinese students were inclined to the theorist LS, the quantitative analysis on Indonesian, Chinese and French students expressed significance of theorist LS; confirming that the theorist dimension deserves no recognition of this research. Additionally, Charlesworth's LS instrument needs empirical retesting to reassess its influence on LOs in HE multicultural-classrooms context, since scholars reflect contradicting effects of LS on LO with minimal understanding of the role of LS on LO such context. Accordingly, the three propositions, P3, P4, and P5: Fig. 1 are

P3: Activist students LS plays a positively significant role on LOs.
P4: Reflector students LS plays a positively significant role on LOs.
P5: Pragmatic Students LS plays a positively significant role on LOs.

As depicted in Fig. 1's conceptual model with five propositions, the holistic integration of IoT, LSs and LOs is inspired through the e-learning initiatives undertaken by

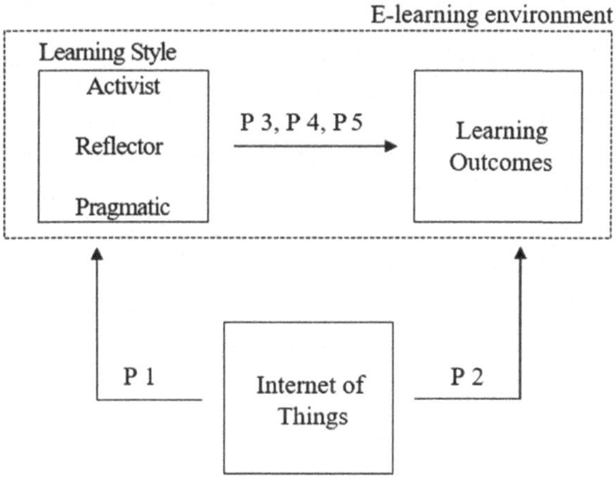

Fig. 1. Research model

the private and public HEIs of the Kingdom of Bahrain. Instructors aligned T&L with students' IoT realize they receive higher scores in student evaluations.

3 Discussion and Conclusion

The aim of this literature review study is to comprehend what perceived benefits could IoT pose on an e-learning environment so that instructors and learners will be aided for improving their LSs to better achieve their LOs. There is a theoretical implication pertaining to this research: i.e., the Fig. 1 holistic integration of IoT, LSs and LOs deepens the under-researched understanding of LOs for solving the declining HEI T&L quality; hence addresses the lack of consensus and contradictions amongst researchers pertains to factors affecting LOs and HEI T&L quality, an uncertain and inconclusive knowledge (Huang et al. 2018). Our research adds value to knowledge towards the education sector since this sector was reported slow at adopting ICT (Sultana and Haque 2018; Huang et al. 2018; Reich et al. 2016; Canfield et al. 2015; Tong et al. 2015; Caspersen et al. 2017a, b; Hill 2012) and scant research understood the extent to which IoT affects LOs and hence T&L quality; through LOs. This research studies the relation between students' perceived use of IoT, and its effect on LS and LOs; as it was motivated to address the empirically unassessed claim of past scholarly stating that LOs do not account for LS (Čolaković and Hadžialićb 2018; Jesus 2018; Mircea et al. 2017; Miraz et al. 2015); should require empirical evidencing. This scarcity is also felt across mature and emergent technologies supporting LS and LOs such as cloud computing, where the focus is on mainly on the research of the application of cloud computing in e-learning environments (El Mhouti et al. 2018). Furthermore, there is a practical implication of the Fig. 1 model. If this model is empirically assessed, within a HEI context, instructors will have a motivating confirmation of how important IoT us when it comes to facilitating LS and LOs. Instructors will be able to comfortable adopting IoT to enhance their curricula which in turn will be more individualized to learners than simply being customized to the whole course. The same advantage goes also for LOs which will not only be centered to a course but individually each learner. The main limitation that we discovered was from our thorough literature review. There is scant literature which even discusses the affect of IoT on e-learning environment, let alone the affect of IoT on LS and LOs. Future research could empirically assess the Fig. 1 model cross-sectionally and longitudinally, and the target population should be learnings and instructors to attain a holistic comprehension of how this model can facilitate the improvement of the T&L quality.

References

Abdel-Basset, M., Manogaran, G., Mohamed, M., Rushdy, E.: Internet of Things in smart education environment: supportive framework in the decision-making process. Concurr. Comput. Pract. Exp. **31**(10) (2018). https://doi.org/10.1002/cpe.4515

Abualkishik, A., Omar, K.: Framework for translating the Holy Quran and its reciting rules to Braille code. In: 2013 International Conference on Research and Innovation in Information Systems (ICRIIS), Kuala Lumpur, Malaysia. IEEE (2013)

Ager, L.: Report says IoT will generate revenue exceeding US$300billion by 2020 (2019). From DevOps Online.co.uk: https://www.devopsonline.co.uk/iot-installed-base-grow-26-billion-units-2020/. Accessed 30 Aug 2019

Akalu, G.A.: Higher education 'massification' and challenges to the professoriate: do academics' conceptions of quality matter? Qual. High. Educ. **22**(3), 260–276 (2016)

Arenas-Gaitán, J., Ramírez-Correa, P.E., Rondán-Cataluña, F.: Cross cultural analysis of the use and perceptions of web based learning systems. Comput. Educ. **57**(2), 1762–1774 (2011)

Ashton, K.: That 'Internet of Things' thing. RFID J. **22**(7), 97–114 (2009a)

Ashton, K.: That 'Internet of Things' thing (2009b). From RFiD Journal: https://www.rfidjournal.com/articles/view?4986. Accessed 30 Aug 2019

Bagheri, M., Movahed, S.H.: The effect of the Internet of Things (IoT) on education business model. In: 2016 12th International Conference on Signal-Image Technology & Internet-Based Systems (SITIS). IEEE (2016)

Bodura, G., Gumus, S., Gursoy, N.G.: Perceptions of Turkish health professional students toward the effects of the Internet of Things (IOT) technology in the future. Nurse Educ. Today **79**, 98–104 (2019)

Budd, J.: On this day in… 1991: the day the World Wide Web shook the channel, 6 August 2019. From Channel Partner Insight: https://www.channelpartnerinsight.com/channel-partner-insight/feature/3080110/on-this-day-in-1991-the-day-the-world-wide-web-was-born. Accessed 30 Aug 2019

Burd, B., Barker, L., Divitini, M., Perez, F.A., Russell, I., Siever, B., Tudor, L.: Courses, content, and tools for Internet of Things in computer science education. In: ITiCSE- WGR 2017 Proceedings of the 2017 ITiCSE Conference on Working Group Reports, Bologna, Italy, pp. 125–139. ACM Digital Library (2017)

Canfield, M.L., Kivisalu, T.M., Karr, C.V.: The use of course grades in the assessment of student learning outcomes for general education. SAGE Open **5**(4), 1–13 (2015)

Caspersen, J., Frølich, N., Muller, J.: Higher education learning outcomes – ambiguity and change in higher education. Eur. J. Educ. **52**(1), 8–19 (2017a)

Caspersen, J., Smeby, J.-C., Aamodt, O.P.: Measuring learning outcomes. Eur. J. Educ. Res. Dev. Policy (2017b). https://doi.org/10.1111/ejed.12205

Charlesworth, Z.M.: Learning styles across cultures: suggestions for educators. Educ. Train. **50** (2), 115–127 (2008)

Cheng, H.-C., Liao, W.-W.: Establishing an lifelong learning environment using IOT and learning analytics. In: 2012 14th International Conference on Advanced Communication Technology (ICACT). IEEE (2012)

Chin, J., Callaghan, V.: Educational living labs: a novel internet-of-things based approach to teaching and research. In: 2013 9th International Conference on Intelligent Environments. Athens, Greece. IEEE (2013). https://doi.org/10.1109/ie.2013.48

Cisco Consulting Services and Cisco EMEAR Education Team: Education and the Internet of Everything, October 2013. From CISCO: https://www.cisco.com/c/dam/en_us/solutions/industries/docs/education/education_internet.pdf. Accessed 30 Aug 2019

Čolaković, A., Hadžialićb, M.: Internet of Things (IoT): a review of enabling technologies, challenges, and open research issues. Comput. Netw. **144**, 17–39 (2018)

Davalos, A.G., Marquez, J., Villanueva, J., Solarte, Z.: IoT in education: integration of objects with virtual academic communities. In: Kacprzyk, J. (ed.) New Advances in Information Systems and Technologies. Advances in Intelligent Systems and Computing. Springer, Cham (2016)

DiBartola, L.M., Miller, M.K., Turley, C.L.: Do learning style and learning environment effect learning outcomes? J. Allied Health **30**(2), 112–115 (2001)

El Mhouti, A., Erradi, M., Nasseh, A.: Using cloud computing services in e-learning process: Benefits and challenges. Educ. Inf. Technol. **23**(2), 893–909 (2018). https://doi.org/10.1007/s10639-017-9642-x

Elyamany, H.F., AlKhairi, A.H.: IoT-academia architecture: a profound approach. In: 2015 IEEE/ACIS 16th International Conference on Software Engineering, Artificial Intelligence, Networking and Parallel/Distributed Computing (SNPD), Takamatsu, Japan. IEEE (2015). https://doi.org/10.1109/snpd.2015.7176275

Ericsson: Internet of Things forecast (2019). From The connected future: https://www.ericsson.com/en/mobility-report/internet-of-things-forecast. Accessed 5 Sept 2019

Farhan, M., Jabbar, S., Aslam, M., Hammoudeh, M., Ahmad, M., Khalid, S., Khan, M., Han, K.: IoT-based students interaction framework using attention-scoring assessment in eLearning. Future Gener. Comput. Syst. **79**(3), 909–919 (2018)

Fatmi, M., Hartling, L., Hillier, T., Campbell, S., Oswald, A.E.: The effectiveness of team-based learning on learning outcomes in health professions education: BEME Guide No. 30. Med. Teach. **35**(12), e1608–e1624 (2013)

Fiselier, E.S., Longhurst, J.W., Gough, G.K.: Exploring the current position of ESD in UK higher education institutions. Int. J. Sustain. High. Educ. **19**(2), 393–412 (2018)

Garrick, B.G., Pende, D.: The impact of national agenda on a local education authority's website: a visual semiotic analysis. Camb. J. Educ. **44**(3), 299–317 (2014)

Factors Influencing Electric Vehicles Adoption in Bahrain: Proposed Research

Abdulrahman Sharida, Mukhtar Al-Hashimi, and Allam Hamdan(✉)

College of Business and Finance, Ahlia University, Manama, Bahrain
allamh3@hotmail.com

Abstract. This study aims to elucidate the prospect of moving toward electric vehicles in Bahrain market which became the market segment in automobiles. Not only for economical reason, also for environmental purposes. This research paper will review pros and cons of electric vehicles and how different stakeholder perceive the value and adoption of electric vehicles, especially it is a new product for Bahrain market. The main challenge is how to persuade supplier, investors and potential customers to move from conventional to electric vehicles. Not only that, this research paper will spread awareness about the future of automobiles industry. Electric vehicles are known with its smart charging as an alternative of fuel and gas. However, there are hybrid cars that are use both an electric motor and a traditional gasoline engine. Nevertheless, involvement of all stakeholders and focus on their engagement need to be taken in account in order to develop the innovation process.

Keywords: Electric vehicles · Market demand · Plug-in hybrid electric vehicles

1 Introduction

Transportation is considered one of the four pillars of urban life which aims to facilitating inhabitants' movement among different places. Various types of transportation are implemented all around the world and it vary from a country to another. Hundred years ago, people used to move goods and themselves by animals where vehicles not invented yet. After cars inventions, a new market was created as well as a new supply-and-demand model. Going back in time, the first concept of automobiles was in 1478 by Leonardo da Vinci which was about self-propelled car (A vehicle without seats). However, it remained as a sketch on paper and never actually made. In 1769, da Vinci concept was transformed to a reality by Nicolas-Joseph Cugnot who built the first self-propelled car in France. Actually, it was built as a tractor for the French army. After 20 years, the first U.S patent was received by Oliver Evans for steam-powered land vehicle. The timeline and achievements go on including patents and building vehicles such internal combustion engine, non-rechargeable primary power cells, gasoline engine, and until the vehicles we see today. Building automobiles was not exclusive to just building them. It was about identifying targeted market and customers and how to create demand and supply for particular markets. Electric Vehicles (EVs) are promoted as a viable near-term vehicle technology to reduce dependence on fossil fuels. This study will focus on

A.-E. Hassanien et al. (Eds.): AICV 2020, AISC 1153, pp. 792–799, 2020.
https://doi.org/10.1007/978-3-030-44289-7_74

Bahrain market and to adopt and supply EVs in Bahrain and discuss the difference between Conventional Vehicles (CVs) and EVs.

In spite that EV are considered a new opportunity to create a new market demand, there are some obstacles in adopting EV in Bahrain market such as lack of awareness and knowledge regarding the perceived value of EVs and what are the benefits of owning an EV. In other words, the fear of change from conventional to EVs. Since EVs are not commonly used in Bahrain and the lack of previous researches, there will be some difficulties to obtain required data in order to test this research hypothesis.

Electric vehicles are well-known in some countries where people tried them either Battery Electric Vehicles (BEV) or Hybrid Electric Vehicles (HEV) which combines electric battery and fuel. However, we still don't know what is the level of awareness and value perceptions of different consumers in Bahrain regarding EVs. Therefore, this research will aim to: Identifying and measuring the factors that affect consumers' purchasing decision to EVs. Enhance consumers' awareness about EVs.

Consumers should be aware about the benefits of EVs and how those vehicles work and maintain. The feedback from consumers will be analyzed and focus on what attracts consumers to purchase EVs and what are the barriers obstruct them to do so.

This study aims to evaluate the impact of adopting EV in Bahrain in addition to create a new market demand beside to value perspective of different stakeholders such as governmental entities, automotive industry and consumers. Nevertheless, analyzing consumers' behavior which will play a primary role in demanding EV which will influence the supply of automotive industry. Taking into consideration the small area of Bahrain and how effective of EV to be used in Bahrain. Moreover, discussing what are the pros and cons of EV and addressing sustainable product development which has which is aligned with Bahrain vision 2030. With increasing oil prices in Bahrain, consumers' behavior had been changed in purchasing new and even second hand vehicles which consume less fuel. Therefore, EV will be a good alternative. This study also will educate and spread awareness on EV benefits and efficiency.

2 Literature Review

In light of fuel and gas depletion and increasing prices of oil and gas, Electric Vehicles (EV) became one of the recent subject and concern of many countries as they look at it as an alternative of conventional vehicles. Most of researches were conducted in North America and Europe. However, few researches included Gulf Cooperation Council (GCC). The aim of this literature review is to discuss different researchers' results and opinions in adopting EVs in addition to its impact on markets and economy and how different stakeholders perceive the value of EVs. Nevertheless, considering barriers and obstacles of the adoption is also a part of this study. The concentration was about spreading awareness about EV and what are different factors that influence consumers to move toward EV instead of conventional vehicles. Currently, Hybrid Electric Vehicle (HEV) which is a combination of electric and conventional fuel vehicles is more common and used than EV. Moreover, the studies illuminate the impact of EV on environment as it is considered as a green solution and it can be a competitive advantage to manufacturers to invest in electric automobile market.

Many countries are developing their strategies to replace fuel with energy in automobile and move toward more efficient energy usage, especially in transportation sector (Alzahrani et al. 2019). Saudi Aramco CEO has stated that “domestic liquids demand was on pace to reach more than 8 million barrels a day of oil equivalent by 2030 if there were no improvements in energy efficiency” (EIA 2014).

When we hear the term “EV”, first thing comes to mind is the fuel consumptions. It is necessary to have both fuel and electric battery (HEV) to guarantee long- distance driving and fuel will be as backup in case of battery degradation (Thomas 2009). However, adopting Plug-in Hybrid Electric (PHEV) is better than HEV as it is an improved version of HEV in terms of battery capacity and plug-in charger as it rechargeable from the grid (Dimatulac 2016).

However, (Poppink, Ruben) has illuminated the difficulty of distributing electric grid network considered as one of primary challenges of implementing EV while it is totally dependent on electricity beside the quality and sustainability of grids which leads to uncertainty of perceived value and benefits. It is also the most complicated in legal and technical wise. In spite that Vehicle to Grid (V2G) has largest possible benefits, smart charging is a better choice because it could be charged faster with less resources, implemented easily and no significant drawbacks (Poppink 2016).

When compared the cost of conventional and electric vehicles, the initial cost of EVs are higher, and getting higher based on battery size or vehicle range. However, electricity is cheaper than fuel and lower emissions. “Better fuel economy, less expensive to run and lower emissions than similar HEVs and conventional vehicles” (Egbue and Long 2012).

The battery life is another concern for potential consumers who consider the cost of maintenance, service and battery repair and replacement. Based on prior reports, Joseph S. Krupa explained that battery-related concerns are barriers of adopting EV and PHEV (Krupa et al. 2014). The third type is battery electric vehicle (BEV) which is all-electric and has a large capacity battery compared to HEV and PHEV that will allow this type of vehicles to be driven for longer range (Rezvani et al. 2015). Some researchers consider EVs as BEV because they are both all-electric (Rezvani et al. 2015).

Schuitema et al. (2013), illustrate that even with electric grids implementation, drivers should plan their trips before they start driving their EVs because they have to charge their vehicle while it’s not in use. Moreover, the number and capacity of required charging stations is another problem which will cause traffic conjunction plus they should be close enough within allowed cruising range (Gagarina and Corcoranb 2018). It is necessary to have a enough number of charging stations distributed on the road and determining appropriate stations locations is challengeable. “Driving range has been found to be a significant factor that determines a consumer’s adoption of a PHEV/EV” (Adnan et al. 2017).

Due to increasing concerns about the environment, electric and hybrid vehicles are playing a primary role in reducing pollution. Therefore, energy and environment reasons can influence consumers’ purchasing decisions of EVs. “Non-financial reasons, especially those associated with environment and energy can influence consumers’ decisions to purchase an EV” (Egbue and Long 2012). However, issues related to energy which impact climate change (global warming) are on top of environmental and political debates (Krupa et al. 2014).

"Across the globe, issues related to energy, its sources, uses, and impacts on climate change are at the forefront of political and environmental debates (e.g., the 2012 United Nations Climate Change Conference at Doha, http://unfccc.int)" (Krupa et al. 2014).

Some countries prefer to use renewable energy such as solar and wind energies as a source in their electric grids. Nevertheless, some cities planning to have smart grid technologies in order to make EV charging more efficient and convenient (Luo 2017).

Before introducing any new product to the market, investors must understand their consumers' behavior toward introduced product. Some researchers defined levels and classifications of behavior to make it easier for them to analyze selected consumers. Others, for instance Moons and De Pelsmacker (2012) who have defined three emotional levels: visceral (Bases on visual attributes as design, size and style, behavioral (emotions from using and experiencing driving EVs) and reflective (self-image of driving EVs). (Rezvani et al. 2015). Moreover, lack of understanding, awareness and familiarity with the idea of owning an EV are related to consumers' behavior. Different types and aspects of vehicles may lead to a confusion which is considered as a constraint to adopt EV and opportunity to spread awareness and introduce EVs by investors to attract potential customers.

"The majority of new vehicle buyers have little or no familiarity with the idea of a PHEV, and may erroneously believe that existing HEVs can perform the same basic function as a PHEV, i.e., have the ability to be refueled by gasoline and to be plugged into an electrical outlet (Axsen and Kurani 2008)" (Kurani and Turrentine 2011).

In addition to lack of awareness, the information delivery and marketing instruments and sources is another concern of educating people about EVs, especially those who didn't experience driving EV. Through a personal experience, consumers will have positive intension and more awareness into EV (Dudenhöffer 2013).

"Analysis reveals that consumers need more knowledge to make informed EV purchase decisions, and they need information from objective and trustworthy sources, including word of mouth. Consumers are currently being informed primarily via advertising and media coverage. Further, consumers have taken little action to find information about EVs, except for experienced users. Ongoing lack of relevant, objective information about EV performance and cost is likely to limit adoption of these vehicles" (Larson et al. 2015). Until 2016 many studies were conducted in Canada to develop a clear awareness of different factors that influence consumers' purchasing decision in owning EV in the household context (Dimatulac 2016). Some researchers stated that majority of people in United States find financial benefits is more important than environmental benefits. "Similar to the works of others (Ozaki and Sevastyanova 2011; Caperello and Kurani 2012; Graham-Rowe et al. 2012), more AMT (Amazon Mechanical Turk) participants stated financial benefits were important than those who stated environmental benefits were important. However, although only 55.1% of participants stated that reducing GHG emissions was an important factor, our results show that those who are most concerned about climate change have 44.4 times greater odds of being willing to consider purchasing a PHEV than those least concerned." (Krupa et al. 2014).

In 2007, Potoglou and Kanarolglou (2007) stated that if consumers have to choose between common combustion and an alternative fueled vehicle, they prefer conventional vehicles because of short driving range of EV (Dudenhöffer 2013).

A survey was conducted in U.S. and 55% of respondents were familiar with EV, almost 30% of respondents heard about it but not really know what it is, and approximately 15% indicate that they didn't hear about it (Roth 2010).

In 2017, consumers became aware of greenhouse gas emission impact on environment as well as. As consumers are becoming more aware of the environmental impact of the greenhouse gas emissions where promotions and government incentives improve the demand on EV which has been growing since then. According to Navigant Research, they expected global demand of light duty EV market and lithium ion batteries of light duty fleet to grow to 6.4 million vehicle sales and $24.1 billion in 2023 respectively (Luo 2017).

Some countries have been started already in developing their strategies to adopt EVs as they are projected a positive impact on country's market and economy. In 2015, Business Strategy and the Environment Journal published Toyota, Volkswagen and General Motors some of business strategies toward EVs:

"(General Motors former vice president, 2012) Toyota explained that it wants to exploit the gradual electrification of the ICEV, enjoying knowledge spill-overs between each step (Toyota, 2012). Volkswagen wants to adopt a 'slow follower EV strategy' and is waiting for the system to materialize before it attempts to exploit EVs (Volkswagen's Managers Future Powertrains Department, 2012). General Motors wanted to mimic the success of the Prius using plug-in hybrid electric vehicles; it leveraged its EV asset position to introduce this hybrid car to the market".

In Saudi Arabia, the government set some initiatives to encourage and promote the adoption of alternate fuel such electric power and natural gas in addition to infrastructure development which will influence commercial vehicles market during 2016–2021 (York 2017). In Malaysia it was projected to increase in PHEVs and EVs overall sales mainly because of increasing awareness toward PHEVs and EVs and increasing popularity of hybrid vehicles. "The overall sales of PHEVs/EVs in Malaysia are projected to rise in the next few years, mainly because of the increasing popularity of hybrid vehicles, increasing awareness among consumers towards PHEV/EV extended incentives of tax breaks by the government of Malaysia, and the latest model promoted by vehicle manufacturing companies" (Adnan et al. 2017).

Manufacturing companies shouldn't focus on business and financial results as they should develop environmental strategies by turning environmental issues to an opportunity to implement green solutions (James 2011). This opportunity will grant the company a competitive advantage among other competitors who didn't turn environmental issues to an opportunity. Oil crises was the main reason to push governments to conduct researches to find alternatives fuels of oil which was the start of EVs development. "The oil crisis of 1973–4 lead to substantial government funding of research on alternative fuels, including the Hybrid Electric Vehicle Act of 1976 which resulted in much of the electric vehicle technology developments that emerged during the 1990s". (Turrentine and Kurani 2011). On one hand, different strategies developed by different countries where each country has its own policies and politics to improve its economy by having new markets such as EVs market. The government play a primary role in granting incentives in order to encourage investors, manufacturers and citizens to adopt EVs. "There is a need for governments to straighten out more incentives for citizens to encourage mass adoption of PHEVs/EVs. Other policies that need to be improved for the consumers' benefits are subsidies and the increased driving range during production which help to increase adoption rates" (Adnan et al. 2017). On the other hand, relying on advocacy influence in government regulation for green solutions, it may not be the most optimum strategy for maximizing returns in terms of environment and social outcomes because corporates interest usually mediates their influence (Arvizu 2010).

Conceptual Framework
There are many factors that influence consumers' purchasing decision to EVs. For instance, vehicles factors which includes purchasing price, driving range, charging time and quality of EVs. Also, personal factors such as age, personal experience, life style, income and occupation which all related to consumers' personal preferences in selecting EVs rather than conventional cars. Moreover, social influence such as word of mouth and media have a major impact on consumers' awareness.

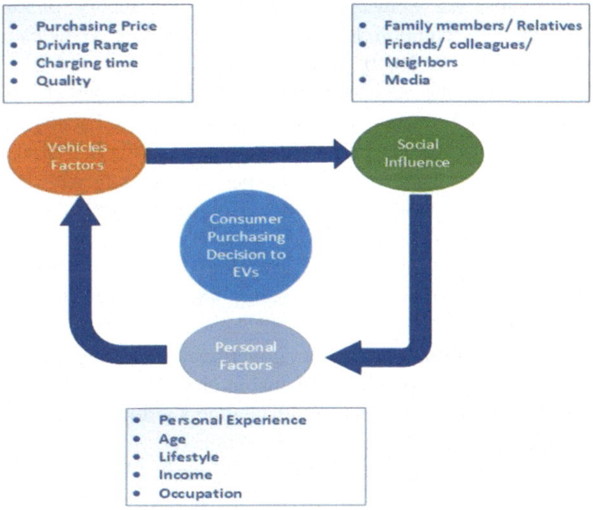

3 Conclusion

Finding an alternate fuel is not a new topic as many researches were conducted in the past where most of them conducted in U.S., Canada and Europe. Few articles covered Asia, especially middles east. "HEVs are not yet available in the Saudi market. Moreover, relevant consumer behavior research on Saudi Arabia is limited" (Alzahrani et al. 2019). However, there is no research conducted about Bahrain strategies to adopt EVs or even PHEV as it is necessary to review previous studies to Bahrain EVs market, consumer behavior and value perspective by different stakeholders. All reviewed studies illustrated common barriers to adopt EVs which is short driving range because of battery capacity, lack of awareness and knowledge of owning an EV and its impact on environment where there are some people who still prefer conventional vehicles regardless the impact of conventional vehicles emission impact on environment and the high prices of oil. "Despite the advantages of electric vehicles, their relatively limited cruising range (in comparison to traditional diesel/petrol vehicles) and significant battery charging time often provide major challenges to their usage" (Egbue and Long 2012). Finally, most literature reviews are discussing engineering and technical perspective of EV rather than business perspective.

References

Adnan, N., Nordin, S.M., Rahman, I., Amini, M.H.: A market modeling review study on predicting Malaysian consumer behavior towards widespread adoption of PHEV/EV. Environ. Sci. Pollut. Res. **24**(22), 17955–17975 (2017). https://doi.org/10.1007/s11356-017-9153-8

Alzahrani, K., Hall-Phillips, A., Zeng, A.Z.: Applying the theory of reasoned action to understanding consumers' intention to adopt hybrid electric vehicles in Saudi Arabia. Transportation **46**(1), 199–215 (2019). https://doi.org/10.1007/s11116-017-9801-3

Dimatulac, T.: Adoption of electric vehicles within the rental fleet market in Canada: modeling factors and determinants by Terence Dimatulac. Submitted to the Faculty of Graduate Studies Through the Department of Civil and Environmental Engineering in Partial Fulfillment (2016)

Dudenhöffer, K.: Why electric vehicles failed: an experimental study with PLS approach based on the technology acceptance model. J. Manage. Control **24**(2), 95–124 (2013). https://doi.org/10.1007/s00187-013-0174-2

Egbue, O., Long, S.: Barriers to widespread adoption of electric vehicles: an analysis of consumer attitudes and perceptions. Energy Policy **48**, 717–729 (2012). https://doi.org/10.1016/j.enpol.2012.06.009

James, D.R.: The perceived influence of the government, consumers, manufacturers, and sustainability on automotive aftermarket sustainable product development. ProQuest Dissertations and Theses, 293 (2011). http://proxy.lib.chalmers.se/login?url=http://search.proquest.com/docview/861341277?accountid=10041 LA - English

Roth, J.: INstitute for PUblic POlicy of RIce UNiversity NEw ENergy TEchnologies: Development (2010)

Krupa, J.S., Rizzo, D.M., Eppstein, M.J., Brad Lanute, D., Gaalema, D.E., Lakkaraju, K., Warrender, C.E.: Analysis of a consumer survey on plug-in hybrid electric vehicles. Transp. Res. Part A Policy Pract. **64**, 14–31 (2014). https://doi.org/10.1016/j.tra.2014.02.019

Kurani, K.S., Turrentine, T.S.: Interpersonal influence within car buyers' social networks: observing consumer assessment of plug-in hybrid electric vehicles (PHEVs) and the spread of pro-societal values by Jonn Axsen. Submitted in partial satisfaction of the requirements for the degree (2011). Transportation Research Part D: Transport and Environment

Larson, P.D., Viáfara, J., Parsons, R.V., Elias, A.: Consumer attitudes about electric cars: pricing analysis and policy implications. Transp. Res. Part A Policy Pract. **69**, 299–314 (2015). https://doi.org/10.1016/j.tra.2014.09.002

Luo, C.: Engineering and economic analysis for electric vehicle charging infrastructure — placement, pricing, and market design. A Dissertation Submitted to the Graduate School of the University of Notre Dame in Partial Fulfillment of the Requirements for the Degree, October 2017

Poppink, R.: Electric vehicle integration (2016)

Rezvani, Z., Jansson, J., Bodin, J.: Advances in consumer electric vehicle adoption research: a review and research agenda. Transp. Res. Part D Transp. Environ. **34**, 122–136 (2015). https://doi.org/10.1016/j.trd.2014.10.010

Thomas, C.E.: Fuel cell and battery electric vehicles compared. Int. J. Hydrogen Energy **34**(15), 6005–6020 (2009). https://doi.org/10.1016/j.ijhydene.2009.06.003

York, P.N.: Saudi Arabia Commercial Vehicles (LCV, Bus, MCV and HCV) Market Forecast & Opportunities, 2021 - Research and Markets. 2017–2019 (2017)

Intelligent Monitoring Technology of the Substation Based on the Micro System

Peng Yin(✉), Hailong Li, Jinhui Liu, Xiaolong Liu, and Zhaohua Wang

Department of Construction, State Grid Liaoning Electric Power Supply Co. Ltd., Shenyang, China
yinpeng168@126.com

Abstract. Substation station power consumption is larger. A provincial power grid is likely to have thousands of substations. The sum of all substation power is a big number. Therefore, a micro-network system is formed in the substation, making full use of renewable energy, reducing the loss of traditional energy, vigorously promoting China's economic restructuring and energy structure optimization, reducing China's energy supply pressure, and reducing the total emissions of greenhouse gases and major pollutants such as carbon dioxide so as to accelerate the realization of energy conservation and emission reduction targets and gain important environmental benefits, economic and social benefits. The micro grid system continues to provide power load after the power grid is cut off, and has the effect of uninterrupted power supply, improving the reliability of the power supply occupied by the substation.

Keywords: Micro system · Intelligent monitoring · Energy saving

1 Introduction

In under the double pressure of energy security and environmental protection, renewable energy sources such as photovoltaic, wind power, large-scale development are to realize the energy structure optimization and adjustment of important way to achieve low carbon development. China has the photovoltaic and wind power industry as a national energy science and technology twelfth five-year plan in key areas. Mature technology has the conditions to develop large scale photovoltaic power generation and wind power. In the world has been the development and application of high speed, and mass according to the global wind energy council (GWEC) installed according to data released by the global wind power market. Global new wind power installed capacity in 2011 41 gw, the accumulative total installed wind 238 GMW, year growth of 21%; China's wind power cumulative installed capacity of about 62 gw, the highest in the world. The photovoltaic power generation in China is got rapid development and the installed ratio significantly increase compared with Germany and Spain [1]. Figure 1 shows the interconnected logical structure of the embedded technology based monitoring and diagnosis system.

A.-E. Hassanien et al. (Eds.): AICV 2020, AISC 1153, pp. 800–807, 2020.
https://doi.org/10.1007/978-3-030-44289-7_75

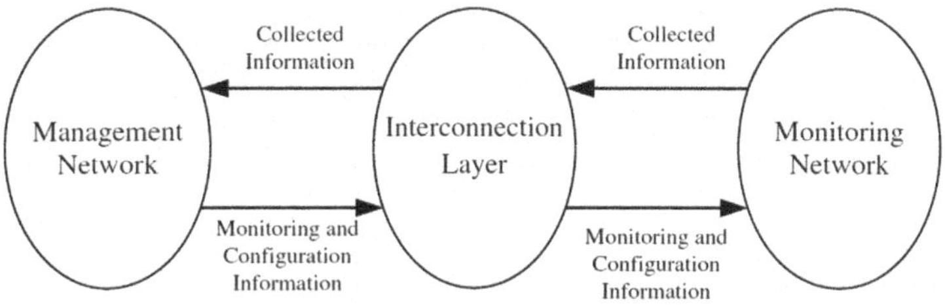

Fig. 1. The interconnected logical structure of the embedded technology based monitoring and diagnosis system

2 Domestic and Overseas Advance

Micro network concept was first proposed by the United States, its micro network theory research results have been passed by platform in the laboratory test, and in North America set up its first piconets demonstration project [2]. The focus of the US. piconets future research mainly concentrated in the meet the requirements of a variety of power quality, improve the reliability of power supply, reduce the cost and realize intelligent [3]. Europe's study of piconets is becoming more and more attention, especially in recent years was carried out by micro network of cooperation between countries and research work [4]. Europe's future research emphasis focused on advanced control strategy and related standards. Japan's domestic energy under the background of increasingly scarce, load increasing, also launched a micro network research, but its development goal mainly locate in the diversification of energy supply, reduce pollution, satisfy the user's individualized demand for electricity [5]. The micro network research in China temporarily or focuses on the study of distributed generation interconnection [6] (Figs. 2 and 3).

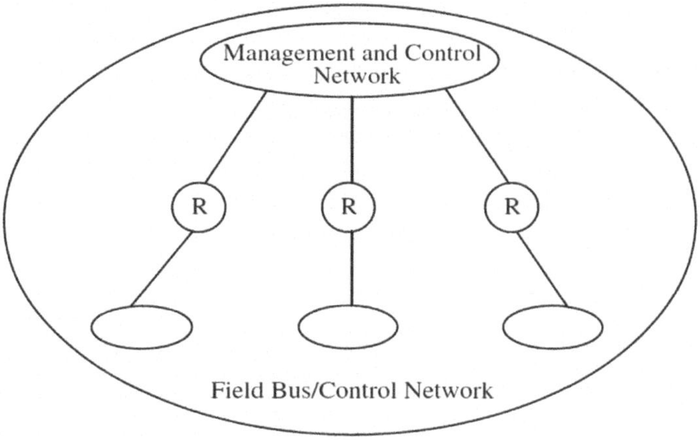

Fig. 2. Group networking of management and control

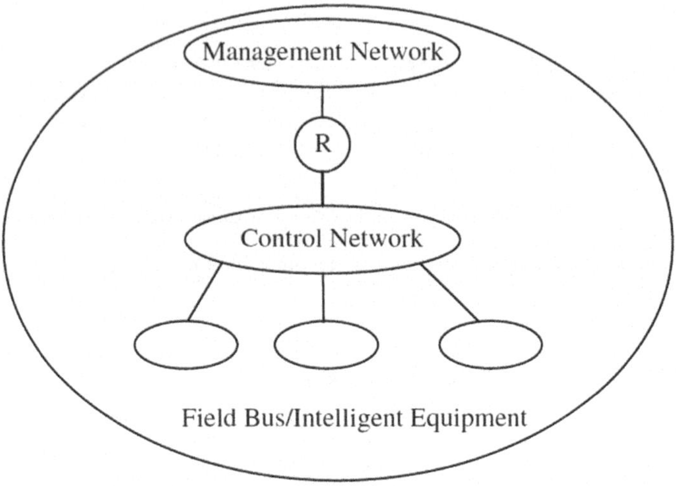

Fig. 3. Separate networking of management and control

3 Related Project Theory Basis

The rapid development of storage technology effectively cracked wind power and photovoltaic output time-varying characteristic and power inherent contradiction between the demand and at the same time to quell wind power and photovoltaic output volatility, tracking, wind power and photovoltaic program output, improve the ability of mass acceptance of wind power and photovoltaic (pv) grid technology provides an efficient and convenient [7]. However, at this stage in the application of energy storage planning, construction and operation of theory and application technology research also has yet to be breakthrough, in energy storage planning, construction and operation is often difficult and wind power and photovoltaic and grid system planning, construction and operation of coordination, energy storage system in the process of power grid operation flexible "identity" effective play to the constrained conversion characteristic advantage, limit the function of energy storage system, wind power and photovoltaic (pv) grid power cannot adapt to the needs of the development of large-scale wind power and photovoltaic industry. But wind power, photovoltaic and joint operation technology in the application of micro network storage system can advance, gradually developed and expanded to a wide range of joint operation [8] (Fig. 4).

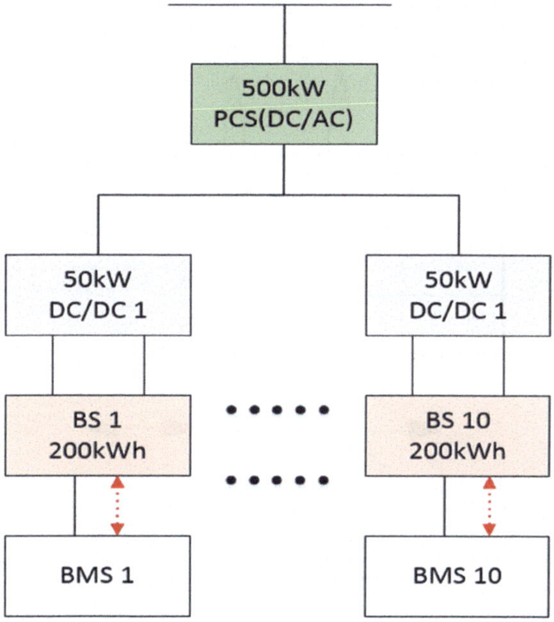

Fig. 4. Energy storage unit figure

The following all kinds of supporting intelligent substation technology development theory for the application of smart step-up substation in wind power technology provides the theory basis of reality:

3.1 Embedded Processing Technology

With the development of the microprocessor and embedded real-time processing technology and progress, with embedded computer technology as the core of the embedded system has become can satisfy the function, reliability, cost, volume, power consumption and so on strict comprehensive performance of computer system, and has a good interactive user interface, information acquisition speed, range and information processing and transmission capacity were greatly improved, provides the equipment function extension [9]. Embedded the processing capacity of processing technology of high speed, wide range of design, superior control functions, easy to upgrade maintenance features, for an equipment state monitoring system of data acquisition analysis provides a foundation to support real time online [10] (Fig. 5).

$$\lambda_2 - \lambda_1 = \mathrm{b}(t_2 - t_1) \tag{1}$$

$$\frac{\lambda_2 - \lambda_1}{\lambda_3 - \lambda_2} = \frac{t_2 - t_1}{t_3 - t_1} \tag{2}$$

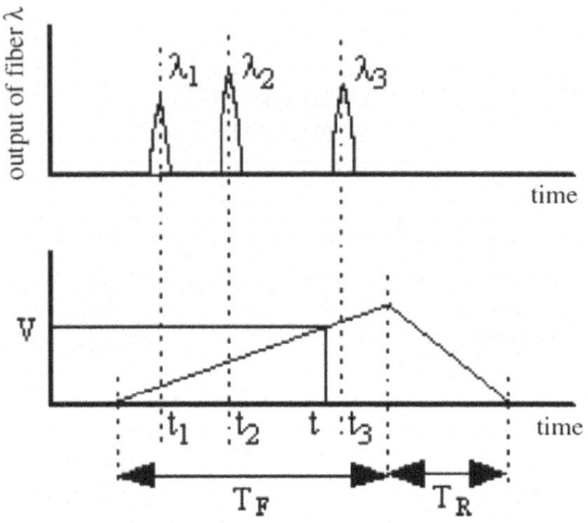

Fig. 5. Relationship between tuning voltage and the wavelength of reflect

3.2 The Domestic Research of IEC 61850 Standard

Intelligent substation construction of IEC61850 standard as a support, and high pressure equipment intelligent substation automation the only international standard, compared with the traditional IEC60870-5-103 standard, it is not only a simple communication protocols, it to server model, logical device model, data model and logic node object model is commonly used in substation automation system equipment is established the unified data model, in order to meet the requirement of interoperability [11]. Based on the unified modeling of objects in the substation automation system, using object-oriented technology and independent of the network structure of abstract communication service interface, realize the seamless connection between different manufacturers of equipment, so as to greatly improve the level of substation automation technology and the safe and stable operation [12]. Domestic each big manufacturers began to focus on IEC61850 standard in 2001, in 2004 entered the stage of substantial research and development of IEC61850 related products, in 2005 began to be substation based on IEC61850 standard is put into practice, after many years of technical accumulation, the current domestic application of IEC61850 has mature technology and its related products, IEC61850 standard to guide the design of the substation automation, development, engineering, maintenance, and other fields [13, 14].

3.3 Intelligent Equipment Standard System Preliminary Established at a Time

Intelligent substation with technical guidance "," guidelines for the high voltage equipment intelligent technology such as promulgated a series of standards, rules and its application in the smart grid pilot project, intelligent, have made it clear that the concept of a device, a smart devices have preliminarily set up the structure of the system, through in an equipment (transformer, switch equipment, etc.) on the outer or embedded condition monitoring sensors, combined with the on-site installation of intelligent components, implement a intelligent equipment [15, 16].

3.4 The Further Development of the Optical Fiber Communication Technology

IEC61850 as widely used in intelligent substation automation seamless communication system is the only international standard, the popularization and application of technical standards is dependent on the communication the development of network technology, communication technology to build intelligent substation based on IEC61850 is very important to the realization of the network architecture, the fiber optic network communication in the construction of digital substation and intelligent substation has gained wide application for many years, domestic digital substation, the optical fiber network in intelligent substation and stable operation of the practical experience shows that high speed optical fiber network for the construction of a highly intelligent device based on IEC61850 communication protocols of the intelligent substation has theoretical and practical basis [17, 18].

4 The Project Plan as a Whole

For the implementation of energy-saving, environmental protection, low carbon requirements, the project will be a photovoltaic power generation, wind power and other new energy source, through the network access station power system, for application of renewable energy in the smart substation experiment beneficial attempt [19, 20]. Standing with micro network refers to the distributed power (solar power, wind power), energy storage device, energy conversion device and monitoring, protection device assembled small power distribution system, is a can realize self control, protection and management of autonomous system, which can run with external power grid interconnection, can also be isolated operation. According to meteorological data, substation site is located in sunshine time is more than 2000 h, the attachment how do wind turbines in operation, is a place where abundant solar and wind power. In order to give full play to the advantages of rich site solar and wind energy resources, comprehensive utilization of the states stand design integrated control within the station building about 400 m^2 area roof building 20 kW photovoltaic power generation system, using the station space to install 150 kW wind turbine, with a total installed capacity of about 70 kW, at the same time using the station space to install 1 set of 60 kW/300 kWh energy storage system, formed by solar energy, wind power generation system and network system is composed of battery energy storage system [21] (Fig. 6).

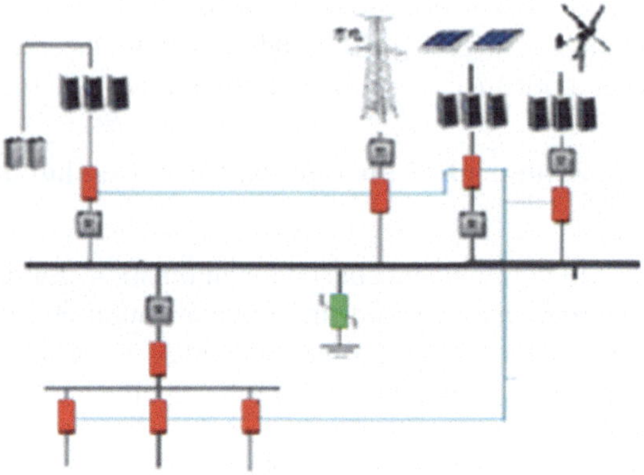

Fig. 6. tMicro network system schematic diagram

5 Conclusion

This article is mainly a photovoltaic power generation, wind power and other new energy through the network system access to 380 v station power system, as a station with power supplement, reducing station electricity from the proportion of the electricity system, at the same time in total loss of electricity as the backup power, guarantee the important load of the continuous power supply, improve the safety and reliability of the power system, for renewable energy application in substation to realize the beneficial attempt.

References

1. Li, R.S., Li, Y.B.: Intelligent substation function architecture and design principles. Power Syst. Prot. Control **21**, 24–27 (2010). (in Chinese)
2. Cheng, J.Y.: Try to talk about intelligent substation function architecture and design principles. Sci. Technol. Innov. Appl. **25**, 15–18 (2015). (in Chinese)
3. Xu, Y.: Intelligent substation function architecture. Electr. Technol. **7**, 87–90 (2013). (in Chinese)
4. Xu, J.: Shallow of intelligent substation design principle. Power Supply Technol. Appl. **6**, 35–40 (2012). (in Chinese)
5. Zhao, B.: Key techniques of the micro grid. Electr. Eng. Technol. **9**, 45–47 (2015). (in Chinese)
6. Shu, L., Zhang, J.H.: The micro grid related issues and technical research. Power Syst. Prot. Control **19**, 235–239 (2010). (in Chinese)
7. Li, J., Wang, B., Dong, X.Z., et al.: The feasibility study of advanced function of merging unit in intelligent distribution substation. In: International Conference on Advanced Power System Automation and Protection, pp. 2341–2347. IEEE (2012)
8. Liu, Y., Gao, H., Wei, X., et al.: Study on new type of IED with integrated functions in intelligent substation. In: International Conference on Advanced Power System Automation and Protection, pp. 161–165. IEEE (2012)

9. Kim, Y., Lee, S., Dollmann, M., Geierhos, M.: Semantic annotation of software requirements with language frame. Int. J. Softw. Eng. Smart Device **4**(2), 1–6 (2017)
10. Kumar, Y.: Identification of malicious packet dropping in ad-hoc networks. Int. J. Wirel. Mob. Commun. Ind. Syst. **3**(1), 13–20 (2016)
11. Mehan, N., Bagwarik, A.: Improved systematic study to show the impact of CFO one the presentation of LTE uplink. Int. J. Wirel. Mob. Commun. Ind. Syst. **4**(1), 57–66 (2017)
12. Park, D.J.: Implementation of 3PL service providers' efficient reverse logistics management system. Int. J. Adv. Sci. Technol. **119**, 67–78 (2018)
13. Fiaidhi, J.: BioTech: the new game changer in science and technology: a summary paper. Int. J. Adv. Sci. Technol. **122**, 31–34 (2019)
14. Baek, S.H., Jang, J.W.: Design of tire surface check system with the laser sensor. Int. J. Control Autom. **8**(3), 9–14 (2015)
15. Md. Shafi, A.A., Aneek, R.H., Bahar, A.N.: Universal reversible gate in quantum-dot cellular automata (QCA): a multilayer design paradigm. Int. J. Grid Distrib. Comput. **10**(1), 43–50 (2017)
16. Park, J.K., Choi, S.Y.: Studying security weaknesses of Android system. Int. J. Secur. Appl. **9**(3), 7–12 (2015)
17. Chuang, P.J., Chen, H.L.: An efficient video content delivery strategy for radio access network environments. Int. J. Future Gener. Commun. Netw. **8**(2), 9–20 (2015)
18. Ullah, Z., Mumtaz, I., Khan, M.S.: Analysis of 3D face modeling. Int. J. Sig. Process. **8**(11), 7–14 (2015)
19. Wang, X., Qin, H.: Implementation of fall detection system based on data fusion technology. Int. J. u- and e- Serv. Sci. Technol. **9**(4), 1–8 (2016)
20. Hu, Y.: Study and design a dynamic e-business application framework based on Java EE and web service. Int. J. Database Theory Appl. **9**(11), 37–48 (2016)
21. Patgar, T.P., Shankaraiah: The impact of hybrid data fusion based on probabilistic detection identification model for intelligent rail communication highway. Int. J. Sens. Appl. Control Syst. **4**(2), 9–20 (2016)

Remote Collaboration in a Complex Environment

Huang Zou[1] and Hao Tang[2(✉)]

[1] Rensselaer Polytechnic Institute, Troy, NY 12180, USA
[2] City University of New York (BMCC), New York 10007, USA
htang@bmcc.cuny.edu

Abstract. A remote collaboration system simplifies a collaboration task among multiple users at a distance. It allows local users see the real world with virtual guidance/annotations (provided by remote experts) seamlessly rendering over target objects in the real world, so they can easily follow the remote guidance to complete complicated tasks. An effective collaboration system often utilizes (1) computer vision algorithms (e.g. 3D reconstruction, and visual tracking and localization algorithms) to analyze objects and scenes in the real world, as well as (2) augmented reality (AR) to accurately superimpose virtual guidance/annotations over the target object in a real environment. However, some complex working environments with deformable linear objects (DLO), e.g. wire harness, prevents remote experts from easily annotating objects in the video, virtual scenes, or clear non-verbal communication due to the difficulty of DLO tracking. In this project, we proposed a remote guidance system, which is able to identify and track multiple DLOs, starting with minimal user input, so it allows remote experts to provide local users with simple and clear instructions/guidance in a complex environment.

Keywords: Computer vision · Augmented Reality · Remote collaboration · Deformable linear objects · Visual tracking

1 Introduction

Nowadays ubiquitous smartphones play an important role in we communications, yet the interactions are limited to verbal or video contents, which sometimes are insufficient. Examples of such tasks may include a customer calling a call center to help troubleshoot a PC, a novice technician calling an expert for guidance in repairing a complex device (e.g., a car engine), or a medical expert guiding a local doctor through a complex procedure. Augmented Reality (AR) is widely adopted in remote collaboration solutions. AR allows local users see the real world, with virtual guidance/annotations (from remote experts) seamlessly superimposed over objects in the real world. Both research prototypes and commercial software apply AR in remote collaboration tasks. However, most of the existing solutions only allow remote collaboration in a simple working environment. It's very difficult to provide precise annotations in a complex environment with deformable linear objects (DLO), e.g. wire harness and cables (Fig. 1), because of

A.-E. Hassanien et al. (Eds.): AICV 2020, AISC 1153, pp. 808–818, 2020.
https://doi.org/10.1007/978-3-030-44289-7_76

complex background and occlusion (e.g. occluded by users) and self-occlusions (e.g. a cable crossing itself), as well as the difficulty of precise wire manipulation guidance.

Compared to rigid objects, it's much harder for the experts to accurately annotate DLO in the video, virtual scenes and clear non-verbal communication, especially the DLO deforms when the local user manipulates it.

In this paper, we proposed a collaboration system, which eases the remote guidance in complex environments with multiple DLOs (e.g. wires and cables). The paper includes two main contributions:

- Propose a visual tracking algorithm to track multiple DLOs in a video sequence.
- Implement a remote guidance system, which incorporates an effective user interface and the DLO tracking algorithm to ease remote guidance between multiple users at a distance.

In the following literature review, we include the study of literature in both remote collaboration and tracking DLO in a complex environment.

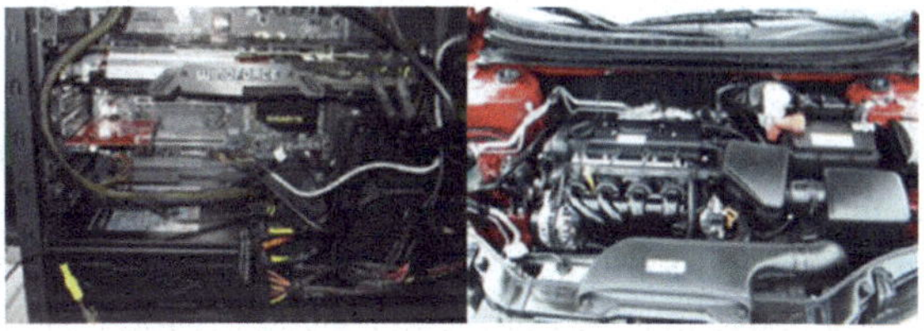

Fig. 1. Examples of complex local environments, with many wires or thin objects.

1.1 Related Work

We summarize related work in two different categories: AR based remote collaboration and tracking deformable linear object (DLO).

1.1.1 AR Based Remote Collaboration

Early work in remote collaboration used video-mediated dialogs [1], or monitors or projectors to overlay remote helpers' gestures (pointers, sketches and hands) on top of the local video stream to be sent back and shown to the local users on a display [2–4]. However, these approaches may cause a fractured ecology in which the worker needs to split attention between the task and the external display [14]. The ReMote system [12, 13] is a remote assistance system that allows a remote expert to assist and guide in real time a local maintenance. The remote expert uses their own hands (hand gestures over a live video of the workspace) as well as annotations on still images to demonstrate how the repair/maintenance task needs to be performed. Their system setup is often complicated and requires many sensors and devices.

AR allows users to see the real world (Fig. 2b shows an example: Pokémon GO AR Game), with virtual objects seamlessly superimposed over the real world [6] and it allows co-located users can see shared 3D virtual objects that they interact with, or a remote helper can annotate the live video view of a local worker, enabling them to collaborate at a distance. Gurevich et al. [5] and Adcock et al. [7] proposed AR-base collaboration systems, which enable a remote helper guide a local user in complex physical tasks. The system can stick the annotations to the original positions in 3D world and it solves the above issue of split attention. [9] captures video of local workspace and send it to a VR-based head-mounted displays worn by the remote user. Wong [8] used AR-based smartglasses (Fig. 2a) for both local and remote users for collaboration.

Fig. 2. (a) AR-based HMD: Microsoft HoloLens and Epson BT-300; (b) a mobile Pokémon GO AR game.

A 3D model was reconstructed by a depth sensor and it's often rendered on a VR HMD worn by the remote user, who can view and annotate the 3D scene from a different view. More studies followed the idea to provide 3D model of local scene [10–13], used independent depth sensors or HMD with embedded depth sensor (HoloLens) for the local user. However, it's difficult to directly reconstruct accurate 3D model of DLOs (e.g. cables and wires) using either 2D sensors (cameras) or even 3D depth sensors, because the linear objects are often thin and lack textures hence difficult to match. In addition, the 3D reconstruction problem becomes more difficult since there are often clutter, occlusion, and self-occlusion in a complex environment. Therefore, it's infeasible to reconstruct 3D of DLOs in remote collaboration task, and then we still rely on a traditional 2D collaboration platform.

1.1.2 Track DLO

To successfully conduct a remote collaboration task with non-verbal communication, a remote expert is often required to annotate a DLO. How to render the annotation of a DLO in the local view if the perspective of the local user (smartphone or smartglass) changes or even the DLO deforms or relocates. These remain challenging problems since DLO needs to be reliably and accurately tracked over a video sequence to ensure the annotation of the DLO can be correctly rendered over the target DLO in the real local view.

In robotics field, plenty of manipulation tasks deal with objects that can be modeled as non-rigid linear structures. There has been some research concerned with the problem of handling wire-like deformable objects using robot arms. In surgical operations like suturing, grasp and knot-tying is a very important and repetitive sub-task [15–17]. Similar knot-tying and path planning procedures, like e.g. knots untangling, are also relevant to contexts like service, collaborative and rescue robotics [18–21]. As for industrial scenarios, one of the hardest tasks involving flexible linear objects is wire routing in assembly processes [22–25].

The general approach to track deformable objects [26–30] is to divide the object into parts. Local connectivity is typically applied among different parts of the object and a trade-off of the visual and geometric agreement is then optimized. However, it's not possible to decompose a DLO into multiple parts. Liu et al. [31] proposed deformable convolution that can both model the appearance of the deformable object as well as be incorporated into the CNN-based framework and trained in an end-to-end manner. The approach reached state-of-the-art performance on public tracking benchmarks, but it's unclear that the method can track DLO since there are few examples of DLO in these benchmarks.

It's a difficult problem to directly track DLOs in a video. Therefore, in this project, we start tracking a few seed points on the DLO and then extend to the entire DLO from the seed points. In the following, we first describe our proposed system design in Sect. 2, and then we present our implementation and preliminary experiments in Sect. 3, before we conclude the paper in Sect. 4.

2 System Design and Implementation

2.1 System Framework

The proposed collaboration platform is a multi-user system with cloud storage. A user on the client side uses a mobile AR system (e.g. a smartphone) to capture a video of local environment and then uploads it onto cloud storage (e.g. Google firebase platform). Users on the remote side query the client's video available on the cloud, allowing remote users to view client's environment on a smartphone. The interface on the remote side also allows users annotate objects and add virtual annotation/landmarks in the client's video, either to guide the client user or to collaborate with client user to complete a complex task.

Figure 3 shows that the remote user tries to guide a local user prototyping electronics on a breadboard. The local user first captures a video of the breadboard with complicated wire connections and send it to the remote users through a cloud server. The remote user can watch the live video of the breadboard and demonstrate the correct wire manipulations. However, it's not an easy task if the remote guidance is done via a simple video sharing with annotations, because the electronics can be very complicated with dozens (or more) wires connecting on the breadboard. Simple verbal communication and video annotation may not be effective particularly when the video includes severe occlusion (by the local users) and self-occlusion (between wires). The remote user needs to mark one wire in the video or a still image at a time and demonstrate how the wire needs to connect to the breadboard, and then the system should continue tracking the wire over the video on both local and remote sides. It requires accurate and robust visual tracking algorithm even if the local user moves the perspective or the wire is temporarily occluded. In the next section, we will describe the algorithm of tracking DLOs.

2.2 Detect and Trace a DLO

Tracking DLOs robustly is the key component of the entire collaboration system, since the local user may be misled or get lost if the DLOs are incorrectly tracked. However, the existing computer vision algorithm does not offer a robust tracking approach on DLOs. in this paper, we extended the algorithm in [32] and develop a robust approach to track multiple DLOs. The basic idea is simple: though it's not easy to directly track DLOs, we can reliably track the endpoints of DLOs (green boxes at the client side in Fig. 3) and then trace the DLOs from two endpoints.

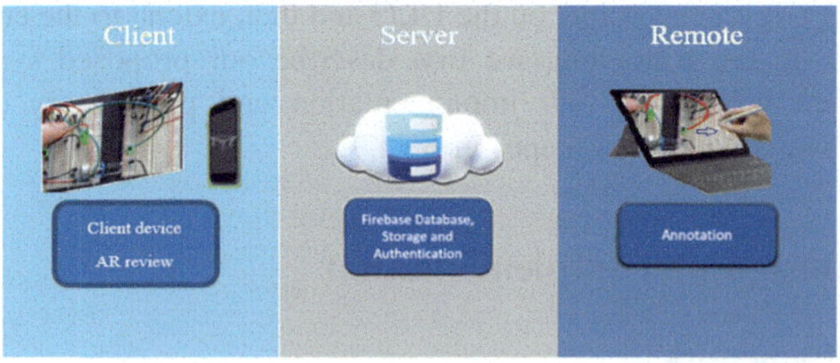

Fig. 3. The overview of the proposed remote collaboration system, including client, server and remote modules. A proposed DLO tracking algorithm allows the remote user easily provide wire/cable manipulation instructions to guide the local user to complete a complex task. Given the two endpoints of a DLO in one video frame, we apply a correlation tracking algorithm [33], which can handle large scale variations in complex image sequences. The algorithm utilizes a scale pyramid to accurately estimate the scale of an object after the optimal translation was found. This breakthrough allows to track objects with changes in both (1) translation and (2) scaling throughout a video stream, and additionally the track algorithm can perform in real-time.

With the endpoints of a DLO being robustly tracked, we can trace each DLO from two endpoints using a superpixel segmentation approach [34]. The algorithm includes the following steps:

1. Superpixel segmentation: a frame with endpoints labeled by the remote expert is first segmented to superpixels using an unsupervised method [34], hence an adjacency graph (Fig. 4b) has been created and the connectivity of neighborhood superpixels can be obtained.
2. Random walk: starting from a superpixel, a walk is defined as one step extension from the current superpixel to its adjacent superpixel according to an optimization function, which will be described in the Sect. 2.2.2. Figure 4b show the random walk algorithm is processed iteratively from two seed superpixels until they meet at one superpixel.

2.2.1 Superpixel Segmentation

Using superpixel segmentation, we can convert an image from a pixel-grid representation to a superpixel representation which holds the following desirable properties:

- Perceptually meaningful: each superpixel is perceptually consistent, i.e. pixels in a superpixel share similar color and texture. Most structure still remains. In our paper, the boundary of DLO can still be conserved after superpixel segmentation.
- Computationally and Representationally efficient: it reduces the complexity of images from several million of pixels to several hundred superpixels. The pairwise constraints exist only between adjacent superpixels. This property is important since the computation of random walk between two endpoints in the graph domain can be much more efficient than a walk in the pixel-grid domain.

In this paper, the superpixels in an image obtained from a segmentation algorithm is converted to a graph representation. Each superpixel is represented as a node and the connection between two adjacent superpixels represents an edge. Figure 4b shows the result after a superpixel segmentation.

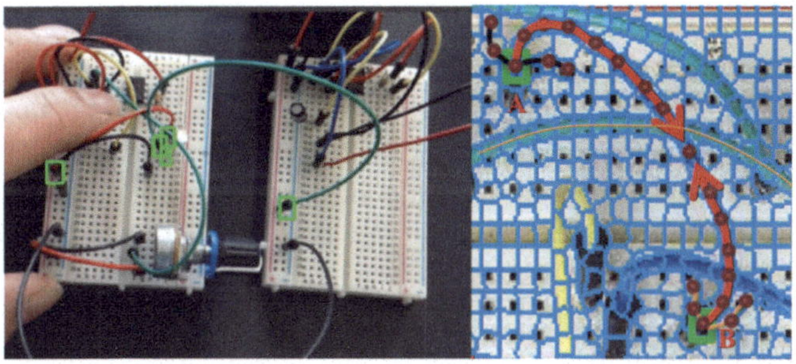

(a) (b)

Fig. 4. (a) A few endpoints of several wires are labeled in one video frame by the remote user, and tracked robustly over video sequence, with possible changes of translation, perspective and scale; (b) The random walk algorithm traces a DLO from two endpoints A and B and both walk finally meets at the middle of DLO. Note: each dark red circle represents one superpixel. The brown line represents another DLO and it occludes the red one in the middle. The random walk can move between any two adjacent superpixels. The optimal path (iterative walk) should minimize the energy function defined in Sect. 2.2.2.

2.2.2 Iterative Walk

Assume after superpixel segmentation, each DLO is segmented into multiple connected superpixels, which share similar color and texture. Tracing the DLO turns out to be a problem to recover a path connecting these superpixels between two labeled endpoints. It can be modeled as a problem of iterative walk. Each walk represents one step move between two adjacent superpixels, (ideally both superpixels are the parts of same DLO). The walk is iterated from two endpoints until two walks meet at one superpixel in the middle of the DLO. Figure 4b shows a path is successfully identified to connect two endpoints over adjacent superpixels to recover a wire (DLO).

The iterative walk always starts from a seed superpixel – an endpoint of a DLO, and move to an adjacent superpixel, which belongs to the same DLO. From a superpixel, the algorithm walks to one of several adjacent superpixels, by choosing the optimal one that shares similar color and texture, as well as minimize the curvature changes in the image. We adopt the energy functions defined in [32], which takes into account the following similarity measurement between two superpixels, when extending a DLO:

- Visual similarity: we assume that the color and texture of two neighbor superpixels on the same DLO are similar. Hence, we compare the normalized color histograms (in the HSV color space) of the two adjacent superpixels, by computing the Euclidean distance of two histograms. The one minimizes the color difference is likely to be a correct extension of the current DLO.
- Curvature similarity: giving an assumption that the curvature changes of a DLO smoothly along consecutive superpixels, we measure the curvature of two adjacent superpixels. Compared with the curvature of the current superpixel on the DLO, the optimal superpixel is likely to minimize the curvature change.
- Distance similarity: this similarity enforces the spatial constraint between two superpixels, because the superpixels on a same DLO is likely connected, if there is no occlusion.

After computing the above similarities, we normalize each to the range in [0 1] and then the superpixel minimizing the product of three similarities is selected to extend the current DLO from each side (from one endpoint). The algorithm is repeated from two labeled endpoints, until two paths meet in the middle of DLO and the DLO is fully traced (Fig. 4b).

2.2.3 Occlusion Handling

We run the iterative walk to trace a DLO from two seeding points, by extending a path from one superpixel to one of its neighbor superpixels at a time. However, it may not always work due to occlusion (Fig. 4b, the wire highlighted in brown occludes the target DLO). Therefore, we have to trace a DLO that is partially occluded by another object, the DLO is broken into two segments which may not connected. We won't be able to trace a path to extract the whole DLO if we always extend a walk to direct neighbor superpixel. Hence, we increase the extension range to the 2nd neighbor superpixels, which is a neighbor of one of the current neighbors superpixels, and it can be easily obtained in any adjacency graph.

2.3 Track DLOs

After the remote user marks the two endpoints of a DLO in one video frame, we apply a correlation tracking algorithm [33] to track the endpoints, and the algorithm can handle large scale variations in complex image sequences. When any new frame comes, the two endpoints are firstly tracked and the algorithm in Sect. 2.2 is used to trace the whole DLO in the new frame. The tracking approach is applied and the DLO can be tracked over the video sequence. Note, the tracking algorithm can be utilized to track multiple DLOs after multiple pairs of endpoints of DLOs are labeled by the user.

2.4 Interface of Client User and Remote Expert

An effective user interface is also very important to the success of any remote collaboration project. Our interface allows both local and remote users to establish a simple remote collaboration, with a live video stream of the local environment accessed by the remote user.

- *Remote side*: the interface allows the remote user to pause the live video anytime, and then labels at least two endpoints of each target DLO in the paused video. Given two endpoints, the algorithm can trace the DLO so that the entire DLO is labeled, and the remote user can confirm the tracing result and provide additional annotations or audio instruction to guide the local user to complete the operation task on the DLO. In case, the iterative walk does not produce a correct labeling of the target DLO, the interface allows a manual correction by the remote user.
- Annotation tools: a few simple 2D annotation and drawing features have been implemented to allow users easily annotate and demonstrate guidance, including drawing of a few simple pre-defined shapes, as well as text input.
- Local side: once the remote user has done the DLO tracing and annotation, and the labeled DLO and the annotation are sent back to the local user. Because both users share the same video and annotations, hence the local user can easily follow the annotation to complete the task.

3 Implementation and Preliminary Test

The local platform was implemented on a smartphone, which sends a live video to a cloud server. Once the annotation and audio instruction are sent back from the remote expert, one or more target DLOs can be tracked over the live video and the annotations are superimposed next to the DLOs.

The superpixel algorithm is computation expensive hence we implement it at the remote side, which is running on a PC and the user label and annotation using a regular tablet. Once the remote expert labels the endpoints of target DLOs, the tracing algorithm identify the entire DLO. Meanwhile, the endpoints are tracked on incoming video frames, and the tracing algorithm is iteratively applied to the incoming frames therefore, the target DLOs can be tracked.

The interface allows the remote user specify the bounding box of the DLOs hence the superpixel based segmentation algorithm is only performed within the bounding box to reduce the computation time.

To validate the effectiveness of the proposed algorithm, we have conducted some preliminary experiments. We test if users could use the proposed remote system to complete a remote guidance task in a complex environment, which includes different manipulations of DLOs. The remote user tries to guide a local user prototyping electronics on a breadboard. Two participants are placed in different rooms in each test, one as a local user and another as remote user. The local user follows the remote user's instructions to connect a few jumper wires to some specific locations on a breadboard. The above test serves to simulate a local user requiring assistance from a remote user to

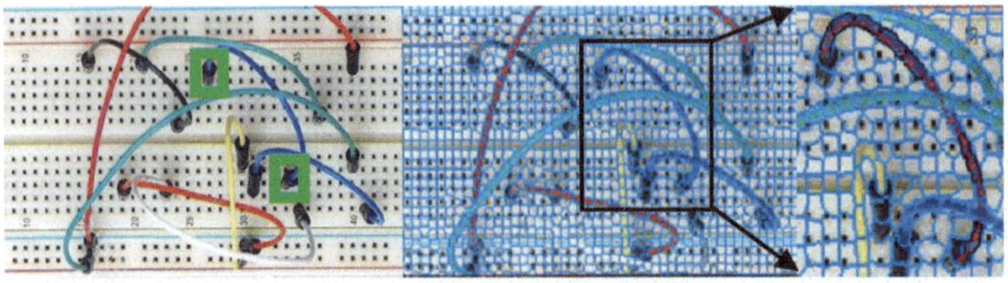

Fig. 5. (a) An input image, (b) The image is segmented into an adjacency graph on superpixels, and two endpoints of a blue wire are carefully marked in red, (c) The output: the blue wire is detected and traced (in red) using an iterative walk.

complete complex tasks. We evaluate the system with a baseline which only uses Skype video chat for collaboration. A few participants are designated as remote users, who are trained to be familiar with the manipulation tool and 2D annotation on a tablet PC system. Others are assigned as local users and use the proposed mobile system. The experiment shows the proposed system improves the collaboration by shortening the completion time and improving the accuracy of annotation. However, we found the system failed if multiple wires with similar colors are close together and self-occlusion exists.

4 Conclusions

In this paper, we propose a remote collaboration system which handles a complex environment including the manipulations of many DLOs. A local user records live video of the local environment, which is available to the remote expert for preview. The remote module allows users to easily identify and track DLOs and add annotation to explain the operation details of DLOs, which is not easy to explain using verbal communication, as well as regular video conference tools.

References

1. Fussell, A.R., Setlock, L.D., Yang, J., Ou, J., Mauer, E., Kramer, A.D.I.: Gestures over video streams to support remote collaboration on physical tasks. HCI **19**, 273–309 (2004)
2. Kirk, D., Fraser, D.S.: Comparing remote gesture technologies for supporting collaborative physical tasks. In: Proceedings of CHI 2006, pp. 1191–1200. ACM Press (2006)
3. Kirk, D., Rodden, T., Fraser, D.S.: Turn it this way: grounding collaborative action with remote gesture. In: Proceedings of CHI 2007, pp. 1039–1048. ACM Press (2007)
4. Gergle, D., Kraut, R.E., Fussell, S.R.: Action as language in a shared visual space. In: Proceedings of CSCW 2004, pp. 487–496. ACM Press (2004)
5. Gurevich, P., Lanir, J., Cohen, B., Stone, R.: TeleAdvisor: a versatile augmented reality tool for remote assistance. In: SIGCHI, p. 6 (2012)
6. Azuma, R.T.: A survey of augmented reality. Presence Teleoperators Virtual Environ. **6**(4), 355–385 (1997)

7. Adcock, M., Gunn, C.: Using projected light for mobile remote guidance. Comput. Support. Coop. Work **24**(6), 591–611 (2015)
8. Wong, K.: HandsOn: a portable system for collaboration on virtual 3D objects using binocular optical head-mounted display. Master thesis, MIT (2015)
9. Tait, M., Billinghurst, M.: The effect of view independence in a collaborative AR system. Comput. Support. Coop. Work **24**(6), 563–589 (2015)
10. Wang, S., et al.: Augmented reality as a telemedicine platform for remote procedural training. Sensors **17**, 2294 (2017)
11. Piumsomboon, T., Lee, G.A., Hart, J.D., Ens, B., Lindeman, R.W., Thomas, B.H., Billinghurst, M.: Mini-Me: an adaptive avatar for mixed reality remote collaboration. In: CHI 2018, Paper 46. ACM, New York (2018)
12. Lee, Y., Masai, K., Kunze, K., Sugimoto, M., Billinghurst, M.: A remote collaboration system with empathy glasses. In: ISMAR (2016)
13. Tecchia, F., Alem, L., Huang, W.: 3D helping hands: a gesture based MR system for remote collaboration. In: Proceedings of ACM VRCAI, pp. 323–328 (2012)
14. O'Neill, J., Castellani, S., Roulland, F., Hairon, N., Juliano, C, Dai, L.: From ethnographic study to mixed reality: a remote collaborative troubleshooting system. In: CSCW11, pp. 225–234 (2011)
15. Javdani, S., Tandon, S., Tang, J., O'Brien, J.F., Abbeel, P.: Modeling and perception of deformable one-dimensional objects. In: 2011 IEEE International Conference on Robotics and Automation (ICRA), pp. 1607–1614. IEEE (2011)
16. Jackson, R.C., Yuan, R., Chow, D.L., Newman, W., Cavusoglu, M.C.: Automatic initialization and dynamic tracking of surgical suture threads. In: ICRA, p. 4710 (2015)
17. Saha, M., Isto, P.: Manipulation planning for deformable linear objects. IEEE Trans. Robot. **23**, 1141–1150 (2007)
18. Lui, W.H., Saxena, A.: Tangled: learning to untangle ropes with RGB-D perception. In: IROS (2013)
19. Nair, A., Chen, D., Agrawal, P., Isola, P., Abbeel, P., Malik, J., Levine, S.: Combining self-supervised learning and imitation for vision-based rope manipulation. In: 2017 IEEE International Conference on Robotics and Automation (ICRA), pp. 2146–2153 (2017)
20. Schulman, J., Lee, A., Ho, J., Abbeel, P.: Tracking deformable objects with point clouds. In: ICRA, pp. 1130–1137 (2013)
21. Hopcroft, J.E., Kearney, J.K., Krat, D.B.: A case study of exible object manipulation. Int. J. Robot. Res. **10**, 41–50 (1991)
22. Remde, A., Henrich, D., Worn, H. Picking-up deformable linear objects with industrial robots (1999)
23. Yue, S., Henrich, D.: Manipulating deformable linear objects: sensor-based fast manipulation during vibration. In: Robotics and Automation, Proceedings. ICRA 2002 (2002)
24. Alvarez, N., Yamazaki, K., Matsubara, T.: An approach to realistic physical simulation of digitally captured deformable linear objects. In: IEEE International Conference on Simulation, Modeling, and Programming for Autonomous Robots (SIMPAR), p. 135. IEEE (2016)
25. Koo, K.M., Jiang, X., Kikuchi, K., Konno, A., Uchiyama, M.: Development of a robot car wiring system. In: 2008 IEEE/ASME International Conference on Advanced Intelligent Mechatronics. AIM 2008, pp. 862–867. IEEE (2008)
26. Sun, X., Cheung, N.-M., Yao, H., Guo, Y.: Non-rigid object tracking via deformable patches using shape-preserved KCF and level sets. In: CVPR (2017)
27. Lukežič, A., Zajc, L.Č., Kristan, M.: Deformable parts correlation filters for robust visual tracking. IEEE Trans. Cybern. **48**, 1849–1861 (2017)

28. Zhong, Y., Jain, A.K., Dubuisson-Jolly, M.-P.: Object tracking using deformable templates. TPAMI **22**, 544–549 (2000)
29. Leymarie, F., Levine, M.D.: Tracking deformable objects in the plane using an active contour model. TPAMI **15**, 617–634 (1993)
30. Dai, J., Qi, H., Xiong, Y., Li, Y., Zhang, G., Hu, H., Wei, Y.: Deformable convolutional networks. In: ICCV (2017)
31. Liu, W., Song, Y., Chen, D., He, S., Yu, Y., Yan, T., Hancke, G., Lau, R.: Deformable object tracking with gated fusion. IEEE Trans. Image Process. **28**, 3766–3777 (2019)
32. De Gregorio, D., Palli, G., Di Stefano, L.: Let's take a walk on superpixels graphs: deformable linear objects segmentation and model estimation. In: ACCV (2018)
33. Danelljan, M., Häger, G., Shahbaz Khan, F., Felsberg, M.: Accurate scale estimation for robust visual tracking. In: BMCV (2014)
34. Achanta, R., Shaji, A., Smith, K., Lucchi, A., Fua, P., Susstrunk, S.: SLIC superpixels compared to state-of-the-art superpixel methods. TPAMI **34**, 2274 (2012)

Home Appliances Control Using Android and Arduino via Bluetooth and GSM Control

David Kwabena Amesimenu[1], Kuo-Chi Chang[1(✉)], Tien-Wen Sung[1], Kai-Chun Chu[2], Yu-Wen Zhou[1], Joram Gakiza[1], Ntawiheba Jean Damour[1], Shoaib Ahmad[1], and Abdalaziz Altayeb Ibrahim Omer[1]

[1] Fujian Provincial Key Laboratory of Big Data Mining and Applications, Fujian University of Technology, Fuzhou, China
albertchangxuite@gmail.com

[2] Department of Business Management, Fujian University of Technology, Fuzhou, China

Abstract. In modern life, home automation can improve the quality of life (QoL), which is an important function and is very convenient. As a result, the popularity of automation has led to the development of "smart homes". Intelligent systems are mainly developed with the help of sensors, which can provide the system with the information necessary to make human-like intelligent decisions. This research design system allows users to remotely monitor and interact via a global system for mobile (GSM) communications and Bluetooth in a system using a graphical user interface (GUI) based on Android applications on their smartphone. The results of this research system focus on how to use GSM and Bluetooth to provide convenient access to the device. This includes the use of people counting systems, electronic door locks, turning on fans at specific temperatures, monitoring home systems, providing security to switch lighting, etc. Function, and can access the smart home and remote access to those wireless sensor networks. In any case, the overall design of this study is successful, it can effectively remotely control all destination systems, and the communication interface between Arduino UNO board, repeater board, GSM, Bluetooth, and smartphone is seamless. This study effectively uses low cost completing the purpose of a smart home is an important contribution to this research.

Keywords: Bluetooth · Smart home automation android · App inventor GSM · SMS · HAS

1 Introduction

The concept of smart home automation has been in full swing since the development of the first home automation network in 1975 (Fig. 1). In terms of technology, the network is based on the communication protocol of electronic devices. In theory, smart home automation uses power transmission for signal transmission and control, and the signal involves radiofrequency of digital data, and according to the X10 industry-standard, maintains the widest "remote" range available today. Past research has also

A.-E. Hassanien et al. (Eds.): AICV 2020, AISC 1153, pp. 819–827, 2020.
https://doi.org/10.1007/978-3-030-44289-7_77

pointed out that a safe and intelligent way to leave is our recent direction, which is the main role, played by automation and Android phones. Nowadays, the automation system needs to control the AC device and wirelessly obtain the current state of the device through the android application over a long distance, that is, open/close wirelessly [1–3].

Fig. 1. The concept of smart home automation

The concept of smart home automation in the system can improve family living standards. This technology is focused on control household appliances like AC, fan, and light, automatically. With the increase of smart devices and people's dependence on their rich attributes, it has become necessary to interconnect a large number of smart devices. Besides, due to advances in wireless technologies such as Bluetooth, GSM, and Wi-Fi, different component device connections have been introduced, but Wi-Fi technology provides an excellent medium for connecting multiple devices on the network. Wi-Fi operates in the internationally recognized 2.4 GHz band, and each connection has its specifications, its functions, and applications. "Home automation" involves the real-time control and monitoring of a large number of household appliances. Of course, home automation can also be defined in many ways. The passed study proposes the overall design structure of a home automation system with low-coast or maintenance and wireless systems. Smart home automation is a modern technology that can transform your house to the extent that it can automate different sets of tasks. This technology is continuously increasing its versatility by integrating modern functions to meet growing demand. The smart home also includes control and automation of household appliances such as washing machines, clothes dryers, ovens, refrigerators, freezers, etc. One of the main purposes of home automation systems is to save electricity. Smart home automation can bring comfortable living and energy management functions to users, and bring more benefits to the elderly and people with disabilities [4–6].

Based on intelligence and automation, the degree of home automation systems varies. It can range from simple remote control of lighting to complex microcontroller-based networks. Remotely monitoring and accessing home appliances and systems is the main characteristic of the home automation system. In an integrated manner, appliances can communicate using home automation systems. Several factors such as convenience, energy efficiency and safety benefits are obtained. Due to expensive costs and exhausting maintenance, most of the present systems are not reasonable for many people. Some systems provide solutions that are not very useful for household applications.

Many protocols were used and easy to maintain to alleviate problems when there are faults [7]. In this project, Bluetooth and GSM are the methods used for home automation. For cost-efficient, flexibility and user-friendly for everyone to be able to use and easy maintenance, the use of Bluetooth and GSM was the main using idea behind this paper.

About related work, the controlling of electrical home appliances with the help of two protocols, that is Bluetooth and GSM are illustrated in this paper. Controlling appliances like lights, fans, AC and the security system in a home with the aid of an Android phone. The android phone app is developed by using MIT APP Inventor and the Arduino sketch is developed by the use of Arduino IDE for control. For user-friendly, cost-efficient and comfortability to use easily, this home automation system was proposed.

For the developments of home automation systems, there have been a lot of proposed methods used. The control of home appliances with the IoT approach is illustrated by some Home Automation Systems. It uses Bluetooth and GSM for interaction between the android mobile application and the appliances under the control of the system [5]. In [4] and [6], Describes how to command and control home gadgets using a mobile phone, people can use this system to control devices in their home from a great distance before they reach home. The user sends an instruction in form of SMS to control a device/appliance from his/her mobile phone to a computer that is connected to the appliance; once the message is received the computer will send the command to a microcontroller for controlling the appliance properly.

2 System Design

1. **System Overview:** The surrounding in a home is automatically controlled with this proposed system in this paper. In Fig. 2 shows the design of the system and some of the components it uses during its operations with the help of Bluetooth and GSM. One of the most successful wireless technologies now is the application of GSM (Global System for Mobile communications) platform. What forms the basis or supports most if the world's mobile phone network is this GSM platform. [7] The GSM has become the world's greatest growing communications technology of all time and the leading global mobile standard, spanning 210 countries. In present days the GSM technology is in use by more than one in five of the world's population – by mid-March 2006 there were over 1.7 billion GSM subscribers, representing approximately 77% of the world's cellular market. The growth of GSM continues unabated with almost 400 million new customers in the last 12

months. The progress hasn't stopped there. Today's GSM platform is living, growing and evolving and already offers an expanded and feature-rich 'family' of voice and multimedia service.

2. **Block Diagram:** In Fig. 2 is the design of the system and some of the components used during its operations with the help of Bluetooth and GSM. Arduino IDE of 8 digital I/O pins is interfaced with 8 channels relay modules for control. NO/NC of these 8 channels Relay modules is also interfaced with the Home Appliances to be controlled. The relay module can only take 5 V supply and the Arduino can also take only 5 V, 2 A supply but the GSM module will take 12 V from the power supply.

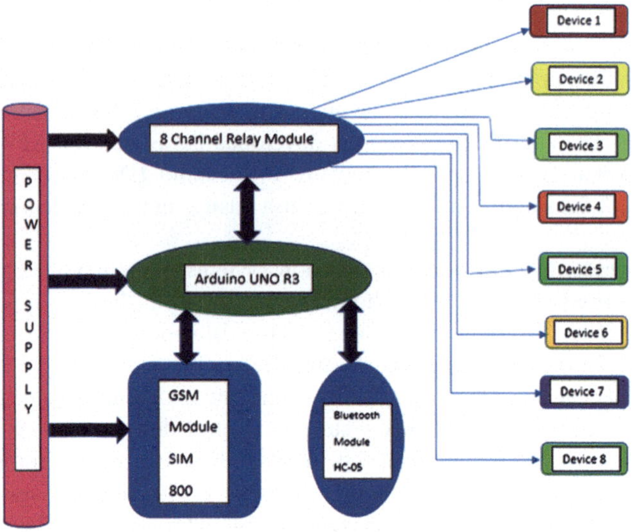

Fig. 2. Block diagram of home automation system

3. **Circuit Diagram:** Using the Bluetooth and the GSM to control the home appliances automatically is shown in Fig. 2. Normally, in designing, the worse scenario is also considered and that is if the system does not work then the traditional way is used to operate. Therefore, the system can be operated in the automatic way and also in the conventional way.
   (A) **Automatic Way:** A click on the Android phone application will automatically control the Home Appliance using Bluetooth and GSM. When the click is made, information is sent through GSM module and Bluetooth to the Arduino and a signal is sent to the relay to put the appliance ON/OFF.
   (B) **Conventional Way:** The electrical appliances are mostly connected to switches through switch board. Anyone who wants to turn ON/OFF an appliance without using the Home Automation System will traditionally use the switch connected to that appliance. Two-way switch is used to integrate the system with the help of a relay connection [8]. Figure 3 shows a relay on a two-way switch.

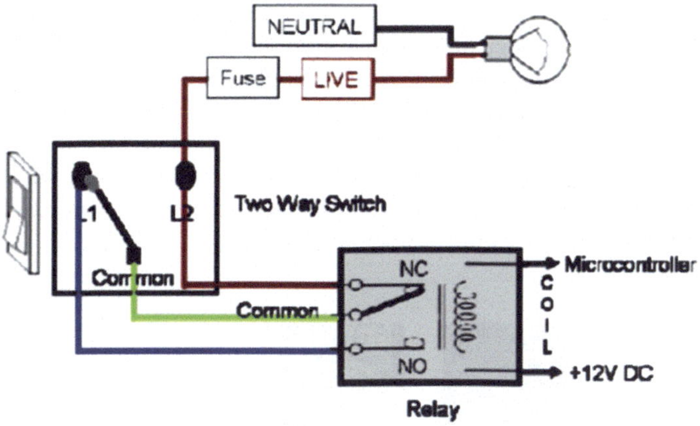

Fig. 3. Relay on a two-way switch

4. System Hardware Design
   (A) **Arduino UNO:** The Arduino Uno is an Atmega 328 based microcontroller board. This microcontroller board has 14 digital I/O pins out of which 8 pins we have used for interfacing home appliances. It also has a USB connection, power jack and a reset button. It is used to decode the signals sent by Bluetooth and GSM and accordingly send the control signal to relay module. Bluetooth module HC-05 and GSM module SIM800 are serially interfaced with Arduino [9]. Figure 4 shows Arduino UNO R3.

Fig. 4. Arduino UNO R3

   (B) **GSM Module:** SIM 800c is a GSM/GPRS module. It is designed for global market [9]. On the Smartphone at the GSM portion, when a particular button is clicked, SMS is sent from the Smartphone to the GSM module SIM800c interfaced with Arduino and Appliances are turned ON/OFF with respect to the information sent. The present status of the Appliance is sent by SMS through GSM module from the Arduino back to the Smartphone.

(C) **Bluetooth:** Bluetooth module used in this project is HC-05. It is easy to use serial port module. It is designed for transparent wireless serial connection setup. The module requires 3.3 V to 5 V power supply [10]. On the Smartphone at the Bluetooth portion, when a particular button is clicked, data is serially sent from the Bluetooth of the Smartphone to the Bluetooth module HC-05 interfaced with Arduino and Appliances are turned ON/OFF with respect to the information sent. The data of the present status of the Appliance is sent back to the Bluetooth of the Smartphone. Figure 5 shows Bluetooth HC-05.

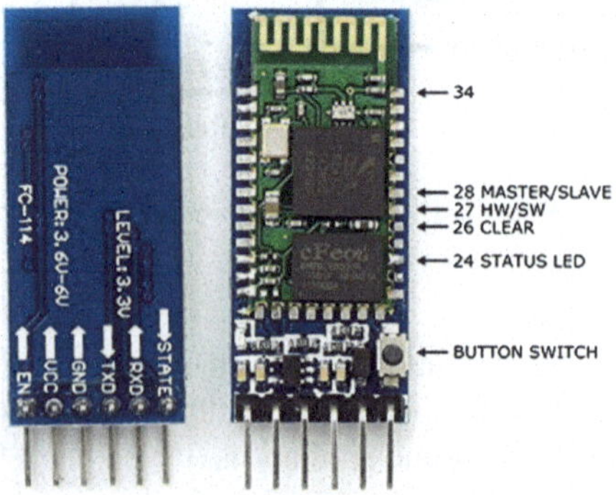

Fig. 5. Bluetooth HC-05

(D) **Android:** Android is operating system for mobile initially developed by Android INC, which is based on Linux kernel. Google bought it in 2005. The android operating system has gone through multiple major releases. The version of android is 7.0 "Nougat" which is released in August 2016 [11]. The current version of Android is 9.0 "Pie". Using the MIT App inventor to create a platform on which Bluetooth and GSM would function perfectly.

(E) **Power supply circuit:** The elementary unit of any electronic system is the power supply which provides required power for system's operation. In this project +5 V supply is used for Arduino and relay module. +12 V power supply is used for GSM module SIM 800 [12].

(F) **Relay Module:** A relay module is used to turn on or turn off home appliances using voltage and/or current much more than Arduino could handle. This is also used for providing isolation between the low voltage circuit on Arduino side and the high voltage circuit side controlling home appliances. Relay module is activated by using external 5 V power supply, which in turn, controls electrical appliances like fans, lights, ovens etc. In this project 8-channel DC 5 V relay module is used. It is equipped with high current relay, AC 250 V 10 A and DC 30 V 10 A. In it each relay needs 15–20 mA driver current [12]. Figure 6 shows the 8- Channel Relay module.

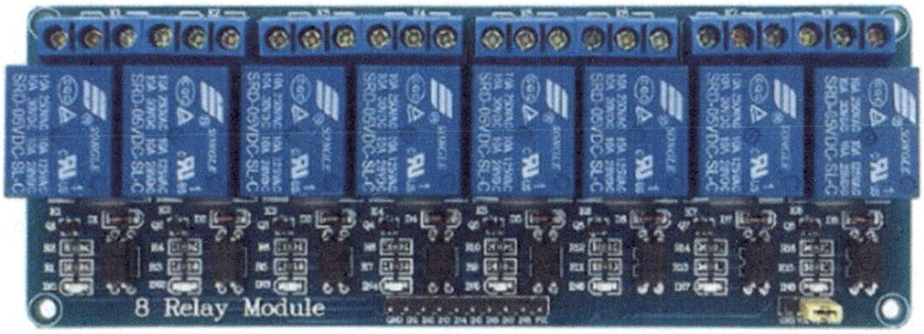

Fig. 6. Shows the 8-channel relay module

(G) **MIT App inventor and android app**

- **MIT App inventor:** The Bluetooth and GSM feature on the Android application was created using the MIT App inventor. It is initially provided by Google. Now it is maintained by Massachusetts Institute of Technology (MIT). It is an open source web application for android [11]. Apps created in app inventor (.apk files) can be ported to any android smart phone.
- **Android App:** When a button is clicked via Bluetooth and GSM on the screen of the Smartphone, data and SMS are sent respectively to Bluetooth module HC-05 interfaced with Arduino board. The relay board is controlled respectively with the data received from the microcontroller board.

3 Performance, Experiment and Results

There are number of popular families of microcontrollers which are used in different applications as per their capability and feasibility to perform the desired task, most common of these are 8051, AVR and PIC microcontrollers. In this propose system we ATMEGA328P from AVR family of microcontrollers was used. Figure 7 shows the experimentation period.

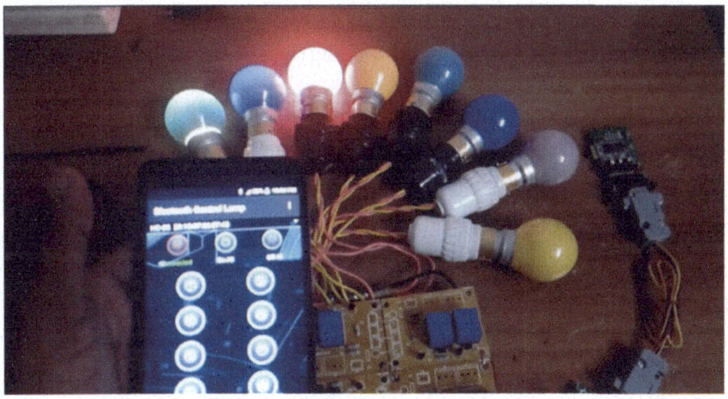

Fig. 7. Experimentation period

The development of the hardware model by connecting the power source required to the Arduino board, the 8 channel Relay module and the GSM module and also interfacing them with the devices was successfully done. It was tried and the application on the Smartphone was able to send SMS to the GSM module SIM 800 and the Arduino was able to decode the information and later send the decoded signal to the Relay to turn ON/OFF the devices connected to it. The advantage of the GSM system is that, anywhere and at any time wherever, the devices connected can be controlled by just sending SMS. Also, the current status of the device is sent back to the Smartphone as SMS. The Bluetooth was also tried by using the Bluetooth of the Smartphone. By operation, data was sent to the Bluetooth module HC-05 and the information was decoded by the Arduino and the decoded signal was sent to the Relay to turn ON/OFF the device connected. The only difference from this to the GSM is that this is with a shorter distance of about 15–20 m.

4 Conclusion and Suggestion

The prime motive and goal of this project is modeling, developing and executing with the use of smartphone to communicate to home appliances via Bluetooth and GSM. The goal of commands has been given on the Smartphone to turn ON/OFF home appliances for convenience's sake. The user can also get the current status of the device from wherever at any time on the Smartphone. Simplicity and economics were considered in the proposed system. Automating the home with features of smartphone was the goal of the Bluetooth system. While using GSM is in the sense of being able to control the home appliances anywhere all over the world. For the GSM, the distance is limited but the Bluetooth is about 15–20 m. Expansion can be made further on the system as a framework to control other appliances like the dishwashing machine, fridges, and television, etc.

References

1. Froiz-Míguez, I., Fernández-Caramés, T.M., Fraga-Lamas, P., Castedo, L.: Design implementation and practical evaluation of an IoT home automation system for fog computing applications based on MQTT and ZigBee-WiFi sensor nodes. Sensors **18**, 2660 (2018). https://doi.org/10.3390/s18082660
2. Singh, H., Pallagani, V., Khandelwal, V., Venkanna, U.: IoT based smart home automation system using sensor node. In: 2018 4th International Conference on Recent Advances in Information Technology (RAIT), Dhanbad, pp. 1–5 (2018). https://doi.org/10.1109/RAIT.2018.8389037
3. Chu, K.C., Horng, D.J., Chang, K.C.: Numerical optimization of the energy consumption for wireless sensor networks based on an improved ant colony algorithm. J. IEEE Access. **7**, 105562–105571 (2019)
4. Beregovskyi, V., Denysyuk, P., Teslyuk, T., Lozynskyi, A.: Development and implementation of the technical accident prevention subsystem for the smart home system. Intell. Syst. Appl. **1**, 1–8 (2018)

5. Kim, H.S., Song, J.-H., Lee, S.: Energy-efficient traffic scheduling in IEEE 802.15.4 for home automation networks. IEEE Trans. Consum. Electron. **53**(2), 369–374 (2007)
6. Ahmad, A.W., Jan, N., Iqbal, S., Lee, C.: Implementation of ZigBee-GSM based home security monitoring and remote control system. In: 2011 IEEE 54th International Midwest Symposium on Circuits and Systems (MWSCAS), Seoul, pp. 1–4 (2011). https://doi.org/10.1109/MWSCAS.2011.6026611
7. Shinde, A., Kanade, S., Jugale, N., Gurav, A., Vatti, R.A., Patwardhan, M.M.: Smart home automation system using ir, bluetooth, GSM and android. In: 2017 Fourth International Conference on Image Information Processing (ICIIP), Shimla, pp. 1–6 (2017). https://doi.org/10.1109/iciip.2017.8313770
8. Rajadurai, S., Nehru, P.P., Selvarasu, R.: Android mobile based home security and device control using GSM. In: 2015 International Conference on Innovations in Information, Embedded and Communication Systems (ICIIECS), Coimbatore, pp. 1–5 (2015). https://doi.org/10.1109/iciiecs.2015.7192980
9. Patel, S.M., Pasha, S.J.: Home automation system (HAS) using android for mobile phone. Int. J. Sci. Eng. Technol. Res. **04**(25), 4844–4849 (2015)
10. Collotta, M., Pau, G., Talty, T., Tonguz, O.K.: Bluetooth 5: a concrete step forward toward the IoT. IEEE Commun. Mag. **56**(7), 125–131 (2018). https://doi.org/10.1109/MCOM.2018.1700053
11. Li, J., Sun, L., Yan, Q., Li, Z., Srisa-an, W., Ye, H.: Significant permission identification for machine-learning-based android malware detection. IEEE Trans. Industr. Inf. **14**(7), 3216–3225 (2018). https://doi.org/10.1109/TII.2017.2789219
12. ElKamchouchi, H., ElShafee, A.: Design and prototype implementation of SMS based home automation system. In: 2012 IEEE International Conference on Electronics Design, Systems and Applications (ICEDSA), Kuala Lumpur, pp. 162–167 (2012). https://doi.org/10.1109/icedsa.2012.6507788

Intelligent Networks

Multi-hop Route Planning Based on Environment Information for Path-Following UAVs

Tien-Wen Sung, Linyun Sun(✉), and Kuo-Chi Chang

College of Information Science and Engineering,
Fujian University of Technology, Fuzhou, China
tienwen.sung@gmail.com, 2191901004@smail.fjut.edu.cn,
albertchangxuite@gmail.com

Abstract. Flight path planning is one of the key research issues in the field of unmanned aerial vehicles (UAVs). There have been various approaches proposed to find or plan an optimal or appropriate path for numerous UAV applications. The obtained paths will be followed as a navigation once the UAV flies. Related works have utilized different methods to find the paths for different situations or considerations in the flight environments. Several bio-inspired algorithms such as PSO, GA, ABC, and ACO as well as the graph-based A* algorithm were usually utilized in the solutions of path planning. In this paper, the concept of Floyd-Warshall algorithm and a grid-based map presenting environmental information are utilized to find an optimal path with minimum hop count. A simulator is also developed for this work. Several simulation results with different grid sizes are illustrated. This study presents a preliminary trial work of grid-based multi-hop route planning for UAVs. Both the grid model and concerned environmental information can be extended for further complex researches.

Keywords: Unmanned aerial vehicles · Path planning · Grids · No-fly zones · Obstacles

1 Introduction

The advances in micro-electro-mechanical systems (MEMS), system-on-chips (SoCs), and communication technologies have promoted the rapid development of unmanned aerial vehicles (UAVs). In addition to military use, UAVs are also widely used in civilian applications and even become consumer electronics. Applications in different civilian domains can be goods delivery, traffic monitoring, aerial photography, fire-fighting, agriculture, etc. [1]. In recent years, the growing of UAVs in various fields has drawn increasing attentions of industrial and academic researchers. There are various topics in the contributed works related to UAVs such as obstacle avoidance, path planning, flight navigation, hardware development, and etc. [2]. Since path planning is one of the key topics for UAVs, this paper focuses on the topic and presents a grid-based path finding study. The study utilizes the concept of Floyd-Warshall algorithm [3] and a grid-based map which provides environmental information to find a shortest

A.-E. Hassanien et al. (Eds.): AICV 2020, AISC 1153, pp. 831–839, 2020.
https://doi.org/10.1007/978-3-030-44289-7_78

path with minimum hop count between source and destination. The model used in this study is simplified but can be extended to more complex ones for further research.

2 Related Works

In many UAV applications, the drones usually need a predefined path to follow and accomplish their tasks [4]. Path planning, therefore, becomes a crucial issue in the field of UAVs. The objective of UAV path planning is to find or compute a flight path between two locations and meet some requirements or constraints. The constraints could be about distance, energy, safety, timeliness, etc. Since UAVs have a limitation of flight height or need to fly at a specific height (e.g., a low-altitude flight mission), the path planning is highly related to the obstacle avoidance, i.e., collision avoidance [5]. Different types, sizes, or flight distances of UAVs, for example fixed-wing drones and micro aerial vehicles (MAVs), may use different approaches for path planning. There are many approaches or algorithms contributed to the path planning for various application scenarios and requirements. The algorithms can be categorized into heuristic and non-heuristic methods [6]. A heuristic approach usually can obtain a near-optimal and time-efficient solution. A non-heuristic approach, on the other hand, usually utilizes mathematical deductions and analyses to obtain the solution which often consumes more computation time. Zhao et al. [7] provided a survey of the computational intelligence (CI) methods used for the UAV path planning. Computational intelligence often uses nature-inspired or bio-inspired algorithms such as genetic algorithm (GA) and particle swarm optimization (PSO) to solve complex problems which mathematical methods do not perform well. Konatowski and Pawłowski [8] described the usages of PSO, Artificial Bee Colony (ABC) and ant colony optimization (ACO) algorithms for UAV path planning. There are also various approaches used, for example Dhulkefl and Durdu [9] utilized the A* algorithm for UAV path planning. To solve those problems that have different details or conditions concerned, one of various approaches can be chosen. This paper describes the trial work of utilizing the concept of Floyd-Warshall algorithm as well as a grid-based map model of environmental information for the UAV path planning. The path with minimum hop count will be determined as an optimal path and no-fly regions and obstacles are avoided.

3 Grid-Based UAV Path Planning

Grid-based concepts are usually utilized to solve problems in various scientific fields. In this study, a two-dimensional grid-based flight map is used to determine an optimal path for UAV flying. The path-finding algorithm performs based on the information given by the grids. Also, the acquired path is expressed in a sequence of grids. Although this grid model is ordinary, it can be further extended to a three-dimensional or more complex one for advanced researches.

3.1 Grid-Based Flight Map

In a flight area there may be obstacles and no-fly zones (NFZ). An obstacle may be a building, tree, etc. whose height is higher than the flying height set by the UAV. A no-fly zone may be an airport or military base as well as their vicinity, for example. Accordingly, this study divides a flight area into three types of grids: no-fly, obstacle, and free-fly grids with type-defined values of T_{NFZ}, T_{OBS} and T_{FRE}, respectively. A given flight area can be described as a two-dimensional matrix $G[i, j]_{m \times n}$ with the values defined in Eq. (1), where m and n depend on the grid size and flight area; g_{ij} represents the corresponding grid; h_{ij} and h_{F} represent obstacle and flying heights, respectively.

$$G[i,j]_{m\times n} = \begin{cases} T_{\mathrm{NFZ}}, & g_{ij} \in \mathrm{NFZ} \\ T_{\mathrm{OBS}}, & g_{ij} \notin \mathrm{NFZ}; h_{ij} \geq h_{\mathrm{F}} \\ T_{\mathrm{FRE}}, & \text{otherwise} \end{cases} \tag{1}$$

As the example shown in Fig. 1, the flight area map with required information is converted into the grids colored with light blue (obstacle), pink (NFZ) and white (free-fly). The path finding will be performed according to the matrix where the information of these grids is stored.

Fig. 1. Dividing flight area into three types of grids.

In regard to the determination of grid size, it is related to the turning radius of the UAV. The radius (r) can be derived from the centripetal acceleration formula and acquired by Eq. (2). It is related to the flying speed (v) and the centripetal acceleration (a_{c}) of the UAV. In case the UAV flies along the grid center line (see Fig. 2), the side length (s_{G}) of the grid should be twice the turning radius, as Eq. (3).

$$r = v^2/a_{\mathrm{c}} \tag{2}$$

$$s_{\mathrm{G}} = 2v^2/a_{\mathrm{c}} \tag{3}$$

It is known that in general the faster the flight speed, the larger the turning radius. Therefore, the flight speed should be bounded to prevent the UAV from colliding with a neighbor grid. The bounded speed (v_B) can be formulated by Eq. (4). Once the flight speed v exceeds v_B, the UAV has a turning radius larger than the grid side length s_G and then a collision or an unlawful entrance may occur while the neighbor grid is an obstacle or a no-fly region, respectively (see Fig. 2).

$$v_B = \sqrt{s_G \cdot a_c} \tag{4}$$

Under the condition of flight speed less than v_B, dividing the flight area into smaller grids can lead to a higher path accuracy. However, the computation time for the grid-based path finding will increase. That is a tradeoff between accuracy and computation time to determine the grid size.

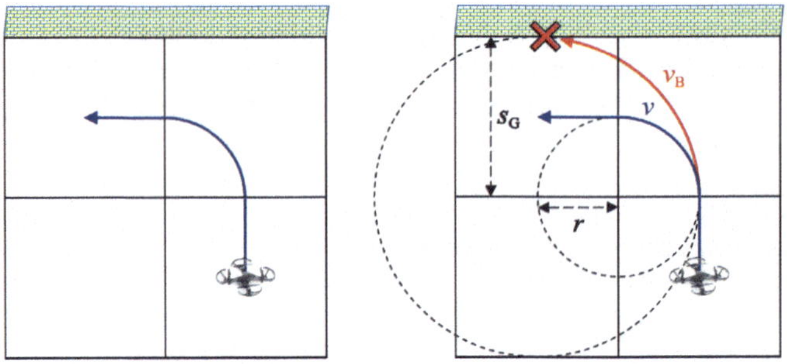

Fig. 2. Grid size and turning radius.

3.2 Grid-to-Grid Distance

Information of the distances between grids is an essential basis for optimal path finding. This study defines the distance between two grids by the grid hop count instead of the actual straight-line distance. Firstly, $f_A(g_{ij}, g_{pq})$ is defined as in Eq. (5) which indicates whether the two grids g_{ij} and g_{pq} are adjacent or not, where $d_E(g_{ij}, g_{pq})$ represents the Euclidean distance between the centers of g_{ij} and g_{pq}. Here two adjacent grids mean that they have a common edge. Moreover, if one of the paired grids belongs to a no-fly zone or obstacle, they are defined as non-adjacent.

$$f_A\left(g_{ij}, g_{pq}\right) = \begin{cases} 1, & d_E\left(g_{ij}, g_{pq}\right) = s_G; G[i,j] = T_{FRE}; G[p,q] = T_{FRE} \\ 0, & \text{otherwise} \end{cases} \tag{5}$$

A matrix $H[i, j, p, q]_{m \times n \times m \times n}$ is defined to indicates the minimum hop count between g_{ij} and g_{pq}. Each element of H is initialized with the value defined in Eq. (6).

$$H[i,j,p,q]_{m\times n\times m\times n} = \begin{cases} 0, & i = p; j = q \\ 1, & f_{\mathrm{A}}(g_{ij}, g_{pq}) = 1 \\ \infty, & \text{otherwise} \end{cases} \tag{6}$$

By utilizing the concept of Floyd-Warshall Algorithm, the final element values in H can be obtained with Eq. (7).

$$\begin{gathered} \forall(g_{ij}, g_{pq}) \text{ where } 1 \leq i,p \leq m \text{ and } 1 \leq j,\, q \leq n \\ H[i,j,p,q] = \min_{1 \leq k_1 \leq m, 1 \leq k_2 \leq n} (H[i,j,p,q], H[i,j,k_1,k_2] + H[k_1,k_2,p,q]) \end{gathered} \tag{7}$$

After the matrix H is obtained with the final values of its elements, it is easy to get the information of minimum hop counts between all pairs of grids. The minimum hop count implies a shortest path between the pair of grids.

3.3 Shortest Path

Although the method used in previous section can obtain the minimum hop count between any pair of grids, it does not provide the path details. Accordingly, an additional procedure of finding path details should be performed in the process of determining minimum hop count. For this purpose, two matrices $X[i, j, p, q]_{m\times n\times m\times n}$ and $Y[i, j, p, q]_{m\times n\times m\times n}$ are defined to respectively represent the horizontal and vertical index numbers of the next hop grid in the path from g_{ij} to g_{pq}. For each grid pair (g_{ij}, g_{pq}), the corresponding elements $X[i, j, p, q]$ and $Y[i, j, p, q]$ are respectively initialized with values of p and q if $f_{\mathrm{A}}(g_{ij}, g_{pq}) = 1$. After the initialization of the matrices, the elements of X and Y will update their values whenever the corresponding hop count in H is updated. Algorithm 1 shown below is used to find the details of shortest (minimum hop count) paths. Here the hop count calculation is omitted from the algorithm. The algorithm is also based on the concept of Floyd-Warshall Algorithm.

Algorithm 1. Determine the shortest paths between all pairs of grids

```
for each (g_ij, g_pq) where 1 ≤ i, p ≤ m and 1 ≤ j, q ≤ n
    if f_A(g_ij, g_pq) = 1
        X [i, j, p, q] = p
        Y [i, j, p, q] = q
    end if
end for
for k1 = 1 to m
    for k2 = 1 to n
        for each (g_ij, g_pq) where 1 ≤ i, p ≤ m and 1 ≤ j, q ≤ n
            if H [i, j, p, q] > H [i, j, k1, k2] + H [k1, k2, p, q]
                X [i, j, p, q] = k1
                Y [i, j, p, q] = k2
            end if
        end for
    end for
end for
```

After performing the algorithm, X and Y have acquired the information of shortest paths between all pairs of grids. The next hop of the minimum hop count path from g_{ij} to g_{pq} will be g_{xy}, where $x = X[i, j, p, q]$ and $y = Y[i, j, p, q]$. Repeat this way for finding the further next hop. After all the hops on the path are traced, the full path can be obtained.

4 Simulations

4.1 Preliminary

According to the method described in Sect. 3, a simple GUI-based simulator was developed to complete the details and verify the path finding results. Figure 3 shows the simulator. It is assumed that the grid sizes selected in the simulations meet the condition of no collision will occur when the UAV turns. In real applications, the grid size needs an estimation with the flight speed and centripetal acceleration driven by the propulsion system and flight control system. The systems are quite different in various types of UAVs such as fixed-wing, single-rotor, or multi-rotor drones.

In the simulator, the flight area is scaled by using one pixel to represent one meter. The simulated grid sizes are 30 m × 30 m, 20 m × 20 m, 15 m × 15 m, and 12 m 12 m, forming 260, 600, 1040, and 1650 grids, respectively. As mentioned in previous sections, there are three types of grids defined: free grid, obstacle grid, and forbidden grid (no-fly grid). The number of different types of grids can be determined by users. In addition, the locations of grids are determined randomly in the simulator, hence it is possible that a too large number of obstacle and no-fly grids results in no available path. That is, an infinity of hop count between the source and destination grids.

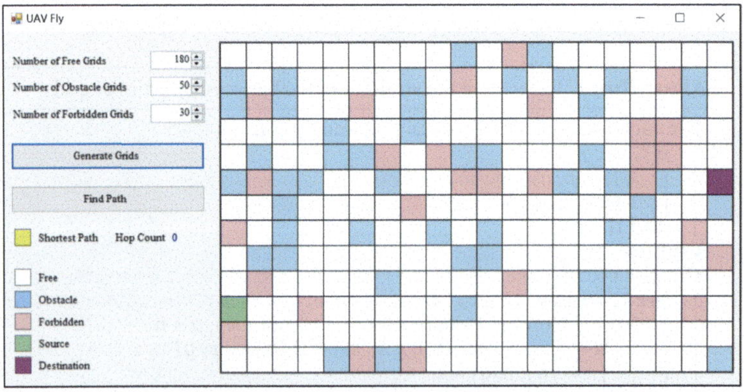

Fig. 3. The developed GUI-based simulator.

4.2 Results

Following Figs. 4, 5, 6 and 7 show the path finding results for four different cases. The main differences are the grid size and the numbers of free, obstacle, and no-fly grids. The yellow grids in sequence indicate the shortest path with minimum hop count. It is obvious that smaller grids can make UAVs fly on a more precise path.

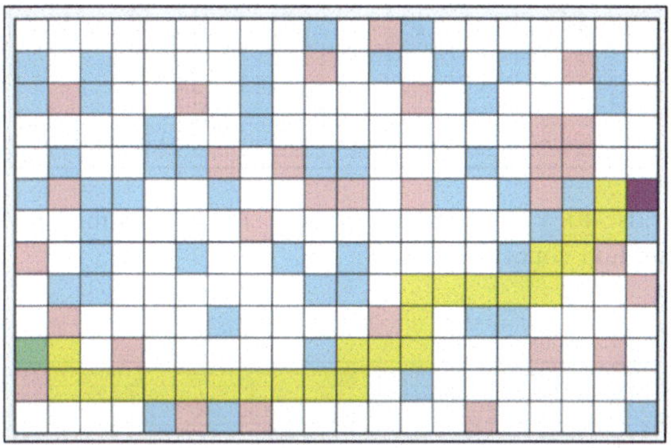

Fig. 4. Number of grids: 20 × 13 = 260, grid size: 30 m × 30 m, Free/Obstacle/NFZ grids: 180/50/30, minimum hop count: 26

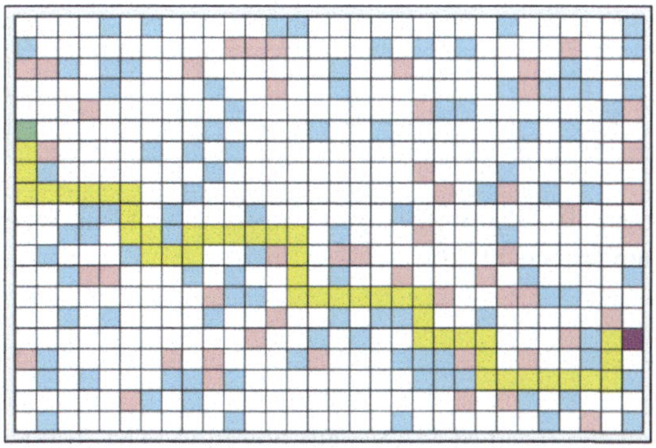

Fig. 5. Number of grids: 30 × 20 = 600, grid size: 20 m × 20 m, Free/Obstacle/NFZ grids: 450/100/50, minimum hop count: 45

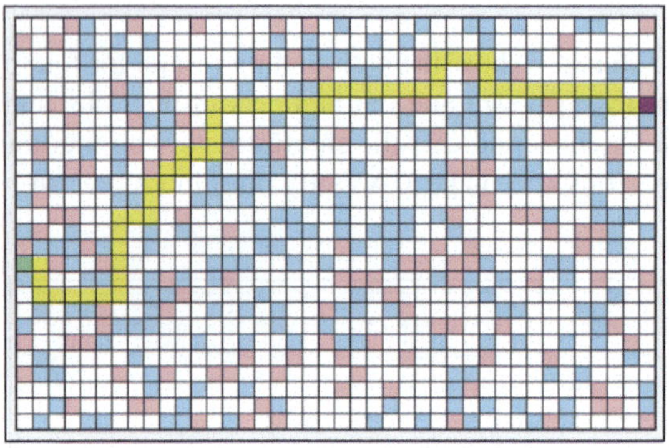

Fig. 6. Number of grids: 40 × 26 = 1040, grid size: 15 m × 15 m, Free/Obstacle/NFZ grids: 680/230/130, minimum hop count: 59

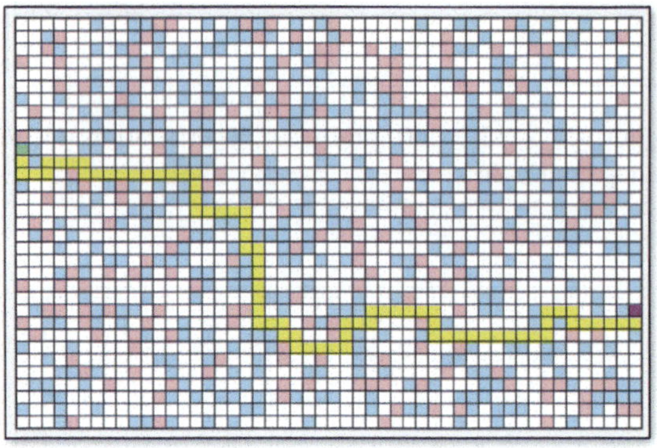

Fig. 7. Number of grids: 50 × 33 = 1650, grid size: 12 m × 12 m, Free/Obstacle/NFZ grids: 1050/400/200, minimum hop count: 76

5 Conclusion

This study focuses on the topic of UAV path planning. The flight area is divided into grids. The grid size could be predefined according to the flight speed and centripetal acceleration of the UAV. In addition, there are three types of grids defined in this study, which are obstacle, no-fly zone, and free-to-fly grids. Based on the environment information of the flight area, the concept of Floyd-Warshall algorithm is used to find the optimal path determined by minimum hop count instead of straight-line distance. A simulator was developed for the experiments of the method. The simulation results illustrate that the method performs well and the smaller grid size will bring a more precise path. This study is a preliminary trial work and a basis for the advanced researches.

In the future works, more complex grid model could be employed and more complex environmental information could be introduced, even the algorithm could be modified and improved for addressing the problems of UAV path planning.

Acknowledgement. This work is supported by the Fujian Provincial Natural Science Foundation in China (Project Number: 2017J01730) and the Education Department of Fujian Province (Project Number: GY-Z19005).

References

1. Hassanalian, M., Abdelkefi, A.: Classifications, applications, and design challenges of drones: a review. Prog. Aerosp. Sci. **91**, 99–131 (2017)
2. Liew, C.F., DeLatte, D., Takeishi, N., Yairi, T.: Recent Developments in aerial robotics: a survey and prototypes overview, pp. 1–14. arXiv:1711.10085v2 (2017)
3. Rosen, K.H.: Discrete Mathematics and Its Applications, 8th edn. McGraw-Hill, New York (2019)
4. Sujit, P.B., Saripalli, S., Sousa, J.B.: Unmanned aerial vehicle path following: a survey and analysis of algorithms for fixed-wing unmanned aerial vehicles. IEEE Control Syst. Mag. **34** (1), 42–59 (2014)
5. Lin, Y., Saripalli, S.: Sampling-based path planning for UAV collision avoidance. IEEE Trans. Intell. Transp. Syst. **18**(11), 3179–3192 (2017)
6. Radmanesh, M., Kumar, M., Guentert, P.H., Sarim, M.: Overview of path-planning and obstacle avoidance algorithms for UAVs: a comparative study. Unmanned Syst. **6**(2), 95–118 (2018)
7. Zhao, Y., Zheng, Z., Liu, Y.: Survey on computational-intelligence-based UAV path planning. Knowl. Based Syst. **158**(15), 54–64 (2018)
8. Konatowski, S., Pawłowski, P.: Ant colony optimization algorithm for UAV path planning. In: International Conference on Advanced Trends in Radioelecrtronics, Telecommunications and Computer Engineering 2018, pp. 177–182. IEEE (2018)
9. Dhulkefl, E.J., Durdu, A.: Path planning algorithms for unmanned aerial vehicles. Int. J. Trend Sci. Res. Dev. **3**(4), 359–362 (2019)

Experimental Comparison of Different Feature Detection Algorithms for UAV Obstacle Avoidance

Tien-Wen Sung, Baohua Zhao(✉), and Kuo-Chi Chang

College of Information Science and Engineering,
Fujian University of Technology, Fuzhou, China
tienwen.sung@gmail.com, 2191901008@smail.fjut.edu.cn,
albertchangxuite@gmail.com

Abstract. For small and low-cost UAVs, real-time autonomous obstacle avoidance is a challenging problem. Due to the weight limit of carrying additional sensors such as radar, laser, and etc., vision-based autonomous obstacle avoidance has become a popular trend for small drones. In the past, several studies have used vision-based approaches for the three-dimensional reconstruction or depth estimation in obstacle avoidance. In this paper, the concepts of emulating human behavior to avoid an obstacle is used in the experiments. Two photos can be compared by detecting and matching the features with the principle of size expansion when an obstacle is approaching, and the matched features can construct a convex hull. This convex hull is treated as an obstacle which needs to be avoided. This method is fast and effective. In this study, several feature detection and matching algorithms are compared by experiments, and constructed convex hulls are used to compare the accuracy of obstacle detection. The larger the overlapped region between the convex hull and real object, the better the obstacle detection accuracy. This study analyzes and summarizes the experimental results to find which algorithm could obtain better accuracy and computation time, i.e. which could be more appropriate to utilized in UAV vision-based obstacle avoidance.

Keywords: UAV · Obstacle detection · Image features detection · Obstacle avoidance

1 Introduction

With the development of microelectronics technology and the improvement of computing power in the 21st century, the use of UAV is becoming more and more popular, not only for military use, but also for our daily life and works. In recent years, as the demand for low-altitude flight applications gradually increases, small drones are becoming more and more popular. However, in some complex environments, due to the limitation of traditional sensor communication and perception, UAV cannot accurately perceive the surrounding environment and can operate normally. Among them, it is particularly important to achieve autonomous obstacle avoidance.

A.-E. Hassanien et al. (Eds.): AICV 2020, AISC 1153, pp. 840–849, 2020.
https://doi.org/10.1007/978-3-030-44289-7_79

Previously, researchers paid more attention to the use of laser or radar in automatic navigation of UAV, but because the weight of laser or radar is too large, it is not suitable for small UAV. With the rapid development of computer vision, the vision sensor which can be used in vision method has shown great advantages in the navigation of UAV because of its lighter and cheaper characteristics. In order to achieve autonomous navigation, the vision based navigation system needs to accurately know the state information of the UAV, which in turn controls the UAV, which has been realized in [1, 2].

In this paper, the method of dimension expansion proposed by Abdulla al Kaff and others is used to detect obstacles, and the effectiveness of this method is tested by a variety of algorithms. In order to more accurately avoid obstacles, it is necessary to capture the contour information of obstacles. The obstacle detection accuracy is presented and the experimental results under various algorithms are compared.

The rest of this paper is arranged as follows: the second section introduces the related work of obstacle detection, the third section introduces the specific methods of obstacle detection experimented in this paper, the fourth section discusses the specific situation of the experiment, and the fifth section is a summary of this paper.

2 Related Work

In the autonomous navigation system of UAV, especially for the complex environment, obstacle avoidance is particularly important. Regarding to this challenging problem, many different vision-based methods have been proposed to solve the problem. In [3], a method based on texture and color change is proposed to detect indoor obstacles, but this method has obvious shortcomings. It is only suitable for indoor environment with detailed texture. Mori [4] proposed a feature detection and template matching algorithm to detect the change of the size of obstacles, but in the experimental part, only tree obstacles are detected, and there are no other shapes of obstacles. Stereo vision is one of the most popular obstacle detection methods [5, 6], but stereo vision needs to use binocular camera or stereo camera, and it does not perform well in areas with less texture. There is a method to detect obstacles by using ranging equipment for UAV [7]. These ranging devices, such as Kinect [8], Xtion pro, are suitable for low light indoor environment, but they are too heavy for small low-cost UAVs. There are also some related works used optical flow to detect obstacles. For example, in [9], the use of translational optical flow method to control the UAV system is proposed. The system can avoid obstacles and keep a safe distance from the ceiling and the ground. Green et al. [10] proposed a method based on optical flow to simulate the flight characteristics of insects to avoid side obstacles. However, many optical flow methods cannot detect the front obstacles. Abdulla et al. Proposed the method of SIFT algorithm to extract feature points and detect convex hull to detect obstacles [11, 12]. The method is only suitable for the environment with obvious texture. Moreover, the convex hull detected by this method cannot accurately display the contour information of obstacles, and cannot avoid obstacles accurately.

The work in this paper use different feature detection and matching algorithms to detect obstacle with constructed convex hull. In order to evaluate the accuracy of the algorithm, the detection effect of each algorithm is compared and analyzed.

3 Method

This method is divided into two parts. The first part is the obstacle detection part, which uses nearly two consecutive pictures when the UAV approaches the obstacle, to detect and match the features of these two pictures, and detect the convex hull of the feature points, through comparing the ratio of these features of the two images to determine whether they are obstacles.

The second part is the accuracy measurement part, which shows the similarity between the detected convex hull and the actual obstacle area. In this paper, the experimental environment is indoor white background, and the obstacles are other colors. The area of the obstacles in the photo can be detected by two value threshold segmentation principle. The accuracy can be calculated by comparing the size of the convex hull and the real object in the photo. The accuracy measurement is to avoid obstacles more accurately in the next actual obstacle avoidance part of UAV.

First, take two photos approximately one meter from an obstacle. the orb, sift and surf are used to extract feature points respectively, and then the Brute Force matching method is used to match feature points, or the FLANN matching method can be used to match feature points. As shown in Fig. 1, the effect map of orb detecting feature points and brute force matching. In particular, when FLANN matching is used, only sift and surf can be used to extract feature points.

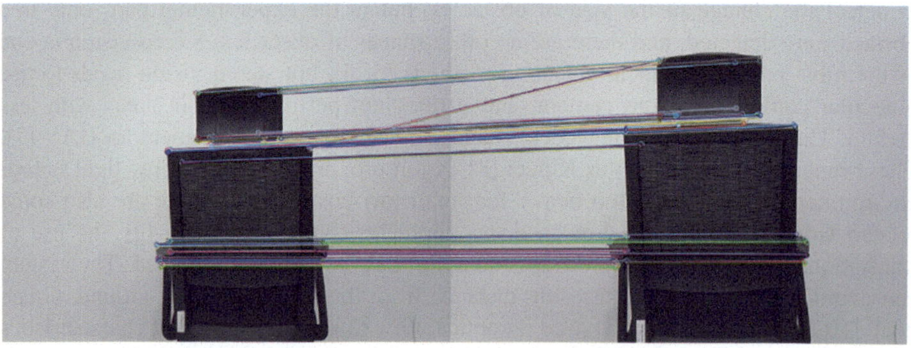

Fig. 1. The original map detected by ORB and Brute Force.

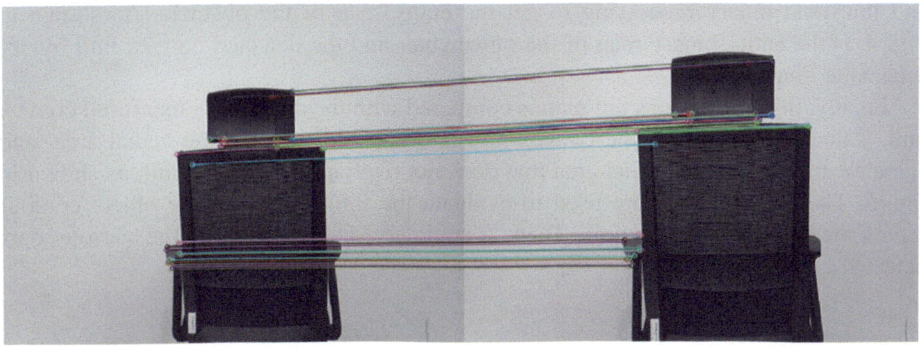

Fig. 2. The better map detected by ORB and Brute Force.

Then we use the empirical value of the distance of the matched feature points to screen out the closer (better) feature points to ensure the accuracy of the matching and the consistency of the convex hull, especially the empirical value of the distance of each method is different. The screening results are shown in Fig. 2.

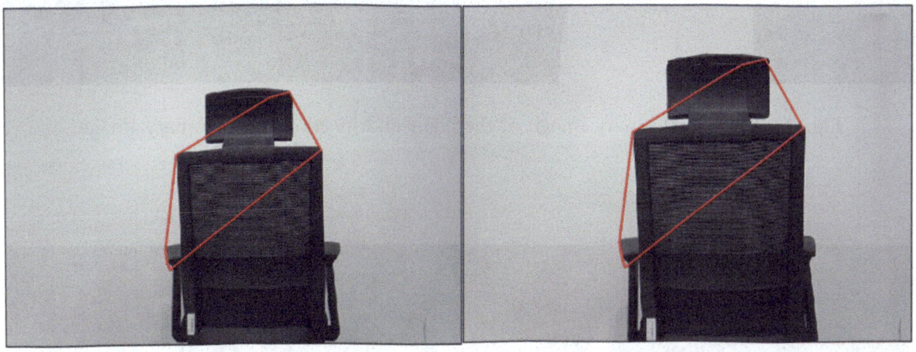

Fig. 3. Convex hull detected by feature point inspection

Calculate the sum of the area of the feature points screened from each graph, calculate the ratio of the two graphs, detect the convex hull of the screened feature points and calculate its area, and still calculate the convex hull ratio of the two graphs. When the area ratio of the feature points and the convex hull area ratio are respectively greater than a certain empirical value, obstacle avoidance can be implemented. The convex hull effect diagram is shown in Fig. 3.

The second part is the accuracy measurement part. Because the photos closer to the obstacle position can better reflect the real-time information of the obstacle, the accuracy of the next photo taken is used to calculate the accuracy. The accuracy measurement first needs to measure the area of the obstacle entity in the photo. In order to facilitate the measurement, this experiment uses the white wall as the background, and threshold binary processing to get the entity map of the obstacle. As shown in Fig. 4, is the entity binary map of the office chair and the detected convex hull binary schematic diagram.

Finally, the area of the solid map is compared with the area of the measured convex hull. Calculate the area of the convex hull that does not overlap the actual area, and calculate the area of the actual area that does not overlap the convex hull, as shown in Fig. 5. These two photos are used to evaluate the information of the object contour represented by the convex hull. Then, a summary and analysis of the various data detected.

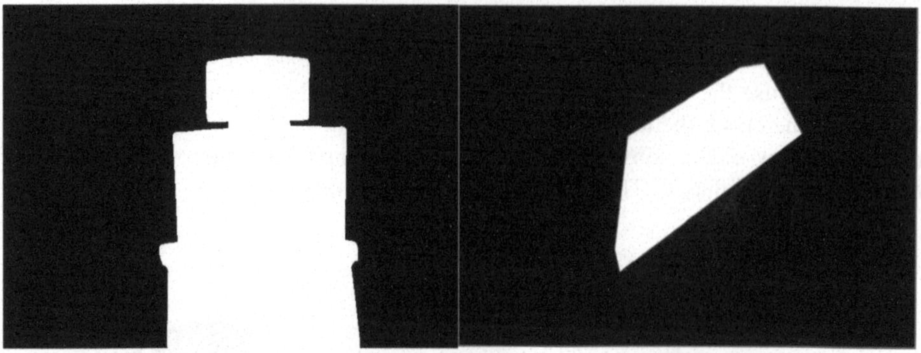

Fig. 4. The actual binary image of the chair and its convex hull binary image

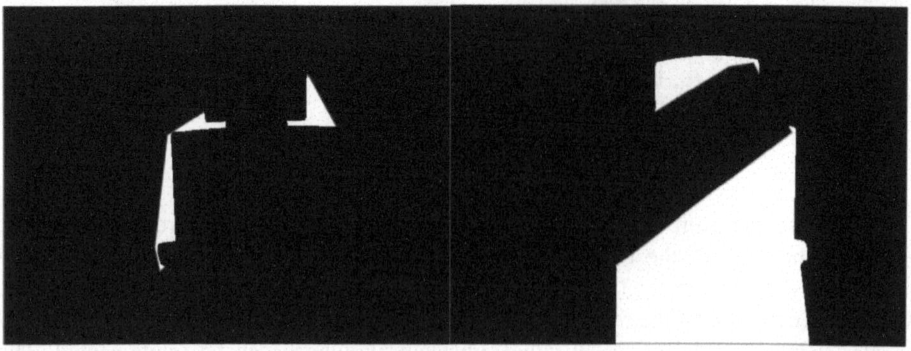

Fig. 5. The convex hull non-overlapping binary diagram and the actual non-overlapping binary diagram

4 Experimental Results

In this experiment, 640 * 480 pixels pictures are used. The pictures are taken indoors in the daytime. In order to show the characteristics and effectiveness of the detection algorithm, the background of the obstacles is white. The platform to execute the algorithm is the computer with Intel i5-4210 m processor at 2.60 GHz.

This experiment mainly tests the effect of different feature detection algorithms in obstacle avoidance of UAV. In this experiment, there are five algorithms, including orb detection algorithm sift detection algorithm and surf detection algorithm under brute force matching. and sift detection algorithm and surf detection algorithm under FLANN matching. The obstacles in the photos used are people, office chairs, folding beds, which are simulated as two consecutive frames, and the shooting distance is about 1 m. The measured data are feature points (front), feature points (back), convex hull area ratio, feature point area ratio, time used, proportion of convex hull not overlapping, proportion of physical objects not overlapping.

The feature points (front) represent the matched original feature point pairs, and the feature points (back) represent the more excellent feature point pairs screened by experience value. The area ratio of the convex hull represents the area ratio of the convex hull of the next photo to that of the previous photo. The area ratio of the feature points is also the ratio of the area of the feature points screened from the next photo and the area sum of the feature points screened from the previous photo. The time used represents the time consumed by the algorithm. The proportion of non-overlapping convex hull represents the ratio of the area of non-overlapping convex hull to the total area of convex hull. The actual area ratio is the same thing.

The experimental data of folding bed, human body and office chair as obstacles are shown in Tables 1, 2 and 3.

Table 1. Folding bed as an obstacle to the experimental table.

Feature detection algorithms	Original feature points	Excellent feature points	Convex hull area ratio	Area ratio of characteristic points	The time used	Convex hull non-overlapping proportion	Actual non-overlapping proportion
orb(bf)	268	73	1.4788	1.51	0.5649	46%	24%
sift(bf)	172	21	1.5359	1.551	0.1656	64%	25%
surf(bf)	53	27	1.4973	1.7845	0.6676	37%	18%
sift(flann)	330	44	1.4633	1.4834	0.1729	39%	22%
surf(flann)	129	24	1.373	1.533	0.6784	56%	23%

Table 2. Human body as an obstacle to the experimental table.

Feature detection algorithms	Original feature points	Excellent feature points	Convex hull area ratio	Area ratio of characteristic points	The time used	Convex hull non-overlapping proportion	Actual non-overlapping proportion
orb(bf)	281	26	1.6077	1.5028	0.5076	52%	0%
sift(bf)	148	24	1.3566	1.8365	0.1520	46%	0%
surf(bf)	36	20	1.8848	1.7647	0.6408	31%	1%
sift(flann)	237	33	1.4068	1.8961	0.1582	45%	0%
surf(flann)	59	9	1.6246	1.7933	0.6005	48%	0%

Table 3. Office chair should be placed above the tables.

Feature detection algorithms	Original feature points	Excellent feature points	Convex hull area ratio	Area ratio of characteristic points	The time used	Convex hull non-overlapping proportion	Actual non-overlapping proportion
orb(bf)	215	47	1.3752	1.5287	0.5429	65%	14%
sift(bf)	39	5	1.1958	1.5798	0.1444	78%	3%
surf(bf)	29	16	1.4963	1.3055	0.6032	29%	5%
sift(flann)	68	12	1.4617	1.4978	0.1416	67%	7%
surf(flann)	40	15	1.5016	1.2860	0.6316	43%	5%

When detecting the above three sets of data, special attention should be paid to the empirical value of the screening conditions used if the same algorithm is used to detect different obstacles. Through observation and analysis, it can be concluded that after the empirical value screening, the feature point pairs become more accurate, then the calculated convex hull area ratio and the feature point area ratio are more reliable.

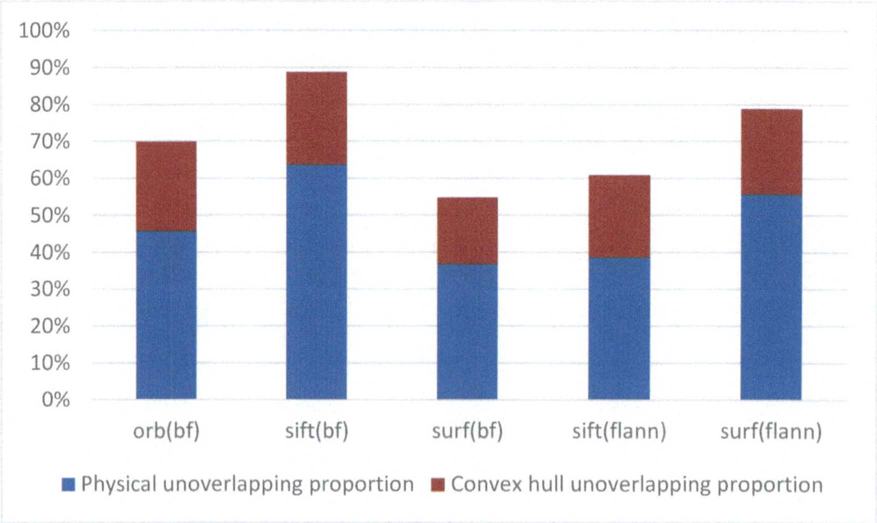

Fig. 6. Folding bed accuracy.

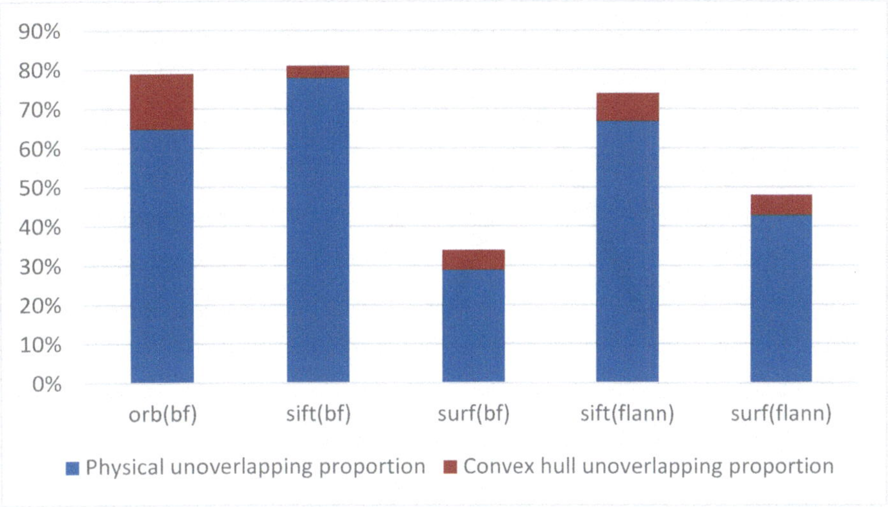

Fig. 7. Office chair accuracy.

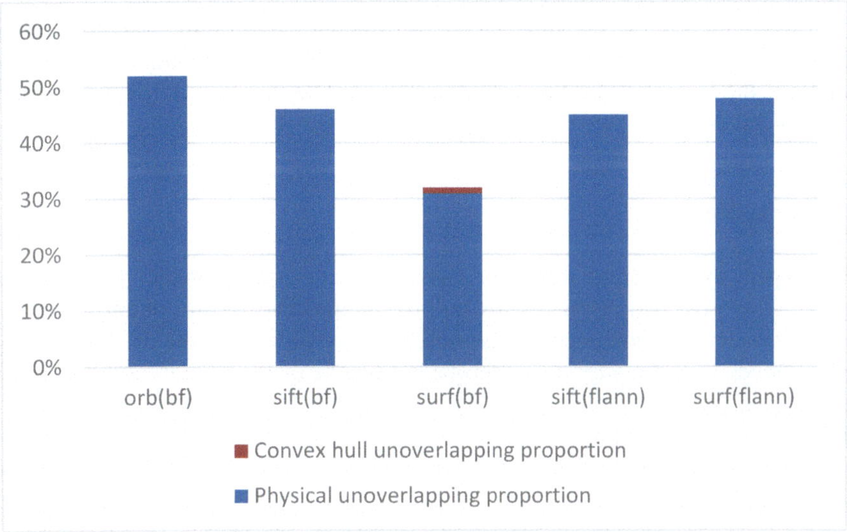

Fig. 8. Human body accuracy.

As shown in Figs. 6, 7 and 8, these three accuracy histograms compare the accuracy relationship of different algorithms under different obstacles. The surf algorithm under brute force matching has the highest accuracy, indicating that the detected convex hull is closer to the real obstacles. When replacing the obstacles with convex hull to carry out the actual obstacle avoidance section, the contour and size information of the obstacles is more accurate.

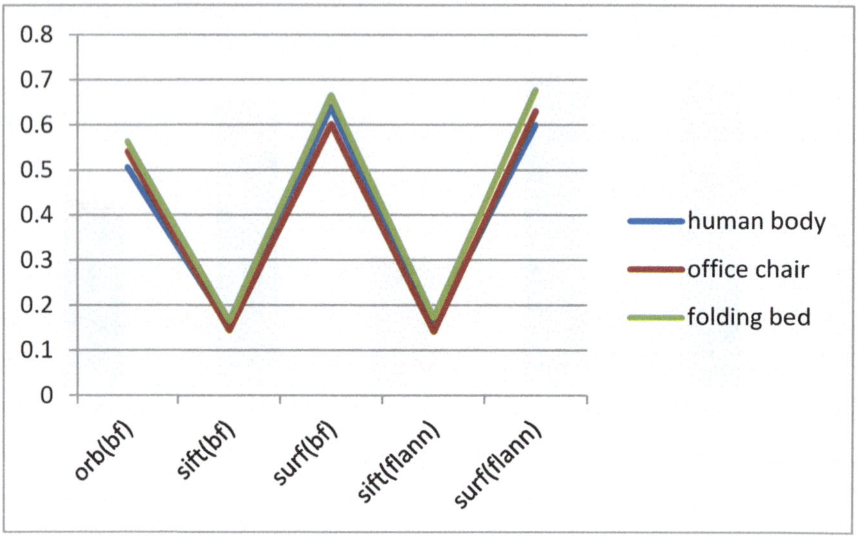

Fig. 9. Algorithm time.

As shown in Fig. 9, the time taken by different algorithms to detect different obstacles is compared. It can be seen from the table that the time used by SIFT algorithm is the least, followed by orb algorithm, but the time used by orb algorithm is nearly four times that of SIFT algorithm.

5 Conclusion

In this paper, five different detection algorithms are used to detect the features of obstacles. The most peripheral convex hull of the matched feature points is extracted, and the obstacle detection accuracy is presented by calculating the area of the object and the convex hull. Finally, the experimental results of the five algorithms are compared and analyzed.

In the future, the main work is to use the algorithm to capture more effective obstacle contour information, and apply the algorithm to the UAV to avoid obstacles more accurately.

Acknowledgement. This work is supported by the Fujian Provincial Natural Science Foundation in China (Project Number: 2017J01730) and the Education Department of Fujian Province (Project Number: GY-Z19005).

References

1. Potena, C., Nardi, D., Pretto, A.: Joint vision-based navigation, control and obstacle avoidance for UAVs in dynamic environments. In: Proceedings of the 2019 European Conference on Mobile Robots (ECMR) (2019)
2. Lu, Y., Xue, Z., Xia, G.-S., Zhang, L.: A survey on vision-based UAV navigation. Geo-spatial Inf. Sci. **21**(1), 21–32 (2018)
3. De Croon, G., de Weerdt, E., De Wagter, C., Remes, B.: The appearance variation cue for obstacle avoidance. In: Proceedings of the 2010 IEEE International Conference on Robotics and Biomimetics (ROBIO), Tianjin, China, 14–18 December 2010, pp. 1606–1611 (2010)
4. Mori, T., Scherer, S.: First results in detecting and avoiding frontal obstacles from a monocular camera for micro unmanned aerial vehicles. In: Proceedings of the 2013 IEEE International Conference on Robotics and Automation (ICRA), Karlsruhe, Germany, 6–10 May 2013, pp. 1750–1757 (2013)
5. Hrabar, S.: 3D path planning and stereo-based obstacle avoidance for rotorcraft UAVs. In: IROS (2008)
6. Na, I., Han, N.S., Jeong, H.: Stereo-based road obstacle detection and tracking. In: ICACT (2011)
7. Merz, T., Kendoul, F.: Beyond visual range obstacle avoidance and infrastructure inspection by an autonomous helicopter. In: International Conference on Intelligent Robots and Systems (IROS), p. 187217 (2011)
8. Bachrach, A., Prentice, S., He, R., Roy, N.: Range - robust autonomous navigation in GPS-denied environments. J. Field Robot. **28**(5), 644666 (2011)
9. Beyeler, A., Zufferey, J.-C., Floreano, D.: 3D vision-based navigation for indoor microflflyers. In: Proceedings 2007 IEEE International Conference on Robotics and Automation, pp. 1336–1341. IEEE (2007)
10. Green, W., Oh, P.: Optic-flow-based collision avoidance. IEEE Robot. Autom. Mag. **15**(1), 96–103 (2008)
11. Al-Kaff, A., Meng, Q.: Monocular vision-based obstacle detection/avoidance for unmanned aerial vehicles. In: 2016 IEEE Intelligent Vehicles Symposium (IV) Gothenburg, Sweden, 19–22 June 2016 (2016)
12. Al-Kaff, A., Garcí, F.: Obstacle detection and avoidance system based on monocular camera and size expansion algorithm for UAVs. Sensors **17**, 1061 (2017)

Gene Regulatory Network Construction Parallel Technique Based on Network Component Analysis

Dina Elsayad(✉), Safwat Hamad, Howida A. Shedeed, and Mohamed F. Tolba

Faculty of Computer and Information Sciences, Ain Shams University, Cairo, Egypt
{dina.elsayad, shamad, Dr_Howida}@cis.asu.edu.eg, fahmytolba@gmail.com

Abstract. Construction of a gene regulatory network is a vital process for understanding the gene functions and gene influences on the other genes. Furthermore, gene regulatory network analysis is a promising method for demonstrating the topological order of genes interactions. One technique for constructing a gene regulatory network is FastNCA that is based on network component analysis methodology. Although FastNCA is widely used to construct the gene regulatory network of cancer diseases associated genes, it is a time-consuming and computational intensive technique. As a result, this paper presents an enhanced parallel implementation of FastNCA that uses distributed parallelism methodology to enhance the performance of FastNCA. Different gene datasets are used to evaluate the performance of the proposed algorithm. The experimental results demonstrate that the proposed algorithm outperforms FastNCA. Where the achieved speedup is up to 250 on 256 processing nodes.

Keywords: Network Component Analysis · Parallel computing · Regulatory network · Genes · High-Performance Computing

1 Introduction

Gene regulatory network (GRN) construction is one of the genetic data analysis tasks [1]. Gene regulatory network [2] inference techniques mainly focus on studying the topological organization of genes' interactions. Furthermore, the analysis of the genetic regulation network investigates how the genes' influence on the other genes. The standard visualization of gene regulatory network's main elements is a set of nodes and a set of edges. Where the genes are represented as a node and the edges represent genes regulatory interactions.

The gene regulatory network techniques can be classified into five categories: the regression based techniques [3–7], reverse engineering techniques [8–13], mutual information based techniques [14–19], graph based approaches [20, 21] and component analysis techniques [22–33]. The component analysis technique is further categorized into four subcategories. These subcategories are: Principal Component Analysis (PCA) [22], Singular value decomposition (SVD) [23], Independent component

A.-E. Hassanien et al. (Eds.): AICV 2020, AISC 1153, pp. 850–857, 2020.
https://doi.org/10.1007/978-3-030-44289-7_80

analysis (ICA) [24–26] and Network component analysis (NCA). This paper focuses on network component analysis approaches.

The Principal Component Analysis (PCA) techniques assume that the data is orthogonal which is not applicable for the genetic data. While Singular value decomposition (SVD) suffers from the noise sensitivity that leads to unmeaning solutions. On the other hand, the multi-local solutions are the drawback of Independent component analysis (ICA) techniques. Therefore, the Network Component Analysis (NCA) techniques presented to avoid the limitations and drawbacks of the Principal Component Analysis (PCA), Independent component analysis (ICA) and Singular value decomposition (SVD) techniques. One of the widely used Network Component Analysis (NCA) techniques for genetic data analysis [34, 35] is a Fast Network Component Analysis (FastNCA) algorithm [28]. Like the other recent approaches for gene regulatory network inference, FastNCA is a time-consuming and intensively computational technique. As a result, parallel techniques are vital. This paper presents an enhanced parallel algorithm that is based on the FastNCA technique. The proposed algorithm named ePFastNCA (Enhanced Parallel implementation for Fast Network Component Analysis technique).

The rest of that paper is arranged as follows; the following Sect. 2 presents the Network Component Analysis (NCA) techniques. While the Sect. 3 illustrates the proposed algorithm (ePFastNCA). Moreover, Sect. 4 provides implementations and experimental results. Finally, the conclusion is discussed in Sect. 5.

2 Network Component Analysis (NCA) Techniques

As stated earlier, one of the gene regulatory network constructions techniques is the Network Component Analysis (NCA) approach. To avoid the noise sensitivity drawback of the other techniques, the NCA takes the data noise into consideration. The mathematical model of NCA is expressed in (1). Where the Γ matrix represents the noise in data. While the gene expression data is represented by [Y] matrix of size $(M \times K)$. Moreover, the connectivity matrix is represented by [A] matrix of size $(M \times N)$. On the other hand, [S] represents the transcription factor (TF) activity matrix of size $(N \times K)$. Where M is a symbol for the number of the gene, N represents the number of transcription factor and K symbolize the number of experiments or samples in the dataset.

$$Y = AS + \Gamma \tag{1}$$

To overcome the multi-local solution drawback of the other techniques and ensure a unique solution, the NCA technique must meet the following criteria in the dataset. The first criterion is the full column rank property of [A] connectivity matrix. Furthermore, the transcription factor activity matrix [S] must have full row rank. While [A] matrix must keep the full column rank property even if one of its columns is deleted. Fast Network Component Analysis (FastNCA) algorithm [28] is one of the approaches to solve this model of NCA.

Figure one illustrates the FastNCA technique. Like the other network component analysis methodologies, the FastNCA is time-consuming and computationally intensive technique. As a result, Parallel Fast Network Component Analysis technique (PFastNCA) [30] was presented. PFastNCA enhances the performance of FastNCA using the distributed parallelism technique [36]. Where the data is divided into smaller chunks each chunk of size M/P where M is the number of the data item and P is the number of processing nodes. In other words, each processing node handles a smaller portion of the dataset of size M/P data items. Moreover, the processing nodes are communicated by sending and receiving messages. The next section demonstrates the proposed algorithm ePFastNCA (Enhanced Parallel implementation for Fast Network Component Analysis technique) that enhances the performance of FastNCA using distributed parallelism methodology as well as decreasing the sequential initialization task time (Figs. 1, 2 and 3).

Input	[Y] : Data matrix
Output	Gene Regulatory Network
Step 1:	Perform rank-L EYM (Eckart–Young–Mirsky) approximation of Y by SVD
Step 2:	Let W = UL (left singular vector)
Step 3:	Estimate connectivity matrix [A]
Step 3.1:	For i=1 to N
Step 3.1.1:	Re-order rows of W such that the ith column of X has the form ai = $\begin{bmatrix} \tilde{a_i} \\ 0 \end{bmatrix}$
Step 3.1.2:	Partition W = $\begin{bmatrix} W_c \\ W_r \end{bmatrix}$
Step 3.1.4:	Get [V_0] the right singular vectors by applying SVD to W_r
Step 3.1.5:	Compute $\tilde{a_i}$ = the left singular vector of $W_c\ V_0\ V_0^T$
Step 4:	Estimate the [S] matrix

Fig. 1. FastNCA technique

Input	[Y] : Data matrix
Output	Gene Regulatory Network
Step 1	Perform rank-L EYM (Eckart–Young–Mirsky) approximation of *Y* by SVD
Step 2:	Let W = U_L (left singular vector)
Step 3:	Estimate connectivity matrix [*A*] in parallel
Step 3.1:	Send chunk of the data for each processing node
Step 3.2:	Collect data from the processing nodes
Step 4:	Estimate the [*S*] matrix

Fig. 2. PFastNCA – master node

Input	[W]: Left singular vector matrix
Output	Connectivity submatrix [X]
Step 1:	Parallel for i=1 to N
	Re-order rows of W such that the ith column of X has the form $a_i = \begin{bmatrix} \tilde{a_i} \\ 0 \end{bmatrix}$
	Partition $W = \begin{bmatrix} W_c \\ W_r \end{bmatrix}$
	Get [V_0] the right singular vectors by applying SVD to W_r
	Compute $\tilde{a_i}$ = the left singular vector of $W_c V_0 V_0^T$
Step 2:	Send connectivity submatrix [X] to the master node

Fig. 3. PFastNCA – worker node

3 Proposed Algorithm (ePFastNCA)

FastNCA is one of network component analysis based techniques for gene regulatory network analysis. However, FastNCA is widely used in biological network construction, it is both a time-consuming and intensive computation technique. Therefore, this paper presents an Enhanced Parallel implementation for the Fast Network Component Analysis technique (ePFastNCA) (illustrated in Figs. 4 and 5) that improves the performance of FastNCA using distributed parallelism methodology and decreases the computational running time of initialization task.

Input	[Y] : Data matrix
Output	Gene Regulatory Network
Step 1:	Divide [Y] into chunks
Step 2:	Send chunk of the data for each processing node
Step 3:	Collect data from the processing nodes
Step 4:	Estimate the [S] matrix

Fig. 4. ePFastNCA – master node

Input	[Y] : Data matrix
Output	Connectivity submatrix [X]
Step 1:	Perform rank-L EYM (Eckart–Young–Mirsky) approximation of Y by SVD
Step 2:	Let W = UL (left singular vector)
Step 3:	Parallel for i=1 to N
	Re-order rows of W such that the ith column of X has the form $a_i = \begin{bmatrix} \tilde{a_i} \\ 0 \end{bmatrix}$
	Partition $W = \begin{bmatrix} W_c \\ W_r \end{bmatrix}$
	Get [V_0] the right singular vectors by applying SVD to W_r
	Compute $\tilde{a_i}$ = the left singular vector of $W_c V_0 V_0^T$
Step 4:	Send connectivity submatrix [X] to the master node

Fig. 5. ePFastNCA – worker node

4 Implementation and Results

Some experiments are performed to evaluate the performance of the proposed algorithm ePFastNCA against the sequential algorithm FastNCA and its parallel version PFastNCA. The ePFastNCA and PFastNCA is implemented using C++ with MPI. While FastNA is implemented using C++. The experiments were conducted on the Bibliotheca Alexandrina High-Performance Computing cluster, where each processing node is Intel ® Xeon® CPU E5-2680 v3 @ 2.50 GHz. The performance of the algorithms is evaluated using six gene datasets that are available from the GEO database through their accession numbers. Table 1 demonstrates the size (number of genes x number of samples) and the accession number of each dataset.

Table 1. Accession number and size of used datasets

Dataset accession no.	Size
GSE30052	10928 × 49
GSE17172	12625 × 18
GSE13009	22277 × 69
GSE2034	22283 × 255
GSE9195	54675 × 78
GSE6532	54675 × 88

Figure 6 shows the speedup of PFastNCA when running on a different number of processing nodes using the six datasets. As depicted in the figure the speedup of PFastNA is decreased after using 32 processing nodes. On the other hand, as illustrated in Fig. 7 the speedup of ePFastNCA is proportionally related to the number of the used processing nodes. Where the speedup reached 247.429 on 256 processing nodes. These

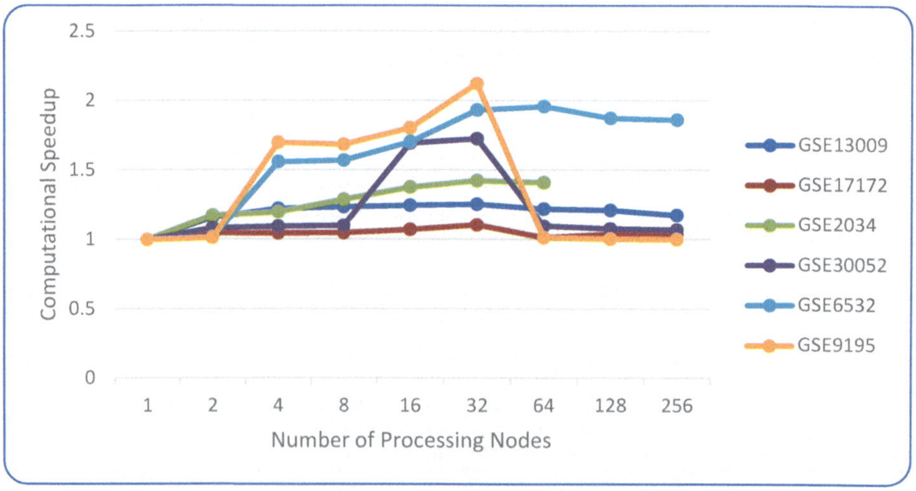

Fig. 6. Speedup of PFastNCA

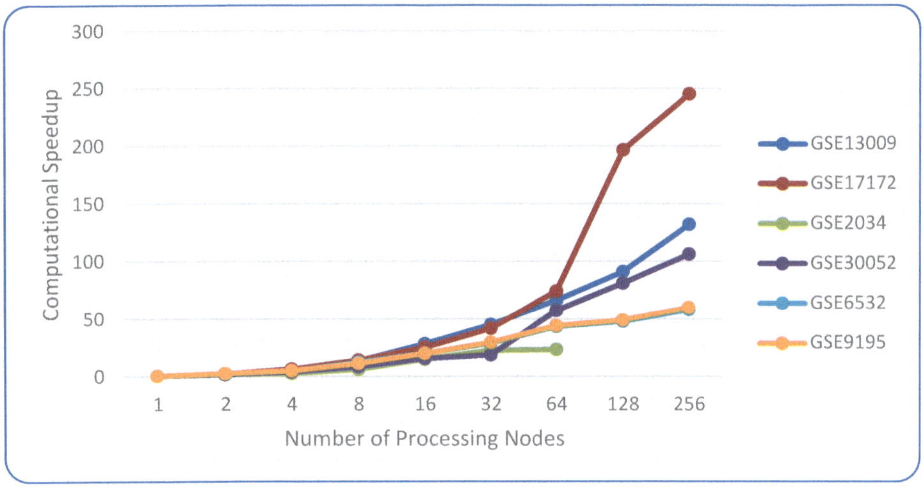

Fig. 7. Speedup of ePFastNCA

results indicate that ePFastNCA successfully improved the performance of PFastNCA and its speedup is incomparable with the speedup of PFastNCA.

5 Conclusion

Gene regulatory network construction is one of the genetic data analysis tasks. The primary goal of gene regulatory network analysis is demonstrating the genes' interactions topological order as well as studying the effects of one gene on the other genes. FastNCA is a technique for gene regulatory analysis that is based on the Network Component analysis methodology. As the other Network Component analysis algorithms, FastNCA is time consuming and computational intensive technique. As a result, PFastNCA was presented as a parallel version of FastNCA. Nevertheless, unfortunately, the excremental results indicated that the speedup of PFastNCA is deprecated after using 32 processing nodes. Therefore, this paper presented a more enhanced implementation for FastNCA that is named ePFastNCA (Enhanced Parallel implementation for Fast Network Component Analysis technique). ePFastNCA improves the performance of FastNCA using distributed parallelism methodology where that data is divided into smaller chunks each chunk of size M/P. Moreover, ePFastNCA decreased the computation running time of the initialization task. The experimental result indicated that ePFastNCA is outperformed PFastNCA where its achieved speedup 45.313 instead of 1.697 for PFastNCA on 32 processing node. Moreover, the speedup of ePFastNCA reached 247.429 on 256 processing nodes.

References

1. Aluru, S.: Handbook of Computational Molecular Biology. CRC Press, Boca Raton (2006)
2. Bolouri, H.: Computational Modeling of Gene Regulatory Networks a Primer. World Scientific Publishing Company, Singapore (2008)
3. Pirgazi, J., Khanteymoori, A.R.: A robust gene regulatory network inference method base on Kalman filter and linear regression. PLoS One **13**(7), e0200094 (2018)
4. Lam, K.Y., Westrick, Z.M., Muller, C.L., Christiaen, L., Bonneau, R.: Fused regression for multi-source gene regulatory network inference. PLoS Comput. Biol. **12**(12), e1005157 (2016)
5. Omranian, N., Eloundou-Mbebi, J.M.O., Mueller-Roeber, B., Nikoloski, Z.: Gene regulatory network inference using fused LASSO on multiple data sets. Sci. Rep. **6**, 20533 (2016)
6. Guerrier, S., Mili, N., Molinari, R., Orso, S., Avella-Medina, M., Ma, Y.: A predictive based regression algorithm for gene network selection. Front. Genet. **7**, 97 (2016)
7. Gregoretti, F., Belcastro, V., Di Bernardo, D., Oliva, G.: A parallel implementation of the network identification by multiple regression (NIR) algorithm to reverse-engineer regulatory gene networks. PLoS One **5**(4), e10179–e10183 (2010)
8. Jostins, L., Jaeger, J.: Reverse engineering a gene network using an asynchronous parallel evolution strategy. BMC Syst. Biol. **4**(1), 17–33 (2010)
9. Klinger, B., Bluthgen, N.: Reverse engineering gene regulatory networks by modular response analysis-a benchmark. Essays Biochem. **62**(4), 535–547 (2018)
10. Perkins, M., Daniels, K.: Visualizing dynamic gene interactions to reverse engineer gene regulatory networks using topological data analysis. In: 2017 21st International Conference Information Visualisation (IV) (2017)
11. Liu, Z.-P.: Reverse engineering of genome-wide gene regulatory networks from gene expression data. Curr. Genomics **16**(1), 3–22 (2015)
12. de Souza, M.C., Higa, C.H.A.: Reverse engineering of gene regulatory networks combining dynamic Bayesian networks and prior biological knowledge. In: International Conference on Computational Science and Its Applications (2018)
13. Villaverde, A.F., Banga, J.R.: Reverse engineering and identification in systems biology: strategies, perspectives and challenges. J. Roy. Soc. Interface **11**(91), 1–16 (2014)
14. Sales, G., Romualdi, C.: parmigene—A parallel R package for mutual information estimation and gene network reconstruction. Bioinformatics **27**(13), 1876–1877 (2011)
15. Shi, H., Schmidt, B., Liu, W., Muller-Wittig, W.: Parallel mutual information estimation for inferring gene regulatory networks on GPUs. BMC Res. Notes **4**(1), 189–194 (2011)
16. Zhang, X., Zhao, X.-M., He, K., Lu, L., Cao, Y., Liu, J., Hao, J.-K., Liu, Z.-P., Chen, L.: Inferring gene regulatory networks from gene expression data by path consistency algorithm based on conditional mutual information. Bioinformatics **28**(1), 98–104 (2011)
17. Meyer, P.E., Lafitte, F., Bontempi, G.: minet: AR/Bioconductor package for inferring large transcriptional networks using mutual information. BMC Bioinf. **9**(1), 461 (2008)
18. Lachmann, A., Giorgi, F.M., Lopez, G., Califano, A.: ARACNe-AP: gene network reverse engineering through adaptive partitioning inference of mutual information. Bioinformatics **32**(14), 2233–2235 (2016)
19. Barman, S., Kwon, Y.-K.: A novel mutual information-based Boolean network inference method from time-series gene expression data. PLoS One **12**(2), e0171097 (2017)
20. Yan, X., Mehan, M.R., Huang, Y., Waterman, M.S., Yu, P.S., Zhou, X.J.: A graph-based approach to systematically reconstruct human transcriptional regulatory modules. Bioinformatics **23**(13), i577–i586 (2007)

21. Jiang, H., Turki, T., Zhang, S., Wang, J.T.L.: Reverse engineering gene regulatory networks using graph mining. In: International Conference on Machine Learning and Data Mining in Pattern Recognition (2018)
22. Raychaudhuri, S., Stuart, J.M., Altman, R.B.: Principal components analysis to summarize microarray experiments: application to sporulation time series. In: Pacific Symposium on Biocomputing, pp. 455–466. NIH Public Access (2000)
23. Holter, N.S., Mitra, M., Maritan, A., Cieplak, M., Banavar, J.R., Fedoroff, N.V.: Fundamental patterns underlying gene expression profiles: simplicity from complexity. Proc. Nat. Acad. Sci. **97**(15), 8409–8414 (2000)
24. Hyvarinen, A., Karhunen, J., Oja, E.: Independent Component Analysis. Wiley, Hoboken (2001)
25. Aapo, H.: Fast and robust fixed-point algorithms for independent component analysis. IEEE Trans. Neural Netw. **10**(3), 626–634 (1999)
26. Liebermeister, W.: Linear modes of gene expression determined by independent component analysis. Bioinformatics **18**(1), 51–60 (2002)
27. Liao, J.C., Boscolo, R., Yang, Y.-L., Tran, L.M., Sabatti, C., Roychowdhury, V.P.: Network component analysis: reconstruction of regulatory signals in biological systems. In: Proceedings of the National Academy of Sciences (2003)
28. Chang, C., Ding, Z., Hung, Y.S., Fung, P.C.W.: Fast network component analysis (FastNCA) for gene regulatory network reconstruction from microarray data. Bioinformatics **24**(11), 1349–1358 (2008)
29. Jayavelu, N.D., Aasgaard, L.S., Bar, N.: Iterative sub-network component analysis enables reconstruction of large scale genetic networks. BMC Bioinf. **16**(1), 366 (2015)
30. Elsayad, D., Ali, A., Shedeed, H.A., Tolba, M.F.: PFastNCA: parallel fast network component analysis for gene regulatory network. In: International Conference on Advanced Machine Learning Technologies and Applications (2018)
31. Shi, Q., Zhang, C., Guo, W., Zeng, T., Lu, L., Jiang, Z., Wang, Z., Liu, J., Chen, L.: Local network component analysis for quantifying transcription factor activities. Methods **124**, 25–35 (2017)
32. Noor, A., Ahmad, A., Serpedin, E., Nounou, M., Nounou, H.: ROBNCA: robust network component analysis for recovering transcription factor activities. Bioinformatics **29**(19), 2410 (2013)
33. Noor, A., Ahmad, A., Serpedin, E.: SparseNCA: sparse network component analysis for recovering transcription factor activities with incomplete prior information. IEEE/ACM Trans. Comput. Biol. Bioinf. **15**(2), 387–395 (2018)
34. Wei, K., Qianni, S., Shuaiqun, W.: Transcriptional regulation analysis of breast cancer based on FastNCA algorithm. Genomics Appl. Biol. **8**, 78 (2018)
35. Sun, Q., Kong, W., Mou, X., Wang, S.: Transcriptional regulation analysis of Alzheimer's disease based on FastNCA algorithm. Curr. Bioinf. **14**(8), 771–782 (2019)
36. Kumar, V.: Introduction to Parallel Computing. Addison-Wesley Longman Publishing Co., Inc., Boston (2002)

Enhanced Security of Home Registration in the Hierarchical Mobile IPv6 Protocol for IoT Applications

Sultan Saleh Alkhliwi(✉)

Faculty of Science, Northern Border University, Arar, Kingdom of Saudi Arabia
salkhliwi@nbu.edu.sa

Abstract. Today, Internet of Things (IoT) is enabling a new generation of dynamic ecosystems in environments such as smart cities and hospitals. Dynamic ecosystems require mobility management to track the current locations of mobile nodes (MNs) around the network without service interruption. To do so, the Hierarchical Mobile IPv6 (HMIPv6) protocol has been chosen, which minimizes the required signalling by introducing a new entity called the mobile anchor point (MAP) to act as a local home agent for all visiting MNs in a specific domain. In the HMIPv6 protocol, MNs typically use the home registration process to inform the home agent about the MN's current location while it is away from its home link and hands it off to the MAP domain. The HMIPv6 protocol protects home registration against outside attacks. However, it fails to prevent the MN from lying about its current location, which can redirect data to the address of a victim. This paper addresses this limitation by proposing the Enhanced Home Agent Registration protocol. This protocol can support the location authentication of MNs to their home agents in the HMIPv6 protocol. This protocol combines two ideas. Firstly, it extends the functions of the MAP to register the regional care-of address of the MN at the home agent on behalf of all MNs. Secondly, it applies a reachability test for a synchronised regional care-of address. To investigate the efficiency, efficacy, and performance of the home agent registration protocol in terms of the home agent registration delay and signalling overhead, simulation is used based on OPNET™ Modeler 14.5.

Keywords: Internet of Things (IoT) · Hierarchical Mobile IPv6 protocol · Home registration · Secure key · OPNET simulation

1 Introduction

The potential of the future Internet is not limited to smart phones. Internet of Things (IoT) represents the emerging area of the future Internet. Because of, the major goal of the IoT is to collect information from the real-world entities and event. Nowadays, there are numbers of entities may around in the real-world environments [1]. Thus, an efficient mobility management protocol is required for IoT solutions. To facilitate IoT abilities to locate a mobile node (MN) point of attachment for delivering data packets and to maintain an MN's connection as it continues to change its point of attachment in the domain, the Engineering Task Force (IETF) proposed the Mobile IPv6 (MIPv6)

A.-E. Hassanien et al. (Eds.): AICV 2020, AISC 1153, pp. 858–869, 2020.
https://doi.org/10.1007/978-3-030-44289-7_81

protocol [3] to both permit this roaming connectivity and reduce the required signalling with the Hierarchical Mobile IPv6 (HMIPv6) protocol [4]. Thus, the HMIPv6 protocol has been chosen to support mobility location management for IoT. The HMIPv6 is a lightweight protocol (compared with others) for the following reasons: (1) It supports hierarchical tunnelling approaches that are flexible, modular and scalable for supporting IP-based macro and micro mobility for WMNs. (2) It introduces a new node agent called the mobile anchor point (MAP) to act as a local home agent for the MNs. Thus, when an MN changes its attached point in the same domain, it will update its location with the MAP instead of the home link, as defined in the MIPv6 protocol [2]. As a result, the HMIPv6 protocol requires minimal bandwidth and computational resources, as well as reduces registration delays when tracking the current location of the MNs [3].

The HMIPv6 protocol has two goals for MNs: (1) allow them to always be reachable at one address, such as a home of address (HoA), regardless of the MN location; and (2) enable them to maintain ongoing connections with other correspondent nodes (CNs) across different locations.

When an MN does a handoff to the MAP domain, the MN configures a local care-of address (LCoA) from the current attachment access router (AR) and regional care-of address (RCoA) from the MAP option. The MN then performs the MAP registration process to registers to its LCoA and RCoA by using a local binding update (LBU) message. To allow the MN to receive packets that are destined for its HoA while it is away from the home link, MN must register its RCoA with an HA on its home link. The MN initiates the HA registration by sending a binding update (BU) message, which contains the HoA and RCoA of the MN, to the HA directly. The HA stores this binding in its binding cache entry. By using this binding, the HA intercepts and forwards the data packet that is destined for the HoA to the bound RCoA.

Providing security for HA registrations is an essential part of the HMIPv6 protocol to prevent various attacks, such as a malicious MN flooding attack and denial-of-service (DoS) attack. The current specification of the HMIPv6 protocol uses the Authentication Header (AH) of Internet Protocol security (IPSec) to protect the BU message and sequence numbers. It is used in this message to prevent any replay attacks. By using both IPSec and the sequence numbers, the home registration is protected against outside attacks. Attackers cannot send spoofed or replay-with BU messages in place of the MN. Furthermore, they can partially prevent an insider attack. That is, the MN cannot send a BU on behalf of another MN that uses the same HA. Although, IPSec HA and sequence number are unsuccessful at preventing the MN from lying about its current location, the MN might send a fake RCoA to the HA and thus redirect data to this address.

The [4] has been proposed to enhance to support the location authentication of MNs in the basic operation of the home registration at the MIPv6 protocol. It bullied in two ideas:

1. Enhancing the current version of the CGA-based technique [5] to cryptographically generate (by the MN itself) and verify (by the MN's HA) the MN's Care-of address (CoA) by using the secret key that is shared between the MN and the HA as an input value instead of the MN's public key.

2. Applying a novel idea to perform the synchronised CoA reachability test to make sure the HA that the MN is connection at the claimed CoA.

Indeed, this scheme increased security for the home registration in the MIPv6 protocol. As the MN (owner address) and the HA (verifier address) are not required to generate and verify the digital signature, respectively. Furthermore, the scheme removes the require embedding the MN's public key and signature in the BU's message, as used in the basic operation of the CGA-based technique, to verify the authenticity of the CoA. As a result, the proposed scheme able to protect both the MN and the HA against DoS attacks and reduces the signalling overhead. In addition, the HA is able to verify the reachability of the MN at the CoA; thus, it prevents a malicious MN from launching a flooding attack. However, this scheme suffers from the following limitations. (1) The proposed scheme needs the MN to cryptographically generate its own CoA, which requires two extra messages to verify the reachability of the MN at the claimed CoA. Therefore, the MN could be exposed to a DoS attack because the MN is a mobile entity with limited computational power. (2) The MN wants to receive the packet from its CN as soon as the handoff to a foreign network is complete, but the proposed scheme has introduced a longer delay compared with the basic operation that is defined in the MIPv6 to complete the HA registration, which increases packet loss.

This paper presents the enhancement of the basic home agent registration (B-HAR) operations that are defined [6] to support location authentication of the MNs to their HAs. This concept is called an Enhancement Home Agent Registration (E-HAR) protocol. In an E-HAR protocol, the HA verifies the ownership of the claimed RCoA and confirms the reachability at that RCoA in the MAP domain. In addition, the E-HAR protocol allows the HA and MAP to authenticate each other and share a secret key. As a result, the E-HAR protocol can reduce the likelihood of a malicious MN successfully luring an HA to flood a third party with useless traffic during its placement in the MAP domain by using the HMIPv6 protocol.

The rest of this paper is organised as follows. Section 2 provides an overview of the E-HAR protocol. Section 3 presents the E-HAR protocol in detail. Section 4 evaluates the performance of the E-HAR protocol compared with the B-HAR protocol in terms of the HA registration delay using the OPNET modeller (version 14.5). Section 5 concludes this paper.

2 Overview of the E-HAR Protocol

The E-HAR protocol enhances the B-HAR protocol [6] by combining two ideas. Firstly, it uses a method to register the RCoA of the MN at the HA. This method extends the functions of the MAP to act as a proxy entity for all visiting MNs. By using this method, the HA confirms the ownership of the RCoA of the claimed MN, and the HA and MAP authenticate each other and generate a shared secret key. Secondly, the E-HAR protocol applies a novel synchronised RCoA reachability test to verify the reachability of an MN at the claimed RCoA. The test enables an HA to register and use the RCoA of an MN while synchronically verifying the MN reachability at that RCoA.

2.1 Extending the MAP Functions

The E-HAR protocol extends the MAP functions that are defined in the basic specification of the HMIPv6 protocol [6] to act as a proxy entity to perform the HA registration procedure for all MNs with their respective HAs. To achieve this, the E-HAR protocol defined a new signalling message called the Proxy Binding Update (PBU) message. The MAP uses the PBU message when the MAP registration procedure is completed successfully to register the RCoA of the MN with the MN's HA. However, a lack of a secure relationship between the MAP and HA can lead to an increased risk of attacks, such as DoS attacks and man-in-the-middle (MTIM) attacks. To solve this issue, the E-HAR protocol allows the MAP and HA to authenticate each other and to share a secret key by executing the Diffie-Hellman (DH) key exchange protocol [7] and the digital signature scheme [8]. Here, the MAP has only its own public key/private key pair (PK_{MAP} and SK_{MAP}) that it obtains from a certification authority (CA). The MAP then sends the PBU message that its private key (SK_{MAP}) signs to the HA. The PBU message includes the public key certificate ($Cert_{MAP}$) and the public value of the MAP (g^x) to initiate a DH procedure with the HA. When the HA receives a valid PBU message from the MAP, it generates its own secret value (y) and public value (g^y). It uses the received value (g^x) to compute the DH key (K_{DH}): $K_{DH} = (g^x)^y$. The HA then configures the secret key ($K_{MAP\text{-}HA}$) based on the key K_{DH} to share with the MAP. When the MAP receives a valid response from the HA that includes the public value of the HA (g^y), it computes the secret key ($K_{MAP\text{-}HA}$) like the HA. In this case, the HA and MAP can protect any further mobility-related messages by using the key $K_{MAP\text{-}HA}$. Extending the MAP functions offers these advantages:

1. It allows the HA to confirm that the MN owns the claimed RCoA because this address is sent with a MAP signature;
2. It reduces the computation overhead that is imposed on the MNs and protects the MN against DoS attacks;
3. It allows the HA and MAP to authenticate each other and share a secret key, which they use to prevent on-path attacks, such as a DoS attack or MITM attack.

However, execution of this idea does not guarantee MN reachability at the RCoA, which is vulnerable to malicious MN flooding attacks. In the next section, to solve this issue, the HA confirms MN reachability at the claimed RCoA.

2.2 The Synchronised RCoA Reachability Test

To extend the MAP functions, the RCoA reachability test synchronizes to validate MN reachability at the claimed RCoA. With a synchronised RCoA reachability test, the HA can register and use the RCoA of the MN while synchronically verifying MN reachability at that RCoA. The reachability test uses two additional messages: a Care-of Token Init (CoTI) message and a Care-of Test (CoT) message.

The reachability test begins when the HA receives a valid PBU message from the MAP. In response, the HA sends a CoTI message to the MAP. The HA uses the CoTI message to deliver a fresh CoT key to the MAP. The MAP uses the received token to confirm MN reachability at the RCoA. That is, the MAP sends a CoT message that contains the received token to the HA. Finally, the HA securely sends the binding acknowledgement (BA) message to the MN via the MAP to acknowledge the binding of the RCoA at its binding cache entry, successfully completing the reachability test.

A CoT is a 64-bit number that is computed by using a node key [2]. The node key is only known to the HA and allows the HA to verify that a token in the CoT message is its own. The HA generates a fresh node key at regular intervals and identifies it by an index. The HA computes a fresh CoT: CoT = First (64, HMAC SHA1 (K_{HA}, (HoA|| RCoA|| Seq_{MAP}))). This CoT is based on its active node key and the values of the HoA of the MN, the claimed RCoA of the MN, and the MAP sequence number (Seq_{MAP}) that is received in a valid PBU message from the MAP.

In addition, the reachability test limits the number of valid PBU messages that can be received from the MAP and the amount of traffic that can be sent to an unreachable RCoA. While the HA is running the reachability test for the RCoA of a specific MN (after sending the CoTI message and before receiving the CoT message), the HA is not yet able to validate MN reachability at a claimed RCoA. Therefore, the test allows the HA to limit the number of PBU messages that can be received from the MAP to a maximum attempt value. This value prevents the MAP from bypassing the test by continuously sending PBU messages without engaging the test. Furthermore, the test allows the HA to set the granted binding lifetime between the HoA of the MN and the unreachable RCoA of the MN to a minimum binding lifetime value. This value limits the amount of data that is sent to the unreachable RCoA. Alternatively, when the HA receives a valid CoT message and successfully verifies the CoT value, the test also allows the HA to set the granted binding lifetime between the HoA and RCoA of the MN to a maximum binding lifetime value. This value reduces the MN requirement to extend the binding lifetime at the HA when it is in the MAP domain.

3 E-HAR Protocol in Detail

The E-HAR protocol, as shown in Fig. 1, utilises four messages and five procedures to complete the home agent registration. Each of these messages is given a specific name:

- **PBU:** Proxy Binding Update;
- **CoTI:** Care-of Token Init;
- **CoT:** Care-of Token Test;
- **BA:** Binding Update Acknowledgment.

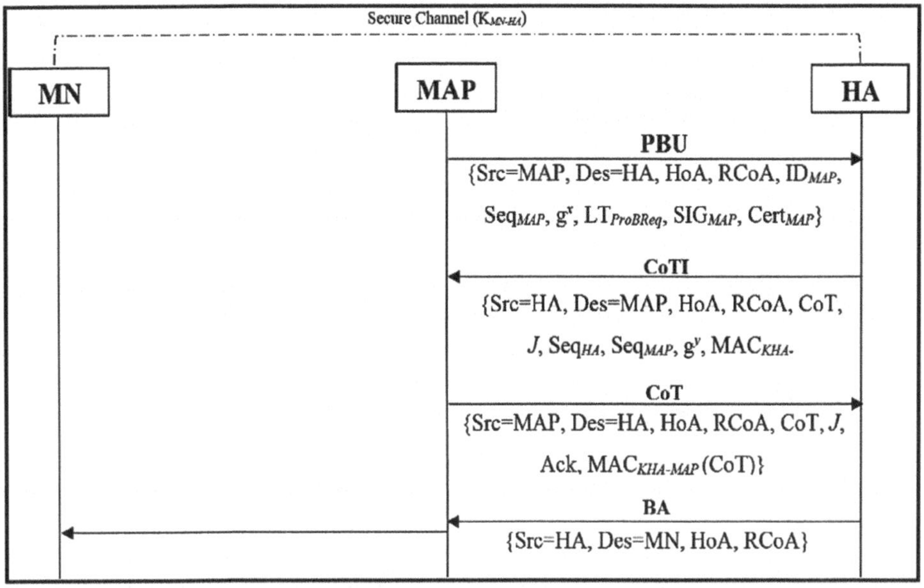

Fig. 1. Detailed view of the E-HAR protocol.

Table 1 lists the notations to be used in the E-HAR protocol description.

Table 1. List of notations

Notation	Indication	Notation	Indication
HA	Home Agent	Seq_X	A sequence number of node X
MAP	Mobile Anchor Point	Ack	An acknowledgement value sent by MAP to HA to send Binding Acknowledgement to MN
MN	Mobile Node	MAC*x*	A keyed hash value used to ensure the integrity and authenticity of the message
HoA	Home of Address	SIG_{MAP}	The MAP sings the PBU message by its private key
RCoA	Regional care-of address	K*x-y*	A shared secret key between two entities
PBU	Proxy Binding Update message	CoT	Care-of keygen token value (64-bits)
CoTI	Care-of Token Init message	K_{DH}	Diffie-Hellman key
CoT	Care-of Test message	$LT_{ProBReq}$	A Proxy binding lifetime request encoded in the PBU message
BA	Binding Acknowledgement	LT_{BGrant}	The granted binding lifetime between the HoA and RCoA of the MN at HA's Binding Cache entry

The E-HAR protocol follows this procedure:

- **Procedure-1:** On behalf of all visiting MNs in the MAP domain, the MAP initiates the E-HAR protocol to bind the HoA and RCoA of the MN with their HAs. To do this, the MAP sends a PBU message. The MAP includes the HoA and RCoA of the MN, an initial sequence number (Seq_{MAP}), a proxy binding lifetime request $LT_{ProBReq}$, the public value (g^x) from the DH method, the MAP signature (SIG_{MAP}) on the message using the private key of the MAP, and the public key certificate of the MAP ($Cert_{MAP}$).
- **Procedure-2:** Upon receipt of the PBU message, the HA verifies the MAP signature (SIG_{MAP}) to confirm the freshness, integrity, and authenticity of the PBU message. If verification fails, the HA discards the message without any further action. Otherwise, the HA performs the following steps:

A The HA checks whether it has a binding cache entry for the HoA that is enclosed in the PBU message. If the HA does not have a binding cache entry, the MAP creates a new binding for the MN as explained in step **'D'**.

B The HA checks the state of the entry. If the state is reachable, that is the previous run of the E-HAR protocol concluded successfully, the process continues to step **'D'**.

C The HA checks the number of attempts in the entry. If the maximum number of attempts is reached, the HA rejects the PBU message without further action. Otherwise, the HA continues to step **'D'**.

D The HA creates or updates a binding cache entry with the claimed RCoA that is enclosed in the PBU message. It sets the entry in an unreachable state and sets the granted binding lifetime (LT_{BGrant}) in the entry between the HoA and RCoA of the MN to a minimum binding lifetime value. This value limits the amount of data that is sent to the unreachable RCoA. The HA also sets the number of attempts to a maximum number of attempts (if an existing entry is created or updated). This value prevents the entity from bypassing the registration and prevents the HA from a DoS attack.

E The HA uses its active node key to generate a fresh CoT. The CoT is returned later to the HA to prevent a replay attack and to confirm the reachability of the MN at the claimed RCoA.

F The HA generates its own secret value y and public value g^y. The HA uses the public value of the MAP (g^x) that is received in the PBU message to compute the DH key (K_{DH}): $K_{DH} = (g^x)^y$. The HA then generates an initial sequence number (Seq_{HA}) and computes the session key ($K_{MAP\text{-}HA}$) from the key K_{DH} and two sequence numbers: Seq_{HA} and Seq_{MAP}.

G The HA sends a CoTI message to the MAP. The HA includes the HoA and RCoA of the MN, the CoT value, Seq_{HA}, Seq_{MAP}, and the public value (g^y) from the DH method. In addition, the HA protects the CoTI message by using the $K_{MAP\text{-}HA}$ key.

- **Procedure-3:** When the MAP receives CoTI message, it checks the value of Seq_{MAP} to verify the freshness of the message. Upon positive verification, like HA, the MAP calculates the DH key (K_{DH}) value to generate the session key ($K_{MAP\text{-}HA}$). The MAP then verifies the integrity and authenticity of the CoTI message by using $K_{MAP\text{-}HA}$ key. If these verifications fail, the MAP discards the message without further action. Otherwise, the MAP stores the value of the $K_{MAP\text{-}HA}$ key that corresponds to the specific HA. Finally, the MAP uses the received CoT to show the presence of the MN at the new RCoA. The MAP sends the CoT message to the HA and requests a longer lifetime in the binding lifetime request field. It also requests a return BA message for the MN. The MAP includes the HoA and RCoA of the MN, the CoT, and the acknowledgement (Ack) in the CoT message. The MAP protects the CoT message by using the $K_{MAP\text{-}HA}$ key.
- **Procedure-4:** Upon receipt of the CoT message, the HA uses the MN HoA as an index to search its binding update cache. The HA then checks the value of the CoT to verify the freshness of the message and MN reachability at the claimed RCoA. When the verification is successful, the HA verifies the integrity and authenticity of the CoT message by using the $K_{MAP\text{-}HA}$ key. If these verifications fail, the HA performs the following actions:

A It discards the message without any further action. Otherwise, the HA resets the number of attempts to zero in its binding cache entry.

B It extends the binding lifetime in the cache entry. It sets the granted binding lifetime (LT_{BGrant}) to a maximum binding lifetime value. This value reduces the required number of MNs to extend the binding lifetime at the HA in the MAP domain, because the HA and MAP achieve mutual authentication and share a secret key ($K_{MAP\text{-}HA}$).

C It sends the BA message to the MN, including the HoA and RCoA of the MN, the HA address, and LT_{BGrant}. Here, the HA and MN have a shared secret key ($K_{MN\text{-}HA}$) before hand-off to the MAP domain. Therefore, the HA encrypts the BA message by using the $K_{MN\text{-}HA}$ key.

- **Procedure-5:** The MAP interrupts the BA message, encapsulates it, and sends it to the MN that is inside the MAP domain by using LCoA of the MN. When the MN receives the BA message, it uses the address of the address as an index to search its binding list entry. The MN then decrypts the BA message and stores the value of LT_{BGrant} for the HA in its binding list entry.

4 Performance Evaluation

This section presents the performance evaluation of the proposed protocol in terms of the HA registration delay and signalling overhead. The HA registration delay is the total amount of time that it takes to complete the HA registration process. The signalling overhead is the total amount of signalling traffic sent and received by all involved entities at each protocol. The control signalling overhead is measured in bits/second OPNET Modeler 14.5 [9] was used to simulate the performance of the proposed protocol under varying network conditions. Specifically, the performance of the proposed protocol is investigated when varying background levels on the network are applied and when varying number of simultaneously roaming MNs are served by the same HA and MAP. The simulation results are collected and then compared to the results when the basic HA registration process is run. The scenario, which is depicted in Fig. 2, consists of a home link domain and MAP domain that are connected to the Internet via a wired link. The home link domain includes a single HA entity, and the MAP domain includes MAP and AR entities. In this scenario, the MN is in the MAP domain and is connected to the Internet via its wirelessly attached AR.

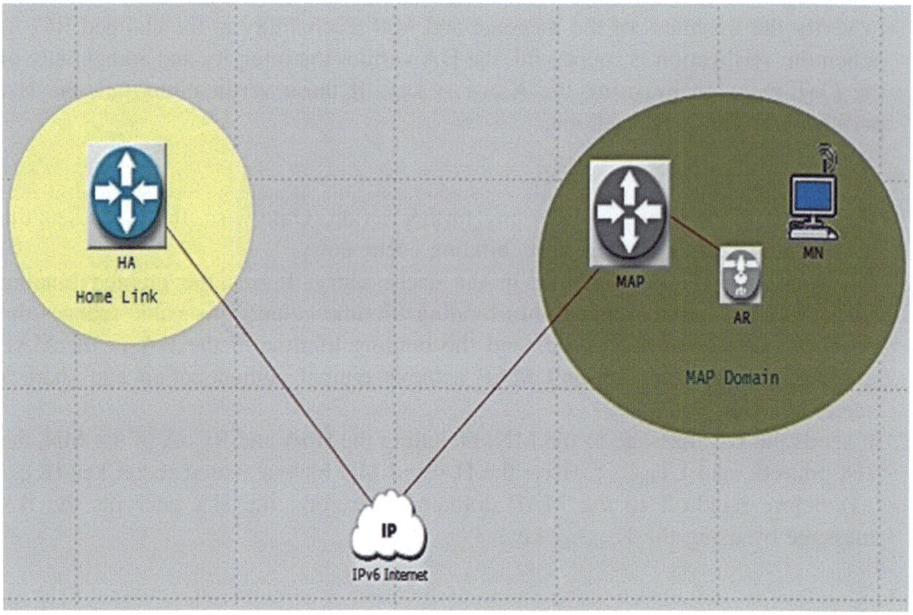

Fig. 2. Simulation scenario

Figure 3 shows the average HA registration delay for the registration process under a different level of background. In this figure, as the background traffic volume increases, the HA and MAP become more congested, and the HA registration delay for registration increases significantly under both protocols. It also shows that, on average, the E-HAR protocol takes around 40% longer than the B-HAR registration protocol to register the RCoA of the MN at the HA. This delay is caused by: (1) two additional messages that are used to register the RCoA of the MN at the HA; (2) HA verification of MN reachability at the claimed RCoA; and (3) the MAP and HA execution of the DH method to compute a shared secret key.

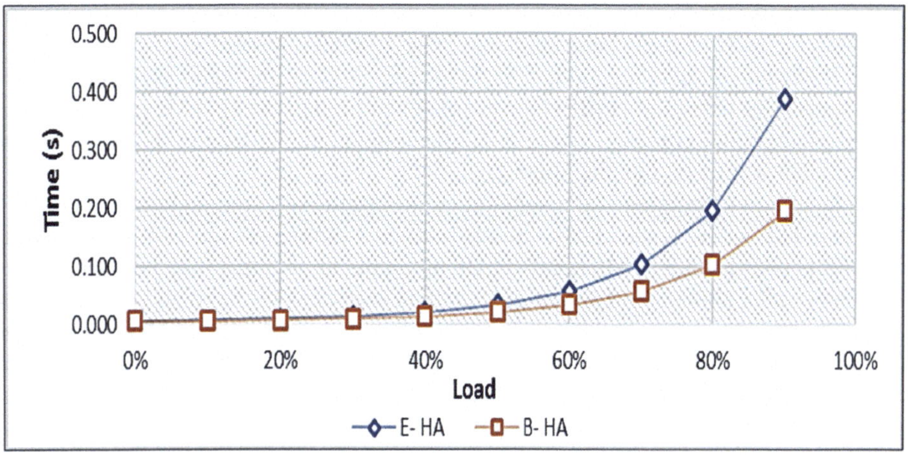

Fig. 3. HA registration delay for E-HAR and B-HAR protocols vs load.

Figure 4 compares the performance of the E-HAR and B-HAR protocols as the number of simultaneously roaming MNs that serve the MAP and HA increases. The E-HAR protocol offers a higher rate of HA registration delay than the B-HAR protocol because the MAP and HAs in the E-HAR protocol must perform more operations during registration. Accordingly, queueing time for the MAP and HA in the E-HAR protocol increases quickly compared to the B-HAR protocol. Therefore, increasing the number of simultaneously roaming MNs led the HA registration delay in the B-HAR protocol to increase less quickly than in the E-HAR protocol.

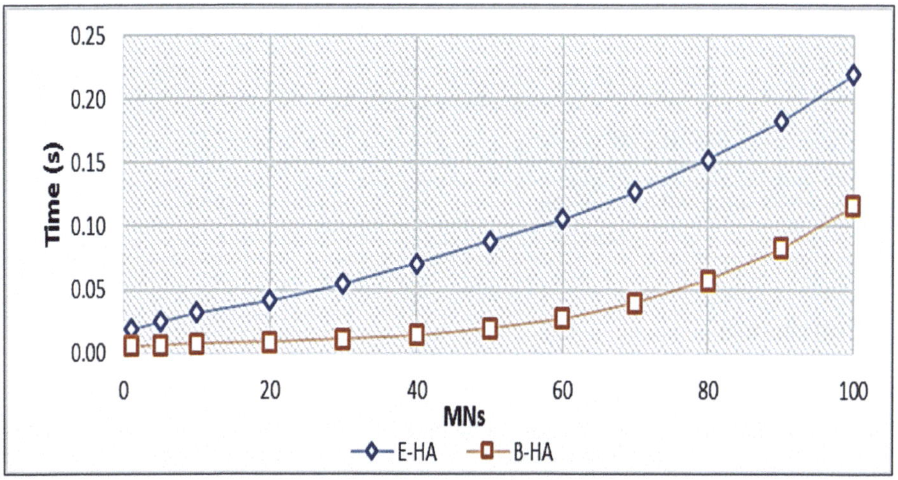

Fig. 4. HA registration delay for E-HAR and B-HAR protocols vs the number of MNs.

Based on the performance evaluation, if the E-HAR and B-HAR protocols are compared for their security and efficiency, the E-HAR protocol is superior in supporting the location authentication of MNs to their HAs during MNs placement in the MAP domain by using the HMIPv6 protocol. Even though the E-HAR protocol offers reduced performance compared to the B-HAR protocol, the E-HAR protocol remains at an acceptable level that satisfies users [10].

Table 2 shows the numerical results for control traffic received and sent on the MN, MAP and HA sides, respectively. In general, any differences in control signalling overhead between the E-HAR and B-HAR protocols at an entity can be attributed to differences in the number and length of signalling messages received and/or sent at that entity. The following observations can be made from this table: (1) At MN, the amount of control traffic received in the E-HAR and B-HAR protocols is identical, but the control traffic sent is introduced in the B-HAR protocol and it is zero bit/second in the E-HAR protocol; (2) The amount of control traffic received and sent at Table 2 shows the numerical results for control traffic received and sent on the MN, MAP and HA sides, respectively. In general, any differences in control signalling overhead between the E-HAR and B-HAR protocols at an entity can be attributed to differences in the number and length of signalling messages received and/or sent at that entity. The following observations can be made from this table: (1) At MN, the amount of control traffic received in the E-HAR and B-HAR protocols is identical, but the control traffic sent is introduced in the B-HAR protocol and it is zero bit/second in the E-HAR protocol; (2) The amount of control traffic received and sent at the MAP is introduced in the E-HAR protocol, while in the B-HAR protocol it is zero bit/second; and (3) The amount of control traffic received at the HA under E-HAR protocol is about 88.9% higher than under B-HAR protocol, and the amount of control traffic sent under E-HAR protocol is about 39.2% higher than under B-HAR protocol.

Table 2. Control traffic received and sent on the MN, MAP and HA sides in E-HAR and B-HAR protocols

Protocols	Control traffic (bit/second)	MN	MAP	HA
E-HAR	Received	1.66	2.32	4.76
	Sent	0	4.02	2.47
B-HAR	Received	1.66	0	1.83
	Sent	1.83	0	1.66

5 Conclusion

This paper presented the design of an E-HAR protocol to register the RCoA of the MN with its HA when the MN is away from its home link and moves to the MAP domain. The E-HAR protocol combines two ideas. Firstly, it extends MAP functions to act as a proxy entity to register the RCoA of the MN at its HA. Secondly, it applies a synchronised RCoA reachability test to ensure the HA that the MN was reachable at the claimed RCoA. Alternatively, the E-HAR protocol allows the MAP and HA to compute the shared secret key. As a result, the E-HAR protocol reduces the likelihood of a malicious MN from successfully luring an HA to flood a third party with useless traffic during its placement in the MAP domain. Furthermore, the evaluation performance, in terms of HA registration delay and signalling overhead, show that the proposed protocol offers higher registration delay under network conditions, and the E-HAR protocol introduces a higher signalling overhead compared with the B-HAR protocol. However, the E-HAR protocol emerges as superior in performance in regard to security and efficiency. Further research is required to identify the optimal values of a maximum binding lifetime; and evaluate the performance of the E-HAR protocol for packet loss.

Acknowledgment. The author gratefully acknowledges the Northern Border University, Kingdom of Saudi Arabia for financial support.

References

1. Jara, A.J., Fernandez, D., Lopez, P., Zamora, M.A., Skarmeta, A.F.: Lightweight MIPv6 with IPSec support. Mob. Inf. Syst. **10**, 7–77 (2014)
2. Johnson, D., Perkins, C., Arkko, J.: Mobility Support in IPv6, RFC 3775 (Proposed Standard) (2004)
3. Alkhliwi, S.: Securing locations of mobile nodes in wireless mesh networks. Int. J. Adv. Comput. Sci. Appl. **9**(11), 70–81 (2018)
4. Elshakankiry, O., Carpenter, A., Zhang, N.: A novel scheme for supporting location authentication of mobile nodes. In: Proceeding of the Second International ICST Conference on Security and Privacy in Mobile Information and Communication Systems, MobiSec 2010, pp. 91–102 (2010)
5. Aura, T.: Cryptographically Generated Addresses (CGA), RFC 3927 (Proposed Standard) (2005)
6. Soliman, H., Castelluccia, C., ElMalki, K., Bellier, L.: Hierarchical Mobile IPv6 (HMIPv6) Mobility Management, RFC 5380 (Proposed Standard) (2008)
7. Stallings, W.: Cryptography and Network Security: Principles and Practice. Prentice Hall, Upper Saddle River (2011)
8. Menezes, A.J., Vanstone, S.A., Van Oorschot, P.C.: Handbook of Applied Cryptography, 1st edn. CRC Press Inc, Boca Raton (1996)
9. Riverbed Modeler. https://www.riverbed.com/gb/products/steelcentral/opnet.html. Accessed 24 Jan 2016
10. Saif, A., Othman, M.: Network load and packet loss optimization during handoff using multi-scan approach. Int. Arab J. Inf. Technol. **8**(1), 16–22 (2011)

Author Index

A.-E. Hassanien et al. (Eds.): AICV 2020, AISC 1153, pp. 871–873, 2020.
https://doi.org/10.1007/978-3-030-44289-7